BIRKHÄUSER

Panos J. Antsaklis
Anthony N. Michel

# A Linear Systems Primer

Birkhäuser
Boston • Basel • Berlin

Panos J. Antsaklis
Department of Electrical Engineering
University of Notre Dame
Notre Dame, IN 46556
U.S.A.

Anthony N. Michel
Department of Electrical Engineering
University of Notre Dame
Notre Dame, IN 46556
U.S.A.

Cover design by Mary Burgess.

Mathematics Subject Classification (2000): 34A30, 34H05, 93-XX, 93-01, 93Axx, 93A30, 93Bxx, 93B03, 93B05, 93B07, 93B10, 93B11, 93B12, 93B15, 93B17, 93B18, 93B20, 93B25, 93B50, 93B55, 93B60, 93Cxx, 93C05, 93C15, 93C35, 93C55, 93C57, 93C62, 93Dxx, 93D05, 93D15, 93D20, 93D25, 93D30

**Library of Congress Control Number:** 2007905134

ISBN-13: 978-0-8176-4460-4    e-ISBN-13: 978-0-8176-4661-5

Printed on acid-free paper.

©2007 Birkhäuser Boston

*www.birkhauser.com*                                                    (MP)

# To our Families

To
*Melinda and our daughter Lily*
*and to my parents*
*Dr. Ioannis and Marina Antsaklis*
  *—Panos J. Antsaklis*

To
*Leone and our children*
*Mary, Kathy, John,*
*Tony, and Pat*
  *—Anthony N. Michel*

# And to our Students

# Preface

## Brief Description

The purpose of this book is to provide an introduction to system theory with emphasis on control theory. It is intended to be the textbook of a typical one-semester course introduction to systems primarily for first-year graduate students in engineering, but also in mathematics, physics, and the rest of the sciences. Prerequisites for such a course include undergraduate-level differential equations and linear algebra, Laplace transforms, and modeling ideas of, say, electric circuits and simple mechanical systems. These topics are typically covered in the usual undergraduate curricula in engineering and sciences. The goal of this text is to provide a clear understanding of the fundamental concepts of systems and control theory, to highlight appropriately the principal results, and to present material sufficiently broad so that the reader will emerge with a clear picture of the dynamical behavior of linear systems and their advantages and limitations.

## Organization and Coverage

This primer covers essential concepts and results in systems and control theory. Since a typical course that uses this book may serve students with different educational experiences, from different disciplines and from different educational systems, the first chapters are intended to build up the understanding of the dynamical behavior of systems as well as provide the necessary mathematical background. Internal and external system descriptions are described in detail, including state variable, impulse response and transfer function, polynomial matrix, and fractional representations. Stability, controllability, observability, and realizations are explained with the emphasis always being on fundamental results. State feedback, state estimation, and eigenvalue assignment are discussed in detail. All stabilizing feedback controllers are also parameterized using polynomial and fractional system representations. The emphasis in this primer is on time-invariant systems, both continuous and discrete time. Although time-varying systems are studied in the first chapter, for a full coverage the reader is encouraged to consult the companion book titled *Linear Systems*[1] that offers detailed descriptions and additional material, including all the proofs of the results presented in this book. In fact, this primer is based on the more complete treatment of *Linear Systems*, which can also serve as a reference for researchers in the field. This primer focuses more on course use of the material, with emphasis on a presentation that is more transparent, without sacrificing rigor, and emphasizes those results that are considered to be fundamental in systems and control and are accepted as important and essential topics of the subject.

## Contents

In a typical course on Linear Systems, the depth of coverage will vary depending on the goals set for the course and the background of the students. We typically cover the material in the first three chapters in about six to seven weeks or about half of the semester; we spend about four to five weeks covering Chapters 4–8 on stability, controllability, and realizations; and we spend the remaining time in the course on state feedback, state estimation, and feedback control presented in Chapters 9–10. This book contains over 175 examples and almost 160 exercises. A Solutions Manual is available to course instructors from the publisher. Answers to selected exercises are given at the end of this book.

By the end of Chapter 3, the students should have gained a good understanding of the role of inputs and initial conditions in the response of systems that are linear and time-invariant and are described by state-variable internal descriptions for both continuous- and discrete-time systems; should have brushed up and acquired background in differential and difference equations, matrix algebra, Laplace and z transforms, vector spaces, and linear transformations; should have gained understanding of linearization and the generality and limitations of the linear models used; should have become familiar with eigenvalues, system modes, and stability of an equilibrium; should have an understanding of external descriptions, impulse responses, and transfer functions; and should have learned how sampled data system descriptions are derived.

Depending on the background of the students, in Chapter 1, one may want to define the initial value problem, discuss examples, briefly discuss existence and uniqueness of solutions of differential equations, identify methods to solve linear differential equations, and derive the state transition matrix. Next, in Chapter 2, one may wish to discuss the system response, introduce the impulse response, and relate it to the state-space descriptions for both continuous- and discrete-time cases. In Chapter 3, one may consider to study in detail the response of the systems to inputs and initial conditions. Note that it is

---

[1] P.J. Antsaklis and A.N. Michel, *Linear Systems*, Birkhäuser, Boston, MA, 2006.

possible to start the coverage of the material with Chapter 3 going back to Chapters 1 and 2 as the need arises.

A convenient way to decide the particular topics from each chapter that need to be covered is by reviewing the Summary and Highlights sections at the end of each chapter.

The Lyapunov stability of an equilibrium and the input/output stability of linear time-invariant systems, along with stability, controllability and observability, are fundamental system properties and are covered in Chapters 4 and 5. Chapter 6 describes useful forms of the state space representations such as the Kalman canonical form and the controller form. They are used in the subsequent chapters to provide insight into the relations between input and output descriptions in Chapter 7. In that chapter the polynomial matrix representation, an alternative internal description, is also introduced. Based on the results of Chapters 5–7, Chapter 8 discusses realizations of transfer functions. Chapter 9 describes state feedback, pole assignment, optimal control, as well as state observers and optimal state estimation. Chapter 10 characterizes all stabilizing controllers and discusses feedback problems using matrix fractional descriptions of the transfer functions.

Depending on the interest and the time constraints, several topics may be omitted completely without loss of continuity. These topics may include, for example, parts of Section 6.4 on controller and observer forms, Section 7.4 on poles and zeros, Section 7.5 on polynomial matrix descriptions, some of the realization algorithms in Section 8.4, sections in Chapter 9 on state feedback and state observers, and all of Chapter 10.

The appendix collects selected results on linear algebra, fields, vector spaces, eigenvectors, the Jordan canonical form, and normed linear spaces, and it addresses numerical analysis issues that arise when computing solutions of equations.

Simulating the behavior of dynamical systems, performing analysis using computational models, and designing systems using digital computers, although not central themes of this book, are certainly encouraged and often required in the examples and in the Exercise sections in each chapter. One could use one of several software packages specifically designed to perform such tasks that come under the label of control systems and signal processing, and work in different operating system environments; or one could also use more general computing languages such as C, which is certainly a more tedious undertaking. Such software packages are readily available commercially and found in many university campuses. In this book we are not endorsing any particular one, but we are encouraging students to make their own informed choices.

## Acknowledgments

We are indebted to our students for their feedback and constructive suggestions during the evolution of this book. We are also grateful to colleagues who provided useful feedback regarding what works best in the classroom in their particular institutions. Special thanks go to Eric Kuehner for his expert preparation of the manuscript. This project would not have been possible without the enthusiastic support of Tom Grasso, Birkhäuser's Computational Sciences and Engineering Editor, who thought that such a companion primer to *Linear Systems* was an excellent idea. We would also like to acknowledge the help of Regina Gorenshteyn, Associate Editor at Birkhäuser.

It was a pleasure writing this book. Our hope is that students enjoy reading it and learn from it. It was written for them.

Notre Dame, IN                                    *Panos J. Antsaklis*
Spring 2007                                       *Anthony N. Michel*

# Contents

**Preface** ........................................................... vii

**1  System Models, Differential Equations, and Initial-Value
   Problems** ........................................................ 1
   1.1  Introduction .................................................. 1
   1.2  Preliminaries ................................................. 6
        1.2.1  Notation ............................................... 7
        1.2.2  Continuous Functions ................................... 7
   1.3  Initial-Value Problems ........................................ 8
        1.3.1  Systems of First-Order Ordinary Differential Equations  9
        1.3.2  Classification of Systems of First-Order Ordinary
               Differential Equations ................................ 10
        1.3.3  $n$th-Order Ordinary Differential Equations ........... 11
   1.4  Examples of Initial-Value Problems ............................ 13
   1.5  Solutions of Initial-Value Problems: Existence, Continuation,
        Uniqueness, and Continuous Dependence on Parameters .......... 17
   1.6  Systems of Linear First-Order Ordinary Differential Equations  20
        1.6.1  Linearization ......................................... 21
        1.6.2  Examples .............................................. 24
   1.7  Linear Systems: Existence, Uniqueness, Continuation, and
        Continuity with Respect to Parameters of Solutions .......... 27
   1.8  Solutions of Linear State Equations ........................... 28
   1.9  Summary and Highlights ........................................ 32
   1.10 Notes ......................................................... 33
   References .......................................................... 33
   Exercises ........................................................... 34

**2  An Introduction to State-Space and Input–Output
   Descriptions of Systems** ........................................ 47
   2.1  Introduction .................................................. 47
   2.2  State-Space Description of Continuous-Time Systems ........... 47

2.3   State-Space Description of Discrete-Time Systems . . . . . . . . . .  50
2.4   Input–Output Description of Systems . . . . . . . . . . . . . . . . . . . . . .  56
      2.4.1   External Description of Systems: General Considerations  56
      2.4.2   Linear Discrete-Time Systems . . . . . . . . . . . . . . . . . . . . . . .  60
      2.4.3   The Dirac Delta Distribution . . . . . . . . . . . . . . . . . . . . . . .  65
      2.4.4   Linear Continuous-Time Systems . . . . . . . . . . . . . . . . . . . . .  68
2.5   Summary and Highlights . . . . . . . . . . . . . . . . . . . . . . . . . . . . . . .  71
2.6   Notes . . . . . . . . . . . . . . . . . . . . . . . . . . . . . . . . . . . . . . . . . . . . . . .  73
References . . . . . . . . . . . . . . . . . . . . . . . . . . . . . . . . . . . . . . . . . . . . . . .  73
Exercises . . . . . . . . . . . . . . . . . . . . . . . . . . . . . . . . . . . . . . . . . . . . . . .  74

3   **Response of Continuous- and Discrete-Time Systems** . . . . . .  77
3.1   Introduction . . . . . . . . . . . . . . . . . . . . . . . . . . . . . . . . . . . . . . . . . .  77
3.2   Solving $\dot{x} = Ax$ and $\dot{x} = Ax + g(t)$: The State Transition
      Matrix $\Phi(t, t_0)$ . . . . . . . . . . . . . . . . . . . . . . . . . . . . . . . . . . . . . .  78
      3.2.1   The Fundamental Matrix . . . . . . . . . . . . . . . . . . . . . . . . . . .  78
      3.2.2   The State Transition Matrix . . . . . . . . . . . . . . . . . . . . . . . .  82
      3.2.3   Nonhomogeneous Equations . . . . . . . . . . . . . . . . . . . . . . . .  84
3.3   The Matrix Exponential $e^{At}$, Modes, and Asymptotic
      Behavior of $\dot{x} = Ax$ . . . . . . . . . . . . . . . . . . . . . . . . . . . . . . . . . .  85
      3.3.1   Properties of $e^{At}$ . . . . . . . . . . . . . . . . . . . . . . . . . . . . . . . .  85
      3.3.2   How to Determine $e^{At}$ . . . . . . . . . . . . . . . . . . . . . . . . . . .  86
      3.3.3   Modes, Asymptotic Behavior, and Stability . . . . . . . . . . .  94
3.4   State Equation and Input–Output Description of
      Continuous-Time Systems . . . . . . . . . . . . . . . . . . . . . . . . . . . . . . 100
      3.4.1   Response of Linear Continuous-Time Systems . . . . . . . . . 100
      3.4.2   Transfer Functions . . . . . . . . . . . . . . . . . . . . . . . . . . . . . . . 102
      3.4.3   Equivalence of State-Space Representations . . . . . . . . . . . 105
3.5   State Equation and Input–Output Description of
      Discrete-Time Systems . . . . . . . . . . . . . . . . . . . . . . . . . . . . . . . . . 108
      3.5.1   Response of Linear Discrete-Time Systems . . . . . . . . . . . . 108
      3.5.2   The Transfer Function and the $z$-Transform . . . . . . . . . . 112
      3.5.3   Equivalence of State-Space Representations . . . . . . . . . . . 115
      3.5.4   Sampled-Data Systems . . . . . . . . . . . . . . . . . . . . . . . . . . . . 116
      3.5.5   Modes, Asymptotic Behavior, and Stability . . . . . . . . . . . 121
3.6   An Important Comment on Notation . . . . . . . . . . . . . . . . . . . . . 126
3.7   Summary and Highlights . . . . . . . . . . . . . . . . . . . . . . . . . . . . . . . 127
3.8   Notes . . . . . . . . . . . . . . . . . . . . . . . . . . . . . . . . . . . . . . . . . . . . . . . 129
References . . . . . . . . . . . . . . . . . . . . . . . . . . . . . . . . . . . . . . . . . . . . . . 130
Exercises . . . . . . . . . . . . . . . . . . . . . . . . . . . . . . . . . . . . . . . . . . . . . . 131

4   **Stability** . . . . . . . . . . . . . . . . . . . . . . . . . . . . . . . . . . . . . . . . . . . . . 141
4.1   Introduction . . . . . . . . . . . . . . . . . . . . . . . . . . . . . . . . . . . . . . . . . . 141
4.2   The Concept of an Equilibrium . . . . . . . . . . . . . . . . . . . . . . . . . . 142
4.3   Qualitative Characterizations of an Equilibrium . . . . . . . . . . . . 144

4.4   Lyapunov Stability of Linear Systems ..................... 148
4.5   The Lyapunov Matrix Equation .......................... 153
4.6   Linearization.......................................... 164
4.7   Input–Output Stability................................. 170
4.8   Discrete-Time Systems ................................. 173
      4.8.1   Preliminaries ................................. 173
      4.8.2   Linear Systems................................ 176
      4.8.3   The Lyapunov Matrix Equation .................... 179
      4.8.4   Linearization ................................. 185
      4.8.5   Input–Output Stability............................ 186
4.9   Summary and Highlights ............................... 188
4.10  Notes ................................................ 189
References ................................................. 190
Exercises .................................................. 191

5   Controllability and Observability:
    Fundamental Results ..................................... 195
    5.1   Introduction ...................................... 195
    5.2   A Brief Introduction to Reachability and Observability ....... 195
          5.2.1   Reachability and Controllability ..................... 196
          5.2.2   Observability and Constructibility ................... 200
          5.2.3   Dual Systems ............................... 203
    5.3   Reachability and Controllability ......................... 204
          5.3.1   Continuous-Time Time-Invariant Systems ........... 205
          5.3.2   Discrete-Time Systems ........................... 213
    5.4   Observability and Constructibility ........................ 218
          5.4.1   Continuous-Time Time-Invariant Systems ........... 219
          5.4.2   Discrete-Time Time-Invariant Systems ............... 225
    5.5   Summary and Highlights .............................. 230
    5.6   Notes ............................................ 232
    References ................................................ 232
    Exercises ................................................. 233

6   Controllability and Observability:
    Special Forms ........................................... 237
    6.1   Introduction ....................................... 237
    6.2   Standard Forms for Uncontrollable and Unobservable Systems  237
          6.2.1   Standard Form for Uncontrollable Systems ........... 238
          6.2.2   Standard Form for Unobservable Systems ............. 241
          6.2.3   Kalman's Decomposition Theorem.................. 244
    6.3   Eigenvalue/Eigenvector Tests for Controllability and
          Observability ...................................... 248
    6.4   Controller and Observer Forms.......................... 250
          6.4.1   Controller Forms ................................ 251
          6.4.2   Observer Forms ................................ 263

6.5   Summary and Highlights ................................. 269
6.6   Notes ................................................ 271
References ................................................ 272
Exercises ................................................ 272

7  **Internal and External Descriptions:**
   **Relations and Properties** ............................. 277
   7.1   Introduction ....................................... 277
   7.2   Relations Between State-Space and Input–Output
         Descriptions ...................................... 277
   7.3   Relations Between Lyapunov and Input–Output Stability ..... 281
   7.4   Poles and Zeros .................................... 282
         7.4.1   Smith and Smith–McMillan Forms ................. 283
         7.4.2   Poles ....................................... 284
         7.4.3   Zeros ....................................... 286
         7.4.4   Relations Between Poles, Zeros, and Eigenvalues of $A$ .. 290
   7.5   Polynomial Matrix and Matrix Fractional Descriptions of
         Systems ........................................... 292
         7.5.1   A Brief Introduction to Polynomial and Fractional
                 Descriptions ................................. 294
         7.5.2   Coprimeness and Common Divisors ............... 298
         7.5.3   Controllability, Observability, and Stability ........... 303
         7.5.4   Poles and Zeros .............................. 304
   7.6   Summary and Highlights ............................. 306
   7.7   Notes .............................................. 308
   References ................................................ 308
   Exercises ................................................ 309

8  **Realization Theory and Algorithms** ..................... 313
   8.1   Introduction ....................................... 313
   8.2   State-Space Realizations of External Descriptions ........... 313
         8.2.1   Continuous-Time Systems ...................... 314
         8.2.2   Discrete-Time Systems ........................ 315
   8.3   Existence and Minimality of Realizations ................. 316
         8.3.1   Existence of Realizations ...................... 316
         8.3.2   Minimality of Realizations ..................... 317
         8.3.3   The Order of Minimal Realizations ............... 321
         8.3.4   Minimality of Realizations: Discrete-Time Systems .... 323
   8.4   Realization Algorithms ............................... 324
         8.4.1   Realizations Using Duality ..................... 324
         8.4.2   Realizations in Controller/Observer Form .......... 326
         8.4.3   Realizations with Matrix $A$ Diagonal .............. 339
         8.4.4   Realizations Using Singular-Value Decomposition ...... 341
   8.5   Polynomial Matrix Realizations ........................ 343
   8.6   Summary and Highlights ............................. 345

8.7   Notes . . . . . . . . . . . . . . . . . . . . . . . . . . . . . . . . . . . . . . . . . . . . 346
References . . . . . . . . . . . . . . . . . . . . . . . . . . . . . . . . . . . . . . . . . . . . . 346
Exercises . . . . . . . . . . . . . . . . . . . . . . . . . . . . . . . . . . . . . . . . . . . . . . 346

**9   State Feedback and State Observers** . . . . . . . . . . . . . . . . . . . . 351
9.1   Introduction . . . . . . . . . . . . . . . . . . . . . . . . . . . . . . . . . . . . . . . 351
9.2   Linear State Feedback . . . . . . . . . . . . . . . . . . . . . . . . . . . . . . . . 352
        9.2.1   Continuous-Time Systems . . . . . . . . . . . . . . . . . . . . . . . 352
        9.2.2   Eigenvalue Assignment . . . . . . . . . . . . . . . . . . . . . . . . . . 355
        9.2.3   The Linear Quadratic Regulator (LQR):
                   Continuous-Time Case . . . . . . . . . . . . . . . . . . . . . . . . . 369
        9.2.4   Input–Output Relations . . . . . . . . . . . . . . . . . . . . . . . . . 372
        9.2.5   Discrete-Time Systems . . . . . . . . . . . . . . . . . . . . . . . . . . 376
        9.2.6   The Linear Quadratic Regulator (LQR): Discrete-Time
                   Case . . . . . . . . . . . . . . . . . . . . . . . . . . . . . . . . . . . . . . . . 377
9.3   Linear State Observers . . . . . . . . . . . . . . . . . . . . . . . . . . . . . . . 378
        9.3.1   Full-Order Observers: Continuous-Time Systems . . . . . . 378
        9.3.2   Reduced-Order Observers: Continuous-Time Systems . . 383
        9.3.3   Optimal State Estimation: Continuous-Time Systems . . 385
        9.3.4   Full-Order Observers: Discrete-Time Systems . . . . . . . . 387
        9.3.5   Reduced-Order Observers: Discrete-Time Systems . . . . . 391
        9.3.6   Optimal State Estimation: Discrete-Time Systems . . . . . 391
9.4   Observer-Based Dynamic Controllers . . . . . . . . . . . . . . . . . . . . . 392
        9.4.1   State-Space Analysis . . . . . . . . . . . . . . . . . . . . . . . . . . . 393
        9.4.2   Transfer Function Analysis . . . . . . . . . . . . . . . . . . . . . . 397
9.5   Summary and Highlights . . . . . . . . . . . . . . . . . . . . . . . . . . . . . . 400
9.6   Notes . . . . . . . . . . . . . . . . . . . . . . . . . . . . . . . . . . . . . . . . . . . . 403
References . . . . . . . . . . . . . . . . . . . . . . . . . . . . . . . . . . . . . . . . . . . . . 404
Exercises . . . . . . . . . . . . . . . . . . . . . . . . . . . . . . . . . . . . . . . . . . . . . . 405

**10  Feedback Control Systems** . . . . . . . . . . . . . . . . . . . . . . . . . . . . . 411
10.1  Introduction . . . . . . . . . . . . . . . . . . . . . . . . . . . . . . . . . . . . . . . 411
10.2  Interconnected Systems . . . . . . . . . . . . . . . . . . . . . . . . . . . . . . 411
        10.2.1  Systems Connected in Parallel and in Series . . . . . . . . . 411
        10.2.2  Systems Connected in Feedback Configuration . . . . . . . . 413
10.3  Parameterization of All Stabilizing Feedback Controllers . . . . . . 422
        10.3.1  Stabilizing Feedback Controllers Using Polynomial
                    MFDs . . . . . . . . . . . . . . . . . . . . . . . . . . . . . . . . . . . . . . 423
        10.3.2  Stabilizing Feedback Controllers Using Proper and
                    Stable MFDs . . . . . . . . . . . . . . . . . . . . . . . . . . . . . . . . 426
10.4  Two Degrees of Freedom Controllers . . . . . . . . . . . . . . . . . . . . . 431
        10.4.1  Internal Stability . . . . . . . . . . . . . . . . . . . . . . . . . . . . . . 432
        10.4.2  Response Maps . . . . . . . . . . . . . . . . . . . . . . . . . . . . . . . 435
        10.4.3  Controller Implementations . . . . . . . . . . . . . . . . . . . . . . 439
        10.4.4  Some Control Problems . . . . . . . . . . . . . . . . . . . . . . . . . 445

10.5 Summary and Highlights ................................. 447
10.6 Notes ................................................. 449
References ................................................. 451
Exercises .................................................. 452

**A  Appendix** ................................................. 455
A.1 Vector Spaces ......................................... 455
    A.1.1 Fields .......................................... 455
    A.1.2 Vector Spaces ................................... 456
A.2 Linear Independence and Bases ......................... 460
    A.2.1 Linear Subspaces ................................ 460
    A.2.2 Linear Independence ............................. 461
    A.2.3 Linear Independence of Functions of Time .......... 462
    A.2.4 Bases ........................................... 463
A.3 Linear Transformations ................................ 464
    A.3.1 Linear Equations ................................ 465
    A.3.2 Representation of Linear Transformations by Matrices . 466
    A.3.3 Solving Linear Algebraic Equations ................ 469
A.4 Equivalence and Similarity .............................. 471
    A.4.1 Change of Bases: Vector Case ..................... 471
    A.4.2 Change of Bases: Matrix Case .................... 472
    A.4.3 Equivalence and Similarity of Matrices ............. 473
A.5 Eigenvalues and Eigenvectors .......................... 474
    A.5.1 Characteristic Polynomial ......................... 474
    A.5.2 The Cayley–Hamilton Theorem and Applications ...... 475
    A.5.3 Minimal Polynomials ............................. 477
A.6 Diagonal and Jordan Canonical Form of Matrices ........... 478
A.7 Normed Linear Spaces ................................. 483
A.8 Some Facts from Matrix Algebra ........................ 486
A.9 Numerical Considerations .............................. 487
    A.9.1 Solving Linear Algebraic Equations ................ 488
    A.9.2 Singular Values and Singular Value Decomposition .... 491
    A.9.3 Least-Squares Problem ........................... 496
A.10 Notes ................................................. 497
References ................................................. 497

**Solutions to Selected Exercises** .............................. 499

**Index** ...................................................... 505

# 1

## System Models, Differential Equations, and Initial-Value Problems

## 1.1 Introduction

The dynamical behavior of systems can be understood by studying their mathematical descriptions. The flight path of an airplane subject to certain engine thrust, rudder and elevator angles, and particular wind conditions, or the behavior of an automobile on cruise control when climbing a certain hill, can be predicted using mathematical descriptions of the pertinent behavior. Mathematical equations, typically differential or difference equations, are used to describe the behavior of processes and to predict their responses to certain inputs. Although computer simulation is an excellent tool for verifying predicted behavior, and thus for enhancing our understanding of processes, it is certainly not an adequate substitute for generating the information captured in a mathematical model, when such a model is available.

This chapter develops mathematical descriptions for linear continuous-time and linear discrete-time finite-dimensional systems. Since such systems are frequently the result of a linearization process of nonlinear systems, or the result of the modeling process of physical systems in which the nonlinear effects have been suppressed or neglected, the origins of these linear systems are frequently nonlinear systems. For this reason, here and in Chapter 4, when we deal with certain qualitative aspects (such as existence, uniqueness, continuation, and continuity with respect to parameters of solutions of system equations, stability of an equilibrium, and so forth), we consider linear as well as nonlinear system models, although the remainder of the book deals exclusively with linear systems.

In this chapter, mathematical models and classification of models are discussed in the remainder of this Introduction, Section 1.1. In Section 1.2, we provide some of the notation used and recall certain facts concerning continuous functions. In Section 1.3 we present the initial-value problem and we give several specific examples in Section 1.4. In Section 1.5 we present results that ensure the existence, continuation, and uniqueness of solutions of initial-value problems and results that ensure that the solutions of inital-value problems depend continuously on initial conditions and system parameters. In this section we also present the Method of Successive Approximations to determine solutions of intial-value problems. The results in Section 1.5 pertain to differential equations that in general are nonlinear. In Section 1.6 we address linearization of such equations and we provide several specific examples.

We utilize the results of Section 1.5 to establish in Section 1.7 conditions for the existence, uniqueness, continuation, and continuity with respect to initial conditions and parameters of solutions of initial-value problems determined by *linear ordinary* differential equations.

In Section 1.8 we determine the solutions of linear ordinary differential equations and introduce for the first time the notions of state and state transition matrix. We also present the variations of constants formula for solving linear nonhomogeneous ordinary differential equations, and we introduce the notions of homogeneous and particular solutions.

Summarizing, the purpose of Sections 1.3 to 1.8 is to provide material dealing with ordinary differential equations and initial-value problems that is essential in the study of continuous-time finite-dimensional systems. This material will enable us to introduce the state-space equations representation of continuous-time finite-dimensional systems. This introduction will be accomplished in the next chapter.

## Physical Processes, Models, and Mathematical Descriptions

A systematic study of (physical) *phenomena* usually begins with a *modeling process*. Examples of models include diagrams of electric circuits consisting of interconnections of resistors, inductors, capacitors, transistors, diodes, voltage or current sources, and so on; mechanical circuits consisting of interconnections of point masses, springs, viscous dampers (dashpots), applied forces, and so on; verbal characterizations of economic and societal systems; among others. Next, appropriate *laws* or *principles* are invoked to generate *equations* that describe the models (e.g., Kirchhoff's current and voltage laws, Newton's laws, conservation laws, and so forth). When using an expression such as "we consider a *system* described by ordinary differential equations," we will have in mind a phenomenon described by an appropriate set of ordinary differential equations (not the description of the physical phenomenon itself).

*A physical process (physical system) will typically give rise to several different models, depending on what questions are being asked.* For instance, in the study of the voltage-current characteristics of a transistor (the physical process), one may utilize a circuit (the model) that is valid at low frequencies or a circuit (a second model) that is valid at high frequencies; alternatively, if semiconductor impurities are of interest, a third model, quite different from the preceding two, is appropriate.

Over the centuries, a great deal of progress has been made in developing mathematical descriptions of physical phenomena (using models of such phenomena). In doing so, we have invoked laws (or principles) of physics, chemistry, biology, economics, and so on, to derive mathematical expressions (usually equations) that characterize the evolution (in time) of the variables of interest. The availability of such mathematical descriptions enables us to make use of the vast resources offered by the many areas of applied and pure mathematics to conduct qualitative and quantitative studies of the behavior of processes. *A given model of a physical process may give rise to several different mathematical descriptions.* For example, when applying Kirchhoff's voltage and current laws to the low-frequency transistor model mentioned earlier, one can derive a set of differential and algebraic equations, a set consisting of only differential equations, or a set of integro-differential equations, and so forth. *This process of mathematical modeling, "from a physical phenomenon to a model to a mathematical description," is essential in science and engineering.* To capture phenomena of interest accurately and in tractable mathematical form is a demanding task, as can be imagined, and requires a thorough understanding of the physical process involved. For this reason, the mathematical description of complex electrical systems, such as power systems, is typically accomplished by electrical engineers, the equations of flight dynamics of an aircraft are derived by aeronautical engineers, the equations of chemical processes are arrived at by chemists and chemical engineers, and the equations that characterize the behavior of economic systems are provided by economists. In most nontrivial cases, this type of modeling process is close to an art form since *a good mathematical description must be detailed enough to accurately describe the phenomena of interest and at the same time simple enough to be amenable to analysis.* Depending on the applications on hand, a given mathematical description of a process may be further simplified before it is used in analysis and especially in design procedures. For example, using the finite element method, one can derive a set of first-order differential equations that describe the motion of a space antenna. Typically, such mathematical descriptions contain hundreds of differential equations. Whereas all these equations are quite useful in simulating the motion of the antenna, a lower order model is more suitable for the control design that, for example, may aim to counteract the effects of certain disturbances. Simpler mathematical models are required mainly because of our inability to deal effectively with hundreds of variables and their interactions. In such simplified mathematical descriptions, only those variables (and their interactions) that have significant effects on the phenomena of interest are included.

A point that cannot be overemphasized is that *the mathematical descriptions we will encounter characterize processes only approximately.* Most often, this is the case because the complexity of physical systems defies exact mathematical formulation. In many other cases, however, it is our own choice that a mathematical description of a given process approximate the actual phenomena by only a certain desired degree of accuracy. As discussed earlier, this is done in the interest of mathematical simplicity. For example, in the description of RLC circuits, one could use nonlinear differential equations that take into consideration parasitic effects in the capacitors; however, most often it suffices to use linear ordinary differential equations with constant coefficients to describe the voltage-current relations of such circuits, since typically such a description provides an adequate approximation and since it is much easier to work with linear rather than nonlinear differential equations.

In this book it will generally be assumed that the mathematical description of a system in question is given. In other words, we assume that the modeling of the process in question has taken place and that equations describing the process are given. Our main objective will be to present a theory of an important class of systems—finite-dimensional linear systems—by studying the equations representing such systems.

## Classification of Systems

For our purposes, a comprehensive classification of systems is not particularly illuminating. However, an enumeration of the more common classes of systems encountered in engineering and science may be quite useful, if for no other reason than to show that the classes of systems considered in this book, although very important, are quite specialized.

As pointed out earlier, the particular set of equations describing a given system will in general depend on the effects one wishes to capture. Thus, one can speak of *lumped parameter* or *finite-dimensional systems* and *distributed parameter* or *infinite-dimensional systems*; *continuous-time* and *discrete-time systems*; *linear* and *nonlinear systems*; *time-varying* and *time-invariant systems*; *deterministic* and *stochastic systems*; appropriate combinations of the above, called *hybrid systems*; and perhaps others.

The appropriate mathematical settings for finite-dimensional systems are finite-dimensional vector spaces, and for infinite-dimensional systems they are most often infinite-dimensional linear spaces. Continuous-time finite-dimensional systems are usually described by ordinary differential equations or certain kinds of integral equations, whereas discrete-time finite-dimensional systems are usually characterized by ordinary difference equations or discrete-time counterparts to those integral equations. Equations used to describe infinite-dimensional systems include partial differential equations, Volterra integro-differential equations, functional differential equations, and so forth. Hybrid system descriptions involve two or more different types of equations. Nondeterministic systems are described by stochastic counterparts to those equations (e.g., Ito differential equations).

In a broader context, not addressed in this book, most of the systems described by the equations enumerated generate *dynamical systems*. It has become customary in the engineering literature to use the term "dynamical system" rather loosely, and it has even been applied to cases where the original definition does not exactly fit. (For a discussion of general dynamical systems, refer, e.g., to Michel et al [5].) We will address in this book dynamical systems determined by ordinary differential equations or ordinary difference equations, considered next.

**Finite-Dimensional Systems**

The dynamical systems we will be concerned with are *continuous-time* and *discrete-time finite-dimensional systems*—primarily *linear systems*. However, since such systems are frequently a consequence of a linearization process, it is important when dealing with fundamental qualitative issues that we have an understanding of the origins of such linear systems. In particular, when dealing with questions of existence and uniqueness of solutions of the equations describing a class of systems, and with stability properties of such systems, we may consider nonlinear models as well.

*Continuous-time finite-dimensional dynamical systems* that we will consider are described by equations of the form

$$\dot{x}_i = f_i(t, x_1, \ldots, x_n, u_1, \ldots, u_m), \qquad i = 1, \ldots, n, \qquad (1.1a)$$

$$y_i = g_i(t, x_1, \ldots, x_n, u_1, \ldots, u_m), \qquad i = 1, \ldots, p, \qquad (1.1b)$$

where $u_i$, $i = 1, \ldots, m$, denote *inputs* or *stimuli*; $y_i$, $i = 1, \ldots, p$, denote *outputs* or *responses*; $x_i$, $i = 1, \ldots, n$, denote *state variables*; $t$ denotes *time*; $\dot{x}_i$ denotes the time derivative of $x_i$; $f_i$, $i = 1, \ldots, n$, are real-valued functions of $1 + n + m$ real variables; and $g_i$, $i = 1, \ldots, p$, are real-valued functions of $1 + n + m$ real variables. A complete description of such systems will usually also require a set of *initial conditions* $x_i(t_0) = x_{i0}$, $i = 1, \ldots, n$, where $t_0$ denotes *initial time*. We will elaborate later on restrictions that need to be imposed on the $f_i, g_i$, and $u_i$ and on the origins of the term "state variables."

Equations (1.1a) and (1.1b) can be represented in vector form as

$$\dot{x} = f(t, x, u), \qquad (1.2a)$$

$$y = g(t, x, u), \qquad (1.2b)$$

where $x$ is the *state vector* with components $x_i$, $u$ is the *input vector* with components $u_i$, $y$ is the *output vector* with components $y_i$, and $f$ and $g$ are vector-valued functions with components $f_i$ and $g_i$, respectively. We call (1.2a) a *state equation* and (1.2b) an *output equation*.

Important special cases of (1.2a) and (1.2b) are the *linear time-varying state equation and output equation* given by

$$\dot{x} = A(t)x + B(t)u, \qquad (1.3a)$$

$$y = C(t)x + D(t)u, \qquad (1.3b)$$

where $A, B, C,$ and $D$ are real $n \times n, n \times m, p \times n$, and $p \times m$ matrices, respectively, whose elements are time-varying. Restrictions on these matrices will be provided later.

*Linear time-invariant state and output equations* given by

$$\dot{x} = Ax + Bu, \qquad (1.4a)$$

$$y = Cx + Du \qquad (1.4b)$$

constitute important special cases of (1.3a) and (1.3b), respectively.

Equations (1.3) and (1.4) may arise in the modeling process, or they may be a consequence of *linearization* of (1.1).

*Discrete-time finite-dimensional dynamical systems* are described by equations of the form

$$x_i(k+1) = f_i(k, x_1(k), \dots, x_n(k), u_1(k), \dots, u_m(k)) \quad i = 1, \dots, n, \quad (1.5a)$$
$$y_i(k) = g_i(k, x_1(k), \dots, x_n(k), u_1(k), \dots, u_m(k)) \quad i = 1, \dots, p, \quad (1.5b)$$

or in vector form,

$$x(k+1) = f(k, x(k), u(k)), \quad (1.6a)$$
$$y(k) = g(k, x(k), u(k)), \quad (1.6b)$$

where $k$ is an integer that denotes *discrete time* and all other symbols are defined as before. A complete description of such systems involves a set of *initial conditions* $x(k_0) = x_{k_0}$, where $k_0$ denotes *initial time*. The corresponding linear time-varying and time-invariant state and output equations are given by

$$x(k+1) = A(k)x(k) + B(k)u(k), \quad (1.7a)$$
$$y(k) = C(k)x(k) + D(k)u(k) \quad (1.7b)$$

and

$$x(k+1) = Ax(k) + Bu(k), \quad (1.8a)$$
$$y(k) = Cx(k) + Du(k), \quad (1.8b)$$

respectively, where all symbols in (1.7) and (1.8) are defined as in (1.3) and (1.4), respectively.

This type of system characterization is called *state-space description* or *state-variable description* or *internal description* of finite-dimensional systems. Another way of describing continuous-time and discrete-time finite-dimensional dynamical systems involves operators that establish a relationship between the system inputs and outputs. Such characterization is called *input–output description* or *external description* of a system. In Chapter 2, we will address both the state-variable description and the input–output description of finite-dimensional systems. Before we can do this, however, we will require some background material concerning ordinary differential equations.

## 1.2 Preliminaries

We will employ a consistent notation and use certain facts from the calculus, analysis, and linear algebra. We will summarize this type of material, as needed, in various sections. This is the first such section.

### 1.2.1 Notation

Let $V$ and $W$ be *sets*. Then $V \cup W, V \cap W, V - W$, and $V \times W$ denote the *union, intersection, difference,* and *Cartesian product* of $V$ and $W$, respectively. If $V$ is a *subset* of $W$, we write $V \subset W$; if $x$ is an *element* of $V$, we write $x \in V$; and if $x$ is not an element of $V$, we write $x \notin V$. We let $V', \partial V, \bar{V}$, and int $V$ denote the *complement, boundary, closure,* and *interior* of $V$, respectively.

Let $\phi$ denote the *empty set*, $R$ the *real numbers*, $R^+ = \{x \in R : x \geq 0\}$ (i.e., $R^+$ denotes the set of nonnegative real numbers), $Z$ the *integers*, and $Z^+ = \{x \in Z : x \geq 0\}$.

We will let $J \subset R$ denote open, closed, or half-open *intervals*. Thus, for $a, b \in R$, $a \leq b$, $J$ may be of the form $J = (a, b) = \{x \in R : a < x < b\}$, $J = [a, b] = \{x \in R : a \leq x \leq b\}$, $J = [a, b) = \{x \in R : a \leq x < b\}$, or $J = (a, b] = \{x \in R : a < x \leq b\}$.

Let $R^n$ denote the real $n$-space. If $x \in R^n$, then

$$x = \begin{bmatrix} x_1 \\ \vdots \\ x_n \end{bmatrix}$$

and $x^T = (x_1, \ldots, x_n)$ denotes the *transpose* of the vector $x$. Also, let $R^{m \times n}$ denote the set of $m \times n$ real matrices. If $A \in R^{m \times n}$, then

$$A = [a_{ij}] = \begin{bmatrix} a_{11} & a_{12} & \cdots & a_{1n} \\ a_{21} & a_{22} & \cdots & a_{2n} \\ & & & \\ a_{m1} & a_{m2} & \cdots & a_{mn} \end{bmatrix}$$

and $A^T = [a_{ji}] \in R^{n \times m}$ denotes the *transpose* of the matrix $A$.

Similarly, we let $C^n$ denote the set of $n$-vectors with complex components and $C^{m \times n}$ denote the set of $m \times n$ matrices with complex elements.

Let $f : V \to W$ denote a *mapping* or *function* from a set $V$ into a set $W$, and denote by $D(f)$ and $R(f)$ the *domain* and the *range* of $f$, respectively. Also, let $f^{-1} : R(f) \to D(f)$, if it exists, denote the *inverse* of $f$.

### 1.2.2 Continuous Functions

First, let $J \subset R$ denote an open interval and consider a function $f : J \to R$. Recall that $f$ is said to be *continuous at the point* $t_0 \in J$ if $\lim_{t \to t_0} f(t) = f(t_0)$ exists; i.e., if for every $\epsilon > 0$ there exists a $\delta > 0$ such that $|f(t) - f(t_0)| < \epsilon$ whenever $|t - t_0| < \delta$ and $t \in J$. The function $f$ is said to be *continuous on* $J$, or simply *continuous*, if it is continuous at each point in $J$.

In the above definition, $\delta$ depends on the choice of $t_0$ and $\epsilon$; i.e., $\delta = \delta(\epsilon, t_0)$. If at *each* $t_0 \in J$ it is true that there is a $\delta > 0$, independent of $t_0$ [i.e., $\delta = \delta(\epsilon)$], such that $|f(t) - f(t_0)| < \epsilon$ whenever $|t - t_0| < \delta$ and $t \in J$, then $f$ is said to be *uniformly continuous* (on $J$).

Let
$$C(J, R) \triangleq \{f : J \to R \mid f \text{ is continuous on } J\}.$$
Now suppose that $J$ contains one or both endpoints. Then continuity is interpreted as being one-sided at these points. For example, if $J = [a, b]$, then $f \in C(J, R)$ will mean that $f \in C((a, b), R)$ and that $\lim_{t \to a^+} f(t) = f(a)$ and $\lim_{t \to b^-} f(t) = f(b)$ exist.

With $k$ any positive integer, and with $J$ an open interval, we will use the notation
$$C^k(J, R) \triangleq \{f : J \to R \mid \text{ the derivative } f^{(j)} \text{ exists on } J \text{ and}$$
$$f^{(j)} \in C(J, R) \text{ for } j = 0, 1, \ldots, k, \text{ where } f^{(0)} \triangleq f\}$$

and we will call $f$ in this case a $C^k$-function. Also, we will call $f$ a *piecewise* $C^k$-function if $f \in C^{k-1}(J, R)$ and $f^{(k-1)}$ has continuous derivatives for all $t \in J$, with the possible exception of a finite set of points where $f^{(k)}$ may have jump discontinuities. As before, when $J$ contains one or both endpoints, then the existence and continuity of derivatives is one-sided at these points.

For any subset $D$ of the $n$-space $R^n$ with nonempty interior, we can define $C(D, R)$ and $C^k(D, R)$ in a similar manner as before. Thus, $f \in C(D, R)$ indicates that at every point $x_0 = (x_{10}, \ldots, x_{n0})^T \in D, \lim_{x \to x_0} f(x) = f(x_0)$ exists, or equivalently, at every $x_0 \in D$ it is true that for every $\epsilon > 0$ there exists a $\delta = \delta(\epsilon, x_0) > 0$ such that $|f(x) - f(x_0)| < \epsilon$ whenever $|x_1 - x_{10}| + \cdots + |x_n - x_{n0}| < \delta$ and $x \in D$. Also, we define $C^k(D, R)$ as

$$C^k(D, R) \triangleq \{f : D \to R \mid \frac{\partial^j f}{\partial x_1^{i_1} \ldots \partial x_n^{i_n}} \in C(D, R), \quad i_1 + \cdots + i_n = j,$$
$$j = 1, \ldots, k, \text{ and } f \in C(D, R)\}$$

(i.e., $i_1, \ldots, i_n$ take on all possible positive integer values such that their sum is $j$). When $D$ contains its boundary (or part of its boundary), then the continuity of $f$ and the existence and continuity of partial derivatives of $f$, $\frac{\partial^j f}{\partial x_1^{i_1} \ldots \partial x_n^{i_n}}$, $i_1 + \cdots + i_n = j$, $j = 1, \ldots, k$, will have to be interpreted in the appropriate way at the boundary points.

Recall that if $K \subset R^n$, $K \neq \phi$, and $K$ is *compact* (i.e., $K$ is closed and bounded), and if $f \in C(K, R)$, then $f$ is uniformly continuous (on $K$) and $f$ attains its maximum and minimum on $K$.

Finally, let $D$ be a subset of $R^n$ with nonempty interior and let $f : D \to R^m$. Then $f = (f_1, \ldots, f_m)^T$ where $f_i : D \to R$, $i = 1, \ldots, m$. We say that $f \in C(D, R^m)$ if $f_i \in C(D, R)$, $i = 1, \ldots, m$, and that for some positive integer $k$, $f \in C^k(D, R^m)$ if $f_i \in C^k(D, R)$, $i = 1, \ldots, m$.

## 1.3 Initial-Value Problems

In this section we make precise the meaning of several concepts that arise in the study of continuous-time finite-dimensional dynamical systems.

### 1.3.1 Systems of First-Order Ordinary Differential Equations

Let $D \subset R^{n+1}$ denote a *domain*, i.e., an open, nonempty, and connected subset of $R^{n+1}$. We call $R^{n+1}$ the $(t, x)$-*space*, and we denote elements of $R^{n+1}$ by $(t, x)$ and elements of $R^n$ by $x = (x_1, \ldots, x_n)^T$. Next, we consider the functions $f_i \in C(D, R)$, $i = 1, \ldots, n$, and if $x_i$ is a function of $t$, we let $x_i^{(n)} = \frac{d^n x_i}{dt^n}$ denote the $n$th derivative of $x_i$ with respect to $t$ (provided that it exists). In particular, when $n = 1$, we usually write

$$x_i^{(1)} = \dot{x}_i = \frac{dx_i}{dt}.$$

We call the system of equations given by

$$\dot{x}_i = f_i(t, x_1, \ldots, x_n), \quad i = 1, \ldots, n, \qquad (1.9)$$

a *system of $n$ first-order ordinary differential equations*. By a *solution* of the system of equations (1.9), we shall mean $n$ continuously differentiable functions $\phi_1, \ldots, \phi_n$ defined on an interval $J = (a, b)$ [i.e., $\phi \in C^1(J, R^n)$] such that $(t, \phi_1(t), \ldots, \phi_n(t)) \in D$ for all $t \in J$ and such that

$$\dot{\phi}_i(t) = f_i(t, \phi_1(t), \ldots, \phi_n(t)), \quad i = 1, \ldots, n,$$

for all $t \in J$.

Next, we let $(t_0, x_{10}, \ldots, x_{n0}) \in D$. Then the *initial-value problem* associated with (1.9) is given by

$$\begin{aligned} \dot{x}_i &= f_i(t, x_1, \ldots, x_n), & i &= 1, \ldots, n, \\ x_i(t_0) &= x_{i0}, & i &= 1, \ldots, n. \end{aligned} \qquad (1.10)$$

A set of functions $\{\phi_1, \ldots, \phi_n\}$ is a *solution* of the initial-value problem (1.10) if $\{\phi_1, \ldots, \phi_n\}$ is a solution of (1.9) on some interval $J$ containing $t_0$ and if $(\phi_1(t_0), \ldots, \phi_n(t_0)) = (x_{10}, \ldots, x_{n0})$.

In Figure 1.1 the solution of a hypothetical initial-value problem is depicted graphically when $n = 1$. Note that $\dot{\phi}(\tau) = f(\tau, \tilde{x}) = \tan \alpha$, where $\alpha$ is the slope of the line $L$ that is tangent to the plot of the curve $\phi(t)$ vs. $t$, at the point $(\tau, \tilde{x})$.

In dealing with systems of equations, we will utilize the vector notation $x = (x_1, \ldots, x_n)^T$, $x_0 = (x_{10}, \ldots, x_{n0})^T$, $\phi = (\phi_1, \ldots, \phi_n)^T$, $f(t, x) = (f_1(t, x_1, \ldots, x_n), \ldots, f_n(t, x_1, \ldots, x_n))^T = (f_1(t, x), \ldots, f_n(t, x))^T$, $\dot{x} = (\dot{x}_1, \ldots, \dot{x}_n)^T$, and $\int_{t_0}^t f(s, \phi(s))ds = [\int_{t_0}^t f_1(s, \phi(s))ds, \ldots, \int_{t_0}^t f_n(s, \phi(s))ds]^T$.

With the above notation we can express the system of first-order ordinary differential equations (1.9) by

$$\dot{x} = f(t, x) \qquad (1.11)$$

and the initial-value problem (1.10) by

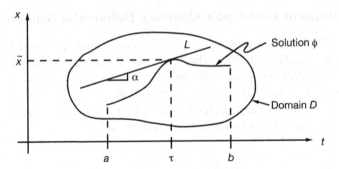

**Figure 1.1.** Solution of an initial-value problem when $n = 1$

$$\dot{x} = f(t, x), \quad x(t_0) = x_0. \tag{1.12}$$

We leave it to the reader to prove that the initial-value problem (1.12) can be equivalently expressed by the *integral equation*

$$\phi(t) = x_0 + \int_{t_0}^{t} f(s, \phi(s))ds, \tag{1.13}$$

where $\phi$ denotes a solution of (1.12).

### 1.3.2 Classification of Systems of First-Order Ordinary Differential Equations

Systems of first-order ordinary differential equations have been classified in many ways. We enumerate here some of the more important cases.

If in (1.11), $f(t, x) \equiv f(x)$ for all $(t, x) \in D$, then

$$\dot{x} = f(x). \tag{1.14}$$

We call (1.14) an *autonomous system* of first-order ordinary differential equations.

If $(t + T, x) \in D$ whenever $(t, x) \in D$ and if $f(t, x) = f(t + T, x)$ for all $(t, x) \in D$, then (1.11) assumes the form

$$\dot{x} = f(t, x) = f(t + T, x). \tag{1.15}$$

We call such an equation a *periodic system* of first-order differential equations with *period* $T$. The smallest $T > 0$ for which (1.15) is true is called the *least period* of this system of equations.

When in (1.11), $f(t, x) = A(t)x$, where $A(t) = [a_{ij}(t)]$ is a real $n \times n$ matrix with elements $a_{ij}$ that are defined and at least piecewise continuous on a $t$-interval $J$, then we have

$$\dot{x} = A(t)x \tag{1.16}$$

and refer to (1.16) as a *linear homogeneous system* of first-order ordinary differential equations.

If for (1.16), $A(t)$ is defined for all real $t$, and if there is a $T > 0$ such that $A(t) = A(t + T)$ for all $t$, then we have

$$\dot{x} = A(t)x = A(t + T)x. \tag{1.17}$$

This system is called a *linear periodic system* of first-order ordinary differential equations.

Next, if in (1.11), $f(t, x) = A(t)x + g(t)$, where $A(t)$ is as defined in (1.16), and $g(t) = [g_1(t), \ldots, g_n(t)]^T$ is a real $n$-vector with elements $g_i$ that are defined and at least piecewise continuous on a $t$-interval $J$, then we have

$$\dot{x} = A(t)x + g(t). \tag{1.18}$$

In this case we speak of a *linear nonhomogeneous system* of first-order ordinary differential equations.

Finally, if in (1.11), $f(t, x) = Ax$, where $A = [a_{ij}] \in R^{n \times n}$, then we have

$$\dot{x} = Ax. \tag{1.19}$$

This type of system is called a *linear, autonomous, homogeneous* system of first-order ordinary differential equations.

### 1.3.3 $n$th-Order Ordinary Differential Equations

Thus far we have been concerned with systems of first-order ordinary differential equations. It is also possible to characterize initial-value problems by means of $n$th-order ordinary differential equations. To this end we let $h$ be a real function that is defined and continuous on a domain $D$ of the real $(t, y, \ldots, y_n)$-space [i.e., $D \subset R^{n+1}$, $D$ is a domain, and $h \in C(D, R)$]. Then

$$y^{(n)} = h(t, y, y^{(1)}, \ldots, y^{(n-1)}) \tag{1.20}$$

is an $n$-*order ordinary differential equation*.

A *solution* of (1.20) is a function $\phi \in C^n(J, R)$ that satisfies $(t, \phi(t), \phi^{(1)}(t), \ldots, \phi^{(n-1)}(t)) \in D$ for all $t \in J$ and

$$\phi^{(n)}(t) = h(t, \phi(t), \phi^{(1)}(t), \ldots, \phi^{(n-1)}(t))$$

for all $t \in J$, where $J = (a, b)$ is a $t$-interval.

Now for a given $(t_0, x_{10}, \ldots, x_{n0}) \in D$, the *initial -value problem* for (1.20) is

$$\begin{aligned} y^{(n)} &= h(t, y, y^{(1)}, \ldots, y^{(n-1)}), \\ y(t_0) &= x_{10}, \ldots, y^{(n-1)}(t_0) = x_{n0}. \end{aligned} \tag{1.21}$$

A function $\phi$ is a *solution* of (1.21) if $\phi$ is a solution of (1.20) on some interval containing $t_0$ and if $\phi(t_0) = x_{10}, \ldots, \phi^{(n-1)}(t_0) = x_{n0}$.

As in the case of systems of first-order ordinary differential equations, we can point to several important special cases. Specifically, we consider equations of the form

$$y^{(n)} + a_{n-1}(t)y^{(n-1)} + \cdots + a_1(t)y^{(1)} + a_0(t)y = g(t), \qquad (1.22)$$

where $a_i \in C(J, R)$, $i = 0, 1, \ldots, n-1$, and $g \in C(J, R)$. We refer to (1.22) as a *linear nonhomogeneous ordinary differential equation of order n*.

If in (1.22) we let $g(t) \equiv 0$, then

$$y^{(n)} + a_{n-1}(t)y^{(n-1)} + \cdots + a_1(t)y^{(1)} + a_0(t)y = 0. \qquad (1.23)$$

We call (1.23) a *linear homogeneous ordinary differential equation of order n*.

If in (1.23) we have $a_i(t) \equiv a_i$, $i = 0, 1, \ldots, n-1$, then

$$y^{(n)} + a_{n-1}y^{(n-1)} + \cdots + a_1y^{(1)} + a_0y = 0, \qquad (1.24)$$

and we call (1.24) a *linear, autonomous, homogeneous ordinary differential equation of order n*.

As in the case of systems of first-order ordinary differential equations, we can define *periodic* and *linear periodic ordinary differential equations of order n* in the obvious way.

It turns out that the theory of $n$th-order ordinary differential equations can be reduced to the theory of a system of $n$ first-order ordinary differential equations. To demonstrate this, we let $y = x_1, y^{(1)} = x_2, \ldots, y^{(n-1)} = x_n$ in (1.21). We now obtain the system of first-order ordinary differential equations

$$
\begin{aligned}
\dot{x}_1 &= x_2 \\
\dot{x}_2 &= x_3 \\
&\vdots \\
\dot{x}_n &= h(t, x_1, \ldots, x_n)
\end{aligned}
\qquad (1.25)
$$

that is defined for all $(t, x_1, \ldots, x_n) \in D$. Assume that $\phi = (\phi_1, \ldots, \phi_n)^T$ is a solution of (1.25) on an interval $J$. Since $\phi_2 = \dot{\phi}_1, \phi_3 = \dot{\phi}_2, \ldots, \phi_n = \phi_1^{(n-1)}$, and since

$$
\begin{aligned}
h(t, \phi_1(t), \ldots, \phi_n(t)) &= h(t, \phi_1(t), \phi_1^{(1)}(t), \ldots, \phi_1^{(n-1)}(t)) \\
&= \phi_1^{(n)}(t),
\end{aligned}
$$

it follows that the first component $\phi_1$ of the vector $\phi$ is a solution of (1.20) on the interval $J$. Conversely, if $\phi_1$ is a solution of (1.20) on $J$, then the vector $(\phi, \phi^{(1)}, \ldots, \phi^{(n-1)})^T$ is clearly a solution of (1.25). Moreover, if $\phi_1(t_0) = x_{10}, \ldots, \phi_1^{(n-1)}(t_0) = x_{n0}$, then the vector $\phi$ satisfies $\phi(t_0) = x_0 = (x_{10}, \ldots, x_{n0})^T$.

## 1.4 Examples of Initial-Value Problems

We now give several specific examples of initial-value problems.

---

*Example 1.1.* The mechanical system of Figure 1.2 consists of two point masses $M_1$ and $M_2$ that are acted upon by viscous damping forces (determined by viscous damping constants $B, B_1$, and $B_2$), spring forces (specified by the spring constants $K, K_1$, and $K_2$), and external forces $f_1$ and $f_2$. The initial displacements of $M_1$ and $M_2$ at $t_0 = 0$ are given by $y_1(0)$ and $y_2(0)$, respectively, and their initial velocities are given by $\dot{y}_1(0)$ and $\dot{y}_2(0)$. The arrows in Figure 1.2 indicate positive directions of displacement for $M_1$ and $M_2$.

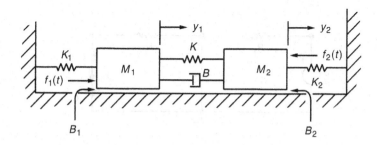

**Figure 1.2.** An example of a mechanical circuit

Newton's second law yields the following coupled second-order ordinary differential equations that describe the motions of the masses in Figure 1.2 (letting $y^{(2)} = d^2y/dt^2 = \ddot{y}$),

$$M_1\ddot{y}_1 + (B + B_1)\dot{y}_1 + (K + K_1)y_1 - B\dot{y}_2 - Ky_2 = f_1(t)$$
$$M_2\ddot{y}_2 + (B + B_2)\dot{y}_2 + (K + K_2)y_2 - B_1\dot{y}_1 - Ky_1 = -f_2(t) \tag{1.26}$$

with initial data $y_1(0), y_2(0), \dot{y}_1(0)$, and $\dot{y}_2(0)$.

Letting $x_1 = y_1, x_2 = \dot{y}_1, x_3 = y_2$, and $x_4 = \dot{y}_2$, we can express (1.26) equivalently by the system of first-order ordinary differential equations

$$
\begin{bmatrix} \dot{x}_1 \\ \dot{x}_2 \\ \dot{x}_3 \\ \dot{x}_4 \end{bmatrix} =
\begin{bmatrix}
0 & 1 & 0 & 0 \\
\frac{-(K_1+K)}{M_1} & \frac{-(B_1+B)}{M_1} & \frac{K}{M_1} & \frac{B}{M_1} \\
0 & 0 & 0 & 1 \\
\frac{K}{M_2} & \frac{B}{M_2} & \frac{-(K+K_2)}{M_2} & \frac{-(B+B_2)}{M_2}
\end{bmatrix}
\begin{bmatrix} x_1 \\ x_2 \\ x_3 \\ x_4 \end{bmatrix}
$$
$$
+ \begin{bmatrix} 0 \\ \frac{1}{M_1}f_1(t) \\ 0 \\ \frac{-1}{M_2}f_2(t) \end{bmatrix} \tag{1.27}
$$

with initial data given by $x(0) = (x_1(0), x_2(0), x_3(0), x_4(0))^T$.

**Example 1.2.** Using the node voltages $v_1, v_2$, and $v_3$ and applying Kirchhoff's current law, we can describe the behavior of the electric circuit given in Figure 1.3 by the system of first-order ordinary differential equations

$$
\begin{bmatrix} \dot{v}_1 \\ \dot{v}_2 \\ \dot{v}_3 \end{bmatrix} = \begin{bmatrix} -\frac{1}{C_1}\left(\frac{1}{R_1}+\frac{1}{R_2}\right) & \frac{1}{R_2 C_1} & 0 \\ -\frac{1}{C_1}\left(\frac{1}{R_1}+\frac{1}{R_2}\right) & -\left(\frac{R_2}{L}-\frac{1}{R_2 C_1}\right) & \frac{R_2}{L} \\ \frac{1}{R_2 C_2} & -\frac{1}{R_2 C_2} & 0 \end{bmatrix} \begin{bmatrix} v_1 \\ v_2 \\ v_3 \end{bmatrix} + \begin{bmatrix} \frac{v}{R_1 C_1} \\ \frac{v}{R_1 C_1} \\ 0 \end{bmatrix}. \quad (1.28)
$$

To complete the description of this circuit, we specify the initial data at $t_0 = 0$, given by $v_1(0), v_2(0)$, and $v_3(0)$.

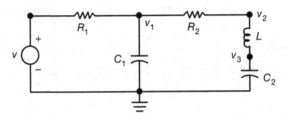

**Figure 1.3.** An example of an electric circuit

**Example 1.3.** Figure 1.4 represents a simplified model of an armature voltage-controlled dc servomotor consisting of a stationary field and a rotating armature and load. We assume that all effects of the field are negligible in the description of this system. The various parameters and variables in Figure 1.4 are $e_a$ = externally applied armature voltage, $i_a$ = armature current, $R_a$ = resistance of the armature winding, $L_a$ = armature winding inductance, $e_m$ = back-emf voltage induced by the rotating armature winding, $B$ = viscous damping due to bearing friction, $J$ = moment of inertia of the armature and load, and $\theta$ = shaft position. The back-emf voltage (with the polarity as shown) is given by

$$e_m = K_\theta \dot{\theta}, \quad (1.29)$$

where $K_\theta > 0$ is a constant, and the torque $T$ generated by the motor is given by

$$T = K_T i_a. \quad (1.30)$$

Application of Newton's second law and Kirchhoff's voltage law yields

$$J\ddot{\theta} + B\dot{\theta} = T(t) \quad (1.31)$$

and

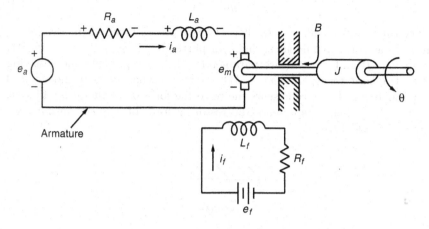

**Figure 1.4.** An example of an electro-mechanical system circuit

$$L_a \frac{di_a}{dt} + R_a i_a + e_m = e_a. \tag{1.32}$$

Combining (1.29) to (1.32) and letting $x_1 = \theta, x_2 = \dot{\theta}$, and $x_3 = i_a$ yields the system of first-order ordinary differential equations

$$\begin{bmatrix} \dot{x}_1 \\ \dot{x}_2 \\ \dot{x}_3 \end{bmatrix} = \begin{bmatrix} 0 & 1 & 0 \\ 0 & -B/J & K_T/J \\ 0 & -K_\theta/L_a & -R_a/L_a \end{bmatrix} \begin{bmatrix} x_1 \\ x_2 \\ x_3 \end{bmatrix} + \begin{bmatrix} 0 \\ 0 \\ e_a/L_a \end{bmatrix}. \tag{1.33}$$

A suitable set of initial data for (1.33) is given by $t_0 = 0$ and $(x_1(0), x_2(0), x_3(0))^T = (\theta(0), \dot{\theta}(0), i_a(0))^T$.

---

**Example 1.4.** A much studied ordinary differential equation is given by

$$\ddot{x} + f(x)\dot{x} + g(x) = 0, \tag{1.34}$$

where $f \in C^1(R, R)$ and $g \in C^1(R, R)$.

When $f(x) \geq 0$ for all $x \in R$ and $xg(x) > 0$ for all $x \neq 0$, then (1.34) is called the *Lienard Equation*. This equation can be used to represent, e.g., RLC circuits with nonlinear circuit elements.

Another important special case of (1.34) is the *van der Pol Equation* given by

$$\ddot{x} - \epsilon(1 - x^2)\dot{x} + x = 0, \tag{1.35}$$

where $\epsilon > 0$ is a parameter. This equation has been used to represent certain electronic oscillators.

If in (1.34), $f(x) \equiv 0$, we obtain

$$\ddot{x} + g(x) = 0. \tag{1.36}$$

When $xg(x) > 0$ for all $x \neq 0$, then (1.36) represents various models of so-called "mass on a nonlinear spring." In particular, if $g(x) = k(1 + a^2x^2)x$, where $k > 0$ and $a^2 > 0$ are parameters, then $g$ represents the restoring force of a *hard spring*. If $g(x) = k(1 - a^2x^2)x$, where $k > 0$ and $a^2 > 0$ are parameters, then $g$ represents the restoring force of a *soft spring*. Finally, if $g(x) = kx$, then $g$ represents the restoring force of a *linear spring*. (See Figures 1.5 and 1.6.)

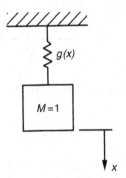

**Figure 1.5.** Mass on a nonlinear spring

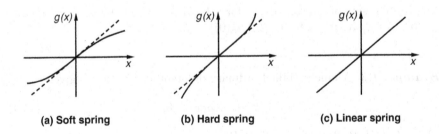

(a) Soft spring          (b) Hard spring          (c) Linear spring

**Figure 1.6.** Mass on a nonlinear spring

For another special case of (1.34), let $f(x) \equiv 0$ and $g(x) = k \sin x$, where $k > 0$ is a parameter. Then (1.34) assumes the form

$$\ddot{x} + k \sin x = 0. \tag{1.37}$$

This equation describes the motion of a point mass moving in a circular path about the axis of rotation normal to a constant gravitational field, as shown in Figure 1.7. The parameter $k$ depends on the radius $l$ of the circular path, the gravitational acceleration $g$, and the mass. The symbol $x$ denotes the angle of deflection measured from the vertical. The present model is called a *simple pendulum*.

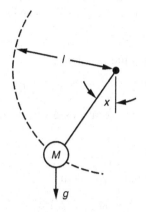

**Figure 1.7.** Model of a simple pendulum

Letting $x_1 = x$ and $x_2 = \dot{x}$, the second-order ordinary differential equation (1.34) can be represented by the system of first-order ordinary differential equations given by

$$\dot{x}_1 = x_2,$$
$$\dot{x}_2 = -f(x_1)x_2 - g(x_1). \tag{1.38}$$

The required initial data for (1.38) are given by $x_1(0)$ and $x_2(0)$.

## 1.5 Solutions of Initial-Value Problems: Existence, Continuation, Uniqueness, and Continuous Dependence on Parameters

The following examples demonstrate that it is necessary to impose restrictions on the right-hand side of equation (1.11) to ensure the existence and uniqueness of solutions of the initial-value problem (1.12).

*Example 1.5.* For the initial-value problem,

$$\dot{x} = g(x), \quad x(0) = 0, \tag{1.39}$$

where $x \in R$, and

$$g(x) = \begin{cases} 1, & x = 0, \\ 0, & x \neq 0, \end{cases}$$

there exists no differentiable function $\phi$ that satisfies (1.39). Hence, *no solution* exists for this initial-value problem (in the sense defined in this chapter).

---

*Example 1.6.* The initial-value problem

$$\dot{x} = x^{1/3}, \quad x(t_0) = 0, \tag{1.40}$$

where $x \in R$, has at least two solutions given by $\phi_1(t) = [\frac{2}{3}(t - t_0)]^{3/2}$ and $\phi_2(t) = 0$ for $t \geq t_0$.

---

*Example 1.7.* The initial-value problem

$$\dot{x} = ax, \quad x(t_0) = x_0, \tag{1.41}$$

where $x \in R$, has a unique solution given by $\phi(t) = e^{a(t-t_0)}x(t_0)$ for $t \geq t_0$.

---

The following result provides a set of sufficient conditions for the *existence of solutions* of initial-value problem (1.12).

**Theorem 1.8.** *Let $f \in C(D, R^n)$. Then for any $(t_0, x_0) \in D$, the initial-value problem (1.12) has a solution defined on $[t_0, t_0 + c)$ for some $c > 0$.* ∎

For a proof of Theorem 1.8, which is called the *Cauchy–Peano Existence Theorem*, refer to [1, Section 1.6].

The next result provides a set of sufficient conditions for the *uniqueness of solutions* for the initial-value problem (1.12).

**Theorem 1.9.** *Let $f \in C(D, R^n)$. Assume that for every compact set $K \subset D$, $f$ satisfies the* Lipschitz condition

$$\| f(t, x) - f(t, y) \| \leq L_K \| x - y \| \tag{1.42}$$

*for all $(t, x), (t, y) \in K$ where $L_K > 0$ is a constant depending only on $K$. Then (1.12) has at most one solution on any interval $[t_0, t_0 + c)$, $c > 0$.* ∎

For a proof of Theorem 1.9, refer to [1, Section 1.8]. In particular, if $f \in C^1(D, R^n)$, then the local Lipschitz condition (1.42) is automatically satisfied.

Now let $\phi$ be a solution of (1.11) on an interval $J$. By a *continuation* or *extension* of $\phi$, we mean an extension $\phi_0$ of $\phi$ to a larger interval $J_0$ in such a way that the extension solves (1.11) on $J_0$. Then $\phi$ is said to be *continued* or *extended* to the larger interval $J_0$. When no such continuation is possible, then $\phi$ is called *noncontinuable*.

***Example 1.10.*** The scalar differential equation

$$\dot{x} = x^2 \tag{1.43}$$

has a solution $\phi(t) = \frac{1}{1-t}$ defined on $J = (-1, 1)$. This solution is continuable to the left to $-\infty$ and is not continuable to the right.

***Example 1.11.*** The differential equation

$$\dot{x} = x^{1/3}, \tag{1.44}$$

where $x \in R$, has a solution $\psi(t) \equiv 0$ on $J = (-\infty, 0)$. This solution is continuable to the right in more than one way. For example, both $\psi_1(t) \equiv 0$ and $\psi_2(t) = (\frac{2t}{3})^{3/2}$ are solutions of (1.44) for $t \geq 0$.

In the next result, $\partial D$ denotes the boundary of a domain $D$ and $\partial J$ denotes the boundary of an interval $J$.

**Theorem 1.12.** *If $f \in C(D, R^n)$ and if $\phi$ is a solution of (1.11) on an open interval $J$, then $\phi$ can be continued to a maximal open interval $J^* \supset J$ in such a way that $(t, \phi(t))$ tends to $\partial D$ as $t \to \partial J^*$ when $\partial D$ is not empty and $|t| + |\phi(t)| \to \infty$ if $\partial D$ is empty. The extended solution $\phi^*$ on $J^*$ is noncontinuable.* ∎

For a proof of Theorem 1.12, refer to [1, Section 1.7].

When $D = J \times R^n$ for some open interval $J$ and $f$ satisfies a Lipschitz condition there (with respect to $x$), we have the following very useful *continuation* result.

**Theorem 1.13.** *Let $f \in C(J \times R^n, R^n)$ for some open interval $J \subset R$ and let $f$ satisfy a Lipschitz condition on $J \times R^n$ (with respect to $x$). Then for any $(t_0, x_0) \in J \times R^n$, the initial-value problem (1.12) has a unique solution that exists on the entire interval $J$.* ∎

For a proof of Theorem 1.13, refer to [1, Section 1.8].

In the next result we address initial-value problems that exhibit dependence on some parameter $\lambda \in G \subset R^m$ given by

$$\dot{x} = f(t, x, \lambda),$$
$$x(\tau) = \xi_\lambda, \tag{1.45}$$

where $f \in C(J \times R^n \times G, R^n)$, $J \subset R$ is an open interval, and $\xi_\lambda$ depends continuously on $\lambda$.

**Theorem 1.14.** *Let $f \in C(J \times R^n \times G, R^n)$, where $J \subset R$ is an open interval and $G \subset R^m$. Assume that for each pair of compact subsets $J_0 \subset J$ and $G_0 \subset G$, there exists a constant $L = L_{J_0, G_0} > 0$ such that for all $(t, \lambda) \in J_0 \times G_0$, $x, y \in R^n$, the Lipschitz condition*

$$\| f(t, x, \lambda) - f(t, y, \lambda) \| \leq L \| x - y \| \tag{1.46}$$

*is true. Then the initial-value problem (1.45) has a unique solution $\phi(t, \tau, \lambda)$, where $\phi \in C(J \times J \times G, R^n)$. Furthermore, if $D$ is a set such that for all $\lambda_0 \in D$ there exists $\epsilon > 0$ such that $\overline{[\lambda_0 - \epsilon, \lambda_0 + \epsilon]} \cap D \subset D$, then $\phi(t, \tau, \lambda) \to \phi(t, \tau_0, \lambda_0)$ uniformly for $t_0 \in J_0$ as $(\tau, \lambda) \to (\tau_0, \lambda_0)$, where $J_0$ is any compact subset of $J$. (Recall that the upper bar denotes closure of a set.)* ■

For a proof of Theorem 1.14, refer to [1, Section 1.9].

Note that Theorem 1.14 applies in the case of Example 1.7 and that the solution $\phi(t)$ of (1.41) depends continuously on the parameter $a$ and the initial conditions $x(t_0) = x_0$.

When Theorem 1.9 is satisfied, it is possible to approximate the unique solutions of the initial-value problem (1.12) arbitrarily closely, using the *method of successive approximations* (also known as *Picard iterations*). Let $f \in C(D, R^n)$, let $K \subset D$ be a compact set, and let $(t_0, x_0) \in K$. *Successive approximations* for (1.12), or equivalently for (1.13), are defined as

$$\phi_0(t) = x_0,$$
$$\phi_{m+1}(t) = x_0 + \int_{t_0}^t f(s, \phi_m(s))ds, \quad m = 0, 1, 2, \ldots \tag{1.47}$$

for $t_0 \leq t \leq t_0 + c$, for some $c > 0$.

**Theorem 1.15.** *If $f \in C(D, R^n)$ and if $f$ is Lipschitz continuous on some compact set $K \subset D$ with constant $L$ (with respect to $x$), then the successive approximations $\phi_m, m = 0, 1, 2, \ldots$ given in (1.47) exist on $[t_0, t_0 + c]$, are continuous there, and converge uniformly, as $m \to \infty$, to the unique solution $\phi$ of (1.12). (Thus, for every $\epsilon > 0$ there exists $N = N(\epsilon)$ such that for all $t \in [t_0, t_0 + c]$, $\| \phi(t) - \phi_m(t) \| < \epsilon$ whenever $m > N(\epsilon)$.)* ■

For the proof of Theorem 1.15, refer to [1, Section 1.8].

## 1.6 Systems of Linear First-Order Ordinary Differential Equations

In this section we will address linear ordinary differential equations of the form

$$\dot{x} = A(t)x + g(t) \tag{1.48}$$

and

$$\dot{x} = A(t)x \tag{1.49}$$

and

$$\dot{x} = Ax + g(t) \tag{1.50}$$

and

$$\dot{x} = Ax, \tag{1.51}$$

where $x \in R^n, A(t) = [a_{ij}(t)] \in C(R, R^{n \times n}), g \in C(R, R^n)$, and $A \in R^{n \times n}$.

Linear equations of the type enumerated above may arise in a natural manner in the modeling process of physical systems (see Section 1.4 for specific examples) or in the process of linearizing equations of the form (1.11) or (1.14) or some other kind of form.

### 1.6.1 Linearization

We consider the system of first-order ordinary differential equations given by

$$\dot{x} = f(t,x), \tag{1.52}$$

where $f : R \times D \to R^n$ and $D \subset R^n$ is some domain.

### Linearization About a Solution $\phi$

If $f \in C^1(R \times D, R^n)$ and if $\phi$ is a given solution of (1.52) defined for all $t \in R$, then we can *linearize* (1.52) about $\phi$ in the following manner. Define $\delta x = x - \phi(t)$ so that

$$
\begin{aligned}
\frac{d(\delta x)}{dt} \triangleq \delta \dot{x} &= f(t,x) - f(t, \phi(t)) \\
&= f(t, \delta x + \phi(t)) - f(t, \phi(t)) \\
&= \frac{\partial f}{\partial x}(t, \phi(t))\delta x + F(t, \delta x), \tag{1.53}
\end{aligned}
$$

where $\frac{\partial f}{\partial x}(t, x)$ denotes the *Jacobian matrix* of $f(t,x) = (f_1(t,x), \ldots, f_n(t,x))^T$ with respect to $x = (x_1, \ldots, x_n)^T$; i.e.,

$$\frac{\partial f}{\partial x}(t,x) = \begin{bmatrix} \frac{\partial f_1}{\partial x_1}(t,x) & \cdots & \frac{\partial f_1}{\partial x_n}(t,x) \\ \vdots & & \vdots \\ \frac{\partial f_n}{\partial x_1}(t,x) & \cdots & \frac{\partial f_n}{\partial x_n}(t,x) \end{bmatrix} \tag{1.54}$$

and

$$F(t, \delta x) \triangleq [f(t, \delta x + \phi(t)) - f(t, \phi(t))] - \frac{\partial f}{\partial x}(t, \phi(t))\delta x. \tag{1.55}$$

It turns out that $F(t, \delta x)$ is $o(\| \delta x \|)$ as $\| \delta x \| \to 0$ uniformly in $t$ on compact subsets of $R$; i.e., for any compact subset $I \subset R$, we have

$$\lim_{\|\delta x\| \to 0} \left( \sup_{t \in I} \frac{\| F(t, \delta x) \|}{\| \delta x \|} \right) = 0.$$

For a proof of this assertion, we refer the reader to [1, Section 1.11]. Letting

$$\frac{\partial f}{\partial x}(t, \phi(t)) = A(t),$$

we obtain from (1.53) the equation

$$\frac{d(\delta x)}{dt} \triangleq \delta \dot{x} = A(t)\delta x + F(t, \delta x). \tag{1.56}$$

Associated with (1.56) we have the linear differential equation

$$\dot{z} = A(t)z, \tag{1.57}$$

called the *linearized equation* of (1.52) about the solution $\phi$.

In applications, the linearization (1.57) of (1.52), about a given solution $\phi$, is frequently used as a means of approximating a nonlinear process by a linear one (in the vicinity of $\phi$). In Chapter 4, where we will study the stability properties of equilibria of (1.52) [which are specific kinds of solutions of (1.52)], we will show under what conditions it makes sense to deduce qualitative properties of a nonlinear process from its linearization.

Of special interest is the case when in (1.52), $f$ is independent of $t$, i.e.,

$$\dot{x} = f(x) \tag{1.58}$$

and $\phi$ is a constant solution of (1.58), say, $\phi(t) = x_0$ for all $t \in R$. Under these conditions we have

$$\frac{d(\delta x)}{dt} \triangleq \delta \dot{x} = A\delta x + F(\delta x), \tag{1.59}$$

where

$$\lim_{\|\delta x\| \to 0} \frac{\| F(\delta x) \|}{\| \delta x \|} = 0 \tag{1.60}$$

and $A$ denotes the Jacobian $\frac{\partial f}{\partial x}(x_0)$. Again, associated with (1.59) we have the linear differential equation

$$\dot{z} = Az,$$

called the *linearized equation* of (1.58) about the solution $\phi(t) \equiv x_0$.

### Linearization About a Solution $\phi$ and an Input $\psi$

We can generalize the above to equations of the form

$$\dot{x} = f(t, x, u), \tag{1.61}$$

where $f : R \times D_1 \times D_2 \to R^n$ and $D_1 \subset R^n, D_2 \subset R^m$ are some domains. If $f \in C^1(R \times D_1 \times D_2, R^n)$ and if $\phi(t)$ is a given solution of (1.61) that we assume to exist for all $t \in R$ and that is determined by the initial condition $x_0$ and the *given specific function* $\psi \in C(R, R^m)$, i.e.,

$$\dot{\phi}(t) = f(t, \phi(t), \psi(t)), \quad t \in R,$$

then we can linearize (1.61) in the following manner. Define $\delta x = x - \phi(t)$ and $\delta u = u - \psi(t)$. Then

$$\begin{aligned}
\frac{d(\delta x)}{dt} = \delta \dot{x} = \dot{x} - \dot{\phi}(t) &= f(t, x, u) - f(t, \phi(t), \psi(t)) \\
&= f(t, \delta x + \phi(t), \delta u + \psi(t)) - f(t, \phi(t), \psi(t)) \\
&= \frac{\partial f}{\partial x}(t, \phi(t), \psi(t))\delta x + \frac{\partial f}{\partial u}(t, \phi(t), \psi(t))\delta u \\
&\quad + F_1(t, \delta x, u) + F_2(t, \delta u),
\end{aligned} \qquad (1.62)$$

where

$$F_1(t, \delta x, u) = f(t, \delta x + \phi(t), u) - f(t, \phi(t), u) - \frac{\partial f}{\partial x}(t, \phi(t), \psi(t))\delta x$$

is $o(\|\delta x\|)$ as $\| \delta x \| \to 0$, uniformly in $t$ on compact subsets of $R$ for fixed $u$ [i.e., for fixed $u$ and for any compact subset $I \subset R$, $\lim\limits_{\|\delta x\| \to 0} \left( \sup_{t \in I} \frac{\|F_1(t, \delta x, u)\|}{\|\delta x\|} \right) = 0$], where

$$F_2(t, \delta u) = f(t, \phi(t), \delta u + \psi(t)) - f(t, \phi(t), \psi(t)) - \frac{\partial f}{\partial u}(t, \phi(t), \psi(t))\delta u$$

is $o(\| \delta u \|)$ as $\| \delta u \| \to 0$, uniformly in $t$ on compact subsets of $R$ [i.e., for any compact subset $I \subset R, \lim_{\|\delta u\| \to 0} \left( \sup_{t \in I} \frac{\|F_2(t, \delta u)\|}{\|\delta u\|} \right) = 0$], and where $\frac{\partial f}{\partial x}(\cdot)$ and $\frac{\partial f}{\partial u}(\cdot)$ denote the Jacobian matrix of $f$ with respect to $x$ and the Jacobian matrix of $f$ with respect to $u$, respectively.

Letting

$$\frac{\partial f}{\partial x}(t, \phi(t), \psi(t)) = A(t) \text{ and } \frac{\partial f}{\partial u}(t, \phi(t), \psi(t)) = B(t),$$

we obtain from (1.62),

$$\frac{d(\delta x)}{dt} = \delta \dot{x} = A(t)\delta x + B(t)\delta u + F_1(t, \delta x, u) + F_2(t, \delta u). \qquad (1.63)$$

Associated with (1.63), we have

$$\dot{z} = A(t)z + B(t)v. \qquad (1.64)$$

We call (1.64) the *linearized equation* of (1.61) about the solution $\phi$ and the input function $\psi$.

As in the case of the linearization of (1.52) by (1.49), the linearization (1.64) of system (1.61) about a given solution $\phi$ and a given input $\psi$ is often used in attempting to capture the qualitative properties of a nonlinear process by a linear process (in the vicinity of $\phi$ and $\psi$). In doing so, great care must be exercised to avoid erroneous conclusions.

The motivation of linearization is of course very obvious: much more is known about linear ordinary differential equations than about nonlinear ones. For example, the explicit forms of the solutions of (1.51) and (1.50) are known; the structures of the solutions of (1.49), (1.48), and (1.64) are known; the qualitative properties of the solutions of linear equations are known; and so forth.

### 1.6.2 Examples

We now consider some specific cases.

---

**Example 1.16.** We consider the *simple pendulum* discussed in Example 1.4 and described by the equation

$$\ddot{x} + k \sin x = 0, \tag{1.65}$$

where $k > 0$ is a constant. Letting $x_1 = x$ and $x_2 = \dot{x}$, (1.65) can be expressed as

$$\dot{x}_1 = x_2,$$
$$\dot{x}_2 = -k \sin x_1. \tag{1.66}$$

It is easily verified that $\phi_1(t) \equiv 0$ and $\phi_2(t) \equiv 0$ is a solution of (1.66). Letting $f_1(x_1, x_2) = x_2$ and $f_2(x_1, x_2) = -k \sin x_1$, the Jacobian of $f(x_1, x_2) = (f_1(x_1, x_2), f_2(x_1, x_2))^T$ evaluated at $(x_1, x_2)^T = (0, 0)^T$ is given by

$$J(0) \triangleq A = \begin{bmatrix} 0 & 1 \\ -k \cos x_1 & 0 \end{bmatrix}_{\begin{bmatrix} x_1 = 0 \\ x_2 = 0 \end{bmatrix}} = \begin{bmatrix} 0 & 1 \\ -k & 0 \end{bmatrix}.$$

The linearized equation of (1.66) about the solution $\phi_1(t) \equiv 0, \phi_2(t) \equiv 0$ is given by

$$\begin{bmatrix} \dot{z}_1 \\ \dot{z}_2 \end{bmatrix} = \begin{bmatrix} 0 & 1 \\ -k & 0 \end{bmatrix} \begin{bmatrix} z_1 \\ z_2 \end{bmatrix}.$$

---

**Example 1.17.** The system of equations

$$\dot{x}_1 = ax_1 - bx_1 x_2 - cx_1^2,$$
$$\dot{x}_2 = dx_2 - ex_1 x_2 - fx_2^2 \tag{1.67}$$

describes the growth of two competing species (e.g., two species of small fish) that prey on each other (e.g., the adult members of one species prey on the young members of the other species, and vice versa). In (1.67) $a, b, c, d, e$, and $f$ are positive parameters and it is assumed that $x_1 \geq 0$ and $x_2 \geq 0$. For (1.67), $\phi_1(t) = \phi_1(t, 0, 0) \equiv 0$ and $\phi_2(t) = \phi_2(t, 0, 0) \equiv 0, t \geq 0$, is a solution of (1.67). A simple computation yields

$$A = \frac{\partial f}{\partial x}(0) = \begin{bmatrix} a & 0 \\ 0 & d \end{bmatrix},$$

and thus the system of equations

$$\begin{bmatrix} \dot{z}_1 \\ \dot{z}_2 \end{bmatrix} = \begin{bmatrix} a & 0 \\ 0 & d \end{bmatrix} \begin{bmatrix} z_1 \\ z_2 \end{bmatrix}$$

constitutes the linearized equation of (1.67) about the solution $\phi_1(t) = 0, \phi_2(t) = 0, t \geq 0$.

---

**Example 1.18.** Consider a unit mass subjected to an inverse square law force field, as depicted in Figure 1.8. In this figure, $r$ denotes radius and $\theta$ denotes angle, and it is assumed that the unit mass (representing, e.g., a satellite) can thrust in the radial and in the tangential directions with thrusts $u_1$ and $u_2$, respectively. The equations that govern this system are given by

$$\ddot{r} = r\dot{\theta}^2 - \frac{k}{r^2} + u_1,$$

$$\ddot{\theta} = \frac{-2\dot{\theta}\dot{r}}{r} + \frac{1}{r}u_2. \qquad (1.68)$$

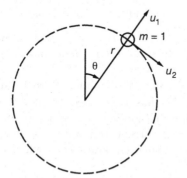

**Figure 1.8.** A unit mass subjected to an inverse square law force field

When $r(0) = r_0, \dot{r}(0) = 0, \theta(0) = \theta_0, \dot{\theta}(0) = \omega_0$, and $u_1(t) \equiv 0, u_2(t) \equiv 0$ for $t \geq 0$, it is easily verified that the system of equations (1.68) has as a solution the circular orbit given by

$$r(t) \equiv r_0 = \text{constant},$$
$$\dot{\theta}(t) = \omega_0 = \text{constant} \tag{1.69}$$

for all $t \geq 0$, which implies that

$$\theta(t) = \omega_0 t + \theta_0, \tag{1.70}$$

where $\omega_0 = (k/r_0^3)^{1/2}$.

If we let $x_1 = r$, $x_2 = \dot{r}$, $x_3 = \theta$, and $x_4 = \dot{\theta}$, the equations of motion (1.68) assume the form

$$\dot{x}_1 = x_2,$$
$$\dot{x}_2 = x_1 x_4^2 - \frac{k}{x_1^2} + u_1,$$
$$\dot{x}_3 = x_4, \tag{1.71}$$
$$\dot{x}_4 = -\frac{2x_2 x_4}{x_1} + \frac{u_2}{x_1}.$$

The linearized equation of (1.71) about the solution (1.70) [with $u_1(t) \equiv 0$, $u_2(t) \equiv 0$] is given by

$$\begin{bmatrix} \dot{z}_1 \\ \dot{z}_2 \\ \dot{z}_3 \\ \dot{z}_4 \end{bmatrix} = \begin{bmatrix} 0 & 1 & 0 & 0 \\ 3\omega_0^2 & 0 & 0 & 2r_0\omega_0 \\ 0 & 0 & 0 & 1 \\ 0 & \frac{-2\omega_0}{r_0} & 0 & 0 \end{bmatrix} \begin{bmatrix} z_1 \\ z_2 \\ z_3 \\ z_4 \end{bmatrix} + \begin{bmatrix} 0 & 0 \\ 1 & 0 \\ 0 & 0 \\ 0 & \frac{1}{r_0} \end{bmatrix} \begin{bmatrix} v_1 \\ v_2 \end{bmatrix}.$$

---

**Example 1.19.** In this example we consider systems described by equations of the form

$$\dot{x} + Af(x) + Bg(x) = u, \tag{1.72}$$

where $x \in R^n$, $A = [a_{ij}] \in R^{n \times n}$, $B = [b_{ij}] \in R^{n \times n}$ with $a_{ii} > 0$, $b_{ii} > 0$, $1 \leq i \leq n$, $f, g \in C^1(R^n, R^n)$, $u \in C(R^+, R^n)$, and $f(x) = 0$, $g(x) = 0$ if and only if $x = 0$.

Equation (1.72) can be used to model a great variety of physical systems. In particular, (1.72) has been used to model a large class of integrated circuits consisting of (nonlinear) transistors and diodes, (linear) capacitors and resistors, and current and voltage sources. (Figure 1.9 gives a symbolic representation of such circuits.) For such circuits, we assume that $f(x) = [f_1(x_1), \ldots, f_n(x_n)]^T$.

If $u(t) = 0$ for all $t \geq 0$, then $\phi_i(t) = 0$, $t \geq 0$, $1 \leq i \leq n$, is a solution of (1.72).

The system of equations (1.72) can be expressed equivalently as

$$\dot{x}_i = -\sum_{j=1}^{n} \left[ a_{ij} \frac{f_j(x_j)}{x_j} + b_{ij} \frac{g_j(x_j)}{x_j} \right] x_j + u_i, \tag{1.73}$$

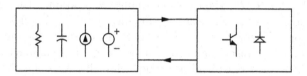

**Figure 1.9.** Integrated circuit

$i = 1, \ldots, n$. The linearized equation of (1.73) about the solution $\phi_i(t) = 0$, and the input $u_i(t) = 0$, $t \geq 0$, $i = 1, \ldots, n$, is given by

$$\dot{z}_i = -\sum_{j=1}^{n} \left[ a_{ij} f_j'(0) + b_{ij} g_j'(0) \right] z_j + v_i, \tag{1.74}$$

where $f_j'(0) = \frac{df_j}{dx_j}(0)$ and $g_j'(0) = \frac{dg_j}{dx_j}(0)$, $i = 1, \ldots, n$.

## 1.7 Linear Systems: Existence, Uniqueness, Continuation, and Continuity with Respect to Parameters of Solutions

In this section we address nonhomogeneous systems of first-order ordinary differential equations given by

$$\dot{x} = A(t)x + g(t), \tag{1.75}$$

where $x \in R^n$, $A(t) = [a_{ij}(t)]$ is a real $n \times n$ matrix and $g$ is a real $n$-vector-valued function.

**Theorem 1.20.** *Suppose that $A \in C(J, R^{n \times n})$ and $g \in C(J, R^n)$, where $J$ is some open interval. Then for any $t_0 \in J$ and any $x_0 \in R^n$, equation (1.75) has a unique solution satisfying $x(t_0) = x_0$. This solution exists on the* entire *interval $J$ and is continuous in $(t, t_0, x_0)$.*

*Proof.* The function $f(t, x) = A(t)x + g(t)$ is continuous in $(t, x)$, and moreover, for any compact subinterval $J_0 \subset J$, there is an $L_0 \geq 0$ such that

$$\| f(t, x) - f(t, y) \|_1 = \| A(t)(x - y) \|_1 \leq \| A(t) \|_1 \| x - y \|_1$$

$$\leq \left( \sum_{i=1}^{n} \max_{1 \leq j \leq n} |a_{ij}(t)| \right) \| x - y \|_1 \leq L_0 \| x - y \|_1$$

for all $(t, x), (t, y) \in J_0 \times R^n$, where $L_0$ is defined in the obvious way. Therefore, $f$ satisfies a Lipschitz condition on $J_0 \times R^n$.

If $(t_0, x_0) \in J_0 \times R^n$, then the continuity of $f$ implies the existence of solutions (Theorem 1.8), whereas the Lipschitz condition implies the uniqueness of solutions (Theorem 1.9). These solutions exist for the entire interval $J_0$ (Theorem 1.13). Since this argument holds for *any* compact subinterval $J_0 \subset J$, the solutions exist and are unique for all $t \in J$. Furthermore, the solutions are continuous with respect to $t_0$ and $x_0$ (Theorem 1.14 modified for the case where $A$ and $g$ do not depend on any parameters $\lambda$).    ∎

For the case when in (1.75) the matrix $A$ and the vector $g$ depend continuously on parameters $\lambda$ and $\mu$, respectively, it is possible to modify Theorem 1.20, and its proof, in the obvious way to show that the unique solutions of the system of equations

$$\dot{x} = A(t, \lambda)x + g(t, \mu) \tag{1.76}$$

are continuous in $\lambda$ and $\mu$ as well. [Assume that $A \in C(J \times R^l, R^{n \times n})$ and $g \in C(J \times R^m, R^n)$ and follow a procedure that is similar to the proof of Theorem 1.20.]

## 1.8 Solutions of Linear State Equations

In this section we determine the specific form of the solutions of systems of linear first-order ordinary differential equations. We will revisit this topic in much greater detail in Chapter 3.

### Homogeneous Equations

We begin by considering linear homogeneous systems

$$\dot{x} = A(t)x, \tag{1.77}$$

where $A \in C(R, R^{n \times n})$. By Theorem 1.20, for every $x_0 \in R^n$, (1.77) has a unique solution that exists for all $t \in R$. We will now use Theorem 1.15 to derive an expression for the solution $\phi(t, t_0, x_0)$ for (1.77) for $t \in R$ with $\phi(t_0, t_0, x_0) = x_0$. In this case the successive approximations given in (1.47) assume the form

$$\phi_0(t, t_0, x_0) = x_0,$$

$$\phi_1(t, t_0, x_0) = x_0 + \int_{t_0}^t A(s)x_0 ds,$$

$$\phi_2(t, t_0, x_0) = x_0 + \int_{t_0}^t A(s)\phi_1(s, t_0, x_0)ds,$$

$$\cdots$$

$$\phi_m(t, t_0, x_0) = x_0 + \int_{t_0}^t A(s)\phi_{m-1}(s, t_0, x_0)ds,$$

or

$$\phi_m(t, t_0, x_0) = x_0 + \int_{t_0}^{t} A(s_1)x_0 ds_1 + \int_{t_0}^{t} A(s_1)\int_{t_0}^{s_1} A(s_2)x_0 ds_2 ds_1 + \cdots$$

$$+ \int_{t_0}^{t} A(s_1)\int_{t_0}^{s_1} A(s_2)\cdots \int_{t_0}^{s_{m-1}} A(s_m)x_0 ds_m \cdots ds_1$$

$$= \left[ I + \int_{t_0}^{t} A(s_1)ds_1 + \int_{t_0}^{t} A(s_1)\int_{t_0}^{s_1} A(s_2)ds_2 ds_1 + \cdots \right.$$

$$\left. + \int_{t_0}^{t} A(s_1)\int_{t_0}^{s_1} A(s_2)\cdots \int_{t_0}^{s_{m-1}} A(s_m)ds_m \cdots ds_1 \right] x_0,$$

$$(1.78)$$

where $I$ denotes the $n \times n$ identity matrix. By Theorem 1.15, the sequence $\{\phi_m\}, m = 0, 1, 2, \ldots$ determined by (1.78) converges uniformly, as $m \to \infty$, to the unique solution $\phi(t, t_0, x_0)$ of (1.77) on compact subsets of $R$. We thus have

$$\phi(t, t_0, x_0) = \Phi(t, t_0)x_0, \qquad (1.79)$$

where

$$\Phi(t, t_0) = I + \int_{t_0}^{t} A(s_1)ds_1 + \int_{t_0}^{t} A(s_1)\int_{t_0}^{s_1} A(s_2)ds_2 ds_1$$

$$+ \int_{t_0}^{t} A(s_1)\int_{t_0}^{s_1} A(s_2)\int_{t_0}^{s_2} A(s_3)ds_3 ds_2 ds_1 + \cdots$$

$$+ \int_{t_0}^{t} A(s_1)\int_{t_0}^{s_1} A(s_2)\ldots \int_{t_0}^{s_{m-1}} A(s_m)ds_m ds_{m-1}\cdots ds_1 + \cdots .$$

$$(1.80)$$

Expression (1.80) is called the *Peano–Baker series*.

From expression (1.80) we immediately note that

$$\Phi(t, t) = I. \qquad (1.81)$$

Furthermore, by differentiating expression (1.80) with respect to time and substituting into (1.77), we obtain that

$$\dot{\Phi}(t, t_0) = A(t)\Phi(t, t_0). \qquad (1.82)$$

From (1.79) it is clear that once the initial data are specified and once the $n \times n$ matrix $\Phi(t, t_0)$ is known, the entire behavior of system (1.77) evolving in time $t$ is known. This has motivated the *state* terminology: $x(t_0) = x_0$ is the *state of the system (1.77) at time* $t_0$, $\phi(t, t_0, x_0)$ is the *state of the system (1.77) at time* $t$, the solution $\phi$ is called the *state vector* of (1.77), the components of $\phi$ are called the *state variables* of (1.77), and the matrix $\Phi(t, t_0)$ that maps $x(t_0)$ into $\phi(t, t_0, x_0)$ is called the *state transition matrix* for (1.77). Also, the vector space containing the state vectors is called the *state space* for (1.77).

We can specialize the preceding discussion to linear systems of equations

$$\dot{x} = Ax. \tag{1.83}$$

In this case the $m$th term in (1.80) assumes the form

$$\int_{t_0}^t A(s_1) \int_{t_0}^{s_1} A(s_2) \int_{t_0}^{s_2} A(s_3) \dots \int_{t_0}^{s_{m-1}} A(s_m) ds_m \cdots ds_1$$

$$= A^m \int_{t_0}^t \int_{t_0}^{s_1} \int_{t_0}^{s_2} \dots \int_{t_0}^{s_{m-1}} 1 ds_m \cdots ds_1 = \frac{A^m(t-t_0)^m}{m!},$$

and expression (1.78) for $\phi_m$ assumes now the form

$$\phi_m(t, t_0, x_0) = \left[ I + \sum_{k=1}^m \frac{A^k(t-t_0)^k}{k!} \right] x_0.$$

We conclude once more from Theorem 1.15 that $\{\phi_m\}$ converges uniformly as $m \to \infty$ to the unique solution $\phi(t, t_0, x_0)$ of (1.83) on compact subsets of $R$. We have

$$\phi(t, t_0, x_0) = \left[ I + \sum_{k=1}^{\infty} \frac{A^k(t-t_0)^k}{k!} \right] x_0$$

$$= \Phi(t, t_0)x_0 \triangleq \Phi(t - t_0)x_0, \tag{1.84}$$

where $\Phi(t - t_0)$ denotes the state transition matrix for (1.83). [Note that by writing $\Phi(t, t_0) = \Phi(t - t_0)$, we have used a slight abuse of notation.] By making the analogy with the scalar $e^a = 1 + \sum_{k=1}^{\infty} \frac{a^k}{k!}$, usage of the notation

$$e^A = I + \sum_{k=1}^{\infty} \frac{A^k}{k!} \tag{1.85}$$

should be clear. We call $e^A$ a *matrix exponential*. In Chapter 3 we will explore several ways of determining $e^A$ for a given $A$.

## Nonhomogeneous Equations

Next, we consider linear nonhomogeneous systems of ordinary differential equations

$$\dot{x} = A(t)x + g(t), \tag{1.86}$$

where $A \in C(R, R^{n \times n})$ and $g \in C(R, R^n)$. Again, by Theorem 1.20, for every $x_0 \in R^n$, (1.86) has a unique solution that exists for all $t \in R$. Instead of *deriving* the complete solution of (1.86) for a given set of initial data

$x(t_0) = x_0$, we will *guess* the solution and verify that it indeed satisfies (1.86). To this end, let us assume that the solution is of the form

$$\phi(t, t_0, x_0) = \Phi(t, t_0)x_0 + \int_{t_0}^{t} \Phi(t, s)g(s)ds, \tag{1.87}$$

where $\Phi(t, t_0)$ denotes the state transition matrix for (1.77).

To show that (1.87) is indeed the solution of (1.86), we first let $t = t_0$. In view of (1.81) and (1.87), we have $\phi(t_0, t_0, x_0) = x_0$. Next, by differentiating both sides of (1.87) and by using (1.81), (1.82), and (1.87), we have

$$\dot{\phi}(t, t_0, x_0) = \dot{\Phi}(t, t_0)x_0 + \Phi(t, t)g(t) + \int_{t_0}^{t} \dot{\Phi}(t, s)g(s)ds$$

$$= A(t)\Phi(t, t_0)x_0 + g(t) + \int_{t_0}^{t} A(t)\Phi(t, s)g(s)ds$$

$$= A(t)[\Phi(t, t_0)x_0 + \int_{t_0}^{t} \Phi(t, s)g(s)ds] + g(t)$$

$$= A(t)\phi(t, t_0, x_0) + g(t);$$

i.e., $\phi(t, t_0, x_0)$ given in (1.87) satisfies (1.86). Therefore, $\phi(t, t_0, x_0)$ is the unique solution of (1.86). Equation (1.87) is called the *variation of constants formula*, which is discussed further in Chapter 3. In the exercise section of Chapter 3 (refer to Exercise 3.13), we ask the reader (with hints) to *derive* the variation of constants formula (1.87), using a *change of variables*.

We note that when $x_0 = 0$, (1.87) reduces to

$$\phi(t, t_0, 0) \triangleq \phi_p(t) = \int_{t_0}^{t} \Phi(t, s)g(s)ds \tag{1.88}$$

and when $x_0 \neq 0$ but $g(t) = 0$ for all $t \in R$, (1.87) reduces to

$$\phi(t, t_0, x_0) \triangleq \phi_h(t) = \Phi(t, t_0)x_0. \tag{1.89}$$

Therefore, the *total solution* of (1.86) may be viewed as consisting of a component that is due to the initial conditions $(t_0, x_0)$ and another component that is due to the *forcing term* $g(t)$. This type of separation is in general possible only in linear systems of differential equations. We call $\phi_p$ a *particular solution* of the nonhomogeneous system (1.86) and $\phi_h$ the *homogeneous solution*.

From (1.87) it is clear that for given initial conditions $x(t_0) = x_0$ and given forcing term $g(t)$, the behavior of system (1.86), summarized by $\phi$, is known for all $t$. Thus, $\phi(t, t_0, x_0)$ specifies the *state vector* of (1.86) at time $t$. The components $\phi_i$ of $\phi$, $i = 1, \ldots, n$, represent the *state variables* for (1.86), and the vector space that contains the state vectors is the *state space* for (1.86).

Before closing this section, it should be pointed out that in applications the matrix $A(t)$ and the vector $g(t)$ in (1.86) may be only *piecewise continuous* rather than continuous, as assumed above [i.e., $A(t)$ and $g(t)$ may have (at most) a finite number of discontinuities over any finite time interval]. In such cases, the derivative of $x$ with respect to $t$ [i.e., the right-hand side in (1.86)] will be discontinuous at a finite number of instants over any finite time interval; however, the state itself, $x$, will still be continuous at these instants [i.e., the solutions of (1.86) will still be continuous over $R$]. In such cases, all the results presented concerning existence, uniqueness, continuation of solutions, and so forth, as well as the explicit expressions of solutions of (1.86), are either still valid or can be modified in the obvious way. For example, should $g(t)$ experience a discontinuity at, say, $t_1 > t_0$, then expression (1.87) will be modified to read as follows:

$$\phi(t, t_0, x_0) = \Phi(t, t_0)x_0 + \int_{t_0}^{t} \Phi(t, s)g(s)ds, \quad t_0 \leq t < t_1, \tag{1.90}$$

$$\phi(t, t_1, x_1) = \Phi(t, t_1)x_1 + \int_{t_1}^{t} \Phi(t, s)g(s)ds, \quad t \geq t_1, \tag{1.91}$$

where $x_1 = \lim_{t \to t_1^-} \phi(t, t_0, x_0)$.

## 1.9 Summary and Highlights

- *Initial-value problem*

$$\dot{x} = f(t, x), \quad x(t_0) = x_0 \tag{1.12}$$

  or

$$\phi(t) = x_0 + \int_{t_0}^{t} f(s, \phi(s))ds, \tag{1.13}$$

  where $\phi(t)$ is a solution of (1.12).
- *Successive approximations*

$$\phi_0(t) = x_0,$$
$$\phi_{m+1}(t) = x_0 + \int_{t_0}^{t} f(s, \phi_m(s))ds, \quad m = 0, 1, 2, \ldots. \tag{1.47}$$

  Under certain conditions (see Theorem 1.15) $\phi_m$, $m = 1, 2$, converges uniformly (on compact sets) as $m \to \infty$ to the unique solution of (1.12).
- *Linearization*
  Given is $\dot{x} = f(t, x)$ and a solution $\phi$. The Jacobian matrix is

$$\frac{\partial f}{\partial x}(t, x) = \begin{bmatrix} \frac{\partial f_1}{\partial x_1}(t, x) & \cdots & \frac{\partial f_1}{\partial x_n}(t, x) \\ \vdots & & \vdots \\ \frac{\partial f_n}{\partial x_1}(t, x) & \cdots & \frac{\partial f_n}{\partial x_n}(t, x) \end{bmatrix}. \tag{1.54}$$

  For $A(t) = \frac{\partial f}{\partial x}(t, \phi(t))$,

$$\dot{z} = A(t)z \tag{1.57}$$

  is the linearized equation about the solution $\phi$.

- *Existence and uniqueness of solutions* of

$$\dot{x} = A(t)x + g(t). \tag{1.75}$$

See Theorem 1.20.
- The *solution* of

$$\dot{x} = A(t)x + g(t), \tag{1.86}$$

with $x(t_0) = x_0$, is given by the variation of constants formula

$$\phi(t, t_0, x_0) = \Phi(t, t_0)x_0 + \int_{t_0}^{t} \Phi(t, s)g(s)ds, \tag{1.87}$$

where the state transition matrix $\Phi(t, t_0)$ is given by

$$\Phi(t, t_0) = I + \int_{t_0}^{t} A(s_1)ds_1 + \int_{t_0}^{t} A(s_1) \int_{t_0}^{s_1} A(s_2)ds_2ds_1 + \cdots \tag{1.80}$$

the Peano–Baker series.
- In the *time-invariant case* $\dot{x} = Ax$,

$$\begin{aligned}
\phi(t, t_0, x_0) &= \left[ I + \sum_{k=1}^{\infty} \frac{A^k(t - t_0)^k}{k!} \right] x_0 \\
&= \Phi(t, t_0)x_0 \triangleq \Phi(t - t_0)x_0 \\
&= e^{A(t-t_0)}x_0,
\end{aligned} \tag{1.84}$$

where

$$e^A = I + \sum_{k=1}^{\infty} \frac{A^k}{k!}. \tag{1.85}$$

## 1.10 Notes

For a classic reference on ordinary differential equations, see Coddington and Levinson [3]. Other excellent sources include Brauer and Nohel [2], Hartman [4], and Simmons [7]. Our treatment of ordinary differential equations in this chapter was greatly influenced by Miller and Michel [6].

## References

1. P.J. Antsaklis and A.N. Michel, *Linear Systems*, Birkhäuser, Boston, MA, 2006.
2. F. Brauer and J.A. Nohel, *Qualitative Theory of Ordinary Differential Equations*, Benjamin, New York, NY, 1969.
3. E.A. Coddington and N. Levinson, *Theory of Ordinary Differential Equations*, McGraw-Hill, New York, NY, 1955.

4. P. Hartman, *Ordinary Differential Equations*, Wiley, New York, NY, 1964.
5. A.N. Michel, K. Wang, and B. Hu, *Qualitative Theory of Dynamical Systems*, 2nd Edition, Marcel Dekker, New York, NY, 2001.
6. R.K. Miller and A.N. Michel, *Ordinary Differential Equations*, Academic Press, New York, NY, 1982.
7. G.F. Simmons, *Differential Equations*, McGraw-Hill, New York, NY, 1972.

## Exercises

**1.1.** (*Hamiltonian dynamical systems*) *Conservative dynamical systems*, also called *Hamiltonian dynamical systems*, are those systems that contain no energy-dissipating elements. Such systems with $n$ degrees of freedom can be characterized by means of a *Hamiltonian function* $H(p,q)$, where $q^T = (q_1, \ldots, q_n)$ denotes $n$ generalized position coordinates and $p^T = (p_1, \ldots, p_n)$ denotes $n$ generalized momentum coordinates. We assume that $H(p,q)$ is of the form

$$H(p,q) = T(q,\dot{q}) + W(q), \tag{1.92}$$

where $T$ denotes the kinetic energy and $W$ denotes the potential energy of the system. These energy terms are obtained from the path-independent line integrals

$$T(q,\dot{q}) = \int_0^{\dot{q}} p(q,\xi)^T d\xi = \int_0^{\dot{q}} \sum_{i=1}^n p_i(q,\xi) d\xi_i, \tag{1.93}$$

$$W(q) = \int_0^q f(\eta)^T d\eta = \int_0^q \sum_{i=1}^n f_i(\eta) d\eta_i, \tag{1.94}$$

where $f_i$, $i = 1, \ldots, n$, denote generalized potential forces.

For the integral (1.93) to be path-independent, it is necessary and sufficient that

$$\frac{\partial p_i(q,\dot{q})}{\partial \dot{q}_j} = \frac{\partial p_j(q,\dot{q})}{\partial \dot{q}_i}, \quad i,j = 1, \ldots, n. \tag{1.95}$$

A similar statement can be made about (1.94).

Conservative dynamical systems are described by the system of $2n$ ordinary differential equations

$$\dot{q}_i = \frac{\partial H}{\partial p_i}(p,q), \quad i = 1, \ldots, n,$$

$$\dot{p}_i = -\frac{\partial H}{\partial q_i}(p,q), \quad i = 1, \ldots, n. \tag{1.96}$$

Note that if we compute the derivative of $H(p,q)$ with respect to $t$ for (1.96) [along the solutions $q_i(t), p_i(t)$, $i = 1, \ldots, n$], then we obtain, by the chain rule,

$$\frac{dH}{dt}(p(t), q(t)) = \sum_{i=1}^{n} \frac{\partial H}{\partial p_i}(p, q)\dot{p}_i + \sum_{i=1}^{n} \frac{\partial H}{\partial q_i}(p, q)\dot{q}_i$$

$$= \sum_{i=1}^{n} -\frac{\partial H}{\partial p_i}(p, q)\frac{\partial H}{\partial q_i}(p, q) + \sum_{i=1}^{n} \frac{\partial H}{\partial q_i}(p, q)\frac{\partial H}{\partial p_i}(p, q)$$

$$= -\sum_{i=1}^{n} \frac{\partial H}{\partial p_i}(p, q)\frac{\partial H}{\partial q_i}(p, q) + \sum_{i=1}^{n} \frac{\partial H}{\partial p_i}(p, q)\frac{\partial H}{\partial q_i}(p, q) \equiv 0.$$

In other words, in a conservative system (1.96), the Hamiltonian, i.e., the total energy, will be constant along the solutions (1.96). This constant is determined by the initial data $(p(0), q(0))$.

(a) In Figure 1.10, $M_1$ and $M_2$ denote point masses; $K_1, K_2, K$ denote spring constants; and $x_1, x_2$ denote the displacements of the masses $M_1$ and $M_2$. Use the Hamiltonian formulation of dynamical systems described above to derive a system of first-order ordinary differential equations that characterize this system. Verify your answer by using Newton's second law of motion to derive the same system of equations. Assume that $x_1(0)$, $\dot{x}_1(0)$, $x_2(0)$, and $\dot{x}_2(0)$ are given.

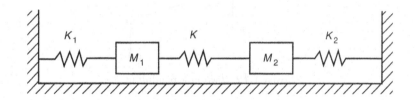

**Figure 1.10.** Example of a conservative dynamical system

(b) In Figure 1.11, a point mass $M$ is moving in a circular path about the axis of rotation normal to a constant gravitational field (this is called the *simple pendulum problem*). Here $l$ is the radius of the circular path, $g$ is the gravitational acceleration, and $\theta$ denotes the angle of deflection measured from the vertical. Use the Hamiltonian formulation of dynamical systems described above to derive a system of first-order ordinary differential equations that characterize this system. Verify your answer by using Newton's second law of motion to derive the same system of equations. Assume that $\theta(0)$ and $\dot{\theta}(0)$ are given.

(c) Determine a system of first-order ordinary differential equations that characterizes the two-link pendulum depicted in Figure 1.12. Assume that $\theta_1(0)$, $\theta_2(0)$, $\dot{\theta}_1(0)$, and $\dot{\theta}_2(0)$ are given.

**1.2.** (*Lagrange's equation*) If a dynamical system contains elements that dissipate energy, such as viscous friction elements in mechanical systems and

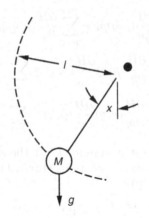

**Figure 1.11.** Simple pendulum

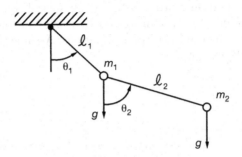

**Figure 1.12.** Two link pendulum

resistors in electric circuits, then we can use *Lagrange's equation* to describe such systems. (In the following, we use some of the same notation used in Exercise 1.1.) For a system with $n$ degrees of freedom, this equation is given by

$$\frac{d}{dt}\left(\frac{\partial L}{\partial \dot{q}_i}(q,\dot{q})\right) - \frac{\partial L}{\partial q}(q,\dot{q}) + \frac{\partial D}{\partial \dot{q}_i}(\dot{q}) = f_i, \quad i = 1,\ldots,n, \qquad (1.97)$$

where $q^T = (q_1,\ldots,q_n)$ denotes the generalized position vector. The function $L(q,\dot{q})$ is called the *Lagrangian* and is defined as

$$L(q,\dot{q}) = T(q,\dot{q}) - W(q),$$

i.e., the difference between the kinetic energy $T$ and the potential energy $W$.

The function $D(\dot{q})$ denotes *Rayleigh's dissipation function*, which we shall assume to be of the form

$$D(\dot{q}) = \frac{1}{2}\sum_{i=1}^{n}\sum_{j=1}^{n}\beta_{ij}\dot{q}_i\dot{q}_j,$$

where $[\beta_{ij}]$ is a positive semidefinite matrix (i.e., $[\beta_{ij}]$ is symmetric and all of its eigenvalues are nonnegative). The dissipation function $D$ represents one-half the rate at which energy is dissipated as heat. It is produced by friction in mechanical systems and by resistance in electric circuits.

Finally, $f_i$ in (1.97) denotes an applied force and includes all external forces associated with the $q_i$ coordinate. The force $f_i$ is defined as being positive when it acts to increase the value of the coordinate $q_i$.

(a) In Figure 1.13, $M_1$ and $M_2$ denote point masses; $K_1, K_2, K$ denote spring constants; $y_1$, $y_2$ denote the displacements of masses $M_1$ and $M_2$, respectively; and $B_1$, $B_2$, $B$ denote viscous damping coefficients. Use the Lagrange formulation of dynamical systems described above to derive two second-order differential equations that characterize this system. Transform these equations into a system of first-order ordinary differential equations. Verify your answer by using Newton's second law of motion to derive the same system equations. Assume that $y_1(0)$, $\dot{y}(0)$, $y_2(0)$, and $\dot{y}(0)$ are given.

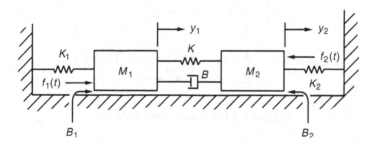

**Figure 1.13.** An example of a mechanical system with energy dissipation

(b) Consider the capacitor microphone depicted in Figure 1.14. Here we have a capacitor constructed from a fixed plate and a moving plate with mass $M$. The moving plate is suspended from the fixed frame by a spring with a spring constant $K$ and has some damping expressed by the damping constant $B$. Sound waves exert an external force $f(t)$ on the moving plate. The output voltage $v_s$, which appears across the resistor $R$, will reproduce electrically the sound-wave patterns that strike the moving plate.

When $f(t) \equiv 0$ there is a charge $q_0$ on the capacitor. This produces a force of attraction between the plates that stretches the spring. When sound waves exert a force on the moving plate, there will be a resulting motion displacement $x$ that is measured from the equilibrium position. The distance between the plates will then be $x_0 - x$, and the charge on the plates will be $q_0 + q$.

When displacements are small, the expression for the capacitance is given approximately by

$$C = \frac{\epsilon A}{x_0 - x}$$

with $C_0 = \epsilon A/x_0$, where $\epsilon > 0$ is the dielectric constant for air and $A$ is the area of the plate.

Use the Lagrange formulation of dynamical systems to derive two second-order ordinary differential equations that characterize this system. Transform these equations into a system of first-order ordinary differential equations. Verify your answer by using Newton's laws of motion and Kirchhoff's voltage/current laws. Assume that $x(0), \dot{x}(0), q(0)$, and $\dot{q}(0)$ are given.

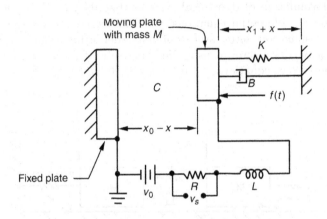

**Figure 1.14.** Capacitor microphone

(c) Use the Lagrange formulation to derive a system of first-order differential equations for the system given in Example 1.3.

**1.3.** Find examples of initial-value problems for which (a) no solutions exist; (b) more than one solution exists; (c) one or more solutions exist, but cannot be continued for all $t \in R$; and (d) unique solutions exist for all $t \in R$.

**1.4.** (*Numerical solution of ordinary differential equations—Euler's method*) An approximation to the solution of the *scalar* initial-value problem

$$\dot{y} = f(t, y), \quad y(t_0) = y_0 \tag{1.98}$$

is given by *Euler's method*,

$$y_{k+1} = y_k + h f(t_k, y_k), \quad k = 0, 1, 2, \ldots, \tag{1.99}$$

where $h = t_{k+1} - t_k$ is the (constant) integration step. The interpretation of this method is that the area below the solution curve is approximated by a sequence of sums of rectangular areas. This method is also called the *forward rectangular rule* (of integration).

(a) Use Euler's method to determine the solution of the initial-value problem

$$\dot{y} = 3y, \quad y(t_0) = 5, \quad t_0 = 0, \quad t_0 \le t \le 10.$$

(b) Use Euler's method to determine the solution of the initial-value problem

$$\ddot{y} = t(\dot{y})^2 - y^2, \quad y(t_0) = 1, \quad \dot{y}(t_0) = 0, \quad t_0 = 0, t_0 \le t \le 10.$$

*Hint:* In both cases, use $h = 0.2$. For part (b), let $y = x_1$, $\dot{x}_1 = x_2$, $\dot{x}_2 = tx_2^2 - x_1^2$, and apply (1.99), appropriately adjusted to the vector case. In both cases, plot $y_k$ vs. $t_k, k = 0, 1, 2, \ldots$.
*Remark:.* Euler's method yields arbitrarily close approximations to the solutions of (1.98), by making $h$ sufficiently small, *assuming infinite (computer) word length*. In practice, however, where truncation errors (quantization) and round-off errors (finite precision operations) are a reality, extremely small values of $h$ may lead to numerical instabilities. Therefore, Euler's method is of limited value as a means of solving initial-value problems numerically.

**1.5.** (*Numerical solution of ordinary differential equations—Runge–Kutta methods*) The Runge–Kutta family of integration methods are among the most widely used techniques to solve initial-value problems (1.98). A simple version is given by

$$y_{i+1} = y_i + k,$$

where

$$k = \frac{1}{6}(k_1 + 2k_2 + 2k_3 + k_4)$$

with

$$k_1 = hf(t_i, y_i),$$
$$k_2 = hf(t_i + \frac{1}{2}h, y_i + \frac{1}{2}k_1),$$
$$k_3 = hf(t_i + \frac{1}{2}h, y_i + \frac{1}{2}k_2),$$
$$k_4 = hf(t_i + h, y_i + k_3),$$

and $t_{i+1} = t_i + h, y(t_0) = y_0$.

The idea of this method is to probe ahead (in time) by one-half or by a whole step $h$ to determine the values of the derivative at several points, and then to form a weighted average.

Runge–Kutta methods can also be applied to higher order ordinary differential equations. For example, after a change of variables, suppose that a second-order differential equation has been changed to a system of two first-order differential equations, say,

$$\begin{align} \dot{x}_1 &= f_1(t, x_1, x_2), \quad x_1(t_0) = x_{10}, \\ \dot{x}_2 &= f_2(t, x_1, x_2), \quad x_2(t_0) = x_{20}. \end{align} \tag{1.100}$$

In solving (1.100), a simple version of the Runge–Kutta method is given by

$$y_{i+1} = y_i + \underline{k},$$

where

$$y_i = (x_{1i}, x_{2i})^T \text{ and } \underline{k} = (k, l)^T$$

with

$$k = \frac{1}{6}(k_1 + 2k_2 + 2k_3 + k_4), \quad l = \frac{1}{6}(l_1 + 2l_2 + 2l_3 + l_4)$$

and

$$k_1 = hf_1(t_i, x_{1i}, x_{2i}), \qquad\qquad\qquad l_1 = hf_2(t_i, x_{1i}, x_{2i}),$$
$$k_2 = hf_1(t_i + \tfrac{1}{2}h, x_{1i} + \tfrac{1}{2}k_1, x_{2i} + \tfrac{1}{2}l_1), \; l_2 = hf_2(t_i + \tfrac{1}{2}h, x_{1i} + \tfrac{1}{2}k_1, x_{2i} + \tfrac{1}{2}l_1),$$
$$k_3 = hf_1(t_i + \tfrac{1}{2}h, x_{1i} + \tfrac{1}{2}k_2, x_{2i} + \tfrac{1}{2}l_2), \; l_3 = hf_2(t_i + \tfrac{1}{2}h, x_{1i} + \tfrac{1}{2}k_2, x_{2i} + \tfrac{1}{2}l_2),$$
$$k_4 = hf_1(t_i + h, x_{1i} + k_3, x_{2i} + l_3), \qquad l_4 = hf_2(t_i + h, x_{1i} + k_3, x_{2i} + l_3).$$

Use the Runge–Kutta method described above to obtain numerical solutions to the initial-value problems given in parts (a) and (b) of Exercise 1.4. Plot your data.

*Remark:* Since Runge–Kutta methods do not use past information, they constitute attractive starting methods for more efficient numerical integration schemes (e.g., predictor–corrector methods) . We note that since there are no built-in accuracy measures in the Runge–Kutta methods, significant computational efforts are frequently expended to achieve a desired accuracy.

**1.6.** (*Numerical solution of ordinary differential equations—Predictor–Corrector methods*) A common predictor–corrector technique for solving initial-value problems determined by ordinary differential equations, such as (1.98), is the *Milne* method, which we now summarize. In this method, $\dot{y}_{i-1}$ denotes the value of the first derivative at time $t_{i-1}$, where $t_i$ is the time for the $i$th iteration step, $\dot{y}_{i-2}$ is similarly defined, and $y_{i+1}$ represents the value of $y$ to be determined. The details of the Milne method are:

1. $y_{i+1,p} = y_{i-3} + \frac{4h}{3}(2\dot{y}_{i-2} - \dot{y}_{i-1} + 2\dot{y}_i)$      (*predictor*)
2. $\dot{y}_{i+1,p} = f(t_{i+1}, y_{i+1,p})$
3. $y_{i+1,c} = y_{i-1} + \frac{h}{3}(\dot{y}_{i-1} + 4\dot{y}_i + \dot{y}_{i+1,p})$      (*corrector*)
4. $\dot{y}_{i+1,c} = f(t_{i+1}, y_{i+1,c})$
5. $y_{i+1,c} = y_{i-1} + \frac{h}{3}(\dot{y}_{i-1} + 4\dot{y}_i + \dot{y}_{i+1,c})$      (*iterating corrector*)

The first step is to obtain a predicted value of $y_{i+1}$ and then to substitute $y_{i+1,p}$ into the given differential equation to obtain a predicted value of $\dot{y}_{i+1,p}$, as indicated in the second equation above. This predicted value, $\dot{y}_{i+1,p}$ is then used in the second equation, the corrector equation, to obtain a corrected value of $y_{i+1}$. The corrected value, $y_{i+1,c}$ is next substituted into the differential equation to obtain an improved value of $\dot{y}_{i+1}$, and so on. If necessary, an iteration process involving the fourth and fifth equations continues until successive values of $y_{i+1}$ differ by less than the value of some desirable tolerance. With $y_{i+1}$ determined to the desired accuracy, the method steps forward one $h$ increment.

A more complicated predictor–corrector method that is more reliable than the Milne method is the *Adams–Bashforth–Moulton* method, the essential equations of which are

$$y_{i+1,p} = y_i + \frac{h}{24}(55\dot{y}_i - 59\dot{y}_{i-1} + 37\dot{y}_{i-2} - 9\dot{y}_{i-3}),$$

$$y_{i+1,c} = y_i + \frac{h}{24}(9\dot{y}_{i+1} + 19\dot{y}_i - 5\dot{y}_{i-1} + \dot{y}_{i-2}),$$

where in the corrector equation, $\dot{y}_{i+1}$ denotes the predicted value.

The application of predictor–corrector methods to systems of first-order ordinary differential equations is straightforward. For example, the application of the Milne method to the second-order system in (1.100) yields from the predictor step

$$x_{k,i+1,p} = x_{k,i-3} + \frac{4h}{3}(2\dot{x}_{k,i-2} - \dot{x}_{k,i-1} + 2\dot{x}_{k,i}), \quad k = 1, 2.$$

Then

$$\dot{x}_{k,i+1,p} = f_k(t_{i+1}, x_{1,i+1,p}, x_{2,i+1,p}), \quad k = 1, 2,$$

and the corrector step assumes the form

$$x_{k,i+1,c} = x_{k,i-1} + \frac{h}{3}(\dot{x}_{k,i-1} + 4\dot{x}_{k,i} + \dot{x}_{k,i+1}), \quad k = 1, 2,$$

and

$$\dot{x}_{k,i+1,c} = f_k(t_{i+1}, x_{1,i+1,c}, x_{2,i+1,c}), \quad k = 1, 2.$$

Use the Milne method and the Adams–Bashforth–Moulton method described above to obtain numerical solutions to the initial-value problems given in parts (a) and (b) of Exercise 1.4. To initiate the algorithm, refer to the Remark in Exercise 1.5.

*Remark.* Derivations and convergence properties of numerical integration schemes, such as those discussed here and in Exercises 1.4 and 1.5, can be found in many of the standard texts on numerical analysis.

**1.7.** Use Theorem 1.15 to solve the initial-value problem $\dot{x} = ax + t$, $x(0) = x_0$ for $t \geq 0$. Here $a \in R$.

**1.8.** Consider the initial-value problem

$$\dot{x} = Ax, \quad x(0) = x_0, \tag{1.101}$$

where $x \in R^2$ and $A \in R^{2 \times 2}$. Let $\lambda_1, \lambda_2$ denote the eigenvalues of $A$; i.e., $\lambda_1$ and $\lambda_2$ are the roots of the equation $\det(A - \lambda I) = 0$, where det denotes determinant, $\lambda$ is a scalar, and $I$ denotes the $2 \times 2$ identity matrix. Make specific choices of $A$ to obtain the following cases:

1. $\lambda_1 > 0, \lambda_2 > 0$, and $\lambda_1 \neq \lambda_2$
2. $\lambda_1 < 0, \lambda_2 < 0$, and $\lambda_1 \neq \lambda_2$
3. $\lambda_1 = \lambda_2 > 0$
4. $\lambda_1 = \lambda_2 < 0$
5. $\lambda_1 > 0, \lambda_2 < 0$
6. $\lambda_1 = \alpha + i\beta, \lambda_2 = \alpha - i\beta, i = \sqrt{-1}, \alpha > 0$
7. $\lambda_1 = \alpha + i\beta, \lambda_2 = \alpha - i\beta, \alpha < 0$
8. $\lambda_1 = i\beta, \lambda_2 = -i\beta$

Using $t$ as a parameter, plot $\phi_2(t, 0, x_0)$ vs. $\phi_1(t, 0, x_0)$ for $0 \leq t \leq t_f$ for every case enumerated above. Here $[\phi_1(t, t_0, x_0), \phi_2(t, t_0, x_0)]^T = \phi(t, t_0, x_0)$ denotes the solution of (1.101). On your plots, indicate increasing time $t$ by means of arrows. Plots of this type are called *trajectories* for (1.101), and sufficiently many plots (using different initial conditions and sufficiently large $t_f$) make up a *phase portrait* for (1.101). Generate a phase portrait for each case given above.

**1.9.** Write two first-order ordinary differential equations for the *van der Pol Equation* (1.35) by choosing $x_1 = x$ and $x_2 = \dot{x}_1$. Determine by simulation *phase portraits* (see Exercise 1.8) for this example for the cases $\epsilon = 0.05$ and $\epsilon = 10$ (refer also to Exercises 1.5 and 1.6 for numerical methods for solving differential equations). The periodic solution to which the trajectories of (1.35) tend to is an example of a *limit cycle*.

**1.10.** Consider a system whose state-space description is given by

$$\dot{x} = -k_1 k_2 \sqrt{x} + k_2 u(t),$$
$$y = k_1 \sqrt{x}.$$

Linearize this system about the nominal solution

$$u_0 \equiv 0, \quad 2\sqrt{x_0(t)} = 2\sqrt{k} - k_1 k_2 t,$$

where $x_0(0) = k$.

**1.11.** For (1.36) consider the *hard, linear, and soft spring models* given by

$$g(x) = k(1 + a^2 x^2)x,$$
$$g(x) = kx,$$
$$g(x) = k(1 - a^2 x^2)x,$$

respectively, where $k > 0$ and $a^2 > 0$. Write two first-order ordinary differential equations for (1.36) by choosing $x_1 = x$ and $x_2 = \dot{x}$. Pick specific values for $k$ and $a^2$. Determine by simulation *phase portraits* (see Exercise 1.8) for this example for the above three cases.

**1.12.** (a) Show that $x^T = (0,0)$ is a solution of the system of equations

$$\dot{x}_1 = x_1^2 + x_2^2 + x_2 \cos x_1,$$
$$\dot{x}_2 = (1 + x_1)x_1 + (1 + x_2)x_2 + x_1 \sin x_2.$$

Linearize this system about the point $x^T = (0,0)$. By means of computer simulations, compare solutions corresponding to different initial conditions in the vicinity of the origin of the above system of equations and its linearization.

(b) Linearize the (bilinear control) system

$$\ddot{x} + (3 + \dot{x}^2)\dot{x} + (1 + x + x^2)u = 0$$

about the solution $x = 0, \dot{x} = 0$, and the input $u(t) \equiv 0$. As in part (a), compare (by means of computer simulations) solutions of the above equation with corresponding solutions of its linearization.

(c) In the circuit given in Figure 1.15, $v_i(t)$ is a voltage source and the nonlinear resistor obeys the relation $i_R = 1.5v_R^3$ [$v_i(t)$ is the circuit input and $v_R(t)$ is the circuit output]. Derive the differential equation for this circuit. Linearize this differential equation for the case when the circuit operates about the point $v_i = 14$.

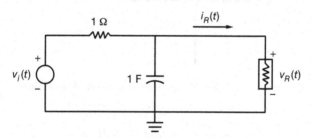

**Figure 1.15.** Nonlinear circuit

**1.13.** (*Inverted pendulum*) The inverted pendulum on a moving carriage subjected to an external force $\mu(t)$ is depicted in Figure 1.16.

The moment of inertia with respect to the center of gravity is $J$, and the coefficient of friction of the carriage (see Figure 1.16) is $F$. From Figure 1.17 we obtain the following equations for the dynamics of this system

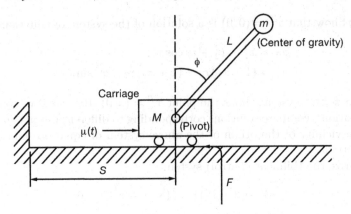

**Figure 1.16.** Inverted pendulum

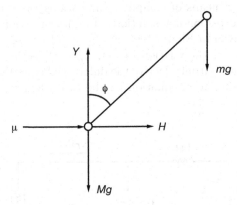

**Figure 1.17.** Force diagram of the inverted pendulum

$$m\frac{d^2}{dt^2}(S + L\sin\phi) \triangleq H, \tag{1.102a}$$

$$m\frac{d^2}{dt^2}(L\cos\phi) \triangleq Y - mg, \tag{1.102b}$$

$$J\frac{d^2\phi}{dt^2} = LY\sin\phi - LH\cos\phi, \tag{1.102c}$$

$$M\frac{d^2S}{dt^2} = \mu(t) - H - F\frac{dS}{dt}. \tag{1.102d}$$

Assuming that $m << M$, (1.102d) reduces to

$$M\frac{d^2S}{dt^2} = \mu(t) - F\frac{dS}{dt}. \tag{1.102e}$$

Eliminating $H$ and $Y$ from (1.102a) to (1.102c), we obtain

$$(J + mL^2)\ddot{\phi} = mgL\sin\phi - mL\ddot{S}\cos\phi. \tag{1.102f}$$

Thus, the system of Figure 1.16 is described by the equations

$$\ddot{\phi} - (g/L')\sin\phi + (1/L')\ddot{S}\cos\phi = 0,$$
$$M\ddot{S} + F\dot{S} = \mu(t), \tag{1.102g}$$

where

$$L' = \frac{J + mL^2}{mL}$$

denotes the effective pendulum length.

Linearize system (1.102g) about $\phi = 0$.

**1.14.** (*Simple pendulum*) A system of first-order ordinary differential equations that characterize the simple pendulum considered in Exercise 1.1b is given by

$$\begin{bmatrix} \dot{x}_1 \\ \dot{x}_2 \end{bmatrix} = \begin{bmatrix} x_2 \\ -\frac{g}{l}\sin x_1 \end{bmatrix},$$

where $x_1 \triangleq \theta$ and $x_2 \triangleq \dot{\theta}$ with $x_1(0) = \theta(0)$ and $x_2(0) = \dot{\theta}(0)$ specified. A linearized model of this system about the solution $x = [0,0]^T$ is given by

$$\begin{bmatrix} \dot{x}_1 \\ \dot{x}_2 \end{bmatrix} = \begin{bmatrix} 0 & 1 \\ -\frac{g}{l} & 0 \end{bmatrix} \begin{bmatrix} x_1 \\ x_2 \end{bmatrix}.$$

Let $g = 10\ (m/sec^2)$ and $l = 1\ (m)$.

(a) For the case when $x(0) = [\theta_0, 0]^T$ with $\theta_0 = \pi/18,\ \pi/12,\ \pi/6$, and $\pi/3$, plot the states for $t \geq 0$ for the nonlinear model.

(b) Repeat (a) for the linear model.

(c) Compare the results in (a) and (b).

# 2

## An Introduction to State-Space and Input–Output Descriptions of Systems

## 2.1 Introduction

State-space representations provide detailed descriptions of the internal behavior of a system, whereas input–output descriptions of systems emphasize external behavior and a system's interaction with this behavior.

In this chapter we address the state-space description of systems, which is an internal description of systems, and the input–output description of systems, also called the external description of systems. We will address continuous-time systems described by ordinary differential equations and discrete-time systems described by ordinary difference equations. We will emphasize linear systems. For such systems, the input–output descriptions involve the convolution integral for the continuous-time case and the convolution sum for the discrete-time case.

This chapter is organized into three parts. In the first of these (Section 2.2), we develop the state-space description of continuous-time systems, whereas in the second part (Section 2.3), we present the state-space representation of discrete-time systems. In the third part (Section 2.4), we address the input–output description of both continuous-time and discrete-time systems. Required background material for this chapter includes certain essentials in ordinary differential equations and linear algebra. This material can be found in Chapter 1 and the appendix, respectively.

## 2.2 State-Space Description of Continuous-Time Systems

Let us consider once more systems described by equations of the form

$$\dot{x} = f(t, x, u), \qquad (2.1a)$$

$$y = g(t, x, u), \qquad (2.1b)$$

where $x \in R^n, y \in R^p, u \in R^m, f : R \times R^n \times R^m \to R^n$, and $g : R \times R^n \times R^m \to R^p$. Here $t$ denotes time and $u$ and $y$ denote system *input* and system *output*, respectively. Equation (2.1a) is called the *state equation*, (2.1b) is called the *output equation*, and (2.1a) and (2.1b) constitute the *state-space description* of continuous-time finite-dimensional systems.

The system input may be a function of $t$ only (i.e., $u : R \to R^m$), or as in the case of *feedback control systems*, it may be a function of $t$ and $x$ (i.e., $u : R \times R^n \to R^m$). In either case, for a *given* (i.e., specified) $u$, we let $f(t, x, u) = F(t, x)$ and rewrite (2.1a) as

$$\dot{x} = F(t, x). \tag{2.2}$$

Now according to Theorems 1.13 and 1.14, if $F \in C(R \times R^n, R^n)$ and if for any compact subinterval $J_0 \subset R$ there is a constant $L_{J_0}$ such that $\| F(t, x) - F(t, \tilde{x}) \| \leq L_{J_0} \| x - \tilde{x} \|$ for all $t \in J_0$ and for all $x, \tilde{x} \in R^n$, then the following statements are true:

1. For any $(t_0, x_0) \in R \times R^n$, (2.2) has a unique solution $\phi(t, t_0, x_0)$ satisfying $\phi(t_0, t_0, x_0) = x_0$ that exists for all $t \in R$.
2. The solution $\phi$ is continuous in $t, t_0$, and $x_0$.
3. If $F$ depends continuously on parameters (say, $\lambda \in R^l$) and if $x_0$ depends continuously on $\lambda$, the solution $\phi$ is continuous in $\lambda$ as well.

Thus, if the above conditions are satisfied, then for a given $t_0, x_0$, and $u$, (2.1a) will have a unique solution that exists for $t \in R$. Therefore, as already discussed in Section 1.8, $\phi(t, t_0, x_0)$ characterizes the *state* of the system at time $t$. Moreover, under these conditions, the system will have a unique *response* for $t \in R$, determined by (2.1b). We usually assume that $g \in C(R \times R^n \times R^m, R^p)$ or that $g \in C^1(R \times R^n \times R^m, R^p)$.

An important special case of (2.1) is systems described by linear time-varying equations of the form

$$\dot{x} = A(t)x + B(t)u, \tag{2.3a}$$
$$y = C(t)x + D(t)u, \tag{2.3b}$$

where $A \in C(R, R^{n \times n}), B \in C(R, R^{n \times m}), C \in C(R, R^{p \times n})$, and $D \in C(R, R^{p \times m})$. Such equations may arise in the modeling process of a physical system, or they may be a consequence of a linearization process, as discussed in Section 1.6.

By applying the results of Section 1.7, we see that for every initial condition $x(t_0) = x_0$ and for every given input $u : R \to R^m$, system (2.3a) possesses a unique solution that exists for all $t \in R$ and that is continuous in $(t, t_0, x_0)$. Moreover, if $A$ and $B$ depend continuously on parameters, say, $\lambda \in R^l$, then the solutions will be continuous in the parameters as well. Indeed, in accordance with (1.87), this solution is given by

$$\phi(t, t_0, x_0) = \Phi(t, t_0)x_0 + \int_{t_0}^{t} \Phi(t, s)B(s)u(s)ds, \tag{2.4}$$

where $\Phi(t, t_0)$ denotes the state transition matrix of the system of equations

$$\dot{x} = A(t)x. \tag{2.5}$$

By using (2.3b) and (2.4) we obtain the *system response* as

$$y(t) = C(t)\Phi(t, t_0)x_0 + C(t) \int_{t_0}^{t} \Phi(t, s)B(s)u(s)ds + D(t)u(t). \tag{2.6}$$

When in (2.3), $A(t) \equiv A, B(t) \equiv B, C(t) \equiv C$, and $D(t) \equiv D$, we have the important linear time-invariant case given by

$$\dot{x} = Ax + Bu, \tag{2.7a}$$
$$y = Cx + Du. \tag{2.7b}$$

In accordance with (1.84), (1.85), (1.87), and (2.4), the solution of (2.7a) is given by

$$\phi(t, t_0, x_0) = e^{A(t-t_0)}x_0 + \int_{t_0}^{t} e^{A(t-s)}Bu(s)ds \tag{2.8}$$

and the response of the system is given by

$$y(t) = Ce^{A(t-t_0)}x_0 + C \int_{t_0}^{t} e^{A(t-s)}Bu(s)ds + Du(t). \tag{2.9}$$

**Linearity**

We have referred to systems described by the linear equations (2.3) [resp., (2.7)] as *linear systems*. In the following discussion, we establish precisely in what sense this linearity is to be understood. To this end, for (2.3) we first let $y_1$ and $y_2$ denote system outputs that correspond to system inputs given by $u_1$ and $u_2$, respectively, *under the condition that* $x_0 = 0$. By invoking (2.6), it is clear that the system output corresponding to the system input $u = \alpha_1 u_1 + \alpha_2 u_2$, where $\alpha_1$ and $\alpha_2$ are real scalars, is given by $y = \alpha_1 y_1 + \alpha_2 y_2$; i.e.,

$$y(t) = C(t) \int_{t_0}^{t} \Phi(t, s)B(s)[\alpha_1 u_1(s) + \alpha_2 u_2(s)]ds + D(t)[\alpha_1 u_1(t) + \alpha_2 u_2(t)]$$

$$= \alpha_1 C(t) \int_{t_0}^{t} \Phi(t, s)B(s)u_1(s)ds + \alpha_2 C(t) \int_{t_0}^{t} \Phi(t, s)B(s)u_2(s)ds$$

$$+ \alpha_1 D(t)u_1(t) + \alpha_2 D(t)u_2(t)$$

$$= \alpha_1 y_1(t) + \alpha_2 y_2(t). \tag{2.10}$$

Next, for (2.3) we let $y_1$ and $y_2$ denote system outputs that correspond to initial conditions $x_0^{(1)}$ and $x_0^{(2)}$, respectively, *under the condition that* $u(t) = 0$

*for all* $t \in R$. Again, by invoking (2.6), it is clear that the system output corresponding to the initial condition $x_0 = \alpha_1 x_0^{(1)} + \alpha_2 x_0^{(2)}$, where $\alpha_1$ and $\alpha_2$ are real scalars, is given by $y = \alpha_1 y_1 + \alpha_2 y_2$; i.e.,

$$
\begin{aligned}
y(t) &= C(t)\Phi(t,t_0)[\alpha_1 x_0^{(1)} + \alpha_2 x_0^{(2)}] \\
&= \alpha_1 C(t)\Phi(t,t_0)x_0^{(1)} + \alpha_2 C(t)\Phi(t,t_0)x_0^{(2)} \\
&= \alpha_1 y_1(t) + \alpha_2 y_2(t).
\end{aligned} \tag{2.11}
$$

Equations (2.10) and (2.11) show that for systems described by the linear equations (2.3) [and, hence, by (2.7)], a *superposition principle* holds in terms of the input $u$ and the corresponding output $y$ of the system under the assumption of zero initial conditions, and in terms of the initial conditions $x_0$ and the corresponding output $y$ under the assumption of zero input. It is important to note, however, that such a superposition principle will in general not hold under conditions that combine nontrivial inputs and nontrivial initial conditions. For example, with $x_0 \neq 0$ given, and with inputs $u_1$ and $u_2$ resulting in corresponding outputs $y_1$ and $y_2$ in (2.3), it does not follow that the input $\alpha_1 u_1 + \alpha_2 u_2$ will result in an output $\alpha_1 y_1 + \alpha_2 y_2$.

## 2.3 State-Space Description of Discrete-Time Systems

### State-Space Representation

The state-space description of discrete-time finite-dimensional dynamical systems is given by equations of the form

$$
x_i(k+1) = f_i(k, x_1(k), \ldots, x_n(k), u_1(k), \ldots, u_m(k)) \quad i = 1, \ldots, n, \tag{2.12a}
$$
$$
y_i(k) = g_i(k, x_1(k), \ldots, x_n(k), u_1(k), \ldots, u_m(k)) \quad i = 1, \ldots, p, \tag{2.12b}
$$

for $k = k_0, k_0+1, \ldots$, where $k_0$ is an integer. (In the following discussion, we let $Z$ denote the set of integers and we let $Z^+$ denote the set of nonnegative integers.) Letting $x(k)^T = (x_1(k), \ldots, x_n(k)), f(\cdot)^T = (f_1(\cdot), \ldots, f_n(\cdot)), u(k)^T = (u_1(k), \ldots, u_m(k)), y(k)^T = (y_1(k), \ldots, y_p(k))$, and $g(\cdot)^T = (g_1(\cdot), \ldots, g_m(\cdot))$, we can rewrite (2.12) more compactly as

$$
x(k+1) = f(k, x(k), u(k)), \tag{2.13a}
$$
$$
y(k) = g(k, x(k), u(k)). \tag{2.13b}
$$

Throughout this section we will assume that $f : Z \times R^n \times R^m \to R^n$ and $g : Z \times R^n \times R^m \to R^p$.

Since $f$ is a function, for given $k_0, x(k_0) = x_0$, and for given $u(k), k = k_0, k_0 + 1, \ldots$, (2.13a) possesses a unique solution $x(k)$ that exists for all $k = k_0, k_0 + 1, \ldots$. Furthermore, under these conditions, $y(k)$ is uniquely defined for $k = k_0, k_0 + 1, \ldots$.

As in the case of continuous-time finite-dimensional systems [see (2.1)], $k_0$ denotes *initial time*, $k$ denotes *time*, $u(k)$ denotes the system *input* (evaluated at time $k$), $y(k)$ denotes the system *output* or system *response* (evaluated at time $k$), $x(k)$ characterizes the *state* (evaluated at time $k$), $x_i(k)$, $i = 1, \ldots, n$, denote the *state variables*, (2.13a) is called the *state equation*, and (2.13b) is called the *output equation*.

A moment's reflection should make it clear that in the case of discrete-time finite-dimensional dynamical systems described by (2.13), questions concerning existence, uniqueness, and continuation of solutions are not an issue, as was the case in continuous-time systems. Furthermore, continuity with respect to initial data $x(k_0) = x_0$, or with respect to system parameters, is not an issue either, provided that $f(\cdot)$ and $g(\cdot)$ have appropriate continuity properties.

In the case of continuous-time systems described by ordinary differential equations [see (2.1)], we allow time $t$ to evolve "forward" and "backward." Note, however, that in the case of discrete-time systems described by (2.13), we restrict the evolution of time $k$ in the forward direction to ensure uniqueness of solutions. (We will revisit this issue in more detail in Chapter 3.)

Special important cases of (2.13) are *linear time-varying systems* given by

$$x(k + 1) = A(k)x(k) + B(k)u(k), \tag{2.14a}$$
$$y(k) = C(k)x(k) + D(k)u(k), \tag{2.14b}$$

where $A : Z \to R^{n \times n}$, $B : Z \to R^{n \times m}$, $C : Z \to R^{p \times n}$, and $D : Z \to R^{p \times m}$. When $A(k) \equiv A, B(k) \equiv B, C(k) \equiv C$, and $D(k) \equiv D$, we have *linear time-invariant systems* given by

$$x(k + 1) = Ax(k) + Bu(k), \tag{2.15a}$$
$$y(k) = Cx(k) + Du(k). \tag{2.15b}$$

As in the case of continuous-time finite-dimensional dynamical systems, many qualitative properties of discrete-time finite-dimensional systems can be studied in terms of *initial-value problems* given by

$$x(k + 1) = f(k, x(k)), \quad x(k_0) = x_0, \tag{2.16}$$

where $x \in R^n$, $f : Z \times R^n \to R^n, k_0 \in Z$, and $k = k_0, k_0 + 1, \cdots$. We call the equation

$$x(k + 1) = f(k, x(k)), \tag{2.17}$$

a *system of first-order ordinary difference equations*. Special important cases of (2.17) include *autonomous systems* described by

$$x(k + 1) = f(x(k)), \tag{2.18}$$

*periodic systems* given by

$$x(k+1) = f(k, x(k)) = f(k+K, x(k)) \qquad (2.19)$$

for fixed $K \in Z^+$ and for all $k \in Z$, *linear homogeneous systems* given by

$$x(k+1) = A(k)x(k), \qquad (2.20)$$

*linear periodic systems* characterized by

$$x(k+1) = A(k)x(k) = A(k+K)x(k) \qquad (2.21)$$

for fixed $K \in Z^+$ and for all $k \in Z$, *linear nonhomogeneous systems*

$$x(k+1) = A(k)x(k) + g(k), \qquad (2.22)$$

and *linear, autonomous, homogeneous systems* characterized by

$$x(k+1) = Ax(k). \qquad (2.23)$$

In these equations all symbols used are defined in the obvious way by making reference to the corresponding systems of ordinary differential equations (see Subsection 1.3.2).

### Difference Equations of Order $n$

Thus far we have addressed systems of first-order difference equations. As in the continuous-time case, it is also possible to characterize initial-value problems by $n$th-order ordinary difference equations, say,

$$y(k+n) = h(k, y(k), y(k+1), \dots, y(k+n-1)), \qquad (2.24)$$

where $h : Z \times R^n \to R, n \in Z^+, k = k_0, k_0 + 1, \dots$. By specifying an *initial time* $k_0 \in Z$ and by specifying $y(k_0), y(k_0 + 1), \dots, y(k_0 + n - 1)$, we again have an initial-value problem given by

$$y(k+n) = h(k, y(k), y(k+1), \dots, y(k+n-1)),$$
$$y(k_0) = x_{10}, \dots, y(k_0 + n - 1) = x_{n0}. \qquad (2.25)$$

We call (2.24) an *$n$th-order ordinary difference equation*, and we note once more that in the case of initial-value problems described by such equations, there are no difficult issues involving the existence, uniqueness, and continuation of solutions.

We can reduce the study of (2.25) to the study of initial-value problems determined by systems of first-order ordinary difference equations. To accomplish this, we let in (2.25) $y(k) = x_1(k), y(k+1) = x_2(k), \dots, y(k+n-1) = x_n(k)$. We now obtain the system of first-order ordinary difference equations

$$x_1(k+1) = x_2(k),$$
$$\dots$$
$$x_{n-1}(k+1) = x_n(k),$$
$$x_n(k+1) = h(k, x_1(k), \dots, x_n(k)). \qquad (2.26)$$

Equations (2.26), together with the initial data $x_0^T = (x_{10}, \ldots, x_{n0})$, are equivalent to the initial-value problem (2.25) in the sense that these two problems will generate identical solutions [and in the sense that the transformation of (2.25) into (2.26) can be reversed unambiguously and uniquely].

As in the case of systems of first-order ordinary difference equations, we can point to several important special cases of $n$th-order ordinary difference equations, including equations of the form

$$y(k+n) + a_{n-1}(k)y(k+n-1) + \cdots + a_1(k)y(k+1) + a_0(k)y(k) = g(k), \quad (2.27)$$
$$y(k+n) + a_{n-1}(k)y(k+n-1) + \cdots + a_1(k)y(k+1) + a_0(k)y(k) = 0, \quad (2.28)$$

and

$$y(k+n) + a_{n-1}y(k+n-1) + \cdots + a_1y(k+1) + a_0y(k) = 0. \quad (2.29)$$

We call (2.27) a *linear nonhomogeneous ordinary difference equation of order $n$*, we call (2.28) a *linear homogeneous ordinary difference equation of order $n$*, and we call (2.29) a *linear, autonomous, homogeneous ordinary difference equation of order $n$*. As in the case of systems of first-order ordinary difference equations, we can define *periodic* and *linear periodic ordinary difference equations of order $n$* in the obvious way.

### Solutions of State Equations

Returning now to linear homogeneous systems

$$x(k+1) = A(k)x(k), \quad (2.30)$$

we observe that

$$x(k+2) = A(k+1)x(k+1) = A(k+1)A(k)x(k)$$
$$\cdots$$
$$x(n) = A(n-1)A(n-2)\cdots A(k+1)A(k)x(k)$$
$$= \prod_{j=k}^{n-1} A(j)x(k);$$

i.e., the state of the system at time $n$ is related to the state at time $k$ by means of the $n \times n$ matrix $\prod_{j=k}^{n-1} A(j)$ (as can easily be proved by induction). This suggests that the *state transition matrix* for (2.30) is given by

$$\Phi(n, k) = \prod_{j=k}^{n-1} A(j), \quad n > k, \quad (2.31)$$

and that

$$\Phi(k, k) = I. \quad (2.32)$$

As in the continuous-time case, the solution to the initial-value problem

$$x(k+1) = A(k)x(k)$$
$$x(k_0) = x_{k_0}, \quad k_0 \in Z, \tag{2.33}$$

is now given by

$$x(n) = \Phi(n, k_0)x_{k_0} = \prod_{j=k_0}^{n-1} A(j)x_{k_0}, n > k_0. \tag{2.34}$$

Continuing, let us next consider initial-value problems determined by linear nonhomogeneous systems (2.22),

$$x(k+1) = A(k)x(k) + g(k),$$
$$x(k_0) = x_{k_0}. \tag{2.35}$$

Then

$$\begin{aligned}
x(k_0+1) &= A(k_0)x(k_0) + g(k_0), \\
x(k_0+2) &= A(k_0+1)x(k_0+1) + g(k_0+1) \\
&= A(k_0+1)A(k_0)x(k_0) + A(k_0+1)g(k_0) + g(k_0+1), \\
x(k_0+3) &= A(k_0+2)x(k_0+2) + g(k_0+2) \\
&= A(k_0+2)A(k_0+1)A(k_0)x(k_0) + A(k_0+2)A(k_0+1)g(k_0) \\
&\quad + A(k_0+2)g(k_0+1) + g(k_0+2) \\
&= \Phi(k_0+3, k_0)x_{k_0} + \Phi(k_0+3, k_0+1)g(k_0) \\
&\quad + \Phi(k_0+3, k_0+2)g(k_0+1) + \Phi(k_0+3, k_0+3)g(k_0+2),
\end{aligned}$$

and so forth. For $k \geq k_0 + 1$, we easily obtain the expression for the solution of (2.35) as

$$x(k) = \Phi(k, k_0)x_{k_0} + \sum_{j=k_0}^{k-1} \Phi(k, j+1)g(j). \tag{2.36}$$

In the time-invariant case

$$x(k+1) = Ax(k) + g(k),$$
$$x(k_0) = x_{k_0}, \tag{2.37}$$

the solution is again given by (2.36) where now the state transition matrix

$$\Phi(k, k_0) = A^{k-k_0}, \quad k \geq k_0, \tag{2.38}$$

in view of (2.31) and (2.32). The solution of (2.37) is then

$$x(k) = A^{k-k_0}x_{k_0} + \sum_{j=k_0}^{k-1} A^{k-(j+1)}g(j), \quad k > k_0. \tag{2.39}$$

We note that when $x_{k_0} = 0$, (2.36) reduces to

$$x_p(k) = \sum_{j=k_0}^{k-1} \Phi(k, j+1)g(j), \tag{2.40}$$

and when $x_{k_0} \neq 0$ but $g(k) \equiv 0$, then (2.36) reduces to

$$x_h(k) = \Phi(k, k_0)x_{k_0}. \tag{2.41}$$

Therefore, the *total solution* of (2.35) consists of the sum of its *particular solution*, $x_p(k)$, and its *homogeneous solution*, $x_h(k)$.

## System Response

Finally, we observe that in view of (2.14b) and (2.36), the *system response* of the system (2.14), is of the form

$$y(k) = C(k)\Phi(k, k_0)x_{k_0} + C(k)\sum_{j=k_0}^{k-1} \Phi(k, j+1)B(j)u(j)$$

$$+ D(k)u(k), \quad k > k_0, \tag{2.42}$$

and

$$y(k_0) = C(k_0)x_{k_0} + D(k_0)u(k_0). \tag{2.43}$$

In the time-invariant case, in view of (2.39), the system response of the system (2.15) is

$$y(k) = CA^{k-k_0}x_{k_0} + C\sum_{j=k_0}^{k-1} A^{k-(j+1)}B(j)u(j) + Du(k), \quad k > k_0, \tag{2.44}$$

and

$$y(k_0) = Cx_{k_0} + Du(k_0). \tag{2.45}$$

Discrete-time systems, as discussed above, arise in several ways, including the *numerical solution* of ordinary differential equations (see, e.g., our discussion in Exercise 1.4 of *Euler's method*); the representation of *sampled-data systems* at discrete points in time (which will be discussed in further detail in Chapter 3); in the modeling process of systems that are defined only at discrete points in time (e.g., digital computer systems); and so forth.

As a specific example of a discrete-time system we consider a *second-order section digital filter* in *direct form*,

$$x_1(k+1) = x_2(k),$$
$$x_2(k+1) = ax_1(k) + bx_2(k) + u(k), \tag{2.46a}$$
$$y(k) = x_1(k), \tag{2.46b}$$

$k \in Z^+$, where $x_1(k)$ and $x_2(k)$ denote the state variables, $u(k)$ denotes the input, and $y(k)$ denotes the output of the digital filter. We depict system (2.46) in block diagram form in Figure 2.1.

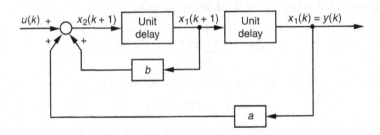

**Figure 2.1.** Second-order section digital filter in direct form

## 2.4 Input–Output Description of Systems

This section consists of four subsections. First we consider rather general aspects of the input–output description of systems. Because of their simplicity, we address the characterization of linear discrete-time systems next. In the third subsection we provide a foundation for the impulse response of linear continuous-time systems. Finally, we address the external description of linear continuous-time systems.

### 2.4.1 External Description of Systems: General Considerations

The state-space representation of systems presupposes knowledge of the *internal structure* of the system. When this structure is unknown, it may still be possible to arrive at a system description—an *external description*—that relates system inputs to system outputs. In linear system theory, a great deal of attention is given to relating the internal description of systems (the state representation) to the external description (the input–output description).

In the present context, we view *system inputs* and *system outputs* as elements of two real vector spaces $U$ and $Y$, respectively, and we view a system as being represented by an operator $T$ that relates elements of $U$ to elements of $Y$. For $u \in U$ and $y \in Y$ we will assume that $u : R \to R^m$ and $y : R \to R^p$ in the case of *continuous-time systems*, and that $u : Z \to R^m$ and $y : Z \to R^p$ in the case of *discrete-time systems*. If $m = p = 1$, we speak of a *single-input/single-output (SISO) system*. Systems for which $m > 1$, $p > 1$, are called *multi-input/multi-output (MIMO) systems*. For continuous-time systems we define vector addition (on $U$) and multiplication of vectors by scalars (on $U$) as

$$(u_1 + u_2)(t) = u_1(t) + u_2(t) \tag{2.47}$$

and

$$(\alpha u)(t) = \alpha u(t) \tag{2.48}$$

for all $u_1, u_2 \in U, \alpha \in R$, and $t \in R$. We similarly define vector addition and multiplication of vectors by scalars on $Y$. Furthermore, for discrete-time systems we define these operations on $U$ and $Y$ analogously. In this case the elements of $U$ and $Y$ are real sequences that we denote, e.g., by $u = \{u_k\}$ or $u = \{u(k)\}$. (It is easily verified that under these rather general conditions, $U$ and $Y$ satisfy all the axioms of a vector space, both for the continuous-time case and the discrete-time case.) In the continuous-time case as well as in the discrete-time case the system is represented by $T : U \to Y$, and we write

$$y = T(u). \tag{2.49}$$

In the subsequent development, we will impose restrictions on the vector spaces $U, Y$, and on the operator $T$, as needed.

*Linearity.* If $T$ is a linear operator, the system is called a *linear system.* In this case we have

$$\begin{aligned} y &= T(\alpha_1 u_1 + \alpha_2 u_2) \\ &= \alpha_1 T(u_1) + \alpha_2 T(u_2) \\ &= \alpha_1 y_1 + \alpha_2 y_2 \end{aligned} \tag{2.50}$$

for all $\alpha_1, \alpha_2 \in R$ and $u_1, u_2 \in U$ where $y_i = T(u_i) \in Y$, $i = 1, 2$, and $y \in Y$. Equation (2.50) represents the well-known *principle of superposition* of linear systems.

*With or Without Memory.* We say that a system is *memoryless,* or *without memory,* if its output for each value of the independent variable ($t$ or $k$) is dependent only on the input evaluated at the same value of the independent variable [e.g., $y(t_1)$ depends only on $u(t_1)$ and $y(k_1)$ depends only on $u(k_1)$]. An example of such a system is the resistor circuit shown in Figure 2.2, where the current $i(t) = u(t)$ denotes the system input at time $t$ and the voltage across the resistor, $v(t) = Ri(t) = y(t)$, denotes the system output at time $t$.

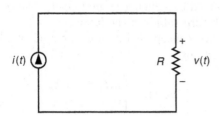

**Figure 2.2.** Resistor circuit

A system that is not memoryless is said to have memory. An example of a continuous-time *system with memory* is the capacitor circuit shown in Figure 2.3, where the current $i(t) = u(t)$ represents the system input at time $t$ and the voltage across the capacitor,

$$y(t) = v(t) = \frac{1}{C} \int_{-\infty}^{t} i(\tau)d\tau,$$

denotes the system output at time $t$. Another example of a continuous-time system with memory is described by the scalar equation

$$y(t) = u(t-1), \quad t \in R,$$

and an example of a discrete-time system with memory is characterized by the scalar equation

$$y(n) = \sum_{k=-\infty}^{n} x(k), \quad n, k \in Z.$$

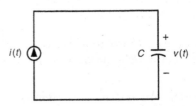

**Figure 2.3.** Capacitor circuit

*Causality.* A system is said to be *causal* if its output at any time, say $t_1$ (or $k_1$), depends only on values of the input evaluated for $t \le t_1$ (for $k \le k_1$). Thus, $y(t_1)$ depends only on $u(t), t \le t_1$ [or $y(k_1)$ depends only on $u(k), k \le k_1$]. Such a system is referred to as being *nonanticipative* since the system output does not anticipate future values of the input.

To make the above concept a bit more precise, we define the function $u_\tau : R \to R^m$ for $u \in U$ by

$$u_\tau(t) = \begin{cases} u(t), & t \le \tau, \\ 0, & t > \tau, \end{cases}$$

and we similarly define the function $y_\tau : R \to R^p$ for $y \in Y$. A system that is represented by the mapping $y = T(u)$ is said to be *causal* if and only if

$$(T(u))_\tau = (T(u_\tau))_\tau \quad \text{for all } \tau \in R, \text{ for all } u \in U.$$

Equivalently, this system is causal if and only if for $u, v \in U$ and $u_\tau = v_\tau$ it is true that

$$(T(u))_\tau = (T(v))_\tau \quad \text{for all } \tau \in R.$$

For example, the discrete-time system described by the scalar equation

$$y(n) = u(n) - u(n+1), \quad n \in Z,$$

is *not causal*. Neither is the continuous-time system characterized by the scalar equation

$$y(t) = x(t+1), \quad t \in R.$$

It should be pointed out that systems that are not causal are by no means useless. For example, causality is *not* of fundamental importance in image-processing applications where the independent variable is not time. Even when time is the independent variable, noncausal systems may play an important role. For example, in the processing of data that have been recorded (such as speech, meteorological data, demographic data, and stock market fluctuations), one is not constrained to processing the data causally. An example of this would be the smoothing of data over a time interval, say, by means of the system

$$y(n) = \frac{1}{2M+1} \sum_{k=-M}^{M} u(n-k).$$

*Time-Invariance.* A system is said to be *time-invariant* if a time shift in the input signal causes a corresponding time shift in the output signal. To make this concept more precise, for fixed $\alpha \in R$, we introduce the *shift operator* $Q_\alpha : U \to U$ as

$$Q_\alpha u(t) = u(t - \alpha), \quad u \in U, t \in R.$$

A system that is represented by the mapping $y = T(u)$ is said to be *time-invariant* if and only if

$$TQ_\alpha(u) = Q_\alpha(T(u)) = Q_\alpha(y)$$

for any $\alpha \in R$ and any $u \in U$. If a system is not time-invariant, it is said to be *time-varying*.

For example, a system described by the relation

$$y(t) = \cos u(t)$$

is time-invariant. To see this, consider the inputs $u_1(t)$ and $u_2(t) = u_1(t - t_0)$. Then

$$y_1(t) = \cos u_1(t), \quad y_2(t) = \cos u_2(t) = \cos u_1(t - t_0)$$

and

$$y_1(t - t_0) = \cos u_1(t - t_0) = y_2(t).$$

As a second example, consider a system described by the relation

$$y(n) = nu(n)$$

and consider two inputs $u_1(n)$ and $u_2(n) = u_1(n - n_0)$. Then

$$y_1(n) = nu_1(n) \quad \text{and} \quad y_2(n) = nu_2(n) = nu_1(n - n_0).$$

However, if we shift the output $y_1(n)$ by $n_0$, we obtain

$$y_1(n - n_0) = (n - n_0)u_1(n - n_0) \neq y_2(n).$$

Therefore, this system is not time-invariant.

### 2.4.2 Linear Discrete-Time Systems

In this subsection we investigate the representation of linear discrete-time systems. We begin our discussion by considering SISO systems.

In the following, we employ the *discrete-time impulse* (or *unit pulse* or *unit sample*), which is defined as

$$\delta(n) = \begin{cases} 0, & n \neq 0, n \in Z, \\ 1, & n = 0. \end{cases} \tag{2.51}$$

Note that if $\{p(n)\}$ denotes the *unit step sequence*, i.e.,

$$p(n) = \begin{cases} 1, & n \geq 0, n \in Z, \\ 0, & n < 0, n \in Z, \end{cases} \tag{2.52}$$

then

$$\delta(n) = p(n) - p(n - 1)$$

and

$$p(n) = \begin{cases} \sum_{k=0}^{\infty} \delta(n - k), & n \geq 0, \\ 0, & n < 0. \end{cases} \tag{2.53}$$

Furthermore, note that an arbitrary sequence $\{x(n)\}$ can be expressed as

$$x(n) = \sum_{k=-\infty}^{\infty} x(k)\delta(n - k). \tag{2.54}$$

We can easily show that a transformation $T : U \rightarrow Y$ determined by the equation

$$y(n) = \sum_{k=-\infty}^{\infty} h(n, k)u(k), \tag{2.55}$$

where $y \triangleq \{y(k)\} \in Y$, $u \triangleq \{u(k)\} \in U$, and $h : Z \times Z \rightarrow R$, is a linear transformation. Also, we note that for (2.55) to make any sense, we need to impose restrictions on $\{h(n, k)\}$ and $\{u(k)\}$. For example, if for every fixed $n, \{h(n, k)\} \in l_2$ and $\{u(k)\} \in l_2 = U$, then it follows from the Hölder Inequality (resp., Schwarz Inequality), see Section A.7, that (2.55) is well defined. There are of course other conditions that one might want to impose on (2.55).

For example, if for every fixed $n$, $\sum_{k=-\infty}^{\infty} |h(n,k)| < \infty$ (i.e., for every fixed $n$, $\{h(n,k)\} \in l_1$) and if $\sup_{k \in Z} |u(k)| < \infty$ (i.e., $\{u(k)\} \in l_\infty$), then (2.55) is also well defined.

We shall now elaborate on the suitability of (2.55) to represent linear discrete-time systems. To this end, we will agree once and for all that, in the ensuing discussion, all assumptions on $\{h(n,k)\}$ and $\{u(k)\}$ are satisfied that ensure that (2.55) is well defined.

We will view $y \in Y$ and $u \in U$ as system outputs and system inputs, respectively, and we will let $T : U \to Y$ denote a linear transformation that relates $u$ to $y$. We first consider the case when $u(k) = 0$ for $k < k_0$, $k$, $k_0 \in Z$. Also, we assume that for $k > n \geq k_0$, the inputs $u(k)$ do not contribute to the system output at time $n$ (i.e., the system is *causal*). Under these assumptions, and in view of the linearity of $T$, and by invoking the representation of signals by (2.54), we obtain for $y = \{y(n)\}$, $n \in Z$, the expression $y(n) = T(\sum_{k=-\infty}^{\infty} u(k)\delta(n-k)) = T(\sum_{k=k_0}^{n} u(k)\delta(n-k)) = \sum_{k=k_0}^{n} u(k)T(\delta(n-k)) = \sum_{k=k_0}^{n} h(n,k)u(k)$, $n \geq k_0$, and $y(n) = 0$, $n < k_0$, where $T(\delta(n-k)) \triangleq (T\delta)(n-k) \triangleq h(n,k)$ represents the response of $T$ to a unit pulse (resp., discrete-time impulse or unit sample) occurring at $n = k$.

When the assumptions in the preceding discussion are no longer valid, then a different argument than the one given above needs to be used to arrive at the system representation. Indeed, for *infinite sums*, the interchanging of the order of the summation operation $\sum$ with the linear transformation $T$ is no longer valid. We refer the reader to a paper by I. W. Sandberg ("A Representation Theorem for Linear Systems," *IEEE Transactions on Circuits and Systems—I*, Vol. 45, No. 5, pp. 578–580, May 1998) for a derivation of the representation of general linear discrete-time systems. In that paper it is shown that an extra term needs to be added to the right-hand side of equation (2.55), even in the representation of *general*, linear, time-invariant, causal, discrete-time systems. [In the proof, the Hahn–Banach Theorem (which is concerned with the extension of bounded linear functionals) is employed and the extra required term is given by $\lim_{l \to \infty} T(\sum_{k=-\infty}^{-c_l-1} u(k)\delta(n-k) + \sum_{k=c_l+1}^{\infty} u(k)\delta(n-k))$ with $c_l \to \infty$ as $l \to \infty$. For a statement and proof of the Hahn–Banach Theorem, refer, e.g., to A. N. Michel and C. J. Herget, *Applied Algebra and Functional Analysis*, Dover, New York, 1993, pp. 367–370.) In that paper it is also pointed out, however, that cases with such extra *nonzero* terms are not necessarily of importance in applications. In particular, if inputs and outputs are defined (to be nonzero) on just the non-negative integers, then for causal systems no additional term is needed (or more specifically, the extra term is zero), as seen in our earlier argument. In any event, *throughout this book we will concern ourselves with linear discrete-time systems that can be represented by equation (2.55)* for the single-input/single-output case (and appropriate generalizations for multi-input/multi-output cases).

Next, suppose that $T$ represents a time-invariant system. This means that if $\{h(n,0)\}$ is the response to $\{\delta(n)\}$, then by time invariance, the response to $\{\delta(n-k)\}$ is simply $\{h(n-k,0)\}$. By a slight abuse of notation, we let $h(n-k,0) \triangleq h(n-k)$. Then (2.55) assumes the form

$$y(n) = \sum_{k=-\infty}^{\infty} u(k)h(n-k). \qquad (2.56)$$

Expression (2.56) is called a *convolution sum* and is written more compactly as

$$y(n) = u(n) * h(n).$$

Now by a substitution of variables, we obtain for (2.56) the alternative expression

$$y(n) = \sum_{k=-\infty}^{\infty} h(k)u(n-k),$$

and therefore, we have

$$y(n) = u(n) * h(n) = h(n) * u(n);$$

i.e., the convolution operation $*$ commutes.

As a specific example, consider a linear, time-invariant, discrete-time system with unit impulse response given by

$$h(n) = \left\{ \begin{array}{ll} a^n, & n \geq 0 \\ 0, & n < 0 \end{array} \right\} = a^n p(n), \quad 0 < a < 1,$$

where $p(n)$ is the unit step sequence given in (2.52). It is an easy matter to show that the response of this system to an input given by

$$u(n) = p(n) - p(n-N)$$

is

$$y(n) = 0, n < 0,$$

$$y(n) = \sum_{k=0}^{n} a^{n-k} = a^n \frac{1 - 1^{-(n+1)}}{1 - a^{-1}} = \frac{1 - a^{n+1}}{1 - a}, \quad 0 \leq n < N,$$

and

$$y(n) = \sum_{k=0}^{N-1} a^{n-k} = a^n \frac{1 - a^{-N}}{1 - a^{-1}} = \frac{a^{n-N+1} - a^{n+1}}{1 - a}, \quad N \leq n.$$

Proceeding, with reference to (2.55) we note that $h(n,k)$ represents the system output at time $n$ due to a $\delta$-function input applied at time $k$. Now if system (2.55) is *causal*, then its output will be identically zero before an input is applied. Hence, a *linear system (2.55) is causal if and only if*

$$h(n, k) = 0 \quad \text{for all} \quad n < k.$$

Therefore, when the system (2.55) is causal, we have in fact

$$y(n) = \sum_{k=-\infty}^{n} h(n, k)u(k). \tag{2.57a}$$

We can rewrite (2.57a) as

$$y(n) = \sum_{k=-\infty}^{k_0-1} h(n, k)u(k) + \sum_{k=k_0}^{n} h(n, k)u(k)$$

$$\triangleq y(k_0 - 1) + \sum_{k=k_0}^{n} h(n, k)u(k). \tag{2.57b}$$

We say that the discrete-time system described by (2.55) is *at rest* at $k = k_0 \in Z$ if $u(k) = 0$ for $k \geq k_0$ implies that $y(k) = 0$ for $k \geq k_0$. Accordingly, if system (2.55) is known to be at rest at $k = k_0$, we have

$$y(n) = \sum_{k=k_0}^{\infty} h(n, k)u(k).$$

Furthermore, if system (2.55) is known to be causal and at rest at $k = k_0$, its input–output description assumes the form [in view of (2.57b)]

$$y(n) = \sum_{k=k_0}^{n} h(n, k)u(k). \tag{2.58}$$

If now, in addition, system (2.55) is also time-invariant, (2.58) becomes

$$y(n) = \sum_{k=k_0}^{n} h(n - k)u(k) = \sum_{k=k_0}^{n} h(k)u(n - k), \tag{2.59}$$

which is a convolution sum. [Note that in (2.59) we have slightly abused the notation for $h(\cdot)$, namely that $h(n - k) = h(n - k, 0)(= h(n, k))$.]

Next, turning to linear, discrete-time, *MIMO systems*, we can generalize (2.55) to

$$y(n) = \sum_{k=-\infty}^{\infty} H(n, k)u(k), \tag{2.60}$$

where $y : Z \to R^p$, $u : Z \to R^m$, and

$$H(n, k) = \begin{bmatrix} h_{11}(n, k) & h_{12}(n, k) & \cdots & h_{1m}(n, k) \\ h_{21}(n, k) & h_{22}(n, k) & \cdots & h_{2m}(n, k) \\ \cdots & \cdots & \cdots & \cdots \\ h_{p1}(n, k) & h_{p2}(n, k) & \cdots & h_{pm}(n, k) \end{bmatrix}, \tag{2.61}$$

where $h_{ij}(n,k)$ represents the system response at time $n$ of the $i$th component of $y$ due to a discrete-time impulse $\delta$ applied at time $k$ at the $j$th component of $u$, whereas the inputs at all other components of $u$ are being held zero. The matrix $H$ is called the *discrete-time unit impulse response matrix* of the system.

Similarly, it follows that the system (2.60) is *causal* if and only if

$$H(n,k) = 0 \quad \text{for all} \quad n < k,$$

and that the input–output description of linear, discrete-time, causal systems is given by

$$y(n) = \sum_{k=-\infty}^{n} H(n,k)u(k). \tag{2.62}$$

A discrete-time system described by (2.60) is said to be *at rest at* $k = k_0 \in Z$ if $u(k) = 0$ for $k \geq k_0$ implies that $y(k) = 0$ for $k \geq k_0$. Accordingly, if system (2.60) is known to be at rest at $k = k_0$, we have

$$y(n) = \sum_{k=k_0}^{\infty} H(n,k)u(k). \tag{2.63}$$

Moreover, if a linear discrete-time system that is at rest at $k_0$ is known to be causal, then its input–output description reduces to

$$y(n) = \sum_{k=k_0}^{n} H(n,k)u(k). \tag{2.64}$$

Finally, as in (2.56), it is easily shown that the unit impulse response $H(n,k)$ of a linear, *time-invariant*, discrete-time MIMO system depends only on the difference of $n$ and $k$; i.e., by a slight abuse of notation we can write

$$H(n,k) = H(n-k,0) \triangleq H(n-k) \tag{2.65}$$

for all $n$ and $k$. Accordingly, linear, time-invariant, causal, discrete-time MIMO systems that are at rest at $k = k_0$ are described by equations of the form

$$y(n) = \sum_{k=k_0}^{n} H(n-k)u(k). \tag{2.66}$$

We conclude by supposing that the system on hand is described by (2.14) under the assumption that $x(k_0) = 0$; i.e., the system is at rest at $k = k_0$. Then, according to (2.42) and (2.43), we obtain

$$H(n,k) = \begin{cases} C(n)\Phi(n,k+1)B(k), & n > k, \\ D(n), & n = k, \\ 0, & n < k. \end{cases} \tag{2.67}$$

Furthermore, for the time-invariant case, we obtain

$$H(n-k) = \begin{cases} CA^{n-(k+1)}B, & n > k, \\ D, & n = k, \\ 0, & n < k. \end{cases} \qquad (2.68)$$

### 2.4.3 The Dirac Delta Distribution

For any linear time-invariant operator $P$ from $C(R, R)$ to itself, we say that $P$ admits an *integral representation* if there exists an integrable function (in the Riemann or Lebesgue sense), $g_p : R \to R$, such that for any $f \in C(R, R)$,

$$(Pf)(x) = (f * g_p)(x) \triangleq \int_{-\infty}^{\infty} f(\tau)g_p(x - \tau)d\tau.$$

We call $g_p$ a *kernel of the integral representation of $P$*.

For the identity operator $I$ [defined by $If = f$ for any $f \in C(R, R)$] an integral representation for which $g_p$ is a function in the usual sense does not exist (see, e.g., Z. Szmydt, *Fourier Transformation and Linear Differential Equations*, D. Reidel Publishing Company, Boston, 1977). However, there exists a sequence of functions $\{\phi_n\}$ such that for any $f \in C(R, R)$,

$$(If)(x) = f(x) = \lim_{n\to\infty} (f * \phi_n)(x). \qquad (2.69)$$

To establish (2.69) we make use of functions $\{\phi_n\}$ given by

$$\phi_n(x) = \begin{cases} n(1 - n|x|), & \text{if } |x| \le \frac{1}{n}, \\ 0, & \text{if } |x| > \frac{1}{n}, \end{cases}$$

$n = 1, 2, 3, \dots$. A plot of $\phi_n$ is depicted in Figure 2.4. In Antsaklis and Michel [1], the following useful property of $\phi_n$ is proved.

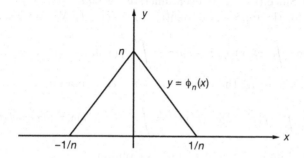

**Figure 2.4.** Generation of $n$ delta distribution

**Lemma 2.1.** *Let $f$ be a continuous real-valued function defined on $R$, and let $\phi_n$ be defined as above (Figure 2.4). Then for any $a \in R$,*

$$\lim_{n \to \infty} \int_{-\infty}^{\infty} f(\tau)\phi_n(a - \tau)d\tau = f(a). \tag{2.70}$$

∎

The above result, when applied to (2.69), now allows us to *define* a *generalized function* $\delta$ (also called a *distribution*) as the kernel of a *formal* or *symbolic* integral representation of the identity operator $I$; i.e.,

$$f(x) = \lim_{n \to \infty} \int_{-\infty}^{\infty} f(\tau)\phi_n(x - \tau)d\tau \tag{2.71}$$

$$\triangleq \int_{-\infty}^{\infty} f(\tau)\delta(x - \tau)d\tau \tag{2.72}$$

$$= f * \delta(x). \tag{2.73}$$

It is emphasized that the expression (2.72) is not an integral at all (in the Riemann or Lebesgue sense) but only a symbolic representation. The generalized function $\delta$ is called the *unit impulse* or the *Dirac delta distribution*.

In applications we frequently encounter functions $f \in C(R^+, R)$. If we extend $f$ to be defined on all of $R$ by letting $f(x) = 0$ for $x < 0$, then (2.70) becomes

$$\lim_{n \to \infty} \int_0^{\infty} f(\tau)\phi_n(a - \tau)d\tau = f(a) \tag{2.74}$$

for any $a > 0$, where we have used the fact that in the proof of Lemma 2.1, we need $f$ to be continuous only in a neighborhood of $a$ (refer to [1]). Therefore, for $f \in C(R^+, R)$, (2.71) to (2.74) yield

$$\lim_{n \to \infty} \int_0^{\infty} f(\tau)\phi_n(t - \tau)d\tau \triangleq \int_0^{\infty} f(\tau)\delta(t - \tau)d\tau = f(t) \tag{2.75}$$

for any $t > 0$. Since the $\phi_n$ are even functions, we have $\phi_n(t - \tau) = \phi_n(\tau - t)$, which allows for the representation $\delta(t - \tau) = \delta(\tau - t)$. We obtain from (2.75) that

$$\lim_{n \to \infty} \int_0^{\infty} f(\tau)\phi_n(\tau - t)d\tau \triangleq \int_0^{\infty} f(\tau)\delta(\tau - t)d\tau = f(t)$$

for any $t > 0$. Changing the variable $\tau' = \tau - t$, we obtain

$$\lim_{n \to \infty} \int_{-t}^{\infty} f(\tau' + t)\phi_n(\tau')d\tau' \triangleq \int_{-t}^{\infty} f(\tau' + t)\delta(\tau')d\tau' = f(t)$$

for any $t > 0$. Taking the limit $t \to 0^+$, we obtain

$$\lim_{n \to \infty} \int_{0^-}^{\infty} f(\tau' + t)\phi_n(\tau')d\tau' \triangleq \int_{0^-}^{\infty} f(\tau')\delta(\tau')d\tau' = f(0), \tag{2.76}$$

where $\int_{0-}^{\infty} f(\tau')\delta(\tau')d\tau'$ is not an integral but a symbolic representation of $\lim_{n\to\infty} \int_{0-}^{\infty} f(\tau'+t)\phi_n(\tau')d\tau'$.

Now let $s$ denote a complex variable. If in (2.75) and (2.76) we let $f(\tau) = e^{-s\tau}, \tau > 0$, then we obtain the *Laplace transform*

$$\lim_{n\to\infty} \int_{0-}^{\infty} e^{-s\tau}\phi_n(\tau)d\tau \triangleq \int_{0-}^{\infty} e^{-s\tau}\delta(\tau)d\tau = 1. \tag{2.77}$$

Symbolically we denote (2.77) by

$$\mathcal{L}(\delta) = 1, \tag{2.78}$$

and we say that the Laplace transform of the unit impulse function or the Dirac delta distribution is equal to one.

Next, we point out another important property of $\delta$. Consider a (time-invariant) operator $P$ and assume that $P$ admits an integral representation with kernel $g_P$. If in (2.75) we let $f = g_P$, we have

$$\lim_{n\to\infty} (P\phi_n)(t) = g_P(t), \tag{2.79}$$

and we write this (symbolically) as

$$P\delta = g_P. \tag{2.80}$$

*This shows that the impulse response of a linear, time-invariant, continuous-time system with integral representation is equal to the kernel of the integral representation of the system.*

Next, for any linear time-varying operator $P$ from $C(R,R)$ to itself, we say that $P$ admits an *integral representation* if there exists an integrable function (in the Riemann or Lebesgue sense), $g_P : R \times R \to R$, such that for any $f \in C(R,R)$,

$$(Pf)(\eta) = \int_{-\infty}^{\infty} f(\tau)g_P(\eta,\tau)d\tau. \tag{2.81}$$

Again, we call $g_P$ a *kernel of the integral representation of $P$*. It turns out that *the impulse response of a linear, time-varying, continuous-time system with integral representation is again equal to the kernel of the integral representation of the system.* To see this, we first observe that if $h \in C(R \times R, R)$, and if in Lemma 2.1 we replace $f \in C(R,R)$ by $h$, then all the ensuing relationships still hold, with obvious modifications. In particular, as in (2.71), we have for all $t \in R$,

$$\lim_{n\to\infty} \int_{-\infty}^{\infty} h(t,\tau)\phi_n(\eta-\tau)d\tau \triangleq \int_{-\infty}^{\infty} h(t,\tau)\delta(\eta-\tau)d\tau = h(t,\eta). \tag{2.82}$$

Also, as in (2.75), we have

$$\lim_{n\to\infty} \int_{0}^{\infty} h(t,\tau)\phi_n(\eta-\tau)d\tau \triangleq \int_{0}^{\infty} h(t,\tau)\delta(\eta-\tau)d\tau = h(t,\eta) \tag{2.83}$$

for $\eta > 0$.

Now let $h(t, \tau) = g_P(t, \tau)$. Then (2.82) yields

$$\lim_{n \to \infty} \int_{-\infty}^{\infty} g_P(t, \tau)\phi_n(\eta - \tau)d\tau \triangleq \int_{-\infty}^{\infty} g_P(t, \tau)\delta(\eta - \tau)d\tau = g_P(t, \eta), \quad (2.84)$$

which establishes our assertion. The common interpretation of (2.84) is that $g_P(t, \eta)$ represents the response of the system at time $t$ due to an impulse applied at time $\eta$.

### 2.4.4 Linear Continuous-Time Systems

We let $P$ denote a linear time-varying operator from $C(R, R^m) \triangleq U$ to $C(R, R^p) = Y$, and we assume that $P$ admits an *integral representation* given by

$$y(t) = (Pu)(t) = \int_{-\infty}^{\infty} H_P(t, \tau)u(\tau)d\tau, \quad (2.85)$$

where $H_P : R \times R \to R^{p \times m}, u \in U$, and $y \in Y$ and where $H_P$ is assumed to be integrable. This means that each element of $H_P$, $h_{P_{ij}} : R \times R \to R$ is integrable (in the Riemann or Lebesgue sense).

Now let $y_1$ and $y_2$ denote the response of system (2.85) corresponding to the input $u_1$ and $u_2$, respectively, let $\alpha_1$ and $\alpha_2$ be real scalars, and let $y$ denote the response of system (2.85) corresponding to the input $\alpha_1 u_1 + \alpha_2 u_2 = u$. Then

$$y = P(u) = P(\alpha_1 u_1 + \alpha_2 u_2) = \int_{-\infty}^{\infty} H_P(t, \tau)[\alpha_1 u_1(\tau) + \alpha_2 u_2(\tau)]d\tau$$

$$= \alpha_1 \int_{-\infty}^{\infty} H_P(t, \tau)u_1(\tau)d\tau + \alpha_2 \int_{-\infty}^{\infty} H_P(t, \tau)u_2(\tau)d\tau$$

$$= \alpha_1 P(u_1) + \alpha_2 P(u_2) = \alpha_1 y_1 + \alpha_2 y_2, \quad (2.86)$$

which shows that system (2.85) is indeed a *linear* system in the sense defined in (2.50).

Next, we let all components of $u(\tau)$ in (2.85) be zero, except for the $j$th component. Then the $i$th component of $y(t)$ in (2.85) assumes the form

$$y_i(t) = \int_{-\infty}^{\infty} h_{P_{ij}}(t, \tau)u_j(\tau)d\tau. \quad (2.87)$$

According to the results of the previous subsection [see (2.84)], $h_{P_{ij}}(t, \tau)$ denotes the response of the $i$th component of the output of system (2.85), measured at time $t$, due to an impulse applied to the $j$th component of the input of system (2.85), applied at time $\tau$, whereas all of the remaining components of the input are zero. Therefore, we call $H_P(t, \tau) = [h_{P_{ij}}(t, \tau)]$ the *impulse response matrix* of system (2.85).

Now suppose that it is known that system (2.85) is *causal*. Then its output will be identically zero before an input is applied. It follows that system (2.85) is causal if and only if

$$H_P(t, \tau) = 0 \quad \text{for all } t < \tau.$$

Therefore, when system (2.85) is causal, we have in fact that

$$y(t) = \int_{-\infty}^{t} H_P(t, \tau) u(\tau) d\tau. \tag{2.88}$$

We can rewrite (2.88) as

$$y(t) = \int_{-\infty}^{t_0} H_P(t, \tau) u(\tau) d\tau + \int_{t_0}^{t} H_P(t, \tau) u(\tau) d\tau$$

$$\triangleq y(t_0) + \int_{t_0}^{t} H_P(t, \tau) u(\tau) d\tau. \tag{2.89}$$

We say that the continuous-time system (2.85) is *at rest at* $t = t_0$ if $u(t) = 0$ for $t \geq t_0$ implies that $y(t) = 0$ for $t \geq t_0$. Note that our problem formulation mandates that the system be at rest at $t_0 = -\infty$. Also, note that if a system (2.85) is known to be causal and to be at rest at $t = t_0$, then according to (2.89) we have

$$y(t) = \int_{t_0}^{t} H_P(t, \tau) u(\tau) d\tau. \tag{2.90}$$

Next, suppose that it is known that the system (2.85) is *time-invariant*. This means that if in (2.87) $h_{P_{ij}}(t, \tau)$ is the response $y_i$ at time $t$ due to an impulse applied at time $\tau$ at the $j$th component of the input [i.e., $u_j(\tau) = \delta(t)$], with all other input components set to zero, then a $-\tau$ time shift in the input [i.e., $u_j(t - \tau) = \delta(t - \tau)$] will result in a corresponding $-\tau$ time shift in the response, which results in $h_{P_{ij}}(t - \tau, 0)$. Since this argument holds for all $t, \tau \in R$ and for all $i = 1, \ldots, p$, and $j = 1, \ldots, m$, we have $H_P(t, \tau) = H_P(t - \tau, 0)$. If we define (using a slight abuse of notation) $H_P(t - \tau, 0) = H_P(t - \tau)$, then (2.85) assumes the form

$$y(t) = \int_{-\infty}^{\infty} H_P(t - \tau) u(\tau) d\tau. \tag{2.91}$$

Note that (2.91) is consistent with the definition of the integral representation of a linear time-invariant operator introduced in the previous subsection.

The right-hand side of (2.91) is the familiar *convolution integral* of $H_P$ and $u$ and is written more compactly as

$$y(t) = (H_P * u)(t). \tag{2.92}$$

We note that since $H_P(t-\tau)$ represents responses at time $t$ due to impulse inputs applied at time $\tau$, then $H_P(t)$ represents responses at time $t$ due to impulse function inputs applied at $\tau = 0$. Therefore, a linear time-invariant system (2.91) is causal if and only if $H_P(t) = 0$ for all $t < 0$.

If it is known that the linear time-invariant system (2.91) is causal and is at rest at $t_0$, then we have

$$y(t) = \int_{t_0}^{t} H_P(t - \tau)u(\tau)d\tau = \int_{t_0}^{t} H_P(\tau)u(t - \tau)d\tau. \tag{2.93}$$

In this case it is customary to choose, without loss of generality, $t_0 = 0$. We thus have

$$y(t) = \int_{0}^{t} H_P(t - \tau)u(\tau)d\tau, \quad t \geq 0. \tag{2.94}$$

If we take the Laplace transform of both sides of (2.94), provided it exists, we obtain

$$\hat{y}(s) = \widehat{H}_P(s)\hat{u}(s), \tag{2.95}$$

where $\hat{y}(s) = [\hat{y}_1(s), \dots, \hat{y}_p(s)]^T$, $\widehat{H}_P(s) = [\hat{h}_{P_{ij}}(s)]$, $\hat{u}(s) = [\hat{u}_1(s), \dots, \hat{u}_m(s)]^T$ where the $\hat{y}_i(s)$, $\hat{u}_j(s)$, and $\hat{h}_{P_{ij}}(s)$ denote the Laplace transforms of $y_i(t)$, $u_j(t)$, and $h_{Pij}(t)$, respectively [see Chapter 3 for more details concerning Laplace transforms]. Consistent with (2.78), we note that $\widehat{H}_P(s)$ represents the Laplace transform of the impulse response matrix $H_P(t)$. We call $\widehat{H}_P(s)$ a *transfer function matrix*.

Now suppose that the input–output relation of a system is specified by the state and output equations (2.3), repeated here as

$$\dot{x} = A(t)x + B(t)u, \tag{2.96a}$$
$$y = C(t)x + D(t)u. \tag{2.96b}$$

If we assume that $x(t_0) = 0$ so that the system is at rest at $t_0 = 0$, we obtain for the response of this system,

$$y(t) = \int_{t_0}^{t} C(t)\Phi(t, \tau)B(\tau)u(\tau)d\tau + D(t)u(t) \tag{2.97}$$

$$= \int_{t_0}^{t} [C(t)\Phi(t, \tau)B(\tau) + D(t)\delta(t - \tau)]u(\tau)d\tau, \tag{2.98}$$

where in (2.98) we have made use of the interpretation of $\delta$ given in Subsection 2.4.3. Comparing (2.98) with (2.90), we conclude that the impulse response matrix for system (2.96) is given by

$$H_P(t, \tau) = \begin{cases} C(t)\Phi(t, \tau)B(\tau) + D(t)\delta(t - \tau), & t \geq \tau, \\ 0, & t < \tau. \end{cases} \tag{2.99}$$

Finally, for time-invariant systems described by the state and output equations (2.7), repeated here as

$$\dot{x} = Ax + Bu, \tag{2.100a}$$

$$y = Cx + Du, \tag{2.100b}$$

we obtain for the impulse response matrix the expression

$$H_P(t - \tau) = \begin{cases} Ce^{A(t-\tau)}B + D\delta(t - \tau), & t \geq \tau, \\ 0, & t < \tau, \end{cases} \tag{2.101}$$

or, as is more commonly written,

$$H_P(t) = \begin{cases} Ce^{At}B + D\delta(t), & t \geq 0, \\ 0, & t < 0. \end{cases} \tag{2.102}$$

We will pursue the topics of this section further in Chapter 3.

## 2.5 Summary and Highlights

*Internal Descriptions*

- The *response of the time-varying continuous-time system*

$$\dot{x} = A(t)x + B(t)u, \quad y = C(t)x + D(t)u, \tag{2.3}$$

with $x(t_0) = x_0$ is given by

$$y(t) = C(t)\Phi(t, t_0)x_0 + C(t) \int_{t_0}^{t} \Phi(t, s)B(s)u(s)ds + D(t)u(t). \tag{2.6}$$

- The *response of the time-invariant continuous-time system*

$$\dot{x} = Ax + Bu, \quad y = Cx + Du, \tag{2.7}$$

is given by

$$y(t) = Ce^{A(t-t_0)}x_0 + C \int_{t_0}^{t} e^{A(t-s)}Bu(s)ds + Du(t). \tag{2.9}$$

- The *response of the discrete-time system*

$$x(k + 1) = A(k)x(k) + B(k)u(k), \quad y(k) = C(k)x(k) + D(k)u(k), \tag{2.14}$$

with $x(k_0) = x_{k_0}$ is given by

$$y(k) = C(k)\Phi(k, k_0)x_{k_0} + C(k) \sum_{j=k_0}^{k-1} \Phi(k, j+1)B(j)u(j)$$

$$+ D(k)u(k), \quad k > k_0 \tag{2.42}$$

and

$$y(k_0) = C(k_0)x_{k_0} + D(k_0)u(k_0), \tag{2.43}$$

where the state transition matrix

$$\Phi(k, k_0) = \prod_{j=k_0}^{k-1} A(j), \quad k > k_0, \tag{2.31}$$

$$\Phi(k_0, k_0) = I. \tag{2.32}$$

In the time-invariant case

$$x(k+1) = Ax(k) + Bu(k), \quad y(k) = Cx(k) + Du(k), \tag{2.15}$$

with $x(k) = x_{k_0}$, the system response is given by

$$y(k) = CA^{k-k_0}x_{k_0} + C \sum_{j=k_0}^{k-1} A^{k-(j+1)}B(j)u(j) + Du(k), \quad k > k_0, \tag{2.44}$$

and

$$y(k_0) = Cx_{k_0} + Du(k_0). \tag{2.45}$$

*External Descriptions*

- Properties: *Linearity* (2.50); with *memory; causality; time-invariance*
- *The input–output description* of a *linear, discrete-time, causal, time-invariant* system that is at rest at $k = k_0$ is given by

$$y(n) = \sum_{k=k_0}^{n} h(n-k)u(k) = \sum_{k=k_0}^{n} h(k)u(n-k). \tag{2.59}$$

$h(n-k)(= h(n-k, 0))$ is the discrete-time unit impulse response of the system.

- For the *discrete-time, time-invariant system*

$$x(k+1) = Ax(k) + Bu(k), \quad y(k) = Cx(k) + Du(k),$$

the discrete-time unit impulse response (for the MIMO case) is

$$H(n-k) = \begin{cases} CA^{n-(k+1)}B, & n > k, \\ D, & n = k, \\ 0, & n < k. \end{cases} \tag{2.68}$$

- *The unit impulse (Dirac delta distribution)* $\delta(t)$ satisfies

$$\int_a^b f(\tau)\delta(t-\tau)d\tau = f(t),$$

where $a < t < b$ [see (2.75)].
- *The input–output description of a linear, continuous-time, causal, time-invariant system* that is at rest at $t = t_0$ is given by

$$y(t) = \int_{t_0}^t H_P(t-\tau)u(\tau)d\tau = \int_{t_0}^t H_P(\tau)u(t-\tau)d\tau. \tag{2.93}$$

$H_P(t-\tau)(= H_P(t-\tau, 0))$ is the continuous-time unit impulse response of the system.
- For the *time-invariant system*

$$\dot{x} = Ax + Bu \quad y = Cx + Du, \tag{2.100}$$

the continuous-time unit impulse response is

$$H_P(t-\tau) = \begin{cases} Ce^{A(t-\tau)}B + D\delta(t-\tau), & t \geq \tau, \\ 0, & t < \tau. \end{cases} \tag{2.101}$$

## 2.6 Notes

An original standard reference on linear systems is by Zadeh and Desoer [7]. Of the many excellent texts on this subject, the reader may want to refer to Brockett [2], Kailath [5], and Chen [3]. For more recent texts on linear systems, consult, e.g., Rugh [6] and DeCarlo [4]. The presentation in this book relies mostly on the recent text by Antsaklis and Michel [1].

## References

1. P.J. Antsaklis and A.N. Michel, *Linear Systems*, Birkhäuser, Boston, MA, 2006.
2. R.W. Brockett, *Finite Dimensional Linear Systems*, Wiley, New York, NY, 1970.
3. C.T. Chen, *Linear System Theory and Design*, Holt, Rinehart and Winston, New York, NY, 1984.
4. R.A. DeCarlo, *Linear Systems*, Prentice-Hall, Englewood Cliffs, NJ, 1989.
5. T. Kailath, *Linear Systems*, Prentice-Hall, Englewood Cliffs, NJ, 1980.
6. W.J. Rugh, *Linear System Theory, Second Edition*, Prentice-Hall, Englewood Cliffs, NJ, 1996.
7. L.A. Zadeh and C.A. Desoer, *Linear System Theory - The State Space Approach*, McGraw-Hill, New York, NY, 1963.

## Exercises

**2.1.** (a) For the mechanical system given in Exercise 1.2a, we view $f_1$ and $f_2$ as making up the system input vector, and $y_1$ and $y_2$ the system output vector. Determine a state-space description for this system.

(b) For the same mechanical system, we view $f_1 + 5f_2$ as the (scalar-valued) system input and we view $8y_1 + 10y_2$ as the (scalar-valued) system output. Determine a state-space description for this system.

(c) For part (a), determine the input–output description of the system.

(d) For part (b), determine the input–output description of the system.

**2.2.** In Example 1.3, we view $e_a$ and $\theta$ as the system input and output, respectively.

(a) Detemine a state-space representation for this system.

(b) Determine the input–output description of this system.

**2.3.** For the second-order section digital filter in direct form, given in Figure 2.1, determine the input–output description, where $x_1(k)$ and $u(k)$ denote the output and input, respectively.

**2.4.** In the circuit of Figure 2.5, $v_i(t)$ and $v_0(t)$ are voltages (at time $t$) and $R_1$ and $R_2$ are resistors. There is also an ideal diode that acts as a short circuit when $v_i$ is positive and as an open circuit when $v_i$ is negative. We view $v_i$ and $v_0$ as the system input and output, respectively.

(a) Determine an input–output description of this system.

(b) Is this system linear? Is it time-varying or time-invariant? Is it causal? Explain your answers.

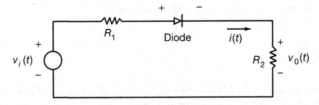

**Figure 2.5.** Diode circuit

**2.5.** We consider the *truncation operator* given by

$$y(t) = T_\tau(u(t))$$

as a system, where $\tau \in R$ is fixed, $u$ and $y$ denote system input and output, respectively, $t$ denotes time, and $T_\tau(\cdot)$ is specified by

$$T_\tau(u(t)) = \begin{cases} u(t) & t \le \tau, \\ 0 & t > \tau. \end{cases}$$

Is this system causal? Is it linear? Is it time-invariant? What is its impulse response?

**2.6.** We consider the *shift operator* given by

$$y(t) = Q_\tau(u(t)) = u(t - \tau)$$

as a system, where $\tau \in R$ is fixed, $u$ and $y$ denote system input and system output, respectively, and $t$ denotes time. Is this system causal? Is it linear? Is it time-invariant? What is its impulse response?

**2.7.** Consider the system whose input–output description is given by

$$y(t) = \min\{u_1(t), u_2(t)\},$$

where $u(t) = [u_1(t), u_2(t)]^T$ denotes the system input and $y(t)$ is the system output. Is this system linear?

**2.8.** Suppose it is known that a linear system has impulse response given by $h(t, \tau) = \exp(-|t - \tau|)$. Is this system causal? Is it time-invariant?

**2.9.** Consider a system with input–output description given by

$$y(k) = 3u(k + 1) + 1, \quad k \in Z,$$

where $y$ and $u$ denote the output and input, respectively (recall that $Z$ denotes the integers). Is this system causal? Is it linear?

**2.10.** Use expression (2.54),

$$x(n) = \sum_{k=-\infty}^{\infty} x(k)\delta(n - k),$$

and $\delta(n) = p(n) - p(n - 1)$ to express the system response $y(n)$ due to any input $u(k)$, as a function of the unit step response of the system [i.e., due to $u(k) = p(k)$].

# 3

# Response of Continuous- and Discrete-Time Systems

## 3.1 Introduction

In system theory it is important to clearly understand how inputs and initial conditions affect the response of a system. Many reasons exist for this. For example, in control theory, it is important to be able to select an input that will cause the system output to satisfy certain properties [e.g., to remain bounded (stability) or to follow a given trajectory (tracking)]. This is in stark contrast to the study of ordinary differential equations, where it is usually assumed that the forcing function (input) is given.

The goal of this chapter is to study the response of linear systems in greater detail than was done in Chapter 2. To this end, solutions of linear ordinary differential equations are reexamined, this time with an emphasis on characterizing all solutions using bases (of the solution vector space) and on determining such solutions. For convenience, certain results from Chapter 2 are repeated. We will find it convenient to treat continuous-time and discrete-time cases separately. Whereas in Chapter 2, certain fundamental issues that include input–output system descriptions, causality, linearity, and time-invariance are emphasized, here we will address in greater detail impulse (and pulse) response and transfer functions for continuous-time systems and discrete-time systems.

In Chapters 1 and 2 we addressed linear as well as nonlinear systems that may be time-varying or time-invariant. We considered this level of generality since this may be mandated during the modeling process of the systems. However, in the analysis and synthesis of such systems, simplified models involving linear time-invariant systems usually suffice. Accordingly, in the remainder of this book, we will emphasize linear, time-invariant continuous-time and discrete-time systems.

In this chapter, in Section 3.2, we further study linear systems of ordinary differential equations with constant coefficients. Specifically, in this section, we develop a general characterization of the solutions of such equations and we study the properties of the solutions by investigating the properties of fundamental matrices and state transition matrices. In Section 3.3 we address several methods of determining the state transition matrix and we study the asymptotic behavior of the solutions of such systems. In Sections 3.4 and 3.5, we further investigate the properties of the state representations and the input–output representations of continuous-time and discrete-time finite-dimensional systems. Specifically, in these sections we study equivalent representations of such systems, we investigate the properties of transfer function matrices, and for the discrete-time case we also address sampled data systems and the asymptotic behavior of the system response of time-invariant systems.

## 3.2 Solving $\dot{x} = Ax$ and $\dot{x} = Ax + g(t)$: The State Transition Matrix $\Phi(t, t_0)$

In this section we consider systems of linear homogeneous ordinary differential equations with constant coefficients.

$$\dot{x} = Ax \tag{3.1}$$

and linear nonhomogeneous ordinary differential equations

$$\dot{x} = Ax + g(t). \tag{3.2}$$

In Theorem 1.20 of Chapter 1 it was shown that these systems of equations, subject to initial conditions $x(t_0) = x_0$, possess unique solutions for every $(t_0, x_0) \in D$, where $D = \{(t, x) : t \in J = (a, b), \ x \in R^n\}$ and where it is assumed that $A \in R^{n \times n}$ and $g \in C(J, R^n)$. These solutions exist over the entire interval $J = (a, b)$, and they depend continuously on the initial conditions. Typically, we will assume that $J = (-\infty, \infty)$. We note that $\phi(t) \equiv 0$, for all $t \in J$, is a solution of (3.1), with $\phi(t_0) = 0$. We call this the *trivial solution*. As in Chapter 1 (refer to Section 1.8), we recall that the preceding statements are also true when $g(t)$ is piecewise continuous on $J$.

In the sequel, we sometimes will encounter the case where $A$ is in Jordan canonical form that may have entries in the complex plane $C$. For this reason, we will allow $D = \{(t, x) : t \in J = (a, b), \ x \in R^n \text{ (or } x \in C^n)\}$ and $A \in R^{n \times n}$ [or $A \in C^{n \times n}$], as needed. For the case of real vectors, the field of scalars for the $x$-space will be the field of real numbers $(F = R)$, whereas for the case of complex vectors, the field of scalars for the $x$-space will be the field of complex numbers $(F = C)$. For the latter case, the theory concerning the existence and uniqueness of solutions for (3.1), as presented in Chapter 1, carries over and can be modified in the obvious way.

### 3.2.1 The Fundamental Matrix

### Solution Space

In our first result we will make use of several facts concerning vector spaces, bases, and linear spaces, which are addressed in the appendix.

**Theorem 3.1.** *The set of solutions of (3.1) on the interval $J$ forms an $n$-dimensional vector space.*

*Proof.* Let $V$ denote the set of all solutions of (3.1) on $J$. Let $\alpha_1, \alpha_2 \in F$ ($F = R$ or $F = C$), and let $\phi_1, \phi_2 \in V$. Then $\alpha_1\phi_1 + \alpha_2\phi_2 \in V$ since $\frac{d}{dt}[\alpha_1\phi_1 + \alpha_2\phi_2] = \alpha_1\frac{d}{dt}\phi_1(t) + \alpha_2\frac{d}{dt}\phi_2(t) = \alpha_1 A\phi_1(t) + \alpha_2 A\phi_2(t) = A[\alpha_1\phi_1(t) + \alpha_2\phi_2(t)]$ for all $t \in J$. [Note that in this time-invariant case, it can be assumed without loss of generality, that $J = (-\infty, \infty)$.] This shows that $V$ is a vector space.

To complete the proof of the theorem, we must show that $V$ is of dimension $n$. To accomplish this, we must find $n$ linearly independent solutions $\phi_1, \ldots, \phi_n$ that span $V$. To this end, we choose a set of $n$ linearly independent vectors $x_0^1, \ldots, x_0^n$ in the $n$-dimensional $x$-space (i.e., in $R^n$ or $C^n$). By the existence results in Chapter 1, if $t_0 \in J$, then there exist $n$ solutions $\phi_1, \ldots, \phi_n$ of (3.1) such that $\phi_1(t_0) = x_0^1, \ldots, \phi_n(t_0) = x_0^n$. We first show that these solutions are linearly independent. If on the contrary, these solutions are linearly dependent, there exist scalars $\alpha_1, \ldots, \alpha_n \in F$, not all zero, such that $\sum_{i=1}^n \alpha_i\phi_i(t) = 0$ for all $t \in J$. This implies in particular that $\sum_{i=1}^n \alpha_i\phi_i(t_0) = \sum_{i=1}^n \alpha_i x_0^i = 0$. But this contradicts the assumption that $\{x_0^1, \ldots, x_0^n\}$ is a linearly independent set. Therefore, the solutions $\phi_1, \ldots, \phi_n$ are linearly independent.

To conclude the proof, we must show that the solutions $\phi_1, \ldots, \phi_n$ span $V$. Let $\phi$ be any solution of (3.1) on the interval $J$ such that $\phi(t_0) = x_0$. Then there exist unique scalars $\alpha_1, \ldots, \alpha_n \in F$ such that

$$x_0 = \sum_{i=1}^n \alpha_i x_0^i,$$

since, by assumption, the vectors $x_0^1, \ldots, x_0^n$ form a basis for the $x$-space. Now

$$\psi = \sum_{i=1}^n \alpha_i \phi_i$$

is a solution of (3.1) on $J$ such that $\psi(t_0) = x_0$. But by the uniqueness results of Chapter 1, we have that

$$\phi = \psi = \sum_{i=1}^n \alpha_i \phi_i.$$

Since $\phi$ was chosen arbitrarily, it follows that $\phi_1, \ldots, \phi_n$ span $V$.  ∎

## Fundamental Matrix and Properties

Theorem 3.1 enables us to make the following definition.

**Definition 3.2.** *A set of $n$ linearly independent solutions of (3.1) on $J$, $\{\phi_1, \ldots, \phi_n\}$, is called a* fundamental set of solutions *of (3.1), and the $n \times n$ matrix*

$$\Phi = [\phi_1, \phi_2, \ldots, \phi_n] = \begin{bmatrix} \phi_{11} & \phi_{12} & \cdots & \phi_{1n} \\ \phi_{21} & \phi_{22} & \cdots & \phi_{2n} \\ \vdots & \vdots & & \vdots \\ \phi_{n1} & \phi_{n2} & \cdots & \phi_{nn} \end{bmatrix}$$

*is called a* fundamental matrix *of (3.1).*     ∎

We note that there are infinitely many different fundamental sets of solutions of (3.1) and, hence, infinitely many different fundamental matrices for (3.1). Clearly $[\phi_1, \phi_2, \ldots, \phi_n]$ is a basis of the solution space. We now study some of the basic properties of a fundamental matrix.

In the next result, $X = [x_{ij}]$ denotes an $n \times n$ matrix, and the derivative of $X$ with respect to $t$ is defined as $\dot{X} = [\dot{x}_{ij}]$. Let $A$ be the $n \times n$ matrix given in (3.1). We call the system of $n^2$ equations

$$\dot{X} = AX \tag{3.3}$$

a *matrix differential equation.*

**Theorem 3.3.** *A fundamental matrix $\Phi$ of (3.1) satisfies the matrix equation (3.3) on the interval J.*

*Proof.* We have

$$\dot{\Phi} = [\dot{\phi}_1, \dot{\phi}_2, \ldots, \dot{\phi}_n] = [A\phi_1, A\phi_2, \ldots, A\phi_n] = A[\phi_1, \phi_2, \ldots, \phi_n] = A\Phi.$$

∎

The next result is called *Abel's formula.*

**Theorem 3.4.** *If $\Phi$ is a solution of the matrix equation (3.3) on an interval J and $\tau$ is any point of J, then*

$$\det \Phi(t) = \det \Phi(\tau) \exp \left[ \int_\tau^t tr\,A ds \right]$$

*for every $t \in J$. [$tr\,A = tr[a_{ij}]$ denotes the trace of A; i.e., $tr\,A = \sum_{j=1}^n a_{jj}$.]*
∎

The proof of Theorem 3.4 is omitted. We refer the reader to [1] for a proof.

Since in Theorem 3.4 $\tau$ is arbitrary, it follows that either $\det \Phi(t) \neq 0$ for all $t \in J$ or $\det \Phi(t) = 0$ for each $t \in J$. The next result provides a test on whether an $n \times n$ matrix $\Phi(t)$ is a fundamental matrix of (3.1).

**Theorem 3.5.** *A solution $\Phi$ of the matrix equation (3.3) is a fundamental matrix of (3.1) if and only if its determinant is nonzero for all $t \in J$.*

*Proof.* If $\Phi = [\phi_1, \phi_2, \ldots, \phi_n]$ is a fundamental matrix for (3.1), then the columns of $\Phi$, $\phi_1, \ldots, \phi_n$, form a linearly independent set. Now let $\phi$ be a nontrivial solution of (3.1). Then by Theorem 3.1 there exist unique scalars $\alpha_1, \ldots, \alpha_n \in F$, not all zero, such that $\phi = \sum_{j=1}^n \alpha_j \phi_j = \Phi a$, where $a^T = (\alpha_1, \ldots, \alpha_n)$. Let $t = \tau \in J$. Then $\phi(\tau) = \Phi(\tau)a$, which is a system of $n$ linear algebraic equations. By construction, this system of equations has a unique solution for any choice of $\phi(\tau)$. Therefore, $\det \Phi(\tau) \neq 0$. It now follows from Theorem 3.4 that $\det \Phi(t) \neq 0$ for any $t \in J$.

Conversely, let $\Phi$ be a solution of (3.3) and assume that $\det \Phi(t) \neq 0$ for all $t \in J$. Then the columns of $\Phi$ are linearly independent for all $t \in J$. Hence, $\Phi$ is a fundamental matrix of (3.1). ∎

It is emphasized that a matrix may have identically zero determinant over some interval, even though its columns are linearly independent. For example, the columns of the matrix

$$\Phi(t) = \begin{bmatrix} 1 & t & t^2 \\ 0 & 1 & t \\ 0 & 0 & 0 \end{bmatrix}$$

are linearly independent, and yet $\det \Phi(t) = 0$ for all $t \in (-\infty, \infty)$. In accordance with Theorem 3.5, the above matrix cannot be a fundamental solution of the matrix equation (3.3) for any matrix $A$.

**Theorem 3.6.** *If $\Phi$ is a fundamental matrix of (3.1) and if $C$ is any nonsingular constant $n \times n$ matrix, then $\Phi C$ is also a fundamental matrix of (3.1). Moreover, if $\Psi$ is any other fundamental matrix of (3.1), then there exists a constant $n \times n$ nonsingular matrix $P$ such that $\Psi = \Phi P$.*

*Proof.* For the matrix $\Phi C$ we have $\frac{d}{dt}(\Phi C) = \dot{\Phi} C = [A\Phi]C = A(\Phi C)$, and therefore, $\Phi C$ is a solution of the matrix equation (3.3). Furthermore, since $\det \Phi(t) \neq 0$ for $t \in J$ and $\det C \neq 0$, it follows that $\det[\Phi(t)C] = [\det \Phi(t)](\det C) \neq 0, t \in J$. By Theorem 3.5, $\Phi C$ is a fundamental matrix.

Next, let $\Psi$ be any other fundamental matrix of (3.1) and consider the product $\Phi^{-1}(t)\Psi$. [Notice that since $\det \Phi(t) \neq 0$ for all $t \in J$, then $\Phi^{-1}(t)$ exists for all $t \in J$.] Also, consider $\Phi\Phi^{-1} = I$ where $I$ denotes the $n \times n$ identity matrix. Differentiating both sides, we obtain $\left(\frac{d}{dt}\Phi\right)\Phi^{-1} + \Phi\left(\frac{d}{dt}\Phi^{-1}\right) = 0$ or $\frac{d}{dt}\Phi^{-1} = -\Phi^{-1}\left(\frac{d}{dt}\Phi\right)\Phi^{-1}$. Therefore, we can compute $\frac{d}{dt}(\Phi^{-1}\Psi) = \Phi^{-1}\left(\frac{d}{dt}\Psi\right) + \left(\frac{d}{dt}\Phi^{-1}\right)\Psi = \Phi^{-1}A\Psi - [\Phi^{-1}\left(\frac{d}{dt}\Phi\right)\Phi^{-1}]\Psi = \Phi^{-1}A\Psi - (\Phi^{-1}A\Phi\Phi^{-1})\Psi = \Phi^{-1}A\Psi - \Phi^{-1}A\Psi = 0$. Hence, $\Phi^{-1}\Psi = P$ or $\Psi = \Phi P$. ∎

*Example 3.7.* It is easily verified that the system of equations

$$\begin{aligned} \dot{x}_1 &= 5x_1 - 2x_2 \\ \dot{x}_2 &= 4x_1 - x_2 \end{aligned} \tag{3.4}$$

has two linearly independent solutions given by $\phi_1(t) = (e^{3t}, e^{3t})^T$, $\phi_2(t) = (e^t, 2e^t)^T$, and therefore, the matrix

$$\Phi(t) = \begin{bmatrix} e^{3t} & e^t \\ e^{3t} & 2e^t \end{bmatrix} \tag{3.5}$$

is a fundamental matrix of (3.4).

Using Theorem 3.6 we can find the particular fundamental matrix $\Psi$ of (3.4) that satisfies the initial condition $\Psi(0) = I$ by using $\Phi(t)$ given in (3.5). We have $\Psi(0) = I = \Phi(0)C$ or $C = \Phi^{-1}(0)$, and therefore,

$$C = \begin{bmatrix} 1 & 1 \\ 1 & 2 \end{bmatrix}^{-1} = \begin{bmatrix} 2 & -1 \\ -1 & 1 \end{bmatrix}$$

and

$$\Psi(t) = \Phi C = \begin{bmatrix} (2e^{3t} - e^t) & (-e^{3t} + e^t) \\ (2e^{3t} - 2e^t) & (-e^{3t} + 2e^t) \end{bmatrix}.$$

---

### 3.2.2 The State Transition Matrix

In Chapter 1 we used the *method of successive approximations* (Theorem 1.15) to prove that for every $(t_0, x_0) \in J \times R^n$,

$$\dot{x} = A(t)x \tag{3.6}$$

possesses a unique solution of the form

$$\phi(t, t_0, x_0) = \Phi(t, t_0)x_0,$$

such that $\phi(t_0, t_0, x_0) = x_0$, which exists for all $t \in J$, where $\Phi(t, t_0)$ is the *state transition matrix* (see Section 1.8). We derived an expression for $\Phi(t, t_0)$ in series form, called the *Peano–Baker series* [see (1.80) of Chapter 1], and we showed that $\Phi(t, t_0)$ is the unique solution of the matrix differential equation

$$\frac{\partial}{\partial t}\Phi(t, t_0) = A(t)\Phi(t, t_0), \tag{3.7}$$

where

$$\Phi(t_0, t_0) = I \text{ for all } t \in J. \tag{3.8}$$

Of course, these results hold for (3.1) as well.

We now provide an alternative formulation of the state transition matrix, and we study some of the properties of such matrices. Even though much of the subsequent discussion applies to system (3.6), we will confine ourselves to system (3.1). In the following definition, we use the natural basis $\{e_1, e_2, \ldots, e_n\}$ (refer to Section A.2).

**Definition 3.8.** *A fundamental matrix $\Phi$ of (3.1) whose columns are determined by the linearly independent solutions $\phi_1, \ldots, \phi_n$ with*

$$\phi_1(t_0) = e_1, \ldots, \phi_n(t_0) = e_n, \quad t_0 \in J,$$

*is called the state transition matrix $\Phi$ for (3.1). Equivalently, if $\Psi$ is any fundamental matrix of (3.1), then the matrix $\Phi$ determined by*

$$\Phi(t, t_0) \triangleq \Psi(t)\Psi^{-1}(t_0) \quad \text{for all} \quad t, t_0 \in J, \tag{3.9}$$

*is said to be the state transition matrix of (3.1).*     ∎

We note that the state transition matrix of (3.1) is *uniquely* determined by the matrix $A$ and is *independent* of the particular choice of the fundamental matrix. To show this, let $\Psi_1$ and $\Psi_2$ be two different fundamental matrices of (3.1). Then by Theorem 3.6 there exists a constant $n \times n$ nonsingular matrix $P$ such that $\Psi_2 = \Psi_1 P$. Now by the definition of state transition matrix, we have $\Phi(t, t_0) = \Psi_2(t)[\Psi_2(t_0)]^{-1} = \Psi_1(t)PP^{-1}[\Psi_1(t_0)]^{-1} = \Psi_1(t)[\Psi_1(t_0)]^{-1}$. This shows that $\Phi(t, t_0)$ is independent of the fundamental matrix chosen.

### Properties of the State Transition Matrix

In the following discussion, we summarize some of the properties of state transition matrix.

**Theorem 3.9.** *Let $t_0 \in J$, let $\phi(t_0) = x_0$, and let $\Phi(t, t_0)$ denote the state transition matrix for (3.1) for all $t \in J$. Then the following statements are true:*

(i)  *$\Phi(t, t_0)$ is the unique solution of the matrix equation $\frac{\partial}{\partial t}\Phi(t, t_0) = A\Phi(t, t_0)$ with $\Phi(t_0, t_0) = I$, the $n \times n$ identity matrix.*
(ii)  *$\Phi(t, t_0)$ is nonsingular for all $t \in J$.*
(iii)  *For any $t, \sigma, \tau \in J$, we have $\Phi(t, \tau) = \Phi(t, \sigma)\Phi(\sigma, \tau)$ (semigroup property).*
(iv)  *$[\Phi(t, t_0)]^{-1} \triangleq \Phi^{-1}(t, t_0) = \Phi(t_0, t)$ for all $t, t_0 \in J$.*
(v)  *The unique solution $\phi(t, t_0, x_0)$ of (3.1), with $\phi(t_0, t_0, x_0) = x_0$ specified, is given by*

$$\phi(t, t_0, x_0) = \Phi(t, t_0)x_0 \text{ for all } t \in J. \tag{3.10}$$

*Proof.* (i)  For any fundamental matrix of (3.1), say $\Psi$, we have, by definition, $\Phi(t, t_0) = \Psi(t)\Psi^{-1}(t_0)$, independent of the choice of $\Psi$. Therefore, $\frac{\partial}{\partial t}\Phi(t, t_0) = \dot{\Psi}(t)\Psi^{-1}(t_0) = A\Psi(t)\Psi^{-1}(t_0) = A\Phi(t, t_0)$. Furthermore, $\Phi(t_0, t_0) = \Psi(t_0)\Psi^{-1}(t_0) = I$.

(ii)  For any fundamental matrix of (3.1) we have that $\det \Psi(t) \neq 0$ for all $t \in J$. Therefore, $\det \Phi(t, t_0) = \det[\Psi(t)\Psi^{-1}(t_0)] = \det \Psi(t) \det \Psi^{-1}(t_0) \neq 0$ for all $t, t_0 \in J$.

(iii) For any fundamental matrix $\Psi$ of (3.1) and for the state transition matrix $\Phi$ of (3.1), we have $\Phi(t,\tau) = \Psi(t)\Psi^{-1}(\tau) = \Psi(t)\Psi^{-1}(\sigma)\Psi(\sigma)\Psi^{-1}(\tau) = \Phi(t,\sigma)\Phi(\sigma,\tau)$ for any $t,\sigma,\tau \in J$.

(iv) Let $\Psi$ be any fundamental matrix of (3.1), and let $\Phi$ be the state transition matrix of (3.1). Then $[\Phi(t,t_0)]^{-1} = [\Psi(t)\Psi(t_0)^{-1}]^{-1} = \Psi(t_0)\Psi^{-1}(t) = \Phi(t_0,t)$ for any $t,t_0 \in J$.

(v) By the results established in Chapter 1, we know that for every $(t_0,x_0) \in D$, (3.1) has a unique solution $\phi(t)$ for all $t \in J$ with $\phi(t_0) = x_0$. To verify (3.10), we note that $\dot{\phi}(t) = \frac{\partial \Phi}{\partial t}(t,t_0)x_0 = A\Phi(t,t_0)x_0 = A\phi(t)$. ∎

In Chapter 1 we pointed out that the state transition matrix $\Phi(t,t_0)$ maps the solution (state) of (3.1) at time $t_0$ to the solution (state) of (3.1) at time $t$. Since there is no restriction on $t$ relative to $t_0$ (i.e., we may have $t < t_0$, $t = t_0$, or $t > t_0$), we can "move forward or backward" in time. Indeed, given the solution (state) of (3.1) at time $t$, we can solve the solution (state) of (3.1) at time $t_0$. Thus, $x(t_0) = x_0 = [\Phi(t,t_0)]^{-1}\phi(t,t_0,x_0) = \Phi(t_0,t)\phi(t,t_0,x_0)$. This *reversibility in time* is possible because $\Phi^{-1}(t,t_0)$ always exists. [In the case of discrete-time systems described by difference equations, this reversibility in time does in general not exist (refer to Section 3.5).]

### 3.2.3 Nonhomogeneous Equations

In Section 1.8, we proved the following result [refer to (1.87) to (1.89)].

**Theorem 3.10.** *Let $t_0 \in J$, let $(t_0,x_0) \in D$, and let $\Phi(t,t_0)$ denote the state transition matrix for (3.1) for all $t \in J$. Then the unique solution $\phi(t,t_0,x_0)$ of (3.2) satisfying $\phi(t_0,t_0,x_0) = x_0$ is given by*

$$\phi(t,t_0,x_0) = \Phi(t,t_0)x_0 + \int_{t_0}^{t} \Phi(t,\eta)g(\eta)d\eta. \tag{3.11}$$

∎

As pointed out in Section 1.8, when $x_0 = 0$, (3.11) reduces to the *zero state response*

$$\phi(t,t_0,0) \triangleq \phi_p(t) = \int_{t_0}^{t} \Phi(t,s)g(s)ds, \tag{3.12}$$

and when $x_0 \neq 0$, but $g(t) \equiv 0$, (3.11) reduces to the *zero input response*

$$\phi(t,t_0,x_0) \triangleq \phi_h(t) = \Phi(t,t_0)x_0 \tag{3.13}$$

and the solution of (3.2) may be viewed as consisting of a component that is due to the initial data $x_0$ and another component that is due to the forcing term $g(t)$. We recall that $\phi_p$ is called a *particular solution* of the nonhomogeneous system (3.2), whereas $\phi_h$ is called the *homogeneous solution*.

## 3.3 The Matrix Exponential $e^{At}$, Modes, and Asymptotic Behavior of $\dot{x} = Ax$

In the time-invariant case $\dot{x} = Ax$, the state transition matrix $\Phi(t, t_0)$ equals the matrix exponential $e^{A(t-t_0)}$, which is studied in the following discussion.

Let $D = \{(t, x) : t \in R, \, x \in R^n\}$. In view of the results of Section 1.8, it follows that for every $(t_0, x_0) \in D$, the unique solution of (3.1) with $x(0) = x_0$ specified is given by

$$\phi(t, t_0, x_0) = \left( I + \sum_{k=1}^{\infty} \frac{A^k (t - t_0)^k}{k!} \right) x_0$$

$$= \Phi(t, t_0) x_0 \triangleq \Phi(t - t_0) x_0 \triangleq e^{A(t-t_0)} x_0, \tag{3.14}$$

where $\Phi(t - t_0) = e^{A(t-t_0)}$ denotes the state transition matrix for (3.1). [By writing $\Phi(t, t_0) = \Phi(t - t_0)$, we are using a slight abuse of notation.]

In arriving at (3.14) we invoked Theorem 1.15 of Chapter 1 in Section 1.5, to show that the sequence $\{\phi_m\}$, where

$$\phi_m(t, t_0, x_0) = \left( I + \sum_{k=1}^{m} \frac{A^k (t - t_0)^k}{k!} \right) x_0 \triangleq S_m(t - t_0) x_0, \tag{3.15}$$

converges uniformly and absolutely as $m \to \infty$ to the unique solution $\phi(t, t_0, x_0)$ of (3.1) given by (3.14) on compact subsets of $R$. In the process of arriving at this result, we also proved the following results.

**Theorem 3.11.** *Let $A$ be a constant $n \times n$ matrix (which may be real or complex), and let $S_m(t)$ denote the partial sum of matrices defined by*

$$S_m(t) = I + \sum_{k=1}^{m} \frac{t^k}{k!} A^k. \tag{3.16}$$

*Then each element of the matrix $S_m(t)$ converges absolutely and uniformly on any finite $t$ interval $(-a, a), a > 0$, as $m \to \infty$. Furthermore, $\dot{S}_m(t) = A S_{m-1}(t) = S_{m-1}(t) A$, and thus, the limit of $S_m(t)$ as $t \to \infty$ is a $C^1$ function on $R$. Moreover, this limit commutes with $A$.* ∎

### 3.3.1 Properties of $e^{At}$

In view of Theorem 3.11, the following definition makes sense (see also Section 1.8).

**Definition 3.12.** *Let $A$ be a constant $n \times n$ matrix (which may be real or complex). We define $e^{At}$ to be the matrix*

$$e^{At} = I + \sum_{k=1}^{\infty} \frac{t^k}{k!} A^k \tag{3.17}$$

*for any $-\infty < t < \infty$, and we call $e^{At}$ a matrix exponential.* ∎

We are now in a position to provide the following characterizations of $e^{At}$.

**Theorem 3.13.** *Let $J = R, t_0 \in J$, and let $A$ be a given constant matrix for (3.1). Then*

(i) $\Phi(t) \triangleq e^{At}$ *is a fundamental matrix for all $t \in J$.*
(ii) *The state transition matrix for (3.1) is given by $\Phi(t, t_0) = e^{A(t-t_0)} \triangleq \Phi(t - t_0), t \in J$.*
(iii) $e^{At_1} e^{At_2} = e^{A(t_1+t_2)}$ *for all $t_1, t_2 \in J$.*
(iv) $A e^{At} = e^{At} A$ *for all $t \in J$.*
(v) $(e^{At})^{-1} = e^{-At}$ *for all $t \in J$.*

*Proof.* By (3.17) and Theorem 3.11 we have that $\frac{d}{dt}[e^{At}] = \lim_{m\to\infty} A S_m(t) = \lim_{m\to\infty} S_m(t) A = A e^{At} = e^{At} A$. Therefore, $\Phi(t) = e^{At}$ is a solution of the matrix equation $\dot{\Phi} = A\Phi$. Next, observe that $\Phi(0) = I$. It follows from Theorem 3.4 that $\det[e^{At}] = e^{trace(At)} \neq 0$ for all $t \in R$. Therefore, by Theorem 3.5 $\Phi(t) = e^{At}$ is a fundamental matrix for $\dot{x} = Ax$. We have proved parts (i) and (iv).

To prove (iii), we note that in view of Theorem 3.9(iii), we have for any $t_1, t_2 \in R$ that $\Phi(t_1, t_2) = \Phi(t_1, 0)\Phi(0, t_2)$. By Theorem 3.9(i) we see that $\Phi(t, t_0)$ solves (3.1) with $\Phi(t_0, t_0) = I$. It was just proved that $\Psi(t) \triangleq e^{A(t-t_0)}$ is also a solution. By uniqueness, it follows that $\Phi(t, t_0) = e^{A(t-t_0)}$. For $t = t_1, t_0 = -t_2$, we therefore obtain $e^{A(t_1+t_2)} = \Phi(t_1, -t_2) = \Phi(t_1)\Phi(-t_2)^{-1}$, and for $t = t_1, t_0 = 0$, we have $\Phi(t_1, 0) = e^{At_1} = \Phi(t_1)$. Also, for $t = 0, t_0 = -t_2$, we obtain $\Phi(0, -t_2) = e^{t_2 A} = \Phi(-t_2)^{-1}$. Therefore, $e^{A(t_1+t_2)} = e^{At_1} e^{At_2}$ for all $t_1, t_2 \in R$.

Finally, to prove (ii), we note that by (iii) we have $\Phi(t, t_0) \triangleq e^{A(t-t_0)} = I + \sum_{k=1}^{\infty} \frac{(t-t_0)^k}{k!} A^k = \Phi(t - t_0)$ is a fundamental matrix for $\dot{x} = Ax$ with $\Phi(t_0, t_0) = I$. Therefore, it is its state transition matrix. ∎

We conclude this section by stating the solution of $\dot{x} = Ax + g(t)$,

$$\phi(t, t_0, x_0) = \Phi(t - t_0)x_0 + \int_{t_0}^{t} \Phi(t - s)g(s)ds$$

$$= e^{A(t-t_0)}x_0 + \int_{t_0}^{t} e^{A(t-s)}g(s)ds$$

$$= e^{A(t-t_0)}x_0 + e^{At}\int_{t_0}^{t} e^{-As}g(s)ds, \qquad (3.18)$$

for all $t \in R$. In arriving at (3.18), we have used expression (1.87) of Chapter 1 and the fact that in this case, $\Phi(t, t_0) = e^{A(t-t_0)}$.

### 3.3.2 How to Determine $e^{At}$

We begin by considering the specific case

$$A = \begin{bmatrix} 0 & \alpha \\ 0 & 0 \end{bmatrix}. \tag{3.19}$$

From (3.17) it follows immediately that

$$e^{At} = I + tA = \begin{bmatrix} 1 & \alpha t \\ 0 & 1 \end{bmatrix}. \tag{3.20}$$

As another example, we consider

$$A = \begin{bmatrix} \lambda_1 & 0 \\ 0 & \lambda_2 \end{bmatrix} \tag{3.21}$$

where $\lambda_1, \lambda_2 \in R$. Again, from (3.17) it follows that

$$
\begin{aligned}
e^{At} &= \begin{bmatrix} 1 + \sum_{k=1}^{\infty} \frac{t^k}{k!} \lambda_1^k & 0 \\ 0 & 1 + \sum_{k=1}^{\infty} \frac{t^k}{k!} \lambda_2^k \end{bmatrix} \\
&= \begin{bmatrix} e^{\lambda_1 t} & 0 \\ 0 & e^{\lambda_2 t} \end{bmatrix}.
\end{aligned} \tag{3.22}
$$

Unfortunately, in general it is much more difficult to evaluate the matrix exponential than the preceding examples suggest. In the following discussion, we consider several methods of evaluating $e^{At}$.

### The Infinite Series Method

In this case we evaluate the partial sum $S_m(t)$ (see Theorem 3.11)

$$S_m(t) = I + \sum_{k=1}^{m} \frac{t^k}{k!} A^k$$

for some fixed $t$, say, $t_1$, and for $m = 1, 2, \ldots$ until no significant changes occur in succeeding sums. This yields the matrix $e^{At_1}$. This method works reasonably well if the smallest and largest real parts of the eigenvalues of $A$ are not widely separated.

In the same spirit as above, we could use any of the vector differential solvers to solve $\dot{x} = Ax$, using the natural basis for $R^n$ as $n$ linearly independent initial conditions [i.e., using as initial conditions the vectors $e_1 = (1, 0, \ldots, 0)^T$, $e_2 = (0, 1, 0, \ldots, 0)^T, \ldots, e_n = (0, \ldots, 0, 1)^T$] and observing that in view of (3.14), the resulting solutions are the columns of $e^{At}$ (with $t_0 = 0$).

---

**Example 3.14.** There are cases when the definition of $e^{At}$ (in series form) directly produces a closed-form expression. This occurs for example when $A^k = 0$ for some $k$. In particular, if all the eigenvalues of $A$ are at the origin, then $A^k = 0$ for some $k \leq n$. In this case, only a finite number of terms in (3.17) will be nonzero and $e^{At}$ can be evaluated in closed form. This was precisely the case in (3.19).

---

### The Similarity Transformation Method

Let us consider the initial-value problem

$$\dot{x} = Ax, \quad x(t_0) = x_0; \tag{3.23}$$

let $P$ be a real $n \times n$ nonsingular matrix, and consider the transformation $x = Py$, or equivalently, $y = P^{-1}x$. Differentiating both sides with respect to $t$, we obtain $\dot{y} = P^{-1}\dot{x} = P^{-1}APy = Jy, y(t_0) = y_0 = P^{-1}x_0$. The solution of the above equation is given by

$$\psi(t, t_0, y_0) = e^{J(t-t_0)}P^{-1}x_0. \tag{3.24}$$

Using (3.24) and $x = Py$, we obtain for the solution of (3.23),

$$\phi(t, t_0, x_0) = Pe^{J(t-t_0)}P^{-1}x_0. \tag{3.25}$$

Now suppose that the similarity transformation $P$ given above has been chosen in such a manner that

$$J = P^{-1}AP \tag{3.26}$$

is in Jordan canonical form (refer to Section A.6). We first consider the case when $A$ has $n$ linearly independent eigenvectors, say, $v_i$, that correspond to the eigenvalues $\lambda_i$ (not necessarily distinct), $i = 1, \ldots, n$. (Necessary and sufficient conditions for this to be the case are given in Sections A.5 and A.6. A sufficient condition for the eigenvectors $v_i$, $i = 1, \ldots, n$, to be linearly independent is that the eigenvalues of $A, \lambda_1, \ldots, \lambda_n$, be distinct.) Then $P$ can be chosen so that $P = [v_1, \ldots, v_n]$ and the matrix $J = P^{-1}AP$ assumes the form

$$J = \begin{bmatrix} \lambda_1 & & 0 \\ & \ddots & \\ 0 & & \lambda_n \end{bmatrix}. \tag{3.27}$$

Using the power series representation

$$e^{Jt} = I + \sum_{k=1}^{\infty} \frac{t^k J^k}{k!}, \tag{3.28}$$

we immediately obtain the expression

$$e^{Jt} = \begin{bmatrix} e^{\lambda_1 t} & & 0 \\ & \ddots & \\ 0 & & e^{\lambda_n t} \end{bmatrix}. \tag{3.29}$$

Accordingly, the solution of the initial-value problem (3.23) is now given by

$$\phi(t, t_0, x_0) = P \begin{bmatrix} e^{\lambda_1(t-t_0)} & & 0 \\ & \ddots & \\ 0 & & e^{\lambda_n(t-t_0)} \end{bmatrix} P^{-1}x_0. \tag{3.30}$$

In the general case when $A$ has repeated eigenvalues, it is no longer possible to diagonalize $A$ (see Section A.6). However, we can generate $n$ linearly independent vectors $v_1, \ldots, v_n$ and an $n \times n$ similarity transformation $P = [v_1, \ldots, v_n]$ that takes $A$ into the Jordan canonical form $J = P^{-1}AP$. Here $J$ is in the block diagonal form given by

$$J = \begin{bmatrix} J_0 & & & 0 \\ & J_1 & & \\ & & \ddots & \\ 0 & & & J_s \end{bmatrix}, \tag{3.31}$$

where $J_0$ is a diagonal matrix with diagonal elements $\lambda_1, \ldots, \lambda_k$ (not necessarily distinct), and each $J_i, i \geq 1$, is an $n_i \times n_i$ matrix of the form

$$J_i = \begin{bmatrix} \lambda_{k+i} & 1 & 0 & \cdots & 0 \\ 0 & \lambda_{k+i} & 1 & \cdots & 0 \\ \vdots & \vdots & \ddots & \ddots & \vdots \\ 0 & 0 & \cdots & \ddots & 1 \\ 0 & 0 & 0 & \cdots & \lambda_{k+i} \end{bmatrix}, \tag{3.32}$$

where $\lambda_{k+i}$ need not be different from $\lambda_{k+j}$ if $i \neq j$, and where $k + n_1 + \cdots + n_s = n$.

Now since for any square block diagonal matrix

$$C = \begin{bmatrix} C_1 & & 0 \\ & \ddots & \\ 0 & & C_l \end{bmatrix}$$

with $C_i, i = 1, \ldots, l$, square, we have that

$$C^k = \begin{bmatrix} C_1^k & & 0 \\ & \ddots & \\ 0 & & C_l^k \end{bmatrix},$$

it follows from the power series representation of $e^{Jt}$ that

$$e^{Jt} = \begin{bmatrix} e^{J_0 t} & & & 0 \\ & e^{J_1 t} & & \\ & & \ddots & \\ 0 & & & e^{J_s t} \end{bmatrix}, \tag{3.33}$$

$t \in R$. As shown earlier, we have

$$e^{J_0 t} = \begin{bmatrix} e^{\lambda_1 t} & & 0 \\ & \ddots & \\ 0 & & e^{\lambda_k t} \end{bmatrix}. \tag{3.34}$$

For $J_i$, $i = 1, \ldots, s$, we have

$$J_i = \lambda_{k+i} I_i + N_i, \tag{3.35}$$

where $I_i$ denotes the $n_i \times n_i$ identity matrix and $N_i$ is the $n_i \times n_i$ nilpotent matrix given by

$$N_i = \begin{bmatrix} 0 & 1 & \cdots & 0 \\ \vdots & \ddots & \ddots & \vdots \\ \vdots & & \ddots & 1 \\ 0 & \cdots & \cdots & 0 \end{bmatrix}. \tag{3.36}$$

Since $\lambda_{k+i} I_i$ and $N_i$ commute, we have that

$$e^{J_i t} = e^{\lambda_{k+i} t} e^{N_i t}. \tag{3.37}$$

Repeated multiplication of $N_i$ by itself results in $N_i^k = 0$ for all $k \geq n_i$. Therefore, the series defining $e^{t N_i}$ terminates, resulting in

$$e^{t J_i} = e^{\lambda_{k+i} t} \begin{bmatrix} 1 & t & \cdots & t^{n_i - 1}/(n_i - 1)! \\ 0 & 1 & \cdots & t^{n_i - 2}/(n_i - 2)! \\ \vdots & \vdots & \ddots & \vdots \\ 0 & 0 & \cdots & 1 \end{bmatrix}, \quad i = 1, \ldots, s. \tag{3.38}$$

It now follows that the solution of (3.23) is given by

$$\phi(t, t_0, x_0) = P \begin{bmatrix} e^{J_0(t-t_0)} & 0 & \cdots & 0 \\ 0 & e^{J_1(t-t_0)} & \cdots & 0 \\ \vdots & \vdots & \ddots & \vdots \\ 0 & 0 & & e^{J_s(t-t_0)} \end{bmatrix} P^{-1} x_0. \tag{3.39}$$

---

**Example 3.15.** In system (3.23), let $A = \begin{bmatrix} -1 & 2 \\ 0 & 1 \end{bmatrix}$. The eigenvalues of $A$ are $\lambda_1 = -1$ and $\lambda_2 = 1$, and corresponding eigenvectors for $A$ are given by $v_1 = (1,0)^T$ and $v_2 = (1,1)^T$, respectively. Then $P = [v_1, v_2] = \begin{bmatrix} 1 & 1 \\ 0 & 1 \end{bmatrix}$, $P^{-1} = \begin{bmatrix} 1 & -1 \\ 0 & 1 \end{bmatrix}$, and $J = P^{-1} A P = \begin{bmatrix} 1 & -1 \\ 0 & 1 \end{bmatrix} \begin{bmatrix} -1 & 2 \\ 0 & 1 \end{bmatrix} \begin{bmatrix} 1 & 1 \\ 0 & 1 \end{bmatrix} = \begin{bmatrix} -1 & 0 \\ 0 & 1 \end{bmatrix} = \begin{bmatrix} \lambda_1 & 0 \\ 0 & \lambda_2 \end{bmatrix}$, as expected. We obtain $e^{At} = Pe^{Jt}P^{-1} = \begin{bmatrix} 1 & 1 \\ 0 & 1 \end{bmatrix} \begin{bmatrix} e^{-t} & 0 \\ 0 & e^t \end{bmatrix} \begin{bmatrix} 1 & -1 \\ 0 & 1 \end{bmatrix} = \begin{bmatrix} e^t & e^t - e^{-t} \\ 0 & e^t \end{bmatrix}$.

---

Suppose next that in (3.23) the matrix $A$ is either in *companion form* or that it has been transformed into this form via some suitable similarity transformation $P$, so that $A = A_c$, where

$$A_c = \begin{bmatrix} 0 & 1 & 0 & \cdots & 0 \\ 0 & 0 & 1 & \cdots & 0 \\ \vdots & \vdots & \vdots & & \vdots \\ 0 & 0 & 0 & \cdots & 1 \\ -a_0 & -a_1 & -a_2 & \cdots & -a_{n-1} \end{bmatrix}. \tag{3.40}$$

Since in this case we have $x_{i+1} = \dot{x}_i$, $i = 1, \ldots, n-1$, it should be clear that in the calculation of $e^{At}$ we need to determine, via some method, only the first row of $e^{At}$. We demonstrate this by means of a specific example.

---

***Example 3.16.*** In system (3.23), assume that $A = A_c = \begin{bmatrix} 0 & 1 \\ -2 & -3 \end{bmatrix}$, which is in companion form. To demonstrate the above observation, let us compute $e^{At}$ by some other method, say diagonalization. The eigenvalues of $A$ are $\lambda_1 = -1$ and $\lambda_2 = -2$, and a set of corresponding eigenvectors is given by $v_1 = (1, -1)^T$ and $v_2 = (1, -2)^T$. We obtain $P = [v_1, v_2] = \begin{bmatrix} 1 & 1 \\ -1 & -2 \end{bmatrix}$, $P^{-1} = \begin{bmatrix} 2 & 1 \\ -1 & -1 \end{bmatrix}$ and $J = P^{-1}A_cP = \begin{bmatrix} -1 & 0 \\ 0 & -2 \end{bmatrix}$, $e^{At} = Pe^{Jt}P^{-1} =$
$\begin{bmatrix} 1 & 1 \\ -1 & -2 \end{bmatrix}\begin{bmatrix} e^{-t} & 0 \\ 0 & e^{-2t} \end{bmatrix}\begin{bmatrix} 2 & 1 \\ -1 & -1 \end{bmatrix} = \begin{bmatrix} (2e^{-t} - e^{-2t}) & (e^{-t} - e^{-2t}) \\ (-2e^{-t} + 2e^{-2t}) & (-e^{-t} + 2e^{-2t}) \end{bmatrix}$. Note that the second row of the above matrix is the derivative of the first row, as expected.

---

## The Cayley–Hamilton Theorem Method

If $\alpha(\lambda) = \det(\lambda I - A)$ is the characteristic polynomial of an $n \times n$ matrix $A$, we have that $\alpha(A) = 0$, in view of the Cayley–Hamilton Theorem; i.e., every $n \times n$ matrix satisfies its characteristic equation (refer to Sections A.5 and A.6). Using this result, along with the series definition of the matrix exponential $e^{At}$, it is easily shown that

$$e^{At} = \sum_{i=0}^{n-1} \alpha_i(t)A^i \tag{3.41}$$

[Refer to Sections A.5 and A.6 for the details on how to determine the terms $\alpha_i(t)$.]

**The Laplace Transform Method**

We assume that the reader is familiar with the basics of the (one-sided) Laplace transform. If $f(t) = [f_1(t), \ldots, f_n(t)]^T$, where $f_i : [0, \infty) \to R$, $i = 1, \ldots, n$, and if each $f_i$ is Laplace transformable, then we define the Laplace transform of the vector $f$ component-wise; i.e., $\hat{f}(s) = [\hat{f}_1(s), \ldots, \hat{f}_n(s)]^T$, where $\hat{f}_i(s) = \mathcal{L}[f_i(t)] \triangleq \int_0^\infty f_i(t)e^{-st}dt$.

We define the Laplace transform of a matrix $C(t) = [c_{ij}(t)]$ similarly. Thus, if each $c_{ij} : [0, \infty) \to R$ and if each $c_{ij}$ is Laplace transformable, then the Laplace transform of $C(t)$ is defined as $\widehat{C}(s) = \mathcal{L}[c_{ij}(t)] = [\mathcal{L}c_{ij}(t)] = [\hat{c}_{ij}(s)]$.

Laplace transforms of some of the common time signals are enumerated in Table 3.1. Also, in Table 3.2 we summarize some of the more important properties of the Laplace transform. In Table 3.1, $\delta(t)$ denotes the *Dirac delta distribution* (see Subsection 2.4.3) and $p(t)$ represents the *unit step function*.

**Table 3.1.** Laplace transforms

| $f(t)(t \geq 0)$ | $\hat{f}(s) = \mathcal{L}[f(t)]$ |
|---|---|
| $\delta(t)$ | $1$ |
| $p(t)$ | $1/s$ |
| $t^k/k!$ | $1/s^{k+1}$ |
| $e^{-at}$ | $1/(s+a)$ |
| $t^k e^{-at}$ | $k!/(s+a)^{k+1}$ |
| $e^{-at}\sin bt$ | $b/[(s+a)^2 + b^2]$ |
| $e^{-at}\cos bt$ | $(s+a)/[(s+a)^2 + b^2]$ |

**Table 3.2.** Laplace transform properties

| | | |
|---|---|---|
| Time different-iation | $df(t)/dt$ | $s\hat{f}(s) - f(0)$ |
| | $d^k f(t)/dt^k$ | $s^k \hat{f}(s) - [s^{k-1}f(0) + \cdots + f^{(k-1)}(0)]$ |
| Frequency shift | $e^{-at}f(t)$ | $\hat{f}(s+a)$ |
| Time shift | $f(t-a)p(t-a), a > 0$ | $e^{-as}\hat{f}(s)$ |
| Scaling | $f(t/\alpha), \alpha > 0$ | $\alpha\hat{f}(\alpha s)$ |
| Convolution | $\int_0^t f(\tau)g(t-\tau)d\tau = f(t) * g(t)$ | $\hat{f}(s)\hat{g}(s)$ |
| Initial value | $\lim_{t \to 0+} f(t) = f(0^+)$ | $\lim_{s \to \infty} s\hat{f}(s)^\dagger$ |
| Final value | $\lim_{t \to \infty} f(t)$ | $\lim_{s \to 0} s\hat{f}(s)^\ddagger$ |

$^\dagger$ If the limit exists.

$^\ddagger$ If $s\hat{f}(s)$ has no singularities on the imaginary axis or in the right half $s$ plane.

Now consider once more the initial-value problem (3.23), letting $t_0 = 0$; i.e.,

$$\dot{x} = Ax, \quad x(0) = x_0. \tag{3.42}$$

Taking the Laplace transform of both sides of $\dot{x} = Ax$, and taking into account the initial condition $x(0) = x_0$, we obtain $s\hat{x}(s) - x_0 = A\hat{x}(s)$, $(sI - A)\hat{x}(s) = x_0$, or

$$\hat{x}(s) = (sI - A)^{-1}x_0. \tag{3.43}$$

It can be shown by analytic continuation that $(sI - A)^{-1}$ exists for all $s$, except at the eigenvalues of $A$. Taking the inverse Laplace transform of (3.43), we obtain the solution

$$\phi(t) = \mathcal{L}^{-1}[(sI - A)^{-1}]x_0 = \Phi(t, 0)x_0 = e^{At}x_0. \tag{3.44}$$

It follows from (3.42) and (3.44) that $\hat{\Phi}(s) = (sI - A)^{-1}$ and that

$$\Phi(t, 0) \triangleq \Phi(t - 0) = \Phi(t) = \mathcal{L}^{-1}[(sI - A)^{-1}] = e^{At}. \tag{3.45}$$

Finally, note that when $t_0 \neq 0$, we can immediately compute $\Phi(t, t_0) = \Phi(t - t_0) = e^{A(t - t_0)}$.

---

***Example 3.17.*** In (3.42), let $A = \begin{bmatrix} -1 & 2 \\ 0 & 1 \end{bmatrix}$. Then

$$(sI - A)^{-1} = \begin{bmatrix} s+1 & -2 \\ 0 & s-1 \end{bmatrix}^{-1} = \begin{bmatrix} \frac{1}{s+1} & \frac{2}{(s+1)(s-1)} \\ 0 & \frac{1}{s-1} \end{bmatrix} = \begin{bmatrix} \frac{1}{s+1} & \left(\frac{1}{s-1} - \frac{1}{s+1}\right) \\ 0 & \frac{1}{s-1} \end{bmatrix}.$$

Using Table 3.1, we obtain $\mathcal{L}^{-1}[(sI - A)^{-1}] = e^{At} = \begin{bmatrix} e^{-t} & (e^t - e^{-t}) \\ 0 & e^t \end{bmatrix}$.

---

Before concluding this subsection, we briefly consider initial-value problems described by

$$\dot{x} = Ax + g(t), \quad x(t_0) = x_0. \tag{3.46}$$

We wish to apply the Laplace transform method discussed above in solving (3.46). To this end we assume $t_0 = 0$ and we take the Laplace transform of both sides of (3.46) to obtain $s\hat{x}(s) - x_0 = A\hat{x}(s) + \hat{g}(s)$, $(sI - A)\hat{x}(s) = x_0 + \hat{g}(s)$, or

$$\begin{aligned} \hat{x}(s) &= (sI - A)^{-1}x_0 + (sI - A)^{-1}\hat{g}(s) \\ &= \hat{\Phi}(s)x_0 + \hat{\Phi}(s)\hat{g}(s) \\ &\triangleq \hat{\phi}_h(s) + \hat{\phi}_p(s). \end{aligned} \tag{3.47}$$

Taking the inverse Laplace transform of both sides of (3.47) and using (3.18) with $t_0 = 0$, we obtain $\phi(t) = \phi_h(t) + \phi_p(t) = \mathcal{L}^{-1}[(sI - A)^{-1}]x_0 + \mathcal{L}^{-1}[(sI - A)^{-1}\hat{g}(s)] = \Phi(t)x_0 + \int_0^t \Phi(t - \eta)g(\eta)d\eta$, where $\phi_h$ denotes the homogeneous solution and $\phi_p$ is the particular solution, as expected.

**Example 3.18.** Consider the initial-value problem given by

$$\dot{x}_1 = -x_1 + x_2,$$
$$\dot{x}_2 = -2x_2 + u(t),$$

with $x_1(0) = -1$, $x_2(0) = 0$, and

$$u(t) = \begin{cases} 1 & \text{for } t > 0, \\ 0 & \text{for } t \leq 0. \end{cases}$$

It is easily verified that in this case

$$\hat{\Phi}(s) = \begin{bmatrix} \frac{1}{s+1} & \left(\frac{1}{s+1} - \frac{1}{s+2}\right) \\ 0 & \frac{1}{s+2} \end{bmatrix},$$

$$\Phi(t) = \begin{bmatrix} e^{-t} & (e^{-t} - e^{-2t}) \\ 0 & e^{-2t} \end{bmatrix},$$

$$\phi_h(t) = \begin{bmatrix} e^{-t} & (e^{-t} - e^{-2t}) \\ 0 & e^{-t} \end{bmatrix}\begin{bmatrix} -1 \\ 0 \end{bmatrix} = \begin{bmatrix} -e^{-t} \\ 0 \end{bmatrix},$$

$$\hat{\phi}_p(s) = \begin{bmatrix} \frac{1}{s+1} & \left(\frac{1}{s+1} - \frac{1}{s+2}\right) \\ 0 & \frac{1}{s+2} \end{bmatrix}\begin{bmatrix} 0 \\ \frac{1}{s} \end{bmatrix} = \begin{bmatrix} \frac{1}{2}\left(\frac{1}{s}\right) + \frac{1}{2}\left(\frac{1}{s+2}\right) - \frac{1}{s+1} \\ \frac{1}{2}\left(\frac{1}{s}\right) - \frac{1}{2}\left(\frac{1}{s+2}\right) \end{bmatrix},$$

$$\phi_p(t) = \begin{bmatrix} \frac{1}{2} + \frac{1}{2}e^{-2t} - e^{-t} \\ \frac{1}{2} - \frac{1}{2}e^{-2t} \end{bmatrix},$$

and

$$\phi(t) = \phi_h(t) + \phi_p(t) = \begin{bmatrix} \frac{1}{2} - 2e^{-t} + \frac{1}{2}e^{-2t} \\ \frac{1}{2} - \frac{1}{2}e^{-2t} \end{bmatrix}.$$

### 3.3.3 Modes, Asymptotic Behavior, and Stability

In this subsection we study the qualitative behavior of the solutions of $\dot{x} = Ax$ by means of the modes of such systems, to be introduced shortly. Although we will not address the stability of systems in detail until Chapter 4, the results here will enable us to give some general stability characterizations for such systems.

### Modes: General Case

We begin by recalling that the unique solution of

$$\dot{x} = Ax, \tag{3.48}$$

satisfying $x(0) = x_0$, is given by

$$\phi(t, 0, x_0) = \Phi(t, 0)x(0) = \Phi(t, 0)x_0 = e^{At}x_0. \qquad (3.49)$$

We also recall that $\det(sI - A) = \prod_{i=1}^{\sigma}(s - \lambda_i)^{n_i}$, where $\lambda_1, \ldots, \lambda_\sigma$ denote the $\sigma$ distinct eigenvalues of $A$, where $\lambda_i$ with $i = 1, \ldots, \sigma$, is assumed to be repeated $n_i$ times (i.e., $n_i$ is the algebraic multiplicity of $\lambda_i$), and $\Sigma_{i=1}^{\sigma} n_i = n$.

To introduce the modes for (3.48), we must show that

$$e^{At} = \sum_{i=1}^{\sigma} \sum_{k=0}^{n_i-1} A_{ik} t^k e^{\lambda_i t}$$

$$= \sum_{i=1}^{\sigma} [A_{i0} e^{\lambda_i t} + A_{i1} t e^{\lambda_i t} + \cdots + A_{i(n_i-1)} t^{n_i-1} e^{\lambda_i t}], \qquad (3.50)$$

where

$$A_{ik} = \frac{1}{k!} \frac{1}{(n_i - 1 - k)!} \lim_{s \to \lambda_i} [[(s - \lambda_i)^{n_i} (sI - A)^{-1}]^{(n_i-1-k)}]. \qquad (3.51)$$

In (3.51), $[\cdot]^{(l)}$ denotes the $l$th derivative with respect to $s$.

Equation (3.50) shows that $e^{At}$ can be expressed as the sum of terms of the form $A_{ik} t^k e^{\lambda_i t}$, where $A_{ik} \in R^{n \times n}$. We call $A_{ik} t^k e^{\lambda_i t}$ a *mode of system (3.48)*. If an eigenvalue $\lambda_i$ is repeated $n_i$ times, there are $n_i$ modes, $A_{ik} t^k e^{\lambda_i t}$, $k = 0, 1, \ldots, n_i - 1$, in $e^{At}$ associated with $\lambda_i$. Accordingly, the solution (3.49) of (3.48) is determined by the $n$ modes of (3.48) corresponding to the $n$ eigenvalues of $A$ and by the initial condition $x(0)$. We note that by selecting $x(0)$ appropriately, modes can be combined or eliminated $[A_{ik} x(0) = 0]$, thus affecting the behavior of $\phi(t, 0, x_0)$.

To verify (3.50) we recall that $e^{At} = \mathcal{L}^{-1}[(sI - A)^{-1}]$ and we make use of the partial fraction expansion method to determine the inverse Laplace transform. As in the scalar case, it can be shown that

$$(sI - A)^{-1} = \sum_{i=1}^{\sigma} \sum_{k=0}^{n_i-1} (k! A_{ik})(s - \lambda_i)^{-(k+1)}, \qquad (3.52)$$

where the $(k! A_{ik})$ are the coefficients of the partial fractions ($k!$ is for scaling). It is known that these coefficients can be evaluated for each $i$ by multiplying both sides of (3.52) by $(s - \lambda_i)^{n_i}$, differentiating $(n_i - 1 - k)$ times with respect to $s$, and then evaluating the resulting expression at $s = \lambda_i$. This yields (3.51). Taking the inverse Laplace transform of (3.52) and using the fact that $\mathcal{L}[t^k e^{\lambda_i t}] = k!(s - \lambda_i)^{-(k+1)}$ (refer to Table 3.1) results in (3.50).

When all $n$ eigenvalues $\lambda_i$ of $A$ are distinct, then $\sigma = n, n_i = 1, i = 1, \ldots, n$, and (3.50) reduces to the expression

$$e^{At} = \sum_{i=1}^{n} A_i e^{\lambda_i t}, \qquad (3.53)$$

where

$$A_i = \lim_{s \to \lambda_i} [(s - \lambda_i)(sI - A)^{-1}]. \tag{3.54}$$

Expression (3.54) can also be derived directly, using a partial fraction expansion of $(sI - A)^{-1}$ given in (3.52).

---

**Example 3.19.** For (3.48) we let $A = \begin{bmatrix} 0 & 1 \\ -4 & -4 \end{bmatrix}$, for which the eigenvalue $\lambda_1 = -2$ is repeated twice; i.e., $n_1 = 2$. Applying (3.50) and (3.51), we obtain

$$e^{At} = A_{10}e^{\lambda_1 t} + A_{11}te^{\lambda_1 t} = \begin{bmatrix} 1 & 0 \\ 0 & 1 \end{bmatrix} e^{-2t} + \begin{bmatrix} 2 & 1 \\ -4 & -2 \end{bmatrix} te^{-2t}.$$

---

**Example 3.20.** For (3.48) we let $A = \begin{bmatrix} 0 & 1 \\ -1 & -1 \end{bmatrix}$, for which the eigenvalues are given by (the complex conjugate pair) $\lambda_1 = -\frac{1}{2} + j\frac{\sqrt{3}}{2}, \lambda_2 = -\frac{1}{2} - j\frac{\sqrt{3}}{2}$. Applying (3.53) and (3.54), we obtain

$$A_1 = \frac{1}{\lambda_1 - \lambda_2} \begin{bmatrix} \lambda_1 + 1 & 1 \\ -1 & \lambda_1 \end{bmatrix} = \frac{1}{j\sqrt{3}} \begin{bmatrix} \frac{1}{2} + j\frac{\sqrt{3}}{2} & 1 \\ -1 & -\frac{1}{2} + j\frac{\sqrt{3}}{2} \end{bmatrix}$$

$$A_2 = \frac{1}{\lambda_2 - \lambda_1} \begin{bmatrix} \lambda_2 + 1 & 1 \\ -1 & \lambda_2 \end{bmatrix} = \frac{1}{-j\sqrt{3}} \begin{bmatrix} \frac{1}{2} - j\frac{\sqrt{3}}{2} & 1 \\ -1 & -\frac{1}{2} - j\frac{\sqrt{3}}{2} \end{bmatrix}$$

[i.e., $A_1 = A_2^*$, where $(\cdot)^*$ denotes the complex conjugate of $(\cdot)$], and

$$e^{At} = A_1 e^{\lambda_1 t} + A_2 e^{\lambda_2 t} = A_1 e^{\lambda_1 t} + A_1^* e^{\lambda_1^* t}$$

$$= 2(Re\, A_1)(Re\, e^{\lambda_1 t}) - 2(Im\, A_1)(Im\, e^{\lambda_1 t})$$

$$= 2e^{-\frac{1}{2}t} \left[ \begin{bmatrix} \frac{1}{2} & 0 \\ 0 & -\frac{1}{2} \end{bmatrix} \cos \frac{\sqrt{3}}{2}t - \begin{bmatrix} -\frac{1}{2\sqrt{3}} & -\frac{1}{\sqrt{3}} \\ \frac{1}{\sqrt{3}} & \frac{1}{2\sqrt{3}} \end{bmatrix} \sin \left( \frac{\sqrt{3}}{2}t \right) \right].$$

The last expression involves only real numbers, as expected, since $A$ and $e^{At}$ are real matrices.

---

**Example 3.21.** For (3.48) we let $A = \begin{bmatrix} 1 & 0 \\ 0 & 1 \end{bmatrix}$, for which the eigenvalue $\lambda_1 = 1$ is repeated twice; i.e., $n_1 = 2$. Applying (3.50) and (3.51), we obtain

$$e^{At} = A_{10}e^{\lambda_1 t} + A_{11}te^{\lambda_1 t} = \begin{bmatrix} 1 & 0 \\ 0 & 1 \end{bmatrix} e^t + \begin{bmatrix} 0 & 0 \\ 0 & 0 \end{bmatrix} te^t = Ie^t.$$

This example shows that not all modes of the system are necessarily present in $e^{At}$. What is present depends in fact on the number and dimensions of the individual blocks of the Jordan canonical form of $A$ corresponding to identical eigenvalues. To illustrate this further, we let for (3.48), $A = \begin{bmatrix} 1 & 1 \\ 0 & 1 \end{bmatrix}$, where the two repeated eigenvalues $\lambda_1 = 1$ belong to the same Jordan block. Then $e^{At} = \begin{bmatrix} 1 & 0 \\ 0 & 1 \end{bmatrix} e^t + \begin{bmatrix} 0 & 1 \\ 0 & 0 \end{bmatrix} te^t$.

---

## Stability of an Equilibrium

In Chapter 4 we will study the *qualitative properties* of linear dynamical systems, including systems described by (3.48). This will be accomplished by studying the *stability properties* of such systems or, more specifically, the *stability properties* of an *equilibrium* of such systems.

If $\phi(t, 0, x_e)$ denotes the solution of system (3.48) with $x(0) = x_e$, then $x_e$ is said to be an *equilibrium* of (3.48) if $\phi(t, 0, x_e) = x_e$ for all $t \geq 0$. Clearly, $x_e = 0$ is an equilibrium of (3.48). In discussing the qualitative properties, it is often customary to speak, somewhat loosely, of the *stability properties of system (3.48)*, rather than the stability properties of the equilibrium $x_e = 0$ of system (3.48).

We will show in Chapter 4 that the following qualitative characterizations of system (3.48) are actually *equivalent* to more fundamental qualitative characterizations of the equilibrium $x_e = 0$ of system (3.48):

1. The system (3.48) is said to be *stable* if all solutions of (3.48) are bounded for all $t \geq 0$ [i.e., for any $\phi(t, 0, x_0) = (\phi_1(t, 0, x_0), \ldots, \phi_n(t, 0, x_0))^T$ of (3.48), there exist constants $M_i$, $i = 1, \ldots, n$ (which in general will depend on the solution on hand) such that $|\phi_i(t, 0, x_0)| < M_i$ for all $t \geq 0$].
2. The system (3.48) is said to be *asymptotically stable* if it is stable and if all solutions of (3.48) tend to the origin as $t$ tends to infinity [i.e., for any solution $\phi(t, 0, x_0) = (\phi_1(t, 0, x_0), \ldots, \phi_n(t, 0, x_0))^T$ of (3.48), we have $\lim_{t \to \infty} \phi_i(t, 0, x_0) = 0$, $i = 1, \ldots, n$].
3. The system (3.48) is said to be *unstable* if it is not stable.

By inspecting the modes of (3.48) given by (3.50), (3.51) and (3.53), (3.54), the following stability criteria for system (3.48) are now evident:

1. The system (3.48) is *asymptotically stable* if and only if all eigenvalues of $A$ have negative real parts (i.e., $Re\lambda_j < 0$, $j = 1, \ldots, n$).
2. The system (3.48) is *stable* if and only if $Re\lambda_j \leq 0$, $j = 1, \ldots, n$, and for all eigenvalues with $Re\lambda_j = 0$ having multiplicity $n_j > 1$, it is true that

$$\lim_{s \to \lambda_j} [(s - \lambda_j)^{n_j} (sI - A)^{-1}]^{(n_j - 1 - k)} = 0, \quad k = 1, \ldots, n_j - 1. \quad (3.55)$$

3. System (3.48) is *unstable* if and only if (2) is not true.

We note in particular that if $Re\lambda_j = 0$ and $n_j > 1$, then there will be modes $A_{jk}t^k$, $k = 0, \ldots, n_j - 1$ that will yield terms in (3.50) whose norm will tend to infinity as $t \to \infty$, unless their coefficients are zero. This shows why the necessary and sufficient conditions for stability of (3.48) include condition (3.55).

---

**Example 3.22.** The systems in Examples 3.19 and 3.20 are asymptotically stable. A system (3.48) with $A = \begin{bmatrix} 0 & 1 \\ 0 & -1 \end{bmatrix}$ is stable, since the eigenvalues of $A$ above are $\lambda_1 = 0$, $\lambda_2 = -1$. A system (3.48) with $A = \begin{bmatrix} -1 & 0 \\ 0 & 1 \end{bmatrix}$ is unstable since the eigenvalues of $A$ are $\lambda_1 = 1$, $\lambda_2 = -1$. The system of Example 3.21 is also unstable.

---

### Modes: Distinct Eigenvalue Case

When the eigenvalues $\lambda_i$ of $A$ are distinct, there is an alternative way to (3.54) of computing the matrix coefficients $A_i$, expressed in terms of the corresponding right and left eigenvectors of $A$. This method offers great insight into questions concerning the presence or absence of modes in the response of a system. Specifically, if $A$ has $n$ distinct eigenvalues $\lambda_i$, then

$$e^{At} = \sum_{i=1}^{n} A_i e^{\lambda_i t}, \tag{3.56}$$

where

$$A_i = v_i \tilde{v}_i, \tag{3.57}$$

where $v_i \in R^n$ and $(\tilde{v}_i)^T \in R^n$ are right and left eigenvectors of $A$ corresponding to the eigenvalue $\lambda_i$, respectively.

To prove the above assertions, we recall that $(\lambda_i I - A)v_i = 0$ and $\tilde{v}_i(\lambda_i I - A) = 0$. If $Q \triangleq [v_1, \ldots, v_n]$, then the $\tilde{v}_i$ are the rows of

$$P = Q^{-1} = \begin{bmatrix} \tilde{v}_1 \\ \vdots \\ \tilde{v}_n \end{bmatrix}.$$

The matrix $Q$ is of course nonsingular, since the eigenvalues $\lambda_i$, $i = 1, \ldots, n$, are by assumption distinct and since the corresponding eigenvectors are linearly independent. Notice that $Q \, \mathrm{diag}[\lambda_1, \ldots, \lambda_n] = AQ$ and that $\mathrm{diag}[\lambda_1, \ldots, \lambda_n]P = PA$. Also, notice that $\tilde{v}_i v_j = \delta_{ij}$, where

$$\delta_{ij} = \begin{cases} 1 & \text{when } i = j, \\ 0 & \text{when } i \neq j. \end{cases}$$

In view of this, we now have $(sI - A)^{-1} = [sI - Q\,\mathrm{diag}[\lambda_1, \ldots, \lambda_n]Q^{-1}]^{-1} = Q[sI - \mathrm{diag}[\lambda_1, \cdots, \lambda_n]]^{-1}Q^{-1} = Q\,\mathrm{diag}[(s - \lambda_1)^{-1}, \ldots, (s - \lambda_n)^{-1}]Q^{-1} = \sum_{i=1}^{n} v_i \tilde{v}_i (s - \lambda_i)^{-1}$. If we take the inverse Laplace transform of the above expression, we obtain (3.56).

If we choose the initial value $x(0)$ for (3.48) to be colinear with an eigenvector $v_j$ of $A$ [i.e., $x(0) = \alpha v_j$ for some real $\alpha \neq 0$], then $e^{\lambda_j t}$ is the only mode that will appear in the solution $\phi$ of (3.48). This can easily be seen from our preceding discussion. In particular if $x(0) = \alpha v_j$, then (3.56) and (3.57) yield

$$\phi(t, 0, x(0)) = e^{At}x(0) = v_1 \tilde{v}_1 x(0)e^{\lambda_1 t} + \cdots + v_n \tilde{v}_n x(0)e^{\lambda_n t} = \alpha v_j e^{\lambda_j t} \quad (3.58)$$

since $\tilde{v}_i v_j = 1$ when $i = j$, and $\tilde{v}_i v_j = 0$ otherwise.

---

**Example 3.23.** In (3.48) we let $A = \begin{bmatrix} -1 & 1 \\ 0 & 1 \end{bmatrix}$. The eigenvalues of $A$ are given

by $\lambda_1 = -1$ and $\lambda_2 = 1$ and $Q = [v_1, v_2] = \begin{bmatrix} 1 & 1 \\ 0 & 2 \end{bmatrix}$, $Q^{-1} = \begin{bmatrix} \tilde{v}_1 \\ \tilde{v}_2 \end{bmatrix} = \begin{bmatrix} 1 & -1/2 \\ 0 & 1/2 \end{bmatrix}$.

Then $e^{At} = v_1 \tilde{v}_1 e^{\lambda_1 t} + v_2 \tilde{v}_2 e^{\lambda_2 t} = \begin{bmatrix} 1 & -1/2 \\ 0 & 0 \end{bmatrix} e^{-t} + \begin{bmatrix} 0 & 1/2 \\ 0 & 1 \end{bmatrix} e^t$. If in particular

we choose $x(0) = \alpha v_1 = (\alpha, 0)^T$, then $\phi(t, 0, x(0)) = e^{At}x(0) = \alpha(1, 0)^T e^{-t}$, which contains only the mode corresponding to the eigenvalue $\lambda_1 = -1$. Thus, for this particular choice of initial vector, the unstable behavior of the system is suppressed.

---

*Remark*

We conclude our discussion of modes and asymptotic behavior by briefly considering systems of linear, nonhomogeneous, ordinary differential equations $\dot{x} = Ax + g(t)$ in (3.2) for the special case where $g(t) = Bu(t)$,

$$\dot{x} = Ax + Bu, \quad (3.59)$$

where $B \in R^{n \times m}$, $u : R \to R^m$, and where it is assumed that the Laplace transform of $u$ exists. Taking the Laplace transform of both sides of (3.59) and rearranging yields

$$\hat{x}(s) = (sI - A)^{-1}x(0) + (sI - A)^{-1}B\hat{u}(s). \quad (3.60)$$

By taking the inverse Laplace transform of (3.60), we see that the solution $\phi$ is the sum of modes that correspond to the singularities or poles of $(sI - A)^{-1}x(0)$ and of $(sI - A)^{-1}B\hat{u}(s)$. If in particular (3.48) is asymptotically stable (i.e., for $\dot{x} = Ax$, $Re\lambda_i < 0$, $i = 1, \ldots, n$) and if $u$ in (3.59) is bounded (i.e., there is an $M$ such that $|u_i(t)| < M$ for all $t \geq 0$, $i = 1, \ldots, m$), then it is easily seen that the solutions of (3.59) are bounded as well. Thus, the fact that the system (3.48) is asymptotically stable has repercussions on the asymptotic behavior of the solution of (3.59). Issues of this type will be addressed in greater detail in Chapter 4.

## 3.4 State Equation and Input–Output Description of Continuous-Time Systems

This section consists of three subsections. We first study the response of linear continuous-time systems. Next, we examine transfer functions of linear time-invariant systems, given the state equations of such systems. Finally, we explore the equivalence of internal representations of systems.

### 3.4.1 Response of Linear Continuous-Time Systems

We consider once more systems described by linear equations of the form

$$\dot{x} = Ax + Bu, \tag{3.61a}$$
$$y = Cx + Du, \tag{3.61b}$$

where $A \in R^{n \times n}$, $B \in R^{n \times m}$, $C \in R^{p \times n}$, $D \in R^{p \times m}$, and $u : R \to R^m$ is assumed to be continuous or piecewise continuous. We recall that in (3.61), $x$ denotes the state vector, $u$ denotes the system input, and $y$ denotes the system output. From Section 2.2 we recall that for given initial conditions $t_0 \in R, x(t_0) = x_0 \in R^n$ and for a given input $u$, the unique solution of (3.61a) is given by

$$\phi(t, t_0, x_0) = \Phi(t, t_0)x_0 + \int_{t_0}^{t} \Phi(t, s)Bu(s)ds \tag{3.62}$$

for $t \in R$, where $\Phi$ denotes the state transition matrix of $A$. Furthermore, by substituting (3.62) into (3.61b), we obtain, for all $t \in R$, the *total system response* given by

$$y(t) = C\Phi(t, t_0)x_0 + C\int_{t_0}^{t} \Phi(t, s)Bu(s)ds + Du(t). \tag{3.63}$$

Recall that the total response (3.63) may be viewed as consisting of the sum of two components, the *zero-input response* given by the term

$$\psi(t, t_0, x_0, 0) = C\Phi(t, t_0)x_0 \tag{3.64}$$

and the *zero-state response* given by the term

$$\rho(t, t_0, 0, u) = C\int_{t_0}^{t} \Phi(t, s)Bu(s)ds + Du(t). \tag{3.65}$$

The cause of the former is the initial condition $x_0$ [and can be obtained from (3.63) by letting $u(t) \equiv 0$], whereas for the latter the cause is the input $u$ [and can be obtained by setting $x_0 = 0$ in (3.63)].

The zero-state response can be used to introduce the *impulse response* of the system (3.61). We recall from Subsection 2.4.3 that by using the Dirac delta distribution $\delta$, we can rewrite (3.63) with $x_0 = 0$ as

$$y(t) = \int_{t_0}^{t} [C\Phi(t,\tau)B + D\delta(t-\tau)]u(\tau)d\tau$$

$$= \int_{t_0}^{t} H(t,\tau)u(\tau)d\tau, \tag{3.66}$$

where $H(t,\tau)$ denotes the impulse response matrix of system (3.61) given by

$$H(t,\tau) = \begin{cases} C\Phi(t,\tau)B + D\delta(t-\tau), & t \geq \tau, \\ 0, & t < \tau. \end{cases} \tag{3.67}$$

Now recall that

$$\Phi(t,t_0) = e^{A(t-t_0)}. \tag{3.68}$$

The solution of (3.61a) is thus given by

$$\phi(t,t_0,x_0) = e^{A(t-t_0)}x_0 + \int_{t_0}^{t} e^{A(t-s)}Bu(s)ds, \tag{3.69}$$

the *total response* of system (3.61) is given by

$$y(t) = Ce^{A(t-t_0)}x_0 + C\int_{t_0}^{t} e^{A(t-s)}Bu(s)ds + Du(t) \tag{3.70}$$

and the *zero-state response* of (3.61), is given by $y(t) = \int_{t_0}^{t}[Ce^{A(t-\tau)}B + D\delta(t-\tau)]u(\tau)d\tau = \int_{t_0}^{t} H(t,\tau)u(\tau)d\tau = \int_{t_0}^{t} H(t-\tau)u(\tau)d\tau$, where the *impulse response matrix H* of system (3.61) is given by

$$H(t-\tau) = \begin{cases} Ce^{A(t-\tau)}B + D\delta(t-\tau), & t \geq \tau, \\ 0, & t < \tau, \end{cases} \tag{3.71}$$

or, as is more commonly written,

$$H(t) = \begin{cases} Ce^{At}B + D\delta(t), & t \geq 0, \\ 0, & t < 0. \end{cases} \tag{3.72}$$

At this point it may be worthwhile to consider some specific cases.

---

**Example 3.24.** In (3.61), let $A = \begin{bmatrix} 0 & 1 \\ 0 & 0 \end{bmatrix}$, $B = \begin{bmatrix} 0 \\ 1 \end{bmatrix}$, $C = [0,1]$, $D = 0$ and consider the case when $t_0 = 0$, $x(0) = (1,-1)^T$, $u$ is the unit step, and $t \geq 0$. We can easily compute the solution of (3.61a) as

$$\phi(t, t_0, x_0) = \phi_h(t, t_0, x_0) + \phi_p(t, t_0, x_0) = \begin{bmatrix} 1 - t \\ -1 \end{bmatrix} + \begin{bmatrix} \frac{1}{2}t^2 \\ t \end{bmatrix}$$

with $t_0 = 0$ and for $t \geq 0$. The total system response $y(t) = Cx(t)$ is given by the sum of the zero-input response and the zero-state response, $y(t, t_0, x_0, u) = \psi(t, t_0, x_0, 0) + \rho(t, t_0, 0, u) = -1 + t$, $t \geq 0$.

---

**Example 3.25.** Consider the time-invariant system given above in Example 3.24. It is easily verified that in the present case

$$\Phi(t) = e^{At} = \begin{bmatrix} 1 & t \\ 0 & 1 \end{bmatrix}.$$

Then $H(t, \tau) = Ce^{A(t-\tau)}B = 1$ for $t \geq \tau$ and $H(t, \tau) = 0$ for $t < \tau$. Thus, the response of this system to an impulse input for zero initial conditions is the unit step.

---

As one might expect, external descriptions of finite-dimensional linear systems are not as complete as internal descriptions of such systems. Indeed, the utility of impulse responses is found in the fact that they represent the input–output relations of a system quite well, assuming that the system is at rest. To describe other dynamic behavior, one needs in general additional information [e.g., the initial state vector (or perhaps the history of the system input since the last time instant when the system was at rest) as well as the internal structure of the system].

Internal descriptions, such as state-space representations, constitute more complete descriptions than external descriptions. However, the latter are simpler to apply than the former. Both types of representations are useful. It is quite straightforward to obtain external descriptions of systems from internal descriptions, as was demonstrated in this section. The reverse process, however, is not quite as straightforward. The process of determining an internal system description from an external description is called *realization* and will be addressed in Chapter 8. The principal issue in system realization is to obtain minimal order internal descriptions that model a given system, avoiding the generation of unnecessary dynamics.

### 3.4.2 Transfer Functions

Next, if as in [(2.95) in Chapter 2], we take the Laplace transform of (3.71), we obtain the input–output relation

$$\hat{y}(s) = \widehat{H}(s)\hat{u}(s). \tag{3.73}$$

We recall from Section 2.4 that $\widehat{H}(s)$ is called the *transfer function matrix* of system (3.61). We can evaluate this matrix in a straightforward manner by first taking the Laplace transform of both sides of (3.61a) and (3.61b) to obtain

$$s\hat{x}(s) - x(0) = A\hat{x}(s) + B\hat{u}(s), \tag{3.74}$$

$$\hat{y}(s) = C\hat{x}(s) + D\hat{u}(s). \tag{3.75}$$

Using (3.74) to solve for $\hat{x}(s)$, we obtain

$$\hat{x}(s) = (sI - A)^{-1}x(0) + (sI - A)^{-1}B\hat{u}(s). \tag{3.76}$$

Substituting (3.76) into (3.75) yields

$$\hat{y}(s) = C(sI - A)^{-1}x(0) + C(sI - A)^{-1}B\hat{u}(s) + D\hat{u}(s) \tag{3.77}$$

and

$$y(t) = \mathcal{L}^{-1}\hat{y}(s) = Ce^{At}x(0) + C\int_0^t e^{A(t-s)}Bu(s)ds + Du(t), \tag{3.78}$$

as expected.

If in (3.77) we let $x(0) = 0$, we obtain the Laplace transform of the zero-state response given by

$$\hat{y}(s) = [C(sI - A)^{-1}B + D]\hat{u}(s)$$

$$= \widehat{H}(s)\hat{u}(s), \tag{3.79}$$

where $\widehat{H}(s)$ denotes the transfer function of system (3.61), given by

$$\widehat{H}(s) = C(sI - A)^{-1}B + D. \tag{3.80}$$

Recalling that $\mathcal{L}[e^{At}] = \Phi(s) = (sI - A)^{-1}$ [refer to (3.45)], we could of course have obtained (3.80) directly by taking the Laplace transform of $H(t)$ given in (3.73).

---

**Example 3.26.** In Example 3.24, let $t_0 = 0$ and $x(0) = 0$. Then

$$\widehat{H}(s) = C(sI - A)^{-1}B + D = [0,1]\begin{bmatrix} s & -1 \\ 0 & s \end{bmatrix}^{-1}\begin{bmatrix} 0 \\ 1 \end{bmatrix}$$

$$= [0,1]\begin{bmatrix} 1/s & 1/s^2 \\ 0 & 1/s \end{bmatrix}\begin{bmatrix} 0 \\ 1 \end{bmatrix} = 1/s$$

and $H(t) = \mathcal{L}^{-1}\widehat{H}(s) = 1$ for $t \geq 0$, as expected (see Example 3.24).

Next, as in Example 3.24, let $x(0) = (1, -1)^T$ and let $u$ be the unit step. Then $\hat{y}(s) = C(sI - A)^{-1}x(0) + \widehat{H}(s)\hat{u}(s) = [0, 1/s](1, -1)^T + (1/s)(1/s) = -1/s + 1/s^2$ and $y(t) = \mathcal{L}^{-1}[\hat{y}(s)] = -1 + t$ for $t \geq 0$, as expected (see Example 3.24).

---

We note that the eigenvalues of the matrix $A$ in Example 3.26 are the roots of the equation $\det(sI - A) = s^2 = 0$, and are given by $s_1 = 0, s_2 = 0$, whereas the transfer function $\widehat{H}(s)$ in this example has only one pole (the zero of its denominator polynomial), located at the origin. It will be shown in Chapter 8 (on realization) that the *poles of the transfer function $\widehat{H}(s)$ (of a SISO system)* are in general a subset of the eigenvalues of $A$. In Chapter 5 we will introduce and study two important system theoretic concepts, called *controllability* and *observability*. We will show in Chapter 8 that the eigenvalues of $A$ are precisely the poles of the transfer function $\widehat{H}(s) = C(sI - A)^{-1}B + D$ if and only if the system (3.61) is observable and controllable. This is demonstrated in the next example.

---

**Example 3.27.** In (3.61), let $A = \begin{bmatrix} 0 & 1 \\ -1 & -2 \end{bmatrix}$, $B = \begin{bmatrix} 0 \\ 1 \end{bmatrix}$, $C = [-3, 3], D = 0$.
The eigenvalues of $A$ are the roots of the equation $\det(sI - A) = s^2 + 2s + 1 = (s+1)^2 = 0$ given by $s_1 = -1, s_2 = -1$, and the transfer function of this SISO system is given by

$$\widehat{H}(s) = C(sI - A)^{-1}B + D = [-3, 3] \begin{bmatrix} s & -1 \\ 1 & s+2 \end{bmatrix}^{-1} \begin{bmatrix} 0 \\ 1 \end{bmatrix}$$

$$= 3[-1, 1] \frac{1}{(s+1)^2} \begin{bmatrix} s+2 & 1 \\ -1 & s \end{bmatrix} \begin{bmatrix} 0 \\ 1 \end{bmatrix} = \frac{3(s-1)}{(s+1)^2},$$

with poles (the zeros of the denominator polynomial) also given by $s_1 = -1, s_2 = -1$.

---

If in Example 3.27 we replace $B = [0, 1]^T$ and $D = 0$ by $B = \begin{bmatrix} 0 & -1/2 \\ 1 & 1/2 \end{bmatrix}$ and $D = [0, 0]$, then we have a multi-input system whose transfer function is given by

$$\widehat{H}(s) = \begin{bmatrix} \dfrac{3(s-1)}{(s+1)^2}, & \dfrac{3}{(s+1)} \end{bmatrix}.$$

The concepts of poles and zeros for MIMO systems (also called multivariable systems) will be introduced in Chapter 7. The determination of the poles of such systems is not as straightforward as in the case of SISO systems. It turns out that in the present case the poles of $\widehat{H}(s)$ are $s_1 = -1, s_2 = -1$, the same as the eigenvalues of $A$.

Before proceeding to our next topic, the equivalence of internal representations, an observation concerning the transfer function $\widehat{H}(s)$ of system (3.61), given by (3.80), $\widehat{H}(s) = C(sI - A)^{-1}B + D$ is in order. Since the numerator matrix polynomial of $(sI - A)^{-1}$ is of degree $(n - 1)$, while its denominator polynomial, the characteristic polynomial $\alpha(s)$ of $A$, is of degree $n$, it is clear that

$$\lim_{s \to \infty} \widehat{H}(s) = D,$$

a real-valued $m \times n$ matrix, and in particular, when the *direct link matrix* $D$ in the output equation (3.61b) is zero, then

$$\lim_{s \to \infty} \widehat{H}(s) = 0,$$

the $m \times n$ matrix with zeros as its entries. In the former case (when $D \neq 0$), $\widehat{H}(s)$ is said to be a *proper transfer function*, whereas in the latter case (when $D = 0$), $\widehat{H}(s)$ is said to be a *strictly proper transfer function*.

When discussing the realization of transfer functions by state-space descriptions (in Chapter 8), we will study the properties of transfer functions in greater detail. In this connection, there are also systems that can be described by models corresponding to transfer functions $\widehat{H}(s)$ that are *not proper*. The differential equation representation of a differentiator (or an inductor) given by $y(t) = (d/dt)u(t)$ is one such example. Indeed, in this case the system cannot be represented by (3.61) and the transfer function, given by $\widehat{H}(s) = s$ is not proper. Such systems will be discussed in Chapter 10.

### 3.4.3 Equivalence of State-Space Representations

In Subsection 3.3.2 it was shown that when a linear, autonomous, homogeneous system of first-order ordinary differential equations $\dot{x} = Ax$ is subjected to an appropriately chosen similarity transformation, the resulting set of equations may be considerably easier to use and may exhibit latent properties of the system of equations. It is therefore natural that we consider a similar course of action in the case of the linear systems (3.61).

We begin by letting

$$\tilde{x} = Px, \tag{3.81}$$

where $P$ is a real, nonsingular matrix (i.e., $P$ is a similarity transformation). Consistent with what has been said thus far, we see that such transformations bring about a *change of basis* for the state space of system (3.61). Application of (3.81) to this system will result, as will be seen, in a system description of the same form as (3.61), but involving different state variables. We will say that the system (3.61), and the system obtained by subjecting (3.61) to the transformation (3.81), constitute *equivalent internal representations* of an underlying system. We will show that equivalent internal representations (of the same system) possess identical external descriptions, as one would expect, by showing that they have identical impulse responses and transfer function matrices. In connection with this discussion, two important notions called *zero-input equivalence* and *zero-state equivalence* of a system will arise in a natural manner.

If we differentiate both sides of (3.81), and if we apply $x = P^{-1}\tilde{x}$ to (3.61), we obtain the equivalent internal representation of (3.61) given by

$$\dot{\tilde{x}} = \tilde{A}\tilde{x} + \tilde{B}u, \tag{3.82a}$$

$$y = \tilde{C}\tilde{x} + \tilde{D}u, \tag{3.82b}$$

where

$$\tilde{A} = PAP^{-1}, \quad \tilde{B} = PB, \quad \tilde{C} = CP^{-1}, \quad \tilde{D} = D \tag{3.83}$$

and where $\tilde{x}$ is given by (3.81). It is now easily verified that the system (3.61) and the system (3.82) have the same external representation. Recall that for (3.61) and for (3.82), we have for the impulse response

$$H(t,\tau) \triangleq H(t-\tau,0) = \begin{cases} Ce^{A(t-\tau)}B + D\delta(t-\tau), & t \ge \tau, \\ 0, & t < \tau, \end{cases} \tag{3.84}$$

and

$$\tilde{H}(t,\tau) \triangleq \tilde{H}(t-\tau,0) = \begin{cases} \tilde{C}e^{\tilde{A}(t-\tau)}\tilde{B} + \tilde{D}\delta(t-\tau), & t \ge \tau, \\ 0, & t < \tau. \end{cases} \tag{3.85}$$

Recalling from Subsection 3.3.2 [see (3.25)] that

$$e^{\tilde{A}(t-\tau)} = Pe^{A(t-\tau)}P^{-1}, \tag{3.86}$$

we obtain from (3.83)–(3.85) that $\tilde{C}e^{\tilde{A}(t-\tau)}\tilde{B}+\tilde{D}\delta(t-\tau)=CP^{-1}Pe^{A(t-\tau)}P^{-1}PB+ D\delta(t-\tau)=Ce^{A(t-\tau)}B+D\delta(t-\tau)$, which proves, in view of (3.84) and (3.85), that

$$\tilde{H}(t,\tau) = H(t,\tau), \tag{3.87}$$

and this in turn shows that

$$\hat{\tilde{H}}(s) = \hat{H}(s). \tag{3.88}$$

This last relationship can also be verified by observing that $\hat{\tilde{H}}(s) = \tilde{C}(sI - \tilde{A})^{-1}\tilde{B} + \tilde{D} = CP^{-1}(sI - PAP^{-1})^{-1}PB + D = CP^{-1}P(sI - A)^{-1}P^{-1}PB + D = C(sI - A)^{-1}B + D = \hat{H}(s)$.

Next, recall that in view of (3.70) we have for (3.61) that

$$y(t) = Ce^{A(t-t_0)}x_0 + \int_{t_0}^t H(t-\tau,0)u(\tau)d\tau$$

$$= \psi(t,t_0,x_0,0) + \rho(t,t_0,0,u) \tag{3.89}$$

and for (3.82) that

$$y(t) = \tilde{C}e^{\tilde{A}(t-t_0)}\tilde{x}_0 + \int_{t_0}^t \tilde{H}(t-\tau,0)u(\tau)d\tau$$

$$= \tilde{\psi}(t,t_0,\tilde{x}_0,0) + \tilde{\rho}(t,t_0,0,u) \tag{3.90}$$

where $\psi$ and $\widetilde{\psi}$ denote the zero-input response of (3.61) and (3.82), respectively, whereas $\rho$ and $\widetilde{\rho}$ denote the zero-state response of (3.61) and (3.82), respectively. The relations (3.89) and (3.90) give rise to the following concepts: Two state-space representations are *zero-state equivalent* if they give rise to the same impulse response (the same external description). Also, two state-space representations are *zero-input equivalent* if for any initial state vector for one representation there exists an initial state vector for the second representation such that the zero-input responses for the two representations are identical.

The following result is now clear: *If two state-space representations are equivalent, then they are both zero-state and zero-input equivalent.* They are clearly zero-state equivalent since $H(t,\tau) = \widetilde{H}(t,\tau)$. Also, in view of (3.89) and (3.90), we have $\widetilde{C}e^{\widetilde{A}(t-t_0)}\widetilde{x}_0 = (CP^{-1})[Pe^{A(t-t_0)}P^{-1}]\widetilde{x}_0 = Ce^{A(t-t_0)}x_0$, where (3.86) was used. Therefore, the two state representations are also zero-input equivalent.

The converse to the above result is in general not true, since there are representations that are both zero-state and zero-input equivalent, yet not equivalent. In Chapter 8, which deals with state-space realizations of transfer functions, we will consider this topic further.

---

**Example 3.28.** System (3.61) with

$$A = \begin{bmatrix} 0 & 1 \\ -2 & -3 \end{bmatrix}, \quad B = \begin{bmatrix} 0 \\ 1 \end{bmatrix}, \quad C = [-1, -5], \quad D = 1$$

has the transfer function

$$H(s) = C(sI - A)^{-1}B + D = \frac{-5s - 1}{s^2 + 3s + 2} + 1 = \frac{(s-1)^2}{(s+1)(s+2)}.$$

Using the similarity transformation

$$P = \begin{bmatrix} 1 & 1 \\ -1 & -2 \end{bmatrix}^{-1} = \begin{bmatrix} 2 & 1 \\ -1 & -1 \end{bmatrix}$$

yields the equivalent representation of the system given by

$$\widetilde{A} = PAP^{-1} = \begin{bmatrix} -1 & 0 \\ 0 & -2 \end{bmatrix}, \quad \widetilde{B} = PB = \begin{bmatrix} 1 \\ -1 \end{bmatrix}, \quad \widetilde{C} = CP^{-1} = [4, 9]$$

and $\widetilde{D} = D = 1$. Note that the columns of $P^{-1}$, given by $[1, -1]^T$ and $[1, -2]^T$, are eigenvectors of $A$ corresponding to the eigenvalues $\lambda_1 = -1, \lambda_2 = -2$ of $A$; that is, $P$ was chosen to diagonalize $A$. Notice that $A$ (which is in companion form) has characteristic polynomial $s^2 + 3s + 2 = (s+1)(s+2)$. Notice also that the eigenvectors given above are of the form $[1, \lambda_i]^T$, $i = 1, 2$. The transfer function of the equivalent representation of the system is now given by

$$\hat{\tilde{H}}(s) = \tilde{C}(sI - \tilde{A})^{-1}\tilde{B} + \tilde{D} = [4, 0]\begin{bmatrix} \frac{1}{s+1} & 0 \\ 0 & \frac{1}{s+2} \end{bmatrix}\begin{bmatrix} 1 \\ -1 \end{bmatrix} + 1$$

$$= \frac{-5s - 1}{(s+1)(s+2)} + 1 = H(s).$$

Finally, it is easily verified that $e^{\tilde{A}t} = Pe^{At}P^{-1}$.

---

From the above discussion it should be clear that systems [of the form (3.61)] described by equivalent representations have identical behavior to the *outside world*, since both their zero-input and zero-state responses are the same. Their states, however, are in general not identical, but are related by the transformation $\tilde{x}(t) = Px(t)$.

## 3.5 State Equation and Input–Output Description of Discrete-Time Systems

In this section, which consists of five subsections, we address the state equation and input–output description of linear discrete-time systems. In the first subsection we study the response of linear time-invariant systems described by the difference equations (2.15) [or (1.8)]. In the second subsection we consider transfer functions for linear time-invariant systems, whereas in the third subsection we address the equivalence of the internal representations of time-invariant linear discrete-time systems [described by (2.15)]. Some of the most important classes of discrete-time systems include linear sampled-data systems that we develop in the fourth subsection. In the final part of this section, we address the modes and asymptotic behavior of linear time-invariant discrete-time systems.

### 3.5.1 Response of Linear Discrete-Time Systems

We consider once again systems described by linear time-invariant equations of the form

$$x(k + 1) = Ax(k) + Bu(k), \tag{3.91a}$$
$$y(k) = Cx(k) + Du(k), \tag{3.91b}$$

where $A \in R^{n \times n}$, $B \in R^{n \times m}$, $C \in R^{p \times n}$, and $D \in R^{p \times m}$. We recall that in (3.91), $x$ denotes the state vector, $u$ denotes the system input, and $y$ denotes the system output. For given initial conditions $k_0 \in Z, x(k_0) = x_{k_0} \in R^n$ and for a given input $u$, equation (3.91a) possesses a unique solution $x(k)$, which is defined for all $k \geq k_0$, and thus, the response $y(k)$ for (3.91b) is also defined for all $k \geq k_0$.

Associated with (3.91a) is the linear autonomous, homogeneous system of equations given by

$$x(k+1) = Ax(k). \tag{3.92}$$

We recall from Section 2.3 that the solution of the initial-value problem

$$x(k+1) = Ax(k), \quad x(k_0) = x_{k_0} \tag{3.93}$$

is given by

$$x(k) = \Phi(k, k_0)x_{k_0} = A^{k-k_0}x_{k_0}, \quad k > k_0, \tag{3.94}$$

where $\Phi(k, k_0)$ denotes the state transition matrix of (3.92) with

$$\Phi(k, k) = I \tag{3.95}$$

[refer to (2.31) to (2.34) in Chapter 2].

Common properties of the state transition matrix $\Phi(k, l)$, such as for example the *semigroup property* (forward in time) given by

$$\Phi(k, l) = \Phi(k, m)\Phi(m, l), \quad k \geq m \geq l,$$

can quite easily be derived from (3.94), (3.95). We caution the reader, however, that not all of the properties of the state transition matrix $\Phi(t, \tau)$ for continuous-time systems $\dot{x} = Ax$ carry over to the discrete-time case (3.92). In particular we recall that if for the continuous-time case we have $t > \tau$, then future values of the state $\phi$ at time $t$ can be obtained from past values of the state $\phi$ at time $\tau$, and vice versa, from the relationships $\phi(t) = \Phi(t, \tau)\phi(\tau)$ and $\phi(\tau) = \Phi^{-1}(t, \tau)\phi(t) = \Phi(\tau, t)\phi(t)$, i.e., for continuous-time systems a principle of *time reversibility exists*. This principle is in general not true for system (3.92), unless $A^{-1}(k)$ exists. The reason for this lies in the fact that $\Phi(k, l)$ will not be nonsingular if $A$ is not nonsingular.

---

**Example 3.29.** In (3.94), let $A = \begin{bmatrix} 1 & 0 \\ 0 & 0 \end{bmatrix}$, $x(0) = \begin{bmatrix} 1 \\ \alpha \end{bmatrix}$, $\alpha \in R$. The initial state $x(0)$ at $k_0 = 0$ for *any* $\alpha \in R$ will map into the state $x(1) = \begin{bmatrix} 1 \\ 0 \end{bmatrix}$. Accordingly, in this case, time reversibility will not apply.

---

**Example 3.30.** In (3.93), let $A = \begin{bmatrix} -1 & 2 \\ 0 & 1 \end{bmatrix}$. In view of (3.94) we have that

$$A^{(k-k_0)} = \begin{bmatrix} (-1)^{(k-k_0)} & 1 - (-1)^{(k-k_0)} \\ 0 & 1 \end{bmatrix}, \quad k \geq k_0;$$ i.e., $A^{(k-k_0)} = A$ when $(k - k_0)$ is odd, and $A^{(k-k_0)} = I$ when $(k - k_0)$ is even. Therefore, given $k_0 = 0$ and $x(0) = \begin{bmatrix} 2 \\ 1 \end{bmatrix}$, then $x(k) = Ax(0) = \begin{bmatrix} 0 \\ 1 \end{bmatrix}$, $k = 1, 3, 5, \ldots$, and

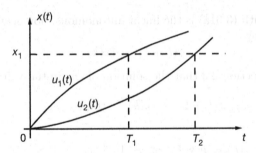

**Figure 3.1.** Plots of states for Example 3.30

$x(k) = Ix(0) = \begin{bmatrix} 2 \\ 1 \end{bmatrix}, k = 2, 4, 6, \ldots$ A plot of the states $x(k) = [x_1(k), x_2(k)]^T$ is given in Figure 3.1.

Continuing, we recall that the solutions of initial-value problems determined by linear nonhomogeneous systems (2.35) are given by expression (2.36). Utilizing (2.36), the solution of (3.91a) for given $x(k_0)$ and $u(k)$ is given as

$$x(k) = \Phi(k, k_0)x(k_0) + \sum_{j=k_0}^{k-1} \Phi(k, j+1)Bu(j), \quad k > k_0. \tag{3.96}$$

This expression in turn can be used to determine the system response for system (3.91) as

$$y(k) = C\Phi(k, k_0)x(k_0)$$
$$+ \sum_{j=k_0}^{k-1} C\Phi(k, j+1)Bu(j) + Du(k), \quad k > k_0,$$
$$y(k_0) = Cx(k_0) + Du(k_0), \tag{3.97}$$

or

$$y(k) = CA^{(k-k_0)}x(k_0) + \sum_{j=k_0}^{k-1} CA^{k-(j+1)}Bu(j) + Du(k), \quad k > k_0,$$
$$y(k_0) = Cx(k_0) + Du(k_0). \tag{3.98}$$

Since the system (3.91) is time-invariant, we can let $k_0 = 0$ without loss of generality to obtain from (3.98) the expression

$$y(k) = CA^k x(0) + \sum_{j=0}^{k-1} CA^{k-(j+1)}Bu(j) + Du(k), \quad k > 0. \tag{3.99}$$

As in the continuous-time case, the *total system response* (3.97) may be viewed as consisting of two components, the *zero-input response*, given by

$$\psi(k) = C\Phi(k, k_0)x(k_0), \quad k > k_0,$$

and the *zero-state response*, given by

$$\left.\begin{aligned}
\rho(k) &= \sum_{j=k_0}^{k-1} C\Phi(k, j+1)Bu(j) + Du(k), \quad k > k_0, \\
\rho(k_0) &= Du(k_0), \quad k = k_0.
\end{aligned}\right\} \tag{3.100}$$

Finally, in view of (2.67), we recall that the (discrete-time) unit impulse response matrix of system (3.91) is given by

$$H(k, l) = \begin{cases} CA^{k-(l+1)}B, & k > l, \\ D, & k = l, \\ 0, & k < l, \end{cases} \tag{3.101}$$

and in particular, when $l = 0$ (i.e., when the pulse is applied at time $l = 0$),

$$H(k, 0) = \begin{cases} CA^{k-1}B, & k > 0, \\ D, & k = 0, \\ 0, & k < 0. \end{cases} \tag{3.102}$$

---

**Example 3.31.** In (3.91), let

$$A = \begin{bmatrix} 0 & 1 \\ 0 & -1 \end{bmatrix}, \quad B = \begin{bmatrix} 0 \\ 1 \end{bmatrix}, \quad C^T = \begin{bmatrix} 1 \\ 0 \end{bmatrix}, \quad D = 0.$$

We first determine $A^k$ by using the Cayley–Hamilton Theorem (refer to Section A.5). To this end we compute the eigenvalues of $A$ as $\lambda_1 = 0, \lambda_2 = -1$, we let $A^k = f(A)$, where $f(s) = s^k$, and we let $g(s) = \alpha_1 s + \alpha_0$. Then $f(\lambda_1) = g(\lambda_1)$, $\alpha_0 = 0$ and $f(\lambda_2) = g(\lambda_2)$, or $(-1)^k = -\alpha_1 + \alpha_0$. Therefore, $A^k = \alpha_1 A + \alpha_0 I = -(-1)^k \begin{bmatrix} 0 & 1 \\ 0 & -1 \end{bmatrix} = \begin{bmatrix} 0 & (-1)^{k-1} \\ 0 & (-1)^k \end{bmatrix}$, $k = 1, 2, \ldots$, or $A^k = \begin{bmatrix} \delta(k) & (-1)^{k-1}p(k-1) \\ 0 & (-1)^k p(k) \end{bmatrix}$, $k = 0, 1, 2, \ldots$, where $A^0 = I$, and where $p(k)$ denotes the unit step given by

$$p(k) = \begin{cases} 1, & k \geq 0, \\ 0, & k < 0. \end{cases}$$

The above expression for $A^k$ is now substituted into (3.98) to determine the response $y(k)$ for $k > 0$ for a given initial condition $x(0)$ and a given input $u(k), k \geq 0$. To determine the unit impulse response, we note that $H(k, 0) = 0$ for $k < 0$ and $k = 0$. When $k > 0$, $H(k, 0) = CA^{k-1}B = (-1)^{k-2}p(k-2)$ for $k > 0$ or $H(k, 0) = 0$ for $k = 1$ and $H(k, 0) = (-1)^{k-2}$ for $k = 2, 3, \ldots$.

---

### 3.5.2 The Transfer Function and the z-Transform

We assume that the reader is familiar with the concept and properties of the *one-sided z-transform* of a real-valued sequence $\{f(k)\}$, given by

$$\mathcal{Z}\{f(k)\} = \hat{f}(z) = \sum_{j=0}^{\infty} z^{-j} f(j). \tag{3.103}$$

An important property of this transform, which is useful in solving difference equations, is given by the relation

$$\mathcal{Z}\{f(k+1)\} = \sum_{j=0}^{\infty} z^{-j} f(j+1) = \sum_{j=1}^{\infty} z^{-(j-1)} f(j)$$

$$= z \left[ \sum_{j=0}^{\infty} z^{-j} f(j) - f(0) \right]$$

$$= z\left[\mathcal{Z}\{f(k)\} - f(0)\right] = z\hat{f}(z) - zf(0). \tag{3.104}$$

If we take the $z$-transform of both sides of (3.91a), we obtain, in view of (3.104), $z\hat{x}(z) - zx(0) = A\hat{x}(z) + B\hat{u}(z)$ or

$$\hat{x}(z) = (zI - A)^{-1} zx(0) + (zI - A)^{-1} B\hat{u}(z). \tag{3.105}$$

Next, by taking the $z$-transform of both sides of (3.91b), and by substituting (3.105) into the resulting expression, we obtain

$$\hat{y}(z) = C(zI - A)^{-1} zx(0) + [C(zI - A)^{-1} B + D]\hat{u}(z). \tag{3.106}$$

The time sequence $\{y(k)\}$ can be recovered from its one-sided $z$-transform $\hat{y}(z)$ by applying the *inverse z-transform*, denoted by $\mathcal{Z}^{-1}[\hat{y}(z)]$.

In Table 3.3 we provide the one-sided $z$-transforms of some of the commonly used sequences, and in Table 3.4 we enumerate some of the more frequently encountered properties of the one-sided $z$-transform.

The *transfer function matrix* $\widehat{H}(z)$ of system (3.91) relates the $z$-transform of the output $y$ to the $z$-transform of the input $u$ under the assumption that $x(0) = 0$. We have

$$\hat{y}(z) = \widehat{H}(z)\hat{u}(z), \tag{3.107}$$

where

$$\widehat{H}(z) = C(zI - A)^{-1} B + D. \tag{3.108}$$

To relate $\widehat{H}(z)$ to the impulse response matrix $H(k,l)$, we notice that $\mathcal{Z}\{\delta(k-l)\} = z^{-l}$, where $\delta$ denotes the *discrete-time impulse* (or *unit pulse* or *unit sample*) defined in (2.51); i.e.,

$$\delta(k - l) = \begin{cases} 1, & k = l, \\ 0, & k \neq l. \end{cases} \tag{3.109}$$

**Table 3.3.** Some commonly used $z$-transforms

| $\{f(k)\}, \quad k \geq 0$ | $\hat{f}(z) = \mathcal{Z}\{f(k)\}$ |
|---|---|
| $\delta(k)$ | $1$ |
| $p(k)$ | $1/(1 - z^{-1})$ |
| $k$ | $z^{-1}/(1 - z^{-1})^2$ |
| $k^2$ | $[z^{-1}(1 + z^{-1})]/(1 - z^{-1})^3$ |
| $a^k$ | $1/(1 - az^{-1})$ |
| $(k+1)a^k$ | $1/(1 - az^{-1})^2$ |
| $[(1/l!)(k+1)\cdots(k+l)]a^k \quad l \geq 1$ | $1/(1 - az^{-1})^{l+1}$ |
| $a\cos\alpha k + b\sin\alpha k$ | $\dfrac{a + z^{-1}(b\sin\alpha - a\cos\alpha)}{1 - 2z^{-1}\cos\alpha + z^{-2}}$ |

**Table 3.4.** Some properties of $z$-transforms

| | $\{f(k)\}, k \geq 0$ | $f(z)$ |
|---|---|---|
| Time shift | $f(k+1)$ | $z\hat{f}(z) - zf(0)$ |
| -Advance | $f(k+l) \quad l \geq 1$ | $z^l \hat{f}(z) - z\sum_{i=1}^{l} z^{l-i} f(i-1)$ |
| Time shift | $f(k-1)$ | $z^{-1}\hat{f}(z) + f(-1)$ |
| -Delay | $f(k-l) \quad l \geq 1$ | $z^{-l}\hat{f}(z) + \sum_{i=1}^{l} z^{-l+i} f(-i)$ |
| Scaling | $a^k f(k)$ | $\hat{f}(z/a)$ |
| | $kf(k)$ | $-z(d/dz)\hat{f}(z)$ |
| Convolution | $\sum_{l=0}^{\infty} f(l)g(k-l) = f(k) * g(k)$ | $\hat{f}(z)\hat{g}(z)$ |
| Initial value | $f(l)$ with $f(k) = 0, \quad k < l$ | $\lim_{z\to\infty} z^l \hat{f}(z)^{\dagger}$ |
| Final value | $\lim_{k\to\infty} f(k)$ | $\lim_{z\to 1}(1 - z^{-1})\hat{f}(z)^{\ddagger}$ |

$^{\dagger}$ If the limit exists.

$^{\ddagger}$ If $(1 - z^{-1})\hat{f}(z)$ has no singularities on or outside the unit circle.

This implies that the $z$-transform of a unit pulse applied at time zero is $\mathcal{Z}\{\delta(k)\} = 1$. It is not difficult to see now that $\{H(k,0)\} = \mathcal{Z}^{-1}[\hat{y}(z)]$, where $\hat{y}(z) = \hat{H}(z)\hat{u}(z)$ with $\hat{u}(z) = 1$. This shows that

$$\mathcal{Z}^{-1}[\hat{H}(z)] = \mathcal{Z}^{-1}[C(zI - A)^{-1}B + D] = \{H(k,0)\}, \tag{3.110}$$

where the unit impulse response matrix $H(k,0)$ is given by (3.102).

The above result can also be derived directly by taking the $z$-transform of $\{H(k,0)\}$ given in (3.102) (prove this). In particular, notice that the $z$-transform of $\{A^{k-1}\}$, $k = 1, 2, \ldots$ is $(zI - A)^{-1}$ since

$$
\begin{aligned}
\mathcal{Z}\{0, A^{k-1}\} &= \sum_{j=1}^{\infty} z^{-j} A^{j-1} = z^{-1} \sum_{j=0}^{\infty} z^{-j} A^j \\
&= z^{-1}(I + z^{-1}A + z^{-2}A^2 + \ldots) \\
&= z^{-1}(I - z^{-1}A)^{-1} = (zI - A)^{-1}.
\end{aligned}
\tag{3.111}
$$

Above, the matrix determined by the expression $(1-\lambda)^{-1} = 1+\lambda+\lambda^2+\cdots$ was used. It is easily shown that the corresponding series involving $A$ converges. Notice also that $\mathcal{Z}\{A^k\}, k = 0, 1, 2, \ldots$ is $z(zI - A)^{-1}$. This fact can be used to show that the inverse $z$-transform of (3.106) yields the time response (3.99), as expected.

We conclude this subsection with a specific example.

---

**Example 3.32.** In system (3.91), we let

$$A = \begin{bmatrix} 0 & 1 \\ 0 & -1 \end{bmatrix}, \quad B = \begin{bmatrix} 0 \\ 1 \end{bmatrix}, \quad C = [1\ 0], \quad D = 0.$$

To verify that $\mathcal{Z}^{-1}[z(zI-A)^{-1}] = A^k$, we compute $z(zI - A)^{-1} = z \begin{bmatrix} z & -1 \\ 0 & z+1 \end{bmatrix}^{-1}$

$$= z \begin{bmatrix} \frac{1}{z} & \frac{1}{z(z+1)} \\ 0 & \frac{1}{z+1} \end{bmatrix} = \begin{bmatrix} 1 & \frac{1}{z+1} \\ 0 & \frac{z}{z+1} \end{bmatrix} \text{ and }$$

$$\mathcal{Z}^{-1}[z(zI - A)^{-1}] = \begin{bmatrix} \delta(k) & (-1)^{k-1}p(k - 1) \\ 0 & (-1)^k p(k) \end{bmatrix}$$

or

$$A^k = \begin{cases} \begin{bmatrix} 1 & 0 \\ 0 & 1 \end{bmatrix}, & \text{when } k = 0, \\ \begin{bmatrix} 0 & (-1)^{k-1} \\ 0 & (-1)^k \end{bmatrix}, & \text{when } k = 1, 2, \ldots, \end{cases}$$

as expected from Example 3.31.

Notice that

$$\mathcal{Z}^{-1}[(zI - A)^{-1}] = \mathcal{Z}^{-1}\left[\begin{bmatrix} 1/z & 1/[z(z + 1)] \\ 0 & 1/(z + 1) \end{bmatrix}\right]$$

$$= \begin{bmatrix} \delta(k - 1)p(k - 1) & \delta(k - 1)p(k - 1) - (-1)^{k-1}p(k - 1) \\ 0 & (-1)^{k-1}p(k - 1) \end{bmatrix}$$

$$= \begin{bmatrix} 0 & 0 \\ 0 & 0 \end{bmatrix} \text{ for } k = 0, \text{ and } \begin{bmatrix} 1 & 0 \\ 0 & 1 \end{bmatrix} \text{ for } k = 1,$$

and

$$\mathcal{Z}^{-1}[(zI - A)^{-1}] = \begin{bmatrix} 0 & -(-1)^{k-1} \\ 0 & (-1)^{k-1} \end{bmatrix} \text{ for } k = 2, 3, \ldots,$$

which is equal to $A^k, k \geq 0$, delayed by one unit; i.e., it is equal to $A^{k-1}, k = 1, 2, \ldots$, as expected.

Next, we consider the system response with $x(0) = 0$ and $u(k) = p(k)$. We have

$$y(k) = \mathcal{Z}^{-1}[\hat{y}(z)] = \mathcal{Z}^{-1}[C(zI - A)^{-1}B \cdot \hat{u}(z)]$$

$$= \mathcal{Z}^{-1}\left[\frac{1}{(z + 1)(z - 1)}\right] = \mathcal{Z}^{-1}\left[\frac{1/2}{z-1} - \frac{1/2}{z+1}\right]$$

$$= \frac{1}{2}[(1)^{k-1} - (-1)^{k-1}]p(k-1)$$

$$= \begin{cases} 0, & k = 0, \\ \frac{1}{2}(1 - (-1)^{k-1}), & k = 1, 2, \ldots, \end{cases}$$

$$= \begin{cases} 0, & k = 0, \\ 0, & k = 1, 3, 5, \ldots, \\ 1, & k = 2, 4, 6, \ldots. \end{cases}$$

Note that if $x(0) = 0$ and $u(k) = \delta(k)$, then

$$y(k) = \mathcal{Z}^{-1}[C(zI - A)^{-1}B] = \mathcal{Z}^{-1}\left[\frac{1}{z(z+1)}\right]$$
$$= \delta(k-1)p(k-1) - (-1)^{k-1}p(k-1)$$
$$= \begin{cases} 0, & k = 0, 1, \\ (-1)^{k-2}, & k = 2, 3, \ldots, \end{cases}$$

which is the unit impulse response of the system (refer to Example 3.31).

---

### 3.5.3 Equivalence of State-Space Representations

Equivalent representations of linear discrete-time systems are defined in a manner analogous to the continuous-time case. For systems (3.91), we let $P$ denote a real nonsingular $n \times n$ matrix and we define

$$\tilde{x}(k) = Px(k). \tag{3.112}$$

Substituting (3.112) into (3.91) yields the equivalent system representation

$$\tilde{x}(k + 1) = \tilde{A}\tilde{x}(k) + \tilde{B}u(k), \tag{3.113a}$$
$$y(k) = \tilde{C}\tilde{x}(k) + \tilde{D}u(k), \tag{3.113b}$$

where

$$\tilde{A} = P^{-1}AP, \quad \tilde{B} = PB, \quad \tilde{C} = CP^{-1}, \quad \tilde{D} = D. \tag{3.114}$$

We note that the terms in (3.114) are identical to corresponding terms obtained for the case of linear continuous-time systems.

We conclude by noting that if $\hat{H}(z)$ and $\hat{\tilde{H}}(z)$ denote the transfer functions of the unit impulse response matrices of system (3.91) and system (3.113), respectively, then it is easily verified that $\hat{H}(z) = \hat{\tilde{H}}(z)$.

### 3.5.4 Sampled-Data Systems

Discrete-time dynamical systems arise in a variety of ways in the modeling process. There are systems that are inherently defined only at discrete points in time, and there are representations of continuous-time systems at discrete points in time. Examples of the former include digital computers and devices (e.g., digital filters) where the behavior of interest of a system is adequately described by values of variables at discrete-time instants (and what happens between the discrete instants of time is quite irrelevant to the problem on hand); inventory systems where only the inventory status at the end of each day (or month) is of interest; economic systems, such as banking, where, e.g., interests are calculated and added to savings accounts at discrete-time intervals only; and so forth. Examples of the latter include simulations of continuous-time processes by means of digital computers, making use of difference equations that approximate the differential equations describing the process in question; feedback control systems that employ digital controllers and give rise to sampled-data systems (as discussed further in the following); and so forth.

In providing a short discussion of sampled-data systems, we make use of the specific class of linear feedback control systems depicted in Figure 3.2. This system may be viewed as an interconnection of a subsystem $S_1$, called the *plant* (the object to be controlled), and a subsystem $S_2$, called the *digital controller*. The plant is described by the equations

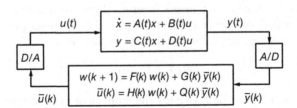

**Figure 3.2.** Digital control system

$$\dot{x} = A(t)x + B(t)u, \tag{3.115a}$$
$$y = C(t)x + D(t)u, \tag{3.115b}$$

where all symbols in (3.115) are defined as in (2.3) and where we assume that $t \geq t_0 \geq 0$.

Since our presentation pertains equally to the time-varying and time-invariant cases, we will first address the more general time-varying case. Next, we specialize our results to the time-invariant case.

The subsystem $S_2$ accepts the continuous-time signal $y(t)$ as its input, and it produces the piecewise continuous-time signal $u(t)$ as its output, where $t \geq t_0$. The continuous-time signal $y$ is converted into a discrete-time signal $\{\bar{y}(k)\}$,

$k \geq k_0 \geq 0$, $k, k_0 \in Z$, by means of an analog-to-digital (A/D) converter and is processed according to a control algorithm given by the difference equations

$$w(k+1) = F(k)w(k) + G(k)\bar{y}(k), \tag{3.116a}$$

$$\bar{u}(k) = H(k)w(k) + Q(k)\bar{y}(k), \tag{3.116b}$$

where the $w(k), \bar{y}(k), \bar{u}(k)$ are real vectors and the $F(k), G(k), H(k)$, and $Q(k)$ are real, time-varying matrices with a consistent set of dimensions. Finally, the discrete-time signal $\{\bar{u}(k)\}$, $k \geq k_0 \geq 0$, is converted into the continuous-time signal $u$ by means of a digital-to-analog (D/A) converter. To simplify our discussion, we assume in the following that $t_0 = k_0$.

An (ideal) A/D converter is a device that has as input a continuous-time signal, in our case $y$, and as output a sequence of real numbers, in our case $\{\bar{y}(k)\}$, $k = k_0, k_0 + 1, \ldots$, determined by the relation

$$\bar{y}(k) = y(t_k). \tag{3.117}$$

In other words, the (ideal) A/D converter is a device that *samples* an input signal, in our case $y(t)$, at times $t_0, t_1, \ldots$ producing the corresponding sequence $\{y(t_0), y(t_1), \ldots\}$.

A *D/A converter* is a device that has as input a discrete-time signal, in our case the sequence $\{\bar{u}(k)\}$, $k = k_0, k_0 + 1, \ldots$, and as output a continuous-time signal, in our case $u$, determined by the relation

$$u(t) = \bar{u}(k), t_k \leq t < t_{k+1}, \quad k = k_0, k_0 + 1, \ldots. \tag{3.118}$$

In other words, the D/A converter is a device that keeps its output constant at the last value of the sequence entered. We also call such a device a *zero-order hold*.

The system of Figure 3.2, as described above, is an example of a *sampled-data system*, since it involves truly *sampled data* (i.e., sampled signals), making use of an *ideal A/D converter*. In practice the digital controller $S_2$ uses *digital signals* as variables. In the scalar case, such signals are represented by real-valued sequences whose numbers belong to a subset of $R$ consisting of a discrete set of points. (In the vector case, the previous statement applies to the components of the vector.) Specifically, in the present case, after the signal $y(t)$ has been sampled, it must be *quantized* (or *digitized*) to yield a *digital signal*, since only such signals are representable in a digital computer. If a computer uses, e.g., 8-bit words, then we can represent $2^8 = 256$ distinct levels for a variable, which determine the signal quantization. By way of a specific example, if we expect in the representation of a function a signal that varies from 9 to 25 volts, we may choose a 0.1-volt quantization step. Then 2.3 and 2.4 volts are represented by two different numbers (quantization levels); however, 2.315, 2.308, and 2.3 are all represented by the bit combination corresponding to 2.3. Quantization is an approximation and for short wordlengths may lead to significant errors. Problems associated with *quantization effects* will not be addressed in this book.

In addition to being a sampled-data system, the system represented by (3.115) to (3.118) constitutes a *hybrid system* as well, since it involves descriptions given by ordinary differential equations and ordinary difference equations. The analysis and synthesis of such systems can be simplified appreciably by replacing the description of subsystem $S_1$ (the plant) by a set of ordinary difference equations, valid only at discrete points in time $t_k, k = 0, 1, 2, \ldots$. [In terms of the blocks of Figure 3.2, this corresponds to considering the plant $S_1$, together with the D/A and A/D devices, to obtain a system with input $\bar{u}(k)$ and output $\bar{y}(k)$, as shown in Figure 3.3.] To accomplish this, we apply the variation of constants formula to (3.115a) to obtain

$$x(t) = \Phi(t, t_k)x(t_k) + \int_{t_k}^{t} \Phi(t, \tau)B(\tau)u(\tau)d\tau, \qquad (3.119)$$

where the notation $\phi(t, t_k, x(t_k)) = x(t)$ has been used. Since the input $u(t)$

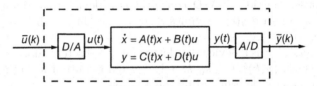

**Figure 3.3.** System described by (3.121) and (3.124)

is the output of the zero-order hold device (the D/A converter), given by (3.118), we obtain from (3.119) the expression

$$x(t_{k+1}) = \Phi(t_{t+1}, t_k)x(t_k) + \left[ \int_{t_k}^{t_{k+1}} \Phi(t_{k+1}, \tau)B(\tau)d\tau \right] u(t_k). \qquad (3.120)$$

Since $\bar{x}(k) \triangleq x(t_k)$ and $\bar{u}(k) \triangleq u(t_k)$, we obtain a discrete-time version of the state equation for the plant, given by

$$\bar{x}(k + 1) = \bar{A}(k)\bar{x}(k) + \bar{B}(k)\bar{u}(k), \qquad (3.121)$$

where

$$\left. \begin{aligned} \bar{A}(k) &\triangleq \Phi(t_{k+1}, t_k), \\ \bar{B}(k) &\triangleq \int_{t_k}^{t_{k+1}} \Phi(t_{k+1}, \tau)B(\tau)d\tau. \end{aligned} \right\} \qquad (3.122)$$

Next, we assume that the output of the plant is sampled at instants $t_k'$ that do not necessarily coincide with the instants $t_k$ at which the input to the plant is adjusted, and we assume that $t_k \leq t_k' < t_{k+1}$. Then (3.115) and (3.119) yield

$$y(t'_k) = C(t'_k)\Phi(t'_k, t_k)x(t_k) + \left[C(t'_k)\int_{t_k}^{t'_k}\Phi(t'_k, \tau)B(\tau)d\tau\right]u(t_k) + D(t'_k)u(t_k).$$

$$(3.123)$$

Defining $\bar{y}(k) \triangleq y(t'_k)$, we obtain from (3.123),

$$\bar{y}(k) = \bar{C}(k)\bar{x}(k) + \bar{D}(k)\bar{u}(k), \qquad (3.124)$$

where

$$\left.\begin{array}{l} \bar{C}(k) \triangleq C(t'_k)\Phi(t'_k, t_k), \\[2mm] \bar{D}(k) \triangleq C(t'_k)\displaystyle\int_{t_k}^{t'_k}\Phi(t'_k, \tau)B(\tau)d\tau + D(t'_k). \end{array}\right\} \qquad (3.125)$$

Summarizing, (3.121) and (3.124) constitute a state-space representation, valid at discrete points in time, of the plant [given by (3.115a)] and including the A/D and D/A devices [given by (3.117) and (3.118), see Figure 3.3]. Furthermore, the entire hybrid system of Figure 3.2, valid at discrete points in time, can now be represented by (3.121), (3.124), and (3.116).

*Time-Invariant System With Constant Sampling Rate*

We now turn to the case of the time-invariant plant, where $A(t) \equiv A, B(t) \equiv B, C(t) \equiv C$, and $D(t) \equiv D$, and we assume that $t_{k+1} - t_k = T$ and $t'_k - t_k = \alpha$ for all $k = 0, 1, 2, \ldots$. Then the expressions given in (3.121), (3.122), (3.124), and (3.125) assume the form

$$\bar{x}(k+1) = \bar{A}\bar{x}(k) + \bar{B}\bar{u}(k), \qquad (3.126a)$$
$$\bar{y}(k) = \bar{C}\bar{x}(k) + \bar{D}\bar{u}(k), \qquad (3.126b)$$

where

$$\left.\begin{array}{ll} \bar{A} = e^{AT}, & \bar{B} = \left(\displaystyle\int_0^T e^{A\tau}d\tau\right)B, \\[4mm] \bar{C} = Ce^{A\alpha}, & \bar{D} = C\left(\displaystyle\int_0^\alpha e^{A\tau}d\tau\right)B + D. \end{array}\right\} \qquad (3.127)$$

If $t'_k = t_k$, or $\alpha = 0$, then $\bar{C} = C$ and $\bar{D} = D$.

In the preceding, $T$ is called the *sampling period* and $1/T$ is called the *sampling rate*. Sampled-data systems are treated in great detail in texts dealing with digital control systems and with digital signal processing.

---

**Example 3.33.** In the control system of Figure 3.2, let

$$A = \begin{bmatrix} 0 & 1 \\ 0 & 0 \end{bmatrix}, \quad B = \begin{bmatrix} 0 \\ 1 \end{bmatrix}, \quad C = [1, 0], \quad D = 0,$$

let $T$ denote the sampling period, and assume that $\alpha = 0$. The discrete-time state-space representation of the plant, preceded by a zero-order hold

(D/A converter) and followed by a sampler [an (ideal) A/D converter], both sampling synchronously at a rate of $1/T$, is given by $\bar{x}(k+1) = \bar{A}\bar{x}(k)+\bar{B}\bar{u}(k)$, $\bar{y}(k) = \bar{C}x(k)$, where

$$\bar{A} = e^{AT} = \sum_{j=1}^{\infty}(T^j/j!)A^j = \begin{bmatrix} 1 & 0 \\ 0 & 1 \end{bmatrix} + \begin{bmatrix} 0 & 1 \\ 0 & 0 \end{bmatrix}T = \begin{bmatrix} 1 & T \\ 0 & 1 \end{bmatrix},$$

$$\bar{B} = \left(\int_0^T e^{A\tau}d\tau\right)B = \left(\int_0^T \begin{bmatrix} 1 & \tau \\ 0 & 1 \end{bmatrix}d\tau\right)\begin{bmatrix} 0 \\ 1 \end{bmatrix}$$

$$= \begin{bmatrix} T & T^2/2 \\ 0 & T \end{bmatrix}\begin{bmatrix} 0 \\ 1 \end{bmatrix} = \begin{bmatrix} T^2/2 \\ T \end{bmatrix},$$

$$\bar{C} = C = [1\ 0].$$

The transfer function (relating $\bar{y}$ to $\bar{u}$ ) is given by

$$\hat{H}(z) = \bar{C}(zI - \bar{A})^{-1}\bar{B}$$

$$= [1\ 0]\begin{bmatrix} z-1 & -T \\ 0 & z-1 \end{bmatrix}^{-1}\begin{bmatrix} T^2/2 \\ T \end{bmatrix}$$

$$= [1\ 0]\begin{bmatrix} 1/(z-1) & T/(z-1)^2 \\ 0 & 1/(z-1) \end{bmatrix}\begin{bmatrix} T^2/2 \\ T \end{bmatrix}$$

$$= \frac{T^2}{2}\frac{(z+1)}{(z-1)^2}.$$

The transfer function of the continuous-time system (continuous-time description of the plant) is determined to be $\hat{H}(s) = C(sI-A)^{-1}B = 1/s^2$, the double integrator.

The behavior of the system between the discrete instants, $t, t_k \leq t < t_{k+1}$, can be determined by using (3.119), letting $x(t_k) = x(k)$ and $u(t_k) = u(k)$.

---

An interesting observation, useful when calculating $\bar{A}$ and $\bar{B}$, is that both can be expressed in terms of a single series. In particular, $\bar{A} = e^{AT} = I+TA+(T^2/2!)A^2+\cdots = I+TA\Psi(T)$, where $\Psi(T) = I+(T/2!)A+(T^2/3!)A^2+\cdots = \sum_{j=0}^{\infty}(T^j/(j+1)!)A^j$. Then $\bar{B} = (\int_0^T e^{A\tau}d\tau) B = (\sum_{j=0}^{\infty}(T^{j+1}/(j+1)!)A^j) B = T\Psi(T)B$. If $\Psi(T)$ is determined first, and then both $\bar{A}$ and $\bar{B}$ can easily be calculated.

---

**Example 3.34.** In Example 3.33, $\Psi(T) = I + TA = \begin{bmatrix} 1 & T \\ 0 & 1 \end{bmatrix}$. Therefore, $\bar{A} = I+TA\Psi(T) = \begin{bmatrix} 1 & T \\ 0 & 1 \end{bmatrix}$ and $\bar{B} = T\Psi(T)B = \begin{bmatrix} T^2/2 \\ T \end{bmatrix}$, as expected.

---

### 3.5.5 Modes, Asymptotic Behavior, and Stability

As in the case of continuous-time systems, we study in this subsection the qualitative behavior of the solutions of linear, autonomous, homogeneous ordinary difference equations

$$x(k+1) = Ax(k) \tag{3.128}$$

in terms of the modes of such systems, where $A \in R^{n \times n}$ and $x(k) \in R^n$ for every $k \in Z^+$. From before, the unique solution of (3.128) satisfying $x(0) = x_0$ is given by

$$\phi(k, 0, x_0) = A^k x_0. \tag{3.129}$$

Let $\lambda_1, \ldots, \lambda_\sigma$, denote the $\sigma$ distinct eigenvalues of $A$, where $\lambda_i$ with $i = 1, \ldots, \sigma$, is assumed to be repeated $n_i$ times so that $\sum_{i=1}^{\sigma} n_i = n$. Then

$$\det(zI - A) = \prod_{i=1}^{\sigma}(z - \lambda_i)^{n_i}. \tag{3.130}$$

To introduce the modes for (3.128), we first derive the expressions

$$
\begin{aligned}
A^k &= \sum_{i=1}^{\sigma}[A_{i0}\lambda_i^k p(k) + \sum_{l=1}^{n_i-1} A_{il}k(k-1)\cdots(k-l+1)\lambda_i^{k-l}p(k-l)] \\
&= \sum_{i=1}^{\sigma}[A_{i0}\lambda_i^k p(k) + A_{i1}k\lambda_i^{k-1}p(k-1) + \cdots \\
&\quad + A_{i(n_i-1)}k(k-1)\cdots(k-n_i+2)\lambda_i^{k-(n_i-1)}p(k-n_i+1)],
\end{aligned} \tag{3.131}
$$

where

$$A_{il} = \frac{1}{l!}\frac{1}{(n_i-1-l)!}\lim_{z \to \lambda_i}\{[(z-\lambda_i)^{n_i}(zI-A)^{-1}]^{(n_i-1-l)}\}. \tag{3.132}$$

In (3.132), $[\cdot]^{(q)}$ denotes the $q$th derivative with respect to $z$, and in (3.131), $p(k)$ denotes the unit step [i.e., $p(k) = 0$ for $k < 0$ and $p(k) = 1$ for $k \geq 0$]. Note that if an eigenvalue $\lambda_i$ of $A$ is zero, then (3.131) must be modified. In this case,

$$\sum_{i=0}^{n_i-1} A_{il}l!\delta(k-l) \tag{3.133}$$

are the terms in (3.131) corresponding to the zero eigenvalue.

To prove (3.131), (3.132), we proceed as in the proof of (3.50), (3.51). We recall that $\{A^k\} = \mathcal{Z}^{-1}[z(zI-A)^{-1}]$ and we use the partial fraction expansion method to determine the $z$-transform. In particular, as in the proof of (3.50), (3.51), we can readily verify that

$$z(zI - A)^{-1} = z \sum_{i=1}^{\sigma} \sum_{l=0}^{n_i-1} (l! A_{il})(z - \lambda_i)^{-(l+1)}, \tag{3.134}$$

where the $A_{il}$ are given in (3.132). We now take the inverse $z$-transform of both sides of (3.134). We first notice that

$$
\begin{aligned}
\mathcal{Z}^{-1}[z(z - \lambda_i)^{-(l+1)}] &= \mathcal{Z}^{-1}[z^{-l} z^{l+1}(z - \lambda_i)^{-(l+1)}] \\
&= \mathcal{Z}^{-1}[z^{-l}(1 - \lambda_i z^{-1})^{-(l+1)}] = f(k - l)p(k - l) \\
&= \begin{cases} f(k - l), & \text{for } k \geq l, \\ 0, & \text{otherwise.} \end{cases}
\end{aligned}
$$

Referring to Tables 3.3 and 3.4 we note that $f(k)p(k) = \mathcal{Z}^{-1}[(1 - \lambda_i z^{-1})^{-(l+1)}] = [\frac{1}{l!}(k+1)\cdots(k+l)]\lambda_i^k$ for $\lambda_i \neq 0$ and $l \geq 1$. Therefore, $\mathcal{Z}^{-1}[l! z(z - \lambda_i)^{-(l+1)}] = l! f(k - l)p(k - l) = k(k - 1)\cdots(k - l + 1)\lambda_i^{k-l}, l \geq 1$. For $l = 0$, we have $\mathcal{Z}^{-1}[(1 - \lambda_i z^{-1})^{-1}] = \lambda_i^k$. This shows that (3.131) is true when $\lambda_i \neq 0$. Finally, if $\lambda_i = 0$, we note that $\mathcal{Z}^{-1}[l! z^{-l}] = l! \delta(k - l)$, which implies (3.133).

Note that one can derive several alternative but equivalent expressions for (3.131) that correspond to different ways of determining the inverse $z$-transform of $z(zI - A)^{-1}$ or of determining $A^k$ via some other methods.

In complete analogy with the continuous-time case, we call the terms $A_{il} k(k - 1)\cdots(k - l + 1)\lambda_i^{k-l}$ the *modes of the system* (3.128). There are $n_i$ modes corresponding to the eigenvalues $\lambda_i, l = 0, \ldots, n_i - 1$, and the system (3.128) has a total of $n$ modes.

It is particularly interesting to study the matrix $A^k, k = 0, 1, 2, \ldots$ using the Jordan canonical form of $A$, i.e., $J = P^{-1}AP$, where the similarity transformation $P$ is constructed by using the generalized eigenvectors of $A$. We recall once more that $J = \text{diag}[J_1, \ldots, J_\sigma] \triangleq \text{diag}[J_i]$ where each $n_i \times n_i$ block $J_i$ corresponds to the eigenvalue $\lambda_i$ and where, in turn, $J_i = \text{diag}[J_{i1}, \ldots, J_{il_i}]$ with $J_{ij}$ being smaller square blocks, the dimensions of which depend on the length of the chains of generalized eigenvectors corresponding to $J_i$ (refer to Subsection 3.3.2). Let $J_{ij}$ denote a typical Jordan canonical form block. We shall investigate the matrix $J_{ij}^k$, since $A^k = P^{-1}J^k P = P^{-1} \text{diag}[J_{ij}^k]P$.

Let

$$
J_{ij} = \begin{bmatrix}
\lambda_i & 1 & 0 & \cdots & 0 \\
0 & \lambda_i & \ddots & & \vdots \\
\vdots & \vdots & \ddots & \ddots & \vdots \\
\vdots & \vdots & & \ddots & 1 \\
0 & 0 & \cdots & \cdots & \lambda_i
\end{bmatrix} = \lambda_i I + N_i, \tag{3.135}
$$

where

$$N_i = \begin{bmatrix} 0 & 1 & 0 & \cdots & 0 \\ 0 & 0 & & & 0 \\ \vdots & \vdots & \ddots & & \vdots \\ \vdots & \vdots & & \ddots & 1 \\ 0 & 0 & \cdots & \cdots & 0 \end{bmatrix}$$

and where we assume that $J_{ij}$ is a $t \times t$ matrix. Then

$$(J_{ij})^k = (\lambda_i I + N_i)^k$$

$$= \lambda_i^k I + k\lambda_i^{k-1} N_i + \frac{k(k-1)}{2!}\lambda_i^{k-2} N_i^2 + \cdots + k\lambda_i N_i^{k-1} + N_i^k. \quad (3.136)$$

Now since $N_i^k = 0$ for $k \geq t$, a typical $t \times t$ Jordan block $J_{ij}$ will generate terms that involve only the scalars $\lambda_i^k, \lambda_i^{k-1}, \ldots, \lambda_i^{k-(t-1)}$. Since the largest possible block associated with the eigenvalue $\lambda_i$ is of dimension $n_i \times n_i$, the expression of $A^k$ in (3.131) should involve at most the terms $\lambda_i^k, \lambda_i^{k-1}, \ldots, \lambda_i^{k-(n_i-1)}$, which it does.

The above enables us to prove the following useful fact: Given $A \in R^{n \times n}$, there exists an integer $k \geq 0$ such that

$$A^k = 0 \quad (3.137)$$

if and only if all the eigenvalues $\lambda_i$ of $A$ are at the origin. Furthermore, the smallest $k$ for which (3.137) holds is equal to the dimension of the largest block $J_{ij}$ of the Jordan canonical form of $A$.

The second part of the above assertion follows readily from (3.136). We ask the reader to prove the first part of the assertion.

We conclude by observing that when all $n$ eigenvalues $\lambda_i$ of $A$ are distinct, then

$$A^k = \sum_{i=1}^{n} A_i \lambda_i^k, \quad k \geq 0, \quad (3.138)$$

where

$$A_i = \lim_{z \to \lambda_i} [(z - \lambda_i)(zI - A)^{-1}]. \quad (3.139)$$

If $\lambda_i = 0$, we use $\delta(k)$, the unit pulse, in place of $\lambda_i^k$ in (3.138). This result is straightforward, in view of (3.131), (3.132).

---

**Example 3.35.** In (3.128) we let $A = \begin{bmatrix} 0 & 1 \\ -\frac{1}{4} & 1 \end{bmatrix}$. The eigenvalues of $A$ are $\lambda_1 = \lambda_2 = \frac{1}{2}$, and therefore, $n_1 = 2$ and $\sigma = 1$. Applying (3.131), (3.132), we obtain

$$A^k = A_{10}\lambda_1^k p(k) + A_{11} k \lambda_1^{k-1} p(k-1)$$

$$= \begin{bmatrix} 1 & 0 \\ 0 & 1 \end{bmatrix} \left(\frac{1}{2}\right)^k p(k) + \begin{bmatrix} -\frac{1}{2} & 1 \\ -\frac{1}{4} & \frac{1}{2} \end{bmatrix} (k) \left(\frac{1}{2}\right)^{k-1} p(k-1).$$

**Example 3.36.** In (3.128) we let $A = \begin{bmatrix} -1 & 2 \\ 0 & 1 \end{bmatrix}$. The eigenvalues of $A$ are $\lambda_1 = -1, \lambda_2 = 1$ (so that $\sigma = 2$). Applying (3.138), (3.139), we obtain

$$A^k = A_{10}\lambda_1^k + A_{20}\lambda_2^k = \begin{bmatrix} 1 & -1 \\ 0 & 0 \end{bmatrix}(-1)^k + \begin{bmatrix} 0 & 1 \\ 0 & 1 \end{bmatrix}, \quad k \geq 0.$$

Note that this same result was obtained by an entirely different method in Example 3.30.

---

**Example 3.37.** In (3.128) we let $A = \begin{bmatrix} 0 & 1 \\ 0 & -1 \end{bmatrix}$. The eigenvalues of $A$ are $\lambda_1 = 0, \lambda_2 = -1$, and $\sigma = 2$. Applying (3.138), (3.139), we obtain

$$A_0 = \lim_{z \to 0}[z(zI - A)^{-1}] = \frac{1}{z+1}\begin{bmatrix} z+1 & 1 \\ 0 & z \end{bmatrix}|_{z=0} = \begin{bmatrix} 1 & 1 \\ 0 & 0 \end{bmatrix}$$

$$A_1 = \lim_{z \to -1}\left[(z+1)\frac{1}{z(z+1)}\begin{bmatrix} z+1 & 1 \\ 0 & z \end{bmatrix}\right] = \begin{bmatrix} 0 & -1 \\ 0 & 1 \end{bmatrix}$$

and

$$A^k = A_0\delta(k) + A_1(-1)^k = \begin{bmatrix} 1 & 1 \\ 0 & 0 \end{bmatrix}\delta(k) + \begin{bmatrix} 0 & -1 \\ 0 & 1 \end{bmatrix}(-1)^k, \quad k \geq 0.$$

---

As in the case of continuous-time systems described by (3.48), various notions of stability of an equilibrium for discrete-time systems described by linear, autonomous, homogeneous ordinary difference equations (3.128) will be studied in detail in Chapter 4. If $\phi(k, 0, x_e)$ denotes the solution of system (3.128) with $x(0) = x_e$, then $x_e$ is said to be an *equilibrium* of (3.128) if $\phi(k, 0, x_e) = x_e$ for all $k \geq 0$. Clearly, $x_e = 0$ is an equilibrium of (3.128). In discussing the qualitative properties, it is customary to speak, somewhat informally, of the stability properties of (3.128), rather than the stability properties of the equilibrium $x_e = 0$ of system (3.128).

The concepts of *stability*, *asymptotic stability*, and *instability* of system (3.128) are now defined in an identical manner as in Subsection 3.3.3 for system (3.48), except that in this case continuous-time $t$ ($t \in R^+$) is replaced by discrete-time $k$ ($k \in Z^+$).

By inspecting the modes of system (3.128) [given by (3.131) and (3.132)], we can readily establish the following stability criteria:

1. The system (3.128) is *asymptotically* stable if and only if all eigenvalues of $A$ are within the unit circle of the complex plane (i.e., $|\lambda_j| < 1$, $j = 1, \ldots, n$).

2. The system (3.128) is *stable* if and only if $|\lambda_j| \leq 1$, $j = 1, \ldots, n$, and for all eigenvalues with $|\lambda_j| = 1$ having multiplicity $n_j > 1$, it is true that

$$\lim_{z \to \lambda_j} [[(z - \lambda_j)^{n_j}(zI - A)^{-1}]^{(n_j - 1 - l)}] = 0 \text{ for } l = 1, \ldots, n_j - 1. \quad (3.140)$$

3. The system (3.128) is *unstable* if and only if (2) is not true.

---

***Example 3.38.*** The system given in Example 3.35 is asymptotically stable. The system given in Example 3.36 is stable. In particular, note that the solution $\phi(k, 0, x(0)) = A^k x(0)$ for Example 3.36 is bounded.

---

When the eigenvalues $\lambda_i$ of $A$ are distinct, then as in the continuous-time case [refer to (3.56), (3.57)], we can readily show that

$$A^k = \sum_{j=1}^{n} A_j \lambda_j^k, \quad A_j = v_j \tilde{v}_j, \quad k \geq 0, \quad (3.141)$$

where the $v_j$ and $\tilde{v}_j$ are right and left eigenvectors of $A$ corresponding to $\lambda_j$, respectively. If $\lambda_j = 0$, we use $\delta(k)$, the unit pulse, in place of $\lambda_j^k$ in (3.141).

In proving (3.141), we use the same approach as in the proof of (3.56), (3.57). We have $A^k = Q \, \text{diag}[\lambda_1^k, \ldots, \lambda_n^k]Q^{-1}$, where the columns of $Q$ are the $n$ right eigenvectors and the rows of $Q^{-1}$ are the $n$ left eigenvectors of $A$.

As in the continuous-time case [system (3.48)], the initial condition $x(0)$ for system (3.128) can be selected to be colinear with the eigenvector $v_i$ to eliminate from the solution of (3.128) all modes except the ones involving $\lambda_i^k$.

---

***Example 3.39.*** As in Example 3.36, we let $A = \begin{bmatrix} -1 & 2 \\ 0 & 1 \end{bmatrix}$. Corresponding to the eigenvalues $\lambda_1 = -1$, $\lambda_2 = 1$, we have the right and left eigenvectors $v_1 = (1, 0)^T, v_2 = (1, 1)^T, \tilde{v}_1 = (1, -1)$, and $\tilde{v}_2 = (0, 1)$. Then

$$A^k = [v_1 \; \tilde{v}_1]\lambda_1^k + [v_2 \; \tilde{v}_2]\lambda_2^k$$
$$= \begin{bmatrix} 1 & -1 \\ 0 & 0 \end{bmatrix}(-1)^k + \begin{bmatrix} 0 & 1 \\ 0 & 1 \end{bmatrix}(1)^k, \quad k \geq 0.$$

Choose $x(0) = \alpha(1, 0)^T = \alpha v_1$ with $\alpha \neq 0$. Then

$$\phi(k, 0, x(0)) = \begin{bmatrix} \alpha \\ 0 \end{bmatrix}(-1)^k,$$

which contains only the mode associated with $\lambda_1 = -1$.

---

We conclude our discussion of modes and asymptotic behavior by briefly considering the state equation

$$x(k+1) = Ax(k) + Bu(k), \tag{3.142}$$

where $x$, $u$, $A$, and $B$ are as defined in (3.91a). Taking the $\mathcal{Z}$-transform of both sides of (3.142) and rearranging yields

$$\tilde{x}(z) = z(zI - A)^{-1}x(0) + (zI - A)^{-1}B\tilde{u}(z). \tag{3.143}$$

By taking the inverse $\mathcal{Z}$-transform of (3.143), we see that the solution $\phi$ of (3.142) is the sum of modes that correspond to the singularities or poles of $z(zI - A)^{-1}x(0)$ and of $(zI - A)^{-1}B\tilde{u}(z)$. If in particular, system (3.128) is asymptotically stable [i.e., for $x(k+1) = Ax(k)$, all eigenvalues $\lambda_j$ of $A$ are such that $|\lambda_j| < 1$, $j = 1, \ldots, n$] and if $u(k)$ in (3.142) is bounded [i.e., there is an $M$ such that $|u_i(k)| < M$ for all $k \geq 0$, $i = 1, \ldots, m$], then it is easily seen that the solutions of (3.142) are bounded as well.

## 3.6 An Important Comment on Notation

Chapters 1–3 are primarily concerned with the basic (qualitative) properties of systems of first-order ordinary differential equations, such as, e.g., the system of equations given by

$$\dot{x} = Ax, \tag{3.144}$$

where $x \in R^n$ and $A \in R^{n \times n}$. In the arguments and proofs to establish various properties for such systems, we highlighted the solutions by using the $\phi$-notation. Thus, the unique solution of (3.144) for a given set of initial data $(t_0, x_0)$ was written as $\phi(t, t_0, x_0)$ with $\phi(t_0, t_0, x_0) = x_0$. A similar notation was used in the case of the equation given by

$$\dot{x} = f(t, x) \tag{3.145}$$

and the equations given by

$$\dot{x} = Ax + Bu, \tag{3.146a}$$
$$y = Cx + Du, \tag{3.146b}$$

where in (3.145) and in (3.146) all symbols are defined as in (1.11) (see Chapter 1) and as in (3.61) of this chapter, respectively.

In the study of control systems such as system (3.61), the center of attention is usually the control input $u$ and the resulting evolution of the system state in the state space and the system output. In the development of control systems theory, the $x$-notation has been adopted to express the solutions of systems. Thus, the solution of (3.61a) is denoted by $x(t)$ [or $x(t, t_0, x_0)$ when $t_0$ and $x_0$ are to be emphasized] and the evolution of the system output $y$

in (3.61b) is denoted by $y(t)$. In all subsequent chapters, except Chapter 4, we will also follow this practice, employing the usual notation utilized in the control systems literature. In Chapter 4, which is concerned with the stability properties of systems, we will use the $\phi$-notation when studying the Lyapunov stability of an equilibrium [such as system (3.144)] and the $x$-notation when investigating the input–output properties of control systems [such as system (3.61)].

## 3.7 Summary and Highlights

*Continuous-Time Systems*

- *The state transition matrix $\Phi(t, t_0)$ of $\dot{x} = Ax$*

$$\Phi(t, t_0) \triangleq \Psi(t)\Psi^{-1}(t_0), \tag{3.9}$$

where $\Psi(t)$ is any fundamental matrix of $\dot{x} = Ax$. See Definitions 3.8 and 3.2 and Theorem 3.9 for properties of $\Phi(t, t_0)$. In the present time-invariant case

$$\Phi(t, t_0) = e^{A(t-t_0)},$$

where

$$e^{At} = I + \sum_{k=1}^{\infty} \frac{t^k A^k}{k!} \tag{3.17}$$

is the matrix exponential. See Theorem 3.13 for properties.
- *Methods to evaluate $e^{At}$.* Via infinite series (3.17) and via similarity transformation

$$e^{At} = P^{-1}e^{Jt}P$$

with $J = P^{-1}AP$ [see (3.25)] where $J$ is diagonal or in Jordan canonical form; via the Cayley–Hamilton Theorem [see (3.41)] and via the Laplace transform, where

$$e^{At} = \mathcal{L}^{-1}[(sI - A)^{-1}], \tag{3.45}$$

or via the system modes [see (3.50)], which simplify to

$$e^{At} = \sum_{i=1}^{n} A_i e^{\lambda_i t} \tag{3.53}$$

when the $n$ eigenvalues of $A$, $\lambda_i$, are distinct. See also (3.56), (3.57).
- *Modes of the system.* $e^{At}$ is expressed in terms of the modes $A_{ik} t^k e^{\lambda_i t}$ in (3.50). The distinct eigenvalue case is found in (3.53), (3.54) and in (3.56), (3.57).
- *The stability of an equilibrium* of $\dot{x} = Ax$ is defined and related to the eigenvalues of $A$ using the expression for $e^{At}$ in terms of the modes.

- Given $\dot{x} = Ax + Bu$,

$$x(t) = e^{At}x(0) + \int_0^t e^{A(t-s)}Bu(s)s$$

is its solution, the *variation of constants formula.*
If in addition $y = Cx + Du$, then the *total response of the system* is

$$y(t) = Ce^{A(t-t_0)}x_0 + C\int_{t_0}^t e^{A(t-s)}Bu(s)ds + Du(t). \tag{3.70}$$

The *impulse response* is

$$H(t) = \begin{cases} Ce^{At}B + D\delta(t), & t \geq \tau, \\ 0, & t < 0, \end{cases} \tag{3.72}$$

and the *transfer function* is

$$\widehat{H}(s) = C(sI - A)^{-1}B + D. \tag{3.80}$$

Note that $H(s) = \mathcal{L}(H(t, 0))$.
- *Equivalent representations*

$$\dot{\tilde{x}} = \widetilde{A}\tilde{x} + \widetilde{B}u,$$
$$y = \widetilde{C}\tilde{x} + \widetilde{D}u, \tag{3.82}$$

where
$$\widetilde{A} = PAP^{-1}, \quad \widetilde{B} = PB, \quad \widetilde{C} = CP^{-1}, \quad \widetilde{D} = D \tag{3.83}$$

is equivalent to $\dot{x} = Ax + Bu$, $y = Cx + Du$.

*Discrete-Time Systems*

- Consider the *discrete-time system*

$$x(k+1) = Ax(k) + Bu(k), \quad y(k) = Cx(k) + Du(k). \tag{3.91}$$

Then

$$y(k) = CA^k x(0) + \sum_{j=0}^{k-1} CA^{k-(j+1)}Bu(j) + Du(k), \quad k > 0. \tag{3.99}$$

The *discrete-time unit impulse response* is

$$H(k, 0) = \begin{cases} CA^{k-1}B & k \geq 0, \\ D & k = 0, \\ 0 & k < 0, \end{cases} \tag{3.102}$$

and the *transfer function* is

$$\widehat{H}(z) = C(zI - A)^{-1}B + D. \tag{3.108}$$

Note that $\widehat{H}(z) = \mathcal{Z}\{H(k, 0)\}$.

- $A^k = \mathcal{Z}^{-1}(z(zI - A)^{-1})$. $A^k$ may also be calculated using the Cayley–Hamilton theorem. Note that when all $n$ eigenvalues of $A$, $\lambda_i$, are distinct then

$$A^k = \sum_{j=0}^{n} A_i \lambda_i^k, \quad k \geq 0, \tag{3.138}$$

$A_i \lambda_i^k$ are the modes of the system.
- The *stability of an equilibrium* of $x(k + 1) = Ax(k)$ is defined and related to the eigenvalues of $A$ using the expressions of $A^k$ in terms of the modes.

*Sampled Data Systems*

- When $\dot{x} = Ax + Bu$, $y = Cx + Du$ is the system in Figure 3.3, the discrete-time description is

$$\bar{x}(k + 1) = \bar{A}\bar{x}(k) + \bar{B}\bar{u}(k),$$
$$\bar{y}(k) = \bar{C}\bar{x}(k) + \bar{D}\bar{u}(k), \tag{3.126}$$

with

$$\bar{A} = e^{AT}, \quad \bar{B} = \left[ \int_0^T e^{A\tau} d\tau \right] B,$$

$$\bar{C} = C, \quad \bar{D} = D, \tag{3.127}$$

where $T$ is the sampling period.

## 3.8 Notes

Our treatment of basic aspects of linear ordinary differential equations in Sections 3.2 and 3.3 follows along lines similar to the development of this subject given in Miller and Michel [8].

State-space and input–output representations of continuous-time systems and discrete-time systems, addressed in Sections 3.4 and 3.5, respectively, are addressed in a variety of textbooks, including Kailath [7], Chen [4], Brockett [3], DeCarlo [5], Rugh [11], and others. For further material on sampled-data systems, refer to Aström and Wittenmark [2] and to the early works on this subject that include Jury [6] and Ragazzini and Franklin [9].

Detailed treatments of the Laplace transform and the $z$-transform, discussed briefly in Sections 3.3 and 3.5, respectively, can be found in numerous texts on signals and linear systems, control systems, and signal processing.

In the presentation of the material in all the sections of this chapter, we have relied principally on Antsaklis and Michel [1].

The state representation of systems received wide acceptance in systems theory beginning in the late 1950s. This was primarily due to the work of R.

E. Kalman and others in filtering theory and quadratic control theory and due to the work of applied mathematicians concerned with the stability theory of dynamical systems. For comments and extensive references on some of the early contributions in these areas, refer to Kailath [7] and Sontag [12]. Of course, differential equations have been used to describe the dynamical behavior of artificial systems for many years. For example, in 1868 J. C. Maxwell presented a complete treatment of the behavior of devices that regulate the steam pressure in steam engines called flyball governors (Watt governors) to explain certain phenomena.

The use of state-space representations in the systems and control area opened the way for the systematic study of systems with multi-inputs and multi-outputs. Since the 1960s an alternative description is also being used to characterize time-invariant MIMO control systems that involves usage of polynomial matrices or differential operators. Some of the original references on this approach include Rosenbrock [10] and Wolovich [13]. This method, which corresponds to system descriptions by means of higher order ordinary differential equations (rather than systems of first-order ordinary differential equations, as is the case in the state-space description), is addressed in Sections 7.5 and 8.5 and in Chapter 10.

# References

1. P.J. Antsaklis and A.N. Michel, *Linear Systems*, Birkhäuser, Boston, MA, 2006.
2. K.J. Aström and B. Wittenmark, *Computer-Controlled Systems. Theory and Design*, Prentice-Hall, Englewood Cliffs, NJ, 1990.
3. R.W. Brockett, *Finite Dimensional Linear Systems*, Wiley, New York, NY, 1970.
4. C.T. Chen, *Linear System Theory and Design*, Holt, Rinehart and Winston, New York, NY, 1984.
5. R.A. DeCarlo, *Linear Systems*, Prentice-Hall, Englewood Cliffs, NJ, 1989.
6. E.I. Jury, *Sampled-Data Control Systems*, Wiley, New York, NY, 1958.
7. T. Kailath, *Linear Systems*, Prentice-Hall, Englewood Cliffs, NJ, 1980.
8. R.K. Miller and A.N. Michel, *Ordinary Differential Equations*, Academic Press, New York, NY, 1982.
9. J.R. Ragazzini and G.F. Franklin, *Sampled-Data Control Systems*, McGraw-Hill, New York, NY, 1958.
10. H.H. Rosenbrock, *State Space and Multivariable Theory*, Wiley, New York, NY, 1970.
11. W.J. Rugh, *Linear System Theory, Second Edition*, Prentice-Hall, Englewood Cliffs, NJ, 1996.
12. E.D. Sontag, *Mathematical Control Theory. Deterministic Finite Dimensional Systems*, TAM 6, Springer-Verlag, New York, NY, 1990.
13. W.A. Wolovich, *Linear Multivariable Systems*, Springer-Verlag, New York, NY, 1974.

## Exercises

For the first 12 exercises, the reader may want to refer to the appendix, which contains appropriate material on matrices and linear algebra.

**3.1.** (a) Let $(V, F) = (R^3, R)$. Determine the representation of $v = (1, 4, 0)^T$ with respect to the basis $v^1 = (1, -1, 0)^T, v^2 = (1, 0, -1)^T$, and $v^3 = (0, 1, 0)^T$.

(b) Let $V = F^3$, and let $F$ be the field of rational functions. Determine the representation of $\tilde{v} = (s+2, 1/s, -2)^T$ with respect to the basis $\{v^1, v^2, v^3\}$ given in (a).

**3.2.** Find the relationship between the two bases $\{v^1, v^2, v^3\}$ and $\{\bar{v}^1, \bar{v}^2, \bar{v}^3\}$ (i.e., find the matrix of $\{\bar{v}^1, \bar{v}^2, \bar{v}^3\}$ with respect to $\{v^1, v^2, v^3\}$) where $v^1 = (2, 1, 0)^T, v^2 = (1, 0, -1)^T, v^3 = (1, 0, 0)^T, \bar{v}^1 = (1, 0, 0)^T, \bar{v}^2 = (0, 1, -1)$, and $\bar{v}^3 = (0, 1, 1)$. Determine the representation of the vector $e_2 = (0, 1, 0)^T$ with respect to both of the above bases.

**3.3.** Let $\alpha \in R$ be fixed. Show that the set of all vectors $(x, \alpha x)^T, x \in R$, determines a vector space of dimension one over $F = R$, where vector addition and multiplication of vectors by scalars is defined in the usual manner. Determine a basis for this space.

**3.4.** Show that the set of all real $n \times n$ matrices with the usual operation of matrix addition and the usual operation of multiplication of matrices by scalars constitutes a vector space over the reals [denoted by $(R^{n \times n}, R)$]. Determine the dimension and a basis for this space. Is the above statement still true if $R^{n \times n}$ is replaced by $R^{m \times n}$, the set of real $m \times n$ matrices? Is the above statement still true if $R^{n \times n}$ is replaced by the set of nonsingular matrices? Justify your answers.

**3.5.** Let $v^1 = (s^2, s)^T$ and $v^2 = (1, 1/s)^T$. Is the set $\{v^1, v^2\}$ linearly independent over the field of rational functions? Is it linearly independent over the field of real numbers?

**3.6.** Determine the rank of the following matrices, carefully specifying the field:

(a) $\begin{bmatrix} j \\ 3j \\ -1 \end{bmatrix}$, (b) $\begin{bmatrix} 1 & 4 & -5 \\ 7 & 0 & 2 \end{bmatrix}$, (c) $\begin{bmatrix} (s+4) & -2 \\ (s^2 - 1) & 6 \\ 0 & 2s + 3 \\ s & -s + 4 \end{bmatrix}$, (d) $\left( \dfrac{s+1}{s^2} \right)$,

where $j = \sqrt{-1}$.

**3.7.** (a) Determine bases for the range and null space of the matrices

$$A_1 = [1\,0\,1], \quad A_2 = \begin{bmatrix} 1 & 1 \\ 0 & 0 \\ 1 & 0 \end{bmatrix}, \quad \text{and} \quad A_3 = \begin{bmatrix} 3 & 2 & 1 \\ 3 & 2 & 1 \\ 3 & 2 & 1 \end{bmatrix}.$$

(b) Characterize all solutions of $A_1 x = 1$ (see Subsection A.3.1).

**3.8.** Show that $e^{(A_1+A_2)t} = e^{A_1 t} e^{A_2 t}$ if $A_1 A_2 = A_2 A_1$.

**3.9.** Show that there exists a similarity transformation matrix $P$ such that

$$PAP^{-1} = A_c = \begin{bmatrix} 0 & 1 & 0 & \cdots & 0 \\ 0 & 0 & 1 & \cdots & 0 \\ \vdots & \vdots & \vdots & & \vdots \\ 0 & 0 & 0 & \cdots & 1 \\ -\alpha_0 & -\alpha_1 & -\alpha_2 & \cdots & -\alpha_{n-1} \end{bmatrix}$$

if and only if there exists a vector $b \in R^n$ such that the rank of $[b, Ab, \ldots, A^{n-1}b]$ is $n$; i.e., $\rho[b, Ab, \ldots, A^{n-1}b] = n$.

**3.10.** Show that if $\lambda_i$ is an eigenvalue of the companion matrix $A_c$ given in Exercise 3.9, then a corresponding eigenvector is $v^i = (1, \lambda_i, \ldots, \lambda_i^{n-1})^T$.

**3.11.** Let $\lambda_i$ be an eigenvalue of a matrix $A$, and let $v^i$ be a corresponding eigenvector. Let $f(\lambda) = \sum_{k=0}^{l} \alpha_k \lambda^k$ be a polynomial with real coefficients. Show that $f(\lambda_i)$ is an eigenvalue of the matrix function $f(A) = \sum_{k=0}^{l} \alpha_k A^k$. Determine an eigenvector corresponding to $f(\lambda_i)$.

**3.12.** For the matrices

$$A_1 = \begin{bmatrix} 1 & 2 & 0 \\ 0 & 0 & 2 \\ 0 & 0 & 1 \end{bmatrix} \quad \text{and} \quad A_2 = \begin{bmatrix} 0 & 1 & 0 & 0 \\ 0 & 0 & 1 & 0 \\ 0 & 0 & 0 & 1 \\ 0 & 0 & 0 & 0 \end{bmatrix},$$

determine the matrices $A_1^{100}, A_2^{100}, e^{A_1 t}$, and $e^{A_2 t}, t \in R$.

**3.13.** For the system

$$\dot{x} = Ax + Bu, \tag{3.147}$$

where all symbols are as defined in (3.61a), derive the *variation of constants formula* (3.11), using the change of variables $z(t) = \Phi(t_0, t)x(t)$.

**3.14.** Show that $\frac{\partial}{\partial \tau} \Phi(t, \tau) = -\Phi(t, \tau)A$ for all $t, \tau \in R$.

**3.15.** The *adjoint equation* of (3.1) is given by

$$\dot{z} = -A^T z. \tag{3.148}$$

Let $\Phi(t, t_0)$ and $\Phi_a(t, t_0)$ denote the state transition matrices of (3.1) and its adjoint equation, respectively. Show that $\Phi_a(t, t_0) = [\Phi(t_0, t)]^T$.

**3.16.** Consider the system described by

$$\dot{x} = Ax + Bu, \quad y = Cx, \tag{3.149}$$

where all symbols are as in (3.61) with $D = 0$, and consider the *adjoint equation* of (3.149), given by

$$\dot{z} = -A^T z + C^T v, \quad w = B^T z. \tag{3.150}$$

(a) Let $H(t, \tau)$ and $H_a(t, \tau)$ denote the impulse response matrices of (3.149) and (3.150), respectively. Show that at the times when the impulse responses are nonzero, they satisfy $H(t, \tau) = H_a(\tau, t)^T$.

(b) Show that $H(s) = -H_a(-s)^T$, where $H(s)$ and $H_a(s)$ are the transfer matrices of (3.149) and (3.150), respectively.

**3.17.** Compute $e^{At}$ for

$$A = \begin{bmatrix} 1 & 4 & 10 \\ 0 & 2 & 0 \\ 0 & 0 & 2 \end{bmatrix}.$$

**3.18.** Given is the matrix

$$A = \begin{bmatrix} 1/2 & -1 & 0 \\ 0 & -1 & 0 \\ 0 & 0 & -2 \end{bmatrix}.$$

(a) Determine $e^{At}$, using the different methods covered in this text. Discuss the advantages and disadvantages of these methods.

(b) For system (3.1) let $A$ be as given. Plot the components of the solution $\phi(t, t_0, x_0)$ when $x_0 = x(0) = (1, 1, 1)^T$ and $x_0 = x(0) = (2/3, 1, 0)^T$. Discuss the differences in these plots, if any.

**3.19.** Show that for $A = \begin{bmatrix} a & b \\ -b & a \end{bmatrix}$, we have $e^{At} = e^{at} \begin{bmatrix} \cos bt & \sin bt \\ -\sin bt & \cos bt \end{bmatrix}$.

**3.20.** Given is the system of equations

$$\begin{bmatrix} \dot{x}_1 \\ \dot{x}_2 \end{bmatrix} = \begin{bmatrix} -1 & 0 \\ 0 & 1 \end{bmatrix} \begin{bmatrix} x_1 \\ x_2 \end{bmatrix} + \begin{bmatrix} 1 \\ 1 \end{bmatrix} u$$

with $x(0) = (1, 0)^T$ and

$$u(t) = p(t) = \begin{cases} 1, & t \geq 0, \\ 0, & \text{elsewhere.} \end{cases}$$

Plot the components of the solution of $\phi$. For different initial conditions $x(0) = (a, b)^T$, investigate the changes in the asymptotic behavior of the solutions.

**3.21.** The system (3.1) with $A = \begin{bmatrix} 0 & 1 \\ -1 & 0 \end{bmatrix}$ is called the *harmonic oscillator* (refer to Chapter 1) because it has periodic solutions $\phi(t) = (\phi_1(t), \phi_2(t))^T$. Simultaneously, for the same values of $t$, plot $\phi_1(t)$ along the horizontal axis and $\phi_2(t)$ along the vertical axis in the $x_1$-$x_2$ plane to obtain a *trajectory* for this system for the specific initial condition $x(0) = x_0 = (x_1(0), x_2(0))^T = (1,1)^T$. In plotting such trajectories, time $t$ is viewed as a parameter, and arrows are used to indicate increasing time. When the horizontal axis corresponds to position and the vertical axis corresponds to velocity, the $x_1$-$x_2$ plane is called the *phase plane* and $\phi_1, \phi_2$ (resp. $x_1, x_2$) are called *phase variables*.

**3.22.** First, determine the solution $\phi$ of $\begin{bmatrix} \dot{x}_1 \\ \dot{x}_2 \end{bmatrix} = \begin{bmatrix} 0 & 1 \\ 1 & 0 \end{bmatrix} \begin{bmatrix} x_1 \\ x_2 \end{bmatrix}$ with $x(0) = (1,1)^T$. Next, determine the solution $\phi$ of the above system for $x(0) = \alpha(1,-1)^T, \alpha \in R, \alpha \neq 0$, and discuss the properties of the two solutions.

**3.23.** In Subsection 3.3.3 it is shown that when the $n$ eigenvalues $\lambda_i$ of a real $n \times n$ matrix $A$ are distinct, then $e^{At} = \sum_{i=1}^{n} A_i e^{\lambda_i t}$ where $A_i = \lim_{s \to \lambda_i} [(s - \lambda_i)(sI - A)^{-1}] = v_i \tilde{v}_i$ [refer to (3.53), (3.54), and (3.57)], where $v_i, \tilde{v}_i$ are the right and left eigenvectors of $A$, respectively, corresponding to the eigenvalue $\lambda_i$. Show that (a) $\sum_{i=1}^{n} A_i = I$, where $I$ denotes the $n \times n$ identity matrix, (b) $AA_i = \lambda_i A_i$, (c) $A_i A = \lambda_i A_i$, (d) $A_i A_j = \delta_{ij} A_i$, where $\delta_{ij} = 1$ if $i = j$ and $\delta_{ij} = 0$ when $i \neq j$.

**3.24.** Consider the system

$$\dot{x} = Ax + Bu, \quad y = Cx, \tag{3.151}$$

where all symbols are defined as in (3.61) with $D = 0$. Let

$$A = \begin{bmatrix} 0 & 1 & 0 & 0 \\ 3 & 0 & 0 & 2 \\ 0 & 0 & 0 & 1 \\ 0 & -2 & 0 & 0 \end{bmatrix}, \quad B = \begin{bmatrix} 0 & 0 \\ 1 & 0 \\ 0 & 0 \\ 0 & 1 \end{bmatrix}, \quad C = [1, 0, 1, 0]. \tag{3.152}$$

(a) Find equivalent representations for system (3.151), (3.152), given by

$$\dot{\tilde{x}} = \tilde{A}\tilde{x} + \tilde{B}u, \quad y = \tilde{C}\tilde{x}, \tag{3.153}$$

where $\tilde{x} = Px$, when $\tilde{A}$ is in (i) the Jordan canonical (or diagonal) form and (ii) the companion form.

(b) Determine the transfer function matrix for this system.

**3.25.** Consider the system (3.61) with $B = 0$.

(a) Let

$$A = \begin{bmatrix} -1 & 1 & 0 \\ 0 & -1 & 0 \\ 0 & 0 & 2 \end{bmatrix} \quad \text{and} \quad C = [1, 1, 1].$$

If possible, select $x(0)$ in such a manner so that $y(t) = te^{-t}, t \geq 0$.

(b) Determine conditions under which it is possible to specify $y(t), t \geq 0$, using only the initial data $x(0)$.

**3.26.** Consider the system given by

$$\begin{bmatrix} \dot{x}_1 \\ \dot{x}_2 \end{bmatrix} = \begin{bmatrix} -1 & 1 \\ -1/2 & 0 \end{bmatrix} \begin{bmatrix} x_1 \\ x_2 \end{bmatrix} + \begin{bmatrix} 0 \\ 1/2 \end{bmatrix} u, \quad y = [1, \ 0] \begin{bmatrix} x_1 \\ x_2 \end{bmatrix}.$$

(a) Determine $x(0)$ so that for $u(t) = e^{-4t}, y(t) = ke^{-4t}$, where $k$ is a real constant. Determine $k$ for the present case. Notice that $y(t)$ does not have any transient components.

(b) Let $u(t) = e^{\alpha t}$. Determine $x(0)$ that will result in $y(t) = ke^{\alpha t}$. Determine the conditions on $\alpha$ for this to be true. What is $k$ in this case?

**3.27.** Consider the system (3.61) with

$$A = \begin{bmatrix} 0 & 0 & 1 & 0 \\ 3 & 0 & -3 & 1 \\ -1 & 1 & 4 & -1 \\ 1 & 0 & -1 & 0 \end{bmatrix}, \quad B = \begin{bmatrix} 0 & 0 \\ 1 & 0 \\ 0 & 1 \\ 0 & 0 \end{bmatrix}, \quad C = \begin{bmatrix} 1 & 0 & 0 & 0 \\ 0 & 0 & 0 & 1 \end{bmatrix}.$$

(a) For $x(0) = [1, 1, 1, 1]^T$ and $u(t) = [1, 1]^T, t \geq 0$, determine the solution $\phi(t, 0, x(0))$ and the output $y(t)$ for this system and plot the components $\phi_i(t, 0, x(0)), i = 1, 2, 3, 4$ and $y_i(t), i = 1, 2$.

(b) Determine the transfer function matrix $H(s)$ for this system.

**3.28.** Consider the system

$$x(k+1) = Ax(k) + Bu(k), \quad y(k) = Cx(k), \tag{3.154}$$

where all symbols are defined as in (3.91) with $D = 0$. Let

$$A = \begin{bmatrix} 1 & 2 \\ 0 & 1 \end{bmatrix}, \quad B = \begin{bmatrix} 2 \\ 3 \end{bmatrix}, \quad C = [1 \ 1],$$

and let $x(0) = 0$ and $u(k) = 1, k \geq 0$.

(a) Determine $\{y(k)\}, k \geq 0$, by working in the (i) time domain and (ii) $z$-transform domain, using the transfer function $H(z)$.

(b) If it is known that when $u(k) = 0$, then $y(0) = y(1) = 1$, can $x(0)$ be uniquely determined? If your answer is affirmative, determine $x(0)$.

**3.29.** Consider $\hat{y}(z) = H(z)\hat{u}(z)$ with transfer function $H(z) = 1/(z + 0.5)$.

(a) Determine and plot the unit pulse response $\{h(k)\}$.

(b) Determine and plot the unit step response.

(c) If

$$u(k) = \begin{cases} 1, & k = 1, 2, \\ 0, & \text{elsewhere,} \end{cases}$$

determine $\{y(k)\}$ for $k = 0, 1, 2, 3$, and 4 via (i) convolution and (ii) the $z$-transform. Plot your answer.

(d) For $u(k)$ given in (c), determine $y(k)$ as $k \to \infty$.

**3.30.** Consider the system (3.91) with $x(0) = x_0$ and $k \geq 0$. Determine conditions under which there exists a sequence of inputs so that the state remains at $x_0$, i.e., so that $x(k) = x_0$ for all $k \geq 0$. How is this input sequence determined? Apply your method to the specific case

$$A = \begin{bmatrix} 2 & 0 \\ 0 & -1 \end{bmatrix}, \quad B = \begin{bmatrix} 1 \\ 1 \end{bmatrix}, \quad x_0 = \begin{bmatrix} -2 \\ 1 \end{bmatrix}.$$

**3.31.** For system (3.92) with $x(0) = x_0$ and $k \geq 0$, it is desired to have the state go to the zero state for any initial condition $x_0$ in at most $n$ steps; i.e., we desire that $x(k) = 0$ for any $x_0 = x(0)$ and for all $k \geq n$.

(a) Derive conditions in terms of the eigenvalues of $A$ under which the above is true. Determine the minimum number of steps under which the above behavior will be true.

(b) For part (a), consider the specific cases

$$A_1 = \begin{bmatrix} 0 & 1 & 0 \\ 0 & 0 & 1 \\ 0 & 0 & 0 \end{bmatrix}, \quad A_2 = \begin{bmatrix} 0 & 1 & 0 \\ 0 & 0 & 0 \\ 0 & 0 & 0 \end{bmatrix}, \quad A_3 = \begin{bmatrix} 0 & 0 & 0 \\ 0 & 0 & 1 \\ 0 & 0 & 0 \end{bmatrix}.$$

*Hint:* Use the Jordan canonical form for $A$. Results of this type are important in *deadbeat control*, where it is desired that a system variable attains some desired value and settles at that value in a finite number of time steps.

**3.32.** Consider a continuous-time system described by the transfer function $H(s) = 4/(s^2 + 2s + 2)$; i.e., $\hat{y}(s) = H(s)\hat{u}(s)$.

(a) Assume that the system is at rest, and assume a unit step input; i.e., $u(t) = 1, t \geq 0, u(t) = 0, t < 0$. Determine and plot $y(t)$ for $t \geq 0$.

(b) Obtain a discrete-time approximation for the above system by following these steps: (i) Determine a *realization* of the form (3.61) of $H(s)$ (see Exercise 3.33); (ii) assuming a sampler and a zero-order hold with sampling period $T$, use (3.151) to obtain a discrete-time system representation

$$\bar{x}(k+1) = \bar{A}\bar{x}(k) + \bar{B}\bar{u}(k), \quad \bar{y}(k) = \bar{C}\bar{x}(k) + \bar{D}\bar{u}(k) \tag{3.155}$$

and determine $\bar{A}, \bar{B}$, and $\bar{C}$ in terms of $T$.

(c) For the unit step input, $u(k) = 1$ for $k \geq 0$ and $u(k) = 0$ for $k < 0$, determine and plot $\bar{y}(k), k \geq 0$, for different values of $T$, assuming the system is at rest. Compare $\bar{y}(k)$ with $y(t)$ obtained in part (a).

(d) Determine for (3.155) the transfer function $\bar{H}(z)$ in terms of $T$. Note that $\bar{H}(z) = \bar{C}(zI - \bar{A})^{-1}\bar{B} + \bar{D}$. It can be shown that $\bar{H}(z) = (1 - z^{-1})\mathcal{Z}\{\mathcal{L}^{-1}[H(s)/s]_{t=kT}\}$. Verify this for the given $H(s)$.

**3.33.** Given a proper rational transfer function matrix $H(s)$, the state-space representation $\{A, B, C, D\}$ is called a *realization of* $H(s)$ if $H(s) = C(sI - A)^{-1}B + D$. Thus, the system (3.61) is a realizations of $H(s)$ if its transfer function matrix is equal to $H(s)$. Realizations of $H(s)$ are studied at length in Chapter 8. When $H(s)$ is scalar, it is straightforward to derive certain realizations, and in the following, we consider one such realization.

Given a proper rational scalar transfer function $H(s)$, let $D \triangleq \lim_{s\to\infty} H(s)$ and let

$$H_{sp}(s) \triangleq H(s) - D = \frac{b_{n-1}s^{n-1} + \cdots + b_1 s + b_0}{s^n + a_{n-1}s^{n-1} + \cdots + a_1 s + a_0},$$

a strictly proper rational function.

(a) Let

$$A = \begin{bmatrix} 0 & 1 & 0 & \cdots & 0 & 0 \\ 0 & 0 & 1 & \cdots & 0 & 0 \\ \cdots & \cdots & \cdots & \cdots & \cdots & \cdots \\ 0 & 0 & 0 & \cdots & 0 & 1 \\ -a_0 & -a_1 & -a_2 & \cdots & -a_{n-2} & -a_{n-1} \end{bmatrix}, \quad B = \begin{bmatrix} 0 \\ 0 \\ \vdots \\ 0 \\ 1 \end{bmatrix}, \quad (3.156)$$

$$C = [b_0\ b_1 \cdots b_{n-1}],$$

and show that $\{A, B, C, D\}$ is indeed a realization of $H(s)$. Also, show that $\{\tilde{A} = A^T, \tilde{B} = C^T, \tilde{C} = B^T, \tilde{D} = D\}$ is a realization of $H(s)$ as well. These two state-space representations are said to be in *controller (companion) form* and in *observer (companion) form*, respectively (refer to Chapter 6).

(b) In particular find realizations in controller and observer form for (i) $H(s) = 1/s^2$, (ii) $H(s) = \omega_n^2/(s^2 + 2\zeta\omega_n s + \omega_n^2)$, and (iii) $H(s) = (s+1)^2/(s-1)^2$.

**3.34.** Assume that $H(s)$ is a $p \times m$ proper rational transfer function matrix. Expand $H(s)$ in a Laurent series about the origin to obtain

$$H(s) = H_0 + H_1 s^{-1} + \cdots + H_k s^{-k} + \cdots = \sum_{k=0}^{\infty} H_k s^{-k}. \quad (3.157)$$

The elements of the sequence $\{H_0, H_1, \ldots, H_k, \ldots\}$ are called the *Markov parameters* of the system. These parameters provide an alternative representation of the transfer function matrix $H(s)$, and they are useful in Realization Theory (refer to Chapter 8).

(a) Show that the impulse response $H(t,0)$ can be expressed as

$$H(t,0) = H_0\delta(t) + \sum_{k=1}^{\infty} H_k(t^{k-1}/(k-1)!). \qquad (3.158)$$

In the following discussion, we assume that the system in question is described by (3.61).

(b) Show that

$$H(s) = D + C(sI - A)^{-1}B = D + \sum_{k=1}^{\infty}[CA^{k-1}B]s^{-k}, \qquad (3.159)$$

which shows that the elements of the sequence $\{D, CB, CAB, ..., CA^{k-1}B, ...\}$ are the Markov parameters of the system; i.e., $H_0 = D$ and $H_k = CA^{k-1}B$, $k = 1, 2, \ldots$.

(c) Show that

$$H(s) = D + \frac{1}{\alpha(s)}C[R_{n-1}s^{n-1} + \cdots + R_1s + R_0]B, \qquad (3.160)$$

where $\alpha(s) = s^n + a_{n-1}s^{n-1} + \cdots + a_1s + a_0 = \det(sI - A)$, the characteristic polynomial of $A$, and $R_{n-1} = I, R_{n-2} = AR_{n-1} + a_{n-1}I = A + a_{n-1}I, \ldots, R_0 = A^{n-1} + a_{n-1}A^{n-2} + \cdots + a_1I$.

Hint: Write $(sI - A)^{-1} = \frac{1}{\alpha(s)}[\text{adjoint}(sI - A)] = \frac{1}{\alpha(s)}[R_{n-1}s^{n-1} + \cdots + R_1s + R_0]$, and equate the coefficients of equal powers of $s$ in the expression

$$\alpha(s)I = (sI - A)[R_{n-1}s^{n-1} + \cdots + R_1s + R_0]. \qquad (3.161)$$

**3.35.** The *frequency response matrix* of a system described by its $p \times m$ transfer function matrix evaluated at $s = j\omega$,

$$H(\omega) \triangleq \widehat{H}(s)|_{s=j\omega},$$

is a very useful means of characterizing a system, since typically it can be determined experimentally, and since control system specifications are frequently expressed in terms of the frequency responses of transfer functions. When the poles of $\widehat{H}(s)$ have negative real parts, the system turns out to be bounded-input/bounded-output (BIBO) stable (refer to Chapter 4). Under these conditions, the frequency response $H(\omega)$ has a clear physical meaning, and this fact can be used to determine $H(\omega)$ experimentally.

(a) Consider a stable SISO system given by $\hat{y}(s) = \widehat{H}(s)\hat{u}(s)$. Show that if $u(t) = k\sin(\omega_0 t + \phi)$ with $k$ constant, then $y(t)$ at steady-state (i.e., after all transients have died out) is given by

$$y_{ss}(t) = k|H(\omega_0)|\sin(\omega_0 t + \phi + \theta(\omega_0)),$$

where $|H(\omega)|$ denotes the magnitude of $H(\omega)$ and $\theta(\omega) = \arg H(\omega)$ is the argument or phase of the complex quantity $H(\omega)$.

From the above it follows that $H(\omega)$ completely characterizes the system response at steady state (of a stable system) to a sinusoidal input. Since $u(t)$ can be expressed in terms of a series of sinusoidal terms via a Fourier series (recall that $u(t)$ is piecewise continuous), $H(\omega)$ characterizes the steady-state response of a stable system to any bounded input $u(t)$. This physical interpretation does not apply when the system is not stable.

(b) For the $p \times m$ transfer function matrix $\widehat{H}(s)$, consider the frequency response matrix $H(\omega)$ and extend the discussion of part (a) above to MIMO systems to give a physical interpretation of $H(\omega)$.

**3.36.** (*Double integrator*)

(a) Plot the response of the double integrator of Example 3.33 to a unit step input.
(b) Consider the discrete-time state-space representation of the double integrator of Example 3.33 for $T = 0.5, 1, 5$ sec and plot the unit step responses. Compare with your results in (a).

**3.37.** (*Spring mass system*) Consider the spring mass system of Example 1.1. For $M_1 = 1$ kg, $M_2 = 1$ kg, $K = 0.091$ N/m, $K_1 = 0.1$ N/m, $K_2 = 0.1$ N/m, $B = 0.0036$ N sec/m, $B_1 = 0.05$ N sec/m, and $B_2 = 0.05$ N sec/m, the state-space representation of the system in (1.27) assumes the form

$$
\begin{bmatrix} \dot{x}_1 \\ \dot{x}_2 \\ \dot{x}_3 \\ \dot{x}_4 \end{bmatrix} = \begin{bmatrix} 0 & 1 & 0 & 0 \\ -0.1910 & -0.0536 & 0.0910 & 0.0036 \\ 0 & 0 & 0 & 1 \\ 0.0910 & 0.0036 & -0.1910 & -0.0536 \end{bmatrix} \begin{bmatrix} x_1 \\ x_2 \\ x_3 \\ x_4 \end{bmatrix} + \begin{bmatrix} 0 & 0 \\ 1 & 0 \\ 0 & 0 \\ 0 & -1 \end{bmatrix} \begin{bmatrix} f_1 \\ f_2 \end{bmatrix},
$$

where $x_1 \triangleq y_1$, $x_2 \triangleq \dot{y}_1$, $x_3 \triangleq y_2$, and $x_4 \triangleq \dot{y}_2$.

(a) Determine the eigenvalues and eigenvectors of the matrix $A$ of the system and express $x(t)$ in terms of the modes and the initial conditions $x(0)$ of the system, assuming that $f_1 = f_2 = 0$.
(b) For $x(0) = [1, 0, -0.5, 0]^T$ and $f_1 = f_2 = 0$, plot the states for $t \geq 0$.
(c) Let $y = Cx$ with $C = \begin{bmatrix} 1 & 0 & 0 & 0 \\ 0 & 1 & 0 & 0 \end{bmatrix}$ denote the output of the system. Determine the transfer function between $y$ and $u \triangleq [f_1, f_2]^T$.
(d) For zero initial conditions, $f_1(t) = \delta(t)$ (the unit impulse), and $f_2(t) = 0$, plot the states for $t \geq 0$ and comment on your results.
(e) It is desirable to explore what happens when the mass ratio $M_2/M_1$ takes on different values. For this, let $M_2 = \alpha M_1$ with $M_1 = 1$ kg and $\alpha = 0.1$, $0.5, 2, 5$. All other parameter values remain the same. Repeat (a) to (d) for the different values of $\alpha$ and discuss your results.

**3.38.** *(Automobile suspension system)* [M.L. James, G.M. Smith, and J.C. Wolford, *Applied Numerical Methods for Digital Computation*, Harper and Row, 1985, p. 667.] Consider the spring mass system in Figure 3.4, which describes part of the suspension system of an automobile. The data for this system are given as

$m_1 = \frac{1}{4} \times$ (mass of the automobile) $= 375$ kg,

$m_2 = $ mass of one wheel $= 30$ kg,

$k_1 = $ spring constant $= 1500$ N/m,

$k_2 = $ linear spring constant of tire $= 6500$ N/m,

$c = $ damping constant of dashpot $= 0, 375, 750,$ and $1125$ N sec/m,

$x_1 = $ displacement of automobile body from equilibrium position m,

$x_3 = $ displacement of wheel from equilibrium position m,

$v = $ velocity of car $= 9, 18, 27,$ or $36$ m/sec.

A linear model $\dot{x} = Ax + Bu$ for this system is given by

$$
\begin{bmatrix} \dot{x}_1 \\ \dot{x}_2 \\ \dot{x}_3 \\ \dot{x}_4 \end{bmatrix} = \begin{bmatrix} 0 & 1 & 0 & 0 \\ -\frac{k_1}{m_1} & -\frac{c}{m_1} & \frac{k_1}{m_1} & \frac{c}{m_1} \\ 0 & 0 & 0 & 1 \\ \frac{k_1}{m_2} & \frac{c}{m_2} & -\frac{k_1+k_2}{m_2} & -\frac{c}{m_2} \end{bmatrix} \begin{bmatrix} x_1 \\ x_2 \\ x_3 \\ x_4 \end{bmatrix} + \begin{bmatrix} 0 \\ 0 \\ 0 \\ \frac{k_2}{m_2} \end{bmatrix} u(t),
$$

where $u(t) = \frac{1}{6} \sin \frac{2\pi vt}{20}$ describes the profile of the roadway.

(a) Determine the eigenvalues of $A$ for all of the above cases.
(b) Plot the states for $t \geq 0$ when the input $u(t) = \frac{1}{6} \sin \frac{2\pi vt}{20}$ and $x(0) = [0,0,0,0]^T$ for all the above cases. Comment on your results.

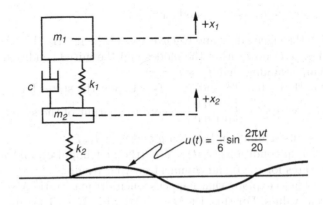

**Figure 3.4.** Model of an automobile suspension system

# 4

---

# Stability

Dynamical systems, either occurring in nature or man made, usually function in some specified mode. The most common such modes are operating points that frequently turn out to be equilibria.

In this chapter we will concern ourselves primarily with the qualitative behavior of equilibria. Most of the time, we will be interested in the asymptotic stability of an equilibrium (operating point), which means that when the state of a given system is displaced (disturbed) from its desired operating point (equilibrium), the expectation is that the state will eventually return to the equilibrium. For example, in the case of an automobile under cruise control, traveling at the desired constant speed of 50 mph (which determines the operating point, or equilibrium condition), perturbations due to hill climbing (hill descending), will result in decreasing (increasing) speeds. In a properly designed cruise control system, it is expected that the car will return to its desired operating speed of 50 mph.

Another qualitative characterization of dynamical systems is the expectation that bounded system inputs will result in bounded system outputs, and that small changes in inputs will result in small changes in outputs. System properties of this type are referred to as input–output stability. Such properties are important for example in tracking systems, where the output of the system is expected to follow a desired input. Frequently, it is possible to establish a connection between the input–output stability properties and the Lyapunov stability properties of an equilibrium. In the case of linear systems, this connection is well understood. This will be addressed in Section 7.3.

## 4.1 Introduction

In this chapter we present a brief introduction to stability theory. We are concerned primarily with linear systems and systems that are a consequence of linearizations of nonlinear systems. As in the other chapters of this book, we consider finite-dimensional continuous-time systems and finite-dimensional discrete-time systems described by systems of first-order ordinary differential equations and systems of first-order ordinary difference equations, respectively.

In Section 4.2 we introduce the concept of equilibrium of dynamical systems described by systems of first-order ordinary differential equations, and in Section 4.3 we give definitions of various types of stability in the sense of Lyapunov (including stability, uniform stability, asymptotic stability, uniform asymptotic stability, exponential stability, and instability).

In Section 4.4 we establish conditions for the various Lyapunov stability and instability types enumerated in Section 4.3 for linear systems $\dot{x} = Ax$. Most of these results are phrased in terms of the properties of the state transition matrix for such systems.

In Section 4.5 we introduce the Second Method of Lyapunov, also called the Direct Method of Lyapunov, to establish necessary and sufficient conditions for various Lyapunov stability types of an equilibrium for linear systems $\dot{x} = Ax$. These results, which are phrased in terms of the system parameters [coefficients of the matrix A], give rise to the Lyapunov matrix equation.

In Section 4.6 we use the Direct Method of Lyapunov in deducing the asymptotic stability and instability of an equilibrium of nonlinear autonomous systems $\dot{x} = Ax + F(x)$ from the stability properties of their linearizations $\dot{w} = Aw$.

In Section 4.7 we establish necessary and sufficient conditions for the input–output stability (more precisely, for the bounded input/bounded output stability) of continuous-time, linear, time-invariant systems. These results involve the system impulse response matrix.

The stability results presented in Sections 4.2 through and including Section 4.7 pertain to continuous-time systems. In Section 4.8 we present analogous stability results for discrete-time systems.

## 4.2 The Concept of an Equilibrium

In this section we concern ourselves with systems of first-order autonomous ordinary differential equations,

$$\dot{x} = f(x), \qquad (4.1)$$

where $x \in R^n$. When discussing global results, we shall assume that $f : R^n \to R^n$, while when considering local results, we may assume that $f : B(h) \to R^n$ for some $h > 0$, where $B(h) = \{x \in R^n : \| x \| < h\}$ and $\| \cdot \|$ denotes a norm on $R^n$. Unless otherwise stated, we shall assume that for every $(t_0, x_0), t_0 \in R^+$, the initial-value problem

$$\dot{x} = f(x), \quad x(t_0) = x_0 \qquad (4.2)$$

possesses a unique solution $\phi(t, t_0, x_0)$ that exists for all $t \geq t_0$ and that depends continuously on the initial data $(t_0, x_0)$. Refer to Section 1.5 for conditions that ensure that (4.2) has these properties. Since (4.1) is time-invariant, we may assume without loss of generality that $t_0 = 0$ and we will denote the solutions of (4.1) by $\phi(t, x_0)$ (rather than $\phi(t, t_0, x_0)$) with $x(0) = x_0$.

**Definition 4.1.** *A point $x_e \in R^n$ is called an* equilibrium point *of (4.1), or simply an* equilibrium *of (4.1), if*

$$f(x_e) = 0.$$

■

We note that

$$\phi(t, x_e) = x_e \quad \text{for all } t \geq 0;$$

i.e., the equilibrium $x_e$ is the unique solution of (4.1) with initial data given by $\phi(0, x_e) = x_e$.

We will usually assume that in a given discussion, unless otherwise stated, the equilibrium of interest is located at the origin of $R^n$. This assumption can be made without loss of generality by noting that if $x_e \neq 0$ is an equilibrium point of (4.1), i.e., $f(x_e) = 0$, then by letting $w = x - x_e$, we obtain the transformed system

$$\dot{w} = F(w) \tag{4.3}$$

with $F(0) = 0$, where

$$F(w) = f(w + x_e). \tag{4.4}$$

Since the above transformation establishes a one-to-one correspondence between the solutions of (4.1) and (4.3), we may assume henceforth that the equilibrium of interest for (4.1) is located at the origin. This equilibrium, $x = 0$, will be referred to as the *trivial solution* of (4.1).

Before concluding this section, it may be fruitful to consider some specific cases.

---

*Example 4.2.* In Example 1.4 we considered the simple pendulum given in Figure 1.7. Letting $x_1 = x$ and $x_2 = \dot{x}$ in (1.37), we obtain the system of equations

$$\dot{x}_1 = x_2,$$
$$\dot{x}_2 = -k \sin x_1, \tag{4.5}$$

where $k > 0$ is a constant. *Physically*, the pendulum has two equilibrium points: one where the mass $M$ is located vertically at the bottom of the figure (i.e., at 6 o'clock) and the other where the mass is located vertically at the top of the figure (i.e., at 12 o'clock). The *model* of this pendulum, however, described by (4.5), has countably infinitely many equilibrium points that are located in $R^2$ at the points $(\pi n, 0)^T, n = 0, \pm 1, \pm 2, \ldots$.

---

***Example 4.3.*** The linear, autonomous, homogenous system of ordinary differential equations

$$\dot{x} = Ax \tag{4.6}$$

has a unique equilibrium that is at the origin if and only if $A$ is nonsingular. Otherwise, (4.6) has nondenumerably many equilibria. [Refer to Chapter 1 for the definitions of symbols in (4.6).]

***Example 4.4.*** Assume that for

$$\dot{x} = f(x), \tag{4.7}$$

$f$ is continuously differentiable with respect to all of its arguments, and let

$$J(x_e) = \left. \frac{\partial f}{\partial x}(x) \right|_{x=x_e},$$

where $\partial f/\partial x$ denotes the $n \times n$ *Jacobian matrix* defined by

$$\frac{\partial f}{\partial x} = \left[ \frac{\partial f_i}{\partial x_j} \right].$$

If $f(x_e) = 0$ and $J(x_e)$ is nonsingular, then $x_e$ is an equilibrium of (4.7).

***Example 4.5.*** The system of ordinary differential equations given by

$$\dot{x}_1 = k + \sin(x_1 + x_2) + x_1,$$
$$\dot{x}_2 = k + \sin(x_1 + x_2) - x_1,$$

with $k > 1$, has no equilibrium points at all.

## 4.3 Qualitative Characterizations of an Equilibrium

In this section we consider several qualitative characterizations that are of fundamental importance in systems theory. These characterizations are concerned with various types of stability properties of an equilibrium and are referred to in the literature as *Lyapunov stability*.

Throughout this section, we consider systems of equations

$$\dot{x} = f(x), \tag{4.8}$$

and we assume that (4.8) possesses an equilibrium at the origin. We thus have $f(0) = 0$.

**Definition 4.6.** *The equilibrium $x = 0$ of (4.8) is said to be* stable *if for every $\epsilon > 0$, there exists a $\delta(\epsilon) > 0$ such that*

$$\| \phi(t, x_0) \| < \epsilon \text{ for all } t \geq 0 \tag{4.9}$$

*whenever*

$$\| x_0 \| < \delta(\epsilon). \tag{4.10}$$

∎

In Definition 4.6, $\| \cdot \|$ denotes any one of the equivalent norms on $R^n$, and (as in Chapters 1 and 2) $\phi(t, x_0)$ denotes the solution of (4.8) with initial condition $x(0) = x_0$. The notation $\delta(\epsilon)$ indicates that $\delta$ depends on the choice of $\epsilon$.

In words, Definition 4.6 states that by choosing the initial points in a sufficiently small spherical neighborhood, when the equilibrium $x = 0$ of (4.8) is stable, we can force the graph of the solution for $t \geq 0$ to lie entirely inside a given cylinder. This is depicted in Figure 4.1 for the case $x \in R^2$.

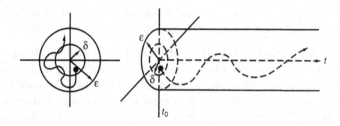

**Figure 4.1.** Stability of an equilibrium

**Definition 4.7.** *The equilibrium $x = 0$ of (4.8) is said to be* asymptotically stable *if*

*(i) it is stable,*
*(ii) there exists an $\eta > 0$ such that $\lim_{t \to \infty} \phi(t, x_0) = 0$ whenever $\| x_0 \| < \eta$.* ∎

The set of all $x_0 \in R^n$ such that $\phi(t, x_0) \to 0$ as $t \to \infty$ is called the *domain of attraction* of the equilibrium $x = 0$ of (4.8). Also, if for (4.8) condition (ii) is true, then the equilibrium $x = 0$ is said to be *attractive*.

**Definition 4.8.** *The equilibrium $x = 0$ of (4.8) is* exponentially stable *if there exists an $\alpha > 0$, and for every $\epsilon > 0$, there exists a $\delta(\epsilon) > 0$, such that*

$$\| \phi(t, x_0) \| \leq \epsilon e^{\alpha t} \text{ for all } t \geq 0$$

*whenever $\|x_0\| < \delta(\epsilon)$.* ∎

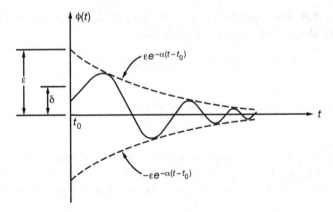

**Figure 4.2.** An exponentially stable equilibrium

Figure 4.2 shows the behavior of a solution in the vicinity of an exponentially stable equilibrium $x = 0$.

**Definition 4.9.** *The equilibrium $x = 0$ of (4.8) is* unstable *if it is not stable. In this case, there exists an $\epsilon > 0$, and a sequence $x_m \to 0$ of initial points and a sequence $\{t_m\}$ such that $\| \phi(t_m, x_m) \| \geq \epsilon$ for all $m, t_m \geq 0$.*  ∎

If $x = 0$ is an unstable equilibrium of (4.8), then it still can happen that all the solutions tend to zero with increasing $t$. This indicates that instability and attractivity of an equilibrium are compatible concepts. We note that the equilibrium $x = 0$ of (4.8) is necessarily unstable if every neighborhood of the origin contains initial conditions corresponding to unbounded solutions (i.e., solutions whose norm grows to infinity on a sequence $t_m \to \infty$). However, it can happen that a system (4.8) with unstable equilibrium $x = 0$ may have only bounded solutions.

The concepts that we have considered thus far pertain to *local* properties of an equilibrium. In the following discussion, we consider *global* characterizations of an equilibrium.

**Definition 4.10.** *The equilibrium $x = 0$ of (4.8) is* asymptotically stable in the large *if it is stable and if every solution of (4.8) tends to zero as $t \to \infty$.*  ∎

When the equilibrium $x = 0$ of (4.8) is asymptotically stable in the large, its domain of attraction is all of $R^n$. Note that in this case, $x = 0$ is the *only* equilibrium of (4.8).

**Definition 4.11.** *The equilibrium $x = 0$ of (4.8) is* exponentially stable in the large *if there exists $\alpha > 0$ and for any $\beta > 0$, there exists $k(\beta) > 0$ such that*

$$\| \phi(t, x_0) \| \leq k(\beta) \| x_0 \| e^{-\alpha t} \quad \text{for all } t \geq 0$$

*whenever $\| x_0 \| < \beta$.*  ∎

We conclude this section with a few specific cases.

The scalar differential equation

$$\dot{x} = 0 \tag{4.11}$$

has for any initial condition $x(0) = x_0$ the solution $\phi(t, x_0) = x_0$; i.e., all solutions are equilibria of (4.11). The trivial solution is stable; however, it is not asymptotically stable.

The scalar differential equation

$$\dot{x} = ax \tag{4.12}$$

has for every $x(0) = x_0$ the solution $\phi(t, x_0) = x_0 e^{at}$, and $x = 0$ is the only equilibrium of (4.12). If $a > 0$, this equilibrium is unstable, and when $a < 0$, this equilibrium is exponentially stable in the large.

As mentioned earlier, a system

$$\dot{x} = f(x) \tag{4.13}$$

can have all solutions approaching an equilibrium, say, $x = 0$, without this equilibrium being asymptotically stable. An example of this type of behavior is given by the *nonlinear* system of equations

$$\dot{x}_1 = \frac{x_1^2(x_2 - x_1) + x_2^5}{(x_1^2 + x_2^2)[1 + (x_1^2 + x_2^2)^2]},$$
$$\dot{x}_2 = \frac{x_2^2(x_2 - 2x_1)}{(x_1^2 + x_2^2)[1 + (x_1^2 + x_2^2)^2]}.$$

For a detailed discussion of this system, refer to [6], pp. 191–194, cited at the end of this chapter.

Before proceeding any further, a few comments are in order concerning the reasons for considering equilibria and their stability properties as well as other types of stability that we will encounter. To this end we consider linear time-invariant systems given by

$$\dot{x} = Ax + Bu, \tag{4.14a}$$
$$y = Cx + Du, \tag{4.14b}$$

where all symbols in (4.14) are defined as in (2.7). The usual qualitative analysis of such systems involves two concepts, *internal stability* and *input–output stability*.

In the case of *internal stability*, the output equation (4.14b) plays no role whatsoever, the system input $u$ is assumed to be identically zero, and the focus of the analysis is concerned with the qualitative behavior of the solutions of linear time-invariant systems

$$\dot{x} = Ax \tag{4.15}$$

near the equilibrium $x = 0$. This is accomplished by making use of the various types of Lyapunov stability concepts introduced in this section. In other words, internal stability of system (4.14) concerns the Lyapunov stability of the equilibrium $x = 0$ of system (4.15).

In the case of *input–output stability*, we view systems as operators determined by (4.14) that relate outputs $y$ to inputs $u$ and the focus of the analysis is concerned with qualitative relations between system inputs and system outputs. We will address this type of stability in Section 4.7.

## 4.4 Lyapunov Stability of Linear Systems

In this section we first study the stability properties of the equilibrium $x = 0$ of linear autonomous homogeneous systems

$$\dot{x} = Ax, \quad t \geq 0. \tag{4.16}$$

Recall that $x = 0$ is always an equilibrium of (4.16) and that $x = 0$ is the only equilibrium of (4.16) if $A$ is nonsingular. Recall also that the solution of (4.16) for $x(0) = x_0$ is given by

$$\phi(t, x_0) = \Phi(t, 0)x_0 = \Phi(t - 0, 0)x_0$$
$$\triangleq \Phi(t)x_0 = e^{At}x_0,$$

where in the preceding equation, a slight abuse of notation has been used.

We first consider some of the basic properties of system (4.16).

**Theorem 4.12.** *The equilibrium $x = 0$ of (4.16) is stable if and only if the solutions of (4.16) are bounded, i.e., if and only if*

$$\sup_{t \geq t_0} \| \Phi(t) \| \triangleq k < \infty,$$

*where $\| \Phi(t) \|$ denotes the matrix norm induced by the vector norm used on $R^n$ and $k$ denotes a constant.*

*Proof.* Assume that the equilibrium $x = 0$ of (4.16) is stable. Then for $\epsilon = 1$ there is a $\delta = \delta(1) > 0$ such that $\| \phi(t, x_0) \| < 1$ for all $t \geq 0$ and all $x_0$ with $\| x_0 \| \leq \delta$. In this case

$$\| \phi(t, x_0) \| = \| \Phi(t)x_0 \| = \| [\Phi(t)(x_0\delta)/ \| x_0 \|] \| (\| x_0 \| /\delta) < \| x_0 \| /\delta$$

for all $x_0 \neq 0$ and all $t \geq 0$. Using the definition of matrix norm [refer to Section A.7], it follows that

$$\| \Phi(t) \| \leq \delta^{-1}, \quad t \geq 0.$$

We have proved that if the equilibrium $x = 0$ of (4.16) is stable, then the solutions of (4.16) are bounded.

Conversely, suppose that all solutions $\phi(t, x_0) = \Phi(t)x_0$ are bounded. Let $\{e_1, \ldots, e_n\}$ denote the natural basis for $n$-space, and let $\| \phi(t, e_j) \| < \beta_j$ for all $t \geq 0$. Then for any vector $x_0 = \sum_{j=1}^n \alpha_j e_j$ we have that

$$\| \phi(t, x_0) \| = \| \sum_{j=1}^n \alpha_j \phi(t, e_j) \| \leq \sum_{j=1}^n |\alpha_j| \beta_j$$

$$\leq (\max_j \beta_j) \sum_{j=1}^n |\alpha_j| \leq k \| x_0 \|$$

for some constant $k > 0$ for $t \geq 0$. For given $\epsilon > 0$, we choose $\delta = \epsilon/k$. Thus, if $\| x_0 \| < \delta$, then $\| \phi(t, x_0) \| < k \| x_0 \| < \epsilon$ for all $t \geq 0$. We have proved that if the solutions of (4.16) are bounded, then the equilibrium $x = 0$ of (4.16) is stable. ∎

**Theorem 4.13.** *The following statements are equivalent.*

*(i)   The equilibrium $x = 0$ of (4.16) is asymptotically stable.*
*(ii)  The equilibrium $x = 0$ of (4.16) is asymptotically stable in the large.*
*(iii)* $\lim_{t \to \infty} \| \Phi(t) \| = 0$.

*Proof.* Assume that statement (i) is true. Then there is an $\eta > 0$ such that when $\| x_0 \| \leq \eta$, then $\phi(t, x_0) \to 0$ as $t \to \infty$. But then we have for any $x_0 \neq 0$ that

$$\phi(t, x_0) = \phi(t, \eta x_0 / \| x_0 \|)(\| x_0 \| /\eta) \to 0$$

as $t \to \infty$. It follows that statement (ii) is true.

Next, assume that statement (ii) is true. For any $\epsilon > 0$, there must exist a $T(\epsilon) > 0$ such that for all $t \geq T(\epsilon)$ we have that $\| \phi(t, x_0) \| = \| \Phi(t)x_0 \| < \epsilon$. To see this, let $\{e_1, \ldots, e_n\}$ be the natural basis for $R^n$. Thus, for some fixed constant $k > 0$, if $x_0 = (\alpha_1, \ldots, \alpha_n)^T$ and if $\| x_0 \| \leq 1$, then $x_0 = \sum_{j=1}^n \alpha_j e_j$ and $\sum_{j=1}^n |\alpha_j| \leq k$. For each $j$, there is a $T_j(\epsilon)$ such that $\| \Phi(t)e_j \| < \epsilon/k$ and $t \geq T_j(\epsilon)$. Define $T(\epsilon) = \max\{T_j(\epsilon) : j = 1, \ldots, n\}$. For $\| x_0 \| \leq 1$ and $t \geq T(\epsilon)$, we have that

$$\| \Phi(t)x_0 \| = \| \sum_{j=1}^n \alpha_j \Phi(t)e_j \| \leq \sum_{j=1}^n |\alpha_j|(\epsilon/k) \leq \epsilon.$$

By the definition of the matrix norm [see the appendix], this means that $\| \Phi(t) \| \leq \epsilon$ for $t \geq T(\epsilon)$. Therefore, statement (iii) is true.

Finally, assume that statement (iii) is true. Then $\| \Phi(t) \|$ is bounded in $t$ for all $t \geq 0$. By Theorem 4.12, the equilibrium $x = 0$ is stable. To prove asymptotic stability, fix $\epsilon > 0$. If $\| x_0 \| < \eta = 1$, then $\| \phi(t, x_0) \| \leq \| \Phi(t) \| \| x_0 \| \to 0$ as $t \to \infty$. Therefore, statement (i) is true. This completes the proof. ∎

**Theorem 4.14.** *The equilibrium $x = 0$ of (4.16) is asymptotically stable if and only if it is exponentially stable.*

*Proof.* The exponential stability of the equilibrium $x = 0$ implies the asymptotic stability of the equilibrium $x = 0$ of systems (4.13) in general and, hence, for systems (4.16) in particular.

Conversely, assume that the equilibrium $x = 0$ of (4.16) is asymptotically stable. Then there is a $\delta > 0$ and a $T > 0$ such that if $\parallel x_0 \parallel \le \delta$, then

$$\parallel \Phi(t + T)x_0 \parallel < \delta/2$$

for all $t \ge 0$. This implies that

$$\parallel \Phi(t + T) \parallel \le \frac{1}{2} \text{ if } t \ge 0. \tag{4.17}$$

From Theorem 3.9 (iii) we have that $\Phi(t - \tau) = \Phi(t - \sigma)\Phi(\sigma - \tau)$ for any $t, \sigma$, and $\tau$. Therefore,

$$\parallel \Phi(t + 2T) \parallel = \parallel \Phi(t + 2T - t - T)\Phi(t + T) \parallel \le \frac{1}{4},$$

in view of (4.17). By induction, we obtain for $t \ge 0$ that

$$\parallel \Phi(t + nT) \parallel \le 2^{-n}. \tag{4.18}$$

Now let $\alpha = (ln2)/T$. Then (4.18) implies that for $0 \le t < T$ we have that

$$\parallel \phi(t + nT, x_0) \parallel \le 2 \parallel x_0 \parallel 2^{-(n+1)} = 2 \parallel x_0 \parallel e^{-\alpha(n+1)T}$$
$$\le 2 \parallel x_0 \parallel e^{-\alpha(t+nT)},$$

which proves the result.    ∎

Even though the preceding results require knowledge of the state transition matrix $\Phi(t)$ of (4.16), they are quite useful in the qualitative analysis of linear systems. In view of the above results, we can state the following equivalent definitions.

The equilibrium $x = 0$ of (4.16) is *stable* if and only if there exists a finite positive constant $\gamma$, such that for any $x_0$, the corresponding solution satisfies the inequality

$$\parallel \phi(t, x_0) \parallel \le \gamma \parallel x_0 \parallel, \quad t \ge 0.$$

Furthermore, in view of the above results, if the equilibrium $x = 0$ of (4.16) is asymptotically stable, then in fact it must be globally asymptotically stable, and exponentially stable in the large. In this case there exist finite constants $\gamma \ge 1$ and $\lambda > 0$ such that

$$\parallel \phi(t, x_0) \parallel \le \gamma e^{-\lambda t} \parallel x_0 \parallel$$

for $t \ge 0$ and $x_0 \in R^n$.

We now continue our investigation of system (4.16) by referring to the discussion in Subsection 3.3.2 [refer to (3.23) to (3.39)] concerning the use of the Jordan canonical form to compute $\exp(At)$. We let $J = P^{-1}AP$ and define $x = Py$. Then (4.16) yields

$$\dot{y} = P^{-1}APy = Jy. \tag{4.19}$$

It is easily verified (the reader is asked to do so in the Exercises section) that the equilibrium $x = 0$ of (4.16) is stable (resp., asymptotically stable or unstable) if and only if $y = 0$ of (4.19) is stable (resp., asymptotically stable or unstable). In view of this, we can assume without loss of generality that the matrix $A$ in (4.16) is in Jordan canonical form, given by

$$A = \text{diag}[J_0, J_1, \ldots, J_s],$$

where

$$J_0 = \text{diag}[\lambda_1, \ldots, \lambda_k] \quad \text{and} \quad J_i = \lambda_{k+i}I_i + N_i$$

for the Jordan blocks $J_1, \ldots, J_s$.

As in (3.33), (3.34), (3.38), and (3.39), we have

$$e^{At} = \begin{bmatrix} e^{J_0 t} & & & 0 \\ & e^{J_1 t} & & \\ & & \ddots & \\ 0 & & & e^{J_s t} \end{bmatrix},$$

where

$$e^{J_0 t} = \text{diag}[e^{\lambda_1 t}, \ldots, e^{\lambda_k t}] \tag{4.20}$$

and

$$e^{J_i t} = e^{\lambda_{k+i} t} \begin{bmatrix} 1 & t & t^2/2 & \cdots & t^{n_i-1}/(n_i-1)! \\ 0 & 1 & t & \cdots & t^{n_i-2}/(n_i-2)! \\ \vdots & \vdots & \vdots & & \vdots \\ 0 & 0 & 0 & \cdots & 1 \end{bmatrix} \tag{4.21}$$

for $i = 1, \ldots, s$.

Now suppose that $Re\lambda_i \leq \beta$ for all $i = 1, \ldots, k$. Then it is clear that $\lim_{t \to \infty}(\| e^{J_0 t} \| /e^{\beta t}) < \infty$, where $\| e^{J_0 t} \|$ is the matrix norm induced by one of the equivalent vector norms defined on $R^n$. We write this as $\| e^{J_0 t} \| = \mathcal{O}(e^{\beta t})$. Similarly, if $\beta = Re\lambda_{k+i}$, then for any $\epsilon > 0$ we have that $\| e^{J_i t} \| = \mathcal{O}(t^{n_i-1}e^{\beta t}) = \mathcal{O}(e^{(\beta+\epsilon)t})$.

From the foregoing it is now clear that $\| e^{At} \| \leq K$ for some $K > 0$ if and only if all eigenvalues of $A$ have nonpositive real parts, and the eigenvalues with zero real part occur in the Jordan form only in $J_0$ and not in any of the Jordan blocks $J_i$, $1 \leq i \leq s$. Hence, by Theorem 4.12, the equilibrium $x = 0$ of (4.16) is under these conditions stable.

Now suppose that all eigenvalues of $A$ have negative real parts. From the preceding discussion it is clear that there is a constant $K > 0$ and an $\alpha > 0$ such that $\| e^{At} \| \leq K e^{-\alpha t}$, and therefore, $\| \phi(t, x_0) \| \leq K e^{-\alpha t} \| x_0 \|$ for all $t \geq 0$ and for all $x_0 \in R^n$. It follows that the equilibrium $x = 0$ is asymptotically stable in the large, in fact exponentially stable in the large. Conversely, assume that there is an eigenvalue $\lambda_i$ with a nonnegative real part. Then either one term in (4.20) does not tend to zero, or else a term in (4.21) is unbounded as $t \to \infty$. In either case, $e^{At} x(0)$ will not tend to zero when the initial condition $x(0) = x_0$ is properly chosen. Hence, the equilibrium $x = 0$ of (4.16) cannot be asymptotically stable (and, hence, it cannot be exponentially stable).

Summarizing the above, we have proved the following result.

**Theorem 4.15.** *The equilibrium $x = 0$ of (4.16) is* stable, *if and only if all eigenvalues of $A$ have nonpositive real parts, and every eigenvalue with zero real part has an associated Jordan block of order one. The equilibrium $x = 0$ of (4.16) is* asymptotically stable in the large, *in fact exponentially stable in the large, if and only if all eigenvalues of $A$ have negative real parts.* ∎

A direct consequence of the above result is that the equilibrium $x = 0$ of (4.16) is *unstable* if and only if at least one of the eigenvalues of $A$ has either positive real part or has zero real part that is associated with a Jordan block of order greater than one.

At this point, it may be appropriate to take note of certain conventions concerning matrices that are used in the literature. It should be noted that some of these are not entirely consistent with the terminology used in Theorem 4.15. Specifically, a real $n \times n$ matrix $A$ is called *stable* or a *Hurwitz matrix* if all its eigenvalues have negative real parts. If at least one of the eigenvalues has a positive real part, then $A$ is called *unstable*. A matrix $A$, which is neither stable nor unstable, is called *critical*, and the eigenvalues with zero real parts are called *critical eigenvalues*.

We conclude our discussion concerning the stability of (4.16) by noting that the results given above can also be obtained by directly using the facts established in Subsection 3.3.3, concerning modes and asymptotic behavior of time-invariant systems.

---

*Example 4.16.* We consider the system (4.16) with

$$A = \begin{bmatrix} 0 & 1 \\ -1 & 0 \end{bmatrix}.$$

The eigenvalues of $A$ are $\lambda_1, \lambda_2 = \pm j$. According to Theorem 4.15, the equilibrium $x = 0$ of this system is stable. This can also be verified by computing the solution of this system for a given set of initial data $x(0)^T = (x_1(0), x_2(0))$,

$$\phi_1(t, x_0) = x_1(0) \cos t + x_2(0) \sin t,$$
$$\phi_2(t, x_0) = -x_1(0) \sin t + x_2(0) \cos t,$$

$t \geq 0$, and then applying Definition 4.6.

---

**Example 4.17.** We consider the system (4.16) with

$$A = \begin{bmatrix} 0 & 1 \\ 0 & 0 \end{bmatrix}.$$

The eigenvalues of $A$ are $\lambda_1 = 0, \lambda_2 = 0$. According to Theorem 4.15, the equilibrium $x = 0$ of this system is unstable. This can also be verified by computing the solution of this system for a given set of initial data $x(0)^T = (x_1(0), x_2(0))$,

$$\phi_1(t, x_0) = x_1(0) + x_2(0)t,$$
$$\phi_2(t, x_0) = x_2(0),$$

$t \geq 0$, and then applying Definition 4.9. (Note that in this example, the entire $x_1$-axis consists of equilibria.)

---

**Example 4.18.** We consider the system (4.16) with

$$A = \begin{bmatrix} 2.8 & 9.6 \\ 9.6 & -2.8 \end{bmatrix}.$$

The eigenvalues of $A$ are $\lambda_1, \lambda_2 = \pm 10$. According to Theorem 4.15, the equilibrium $x = 0$ of this system is unstable.

---

**Example 4.19.** We consider the system (4.16) with

$$A = \begin{bmatrix} -1 & 0 \\ -1 & -2 \end{bmatrix}.$$

The eigenvalues of $A$ are $\lambda_1, \lambda_2 = -1, -2$. According to Theorem 4.15, the equilibrium $x = 0$ of this system is exponentially stable.

---

## 4.5 The Lyapunov Matrix Equation

In Section 4.4 we established a variety of stability results that require explicit knowledge of the solutions of (4.16). In this section we will develop stability criteria for (4.16) with *arbitrary* matrix $A$. In doing so, we will employ *Lyapunov's Second Method* (also called *Lyapunov's Direct Method*) for the case of linear systems (4.16). This method utilizes auxiliary real-valued functions

$v(x)$, called *Lyapunov functions*, that may be viewed as *generalized energy functions* or *generalized distance functions* (from the equilibrium $x = 0$), and the stability properties are then deduced directly from the properties of $v(x)$ and its time derivative $\dot{v}(x)$, evaluated along the solutions of (4.16).

A logical choice of Lyapunov function is $v(x) = x^T x = \| x \|^2$, which represents the square of the Euclidean distance of the state from the equilibrium $x = 0$ of (4.16). The stability properties of the equilibrium are then determined by examining the properties of $\dot{v}(x)$, the time derivative of $v(x)$ along the solutions of (4.16), which we repeat here,

$$\dot{x} = Ax. \tag{4.22}$$

This derivative can be determined *without explicitly solving for the solutions of (4.22)* by noting that

$$\dot{v}(x) = \dot{x}^T x + x^T \dot{x} = (Ax)^T x + x^T (Ax)$$
$$= x^T (A^T + A)x.$$

If the matrix $A$ is such that $\dot{v}(x)$ is negative for all $x \neq 0$, then it is reasonable to expect that the distance of the state of (4.22) from $x = 0$ will decrease with increasing time, and that the state will therefore tend to the equilibrium $x = 0$ of (4.22) with increasing time $t$.

It turns out that the Lyapunov function used in the above discussion is not sufficiently flexible. In the following discussion, we will employ as a "generalized distance function" the quadratic form given by

$$v(x) = x^T P x, \quad P = P^T, \tag{4.23}$$

where $P$ is a real $n \times n$ matrix. The time derivative of $v(x)$ along the solutions of (4.22) is determined as

$$\dot{v}(x) = \dot{x}^T P x + x^T P \dot{x} = x^T A^T P x + x^T P A x$$
$$= x^T (A^T P + P A)x;$$

i.e.,

$$\dot{v} = x^T C x, \tag{4.24}$$

where

$$C = A^T P + P A. \tag{4.25}$$

Note that $C$ is real and $C^T = C$. The system of equations given in (4.25) is called the *Lyapunov Matrix Equation*.

We recall that since $P$ is real and symmetric, all its eigenvalues are real. Also, we recall that $P$ is said to be *positive definite* (resp., *positive semidefinite*) if all its eigenvalues are positive (resp., nonnegative), and it is called *indefinite* if $P$ has eigenvalues of opposite sign. The concepts of *negative definite* and *negative semidefinite* (for $P$) are similarly defined. Furthermore, we recall that the *function* $v(x)$ given in (4.23) is said to be *positive definite, positive semidefinite, indefinite*, and so forth, if $P$ has the corresponding definiteness properties.

Instead of solving for the eigenvalues of a real symmetric matrix to determine its definiteness properties, there are more efficient and direct methods of accomplishing this. We now digress to discuss some of these.

Let $G = [g_{ij}]$ be a real $n \times n$ matrix (not necessarily symmetric). Recall that the *minors* of $G$ are the matrix itself and the matrix obtained by removing successively a row and a column. The *principal minors* of $G$ are $G$ itself and the matrices obtained by successively removing an $i$th row and an $i$th column, and the *leading principal minors* of $G$ are $G$ itself and the minors obtained by successively removing the last row and the last column. For example, if $G = [g_{ij}] \in R^{3 \times 3}$, then the principal minors are

$$\begin{bmatrix} g_{11} & g_{12} & g_{13} \\ g_{21} & g_{22} & g_{23} \\ g_{31} & g_{32} & g_{33} \end{bmatrix}, \quad \begin{bmatrix} g_{11} & g_{12} \\ g_{21} & g_{22} \end{bmatrix}, \quad [g_{11}],$$

$$\begin{bmatrix} g_{11} & g_{13} \\ g_{31} & g_{33} \end{bmatrix}, \quad \begin{bmatrix} g_{22} & g_{23} \\ g_{32} & g_{33} \end{bmatrix}, \quad [g_{22}], \quad [g_{33}].$$

The first three matrices above are the leading principal minors of $G$. On the other hand, the matrix

$$\begin{bmatrix} g_{21} & g_{22} \\ g_{31} & g_{32} \end{bmatrix}$$

is a minor but not a principal minor.

The following results, due to Sylvester, allow efficient determination of the definiteness properties of a *real, symmetric* matrix.

**Proposition 4.20.** *(i) A real symmetric matrix $P = [p_{ij}] \in R^{n \times n}$ is* positive definite *if and only if the determinants of its leading principal minors are positive, i.e., if and only if*

$$p_{11} > 0, \quad \det \begin{bmatrix} p_{11} & p_{12} \\ p_{12} & p_{22} \end{bmatrix} > 0, \ldots, \det P > 0.$$

*(ii) A real symmetric matrix $P$ is* positive semidefinite *if and only if the determinants of* all *of its principal minors are nonnegative.* ∎

Still digressing, we consider next the quadratic form

$$v(w) = w^T G w, \quad G = G^T,$$

where $G \in R^{n \times n}$. Now recall that there exists an orthogonal matrix $Q$ such that the matrix $P$ defined by

$$P = Q^{-1} G Q = Q^T G Q$$

is diagonal. Therefore, if we let $w = Qx$, then

$$v(Qx) \triangleq v(x) = x^T Q^T G Q x = x^T P x,$$

where $P$ is in the form given by

$$P = \text{diag}[\Lambda_i] \quad i = 1, \ldots, p,$$

where $\Lambda_i = \text{diag} \lambda_i$. From this, we immediately obtain the following useful result.

**Proposition 4.21.** *Let $P = P^T \in R^{n \times n}$, let $\lambda_M(P)$ and $\lambda_m(P)$ denote the largest and smallest eigenvalues of $P$, respectively, and let $\| \cdot \|$ denote the Euclidean norm. Then*

$$\lambda_m(P) \| x \|^2 \leq v(x) = x^T P x \leq \lambda_M(P) \| x \|^2 \tag{4.26}$$

*for all $x \in R^n$ (refer to [1]).* ∎

Let $c_1 \triangleq \lambda_m(P)$ and $c_2 = \lambda_M(P)$. Clearly, $v(x)$ is positive definite if and only if $c_2 \geq c_1 > 0$, $v(x)$ is positive semidefinite if and only if $c_2 \geq c_1 \geq 0$, $v(x)$ is indefinite if and only if $c_2 > 0, c_1 < 0$, and so forth.

We are now in a position to prove several results.

**Theorem 4.22.** *The equilibrium $x = 0$ of (4.22) is stable if there exists a real, symmetric, and positive definite $n \times n$ matrix $P$ such that the matrix $C$ given in (4.25) is negative semidefinite.*

*Proof.* Along any solution $\phi(t, x_0) \triangleq \phi(t)$ of (4.22) with $\phi(0, x_0) = \phi(0) = x_0$, we have

$$\phi(t)^T P \phi(t) = x_0^T P x_0 + \int_0^t \frac{d}{d\eta} \phi(\eta)^T P \phi(\eta) d\eta = x_0^T P x_0 + \int_0^t \phi(\eta)^T C \phi(\eta) d\eta$$

for all $t \geq 0$. Since $P$ is positive definite and $C$ is negative semidefinite, we have

$$\phi(t)^T P \phi(t) - x_0^T P x_0 \leq 0$$

for all $t \geq 0$, and there exist $c_2 \geq c_1 > 0$ such that

$$c_1 \| \phi(t) \|^2 \leq \phi(t)^T P \phi(t) \leq x_0^T P x_0 \leq c_2 \| x_0 \|^2$$

for all $t \geq 0$. It follows that

$$\| \phi(t) \| \leq (c_2/c_1)^{1/2} \| x_0 \|$$

for all $t \geq 0$ and for any $x_0 \in R^n$. Therefore, the equilibrium $x = 0$ of (4.22) is stable (refer to Theorem 4.12). ∎

**Example 4.23.** For the system given in Example 4.16 we choose $P = I$, and we compute

$$C = A^T P + PA = A^T + A = 0.$$

According to Theorem 4.22, the equilibrium $x = 0$ of this system is stable (as expected from Example 4.16).

---

**Theorem 4.24.** *The equilibrium $x = 0$ of (4.22) is* exponentially stable in the large *if there exists a real, symmetric, and positive definite $n \times n$ matrix $P$ such that the matrix $C$ given in (4.25) is negative definite.*

*Proof.* We let $\phi(t, x_0) \triangleq \phi(t)$ denote an arbitrary solution of (4.22) with $\phi(0) = x_0$. In view of the hypotheses of the theorem, there exist constants $c_2 \geq c_1 > 0$ and $c_3 \geq c_4 > 0$ such that

$$c_1 \parallel \phi(t) \parallel^2 \leq v(\phi(t)) = \phi(t)^T P\phi(t) \leq c_2 \parallel \phi(t) \parallel^2$$

and

$$-c_3 \parallel \phi(t) \parallel^2 \leq \dot{v}(\phi(t)) = \phi(t)^T C\phi(t) \leq -c_4 \parallel \phi(t) \parallel^2$$

for all $t \geq 0$ and for any $x_0 \in R^n$. Then

$$\dot{v}(\phi(t)) = \frac{d}{dt}[\phi(t)^T P\phi(t)] \leq (-c_4/c_2)\phi(t)^T P\phi(t)$$
$$= (-c_4/c_2)v(\phi(t))$$

for all $t \geq t_0$. This implies, after multiplication by the appropriate integrating factor, and integrating from 0 to $t$, that

$$v(\phi(t)) = \phi(t)^T P\phi(t) \leq x_0^T P x_0 e^{-(c_4/c_2)t}$$

or

$$c_1 \parallel \phi(t) \parallel^2 \leq \phi(t)^T P\phi(t) \leq c_2 \parallel x_0 \parallel^2 e^{-(c_4/c_2)t}$$

or

$$\parallel \phi(t) \parallel \leq (c_2/c_1)^{1/2} \parallel x_0 \parallel e^{-\frac{1}{2}(c_4/c_2)t}, \quad t \geq 0.$$

This inequality holds for all $x_0 \in R^n$. Therefore, the equilibrium $x = 0$ of (4.22) is exponentially stable in the large (refer to Sections 4.3 and 4.4). ∎

In Figure 4.3 we provide an interpretation of Theorem 4.24 for the two-dimensional case ($n = 2$). The curves $C_i$, called *level curves*, depict loci where $v(x)$ is constant; i.e., $C_i = \{x \in R^2 : v(x) = x^T Px = c_i\}, i = 0, 1, 2, 3, \ldots$. When the hypotheses of Theorem 4.24 are satisfied, trajectories determined by (4.22) penetrate level curves corresponding to decreasing values of $c_i$ as $t$ increases, tending to the origin as $t$ becomes arbitrarily large.

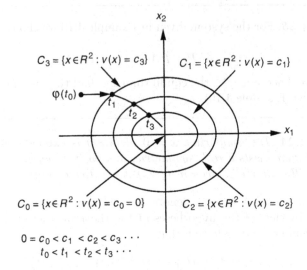

$$0 = c_0 < c_1 < c_2 < c_3 \cdots$$
$$t_0 < t_1 < t_2 < t_3 \cdots$$

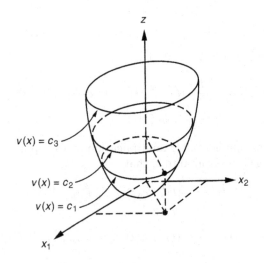

**Figure 4.3.** Asymptotic stability

---

***Example 4.25.*** For the system given in Example 4.19, we choose

$$P = \begin{bmatrix} 1 & 0 \\ 0 & 0.5 \end{bmatrix},$$

and we compute the matrix

$$C = A^T P + PA = \begin{bmatrix} -2 & 0 \\ 0 & -2 \end{bmatrix}.$$

According to Theorem 4.24, the equilibrium $x = 0$ of this system is exponentially stable in the large (as expected from Example 4.19).

---

**Theorem 4.26.** *The equilibrium $x = 0$ of (4.22) is* unstable *if there exists a real, symmetric $n \times n$ matrix $P$ that is either negative definite or indefinite such that the matrix $C$ given in (4.25) is negative definite.*

*Proof.* We first assume that $P$ is indefinite. Then $P$ possesses eigenvalues of either sign, and every neighborhood of the origin contains points where the function

$$v(x) = x^T P x$$

is positive and negative. Consider the neighborhood

$$B(\epsilon) = \{x \in R^n : \| x \| < \epsilon\},$$

where $\| \cdot \|$ denotes the Euclidean norm, and let

$$G = \{x \in B(\epsilon) : v(x) < 0\}.$$

On the boundary of $G$ we have either $\| x \| = \epsilon$ or $v(x) = 0$. In particular, note that the origin $x = 0$ is on the boundary of $G$. Now, since the matrix $C$ is negative definite, there exist constants $c_3 > c_4 > 0$ such that

$$-c_3 \| x \|^2 \leq x^T C x = \dot{v}(x) \leq -c_4 \| x \|^2$$

for all $x \in R^n$. Let $\phi(t, x_0) \triangleq \phi(t)$ and let $x_0 = \phi(0) \in G$. Then $v(x_0) = -a < 0$. The solution $\phi(t)$ starting at $x_0$ must leave the set $G$. To see this, note that as long as $\phi(t) \in G, v(\phi(t)) \leq -a$ since $\dot{v}(x) < 0$ in $G$. Let $-c = \sup\{\dot{v}(x) : x \in G$ and $v(x) \leq -a\}$.

Then $c > 0$ and

$$v(\phi(t)) = v(x_0) + \int_0^t \dot{v}(\phi(s))ds \leq -a - \int_0^t cds$$
$$= -a - tc, t \geq t_0.$$

This inequality shows that $\phi(t)$ must escape the set $G$ (in finite time) because $v(x)$ is bounded from below on $G$. But $\phi(t)$ cannot leave $G$ through the surface determined by $v(x) = 0$ since $v(\phi(t)) \leq -a$. Hence, it must leave $G$ through the sphere determined by $\| x \| = \epsilon$. Since the above argument holds for arbitrarily small $\epsilon > 0$, it follows that the origin $x = 0$ of (4.22) is unstable.

Next, we assume that $P$ is negative definite. Then $G$ as defined is all of $B(\epsilon)$. The proof proceeds as above. ∎

The proof of Theorem 4.26 shows that for $\epsilon > 0$ sufficiently small when $P$ is negative definite, *all* solutions $\phi(t)$ of (4.22) with initial conditions $x_0 \in B(\epsilon)$ will tend away from the origin. This constitutes a severe case of instability, called *complete instability*.

---

**Example 4.27.** For the system given in Example 4.18, we choose

$$P = \begin{bmatrix} -0.28 & -0.96 \\ -0.96 & 0.28 \end{bmatrix},$$

and we compute the matrix

$$C = A^T P + PA = \begin{bmatrix} -20 & 0 \\ 0 & -20 \end{bmatrix}.$$

The eigenvalues of $P$ are $\pm 1$. According to Theorem 4.26, the equilibrium $x = 0$ of this system is unstable (as expected from Example 4.18).

---

In applying the results derived thus far in this section, we start by choosing (guessing) a matrix $P$ having certain desired properties. Next, we solve for the matrix $C$, using (4.25). If $C$ possesses certain desired properties (i.e., it is negative definite), we draw appropriate conclusions by applying one of the preceding theorems of this section; if not, we need to choose another matrix $P$. This points to the principal shortcoming of Lyapunov's Direct Method, when applied to general systems. However, in the *special case* of linear systems described by (4.22), it is possible to *construct* Lyapunov functions of the form $v(x) = x^T P x$ in a *systematic* manner. In doing so, one first chooses the matrix $C$ in (4.25) (having desired properties), and then one solves (4.25) for $P$. Conclusions are then drawn by applying the appropriate results of this section. In applying this construction procedure, we need to know conditions under which (4.25) possesses a (unique) solution $P$ for a given $C$. We will address this topic next.

We consider the quadratic form

$$v(x) = x^T P x, \quad P = P^T, \tag{4.27}$$

and the time derivative of $v(x)$ along the solutions of (4.22), given by

$$\dot{v}(x) = x^T C x, \quad C = C^T, \tag{4.28}$$

where

$$C = A^T P + PA \tag{4.29}$$

and where all symbols are as defined in (4.23) to (4.25). Our objective is to determine the as yet unknown matrix $P$ in such a way that $\dot{v}(x)$ becomes a preassigned negative definite quadratic form, i.e., in such a way that $C$ is a preassigned negative definite matrix.

Equation (4.29) constitutes a system of $n(n+1)/2$ linear equations. We need to determine under what conditions we can solve for the $n(n+1)/2$ elements, $p_{ik}$, given $C$ and $A$. To this end, we choose a similarity transformation $Q$ such that

$$QAQ^{-1} = \bar{A}, \tag{4.30}$$

or equivalently,

$$A = Q^{-1}\bar{A}Q, \tag{4.31}$$

where $\bar{A}$ is similar to $A$ and $Q$ is a real $n \times n$ nonsingular matrix. From (4.31) and (4.29) we obtain

$$(\bar{A})^T(Q^{-1})^T PQ^{-1} + (Q^{-1})^T PQ^{-1}\bar{A} = (Q^{-1})^T CQ^{-1} \tag{4.32}$$

or

$$(\bar{A})^T \bar{P} + \bar{P}\bar{A} = \bar{C}, \quad \bar{P} = (Q^{-1})^T PQ^{-1}, \quad \bar{C} = (Q^{-1})^T CQ^{-1}. \tag{4.33}$$

In (4.33), $P$ and $C$ are subjected to a congruence transformation and $\bar{P}$ and $\bar{C}$ have the same definiteness properties as $P$ and $C$, respectively. Since every real $n \times n$ matrix can be triangularized (refer to [1]), we can choose $Q$ in such a fashion that $\bar{A} = [\bar{a}_{ij}]$ is *triangular*; i.e., $\bar{a}_{ij} = 0$ for $i > j$. Note that in this case the eigenvalues of $A$, $\lambda_1, \ldots, \lambda_n$, appear in the main diagonal of $\bar{A}$. To simplify our notation, we rewrite (4.33) in the form (4.29) by dropping the bars, i.e.,

$$A^T P + PA = C, \quad C = C^T, \tag{4.34}$$

and *we assume that* $A = [a_{ij}]$ *has been triangularized*; i.e., $a_{ij} = 0$ for $i > j$. Since the eigenvalues $\lambda_1, \ldots, \lambda_n$ appear in the diagonal of $A$, we can rewrite (4.34) as

$$2\lambda_1 p_{11} = c_{11}$$
$$a_{12}p_{11} + (\lambda_1 + \lambda_2)p_{12} = c_{12} \tag{4.35}$$
$$\cdots\cdots\cdots\cdots\cdots\cdots\cdots\cdots\cdots$$

Note that $\lambda_1$ may be a complex number; in which case, $c_{11}$ will also be complex. Since this system of equations is triangular, and since its determinant is equal to

$$2^n \lambda_1 \ldots \lambda_n \prod_{i<j}(\lambda_i + \lambda_j), \tag{4.36}$$

the matrix $P$ can be determined (uniquely) if and only if this determinant is not zero. This is true when all eigenvalues of $A$ are nonzero and no two of them are such that $\lambda_i + \lambda_j = 0$. This condition is not affected by a similarity transformation and is therefore also valid for the original system of equations (4.29).

We summarize the above discussion in the following lemma.

**Lemma 4.28.** *Let $A \in R^{n \times n}$ and let $\lambda_1, \ldots, \lambda_n$ denote the (not necessarily distinct) eigenvalues of $A$. Then (4.34) has a unique solution for $P$ corresponding to each $C \in R^{n \times n}$ if and only if*

$$\lambda_i \neq 0, \lambda_i + \lambda_j \neq 0 \text{ for all } i, j. \tag{4.37}$$

∎

To construct $v(x)$, we must still check the definiteness of $P$. This can be done in a purely algebraic way; however, in the present case, it is much easier to apply the results of this section and argue as follows:

(a) If all the eigenvalues $\lambda_i$ of $A$ have negative real parts, then the equilibrium $x = 0$ of (4.22) is exponentially stable in the large, and if $C$ in (4.29) is negative definite, then $P$ must be positive definite. To prove this, we note that if $P$ is not positive definite, then for $\delta > 0$ and sufficiently small, $(P - \delta I)$ has at least one negative eigenvalue, whereas the function $v(x) = x^T(P - \delta I)x$ has a negative definite derivative; i.e.,

$$v_{(L)}^1(x) = x^T[C - \delta(A + A^T)]x < 0$$

for all $x \neq 0$. By Theorem 4.26, the equilibrium $x = 0$ of (4.22) is unstable. We have arrived at a contradiction. Therefore, $P$ must be positive definite.

(b) If $A$ has eigenvalues with positive real parts and no eigenvalues with zero real parts, we can use a similarity transformation $x = Qy$ in such a way that $Q^{-1}AQ$ is a block diagonal matrix of the form diag$[A_1, A_2]$, where all the eigenvalues of $A_1$ have positive real parts, whereas all eigenvalues of $A_2$ have negative real parts (refer to [1]). (If $A$ does not have any eigenvalues with negative real parts, then we take $A = A_1$). By the result established in (a), noting that all eigenvalues of $-A_1$ have negative real parts, given any negative definite matrices $C_1$ and $C_2$, there exist positive definite matrices $P_1$ and $P_2$ such that

$$(-A_1^T)P_1 + P_1(-A_1) = C_1, \quad A_2^T P_2 + P_2 A_2 = C_2.$$

Then $w(y) = y^T P y$, with $P = $ diag$[-P_1, P_2]$ is a Lyapunov function for the system $\dot{y} = Q^{-1}AQy$ (and, hence, for the system $\dot{x} = Ax$), which satisfies the hypotheses of Theorem 4.26. Therefore, the equilibrium $x = 0$ of system (4.22) is unstable. If $A$ does not have any eigenvalues with negative real parts, then the equilibrium $x = 0$ of system (4.22) is *completely unstable.*]

In the above proof, we did not invoke Lemma 4.28. We note, however, that if additionally, (4.37) is true, then we can construct the Lyapunov function for (4.22) in a systematic manner.

Summarizing the above discussion, we can now state the main result of this subsection.

**Theorem 4.29.** *Assume that the matrix $A$ [for system (4.22)] has no eigenvalues with real part equal to zero. If all the eigenvalues of $A$ have negative real parts, or if at least one of the eigenvalues of $A$ has a positive real part, then there exists a quadratic Lyapunov function*

$$v(x) = x^T P x, \quad P = P^T,$$

*whose derivative along the solutions of (4.22) is definite (i.e., it is either negative definite or positive definite).*  ∎

This result shows that when $A$ is a stable matrix (i.e., all the eigenvalues of $A$ have negative real parts), then for system (4.22) the conditions of Theorem 4.24 are also necessary conditions for exponential stability in the large. Moreover, in the case when the matrix $A$ has at least one eigenvalue with positive real part and no eigenvalues on the imaginary axis, then the conditions of Theorem 4.26 are also necessary conditions for instability.

---

**Example 4.30.** We consider the system (4.22) with

$$A = \begin{bmatrix} 0 & 1 \\ -1 & 0 \end{bmatrix}.$$

The eigenvalues of $A$ are $\lambda_1, \lambda_2 = \pm j$, and therefore condition (4.37) is violated. According to Lemma 4.28, the Lyapunov matrix equation

$$A^T P + P A = C$$

does not possess a unique solution for a given $C$. We now verify this for two specific cases.

(i) When $C = 0$, we obtain

$$\begin{bmatrix} 0 & -1 \\ 1 & 0 \end{bmatrix} \begin{bmatrix} p_{11} & p_{12} \\ p_{12} & p_{22} \end{bmatrix} + \begin{bmatrix} p_{11} & p_{12} \\ p_{12} & p_{22} \end{bmatrix} \begin{bmatrix} 0 & 1 \\ -1 & 0 \end{bmatrix} = \begin{bmatrix} -2p_{12} & p_{11} - p_{22} \\ p_{11} - p_{22} & 2p_{12} \end{bmatrix}$$

$$= \begin{bmatrix} 0 & 0 \\ 0 & 0 \end{bmatrix},$$

or $p_{12} = 0$ and $p_{11} = p_{22}$. Therefore, for any $a \in R$, the matrix $P = aI$ is a solution of the Lyapunov matrix equation. In other words, for $C = 0$, the Lyapunov matrix equation has in this example denumerably many solutions.

(ii) When $C = -2I$, we obtain

$$\begin{bmatrix} -2p_{12} & p_{11} - p_{22} \\ p_{11} - p_{22} & 2p_{12} \end{bmatrix} = \begin{bmatrix} -2 & 0 \\ 0 & -2 \end{bmatrix},$$

or $p_{11} = p_{22}$ and $p_{12} = 1$ and $p_{12} = -1$, which is impossible. Therefore, for $C = -2I$, the Lyapunov matrix equation has in this example no solutions at all.

---

It turns out that if all the eigenvalues of matrix $A$ have negative real parts, then we can compute $P$ in (4.29) explicitly.

**Theorem 4.31.** *If all eigenvalues of a real $n \times n$ matrix $A$ have negative real parts, then for each matrix $C \in R^{n \times n}$, the unique solution of (4.29) is given by*

$$P = \int_0^\infty e^{A^T t}(-C)e^{At}dt. \tag{4.38}$$

*Proof.* If all eigenvalues of $A$ have negative real parts, then (4.37) is satisfied and therefore (4.29) has a unique solution for every $C \in R^{n \times n}$. To verify that (4.38) is indeed this solution, we first note that the right-hand side of (4.38) is well defined, since all eigenvalues of $A$ have negative real parts. Substituting the right-hand side of (4.38) for $P$ into (4.29), we obtain

$$
\begin{aligned}
A^T P + PA &= \int_0^\infty A^T e^{A^T t}(-C)e^{At}dt + \int_0^\infty e^{A^T t}(-C)e^{At}A\,dt \\
&= \int_0^\infty \frac{d}{dt}[e^{A^T t}(-C)e^{At}]dt \\
&= e^{A^T t}(-C)e^{At}\Big|_0^\infty = C,
\end{aligned}
$$

which proves the theorem.    ∎

## 4.6 Linearization

In this section we consider *nonlinear, finite-dimensional, continuous-time* dynamical systems described by equations of the form

$$\dot{w} = f(w), \tag{4.39}$$

where $f \in C^1(R^n, R^n)$. We assume that $w = 0$ is an equilibrium of (4.39). In accordance with Subsection 1.6.1, we linearize system (4.39) about the origin to obtain

$$\dot{x} = Ax + F(x), \tag{4.40}$$

$x \in R^n$, where $F \in C(R^n, R^n)$ and where $A$ denotes the Jacobian of $f(w)$ evaluated at $w = 0$, given by

$$A = \frac{\partial f}{\partial w}(0), \tag{4.41}$$

and where

$$F(x) = o(\| x \|) \quad \text{as} \quad \| x \| \to 0. \tag{4.42}$$

Associated with (4.40) is the *linearization* of (4.39), given by

$$\dot{y} = Ay. \tag{4.43}$$

In the following discussion, we use the results of Section 4.5 to establish criteria that allow us to deduce the stability properties of the equilibrium $w = 0$ of the nonlinear system (4.39) from the stability properties of the equilibrium $y = 0$ of the linear system (4.43).

**Theorem 4.32.** *Let $A \in R^{n \times n}$ be a Hurwitz matrix (i.e., all of its eigevnalues have negative real parts), let $F \in C(R^n, R^n)$, and assume that (4.42) holds. Then the equilibrium $x = 0$ of (4.40) [and, hence, of (4.39)] is* exponentially stable.

*Proof.* Theorem 4.29 applies to (4.43) since all the eigenvalues of $A$ have negative real parts. In view of that theorem (and the comments following Lemma 4.28), there exists a symmetric, real, positive definite $n \times n$ matrix $P$ such that

$$PA + A^T P = C, \qquad (4.44)$$

where $C$ is negative definite. Consider the Lyapunov function

$$v(x) = x^T P x. \qquad (4.45)$$

The derivative of $v$ with respect to $t$ along the solutions of (4.40) is given by

$$\begin{aligned}
\dot{v}(x) &= \dot{x}^T P x + x^T P \dot{x} \\
&= (Ax + F(x))^T P x + x^T P (Ax + F(x)) \\
&= x^T C x + 2x^T P F(x).
\end{aligned} \qquad (4.46)$$

Now choose $\gamma < 0$ such that $x^T C x \leq 3\gamma \parallel x \parallel^2$ for all $x \in R^n$. Since it is assumed that (4.42) holds, there is a $\delta > 0$ such that if $\parallel x \parallel \leq \delta$, then $\parallel PF(x) \parallel \leq -\gamma \parallel x \parallel$ for all $x \in \overline{B(\delta)} = \{x \in R^n : \parallel x \parallel \leq \delta\}$. Therefore, for all $x \in B(\delta)$, we obtain, in view of (4.46), the estimate

$$\dot{v}(x) \leq 3\gamma \parallel x \parallel^2 -2\gamma \parallel x \parallel^2 = \gamma \parallel x \parallel^2 . \qquad (4.47)$$

Now let $\alpha = \min_{\parallel x \parallel = \delta} v(x)$. Then $\alpha > 0$ (since $P$ is positive definite). Take $\lambda \in (0, \alpha)$, and let

$$C_\lambda = \{x \in B(\delta) = \{x \in R^n : \parallel x \parallel < \delta\} : v(x) \leq \lambda\}. \qquad (4.48)$$

Then $C_\lambda \subset B(\delta)$. [This can be shown by contradiction. Suppose that $C_\lambda$ is not entirely inside $B(\delta)$. Then there is a point $\bar{x} \in C_\lambda$ that lies on the boundary of $B(\delta)$. At this point, $v(\bar{x}) \geq \alpha > \lambda$. We have thus arrived at a contradiction.] The set $C_\lambda$ has the property that any solution of (4.40) starting in $C_\lambda$ at $t = 0$ will stay in $C_\lambda$ for all $t \geq 0$. To see this, we let $\phi(t, x_0) \triangleq \phi(t)$ and we recall that $\dot{v}(x) \leq \gamma \parallel x \parallel^2, \gamma < 0, x \in B(\delta) \supset C_\lambda$. Then $\dot{v}(\phi(t)) \leq 0$ implies that $v(\phi(t)) \leq v(x_0) \leq \lambda$ for all $t \geq t_0 \geq 0$. Therefore, $\phi(t) \in C_\lambda$ for all $t \geq t_0 \geq 0$.

We now proceed in a similar manner as in the proof of Theorem 4.24 to complete this proof. In doing so, we first obtain the estimate

$$\dot{v}(\phi(t)) \le (\gamma/c_2)v(\phi(t)), \tag{4.49}$$

where $\gamma$ is given in (4.47) and $c_2$ is determined by the relation

$$c_1 \parallel x \parallel^2 \le v(x) = x^T P x \le c_2 \parallel x \parallel^2 . \tag{4.50}$$

Following now in an identical manner as was done in the proof of Theorem 4.22, we have

$$\parallel \phi(t) \parallel \le (c_2/c_1)^{\frac{1}{2}} \parallel x_0 \parallel e^{\frac{1}{2}(\gamma/c_2)t}, \quad t \ge 0, \tag{4.51}$$

whenever $x_0 \in B(r')$, where $r'$ has been chosen sufficiently small so that $B(r') \subset C_\lambda$. This proves that the equilibrium $x = 0$ of (4.40) is exponentially stable. ∎

It is important to recognize that Theorem 4.32 is a *local result* that yields sufficient conditions for the exponential stability of the equilibrium $x = 0$ of (4.40); it does not yield conditions for exponential stability in the large. The proof of Theorem 4.32, however, enables us to determine an estimate of the domain of attraction of the equilibrium $x = 0$ of (4.39), involving the following steps:

1. Determine an equilibrium, $x_e$, of (4.39) and transform (4.39) to a new system that translates $x_e$ to the origin $x = 0$ (refer to Section 4.2).
2. Linearize (4.39) about the origin and determine $F(x), A$, and the eigenvalues of $A$.
3. If all eigenvalues of $A$ have negative real parts, choose a negative definite matrix $C$ and solve the Lyapunov matrix equation

$$C = A^T P + PA.$$

4. Determine the Lyapunov function

$$v(x) = x^T P x.$$

5. Compute the derivative of $v$ along the solutions of (4.40), given by

$$\dot{v}(x) = x^T C x + 2x^T P F(x).$$

6. Determine $\delta > 0$ such that $\dot{v}(x) < 0$ for all $x \in B(\delta) - \{0\}$.
7. Determine the largest $\lambda = \lambda_M$ such that $C_{\lambda_M} \subset B(\delta)$, where

$$C_\lambda = \{x \in R^n : v(x) < \lambda\}.$$

8. $C_{\lambda_M}$ is a subset of the domain of attraction of the equilibrium $x = 0$ of (4.40) and, hence, of (4.39).

The above procedure may be repeated for different choices of matrix $C$ given in step (3), resulting in different matrices $P_i$, which in turn may result in different estimates for the domain of attraction, $C^i_{\lambda_M}$, $i \in \Lambda$, where $\Lambda$ is an index set. The union of the sets $C^i_{\lambda_M} \triangleq D_i$, $D = \cup_i D_i$, is also a subset of the domain of attraction of the equilibrium $x = 0$ of (4.39).

**Theorem 4.33.** *Assume that $A$ is a real $n \times n$ matrix that has at least one eigenvalue with positive real part and no eigenvalue with zero real part. Let $F \in C(R^n, R^n)$, and assume that (4.42) holds. Then the equilibrium $x = 0$ of (4.40) [and, hence, of (4.39)] is unstable.*

*Proof.* We use Theorem 4.29 to choose a real, symmetric $n \times n$ matrix $P$ such that the matrix $PA + A^T P = C$ is negative definite. The matrix $P$ is not positive definite, or even positive semidefinite (refer to the comments following Lemma 4.28). Hence, the function $v(x) = x^T P x$ is negative at some points arbitrarily close to the origin. The derivative of $v(x)$ with respect to $t$ along the solutions of (4.40) is given by (4.46). As in the proof of Theorem 4.32, we can choose a $\gamma < 0$ such that $x^T C x \leq 3\gamma \parallel x \parallel^2$ for all $x \in R^n$, and in view of (4.42) we can choose a $\delta > 0$ such that $\parallel PF(x) \parallel \leq -\gamma \parallel x \parallel$ for all $x \in B(\delta)$. Therefore, for all $x \in B(\delta)$, we obtain that

$$\dot{v}(x) \leq 3\gamma \parallel x \parallel^2 -2\gamma \parallel x \parallel^2 = \gamma \parallel x \parallel^2 .$$

Now let

$$G = \{x \in B(\delta) : v(x) < 0\}.$$

The boundary of $G$ is made up of points where $v(x) = 0$ and where $\parallel x \parallel = \delta$. Note in particular that the equilibrium $x = 0$ of (4.40) is in the boundary of $G$. Now following an identical procedure as in the proof of Theorem 4.26, we show that any solution $\phi(t)$ of (4.40) with $\phi(0) = x_0 \in G$ must escape $G$ in finite time through the surface determined by $\parallel x \parallel = \delta$. Since the above argument holds for arbitrarily small $\delta > 0$, it follows that the origin $x = 0$ of (4.40) is unstable. ∎

Before concluding this section, we consider a few specific cases.

---

*Example 4.34.* The *Lienard Equation* is given by

$$\ddot{w} + f(w)\dot{w} + w = 0, \tag{4.52}$$

where $f \in C^1(R, R)$ with $f(0) > 0$. Letting $x_1 = w$ and $x_2 = \dot{w}$, we obtain

$$\begin{aligned} \dot{x}_1 &= x_2, \\ \dot{x}_2 &= -x_1 - f(x_1)x_2. \end{aligned} \tag{4.53}$$

Let $x^T = (x_1, x_2)$, $f(x)^T = (f_1(x), f_2(x))$, and let

$$J(0) = A = \begin{bmatrix} \frac{\partial f_1}{\partial x_1}(0) & \frac{\partial f_1}{\partial x_2}(0) \\ \frac{\partial f_2}{\partial x_1}(0) & \frac{\partial f_2}{\partial x_2}(0) \end{bmatrix} = \begin{bmatrix} 0 & 1 \\ -1 & -f(0) \end{bmatrix}.$$

Then

$$\dot{x} = Ax + [f(x) - Ax] = Ax + F(x),$$

where

$$F(x) = \begin{bmatrix} 0 \\ [f(0) - f(x_1)] x_2 \end{bmatrix}.$$

The origin $x = 0$ is clearly an equilibrium of (4.52) and hence of (4.53). The eigenvalues of $A$ are given by

$$\lambda_1, \lambda_2 = \frac{-f(0) \pm \sqrt{f(0)^2 - 4}}{2},$$

and therefore, $A$ is a Hurwitz matrix. Also, (4.42) holds. Therefore, all the conditions of Theorem 4.32 are satisfied. We conclude that the equilibrium $x = 0$ of (4.53) is *exponentially stable*.

---

**Example 4.35.** We consider the system given by

$$\begin{aligned} \dot{x}_1 &= -x_1 + x_1(x_1^2 + x_2^2), \\ \dot{x}_2 &= -x_2 + x_2(x_1^2 + x_2^2). \end{aligned} \tag{4.54}$$

The origin is clearly an equilibrium of (4.54). Also, the system is already in the form (4.40) with

$$A = \begin{bmatrix} -1 & 0 \\ 0 & -1 \end{bmatrix}, \quad F(x) = \begin{bmatrix} x_1(x_1^2 + x_2^2) \\ x_2(x_1^2 + x_2^2) \end{bmatrix},$$

and condition (4.42) is clearly satisfied. The eigenvalues of $A$ are $\lambda_1 = -1, \lambda_2 = -1$. Therefore, all conditions of Theorem 4.32 are satisfied and we conclude that the equilibrium $x^T = (x_1, x_2) = 0$ is *exponentially stable*; however, we cannot conclude that this equilibrium is exponentially stable in the large. Accordingly, we seek to determine an estimate for the domain of attraction of this equilibrium.

We choose $C = -I$ (where $I \in R^{2 \times 2}$ denotes the identity matrix), and we solve the matrix equation $A^T P + PA = C$ to obtain $P = (1/2)I$, and therefore,

$$v(x_1, x_2) = x^T P x = \frac{1}{2}(x_1^2 + x_2^2).$$

Along the solutions of (4.54) we obtain

$$\begin{aligned} \dot{v}(x_1, x_2) &= x^T C x + 2x^T P F(x) \\ &= -(x_1^2 + x_2^2) + (x_1^2 + x_2^2)^2. \end{aligned}$$

Clearly, $\dot{v}(x_1, x_2) < 0$ when $(x_1, x_2) \neq (0, 0)$ and $x_1^2 + x_2^2 < 1$. In the language of the proof of Theorem 4.32, we can therefore choose $\delta = 1$.

Now let

$$C_{1/2} = \{x \in R^2 : v(x_1, x_2) = \frac{1}{2}(x_1^2 + x_2^2) < \frac{1}{2}\}.$$

Then clearly, $C_{1/2} \subset B(\delta), \delta = 1$, in fact $C_{1/2} = B(\delta)$. Therefore, the set $\{x \in R^2 : x_1^2 + x_2^2 < 1\}$ is a subset of the domain of attraction of the equilibrium $(x_1, x_2)^T = 0$ of system (4.54).

---

**Example 4.36.** The differential equation governing the motion of a pendulum is given by

$$\ddot{\theta} + a \sin \theta = 0, \tag{4.55}$$

where $a > 0$ is a constant (refer to Chapter 1). Letting $\theta = x_1$ and $\dot{\theta} = x_2$, we obtain the system description

$$\begin{aligned}
\dot{x}_1 &= x_2, \\
\dot{x}_2 &= -a \sin x_1.
\end{aligned} \tag{4.56}$$

The points $x_e^{(1)} = (0, 0)^T$ and $x_e^{(2)} = (\pi, 0)^T$ are equilibria of (4.56).

(i)  Linearizing (4.56) about the equilibrium $x_e^{(1)}$, we put (4.56) into the form (4.40) with

$$A = \begin{bmatrix} 0 & 1 \\ -a & 0 \end{bmatrix}.$$

The eigenvalues of $A$ are $\lambda_1, \lambda_2 = \pm j\sqrt{a}$. Therefore, the results of this section (Theorem 4.32 and 4.33) are not applicable in the present case.

(ii)  In (4.56), we let $y_1 = x_1 - \pi$ and $y_2 = x_2$. Then (4.56) assumes the form

$$\begin{aligned}
\dot{y}_1 &= y_2, \\
\dot{y}_2 &= -a \sin(y_1 + \pi).
\end{aligned} \tag{4.57}$$

The point $(y_1, y_2)^T = (0, 0)^T$ is clearly an equilibrium of system (4.57). Linearizing about this equilibrium, we put (4.57) into the form (4.40), where

$$A = \begin{bmatrix} 0 & 1 \\ a & 0 \end{bmatrix}, \quad F(y_1, y_2) = \begin{bmatrix} 0 \\ -a(\sin(y_1 + \pi) + y_1) \end{bmatrix}.$$

The eigenvalues of $A$ are $\lambda_1, \lambda_2 = a, -a$. All conditions of Theorem 4.33 are satisfied, and we conclude that the equilibrium $x_e^{(2)} = (\pi, 0)^T$ of system (4.56) is *unstable*.

---

## 4.7 Input–Output Stability

We now turn our attention to systems described by the state equations

$$\dot{x} = Ax + Bu,$$
$$y = Cx + Du, \tag{4.58}$$

where $A \in R^{n \times n}, B \in R^{n \times m}, C \in R^{p \times n}$, and $D \in R^{p \times m}$. In the preceding sections of this chapter we investigated the *internal stability* properties of system (4.58) by studying the Lyapunov stability of the trivial solution of the associated system

$$\dot{w} = Aw. \tag{4.59}$$

In this approach, system inputs and system outputs played no role. To account for these, we now consider the *external stability* properties of system (4.58), called *input–output stability*: Every bounded input of a system should produce a bounded output. More specifically, in the present context, we say that system (4.58) is *bounded-input/bounded-output (BIBO) stable*, if for zero initial conditions at $t = 0$, every bounded input defined on $[0, \infty)$ gives rise to a bounded response on $[0, \infty)$.

Matrix $D$ does not affect the BIBO stability of (4.58). Accordingly, we will consider without any loss of generality the case where $D \equiv 0$; i.e., throughout this section we will concern ourselves with systems described by equations of the form

$$\dot{x} = Ax + Bu,$$
$$y = Cx. \tag{4.60}$$

We will say that the system (4.60) is *BIBO stable* if there exists a constant $c > 0$ such that the conditions

$$x(0) = 0,$$
$$\|u(t)\| \leq 1, \quad t \geq 0,$$

imply that $\| y(t) \| \leq c$ for all $t \geq 0$. (The symbol $\| \cdot \|$ denotes the Euclidean norm.)

Recall that for system (4.60) the impulse response matrix is given by

$$H(t) = Ce^{At}B, \quad t \geq 0,$$
$$= 0, \qquad\quad t < 0, \tag{4.61}$$

and the transfer function matrix is given by

$$\widehat{H}(s) = C(sI - A)^{-1}B. \tag{4.62}$$

**Theorem 4.37.** *The system (4.60) is BIBO stable if and only if there exists a finite constant $L > 0$ such that for all $t$,*

$$\int_0^t \| H(t - \tau) \| \, d\tau \leq L. \tag{4.63}$$

*Proof.* The first part of the proof of Theorem 4.37 (sufficiency) is straightforward. Indeed, if $\| u(t) \| \leq 1$ for all $t \geq 0$ and if (4.63) is true, then we have for all $t \geq 0$ that

$$\| y(t) \| = \left\| \int_0^t H(t - \tau)u(\tau)d\tau \right\|$$

$$\leq \int_0^t \| H(t - \tau)u(\tau) \| \, d\tau$$

$$\leq \int_0^t \| H(t - \tau) \| \, \| u(\tau) \| \, d\tau$$

$$\leq \int_0^t \| H(t - \tau) \| \, d\tau \leq L.$$

Therefore, system (4.60) is BIBO stable.

In proving the second part of Theorem 4.37 (necessity), we simplify matters by first considering in (4.60) the single-variable case ($n = 1$) with the input–output description given by

$$y(t) = \int_0^t h(t - \tau)u(\tau)d\tau. \tag{4.64}$$

For purposes of contradiction, we assume that the system is BIBO stable, but no finite $L$ exists such that (4.63) is satisfied. Another way of stating this is that for *every finite* $L$, there exists $t_1 = t_1(L), t_1 > 0$, such that

$$\int_0^{t_1} |h(t_1, \tau)|d\tau > L.$$

We now choose in particular the input given by

$$u(t) = \begin{cases} +1 & \text{if } h(t - \tau) > 0, \\ 0 & \text{if } h(t - \tau) = 0, \\ -1 & \text{if } h(t - \tau) < 0, \end{cases} \tag{4.65}$$

$0 \leq t \leq t_1$. Clearly, $|u(t)| \leq 1$ for all $t \geq 0$. The output of the system at $t = t_1$ due to the above input, however, is

$$y(t_1) = \int_0^{t_1} h(t_1 - \tau)u(\tau)d\tau = \int_0^{t_1} |h(t_1 - \tau)|d\tau > L,$$

which contradicts the assumption that the system is BIBO stable.

The above can now be generalized to the multivariable case. In doing so, we apply the single-variable result to every possible pair of input and output vector components, we make use of the fact that the sum of a finite number of bounded sums will be bounded, and we recall that a vector is bounded if and only if each of its components is bounded. We leave the details to the reader. ∎

In the preceding argument we made the tacit assumption that $u$ is continuous, or piecewise continuous. However, our particular choice of $u$ may involve nondenumerably many switchings (discontinuities) over a given finite-time interval. In such cases, $u$ is no longer piecewise continuous; however, it is measurable (in the Lebesgue sense). This generalization can be handled, although in a broader mathematical setting that we do not wish to pursue here. The interested reader may want to refer, e.g., to the books by Desoer and Vidyasagar [5], Michel and Miller [13], and Vidyasagar [20] and the papers by Sandberg [17] to [19] and Zames [21], [22] for further details.

From Theorem 4.37 and from (4.61) it follows readily that a necessary and sufficient condition for the BIBO stability of system (4.60) is the condition

$$\int_0^\infty \| H(t) \| \, dt < \infty. \tag{4.66}$$

**Corollary 4.38.** *Assume that the equilibrium $w = 0$ of (4.59) is exponentially stable. Then system (4.60) is* BIBO *stable.*

*Proof.* Under the hypotheses of the corollary, we have

$$\| \int_0^t H(t-\tau)d\tau \| \leq \int_0^t \| H(t-\tau) \| \, d\tau$$

$$= \int_0^t \| C\Phi(t-\tau)B \| \, d\tau \leq \| C \| \| B \| \int_0^t \| \Phi(t-\tau) \| \, d\tau.$$

Since the equilibrium $w = 0$ of (4.59) is exponentially stable, there exist $\delta > 0$, $\lambda > 0$ such that $\| \Phi(t,\tau) \| \leq \delta e^{-\lambda(t-\tau)}, t \geq \tau$. Therefore,

$$\int_0^t \| H(t-\tau) \| \, d\tau \leq \int_0^t \| C \| \| B \| \delta e^{-\lambda(t-\tau)} d\tau$$

$$\leq (\| C \| \| B \| \delta)/\lambda \triangleq L$$

for all $\tau, t$ with $t \geq \tau$. It now follows from Theorem 4.37 that system (4.60) is BIBO stable. ∎

In Section 7.3 we will establish a connection between the BIBO stability of (4.60) and the exponential stability of the trivial solution of (4.59).

Next, we recall that a complex number $s_p$ is a *pole* of $\widehat{H}(s) = [\hat{h}_{ij}(s)]$ if for some pair $(i,j)$, we have $|\hat{h}_{ij}(s_p)| = \infty$. If each entry of $\widehat{H}(s)$ has only poles with negative real values, then, as shown in Chapter 3, each entry of $H(t) = [h_{ij}(t)]$ has a sum of exponentials with exponents with real part negative. It follows that the integral

$$\int_0^\infty \| H(t) \| \, dt$$

is finite, and any realization of $\widehat{H}(s)$ will result in a system that is BIBO stable.

Now conversely, if

$$\int_0^\infty \| H(t) \| \, dt$$

is finite, then the exponential terms in any entry of $H(t)$ must have negative real parts. But then every entry of $\widehat{H}(s)$ has poles whose real parts are negative.

We have proved the following result.

**Theorem 4.39.** *The system (4.60) is BIBO stable if and only if all poles of the transfer function $\widehat{H}(s)$ given in (4.62) have only poles with negative real parts.* ∎

---

**Example 4.40.** A system with $H(s) = 1/s$ is not BIBO stable. To see this consider a step input. The response is then given by $y(t) = t$, $t \geq 0$, which is not bounded.

---

## 4.8 Discrete-Time Systems

In this section we address the Lyapunov stability of an equilibrium of discrete-time systems (internal stability) and the input–output stability of discrete-time systems (external stability). We establish results for discrete-time systems that are analogous to practically all the stability results that we presented for continuous-time systems.

This section is organized into five subsections. In the first subsection we provide essential preliminary material. In the second and third subsections we establish results for the stability, instability, asymptotic stability, and exponential stability of an equilibrium and boundedness of solutions of systems described by linear autonomous ordinary difference equations. These results are used to develop Lyapunov stability results for linearizations of nonlinear systems described by ordinary difference equations in the fourth subsection. In the last subsection we present results for the input–output stability of linear time-invariant discrete-time systems.

### 4.8.1 Preliminaries

We concern ourselves here with finite-dimensional discrete-time systems described by difference equations of the form

$$\begin{aligned}
x(k + 1) &= Ax(k) + Bu(k), \\
y(k) &= Cx(k),
\end{aligned}$$

(4.67)

where $A \in R^{n \times n}, B \in R^{n \times m}, C \in R^{p \times n}, k \geq k_0$, and $k, k_0 \in Z^+$. Since (4.67) is time-invariant, we will assume without loss of generality that $k_0 = 0$, and thus, $x : Z^+ \to R^n, y : Z^+ \to R^p$, and $u : Z^+ \to R^m$.

The internal dynamics of (4.67) under conditions of no input are described by equations of the form

$$x(k+1) = Ax(k). \tag{4.68}$$

Such equations may arise in the modeling process, or they may be the consequence of the linearization of nonlinear systems described by equations of the form

$$x(k+1) = g(x(k)), \tag{4.69}$$

where $g : R^n \to R^n$. For example, if $g \in C^1(R^n, R^n)$, then in linearizing (4.69) about, e.g., $x = 0$, we obtain

$$x(k+1) = Ax(k) + f(x(k)), \tag{4.70}$$

where $A = \frac{\partial f}{\partial x}(x)\Big|_{x=0}$ and where $f : R^n \to R^n$ is $o(||x||)$ as a norm of $x$ (e.g., the Euclidean norm) approaches zero. Recall that this means that given $\epsilon > 0$, there is a $\delta > 0$ such that $|| f(x) || < \epsilon || x ||$ for all $|| x || < \delta$.

As in Section 4.7, we will study the *external qualitative properties* of system (4.67) by means of the BIBO stability of such systems. Consistent with the definition of input–output stability of continuous-time systems, we will say that the system (4.67) is *BIBO stable* if there exists a constant $L > 0$ such that the conditions

$$x(0) = 0,$$
$$|| u(k) || \leq 1, \quad k \geq 0,$$

imply that $|| y(k) || \leq L$ for all $k \geq 0$.

We will study the *internal qualitative properties* of system (4.67) by studying the *Lyapunov stability* properties of an equilibrium of (4.68).

Since system (4.69) is time-invariant, we will assume without loss of generality that $k_0 = 0$. As in Chapters 1 and 2, we will denote for a given set of initial data $x(0) = x_0$ the solution of (4.69) by $\phi(k, x_0)$. When $x_0$ is understood or of no importance, we will frequently write $\phi(k)$ in place of $\phi(k, x_0)$. Recall that for system (4.69) [as well as systems (4.67), (4.68), and (4.70)], there are no particular difficulties concerning the existence and uniqueness of solutions, and furthermore, as long as $g$ in (4.69) is continuous, the solutions will be continuous with respect to initial data. Recall also that in contrast to systems described by ordinary differential equations, the solutions of systems described by ordinary difference equations [such as (4.69)] exist only in the forward direction of time ($k \geq 0$).

We say that $x_e \in R^n$ is an *equilibrium* of system (4.69) if $\phi(k, x_e) \equiv x_e$ for all $k \geq 0$, or equivalently,

$$g(x_e) = x_e. \tag{4.71}$$

As in the continuous-time case, we will assume without loss of generality that the equilibrium of interest will be the origin; i.e., $x_e = 0$. If this is not the case, then we can always transform (similarly as in the continuous-time case) system (4.69) into a system of equations that has an equilibrium at the origin.

---

***Example 4.41.*** The system described by the equation

$$x(k + 1) = x(k)[x(k) - 1]$$

has two equilibria, one at $x_{e1} = 0$ and another at $x_{e2} = 1$.

---

***Example 4.42.*** The system described by the equations

$$x_1(k + 1) = x_2(k),$$
$$x_2(k + 1) = -x_1(k)$$

has an equilibrium at $x_e^T = (0, 0)$.

---

Throughout this section we will assume that the function $g$ in (4.69) is continuous, or if required, continuously differentiable. The various definitions of Lyapunov stability of the equilibrium $x = 0$ of system (4.69) are essentially identical to the corresponding definitions of Lyapunov stability of an equilibrium of continuous-time systems described by ordinary differential equations, replacing $t \in R^+$ by $k \in Z^+$. We will concern ourselves with stability, instability, asymptotic stability, and exponential stability of the equilibrium $x = 0$ of (4.69).

We say that the equilibrium $x = 0$ of (4.69) is *stable* if for every $\epsilon > 0$ there exists a $\delta = \delta(\epsilon) > 0$ such that $\| \phi(k, x_0) \| < \epsilon$ for all $k \geq 0$ whenever $\| x_0 \| < \delta$. If the equilibrium $x = 0$ of (4.69) is not stable, it is said to be *unstable*. We say that the equilibrium $x = 0$ of (4.69) is *asymptotically stable* if (i) it is stable and (ii) there exists an $\eta > 0$ such that if $\| x_0 \| < \eta$, then $\lim_{k \to \infty} \| \phi(k, x_0) \| = 0$. If the equilibrium $x = 0$ satisfies property (ii), it is said to be *attractive*, and we call the set of all $x_0 \in R^n$ for which $x = 0$ is attractive the *domain of attraction* of this equilibrium. If $x = 0$ is asymptotically stable and if its domain of attraction is all of $R^n$, then it is said to be *asymptotically stable in the large* or *globally asymptotically stable*. We say that the equililbrium $x = 0$ of (4.69) is *exponentially stable* if there exists an $\alpha > 0$ and for every $\epsilon > 0$, there exists a $\delta(\epsilon) > 0$, such that $\| \phi(k, x_0) \| \leq \epsilon e^{-\alpha k}$ for all $k \geq 0$ whenever $\| x_0 \| < \delta(\epsilon)$. The equilibrium $x = 0$ of (4.69) is *exponentially stable in the large* if there exists $\alpha > 0$ and for any $\beta > 0$, there exists $k(\beta) > 0$ such that $\| \phi(t, x_0) \| \leq k(\beta) \| x_0 \| e^{-\alpha k}$ for all $k > 0$ whenever $\| x_0 \| < \beta$. Finally, we say that a solution of (4.69) through $x_0$ is *bounded* if there is a constant $M$ such that $\| \phi(k, x_0) \| \leq M$ for all $k \geq 0$.

## 4.8.2 Linear Systems

In proving some of the results of this section, we require a result for system (4.68) that is analogous to Theorem 3.1. As in the proof of that theorem, we note that the linear combination of solutions of system (4.68) is also a solution of system (4.68), and hence, the set of solutions $\{\phi : Z^+ \times R^n \to R^n\}$ constitutes a vector space (over $F = R$ or $F = C$). The dimension of this vector space is $n$. To show this, we choose a set of linearly independent vectors $x_0^1, \ldots, x_0^n$ in the $n$-dimensional $x$-space ($R^n$ or $C^n$) and we show, in an identical manner as in the proof of Theorem 3.1, that the set of solutions $\phi(k, x_0^i), i = 1, \ldots, n$, is linearly independent and spans the set of solutions of system (4.68). (We ask the reader in the Exercise section to provide the details of the proof of the above assertions.) This yields the following result.

**Theorem 4.43.** *The set of solutions of system (4.68) over the time interval $Z^+$ forms an $n$-dimensional vector space.* ∎

Incidentally, if in particular we choose $\phi(k, e^i), i = 1, \ldots, n$, where $e^i, i = 1, \ldots, n$, denotes the natural basis for $R^n$, and if we let $\Phi(k, k_0 = 0) \triangleq \Phi(k) = [\phi(k, e^1), \ldots, \phi(k, e^n)]$, then it is easily verified that the $n \times n$ matrix $\Phi(k)$ satisfies the matrix equation

$$\Phi(k + 1) = A\Phi(k), \quad \Phi(0) = I,$$

and that $\Phi(k) = A^k, k \geq 0$ [i.e., $\Phi(k)$ is the state transition matrix for system (4.68)].

**Theorem 4.44.** *The equilibrium $x = 0$ of system (4.68) is stable if and only if the solutions of (4.68) are bounded.*

*Proof.* Assume that the equilibrium $x = 0$ of (4.68) is stable. Then for $\epsilon = 1$ there is a $\delta > 0$ such that $\| \phi(k, x_0) \| < 1$ for all $k \geq 0$ and all $\| x_0 \| \leq \delta$. In this case

$$\| \phi(k, x_0) \| = \| A^k x_0 \| = \| A^k x_0 \delta / \| x_0 \| \| (\| x_0 \| / \delta) < \| x_0 \| / \delta$$

for all $x_0 \neq 0$ and all $k \geq 0$. Using the definition of matrix norm [refer to Section A.7] it follows that $\| A^k \| \leq \delta^{-1}, k \geq 0$. We have proved that if the equilibrium $x = 0$ of (4.68) is stable, then the solutions of (4.68) are bounded.

Conversely, suppose that all solutions $\phi(k, x_0) = A^k x_0$ are bounded. Let $\{e^1, \ldots, e^n\}$ denote the natural basis for $n$-space and let $\| \phi(k, e^j) \| < \beta_j$ for all $k \geq 0$. Then for any vector $x_0 = \sum_{j=1}^n \alpha_j e^j$ we have that

$$\| \phi(k, x_0) \| = \| \sum_{j=1}^n \alpha_j \phi(k, e^j) \| \leq \sum_{j=1}^n |\alpha_j| \beta_j \leq (\max_j \beta_j) \sum_{j=1}^n |\alpha_j|$$

$$\leq c \| x_0 \|, \quad k \geq 0,$$

for some constant $c$. For given $\epsilon > 0$, we choose $\delta = \epsilon/c$. Then, if $\| x_0 \| < \delta$, we have $\| \phi(k, x_0) \| < c \| x_0 \| < \epsilon$ for all $k \geq 0$. We have proved that if the solutions of (4.68) are bounded, then the equilibrium $x = 0$ of (4.68) is stable. ∎

**Theorem 4.45.** *The following statements are equivalent:*

*(i)   The equilibrium $x = 0$ of (4.68) is asymptotically stable,*
*(ii)  The equilibrium $x = 0$ of (4.68) is asymptotically stable in the large,*
*(iii) $\lim_{k \to \infty} \| A^k \| = 0$.*

*Proof.* Assume that statement (i) is true. Then there is an $\eta > 0$ such that when $\| x_0 \| \leq \eta$, then $\phi(k, x_0) \to 0$ as $k \to \infty$. But then we have for *any* $x_0 \neq 0$ that

$$\phi(k, x_0) = A^k x_0 = [A^k(\eta x_0 / \| x_0 \|)] \| x_0 \| / \eta \to 0 \text{ as } k \to \infty.$$

It follows that statement (ii) is true.

Next, assume that statement (ii) is true. Then for any $\epsilon > 0$ there must exist a $K = K(\epsilon)$ such that for all $k \geq K$ we have that $\| \phi(k, x_0) \| = \| A^k x_0 \| < \epsilon$. To see this, let $\{e^1, \ldots, e^n\}$ be the natural basis for $R^n$. Thus, for a fixed constant $c > 0$, if $x_0 = (\alpha_1, \ldots, \alpha_n)^T$ and if $\| x_0 \| \leq 1$, then $x_0 = \sum_{j=1}^{n} \alpha_j e^j$ and $\sum_{j=1}^{n} |\alpha_j| \leq c$. For each $j$ there is a $K_j = K_j(\epsilon)$ such that $\| A^k e^j \| < \epsilon/c$ for $k \geq K_j$. Define $K = K(\epsilon) = \max\{K_j(\epsilon) : j = 1, \ldots, n\}$. For $\| x_0 \| \leq 1$ and $k \geq K$ we have that

$$\| A^k x_0 \| = \| \sum_{j=1}^{n} \alpha_j A^k e^j \| \leq \sum_{j=1}^{n} |\alpha_j|(\epsilon/c) \leq \epsilon.$$

By the definition of matrix norm [see Section A.7], this means that $\| A^k \| \leq \epsilon$ for $k > K$. Therefore, statement (iii) is true.

Finally, assume that statement (iii) is true. Then $\| A^k \|$ is bounded for all $k \geq 0$. By Theorem 4.44, the equilibrium $x = 0$ is stable. To prove asymptotic stability, fix $\epsilon > 0$. If $\| x_0 \| < \eta = 1$, then $\| \phi(k, x_0) \| \leq \| A^k \| \| x_0 \| \to 0$ as $k \to \infty$. Therefore, statement (i) is true. This completes the proof. ∎

**Theorem 4.46.** *The equilibrium $x = 0$ of (4.68) is asymptotically stable if and only if it is exponentially stable.*

*Proof.* The exponential stability of the equilibrium $x = 0$ implies the asymptotic stability of the equilibrium $x = 0$ of systems (4.69) in general and, hence, for systems (4.68) in particular.

Conversely, assume that the equilibrium $x = 0$ of (4.68) is asymptotically stable. Then there is a $\delta > 0$ and a $K > 0$ such that if $\| x_0 \| \leq \delta$, then

$$\| \Phi(k + K)x_0 \| \leq \frac{\delta}{2}$$

for all $k \geq 0$. This implies that

$$\| \Phi(k + K) \| \leq \frac{1}{2} \quad \text{if } k \geq 0. \tag{4.72}$$

From Section 3.5.1 we have that $\Phi(k - l) = \Phi(k - s)\Phi(s - l)$ for any $k, l, s$. Therefore,

$$\| \Phi(k + 2K) \| = \| \Phi[(k + 2K) - (k + K)]\Phi(k + K) \| \leq \frac{1}{4}$$

in view of (4.72). By induction we obtain for $k \geq 0$ that

$$\| \Phi(k + nK) \| \leq 2^{-n}. \tag{4.73}$$

Let $\alpha = \frac{(\ln 2)}{K}$. Then (4.73) implies that for $0 \leq k < K$ we have that

$$\| (k + nK, x_0) \| \leq 2 \| x_0 \| \, 2^{-(n+1)}$$
$$= 2 \| x_0 \| \, e^{-\alpha(n+1)K}$$
$$\leq 2 \| x_0 \| \, e^{-\alpha(k+nK)},$$

which proves the result.    ∎

To arrive at the next result, we make reference to the results of Subsection 3.5.5. Specifically, by inspecting the expressions for the modes of system (4.68) given in (3.131) and (3.132), or by utilizing the Jordan canonical form of $A$ [refer to (3.135) and (3.136)], the following result is evident.

**Theorem 4.47.** *(i)   The equilibrium $x = 0$ of system (4.68) is asymptotically stable if and only if all eigenvalues of $A$ are within the unit circle of the complex plane (i.e., if $\lambda_1, \ldots, \lambda_n$ denote the eigenvalues of $A$, then $|\lambda_j| < 1, j = 1, \ldots, n$). In this case we say that the matrix $A$ is Schur stable, or simply, the matrix $A$ is stable.*
*(ii)  The equilibrium $x = 0$ of system (4.68) is stable if and only if $|\lambda_j| \leq 1, j = 1, \ldots, n$, and for each eigenvalue with $|\lambda_j| = 1$ having multiplicity $n_j > 1$, it is true that*

$$\lim_{z \to \lambda_j} \left\{ \frac{d^{n_j - 1 - l}}{dz^{n_j - 1 - l}} [(z - \lambda_j)^{n_j} (zI - A)^{-1}] \right\} = 0, \quad l = 1, \ldots, n_j - 1.$$

*(iii) The equilibrium $x = 0$ of system (4.68) is unstable if and only if the conditions in (ii) above are not true.*    ∎

Alternatively, it is evident that the equilibrium $x = 0$ of system (4.68) is *stable* if and only if all eigenvalues of $A$ are within or on the unit circle of the complex plane, and every eigenvalue that is on the unit circle has an associated Jordan block of order 1.

**Example 4.48.** (i)   For the system in Example 4.42 we have

$$A = \begin{bmatrix} 0 & 1 \\ -1 & 0 \end{bmatrix}.$$

The eigenvalues of $A$ are $\lambda_1, \lambda_2 = \pm\sqrt{-1}$. According to Theorem 4.47, the equilibrium $x = 0$ of the system is stable, and according to Theorem 4.44 the matrix $A^k$ is bounded for all $k \geq 0$.

(ii)   For system (4.68) let

$$A = \begin{bmatrix} 0 & -1/2 \\ -1 & 0 \end{bmatrix}.$$

The eigenvalues of $A$ are $\lambda_1, \lambda_2 = \pm 1/\sqrt{2}$. According to Theorem 4.47, the equilibrium $x = 0$ of the system is asymptotically stable, and according to Theorem 4.45, $\lim_{k \to \infty} A^k = 0$.

(iii)   For system (4.68) let

$$A = \begin{bmatrix} 0 & -1/2 \\ -3 & 0 \end{bmatrix}.$$

The eigenvalues of $A$ are $\lambda_1, \lambda_2 = \pm\sqrt{3/2}$. According to Theorem 4.47, the equilibrium $x = 0$ of the system is unstable, and according to Theorem 4.44, the matrix $A^k$ is not bounded with increasing $k$.

(iv)   For system (4.68) let

$$A = \begin{bmatrix} 1 & 1 \\ 0 & 1 \end{bmatrix}.$$

The matrix $A$ is a Jordan block of order 2 for the eigenvalue $\lambda = 1$. Accordingly, the equilibrium $x = 0$ of the system is unstable (refer to the remark following Theorem 4.47) and the matrix $A^k$ is unbounded with increasing $k$.

### 4.8.3 The Lyapunov Matrix Equation

In this subsection we obtain another characterization of stable matrices by means of the *Lyapunov matrix equation*.

Returning to system (4.68) we choose as a Lyapunov function

$$v(x) = x^T B x, B = B^T, \tag{4.74}$$

and we evaluate the first forward difference of $v$ along the solutions of (4.68) as

$$Dv(x(k)) \triangleq v(x(k+1)) - v(x(k)) = x(k+1)^T Bx(k+1) - x(k)^T Bx(k)$$
$$= x(k)^T A^T BAx(k) - x(k)^T Bx(k)$$
$$= x(k)^T (A^T BA - B)x(k),$$

and therefore,

$$Dv(x) = x^T (A^T BA - B)x \triangleq -x^T Cx,$$

where

$$A^T BA - B = C, \quad C^T = C. \tag{4.75}$$

**Theorem 4.49.** *(i)  The equilibrium $x = 0$ of system (4.68) is stable if there exists a real, symmetric, and positive definite matrix $B$ such that the matrix $C$ given in (4.75) is negative semidefinite.*

*(ii)  The equilibrium $x = 0$ of system (4.68) is asymptotically stable in the large if there exists a real, symmetric, and positive definite matrix $B$ such that the matrix $C$ given in (4.75) is negative definite.*

*(iii)  The equilibrium $x = 0$ of system (4.68) is unstable if there exists a real, symmetric matrix $B$ that is either negative definite or indefinite such that the matrix $C$ given in (4.75) is negative definite.*  ∎

In proving Theorem 4.49 one can follow a similar approach as in the proofs of Theorems 4.22, 4.24 and 4.26. We leave the details to the reader as an exercise.

In applying Theorem 4.49, we start by choosing (guessing) a matrix $B$ having certain desired properties and we then solve for the matrix $C$, using equation (4.75). If $C$ possesses certain desired properties (i.e., it is negative definite), we can draw appropriate conclusions by applying one of the results given in Theorem 4.49; if not, we need to choose another matrix $B$. This approach is not very satisfactory, and in the following we will derive results that will allow us (as in the case of continuous-time systems) to *construct* Lyapunov functions of the form $v(x) = x^T Bx$ in a systematic manner. In doing so, we first choose a matrix $C$ in (4.75) that is either negative definite or positive definite, and then we solve (4.75) for $B$. Conclusions are then made by applying Theorem 4.49. In applying this construction procedure, we need to know conditions under which (4.75) possesses a (unique) solution $B$ for *any* definite (i.e., positive or negative definite) matrix $C$. We will address this issue next.

We first show that if $A$ is stable, i.e., if all eigenvalues of matrix $A$ [in system (4.68)] are inside the unit circle of the complex plane, then we can compute $B$ in (4.75) explicitly. To show this, we assume that in (4.75) $C$ is a given matrix and that $A$ is stable. Then

$$(A^T)^{k+1} BA^{k+1} - (A^T)^k BA^k = (A^T)^k CA^k,$$

and summing from $k = 0$ to $l$ yields

$$A^T BA - B + (A^T)^2 BA^2 - A^T BA + \cdots + (A^T)^{l+1} BA^{l+1} - (A^T)^l BA^l = \sum_{k=0}^{l} (A^T)^k CA^k$$

or

$$(A^T)^{l+1} BA^{l+1} - B = \sum_{k=0}^{l} (A^T)^k CA^k.$$

Letting $l \to \infty$, we obtain

$$B = -\sum_{k=0}^{\infty} (A^T)^k CA^k. \tag{4.76}$$

It is easy to verify that (4.76) is a solution of (4.75). We have

$$-A^T \left[ \sum_{k=0}^{\infty} (A^T)^k CA^k \right] A + \sum_{k=0}^{\infty} (A^T)^k CA^k = C$$

or

$$-A^T CA + C - (A^T)^2 CA^2 + A^T CA - (A^T)^3 CA^3 + (A^T)^2 CA^2 - \cdots = C.$$

Therefore (4.76) is a solution of (4.75). Furthermore, if $C$ is negative definite, then $B$ is positive definite.

Combining the above with Theorem 4.49(ii) we have the following result.

**Theorem 4.50.** *If there is a positive definite and symmetric matrix $B$ and a negative definite and symmetric matrix $C$ satisfying (4.75), then the matrix $A$ is stable. Conversely, if $A$ is stable, then, given any symmetric matrix $C$, (4.75) has a unique solution, and if $C$ is negative definite, then $B$ is positive definite.* ∎

Next, we determine conditions under which the system of equations (4.75) has a (unique) solution $B = B^T \in R^{n \times n}$ for a given matrix $C = C^T \in R^{n \times n}$. To accomplish this, we consider the more general equation

$$A_1 X A_2 - X = C, \tag{4.77}$$

where $A_1 \in R^{m \times m}, A_2 \in R^{n \times n}$, and $X$ and $C$ are $m \times n$ matrices.

**Lemma 4.51.** *Let $A_1 \in R^{m \times m}$ and $A_2 \in R^{n \times n}$. Then (4.77) has a unique solution $X \in R^{m \times n}$ for a given $C \in R^{m \times n}$ if and only if no eigenvalue of $A_1$ is a reciprocal of an eigenvalue of $A_2$.*

*Proof.* We need to show that the condition on $A_1$ and $A_2$ is equivalent to the condition that $A_1 X A_2 = X$ implies $X = 0$. Once we have proved that $A_1 X A_2 = X$ has the unique solution $X = 0$, then it can be shown that (4.77) has a unique solution for every $C$, since (4.77) is a linear equation.

Assume first that the condition on $A_1$ and $A_2$ is satisfied. Now $A_1 X A_2 = X$ implies that $A_1^{k-j} X A_2^{k-j} = X$ and

$$A_1^j X = A_1^k X A_2^{k-j} \quad \text{for } k \geq j \geq 0.$$

Now for a polynomial of degree $k$,

$$p(\lambda) = \sum_{j=0}^{k} a_j \lambda^j,$$

we define the polynomial of degree $k$,

$$p^*(\lambda) = \sum_{j=0}^{k} a_j \lambda^{k-j} = \lambda^k p(1/\lambda),$$

from which it follows that

$$p(A_1) X = A_1^k X p^*(A_2).$$

Now let $\phi_i(\lambda)$ be the characteristic polynomial of $A_i, i = 1, 2$. Since $\phi_1(\lambda)$ and $\phi_2^*(\lambda)$ are relatively prime, there are polynomials $p(\lambda)$ and $q(\lambda)$ such that

$$p(\lambda)\phi_1(\lambda) + q(\lambda)\phi_2^*(\lambda) = 1.$$

Now define $\phi(\lambda) = q(\lambda)\phi_2^*(\lambda)$ and note that $\phi^*(\lambda) = q^*(\lambda)\phi_2(\lambda)$. It follows that $\phi^*(A_2) = 0$ and $\phi(A_1) = I$. From this it follows that $A_1 X A_2 = X$ implies $X = 0$.

To prove the converse, we assume that $\lambda$ is an eigenvalue of $A_1$ and $\lambda^{-1}$ is an eigenvalue of $A_2$ (and, hence, is also an eigenvalue of $A_2^T$). Let $A_1 x^1 = \lambda x^1$ and $A_2^T x^2 = \lambda^{-1} x^2, x^1 \neq 0$ and $x^2 \neq 0$. Define $X = (x_1^2 x^1, x_2^2 x^1, \dots, x_n^2 x^1)$. Then $X \neq 0$ and $A_1 X A_2 = X$. ∎

To construct $v(x)$ by using Lemma 4.51, we must still check the definiteness of $B$. To accomplish this, we use Theorem 4.49.

1. If all eigenvalue of $A$ [for system (4.68)] are inside the unit circle of the complex plane, then no reciprocal of an eigenvalue of $A$ is an eigenvalue, and Lemma 4.51 gives another way of showing that (4.75) has a unique solution $B$ for each $C$ if $A$ is stable. If $C$ is negative definite, then $B$ is positive definite. This can be shown as was done for the case of linear ordinary differential equations.

2. Suppose that at least one of the eigenvalues of $A$ is outside the unit circle in the complex plane and that $A$ has no eigenvalues on the unit circle. As in the case of linear differential equations (4.22) (Section 4.5), we use a similarity transformation $x = Qy$ in such a way that $Q^{-1}AQ = \text{diag}[A_1, A_2]$, where all eigenvalues of $A_1$ are outside the unit circle while all eigenvalues

of $A_2$ are within the unit circle. We then proceed identically as in the case of linear differential equations to show that under the present assumptions there exists for system (4.68) a Lyapunov function that satisfies the hypotheses of Theorem 4.49(iii). Therefore, the equilibrium $x = 0$ of system (4.68) is unstable. If $A$ does not have any eigenvalues within the unit circle, then the equilibrium $x = 0$ of (4.68) is completely unstable. In this proof, Lemma 4.51 has not been invoked. If additionally, the hypotheses of Lemma 4.51 are true (i.e., no reciprocal of an eigenvalue of $A$ is an eigenvalue of $A$), then we can construct the Lyapunov function for system (4.68) in a systematic manner.

Summarizing the above discussion, we have proved the following result.

**Theorem 4.52.** *Assume that the matrix $A$ for system (4.68) has no eigenvalues on the unit circle in the complex plane. If all the eigenvalues of the matrix $A$ are within the unit circle of the complex plane, or if at least one eigenvalue is outside the unit circle of the complex plane, then there exists a Lyapunov function of the form $v(x) = x^T Bx, B = B^T$, whose first forward difference along the solutions of system (4.68) is definite (i.e., it is either negative definite or positive definite).* ■

Theorem 4.52 shows that when all the eigenvalues of $A$ are within the unit circle, then for system (4.68), the conditions of Theorem 4.49(ii) are also necessary conditions for exponential stability in the large. Furthermore, when at least one eigenvalue of $A$ is outside the unit circle and no eigenvalues are on the unit circle, then the conditions of Theorem 4.49(iii) are also necessary conditions for instability.

We conclude this subsection with some specific examples.

---

**Example 4.53.** (i)  For system (4.68), let

$$A = \begin{bmatrix} 0 & 1 \\ -1 & 0 \end{bmatrix}.$$

Let $B = I$, which is positive definite. From (4.75) we obtain

$$C = A^T A - I = \begin{bmatrix} 0 & -1 \\ 1 & 0 \end{bmatrix} \begin{bmatrix} 0 & 1 \\ -1 & 0 \end{bmatrix} - \begin{bmatrix} 1 & 0 \\ 0 & 1 \end{bmatrix} = \begin{bmatrix} 0 & 0 \\ 0 & 0 \end{bmatrix}.$$

It follows from Theorem 4.49(i) that the equilibrium $x = 0$ of this system is stable. This is the same conclusion that was made in Example 4.48.
(ii)  For system (4.68), let

$$A = \begin{bmatrix} 0 & -\frac{1}{2} \\ -1 & 0 \end{bmatrix}.$$

Choose

$$B = \begin{bmatrix} \frac{8}{3} & 0 \\ 0 & \frac{5}{3} \end{bmatrix},$$

which is positive definite. From (4.75) we obtain

$$C = A^T B A - B = \begin{bmatrix} 0 & -1 \\ -\frac{1}{2} & 0 \end{bmatrix} \begin{bmatrix} \frac{8}{3} & 0 \\ 0 & \frac{5}{3} \end{bmatrix} \begin{bmatrix} 0 & -\frac{1}{2} \\ -1 & 0 \end{bmatrix} - \begin{bmatrix} \frac{8}{3} & 0 \\ 0 & \frac{5}{3} \end{bmatrix} = \begin{bmatrix} -1 & 0 \\ 0 & -1 \end{bmatrix},$$

which is negative definite. It follows from Theorem 4.49(ii) that the equilibrium $x = 0$ of this system is asymptotically stable in the large. This is the same conclusion that was made in Example 4.48(ii).

(iii) For system (4.68), let

$$A = \begin{bmatrix} 0 & -\frac{1}{2} \\ -3 & 0 \end{bmatrix}.$$

Choose

$$C = \begin{bmatrix} -1 & 0 \\ 0 & -1 \end{bmatrix},$$

which is negative definite. From (4.75) we obtain

$$C = A^T B A - B = \begin{bmatrix} 0 & -3 \\ -\frac{1}{2} & 0 \end{bmatrix} \begin{bmatrix} b_{11} & b_{12} \\ b_{12} & b_{22} \end{bmatrix} \begin{bmatrix} 0 & -\frac{1}{2} \\ -3 & 0 \end{bmatrix} - \begin{bmatrix} b_{11} & b_{12} \\ b_{12} & b_{22} \end{bmatrix}$$

or

$$\begin{bmatrix} (9b_{22} - b_{11}) & \frac{1}{2}b_{12} \\ \frac{1}{2}b_{12} & (\frac{1}{4}b_{11} - b_{22}) \end{bmatrix} = \begin{bmatrix} -1 & 0 \\ 0 & -1 \end{bmatrix},$$

which yields

$$B = \begin{bmatrix} -8 & 0 \\ 0 & -1 \end{bmatrix},$$

which is also negative definite. It follows from Theorem 4.49(iii) that the equilibrium $x = 0$ of this system is unstable. This conclusion is consistent with the conclusion made in Example 4.48(iii).

(iv) For system (4.68), let

$$A = \begin{bmatrix} \frac{1}{3} & 1 \\ 0 & 3 \end{bmatrix}.$$

The eigenvalues of $A$ are $\lambda_1 = \frac{1}{3}$ and $\lambda_2 = 3$. According to Lemma 4.51, for a given $C$, (4.77) does *not* have a unique solution in this case since $\lambda_1 = 1/\lambda_2$. For purposes of illustration, we choose $C = -I$. Then

$$-I = A^T B A - B = \begin{bmatrix} \frac{1}{3} & 0 \\ 1 & 3 \end{bmatrix} \begin{bmatrix} b_{11} & b_{12} \\ b_{12} & b_{22} \end{bmatrix} \begin{bmatrix} \frac{1}{3} & 1 \\ 0 & 3 \end{bmatrix} = \begin{bmatrix} b_{11} & b_{12} \\ b_{12} & b_{22} \end{bmatrix}$$

or

$$\begin{bmatrix} -\frac{8}{9}b_{11} & \frac{1}{3}b_{11} \\ \frac{1}{3}b_{11} & b_{11} + 6b_{12} + 8b_{22} \end{bmatrix} = \begin{bmatrix} -1 & 0 \\ 0 & -1 \end{bmatrix},$$

which shows that for $C = -I$, (4.77) does not have any solution (for $B$) at all.

### 4.8.4 Linearization

In this subsection we determine conditions under which the stability properties of the equilibrium $w = 0$ of the linear system

$$w(k + 1) = Aw(k) \qquad (4.78)$$

determine the stability properties of the equilibrium $x = 0$ of the nonlinear system

$$x(k + 1) = Ax(k) + f(x(k)), \qquad (4.79)$$

under the assumption that $f(x) = o(\| x \|)$ as $\| x \| \to 0$ (i.e., given $\epsilon > 0$, there exists $\delta > 0$ such that $\| f(x(k)) \| < \epsilon \| x(k) \|$ for all $k \geq 0$ and all $\| x(k) \| < \delta$). [Refer to the discussion concerning (4.68) to (4.70) in Subsection 4.8.1.]

**Theorem 4.54.** *Assume that $f \in C(R^n, R^n)$ and that $f(x)$ is $o(\| x \|)$ as $\| x \| \to 0$. (i) If $A$ is stable (i.e., all the eigenvalues of $A$ are within the unit circle of the complex plane), then the equilibrium $x = 0$ of system (4.79) is asymptotically stable. (ii) If at least one eigenvalue of $A$ is outside the unit circle of the complex plane and no eigenvalue is on the unit circle, then the equilibrium $x = 0$ of system (4.79) is unstable.* ∎

In proving Theorem 4.54 one can follow a similar approach as in the proofs of Theorems 4.32 and 4.33. We leave the details to the reader as an exercise.

Before concluding this subsection, we consider some specific examples.

---

*Example 4.55.* (i) Consider the system

$$x_1(k + 1) = -\frac{1}{2}x_2(k) + x_1(k)^2 + x_2(k)^2,$$
$$x_2(k + 1) = -x_1(k) + x_1(k)^2 + x_2(k)^2. \qquad (4.80)$$

Using the notation of (4.79), we have

$$A = \begin{bmatrix} 0 & -\frac{1}{2} \\ -1 & 0 \end{bmatrix}, \quad f(x_1, x_2) = \begin{bmatrix} x_1^2 + x_2^2 \\ x_1^2 + x_2^2 \end{bmatrix}.$$

The linearization of (4.80) is given by

$$w(k + 1) = Aw(k). \qquad (4.81)$$

From Example 4.48(ii) [and Example 4.53(ii)], it follows that the equilibrium $w = 0$ of (4.81) is asymptotically stable. Furthermore, in the present case $f(x) = o(\| x \|)$ as $\| x \| \to 0$. Therefore, in view of Theorem 4.54, the equilibrium $x = 0$ of system (4.80) is asymptotically stable.

(ii) Consider the system

$$x_1(k+1) = -\frac{1}{2}x_2(k) + x_1(k)^3 + x_2(k)^2,$$
$$x_2(k+1) = -3x_1(k) + x_1^4(k) - x_2(k)^5. \tag{4.82}$$

Using the notation of (4.78) and (4.79), we have in the present case

$$A = \begin{bmatrix} 0 & -\frac{1}{2} \\ -3 & 0 \end{bmatrix}, \quad f(x_1, x_2) = \begin{bmatrix} x_1^3 + x_2^2 \\ x_1^4 - x_2^5 \end{bmatrix}.$$

Since $A$ is unstable [refer to Example 4.53(iii) and Example 4.48(iii)] and since $f(x) = o(\| x \|)$ as $\| x \| \to 0$, it follows from Theorem 4.54 that the equilibrium $x = 0$ of system (4.82) is unstable.

## 4.8.5 Input–Output Stability

We conclude this chapter by considering the input–output stability of discrete-time systems described by equations of the form

$$x(k+1) = Ax(k) + Bu(k),$$
$$y(k) = Cx(k), \tag{4.83}$$

where all matrices and vectors are defined as in (4.67). Throughout this subsection we will assume that $k_0 = 0$, $x(0) = 0$, and $k \geq 0$.

As in the continuous-time case, we say that system (4.83) is *BIBO stable* if there exists a constant $c > 0$ such that the conditions

$$x(0) = 0,$$
$$\| u(k) \| \leq 1, \quad k \geq 0,$$

imply that $\| y(k) \| \leq c$ for all $k \geq 0$.

The results that we will present involve the impulse response matrix of (4.83) given by

$$H(k) = \begin{cases} CA^{k-1}B, & k > 0, \\ 0, & k \leq 0, \end{cases} \tag{4.84}$$

and the transfer function matrix given by

$$\widehat{H}(z) = C(zI - A)^{-1}B. \tag{4.85}$$

Recall that

$$y(n) = \sum_{k=0}^{n} H(n-k)u(k). \tag{4.86}$$

Associated with system (4.83) is the free dynamical system described by the equation

$$p(k+1) = Ap(k). \tag{4.87}$$

**Theorem 4.56.** *The system (4.83) is* BIBO stable *if and only if there exists a constant $L > 0$ such that for all $n \geq 0$,*

$$\sum_{k=0}^{n} \| H(k) \| \leq L. \tag{4.88}$$

■

As in the continous-time case, the first part of the proof of Theorem 4.56 (sufficiency) is straightforward. Specifically, if $\| u(k) \| \leq 1$ for all $k \geq 0$ and if (4.88) is true, then we have for all $n \geq 0$,

$$\| y(n) \| = \| \sum_{k=0}^{n} H(n-k)u(k) \| \leq \sum_{k=0}^{n} \| H(n-k)u(k) \|$$

$$\leq \sum_{k=0}^{n} \| H(n-k) \| \| u(k) \| \leq \sum_{k=0}^{n} \| H(n-k) \| \leq L.$$

Therefore, system (4.83) is BIBO stable.

In proving the second part of Theorem 4.56 (necessity), we simplify matters by first considering in (4.83) the single-variable case ($n = 1$) with the system description given by

$$y(t) = \sum_{k=0}^{t} h(t-k)u(k), \quad t > 0. \tag{4.89}$$

For purposes of contradiction, we assume that the system is BIBO stable, but no finite $L$ exists such that (4.88) is satisfied. Another way of expressing the last assumption is that for *any finite $L$*, there exists $t = k_1(L) \triangleq k_1$ such that

$$\sum_{k=0}^{k_1} |h(k_1 - k)| > L.$$

We now choose in particular the input $u$ given by

$$u(k) = \begin{cases} +1 & \text{if } h(t-k) > 0, \\ 0 & \text{if } h(t-k) = 0, \\ -1 & \text{if } h(t-k) < 0, \end{cases}$$

$0 \leq k \leq k_1$. Clearly, $|u(k)| \leq 1$ for all $k \geq 0$. The output of the system at $t = k_1$ due to the above input, however, is

$$y(k_1) = \sum_{k=0}^{k_1} h(k_1 - k)u(k) = \sum_{k=0}^{k_1} |h(k_1 - k)| > L,$$

which contradicts the assumption that the system is BIBO stable.

The above can now be extended to the multivariable case. In doing so we apply the single-variable result to every possible pair of input and output vector components, we make use of the fact that the sum of a finite number of bounded sums will be bounded, and we note that a vector is bounded if and only if each of its components is bounded. We leave the details to the reader.

Next, as in the case of continuous-time systems, we note that the asymptotic stability of the equilibrium $p = 0$ of system (4.87) implies the BIBO stability of system (4.83) since the sum

$$\| \sum_{k=1}^{\infty} CA^{k-1}B \| \leq \sum_{k=1}^{\infty} \| C \| \| A^{k-1} \| \| B \|$$

is finite.

Next, we recall that a complex number $z_p$ is a *pole* of $\widehat{H}(z) = [\hat{h}_{ij}(z)]$ if for some $(i, j)$ we have $|\hat{h}_{ij}(z_p)| = \infty$. If each entry of $\widehat{H}(z)$ has only poles with modulus (magnitude) less than 1, then, as shown in Chapter 3, each entry of $H(k) = [h_{ij}(k)]$ consists of a sum of convergent terms. It follows that under these conditions the sum

$$\sum_{k=0}^{\infty} \| H(k) \|$$

is finite, and any realization of $\widehat{H}(z)$ will result in a system that is BIBO stable.

Conversely, if

$$\sum_{k=0}^{\infty} \| H(k) \|$$

is finite, then the terms in every entry of $H(k)$ must be convergent. But then every entry of $\widehat{H}(z)$ has poles whose modulus is within the unit circle of the complex plane. We have proved the final result of this section.

**Theorem 4.57.** *The time-invariant system (4.83) is BIBO stable if and only if the poles of the transfer function*

$$\widehat{H}(z) = C(zI - A)^{-1}B$$

*are within the unit circle of the complex plane.* ∎

## 4.9 Summary and Highlights

In this chapter we first addressed the stability of an equilibrium of continuous-time finite-dimensional systems. In doing so, we first introduced the concept of equilibrium and defined several types of stability in the sense of Lyapunov (Sections 4.2 and 4.3). Next, we established several stability conditions of an equilibrium for linear systems $\dot{x} = Ax$, $t \geq 0$ in terms of the state transition matrix in Theorems 4.12–4.14 and in terms of eigenvalues in Theorem 4.15 (Section 4.4). Next, we established various stability conditions that are phrased in terms of the Lyapunov matrix equation (4.25) for system $\dot{x} = Ax$ (Section 4.5). The existence of Lyapunov functions for $\dot{x} = Ax$ of the form $x^T P x$ is established in Theorem 4.29. In Section 4.6 we established conditions under which the asymptotic stability and the instability of an equilibrium for a nonlinear time-invariant system can be deduced via linearization; see Theorems 4.32 and 4.33.

Next, we addressed the input–output stability of time-invariant linear, continuous-time, finite-dimensional systems (Section 4.7). For such systems we established several conditions for bounded input/bounded output stability (BIBO stability); see Theorems 4.37 and 4.39.

The chapter is concluded with Section 4.8, where we addressed the Lyapunov stability and the input–output stability of linear, time-invariant, discrete-time systems. For such systems, we established results that are analogous to the stability results of continuous-time systems. The stability of an equilibrium is expressed in terms of the state transition matrix in Theorem 4.45, in terms of the eigenvalues in Theorem 4.47, and in terms of the Lyapunov Matrix Equation in Theorems 4.49 and 4.50. The existence of Lyapunov functions of the form $x^T P x$ for $x(k + 1) = Ax(k)$ is established in Theorem 4.52. Stability results based on linearization are presented in Theorem 4.54 and for BIBO stability in Theorems 4.56 and 4.57.

## 4.10 Notes

The initial contributions to stability theory that took place toward the end of the nineteenth century are primarily due to physicists and mathematicians (Lyapunov [11]), whereas input–output stability is the brainchild of electrical engineers (Sandberg [17] to [19], Zames [21], [22]). Sources with extensive coverage of Lyapunov stability theory include, e.g., Hahn [6], Khalil [8], LaSalle [9], LaSalle and Lefschetz [10], Michel and Miller [13], Michel et al. [14], Miller and Michel [15], and Vidyasagar [20]. Input–output stability is addressed in great detail in Desoer and Vidyasagar [5], Vidyasagar [20], and Michel and Miller [13]. For a survey that traces many of the important developments of stability in feedback control, refer to Michel [12].

In the context of *linear systems*, sources on both Lyapunov stability and input–output stability can be found in numerous texts, including Antsaklis and Michel [1], Brockett [2], Chen [3], DeCarlo [4], Kailath [7], and Rugh [16]. In developing our presentation, we found the texts by Antsaklis and Michel [1], Brockett [2], Hahn [6], LaSalle [9], and Miller and Michel [15] especially helpful.

In this chapter, we addressed various types of Lyapunov stability and bounded input/bounded output stability of time-invariant systems. In the various stability concepts for such systems, the initial time $t_0$ (resp., $k_0$) plays no significant role, and for this reason, we chose without loss of generality $t_0 = 0$ (resp., $k_0 = 0$). In the case of time-varying systems, this is in general not true, and in defining the various Lyapunov stability concepts and the concept of bounded input/bounded output stability, one has to take into account the effects of initial time. In doing so, we have to distinguish between *uniformity* and *nonuniformity* when defining the various types of Lyapunov stability of an equilibrium and the BIBO stability of a system. For a treatment of the Lyapunov stability and the BIBO stablity of the time-varying counterparts of systems (4.14), (4.15) and (4.67), (4.68), we refer the reader to Chapter 6 in Antsaklis and Michel [1].

We conclude by noting that there are graphical criteria (i.e., frequency domain criteria), such as, the Leonhard–Mikhailov criterion, and algebraic criteria, such as the Routh–Hurwitz criterion and the Schur–Cohn criterion, which yield necessary and sufficient conditions for the asymptotic stability of the equilibrium $x = 0$ for system (4.15) and (4.68). For a presentation of these results, the reader should consult Chapter 6 in Antsaklis and Michel [1] and Michel [12].

# References

1. P.J. Antsaklis and A.N. Michel, *Linear Systems*, Birkhäuser, Boston, MA, 2006.
2. R.W. Brockett, *Finite Dimensional Linear Systems*, Wiley, New York, NY, 1970.
3. C.T. Chen, *Linear System Theory and Design*, Holt, Rinehart and Winston, New York, NY, 1984.
4. R.A. DeCarlo, *Linear Systems*, Prentice-Hall, Englewood Cliffs, NJ, 1989.
5. C.A. Desoer and M. Vidyasagar, *Feedback Systems: Input–Output Properties*, Academic Press, New York, NY, 1975.
6. W. Hahn, *Stability of Motion*, Springer-Verlag, New York, NY, 1967.
7. T. Kailath, *Linear Systems*, Prentice-Hall, Englewood Cliffs, NJ, 1980.
8. H.K. Khalil, *Nonlinear Systems*, Macmillan, New York, NY, 1992.
9. J.P. LaSalle, *The Stability and Control of Discrete Processes*, Springer-Verlag, New York, NY, 1986.
10. J.P. LaSalle and S. Lefschetz, *Stability by Liapunov's Direct Method*, Academic Press, New York, NY, 1961.
11. M.A. Liapounoff, "Problème générale de la stabilité de mouvement," *Ann. Fac. Sci. Toulouse*, Vol. 9, 1907, pp. 203–474. (Translation of a paper published in *Comm. Soc. Math.* Kharkow 1893, reprinted in *Ann. Math. Studies*, Vol. 17, 1949, Princeton, NJ).
12. A.N. Michel, "Stability: the common thread in the evolution of feedback control," *IEEE Control Systems*, Vol. 16, 1996, pp. 50–60.
13. A.N. Michel and R.K. Miller, *Qualitative Analysis of Large Scale Dynamical Systems*, Academic Press, New York, NY, 1977.
14. A.N. Michel, K. Wang, and B. Hu, *Qualitative Theory of Dynamical Systems, Second Edition*, Marcel Dekker, New York, NY, 2001.

15. R.K. Miller and A.N. Michel, *Ordinary Differential Equations*, Academic Press, New York, NY, 1982.

16. W.J. Rugh, *Linear System Theory, Second Edition*, Prentice-Hall, Englewood Cliffs, NJ, 1999.

17. I.W. Sandberg, "On the $L_2$-boundedness of solutions of nonlinear functional equations," *Bell Syst. Tech. J.*, Vol. 43, 1964, pp. 1581–1599.

18. I.W. Sandberg, "A frequency-domain condition for stability of feedback systems containing a single time-varying nonlinear element," *Bell Syst. Tech. J.*, Vol. 43, 1964, pp. 1601–1608.

19. I.W. Sandberg, "Some results on the theory of physical systems governed by nonlinear functional equations," *Bell Syst. Tech. J.*, Vol. 44, 1965, pp. 871–898.

20. M. Vidyasagar, *Nonlinear Systems Analysis*, 2d edition, Prentice Hall, Englewood Cliffs, NJ, 1993.

21. G. Zames, "On the input–output stability of time-varying nonlinear feedback systems, Part I," *IEEE Trans. on Automat. Contr.*, Vol. 11, 1966, pp. 228–238.

22. G. Zames, "On the input–output stability of time-varying nonlinear feedback systems, Part II," *IEEE Trans. on Automat. Contr.*, Vol. 11, 1966, pp. 465–476.

## Exercises

**4.1.** Determine the set of equilibrium points of a system described by the differential equations

$$\dot{x}_1 = x_1 - x_2 + x_3,$$
$$\dot{x}_2 = 2x_1 + 3x_2 + x_3,$$
$$\dot{x}_3 = 3x_1 + 2x_2 + 2x_3.$$

**4.2.** Determine the set of equilibria of a system described by the differential equations

$$\dot{x}_1 = x_2,$$
$$\dot{x}_2 = \begin{cases} x_1 \sin(1/x_1), & \text{when } x_1 \neq 0, \\ 0, & \text{when } x_1 = 0. \end{cases}$$

**4.3.** Determine the equilibrium points and their stability properties of a system described by the ordinary differential equation

$$\dot{x} = x(x - 1) \tag{4.90}$$

by solving (4.90) and then applying the definitions of stability, asymptotic stability, etc.

**4.4.** Prove that the equilibrium $x = 0$ of (4.16) is stable (resp., asymptotically stable or unstable) if and only if $y = 0$ of (4.19) is stable (resp., asymptotically stable or unstable).

**4.5.** Apply Proposition 4.20 to determine the definiteness properties of the matrix $A$ given by

$$A = \begin{bmatrix} 1 & 2 & 1 \\ 2 & 5 & -1 \\ 1 & -1 & 10 \end{bmatrix}.$$

**4.6.** Use Theorem 4.26 to prove that the trivial solution of the system

$$\begin{bmatrix} \dot{x}_1 \\ \dot{x}_2 \end{bmatrix} = \begin{bmatrix} 3 & 4 \\ 2 & 1 \end{bmatrix} \begin{bmatrix} x_1 \\ x_2 \end{bmatrix}$$

is unstable.

**4.7.** Determine the equilibrium points of a system described by the differential equation

$$\dot{x} = -x + x^2,$$

and determine the stability properties of the equilibrium points, if applicable, by using Theorem 4.32 or 4.33.

**4.8.** The system described by the differential equations

$$\begin{aligned} \dot{x}_1 &= x_2 + x_1(x_1^2 + x_2^2), \\ \dot{x}_2 &= -x_1 + x_2(x_1^2 + x_2^2) \end{aligned} \tag{4.91}$$

has an equilibrium at the origin $x^T = (x_1, x_2) = (0,0)$. Show that the trivial solution of the *linearization* of system (4.91) is stable. Prove that the equilibrium $x = 0$ of system (4.91) is unstable. (This example shows that the assumptions on the matrix $A$ in Theorems 4.32 and 4.33 are absolutely essential.)

**4.9.** Use Corollary 4.38 to analyze the stability properties of the system given by

$$\dot{x} = Ax + Bu,$$
$$y = Cx,$$
$$A = \begin{bmatrix} -1 & 0 \\ 1 & -1 \end{bmatrix}, \quad B = \begin{bmatrix} 1 \\ -1 \end{bmatrix}, \quad C = [0,1].$$

**4.10.** Determine all equilibrium points for the discrete-time systems given by

(a)

$$\begin{aligned} x_1(k+1) &= x_2(k) + |x_1(k)|, \\ x_2(k+1) &= -x_1(k) + |x_2(k)|. \end{aligned}$$

(b)

$$x_1(k+1) = x_1(k)x_2(k) - 1,$$
$$x_2(k+1) = 2x_1(k)x_2(k) + 1.$$

**4.11.** Prove Theorem 4.43.

**4.12.** Determine the stability properties of the trivial solution of the discrete-time system given by the equations

$$\begin{bmatrix} x_1(k+1) \\ x_2(k+1) \end{bmatrix} = \begin{bmatrix} \cos\theta & \sin\theta \\ -\sin\theta & \cos\theta \end{bmatrix} \begin{bmatrix} x_1(k) \\ x_2(k) \end{bmatrix}$$

with $\theta$ fixed.

**4.13.** Analyze the stability of the equilibrium $x = 0$ of the system described by the scalar-valued difference equation

$$x(k+1) = \sin[x(k)].$$

**4.14.** Analyze the stability of the equilibrium $x = 0$ of the system described by the difference equations

$$x_1(k+1) = x_1(k) + x_2(k)[x_1(k)^2 + x_2(k)^2],$$
$$x_2(k+1) = x_2(k) - x_1(k)[x_1(k)^2 + x_2(k)^2].$$

**4.15.** Determine a basis of the solution space of the system

$$\begin{bmatrix} x_1(k+1) \\ x_2(k+1) \end{bmatrix} = \begin{bmatrix} 0 & 1 \\ -6 & 5 \end{bmatrix} \begin{bmatrix} x_1(k) \\ x_2(k) \end{bmatrix}.$$

Use your answer in analyzing the stability of the trivial solution of this system.

**4.16.** Let $A \in R^{n \times n}$. Prove that part (iii) of Theorem 4.45 is equivalent to the statement that all eigenvalues of $A$ have modulus less than 1; i.e.,

$$\lim_{k \to \infty} \| A^k \| = 0$$

if and only if for any eigenvalue $\lambda$ of $A$, it is true that $|\lambda| < 1$.

**4.17.** Use Theorem 4.44 to show that the equilibrium $x = 0$ of the system

$$x(k+1) = \begin{bmatrix} 1 & 1 & 1 & \cdots & 1 \\ 0 & 1 & 1 & \cdots & 1 \\ \cdots & \cdots & \cdots & \cdots & \cdots \\ 0 & 0 & 0 & \cdots & 1 \end{bmatrix} x(k)$$

is unstable.

**4.18.** (a) Use Theorem 4.47 to determine the stability of the equilibrium $x = 0$ of the system

$$x(k+1) = \begin{bmatrix} 1 & 1 & -2 \\ 0 & 1 & 3 \\ 0 & 9 & -1 \end{bmatrix} x(k).$$

(b) Use Theorem 4.47 to determine the stability of the equilibrium $x = 0$ of the system

$$x(k+1) = \begin{bmatrix} 1 & 0 & -2 \\ 0 & 1 & 3 \\ 0 & 9 & -1 \end{bmatrix} x(k).$$

**4.19.** Apply Theorems 4.24 and 4.49 to show that if the equilibrium $x = 0$ $(x \in R^n)$ of the system

$$x(k+1) = e^A x(k)$$

is asymptotically stable, then the equilibrium $x = 0$ of the system

$$\dot{x} = Ax$$

is also asymptotically stable.

**4.20.** Apply Theorem 4.49 to show that the trivial solution of the system given by

$$\begin{bmatrix} x_1(k+1) \\ x_2(k+1) \end{bmatrix} = \begin{bmatrix} 0 & 2 \\ 2 & 0 \end{bmatrix} \begin{bmatrix} x_1(k) \\ x_2(k) \end{bmatrix}$$

is unstable.

**4.21.** Determine the stability of the equilibrium $x = 0$ of the scalar-valued system given by

$$x(k+1) = \frac{1}{2}x(k) + \frac{2}{3}\sin x(k).$$

**4.22.** Analyze the stability properties of the discrete-time system given by

$$x(k+1) = x(k) + \frac{1}{2}u(k)$$

$$y(k) = \frac{1}{2}x(k)$$

where $x, y$, and $u$ are scalar-valued variables. Is this system BIBO stable?

**4.23.** Prove Theorem 4.47.

**4.24.** Prove Theorem 4.49 by following a similar approach as was used in the proofs of Theorems 4.22, 4.24, and 4.26.

**4.25.** Prove Theorem 4.54 by following a similar approach as was used in the proofs of Theorems 4.32 and 4.33.

# 5

# Controllability and Observability: Fundamental Results

## 5.1 Introduction

The principal goals of this chapter are to introduce the system properties of controllability and observability (and of reachability and constructibility), which play a central role in the study of state feedback controllers and state observers, and in establishing the relations between internal and external system representations, topics that will be studied in Chapters 7, 8, and 9. State controllability refers to the ability to manipulate the state by applying appropriate inputs (in particular, by steering the state vector from one vector value to any other vector value in finite time). Such is the case, for example, in satellite attitude control, where the satellite must change its orientation. State observability refers to the ability to determine the state vector of the system from knowledge of the input and the corresponding output over some finite time interval. Since it is frequently difficult or impossible to measure the state of a system directly (for example, internal temperatures and pressures in an internal combustion engine), it is very desirable to determine such states by observing the inputs and outputs of the system over some finite time interval.

In Section 5.2, the concepts of reachability and controllability and observability and constructibility are introduced, using discrete-time time-invariant systems. Discrete-time systems are selected for this exposition because the mathematical development is much simpler in this case. In subsection 5.2.3 the concept of duality is also introduced. Reachability and controllability are treated in detail in Section 5.3 and observability and constructibility in Section 5.4 for both continuous-time and discrete-time time-invariant systems.

## 5.2 A Brief Introduction to Reachability and Observability

Reachability and controllability are introduced first, followed by observability and constructibility. These important system concepts are more easily explained in the discrete-time case, and this is the approach taken in this section. Duality is also discussed at the end of the section.

### 5.2.1 Reachability and Controllability

The concepts of *state reachability* (or *controllability-from-the-origin*) and *controllability* (or *controllability-to-the-origin*) are introduced here and are discussed at length in Section 5.3.

In the case of time-invariant systems, a state $x_1$ is called *reachable* if there exists an input that transfers the state of the system $x(t)$ from the zero state to $x_1$ in some finite time $T$. The definition of reachability for the discrete-time case is completely analogous. Figure 5.1 shows that different control inputs $u_1(t)$ and $u_2(t)$ may force the state of a continuous-time system to reach the value $x_1$ from the origin at different finite times $T_1$ and $T_2$, following different paths. Note that reachability refers to the ability of the system to reach $x_1$ from the origin in some finite time; it specifies neither the exact time it takes to achieve this nor the trajectory to be followed.

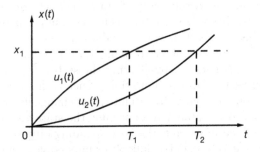

**Figure 5.1.** A reachable state $x_1$

A state $x_0$ is called *controllable* if there exists an input that transfers the state from $x_0$ to the zero state in some finite time $T$. See Figure 5.2. The definition of controllability for the discrete-time case is completely analogous. Similar to reachability, controllability specifies neither the time it takes to achieve the transfer nor the trajectory to be followed.

We note that when particular types of trajectories to be followed are of interest, then one seeks particular control inputs that will achieve such transfers. This leads to various control problem formulations, including the Linear Quadratic (Optimal) Regulator (LQR). The LQR problem is discussed in Chapter 9.

Section 5.3 shows that reachability always implies controllability, but controllability implies reachability only when the state transition matrix $\Phi$ of the system is nonsingular. This is always true for continuous-time systems, but is

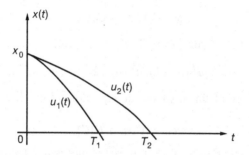

**Figure 5.2.** A controllable state $x_0$

true for discrete-time systems only when the matrix $A$ of the system is non-singular. If the system is state reachable, then there always exists an input that transfers any state $x_0$ to any other state $x_1$ in finite time.

In the time-invariant case, a system is *reachable* (or *controllable-from-the-origin*) if and only if its *controllability matrix* $\mathcal{C}$,

$$\mathcal{C} \triangleq [B, AB, \ldots, A^{n-1}B] \in R^{n \times mn}, \tag{5.1}$$

has full row rank $n$; that is, rank $\mathcal{C} = n$. The matrices $A \in R^{n \times n}$ and $B \in R^{n \times m}$ come from either the continuous-time state equations

$$\dot{x} = Ax + Bu \tag{5.2}$$

or the discrete-time state equations

$$x(k+1) = Ax(k) + Bu(k), \tag{5.3}$$

$k \geq k_0 = 0$. Alternatively, we say that the pair $(A, B)$ is reachable. The matrix $\mathcal{C}$ should perhaps more appropriately be called the "reachability matrix" or the "controllability-from-the-origin matrix." The term "controllability matrix," however, has been in use for some time and is expected to stay in use. Therefore, we shall call $\mathcal{C}$ the "controllability matrix," having in mind the "controllability-from-the-origin matrix."

We shall now discuss reachability and controllability for discrete-time time-invariant systems (5.3).

If the state $x(k)$ in (5.3) is expressed in terms of the initial vector $x(0)$, then (see Subsection 3.5.1)

$$x(k) = A^k x(0) + \sum_{i=0}^{k-1} A^{k-(i+1)} Bu(i) \tag{5.4}$$

for $k > 0$. Rewriting the summation in terms of matrix-vector multiplication, it follows that it is possible to transfer the state from some value $x(0) = x_0$ to some $x_1$ in $n$ steps, that is, $x(n) = x_1$, if there exists an $n$-step input sequence $\{u(0), u(1), \ldots, u(n-1)\}$ that satisfies the equation

$$x_1 - A^n x_0 = C_n U_n, \tag{5.5}$$

where $C_n \triangleq [B, AB, \ldots, A^{n-1}B] = C$ [see (5.1)] and

$$U_n \triangleq [u^T(n-1), u^T(n-2), \ldots, u^T(0)]^T. \tag{5.6}$$

From the theory of linear algebraic equations, (5.5) has a solution $U_n$ if and only if

$$x_1 - A^n x_0 \in \mathcal{R}(C), \tag{5.7}$$

where $\mathcal{R}(C) = \text{range}(C)$. Note that it is not necessary to take more than $n$ steps in the control sequence, since if this transfer cannot be accomplished in $n$ steps, it cannot be accomplished at all. This follows from the Cayley–Hamilton Theorem, in view of which it can be shown that $\mathcal{R}(C_n) = \mathcal{R}(C_k)$ for $k \geq n$. Also note that $\mathcal{R}(C_n)$ includes $\mathcal{R}(C_k)$ for $k < n$ [i.e., $\mathcal{R}(C_n) \supset \mathcal{R}(C_k), k < n$]. (See Exercise 5.1.)

It is now easy to see that the system (5.3) or the pair $(A, B)$ is *reachable* (*controllable-from-the-origin*), if and only if $\text{rank} C = n$, since in this case $\mathcal{R}(C) = R^n$, the entire state space. Note that $x_1 \in \mathcal{R}(C)$ is the condition for a particular state $x_1$ to be reachable from the zero state. Since $\mathcal{R}(C)$ contains all such states, it is called the *reachable subspace* of the system. It is also clear from (5.5) that if the system is reachable, any state $x_0$ can be transferred to any other state $x_1$ in $n$ steps. In addition, the input that accomplishes this transfer is any solution $U_n$ of (5.5). Note that, depending on $x_1$ and $x_0$, this transfer may be accomplished in fewer than $n$ steps (see Section 5.3).

---

***Example 5.1.*** Consider $x(k+1) = Ax(k) + Bu(k)$, where $A = \begin{bmatrix} 0 & 1 \\ 1 & 1 \end{bmatrix}$, $B = \begin{bmatrix} 0 \\ 1 \end{bmatrix}$. Here the controllability (-from-the-origin) matrix $C$ is $C = [B, AB] = \begin{bmatrix} 0 & 1 \\ 1 & 1 \end{bmatrix}$ with $\text{rank} C = 2$. Therefore, the system [or the pair $(A, B)$] is reachable, meaning that any state $x_1$ can be reached from the zero state in a finite number of steps by applying at most $n$ inputs $\{u(0), u(1), \ldots, u(n-1)\}$ (presently, $n = 2$). To see this, let $x_1 = \begin{bmatrix} a \\ b \end{bmatrix}$. Then (5.5) implies that $\begin{bmatrix} a \\ b \end{bmatrix} = \begin{bmatrix} 0 & 1 \\ 1 & 1 \end{bmatrix} \begin{bmatrix} u(1) \\ u(0) \end{bmatrix}$ or $\begin{bmatrix} u(1) \\ u(0) \end{bmatrix} = \begin{bmatrix} -1 & 1 \\ 1 & 0 \end{bmatrix} \begin{bmatrix} a \\ b \end{bmatrix} = \begin{bmatrix} b-a \\ a \end{bmatrix}$. Thus, the control $u(0) = a, u(1) = b - a$ will transfer the state from the origin at $k = 0$ to the state $\begin{bmatrix} a \\ b \end{bmatrix}$ at $k = 2$.

To verify this, we observe that $x(1) = Ax(0) + Bu(0) = \begin{bmatrix} 0 \\ 1 \end{bmatrix} a = \begin{bmatrix} 0 \\ a \end{bmatrix}$ and

$$x(2) = Ax(1) + Bu(1) = \begin{bmatrix} a \\ a \end{bmatrix} + \begin{bmatrix} 0 \\ 1 \end{bmatrix} (b-a) = \begin{bmatrix} a \\ b \end{bmatrix}.$$

Reachability of the system also implies that a state $x_1$ can be reached from any other state $x_0$ in at most $n = 2$ steps. To illustrate this, let

$x(0) = \begin{bmatrix} 1 \\ 1 \end{bmatrix}$. Then (5.5) implies that $x_1 - A^2 x_0 = \begin{bmatrix} a \\ b \end{bmatrix} - \begin{bmatrix} 1 & 1 \\ 1 & 2 \end{bmatrix} \begin{bmatrix} 1 \\ 1 \end{bmatrix} =$
$\begin{bmatrix} a - 2 \\ b - 3 \end{bmatrix} = \begin{bmatrix} 0 & 1 \\ 1 & 1 \end{bmatrix} \begin{bmatrix} u(1) \\ u(0) \end{bmatrix}$. Solving, $\begin{bmatrix} u(1) \\ u(0) \end{bmatrix} = \begin{bmatrix} b - a - 1 \\ a - 2 \end{bmatrix}$, which will drive
the state from $\begin{bmatrix} 1 \\ 1 \end{bmatrix}$ at $k = 0$ to $\begin{bmatrix} a \\ b \end{bmatrix}$ at $k = 2$.

---

Notice that in general the solution $U_n$ of (5.5) is not unique; i.e., many inputs can accomplish the transfer from $x(0) = x_0$ to $x(n) = x_1$, each corresponding to a particular state trajectory. In control problems, particular inputs are frequently selected that, in addition to transferring the state, satisfy additional criteria, such as, e.g., minimization of an appropriate performance index (optimal control).

A system [or the pair $(A, B)$] is *controllable, or controllable-to-the-origin,* when any state $x_0$ can be driven to the zero state in a finite number of steps. From (5.5) we see that a system is controllable when $A^n x_0 \in \mathcal{R}(\mathcal{C})$ for any $x_0$. If rank $A = n$, a system is controllable when rank $\mathcal{C} = n$, i.e., when the reachability condition is satisfied. In this case the $n \times mn$ matrix

$$A^{-n}\mathcal{C} = [A^{-n}B, \ldots, A^{-1}B] \tag{5.8}$$

is of interest and the system is controllable if and only if rank$(A^{-n}\mathcal{C}) =$ rank $\mathcal{C} = n$. If, however, rank $A < n$, then controllability does not imply reachability (see Section 5.3).

---

***Example 5.2.*** The system in Example 5.1 is controllable (-to-the-origin). To see this, we let, $x_1 = 0$ in (5.5) and write $-A^2 x_0 = -\begin{bmatrix} 1 & 1 \\ 1 & 2 \end{bmatrix} \begin{bmatrix} a \\ b \end{bmatrix} =$
$[B, AB] \begin{bmatrix} u(1) \\ u(0) \end{bmatrix} = \begin{bmatrix} 0 & 1 \\ 1 & 1 \end{bmatrix} \begin{bmatrix} u(1) \\ u(0) \end{bmatrix}$, where $x_0 = \begin{bmatrix} a \\ b \end{bmatrix}$. From this we obtain
$\begin{bmatrix} u(1) \\ u(0) \end{bmatrix} = -\begin{bmatrix} -1 & 1 \\ 1 & 0 \end{bmatrix} \begin{bmatrix} 1 & 1 \\ 1 & 2 \end{bmatrix} \begin{bmatrix} a \\ b \end{bmatrix} = \begin{bmatrix} 0 & -1 \\ -1 & -1 \end{bmatrix} \begin{bmatrix} a \\ b \end{bmatrix} = \begin{bmatrix} -b \\ -a - b \end{bmatrix}$, which is the
input that will drive the state from $\begin{bmatrix} a \\ b \end{bmatrix}$ at $k = 0$ to $\begin{bmatrix} 0 \\ 0 \end{bmatrix}$ at $k = 2$.

---

***Example 5.3.*** The system $x(k + 1) = 0$ is controllable since any state, say, $x(0) = \begin{bmatrix} a \\ b \end{bmatrix}$, can be transferred to the zero state in one step. In this system, however, the input $u$ does not affect the state at all! This example shows that reachability is a more useful concept than controllability for discrete-time systems.

---

It should be pointed out that nothing has been said up to now about maintaining the desired system state after reaching it [refer to (5.5)]. Zeroing the input for $k \geq n$, i.e., letting $u(k) = 0$ for $k \geq n$, will not typically work, unless $Ax_1 = x_1$. In general a state starting at $x_1$, will remain at $x_1$ for all $k \geq n$ if and only if there exists an input $u(k), k \geq n$, such that

$$x_1 = Ax_1 + Bu(k), \tag{5.9}$$

that is, if and only if $(I - A)x_1 \in \mathcal{R}(B)$. Clearly, there are states for which this condition may not be satisfied.

### 5.2.2 Observability and Constructibility

In Section 5.4, definitions for state *observability* and *constructibility* are given, and appropriate tests for these concepts are derived. It is shown that observability always implies constructibility, whereas constructibility implies observability only when the state transition matrix $\Phi$ of the system is nonsingular. Whereas this is always true for continuous-time systems, it is true for discrete-time systems only when the matrix $A$ of the system is nonsingular. If a system is state observable, then its present state can be determined from knowledge of the present and future outputs and inputs. Constructibility refers to the ability of determining the present state from present and past outputs and inputs, and as such, it is of greater interest in applications.

In the time-invariant case a system [or a pair $(A, C)$] is observable if and only if its *observability matrix* $\mathcal{O}$, where

$$\mathcal{O} \triangleq \begin{bmatrix} C \\ CA \\ \vdots \\ CA^{n-1} \end{bmatrix} \in R^{pn \times n}, \tag{5.10}$$

has full column rank; i.e., rank $\mathcal{O} = n$. The matrices $A \in R^{n \times n}$ and $C \in R^{p \times n}$ are given by the system description

$$\dot{x} = Ax + Bu, \quad y = Cx + Du \tag{5.11}$$

in the continuous-time case, and by the system description

$$x(k+1) = Ax(k) + Bu(k), \quad y(k) = Cx(k) + Du(k), \tag{5.12}$$

with $k \geq k_0 = 0$, in the discrete-time case.

We shall now briefly discuss observability and constructibility for the discrete-time time-invariant case. As in the case of reachability and controllability, this discussion will provide insight into the underlying concepts and will clarify what these imply for a system.

If the output in (5.12) is expressed in terms of the initial vector $x(0)$, then

$$y(k) = CA^k x(0) + \sum_{i=0}^{k-1} CA^{k-(i+1)} Bu(i) + Du(k) \qquad (5.13)$$

for $k > 0$ (see Section 3.5). This implies that

$$\tilde{y}(k) = CA^k x_0 \qquad (5.14)$$

for $k \geq 0$, where

$$\tilde{y}(k) \triangleq y(k) - \left[ \sum_{i=0}^{k-1} CA^{k-(i+1)} Bu(i) + Du(k) \right]$$

for $k > 0$, $\tilde{y}(0) \triangleq y(0) - Du(0)$, and $x_0 = x(0)$. In (5.14) $x_0$ is to be determined assuming that the system parameters are given and the inputs and outputs are measured. Note that if $u(k) = 0$ for $k \geq 0$, then the problem is simplified, since $\tilde{y}(k) = y(k)$ and since the output is generated only by the initial condition $x_0$. It is clear that the ability of determining $x_0$ from output and input measurements depends only on the matrices $A$ and $C$, since the left-hand side of (5.14) is a known quantity. Now if $x(0) = x_0$ is known, then all $x(k), k \geq 0$, can be determined by means of (5.12). To determine $x_0$, we apply (5.14) for $k = 0, \ldots, n-1$. Then

$$\tilde{Y}_{0,n-1} = \mathcal{O}_n x_0, \qquad (5.15)$$

where $\mathcal{O}_n \triangleq [C^T, (CA)^T, \ldots, (CA^{n-1})^T]^T = \mathcal{O}$ [as in (5.10)] and

$$\tilde{Y}_{0,n-1} \triangleq [\tilde{y}^T(0), \ldots, \tilde{y}^T(n-1)]^T.$$

Now (5.15) always has a solution $x_0$, by construction. A system is observable if the solution $x_0$ is unique, i.e., if it is the only initial condition that, together with the given input sequence, can generate the observed output sequence. From the theory of linear systems of equations, (5.15) has a unique solution $x_0$ if and only if the null space of $\mathcal{O}$ consists of only the zero vector, i.e., Null$(\mathcal{O}) = \mathcal{N}(\mathcal{O}) = \{0\}$, or equivalently, if and only if the only $x \in R^n$ that satisfies

$$\mathcal{O}x = 0 \qquad (5.16)$$

is the zero vector. This is true if and only if rank $\mathcal{O} = n$. Thus, a system is observable if and only if rank $\mathcal{O} = n$. Any nonzero state vector $x \in R^n$ that satisfies (5.16) is said to be an unobservable state, and $\mathcal{N}(\mathcal{O})$ is said to be the *unobservable subspace*. Note that any such $x$ satisfies $CA^k x = 0$ for $k = 0, 1, \ldots, n-1$. If rank $\mathcal{O} < n$, then all vectors $x_0$ that satisfy (5.15) are given by $x_0 = x_{0p} + x_{0h}$, where $x_{0p}$ is a particular solution and $x_{0h}$ is any vector in $\mathcal{N}(\mathcal{O})$. Any of these state vectors, together with the given inputs, could have generated the measured outputs.

To determine $x_0$ from (5.15) it is not necessary to use more than $n$ values for $\tilde{y}(k), k = 0, \ldots, n - 1$, or to observe $y(k)$ for more than $n$ steps in the future. This is true because, in view of the Cayley–Hamilton Theorem, it can be shown that $\mathcal{N}(\mathcal{O}_n) = \mathcal{N}(\mathcal{O}_k)$ for $k \geq n$. Note also that $\mathcal{N}(\mathcal{O}_n)$ is included in $\mathcal{N}(\mathcal{O}_k)$ ($\mathcal{N}(\mathcal{O}_n) \subset \mathcal{N}(\mathcal{O}_k)$) for $k < n$. Therefore, in general, one has to observe the output for $n$ steps (see Exercise 5.1).

---

**Example 5.4.** Consider the system $x(k + 1) = Ax(k), y(k) = Cx(k)$, where $A = \begin{bmatrix} 0 & 1 \\ 1 & 1 \end{bmatrix}$ and $C = [0\ 1]$. Here, $\mathcal{O} = \begin{bmatrix} C \\ CA \end{bmatrix} = \begin{bmatrix} 0 & 1 \\ 1 & 1 \end{bmatrix}$ with rank $\mathcal{O} = 2$. Therefore, the system [or the pair $(A, C)$] is observable. This means that $x(0)$ can uniquely be determined from $n = 2$ output measurements (in the present cases, the input is zero). In fact, in view of (5.15), $\begin{bmatrix} y(0) \\ y(1) \end{bmatrix} = \begin{bmatrix} 0 & 1 \\ 1 & 1 \end{bmatrix} \begin{bmatrix} x_1(0) \\ x_2(0) \end{bmatrix}$ or $\begin{bmatrix} x_1(0) \\ x_2(0) \end{bmatrix} = \begin{bmatrix} -1 & 1 \\ 1 & 0 \end{bmatrix} \begin{bmatrix} y(0) \\ y(1) \end{bmatrix} = \begin{bmatrix} y(1) - y(0) \\ y(0) \end{bmatrix}$.

---

**Example 5.5.** Consider the system $x(k + 1) = Ax(k), y(k) = Cx(k)$, where $A = \begin{bmatrix} 1 & 0 \\ 1 & 1 \end{bmatrix}$ and $C = [1\ 0]$. Here, $\mathcal{O} = \begin{bmatrix} C \\ CA \end{bmatrix} = \begin{bmatrix} 1 & 0 \\ 1 & 0 \end{bmatrix}$ with rank $\mathcal{O} = 1$. Therefore, the system is not observable. Note that a basis for $\mathcal{N}(\mathcal{O})$ is $\left\{ \begin{bmatrix} 0 \\ 1 \end{bmatrix} \right\}$, which in view of (5.16) implies that all state vectors of the form $\begin{bmatrix} 0 \\ c \end{bmatrix}, c \in R$, are unobservable. Relation (5.15) implies that $\begin{bmatrix} y(0) \\ y(1) \end{bmatrix} = \begin{bmatrix} 1 & 0 \\ 1 & 0 \end{bmatrix} \begin{bmatrix} x_1(0) \\ x_2(0) \end{bmatrix}$. For a solution $x(0)$ to exist, as it must, we have that $y(0) = y(1) = a$. Thus, this system will generate an identical output for $k \geq 0$. Accordingly, all $x(0)$ that satisfy (5.15) and can generate this output are given by $\begin{bmatrix} x_1(0) \\ x_2(0) \end{bmatrix} = \begin{bmatrix} a \\ 0 \end{bmatrix} + \begin{bmatrix} 0 \\ c \end{bmatrix} = \begin{bmatrix} a \\ c \end{bmatrix}$, where $c \in R$.

---

In general, a system (5.12) [or a pair $(A, C)$] is *constructible* if the only vector $x$ that satisfies $x = A^k \hat{x}$ with $C\hat{x} = 0$ for every $k \geq 0$ is the zero vector. When $A$ is nonsingular, this condition can be stated more simply, namely, that the system is constructible if the only vector $x$ that satisfies $CA^{-k}x = 0$ for every $k \geq 0$ is the zero vector. Compare this with the condition $CA^k x = 0, k \geq 0$, for $x$ to be an unobservable state; or with the condition that a system is observable if the only vector $x$ that satisfies $CA^k x = 0$ for every $k \geq 0$ is the zero vector. In view of (5.14), the above condition for a system to be constructible is the condition for the existence of a unique solution $x_0$ when past outputs and inputs are used. This, of course, makes sense since constructibility refers to determining the present state from knowledge of past outputs and inputs. Therefore, when $A$ is nonsingular, the system is constructible if and only if the $pn \times n$ matrix

$$\mathcal{O}A^{-n} = \begin{bmatrix} CA^{-n} \\ \vdots \\ CA^{-1} \end{bmatrix} \tag{5.17}$$

has full rank, since in this case the only $x$ that satisfies $CA^{-k}x = 0$ for every $k \geq 0$ is $x = 0$. Note that if the system is observable, then it is also constructible; however, if it is constructible, then it is also observable only when $A$ is nonsingular (see Section 5.3).

---

**Example 5.6.** Consider the (unobservable) system in Example 5.5. Since $A$ is nonsingular, $\mathcal{O}A^{-2} = \begin{bmatrix} 1 & 0 \\ 1 & 0 \end{bmatrix} \begin{bmatrix} 1 & 0 \\ -2 & 1 \end{bmatrix} = \begin{bmatrix} 1 & 0 \\ 1 & 0 \end{bmatrix}$. Since $\operatorname{rank} \mathcal{O}A^{-2} = 1 < 2$, the system [or the pair $(A, C)$] is not constructible. This can also be seen from the relation $CA^{-k}x = 0$, $k \geq 0$, that has nonzero solutions $x$, since $C = [1, 0] = CA^{-1} = CA^{-2} = \cdots = CA^{-k}$ for $k \geq 0$, which implies that any $x = \begin{bmatrix} 0 \\ c \end{bmatrix}$, $c \in R$, is a solution.

---

### 5.2.3 Dual Systems

Consider the system described by

$$\dot{x} = Ax + Bu, \quad y = Cx + Du, \tag{5.18}$$

where $A \in R^{n \times n}, B \in R^{n \times m}, C \in R^{p \times n}$, and $D \in R^{p \times m}$. The *dual system* of (5.18) is defined as the system

$$\dot{x}_D = A_D x_D + B_D u_D, \quad y_D = C_D x_D + D_D u_D, \tag{5.19}$$

where $A_D = A^T, B_D = C^T, C_D = B^T$, and $D_D = D^T$.

**Lemma 5.7.** *System (5.18), denoted by $\{A, B, C, D\}$, is reachable (controllable) if and only if its dual $\{A_D, B_D, C_D, D_D\}$ in (5.19) is observable (constructible), and vice versa.*

*Proof.* System $\{A, B, C, D\}$ is reachable if and only if $\mathcal{C} \triangleq [B, AB, \dots, A^{n-1}B]$ has full rank $n$, and its dual is observable if and only if

$$\mathcal{O}_D \triangleq \begin{bmatrix} B^T \\ B^T A^T \\ \vdots \\ B^T (A^T)^{n-1} \end{bmatrix}$$

has full rank $n$. Since $\mathcal{O}_D^T = \mathcal{C}$, $\{A, B, C, D\}$ is reachable if and only if $\{A_D, B_D, C_D, D_D\}$ is observable. Similarly, $\{A, B, C, D\}$ is observable if and only if $\{A_D, B_D, C_D, D_D\}$ is reachable. Now $\{A, B, C, D\}$ is controllable if and only if its dual is constructible, and vice versa, since it is shown in Sections 5.3 and 5.4, that a continuous-time system is controllable if and only if it is reachable; it is constructible if and only if it is observable.    ∎

For the discrete-time time-invariant case, the dual system is again defined as $A_D = A^T$, $B_D = C^T$, $C_D = B^T$, and $D_D = D^T$. That such a system is reachable if and only if its dual is observable can be shown in exactly the same way as in the proof of Lemma 5.7. That such a system is controllable if and only if its dual is constructible in the case when $A$ is nonsingular is because in this case the system is reachable if and only if it is controllable; and the same holds for observability and constructibility. The proof for the case when $A$ is singular involves the controllable and unconstructible subspaces of a system and its dual. We omit the details. The reader is encouraged to complete this proof after studying Sections 5.3 and 5.4.

Figure 5.3 summarizes the relationships between reachability (observability) and controllability (constructibility) for continuous- and discrete-time systems.

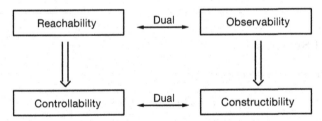

**Figure 5.3.** In continuous-time systems, reachability (observability) always implies and is implied by controllability (constructibility). In discrete-time systems, reachability (observability) always implies but in general is not implied by controllability (constructibility).

## 5.3 Reachability and Controllability

The objective here is to study the important properties of state controllability and reachability when a system is described by a state-space representation. In the previous section, a brief introduction to these concepts was given for discrete-time systems, and it was shown that a system is completely reachable if and only if the controllability (-from-the-origin) matrix $\mathcal{C}$ in (5.1) has full rank $n$ (rank $\mathcal{C} = n$). Furthermore, it was shown that the input sequence necessary to accomplish the transfer can be determined directly from $\mathcal{C}$ by solving a system of linear algebraic equations (5.5). In a similar manner, we would like to derive tests for reachability and controllability and determine the necessary system inputs to accomplish the state transfer for the continuous-time case. We note, however, that whereas the test for reachability can be derived by a number of methods, the appropriate sequence of system inputs to use cannot easily be determined directly from $\mathcal{C}$, as was the case for discrete-time systems. For this reason, we use an approach that utilizes ranges of maps, in particular, the range of an important $n \times n$ matrix—the reachability Gramian. The inputs that accomplish the desired state transfer can be determined directly from this matrix.

### 5.3.1 Continuous-Time Time-Invariant Systems

We consider the state equation

$$\dot{x} = Ax + Bu, \tag{5.20}$$

where $A \in R^{n \times n}, B \in R^{n \times m}$, and $u(t) \in R^m$ is (piecewise) continuous. The state at time $t$ is given by

$$x(t) = \Phi(t, t_0)x(t_0) + \int_{t_0}^{t} \Phi(t, \tau)Bu(\tau)d\tau, \tag{5.21}$$

where $\Phi(t, \tau)$ is the state transition matrix of the system, and $x(t_0) = x_0$ denotes the state at initial time.

Here

$$\Phi(t, \tau) = \Phi(t - \tau, 0) = \exp[(t - \tau)A] = e^{A(t - \tau)}. \tag{5.22}$$

We are interested in using the input to transfer the state from $x_0$ to some other value $x_1$ at some finite time $t_1 > t_0$, [i.e., $x(t_1) = x_1$ in (5.21)]. Because of time invariance, the difference $t_1 - t_0 = T$, rather than the individual times $t_0$ and $t_1$, is important. Accordingly, we can always take $t_0 = 0$ and $t_1 = T$. Equation (5.21) assumes the form

$$x_1 - e^{AT}x_0 = \int_0^T e^{A(T-\tau)}Bu(\tau)d\tau, \tag{5.23}$$

and clearly, there exists $u(t), t \in [0, T]$, that satisfies (5.23) if and only if such transfer of the state is possible. Letting $\hat{x}_1 \triangleq x_1 - e^{AT}x_0$, we note that the $u(t)$ that transfers the state from $x_0$ at time 0 to $x_1$ at time $T$ will also cause the state to reach $\hat{x}_1$ at $T$, starting from the origin at 0 (i.e., $x(0) = 0$).

For the time-invariant system (5.20), we introduce the following concepts.

**Definition 5.8.** (i) A state $x_1$ is reachable if there exists an input $u(t), t \in [0, T]$, that transfers the state $x(t)$ from the origin at $t = 0$ to $x_1$ in some finite time $T$.

*(ii) The set of all reachable states $R_r$ is the reachable subspace of the system $\dot{x} = Ax + Bu$, or of the pair $(A, B)$.*

*(iii) The system $\dot{x} = Ax + Bu$, or the pair $(A, B)$ is (completely state) reachable if every state is reachable, i.e., if $R_r = R^n$.*  ∎

Regarding (ii), note that the set of all reachable states $x_1$ contains the origin and constitutes a linear subspace of the state space $(R^n, R)$.

A reachable state is sometimes also called *controllable-from-the-origin*. Additionally, there are also states defined to be *controllable-to-the-origin* or simply *controllable*; see the definition later in this section.

**Definition 5.9.** *The $n \times n$ reachability Gramian of the time-invariant system $\dot{x} = Ax + Bu$ is*

$$W_r(0, T) \triangleq \int_0^T e^{(T-\tau)A} BB^T e^{(T-\tau)A^T} d\tau. \tag{5.24}$$

∎

Note that $W_r$ is symmetric and positive semidefinite for every $T > 0$; i.e., $W_r = W_r^T$ and $W_r \geq 0$ (show this).

It can now be shown in [1, p. 230, Lemma 3.2.1] that the reachable subspace of the system (5.20) is exactly the range of the reachability Gramian $W_r$ in (5.24). Let the $n \times mn$ *controllability (-from-the-origin) matrix* be

$$\mathcal{C} \triangleq [B, AB, \dots, A^{n-1}B]. \tag{5.25}$$

The range of $W_r(0, T)$, denoted by $\mathcal{R}(W_r(0, T))$, is independent of $T$; i.e., it is the same for any finite $T(> 0)$, and in particular, it is equal to the range of the controllability matrix $\mathcal{C}$. Thus, the reachable subspace $R_r$ of system (5.20), which is the set of all states that can be reached from the origin in finite time, is given by the range of $\mathcal{C}, \mathcal{R}(\mathcal{C})$, or the range of $W_r(0, T), \mathcal{R}(W_r(0, T))$, for some finite (and therefore for any) $T > 0$. This is stated as Lemma 5.10 below; for the proof, see [1, p. 236, Lemma 3.2.10].

**Lemma 5.10.** $\mathcal{R}(W_r(0, T)) = \mathcal{R}(\mathcal{C})$ *for every $T > 0$.*  ∎

---

**Example 5.11.** For the system $\dot{x} = Ax + Bu$ with $A = \begin{bmatrix} 0 & 1 \\ 0 & 0 \end{bmatrix}$ and $B = \begin{bmatrix} 0 \\ 1 \end{bmatrix}$,

we have $e^{At} = \begin{bmatrix} 1 & t \\ 0 & 1 \end{bmatrix}$ and $e^{At}B = \begin{bmatrix} t \\ 1 \end{bmatrix}$. The reachability Gramian is

$W_r(0, T) = \int_0^T \begin{bmatrix} T - \tau \\ 1 \end{bmatrix} [T-\tau, 1] d\tau = \int_0^T \begin{bmatrix} (T-\tau)^2 & T-\tau \\ T-\tau & 1 \end{bmatrix} d\tau = \begin{bmatrix} \frac{1}{3}T^3 & \frac{1}{2}T^2 \\ \frac{1}{2}T^2 & T \end{bmatrix}$.

Since $\det W_r(0, T) = \frac{1}{12}T^4 \neq 0$ for any $T > 0$, $\operatorname{rank} W_r(0, T) = n$ and $(A, B)$

is reachable. Note that $\mathcal{C} = [B, AB] = \begin{bmatrix} 0 & 1 \\ 1 & 0 \end{bmatrix}$ and that $\mathcal{R}(W_r(0, T)) = \mathcal{R}(\mathcal{C}) = R^2$, as expected (Lemma 5.10).

If $B = \begin{bmatrix} 1 \\ 0 \end{bmatrix}$, instead of $\begin{bmatrix} 0 \\ 1 \end{bmatrix}$, then $\mathcal{C} = [B, AB] = \begin{bmatrix} 1 & 0 \\ 0 & 0 \end{bmatrix}$ and $(A, B)$ is not reachable. In this case $e^{At}B = \begin{bmatrix} 1 \\ 0 \end{bmatrix}$ and the reachability matrix is $W_r(0, T) = \int_0^T \begin{bmatrix} 1 & 0 \\ 0 & 0 \end{bmatrix} d\tau = \begin{bmatrix} T & 0 \\ 0 & 0 \end{bmatrix}$. Notice again that $\mathcal{R}(\mathcal{C}) = \mathcal{R}(W_r(0, T))$ for every $T > 0$.

---

The following theorems and corollaries 5.12 to 5.15 contain the main reachability results. Their proofs may be found in [1, p. 237, Chapter 3], starting with Theorem 2.11.

**Theorem 5.12.** *Consider the system $\dot{x} = Ax + Bu$, and let $x(0) = 0$. There exists an input $u$ that transfers the state to $x_1$ in finite time if and only if $x_1 \in \mathcal{R}(\mathcal{C})$, or equivalently, if and only if $x_1 \in \mathcal{R}(W_r(0, T))$ for some finite (and therefore for any) $T$. Thus, the reachable subspace $R_r = \mathcal{R}(\mathcal{C}) = \mathcal{R}(W_r(0, T))$. Furthermore, an appropriate $u$ that will accomplish this transfer in time $T$ is given by*

$$u(t) = B^T e^{A^T(T-t)} \eta_1 \tag{5.26}$$

*with $\eta_1$ such that $W_r(0, T)\eta_1 = x_1$ and $t \in [0, T]$.* ∎

Note that in (5.26) no restrictions are imposed on time $T$, other than $T$ be finite. $T$ can be as small as we wish; i.e., the transfer can be accomplished in a very short time indeed.

**Corollary 5.13.** *The system $\dot{x} = Ax + Bu$, or the pair $(A, B)$, is (completely state) reachable, if and only if*

$$\mathrm{rank}\, \mathcal{C} = n, \tag{5.27}$$

*or equivalently, if and only if*

$$\mathrm{rank}\, W_r(0, T) = n \tag{5.28}$$

*for some finite (and therefore for any) $T$.* ∎

**Theorem 5.14.** *There exists an input $u$ that transfers the state of the system $\dot{x} = Ax + Bu$ from $x_0$ to $x_1$ in some finite time $T$ if and only if*

$$x_1 - e^{AT}x_0 \in \mathcal{R}(\mathcal{C}), \tag{5.29}$$

*or equivalently, if and only if*

$$x_1 - e^{AT}x_0 \in \mathcal{R}(W_r(0, T)). \tag{5.30}$$

*Such an input is given by*

$$u(t) = B^T e^{A^T (T-t)} \eta_1 \tag{5.31}$$

with $t \in [0, T]$, where $\eta_1$ is a solution of

$$W_r(0, T)\eta_1 = x_1 - e^{AT} x_0. \tag{5.32}$$

∎

The above theorem leads to the next result, which establishes the importance of reachability in determining an input $u$ to transfer the state from any $x_0$ to any $x_1$ in finite time.

**Corollary 5.15.** *Let the system $\dot{x} = Ax + Bu$ be (completely state) reachable, or the pair $(A, B)$ be reachable. Then there exists an input that will transfer any state $x_0$ to any other state $x_1$ in some finite time $T$. Such input is given by*

$$u(t) = B^T e^{A^T (T-t)} W_r^{-1}(0, T)[x_1 - e^{AT} x_0] \tag{5.33}$$

*for $t \in [0, T]$.*

∎

There are many different control inputs $u$ that can accomplish the transfer from $x_0$ to $x_1$ in time $T$. It can be shown that the input $u$ given by (5.33) accomplishes this transfer while expending a minimum amount of energy; in fact, $u$ minimizes the cost functional $\int_0^T \| u(\tau) \|^2 d\tau$, where $\| u(t) \| \triangleq [u^T(t)u(t)]^{1/2}$ denotes the Euclidean norm of $u(t)$.

---

**Example 5.16.** The system $\dot{x} = Ax + Bu$ with $A = \begin{bmatrix} 0 & 1 \\ 0 & 0 \end{bmatrix}$ and $B = \begin{bmatrix} 0 \\ 1 \end{bmatrix}$ is reachable (see Example 5.11). A control input $u(t)$ that will transfer any state $x_0$ to any other state $x_1$ in some finite time $T$ is given by (see Corollary 5.15 and Example 5.11)

$$u(t) = B^T e^{A^T (T-t)} W_r^{-1}(0, T)[x_1 - e^{AT} x_0]$$
$$= [T - t, 1] \begin{bmatrix} 12/T^3 & -6/T^2 \\ -6/T^2 & 4/T \end{bmatrix} \begin{bmatrix} x_1 - \begin{bmatrix} 1 & T \\ 0 & 1 \end{bmatrix} x_0 \end{bmatrix}.$$

---

**Example 5.17.** For the (scalar) system $\dot{x} = -ax + bu$, determine $u(t)$ that will transfer the state from $x(0) = x_0$ to the origin in $T$ sec; i.e., $x(T) = 0$.

We shall apply Corollary 5.15. The reachability Gramian is $W_r(0, T) = \int_0^T e^{-(T-\tau)a} bb e^{-(T-\tau)a} d\tau = e^{-2aT} b^2 \int_0^T e^{2a\tau} d\tau = e^{-2aT} b^2 \frac{1}{2a}[e^{2aT} - 1] = \frac{b^2}{2a}[1 - e^{-2aT}]$. (Note [see (5.36) below] that the controllability Gramian is $W_c(0, T) = \frac{b^2}{2a}[e^{2aT} - 1]$.) Now in view of (5.33), we have

$$u(t) = b e^{-(T-t)a} \frac{2a}{b^2} \frac{1}{1 - e^{-2aT}}[-e^{-aT} x_0]$$

$$= -\frac{2a}{b} \frac{e^{-2aT}}{1 - e^{-2aT}} e^{aT} x_0 = -\frac{2a}{b} \frac{1}{e^{2aT} - 1} e^{at} x_0.$$

To verify that this $u(t)$ accomplishes the desired transfer, we compute $x(t) = e^{At}x_0 + \int_0^t e^{A(t-\tau)}Bu(\tau)d\tau = e^{-at}x_0 + \int_0^t e^{-at}e^{a\tau}bu(\tau)d\tau = e^{-at}[x_0 + \int_0^t e^{a\tau}b \times \left(-\frac{2a}{b}\frac{1}{e^{2aT}-1} \times e^{a\tau}\right)d\tau = e^{-at}\left[1 - \frac{e^{2at}-1}{e^{2aT}-1}\right]x_0$. Note that $x(T) = 0$, as desired, and also that $x(0) = x_0$. The above expression shows also that for $t > T$, the state does not remain at the origin. An important point to notice here is that as $T \to 0$, the control magnitude $|u| \to \infty$. Thus, although it is (theoretically) possible to accomplish the desired transfer instantaneously, this will require infinite control magnitude. In general the faster the transfer, the larger the control magnitude required.

---

We now introduce the concept of a controllable state.

**Definition 5.18.** *(i) A state $x_0$ is controllable if there exists an input $u(t)$, $t \in [0, T]$, which transfers the state $x(t)$ from $x_0$ at $t = 0$ to the origin in some finite time $T$.*
*(ii) The set of all controllable states $R_c$, is the controllable subspace of the system $\dot{x} = Ax + Bu$, or of the pair $(A, B)$.*
*(iii) The system $\dot{x} = Ax + Bu$, or the pair $(A, B)$, is (completely state) controllable if every state is controllable, i.e., if $R_c = R^n$.* ∎

We shall now establish the relationship between reachability and controllability for the continuous-time time-invariant systems (5.20).

In view of (5.23), $x_0$ is controllable when there exists $u(t), t \in [0, T]$, so that

$$-e^{AT}x_0 = \int_0^T e^{A(T-\tau)}Bu(\tau)d\tau$$

or when $e^{AT}x_0 \in \mathcal{R}(W_r(0,T))$ [1, p. 230, Lemma 3.2.1], or equivalently, in view of Lemma 5.10, when

$$e^{AT}x_0 \in \mathcal{R}(\mathcal{C}) \tag{5.34}$$

for some finite $T$. Recall that $x_1$ is reachable when

$$x_1 \in \mathcal{R}(\mathcal{C}). \tag{5.35}$$

We require the following result.

**Lemma 5.19.** *If $x \in \mathcal{R}(\mathcal{C})$, then $Ax \in \mathcal{R}(\mathcal{C})$; i.e., the reachable subspace $R_r = \mathcal{R}(\mathcal{C})$ is an $A$-invariant subspace.*

*Proof.* If $x \in \mathcal{R}(\mathcal{C})$, this means that there exists a vector $\alpha$ such that $[B, AB, \ldots, A^{n-1}B]\alpha = x$. Then $Ax = [AB, A^2B, \ldots, A^nB]\alpha$. In view of the Cayley–Hamilton Theorem, $A^n$ can be expressed as a linear combination of $A^{n-1}, \ldots, A, I$, which implies that $Ax = \mathcal{C}\beta$ for some appropriate vector $\beta$. Therefore, $Ax \in \mathcal{R}(\mathcal{C})$. ∎

**Theorem 5.20.** *Consider the system* $\dot{x} = Ax + Bu$.

*(i)   A state* $x$ *is reachable if and only if it is controllable.*
*(ii)  $R_c = R_r$.*
*(iii) The system (2.3), or the pair $(A, B)$, is (completely state) reachable if and only if it is (completely state) controllable.*

*Proof.* (i) Let $x$ be reachable, that is, $x \in \mathcal{R}(\mathcal{C})$. Premultiply $x$ by $e^{AT} = \sum_{k=0}^{\infty}(T^k/k!)A^k$ and notice that, in view of Lemma 5.19, $Ax, A^2x, \ldots, A^kx \in \mathcal{R}(\mathcal{C})$. Therefore, $e^{AT}x \in \mathcal{R}(\mathcal{C})$ for any $T$ that, in view of (5.34), implies that $x$ is also controllable. If now $x$ is controllable, i.e., $e^{AT}x \in \mathcal{R}(\mathcal{C})$, then premultiplying by $e^{-AT}$, the vector $e^{-AT}\left(e^{AT}x\right) = x$ will also be in $\mathcal{R}(\mathcal{C})$. Therefore, $x$ is also reachable. Note that the second part of (i), that controllability implies reachability, is true because the inverse $(e^{AT})^{-1} = e^{-AT}$ does exist. This is in contrast to the discrete-time case where the state transition matrix $\Phi(k, 0)$ is nonsingular if and only if $A$ is nonsingular [nonreversibility of time in discrete-time systems].

Parts (ii) and (iii) of the theorem follow directly from (i).    ∎

The reachability Gramian for the time-invariant case, $W_r(0, T)$, was defined in (5.24). For completeness the controllability Gramian is defined below.

**Definition 5.21.** *The controllability Gramian in the time-invariant case is the $n \times n$ matrix*

$$W_c(0, T) \triangleq \int_0^T e^{-A\tau}BB^T e^{-A^T\tau}d\tau. \tag{5.36}$$

■

We note that

$$W_r(0, T) = e^{AT}W_c(0, T)e^{A^T T},$$

which can be verified directly.

## Additional Criteria for Reachability and Controllability

We first recall the definition of a set of linearly independent functions of time and consider in particular $n$ complex-valued functions $f_i(t)$, $i = 1, \ldots, n$, where $f_i^T(t) \in C^m$. Recall that the set of functions $f_i$, $i = 1, \ldots, n$, is *linearly dependent* on a time interval $[t_1, t_2]$ over the field of complex numbers $C$ if there exist complex numbers $a_i$, $i = 1, \ldots, n$, not all zero, such that

$$a_1 f_1(t) + \cdots + a_n f_n(t) = 0 \quad \text{for all } t \text{ in } [t_1, t_2];$$

otherwise, the set of functions is said to be *linearly independent* on $[t_1, t_2]$ over the field of complex numbers.

It is possible to test linear independence using the *Gram matrix of the functions $f_i$.*

**Lemma 5.22.** *Let $F(t) \in C^{n \times m}$ be a matrix with $f_i(t) \in C^{1 \times m}$ in its ith row. Define the* Gram matrix *of $f_i(t)$, $i = 1, \ldots, n$, by*

$$W(t_1, t_2) \triangleq \int_{t_1}^{t_2} F(t)F^*(t)dt, \tag{5.37}$$

*where $(\cdot)^*$ denotes the complex conjugate transpose. The set $f_i(t)$, $i = 1, \ldots, n$, is linearly independent on $[t_1, t_2]$ over the field of complex numbers if and only if the Gram matrix $W(t_1, t_2)$ is nonsingular, or equivalently, if and only if the Gram determinant $\det W(t_1, t_2) \neq 0$.*

*Proof.* (*Necessity*) Assume the set $f_i, i = 1, \ldots, n$, is linearly independent but $W(t_1, t_2)$ is singular. Then there exists some nonzero $\alpha \in C^{1 \times n}$ so that $\alpha W(t_1, t_2) = 0$, from which $\alpha W(t_1, t_2)\alpha^* = \int_{t_1}^{t_2} (\alpha F(t))(\alpha F(t))^* dt = 0$. Since $(\alpha F(t))(\alpha F(t))^* \geq 0$ for all $t$, this implies that $\alpha F(t) = 0$ for all $t$ in $[t_1, t_2]$, which is a contradiction. Therefore, $W(t_1, t_2)$ is nonsingular.

(*Sufficiency*) Assume that $W(t_1, t_2)$ is nonsingular but the set $f_i, i = 1, \ldots, n$, is linearly dependent. Then there exists some nonzero $\alpha \in C^{1 \times n}$ so that $\alpha F(t) = 0$. Then $\alpha W(t_1, t_2) = \int_{t_1}^{t_2} \alpha F(t)F^*(t)dt = 0$, which is a contradiction. Therefore, the set $f_i$, $i = 1, \ldots, n$, is linearly independent. ∎

We now introduce a number of additional tests for reachability and controllability of time-invariant systems. Some earlier results are also repeated here for convenience.

**Theorem 5.23.** *The system $\dot{x} = Ax + Bu$ is reachable (controllable-from-the-origin)*

(i)   *if and only if*

$$\mathrm{rank}\, W_r(0, T) = n \quad \text{for some finite } T > 0,$$

where

$$W_r(0, T) \triangleq \int_0^T e^{(T-\tau)A} BB^T e^{(T-\tau)A^T} d\tau, \tag{5.38}$$

*the reachability Gramian; or*

(ii)  *if and only if the n rows of*

$$e^{At}B \tag{5.39}$$

*are linearly independent on $[0, \infty)$ over the field of complex numbers; or alternatively, if and only if the n rows of*

$$(sI - A)^{-1}B \tag{5.40}$$

*are linearly independent over the field of complex numbers; or*

(iii) *if and only if*

$$\mathrm{rank}\, \mathcal{C} = n, \tag{5.41}$$

*where $\mathcal{C} \triangleq [B, A, B, \ldots, A^{n-1}B]$, the controllability matrix; or*

*(iv) if and only if*

$$\text{rank}[s_i I - A, B] = n \tag{5.42}$$

*for all complex numbers $s_i$; or alternatively, for $s_i$, $i = 1, \ldots, n$, the eigenvalues of $A$.*

*Proof.* Parts (i) and (ii) were proved in Corollary 5.13.

In part (ii), rank $W_r(0, T) = n$ implies and is implied by the linear independence of the $n$ rows of $e^{(T-t)A}B$ on $[0, T]$ over the field of complex numbers, in view of Lemma 5.22, or by the linear independence of the $n$ rows of $e^{\hat{t}A}B$, where $\hat{t} \triangleq T - t$, on $[0, T]$. Therefore, the system is reachable if and only if the $n$ rows of $e^{At}B$ are linearly independent on $[0, \infty)$ over the field of complex numbers. Note that the time interval can be taken to be $[0, \infty)$ since in $[0, T]$, $T$ can be taken to be any finite positive real number. To prove the second part of (ii), recall that $\mathcal{L}(e^{At}B) = (sI - A)^{-1}B$ and that the Laplace transform is a one-to-one linear operator.

Part (iv) will be proved later in Section 6.3.    ∎

Since reachability implies and is implied by controllability, the criteria developed in the theorem for reachability are typically used to test the controllability of a system as well.

---

**Example 5.24.** For the system $\dot{x} = Ax + Bu$, where $A = \begin{bmatrix} 0 & 1 \\ 0 & 0 \end{bmatrix}$ and $B = \begin{bmatrix} 0 \\ 1 \end{bmatrix}$ (as in Example 5.11), we shall verify Theorem 5.23. The system is reachable since

(i)   the reachability Gramian $W_r(0, T) = \begin{bmatrix} \frac{1}{3}T^3 & \frac{1}{2}T^2 \\ \frac{1}{2}T^2 & T \end{bmatrix}$ has rank $W_r(0, T) = 2 = n$ for any $T > 0$, or since

(ii)  $e^{At}B = \begin{bmatrix} t \\ 1 \end{bmatrix}$ has rows that are linearly independent on $[0, \infty)$ over the field of complex numbers (since $a_1 \times t + a_2 \times 1 = 0$, where $a_1$ and $a_2$ are complex numbers implies that $a_1 = a_2 = 0$). Similarly, the rows of $(sI - A)^{-1}B = \begin{bmatrix} 1/s^2 \\ 1/s \end{bmatrix}$ are linearly independent over the field of complex numbers. Also, since

(iii) $\text{rank}\,\mathcal{C} = \text{rank}[B, AB] = \text{rank}\begin{bmatrix} 0 & 1 \\ 1 & 0 \end{bmatrix} = 2 = n$, or

(iv)  $\text{rank}[s_i I - A, B] = \text{rank}\begin{bmatrix} s_i & -1 & 0 \\ 0 & s_i & 1 \end{bmatrix} = 2 = n$ for $s_i = 0$, $i = 1, 2$, the eigenvalues of $A$.

If $B = \begin{bmatrix} 1 \\ 0 \end{bmatrix}$ in place of $\begin{bmatrix} 0 \\ 1 \end{bmatrix}$, then

(i)  $W_r(0,T) = \begin{bmatrix} T & 0 \\ 0 & 0 \end{bmatrix}$ (see Example 5.11) with rank $W_r(0,T) = 1 < 2 = n$, and

(ii)  $e^{At}B = \begin{bmatrix} 1 \\ 0 \end{bmatrix}$ and $(sI - A)^{-1}B = \begin{bmatrix} 1/s \\ 0 \end{bmatrix}$, neither of which has rows that are linearly independent over the complex numbers. Also,

(iii)  rank $\mathcal{C} = \begin{bmatrix} 1 & 0 \\ 0 & 0 \end{bmatrix} = 1 < 2 = n$, and

(iv)  rank$[s_i I - A, B] = $ rank $\begin{bmatrix} s_i & -1 & 1 \\ 0 & s_i & 0 \end{bmatrix} = 1 < 2 = n$ for $s_i = 0$.

Based on any of the above tests, it is concluded that the system is not reachable.

---

### 5.3.2 Discrete-Time Systems

The response of discrete-time systems was studied in Section 3.5. We consider systems described by equations of the form

$$x(k+1) = Ax(k) + Bu(k), \quad k \geq k_0, \tag{5.43}$$

where $A \in R^{n \times n}$ and $B \in R^{n \times m}$. The state $x(k)$ is given by

$$x(k) = \Phi(k, k_0)x(k_0) + \sum_{i=k_0}^{k-1} \Phi(k, i+1)Bu(i), \tag{5.44}$$

where the state transition matrix is

$$\Phi(k, k_0) = A^{k-k_0}, \quad k \geq k_0. \tag{5.45}$$

Let the state at time $k_0$ be $x_0$. For the state at some time $k_1 > k_0$ to assume the value $x_1$, an input $u$ must exist that satisfies $x(k_1) = x_1$ in (5.44).

For the time-invariant system the elapsed time $k_1 - k_0$ is of interest, and we therefore take $k_0 = 0$ and $k_1 = K$. Recalling that $\Phi(k, 0) = A^k$, for the state $x_1$ to be reached from $x(0) = x_0$ in $K$ steps, i.e., $x(K) = x_1$, an input $u$ must exist that satisfies

$$x_1 = A^K x_0 + \sum_{i=0}^{K-1} A^{K-(i+1)} Bu(i), \tag{5.46}$$

when $K > 0$, or

$$x_1 = A^K x_0 + \mathcal{C}_K U_K, \tag{5.47}$$

where

$$\mathcal{C}_K \triangleq [B, AB, \ldots, A^{K-1}B] \tag{5.48}$$

and

$$U_K \triangleq [u^T(K-1), u^T(K-2), \ldots, u^T(0)]^T. \tag{5.49}$$

The definitions of *reachable state* $x_1$, *reachable subspace* $R_r$, and a *system being (completely state) reachable*, or *the pair (A,B) being reachable*, are the same as in the continuous-time case (see Definition 5.8, and use integer $K$ in place of real time $T$).

To determine the finite input sequence for discrete-time systems that will accomplish a desired state transfer, if such a sequence exists, one does not have to define matrices comparable with the reachability Gramian $W_r$, as in the case for continuous-time systems, but we can work directly with the controllability matrix $\mathcal{C}_n = \mathcal{C}$; see also the introductory discussion in Section 5.2.1. In particular, we have the following result.

**Theorem 5.25.** *Consider the system* $x(k+1) = Ax(k) + Bu(k)$ *given in (5.43), and let* $x(0) = 0$. *There exists an input* $u$ *that transfers the state to* $x_1$ *in finite time if and only if*

$$x_1 \in \mathcal{R}(\mathcal{C}).$$

*In this case,* $x_1$ *is reachable and* $R_r = \mathcal{R}(\mathcal{C})$. *An appropriate input sequence* $\{u(k)\}$, $k = 0, \ldots, n-1$, *that accomplishes this transfer in* $n$ *steps is determined by* $U_n \triangleq [u^T(n-1), u^T(n-2), \ldots, u^T(0)]^T$, *which is a solution to the equation*

$$\mathcal{C}U_n = x_1. \tag{5.50}$$

*Henceforth, with an abuse of language, we will refer to* $U_n$ *as a control sequence, when in fact we actually have in mind* $\{u(k)\}$.

*Proof.* In view of (5.47), $x_1$ can be reached from the origin in $K$ steps if and only if $x_1 = \mathcal{C}_K U_K$ has a solution $U_K$, or if and only if $x_1 \in \mathcal{R}(\mathcal{C}_K)$. Furthermore, all input sequences that accomplish this are solutions to the equation $x_1 = \mathcal{C}_K U_K$. For $x_1$ to be reachable we must have $x_1 \in \mathcal{R}(\mathcal{C}_K)$ for some finite $K$. This range, however, cannot increase beyond the range of $\mathcal{C}_n = \mathcal{C}$; i.e., $\mathcal{R}(\mathcal{C}_K) = \mathcal{R}(\mathcal{C}_n)$ for $K \geq n$ [see Exercise 5.1]. This follows from the Cayley–Hamilton Theorem, which implies that any vector $x$ in $\mathcal{R}(\mathcal{C}_K)$, $K \geq n$, can be expressed as a linear combination of $B, AB, \ldots, A^{n-1}B$. Therefore, $x \in \mathcal{R}(\mathcal{C}_n)$. It is of course possible to have $x_1 \in \mathcal{R}(\mathcal{C}_K)$ with $K < n$, for a particular $x_1$; however, in this case $x_1 \in \mathcal{R}(\mathcal{C}_n)$, since $\mathcal{C}_K$ is a subset of $\mathcal{C}_n$. Thus, $x_1$ is reachable if and only if it is in the range of $\mathcal{C}_n = \mathcal{C}$. Clearly, any $U_n$ that accomplishes the transfer satisfies (5.50). ∎

As pointed out in the above proof, for given $x_1$ we may have $x_1 \in \mathcal{R}(\mathcal{C}_K)$ for some $K < n$. In this case the transfer can be accomplished in fewer than $n$ steps, and appropriate inputs are obtained by solving the equation $\mathcal{C}_K U_K = x_1$.

**Corollary 5.26.** *The system* $x(k + 1) = Ax(k) + Bu(k)$ *in (5.43) is (completely state) reachable, or the pair* $(A, B)$ *is reachable, if and only if*

$$\text{rank}\, \mathcal{C} = n. \tag{5.51}$$

*Proof.* Apply Theorem 5.25, noting that $\mathcal{R}(\mathcal{C}) = R_r = R^n$ if and only if $\text{rank}\, \mathcal{C} = n$. ∎

**Theorem 5.27.** *There exists an input* $u$ *that transfers the state of the system* $x(k + 1) = Ax(k) + Bu(k)$ *in (5.43) from* $x_0$ *to* $x_1$ *in some finite number of steps* $K$, *if and only if*

$$x_1 - A^K x_0 \in \mathcal{R}(\mathcal{C}_K). \tag{5.52}$$

*Such an input sequence* $U_K \triangleq [u^T(K - 1), u^T(K - 2), \dots, u^T(0)]^T$ *is determined by solving the equation*

$$\mathcal{C}_K U_K = x_1 - A^K x_0. \tag{5.53}$$

*Proof.* The proof follows directly from (5.47). ∎

The above theorem leads to the following result that establishes the importance of reachability in determining $u$ to transfer the state from any $x_0$ to any $x_1$ in a finite number of steps.

**Corollary 5.28.** *Let the system* $x(k+1) = Ax(k) + Bu(k)$ *given in (5.43) be (completely state) reachable or the pair* $(A, B)$ *be reachable. Then there exists an input sequence that transfers the state from any* $x_0$ *to any* $x_1$ *in a finite number of steps. Such input is determined by solving Eq. (5.54).*

*Proof.* Consider (5.47). Since $(A, B)$ is reachable, $\text{rank}\, \mathcal{C}_n = \text{rank}\, \mathcal{C} = n$ and $\mathcal{R}(\mathcal{C}) = R^n$. Then

$$\mathcal{C} U_n = x_1 - A^n x_0 \tag{5.54}$$

always has a solution $U_n = [u^T(n - 1), \dots, u^T(0)]^T$ for any $x_0$ and $x_1$. This input sequence transfers the state from $x_0$ to $x_1$ in $n$ steps. ∎

Note that, in view of Theorem 5.27, for particular $x_0$ and $x_1$, the state transfer may be accomplished in $K < n$ steps, using (5.53).

---

*Example 5.29.* Consider the system in Example 5.1, namely, $x(k + 1) = Ax(k) + Bu(k)$, where $A = \begin{bmatrix} 0 & 1 \\ 1 & 1 \end{bmatrix}$ and $B = \begin{bmatrix} 0 \\ 1 \end{bmatrix}$. Since $\text{rank}\, \mathcal{C} = \text{rank}[B, AB] = \text{rank} \begin{bmatrix} 0 & 1 \\ 1 & 1 \end{bmatrix} = 2 = n$, the system is reachable and any state $x_0$ can be transferred to any other state $x_1$ in two steps. Let $x_1 = \begin{bmatrix} a \\ b \end{bmatrix}$, $x_0 = \begin{bmatrix} a_0 \\ b_0 \end{bmatrix}$.

Then (5.54) implies that $\begin{bmatrix} 0 & 1 \\ 1 & 1 \end{bmatrix} \begin{bmatrix} u(1) \\ u(0) \end{bmatrix} = \begin{bmatrix} a \\ b \end{bmatrix} - \begin{bmatrix} 1 & 1 \\ 1 & 2 \end{bmatrix} \begin{bmatrix} a_0 \\ b_0 \end{bmatrix}$ or $\begin{bmatrix} u(1) \\ u(0) \end{bmatrix} = \begin{bmatrix} -1 & 1 \\ 1 & 0 \end{bmatrix} \begin{bmatrix} a \\ b \end{bmatrix} - \begin{bmatrix} 0 & 1 \\ 1 & 1 \end{bmatrix} \begin{bmatrix} a_0 \\ b_0 \end{bmatrix} = \begin{bmatrix} b - 1 - b_0 \\ a - a_0 - b_0 \end{bmatrix}$. This agrees with the results obtained in Example 5.1. In view of (5.53), if $x_1$ and $x_0$ are chosen so that

$$x_1 - Ax_0 = \begin{bmatrix} a \\ b \end{bmatrix} - \begin{bmatrix} 0 & 1 \\ 1 & 1 \end{bmatrix} \begin{bmatrix} a_0 \\ b_0 \end{bmatrix} = \begin{bmatrix} a - b_0 \\ b - a_0 - b_0 \end{bmatrix} \text{ is in the } \mathcal{R}(\mathcal{C}_1) = \mathcal{R}(B) =$$

span $\left\{ \begin{bmatrix} 0 \\ 1 \end{bmatrix} \right\}$, then the state transfer can be achieved in one step. For exam-

ple, if $x_1 = \begin{bmatrix} 1 \\ 3 \end{bmatrix}$ and $x_0 = \begin{bmatrix} 0 \\ 1 \end{bmatrix}$, then $Bu(0) = \begin{bmatrix} 0 \\ 1 \end{bmatrix} u(0) = x_1 - Ax_0 = \begin{bmatrix} 0 \\ 2 \end{bmatrix}$

implies that the transfer from $x_0$ to $x_1$ can be accomplished in this case in $1 < 2 = n$ steps with $u(0) = 2$.

---

**Example 5.30.** Consider the system $x(k + 1) = Ax(k) + Bu(k)$ with $A = \begin{bmatrix} 0 & 1 \\ 0 & 0 \end{bmatrix}$ and $B = \begin{bmatrix} 0 \\ 1 \end{bmatrix}$. Since $\mathcal{C} = [B, AB] = \begin{bmatrix} 0 & 1 \\ 1 & 0 \end{bmatrix}$ has full rank, there exists an input sequence that will transfer the state from any $x(0) = x_0$ to any $x(n) = x_1$ (in $n$ steps), given by (5.54), $U_2 = \begin{bmatrix} u(1) \\ u(0) \end{bmatrix} = \mathcal{C}^{-1}(x_1 - A^2 x_0) = \begin{bmatrix} 0 & 1 \\ 1 & 0 \end{bmatrix} (x_1 - x_0)$. Compare this with Example 5.16, where the continuous-time system had the same system parameters $A$ and $B$.

---

*Additional Criteria for Reachability.* Note that completely analogous results to Theorem 5.23(ii)–(iv) exist for the discrete-time case.

We now turn to the concept of controllability. The definitions of *controllable state* $x_0$, *controllable subspace* $R_c$, and a *system* being *(completely state) controllable*, or *the pair (A,B) being controllable* are similar to the corresponding concepts given in Definition 5.18 for the case of continuous-time systems.

We shall now establish the relationship between reachability and controllability for the discrete-time time-invariant systems $x(k+1) = Ax(k) + Bu(k)$ in (5.43).

Consider (5.46). The state $x_0$ is controllable if it can be steered to the origin $x_1 = 0$ in a finite number of steps $K$. That is, $x_0$ is controllable if and only if

$$-A^K x_0 = \mathcal{C}_K U_K \tag{5.55}$$

for some finite positive integer $K$, or when

$$A^K x_0 \in \mathcal{R}(\mathcal{C}_K) \tag{5.56}$$

for some $K$. Recall that $x_1$ is reachable when

$$x_1 \in \mathcal{R}(\mathcal{C}). \tag{5.57}$$

**Theorem 5.31.** *Consider the system* $x(k + 1) = Ax(k) + Bu(k)$ *in (5.43).*

*(i)  If state $x$ is reachable, then it is controllable.*

*(ii)* $R_r \subset R_c$.

*(iii) If the system is (completely state) reachable, or the pair $(A, B)$ is reachable, then the system is also (completely state) controllable, or the pair $(A, B)$ is controllable.*

Furthermore, if $A$ is nonsingular, then relations *(i)* and *(iii)* become if and only if statements, since controllability also implies reachability, and relation *(ii)* becomes an equality; i.e., $R_c = R_r$.

*Proof.* (i) If $x$ is reachable, then $x \in \mathcal{R}(\mathcal{C})$. In view of Lemma 5.19, $\mathcal{R}(\mathcal{C})$ is an $A$-invariant subspace and so $A^n x \in \mathcal{R}(\mathcal{C})$, which in view of (5.56), implies that $x$ is also controllable. Since $x$ is an arbitrary vector in $R_r$, this implies (ii). If $\mathcal{R}(\mathcal{C}) = R^n$, the whole state space, then $A^n x$ for any $x$ is in $\mathcal{R}(\mathcal{C})$ and so any vector $x$ is also controllable. Thus, reachability implies controllability. Now, if $A$ is nonsingular, then $A^{-n}$ exists. If $x$ is controllable, i.e., $A^n x \in \mathcal{R}(\mathcal{C})$, then $x \in \mathcal{R}(\mathcal{C})$, i.e., $x$ is also reachable. This can be seen by noting that $A^{-n}$ can be written as a power series in terms of $A$, which in view of Lemma 5.19, implies that $A^{-n}(A^n x) = x$ is also in $\mathcal{R}(\mathcal{C})$. ∎

Matrix $A$ being nonsingular is the necessary and sufficient condition for the state transition matrix $\Phi(k, k_0)$ to be nonsingular, which in turn is the condition for *time reversibility* in discrete-time systems. Recall that reversibility in time may not be present in such systems since $\Phi(k, k_0)$ may be singular. In contrast to this, in continuous-time systems, $\Phi(t, t_0)$ is always nonsingular. This causes differences in behavior between continuous- and discrete-time systems and implies that in discrete-time systems controllability may not imply reachability (see Theorem 5.31). Note that, in view of Theorem 5.20, in the case of continuous-time systems, it is not only reachability that always implies controllability, but also vice versa, controllability always implies reachability.

When $A$ is nonsingular, the input that will transfer the state from $x_0$ at $k = 0$ to $x_1 = 0$ in $n$ steps can be determined using (5.54). In particular, one needs to solve

$$[A^{-n}\mathcal{C}]U_n = [A^{-n}B, \ldots, A^{-1}B]U_n = -x_0 \tag{5.58}$$

for $U_n = [u^T(n-1), \ldots, u^T(0)]^T$. Note that $x_0$ is controllable if and only if $-A^n x_0 \in \mathcal{R}(\mathcal{C})$, or if and only if $x_0 \in \mathcal{R}(A^{-n}\mathcal{C})$ for $A$ nonsingular.

Clearly, in the case of controllability (and under the assumption that $A$ is nonsingular), the matrix $A^{-n}\mathcal{C}$ is of interest, instead of $\mathcal{C}$ [see also (5.8)]. In particular, a system is controllable if and only if $\text{rank}(A^{-n}\mathcal{C}) = \text{rank}\,\mathcal{C} = n$.

---

**Example 5.32.** Consider the system $x(k+1) = Ax(k) + Bu(k)$, where $A = \begin{bmatrix} 1 & 1 \\ 0 & 1 \end{bmatrix}$ and $B = \begin{bmatrix} 1 \\ 0 \end{bmatrix}$. Since $\text{rank}\,\mathcal{C} = \text{rank}[B, AB] = \text{rank} \begin{bmatrix} 1 & 1 \\ 0 & 0 \end{bmatrix} = 1 < 2 = n$, this system is not (completely) reachable (controllable-from-the-origin). All reachable states are of the form $\alpha \begin{bmatrix} 1 \\ 0 \end{bmatrix}$, where $\alpha \in R$ since $\left\{ \begin{bmatrix} 1 \\ 0 \end{bmatrix} \right\}$ is a basis for the $\mathcal{R}(\mathcal{C}) = R_r$, the reachability subspace.

In view of (5.56) and the Cayley–Hamilton Theorem, all controllable states $x_0$ satisfy $A^2 x_0 \in \mathcal{R}(\mathcal{C})$; i.e., all controllable states are of the form $\alpha \begin{bmatrix} 1 \\ 0 \end{bmatrix}$, where $\alpha \in R$. This verifies Theorem 5.31 for the case when $A$ is nonsingular. Note that presently $R_r = R_c$.

---

***Example 5.33.*** Consider the system $x(k+1) = Ax(k) + Bu(k)$, where $A = \begin{bmatrix} 0 & 1 \\ 0 & 0 \end{bmatrix}$ and $B = \begin{bmatrix} 1 \\ 0 \end{bmatrix}$. Since $\operatorname{rank}\mathcal{C} = \operatorname{rank}[B, AB] = \operatorname{rank} \begin{bmatrix} 1 & 0 \\ 0 & 0 \end{bmatrix} = 1 < 2 = n$, the system is not (completely) reachable. All reachable states are of the form $\alpha \begin{bmatrix} 1 \\ 0 \end{bmatrix}$, where $\alpha \in R$ since $\left\{ \begin{bmatrix} 1 \\ 0 \end{bmatrix} \right\}$ is a basis for $\mathcal{R}(\mathcal{C}) = R_r$, the reachability subspace.

To determine the controllable subspace $R_c$, consider (5.56) for $K = n$, in view of the Cayley–Hamilton Theorem. Note that $A^{-1}\mathcal{C}$ cannot be used in the present case, since $A$ is singular. Since $A^2 x_0 = \begin{bmatrix} 0 & 0 \\ 0 & 0 \end{bmatrix} x_0 = \begin{bmatrix} 0 \\ 0 \end{bmatrix} \in \mathcal{R}(\mathcal{C})$, any state $x_0$ will be a controllable state; i.e., the system is (completely) controllable and $R_c = R^n$. This verifies Theorem 5.31 and illustrates that controllability does not in general imply reachability.

Note that (5.54) can be used to determine the control sequence that will drive any state $x_0$ to the origin ($x_1 = 0$). In particular,

$$CU_n = \begin{bmatrix} 1 & 0 \\ 0 & 0 \end{bmatrix} \begin{bmatrix} u(1) \\ u(0) \end{bmatrix} = \begin{bmatrix} 0 \\ 0 \end{bmatrix} = -A^2 x_0.$$

Therefore, $u(0) = \alpha$ and $u(1) = 0$, where $\alpha \in R$ will drive any state to the origin. To verify this, we consider $x(1) = Ax(0) + Bu(0) = \begin{bmatrix} 0 & 1 \\ 0 & 0 \end{bmatrix} \begin{bmatrix} x_{01} \\ x_{02} \end{bmatrix} + \begin{bmatrix} 1 \\ 0 \end{bmatrix} \alpha = \begin{bmatrix} x_{02} + \alpha \\ 0 \end{bmatrix}$ and $x(2) = Ax(1) + Bu(1) = \begin{bmatrix} 0 & 1 \\ 0 & 0 \end{bmatrix} \begin{bmatrix} x_{02} + \alpha \\ 0 \end{bmatrix} + \begin{bmatrix} 1 \\ 0 \end{bmatrix} 0 = \begin{bmatrix} 0 \\ 0 \end{bmatrix}$.

---

## 5.4 Observability and Constructibility

In applications, the state of a system is frequently required but not accessible. Under such conditions, the question arises whether it is possible to determine the state by observing the response of the system to some input over some period of time. It turns out that the answer to this question is affirmative if the system is observable. *Observability* refers to the ability of determining the present state $x(t_0)$ from knowledge of current and future system outputs, $y(t)$, and system inputs, $u(t), t \geq t_0$. *Constructibility* refers to the ability of determining the present state $x(t_0)$ from knowledge of current and past system outputs, $y(t)$, and system inputs, $u(t), t \leq t_0$. Observability was briefly addressed in Section 5.2. In this section this concept is formally defined and the (present) state is explicitly determined from input and output measurements.

### 5.4.1 Continuous-Time Time-Invariant Systems

We shall now study observability and constructibility for time-invariant systems described by equations of the form

$$\dot{x} = Ax + Bu, \quad y = Cx + Du, \tag{5.59}$$

where $A \in R^{n \times n}, B \in R^{n \times m}, C \in R^{p \times n}, D \in R^{p \times m}$, and $u(t) \in R^m$ is (piecewise) continuous. As was shown in Section 3.3, the output of this system is given by

$$y(t) = Ce^{At}x(0) + \int_0^t Ce^{A(t-\tau)}Bu(\tau)d\tau + Du(t). \tag{5.60}$$

We recall that the initial time can always be taken to be $t_0 = 0$. We will find it convenient to rewrite (5.60) as

$$\tilde{y}(t) = Ce^{At}x_0, \tag{5.61}$$

where $\tilde{y}(t) \triangleq y(t) - \left[ \int_0^t Ce^{A(t-\tau)}Bu(\tau)d\tau + Du(t) \right]$ and $x_0 = x(0)$.

**Definition 5.34.** *A state $x$ is* unobservable *if the zero-input response of the system (5.59) is zero for every $t \geq 0$, i.e., if*

$$Ce^{At}x = 0 \quad \text{for every } t \geq 0. \tag{5.62}$$

*The set of all unobservable states $x, R_{\bar{o}}$, is called the* unobservable subspace *of (5.59). System (5.59) is* (completely state) observable, *or the pair $(A, C)$ is* observable, *if the only state $x \in R^n$ that is unobservable is $x = 0$, i.e., if $R_{\bar{o}} = \{0\}$.* ∎

Definition 5.34 states that a state is unobservable precisely when it cannot be distinguished as an initial condition at time 0 from the initial condition $x(0) = 0$. This is because in this case the output is the same as if the initial condition were the zero vector. Note that the set of all unobservable states contains the zero vector and it can be shown to be a linear subspace. We now define the observability Gramian.

**Definition 5.35.** *The* observability Gramian *of system (5.59) is the $n \times n$ matrix*

$$W_o(0, T) \triangleq \int_0^T e^{A^T \tau} C^T C e^{A \tau} d\tau. \tag{5.63}$$

■

We note that $W_o$ is symmetric and positive semidefinite for every $T > 0$; i.e., $W_o = W_o^T$ and $W_o \geq 0$ (show this). Recall that the $pn \times n$ *observability matrix*

$$\mathcal{O} \triangleq \begin{bmatrix} C \\ CA \\ \vdots \\ CA^{n-1} \end{bmatrix} \tag{5.64}$$

was defined in Section 5.2.

We now show that the null space of $W_o(0, T)$, denoted by $\mathcal{N}(W_o(0, T))$, is independent of $T$; i.e., it is the same for any $T > 0$, and in particular, it is equal to the null space of the observability matrix $\mathcal{O}$. Thus, the unobservable subspace $R_{\bar{o}}$ of the system is given by the null space of $\mathcal{O}, \mathcal{N}(\mathcal{O})$, or the null space of $W_o(0, T), \mathcal{N}(W_o(0, T))$ for some finite (and therefore for all) $T > 0$.

**Lemma 5.36.** $\mathcal{N}(\mathcal{O}) = \mathcal{N}(W_o(0, T))$ *for every* $T > 0$.

*Proof.* If $x \in \mathcal{N}(\mathcal{O})$, then $\mathcal{O}x = 0$. Thus, $CA^k x = 0$ for all $0 \leq k \leq n-1$, which is also true for every $k > n - 1$, in view of the Cayley–Hamilton Theorem. Then $Ce^{At}x = C[\sum_{k=0}^{\infty}(t^k/k!)A^i]x = 0$ for every finite $t$. Therefore, in view of (5.63), $W_o(0, T)x = 0$ for every $T > 0$; i.e., $x \in \mathcal{N}(W_o(0, T))$ for every $T > 0$. Now let $x \in \mathcal{N}(W_o(0, T))$ for some $T > 0$, so that $x^T W(0, T)x = \int_0^T \| Ce^{A\tau}x \|^2 d\tau = 0$, or $Ce^{At}x = 0$ for every $t \in [0, T]$. Taking derivatives of the last equation with respect to $t$ and evaluating at $t = 0$, we obtain $Cx = CAx = \cdots = CA^k x = 0$ for every $k > 0$. Therefore, $CA^k x = 0$ for every $k \geq 0$, i.e., $\mathcal{O}x = 0$ or $x \in \mathcal{N}(\mathcal{O})$. ■

**Theorem 5.37.** *A state $x$ is unobservable if and only if*

$$x \in \mathcal{N}(\mathcal{O}), \tag{5.65}$$

*or equivalently, if and only if*

$$x \in \mathcal{N}(W_o(0, T)) \tag{5.66}$$

*for some finite (and therefore for all) $T > 0$. Thus, the unobservable subspace $R_{\bar{o}} = \mathcal{N}(\mathcal{O}) = \mathcal{N}(W_o(0, T))$ for some $T > 0$.*

*Proof.* If $x$ is unobservable, (5.62) is satisfied. Taking derivatives with respect to $t$ and evaluating at $t = 0$, we obtain $Cx = CAx = \cdots = CA^k x = 0$ for $k > 0$ or $CA^k x = 0$ for every $k \geq 0$. Therefore, $\mathcal{O}x = 0$ and (5.65) is satisfied.

Assume now that $\mathcal{O}x = 0$; i.e., $CA^k x = 0$ for $0 \le k \le n-1$, which is also true for every $k > n-1$, in view of the Cayley–Hamilton Theorem. Then $Ce^{At}x = C[\Sigma_{k=0}^{\infty}(t^k/k!)A^i]x = 0$ for every finite $t$; i.e., (5.62) is satisfied and $x$ is unobservable. Therefore, $x$ is unobservable if and only if (5.65) is satisfied. In view of Lemma 5.36, (5.66) follows.    ∎

Clearly, $x$ is observable if and only if $\mathcal{O}x \neq 0$ or $W_o(0,T)x \neq 0$ for some $T > 0$.

**Corollary 5.38.** *The system (5.59) is (completely state) observable, or the pair $(A,C)$ is observable, if and only if*

$$\operatorname{rank} \mathcal{O} = n, \tag{5.67}$$

*or equivalently, if and only if*

$$\operatorname{rank} W_o(0,T) = n \tag{5.68}$$

*for some finite (and therefore for all) $T > 0$. If the system is observable, the state $x_0$ at $t = 0$ is given by*

$$x_0 = W_o^{-1}(0,T)\left[\int_0^T e^{A^T \tau} C^T \tilde{y}(\tau)d\tau\right]. \tag{5.69}$$

*Proof.* The system is observable if and only if the only vector that satisfies (5.62) or (5.65) is the zero vector. This is true if and only if the null space is empty, i.e., if and only if (5.67) or (5.68) are true. To determine the state $x_0$ at $t = 0$, given the output and input values over some interval $[0,T]$, we premultiply (5.61) by $e^{A^T \tau} C^T$ and integrate over $[0,T]$ to obtain

$$W_o(0,T)x_0 = \int_0^T e^{A^T \tau} C^T \tilde{y}(\tau)d\tau, \tag{5.70}$$

in view of (5.63). When the system is observable, (5.70) has the unique solution (5.69).    ∎

Note that $T > 0$, the time span over which the input and output are observed, is arbitrary. Intuitively, one would expect in practice to have difficulties in evaluating $x_0$ accurately when $T$ is small, using any numerical method. Note that for very small $T$, $\|W_o(0,T)\|$ can be very small, which can lead to numerical difficulties in solving (5.70). Compare this with the analogous case for reachability, where small $T$ leads in general to large values in control action.

It is clear that if the state at some time $t_0$ is determined, then the state $x(t)$ at any subsequent time is easily determined, given $u(t), t \ge t_0$.

Alternative methods to (5.69) to determine the state of the system when the system is observable are provided in Section 9.3 on state observers.

**Example 5.39.** (i) Consider the system $\dot{x} = Ax, y = Cx$, where $A = \begin{bmatrix} 0 & 1 \\ 0 & 0 \end{bmatrix}$

and $C = [1, \ 0]$. Here $e^{At} = \begin{bmatrix} 1 & t \\ 0 & 1 \end{bmatrix}$ and $Ce^{At} = [1, \ t]$. The observ-

ability Gramian is then $W_o(0, T) = \int_0^T \begin{bmatrix} 1 \\ \tau \end{bmatrix} [1 \ \tau] d\tau = \int_0^T \begin{bmatrix} 1 & \tau \\ \tau & \tau^2 \end{bmatrix} d\tau =$

$\begin{bmatrix} T & \frac{1}{2}T^2 \\ \frac{1}{2}T^2 & \frac{1}{3}T^3 \end{bmatrix}$. Notice that $\det W_o(0, T) = \frac{1}{12}T^4 \neq 0$ for any $T > 0$, i.e.,

rank $W_o(0, T) = 2 = n$ for any $T > 0$, and therefore (Corollary 5.38),
the system is observable. Alternatively, note that the observability ma-

trix $\mathcal{O} = \begin{bmatrix} C \\ CA \end{bmatrix} = \begin{bmatrix} 1 & 0 \\ 0 & 1 \end{bmatrix}$ and rank $\mathcal{O} = 2 = n$. Clearly, in this case

$\mathcal{N}(\mathcal{O}) = \mathcal{N}(W_o(0, T)) = \left\{ \begin{bmatrix} 0 \\ 0 \end{bmatrix} \right\}$, which verifies Lemma 5.36.

(ii) If $A = \begin{bmatrix} 0 & 1 \\ 0 & 0 \end{bmatrix}$, as before, but $C = [0, \ 1]$, in place of $[1, \ 0]$, then $Ce^{At} =$

$[0, \ 1]$ and the observability Gramian is $W_o(0, T) = \int_0^T \begin{bmatrix} 0 \\ 1 \end{bmatrix} [0, \ 1] d\tau =$

$\begin{bmatrix} 0 & 0 \\ 0 & T \end{bmatrix}$. We have rank $W_o(0, T) = 1 < 2 = n$, and the system is not

completely observable. In view of Theorem 5.37, all unobservable states
$x \in \mathcal{N}(W_o(0, T))$ and are therefore of the form $\begin{bmatrix} \alpha \\ 0 \end{bmatrix}, \alpha \in R$. Alter-

natively, the observability matrix $\mathcal{O} = \begin{bmatrix} C \\ CA \end{bmatrix} = \begin{bmatrix} 0 & 1 \\ 0 & 0 \end{bmatrix}$. Note that

$\mathcal{N}(\mathcal{O}) = \mathcal{N}(W_0(0, T)) = \text{span} \left\{ \begin{bmatrix} 1 \\ 0 \end{bmatrix} \right\}$.

Observability utilizes future output measurements to determine the present
state. In (re)constructibility, past output measurements are used. Constructibility is defined in the following, and its relation to observability is determined.

**Definition 5.40.** A state $x$ is unconstructible *if the zero-input response of
the system (5.59) is zero for all $t \leq 0$; i.e.,*

$$Ce^{At}x = 0 \quad \text{for every } t \leq 0. \tag{5.71}$$

*The set of all unconstructible states $x$, $R_{\overline{cn}}$, is called the* unconstructible subspace *of (5.59). The system (5.59) is (completely state) (re)constructible, or
the pair $(A, C)$ is (re)constructible, if the only state $x \in R^n$ that is unconstructible is $x = 0$; i.e., $R_{\overline{cn}} = \{0\}$.*

We shall now establish a relationship between observability and constructibility for the continuous-time time-invariant systems (5.59). Recall that
$x$ is unobservable if and only if

$$Ce^{At}x = 0 \quad \text{for every } t \geq 0. \tag{5.72}$$

**Theorem 5.41.** *Consider the system* $\dot{x} = Ax + Bu, y = Cx + Du$ *given in* (5.59).

(i) *A state* $x$ *is unobservable if and only if it is unconstructible.*
(ii) $R_{\bar{o}} = R_{\overline{cn}}$.
(iii) *The system, or the pair* $(A, C)$, *is (completely state) observable if and only if it is (completely state) (re)constructible.*

*Proof.* (i) If $x$ is unobservable, then $Ce^{At}x = 0$ for every $t \geq 0$. Taking derivatives with respect to $t$ and evaluating at $t = 0$, we obtain $Cx = CAx = \cdots = CA^k x = 0$ for $k > 0$ or $CA^k x = 0$ for every $k \geq 0$. This, in view of $Ce^{At}x = \sum_{k=0}^{\infty}(t^k/k!)CA^k x$, implies that $Ce^{At}x = 0$ for every $t \leq 0$; i.e., $x$ is unconstructible. The converse is proved in a similar manner. Parts (ii) and (iii) of the theorem follow directly from (i). ∎

The observability Gramian for the time-invariant case, $W_o(0, T)$, was defined in (5.63). The constructibility Gramian is now defined.

**Definition 5.42.** *The* constructibility Gramian *of system (5.59) is the* $n \times n$ *matrix*

$$W_{cn}(0, T) \triangleq \int_0^T e^{A^T(\tau - T)}C^T C e^{A(\tau - T)}d\tau. \tag{5.73}$$

∎

Note that

$$W_o(0, T) = e^{A^T T}W_{cn}(0, T)e^{AT}, \tag{5.74}$$

as can be verified directly.

## Additional Criteria for Observability and Constructibility

We shall now use Lemma 5.22 to develop additional tests for observability and constructibility. These are analogous to the corresponding results established for reachability and controllability in Theorem 5.23.

**Theorem 5.43.** *The system* $\dot{x} = Ax + Bu, y = Cx + Du$ *is observable*

(i) *if and only if*

$$\text{rank } W_o(0, T) = n \tag{5.75}$$

*for some finite* $T > 0$, *where* $W_0(0, T) \triangleq \int_0^T e^{A^T \tau}C^T C e^{A\tau}d\tau$, *the observability Gramian, or*

*(ii) if and only if the n columns of*

$$Ce^{At} \tag{5.76}$$

*are linearly independent on $[0, \infty)$ over the field of complex numbers, or alternatively, if and only if the n columns of*

$$C(sI - A)^{-1} \tag{5.77}$$

*are linearly independent over the field of complex numbers, or*
*(iii) if and only if*

$$\text{rank}\,\mathcal{O} = n, \tag{5.78}$$

*where* $\mathcal{O} \triangleq \begin{bmatrix} C \\ CA \\ \vdots \\ CA^{n-1} \end{bmatrix}$, *the observability matrix, or*

*(iv) if and only if*

$$\text{rank}\begin{bmatrix} s_iI - A \\ C \end{bmatrix} = n \tag{5.79}$$

*for all complex numbers $s_i$, or alternatively, for all eigenvalues of A.*

*Proof.* The proof of this theorem is completely analogous to the (dual) results on reachability (Theorem 5.23) and is omitted. ∎

Since it was shown (in Theorem 5.41) that observability implies and is implied by constructibility, the tests developed in the theorem for observability are typically also used to test for constructibility.

---

**Example 5.44.** Consider the system $\dot{x} = Ax, y = Cx$, where $A = \begin{bmatrix} 0 & 1 \\ 0 & 0 \end{bmatrix}$ and $C = [1,\ 0]$, as in Example 5.39(i). We shall verify (i) to (iv) of Theorem 5.43 for this case.

(i)  For the observability Gramian, $W_o(0,T) = \begin{bmatrix} T & \frac{1}{2}T^2 \\ \frac{1}{2}T^2 & \frac{1}{3}T^3 \end{bmatrix}$, we have rank $W_o(0,T) = 2 = n$ for any $T > 0$.

(ii) The columns of $Ce^{At} = [1,\ t]$ are linearly independent on $[0, \infty)$ over the field of complex numbers, since $a_1 \times 1 + a_2 \times t = 0$ implies that the complex numbers $a_1$ and $a_2$ must both be zero. Similarly, the columns of $C(sI - A)^{-1} = [\frac{1}{s}, \frac{1}{s^2}]$ are linearly independent over the field of complex numbers.

(iii) rank $\mathcal{O} = $ rank $\begin{bmatrix} C \\ CA \end{bmatrix} = $ rank $\begin{bmatrix} 1 & 0 \\ 0 & 1 \end{bmatrix} = 2 = n.$

(iv) rank $\begin{bmatrix} s_iI - A \\ C \end{bmatrix} = $ rank $\begin{bmatrix} s_i & -1 \\ 0 & s_i \\ 1 & 0 \end{bmatrix} = 2 = n$ for $s_i = 0$, $i = 1,2$, the eigenvalues of $A$.

Consider again $A = \begin{bmatrix} 0 & 1 \\ 0 & 0 \end{bmatrix}$ but $C = [0, 1]$ [in place of $[1, 0]$, as in Example 5.39(ii)].

The system is not observable for the reasons given below.

(i)  $W_o(0, T) = \begin{bmatrix} 0 & 0 \\ 0 & T \end{bmatrix}$ with rank $W_o(0, T) = 1 < 2 = n$.

(ii)  $Ce^{At} = [0, 1]$ and its columns are not linearly independent. Similarly, the columns of $C(sI - A)^{-1} = [0, \frac{1}{s}]$ are not linearly independent.

(iii)  rank $\mathcal{O} = \text{rank} \begin{bmatrix} C \\ CA \end{bmatrix} = \text{rank} \begin{bmatrix} 0 & 1 \\ 0 & 0 \end{bmatrix} = 1 < 2 = n$.

(iv)  rank $\begin{bmatrix} s_i I - A \\ C \end{bmatrix} = \text{rank} \begin{bmatrix} s_i & -1 \\ 0 & s_i \\ 0 & 1 \end{bmatrix} = 1 < 2 = n$ for $s_i = 0$ an eigenvalue of

$A$.

---

## 5.4.2 Discrete-Time Time-Invariant Systems

We consider systems described by equations of the form

$$x(k + 1) = Ax(k) + Bu(k), \quad y(k) = Cx(k) + Du(k), \quad k \geq k_0, \qquad (5.80)$$

where $A \in R^{n \times n}, C \in R^{n \times m}, C \in R^{p \times n}, D \in R^{p \times m}$. The output $y(k)$ for $k > k_0$ is given by

$$y(k) = C(k)\Phi(k, k_0)x(k_0) + \sum_{i=k_0}^{k-1} C(k)\Phi(k, i+1)B(i)u(i) + D(k)u(k), \qquad (5.81)$$

where the state transition matrix $\Phi(k, k_0)$ is given by

$$\Phi(k, k_0) = A^{k-k_0}, \quad k \geq k_0. \qquad (5.82)$$

Observability and (re)constructibility for discrete-time systems are defined as in the continuous-time case. Observability refers to the ability to uniquely determine the state from knowledge of current and future outputs and inputs, whereas constructibility refers to the ability to determine the state from knowledge of current and past outputs and inputs. Without loss of generality, we take $k_0 = 0$. Then

$$y(k) = CA^k x(0) + \sum_{i=0}^{k-1} CA^{k-(i+1)}Bu(i) + Du(k) \qquad (5.83)$$

for $k > 0$ and $y(0) = Cx(0) + Du(0)$. Rewrite as

$$\tilde{y}(k) = CA^k x_0 \qquad (5.84)$$

for $k \geq 0$, where $\tilde{y}(k) \triangleq y(k) - \left[\sum_{i=0}^{k-1} CA^{k-(i+1)}Bu(i) + Du(k)\right]$ for $k > 0$ and $\tilde{y}(0) \triangleq y(0)$, and $x_0 = x(0)$.

**Definition 5.45.** *A state $x$ is unobservable if the zero-input response of system (5.80) is zero for all $k \geq 0$, i.e., if*

$$CA^k x = 0 \quad \text{for every } k \geq 0. \tag{5.85}$$

*The set of all unobservable states $x$, $R_{\bar{o}}$, is called the* unobservable subspace *of (5.80). The system (5.80) is* (completely state) observable, *or the pair $(A, C)$ is observable, if the only state $x \in R^n$ that is unobservable is $x = 0$, i.e., if $R_{\bar{o}} = \{0\}$.* ∎

The $pn \times n$ *observability matrix* $\mathcal{O}$ was defined in (5.64). Let $\mathcal{N}(\mathcal{O})$ denote the null space of $\mathcal{O}$.

**Theorem 5.46.** *A state $x$ is unobservable if and only if*

$$x \in \mathcal{N}(\mathcal{O}); \tag{5.86}$$

*i.e., the unobservable subspace $R_{\bar{o}} = \mathcal{N}(\mathcal{O})$.*

*Proof.* If $x \in \mathcal{N}(\mathcal{O})$, then $\mathcal{O}x = 0$ or $CA^k x = 0$ for $0 \leq k \leq n - 1$. This statement is also true for $k > n - 1$, in view of the Cayley–Hamilton Theorem. Therefore, (5.85) is satisfied and $x$ is unobservable. Conversely, if $x$ is unobservable, then (5.85) is satisfied and $\mathcal{O}x = 0$. ∎

Clearly, $x$ is observable if and only if $\mathcal{O}x \neq 0$.

**Corollary 5.47.** *The system (5.80) is* (completely state) observable, *or the pair $(A, C)$ is observable, if and only if*

$$\text{rank}\, \mathcal{O} = n. \tag{5.87}$$

*If the system is observable, the state $x_0$ at $k = 0$ can be determined as the unique solution of*

$$[Y_{0,n-1} - M_n U_{0,n-1}] = \mathcal{O}x_0, \tag{5.88}$$

*where*

$$Y_{0,n-1} \triangleq [y^T(0), y^T(1), \ldots, y^T(n-1)]^T \text{ is a } pn \times 1 \text{ matrix,}$$
$$U_{0,n-1} \triangleq [u^T(0), u^T(1), \ldots, u^T(n-1)]^T \text{ is an } mn \times 1 \text{ matrix,}$$

*and $M_n$ is the $pn \times mn$ matrix given by*

$$M_n \triangleq \begin{bmatrix} D & 0 & \cdots & 0 & 0 \\ CB & D & \cdots & 0 & 0 \\ \vdots & \vdots & \ddots & \vdots & \vdots \\ CA^{n-2}B & CA^{n-3}B & \cdots & D & \\ CA^{n-1}B & CA^{n-2}B & \cdots & CB & D \end{bmatrix}.$$

*Proof.* The system is observable if and only if the only vector that satisfies (5.85) is the zero vector. This is true if and only if $\mathcal{N}(\mathcal{O}) = \{0\}$, or if (5.87) is true. To determine the state $x_0$, apply (5.83) for $k = 0, 1, \ldots, n - 1$, and rearrange in a form of a system of linear equations to obtain (5.88).    ∎

The matrix $M_n$ defined above has the special structure of a *Toeplitz* matrix. Note that a matrix $T$ is Toeplitz if its $(i, j)$th entry depends on the value $i - j$; that is, $T$ is "constant along the diagonals."

*Additional Criteria for Observability.* Note that completely analogous results to Theorem 5.43(ii)–(iv) exist for the discrete-time case.

Constructibility refers to the ability to determine uniquely the state $x(0)$ from knowledge of current and past outputs and inputs. This is in contrast to observability, which utilizes future outputs and inputs. The easiest way to define constructibility is by the use of (5.84), where $x(0) = x_0$ is to be determined from past data $\tilde{y}(k)$, $k \leq 0$. Note, however, that for $k \leq 0$, $A^k$ may not exist; in fact, it exists only when $A$ is nonsingular. To avoid making restrictive assumptions, we shall define unconstructible states in a slightly different way than anticipated. Unfortunately, this definition is not very transparent. It turns out that by using this definition, an unconstructible state can be related to an unobservable state in a manner analogous to the way a controllable state was related to a reachable state in Section 5.3 (see also the discussion of duality in Section 5.2).

**Definition 5.48.** *A state $x$ is unconstructible if for every $k \geq 0$, there exists $\hat{x} \in R^n$ such that*

$$x = A^k \hat{x}, \quad C\hat{x} = 0. \tag{5.89}$$

*The set of all unconstructible states, $R_{\overline{cn}}$, is called the* unconstructible subspace. *The system (5.80) is* (completely state) constructible, *or the pair $(A, C)$ is constructible, if the only state $x \in R^n$ that is unconstructible is $x = 0$, i.e., if $R_{\overline{cn}} = \{0\}$.*    ∎

Note that if $A$ is nonsingular, then (5.89) simply states that $x$ is unconstructible if $CA^{-k}x = 0$ for every $k \geq 0$ (compare this with Definition 5.45 of an unobservable state).

The results that can be derived for constructibility are simply dual to the results on controllability. They are presented briefly below, but first, a technical result must be established.

**Lemma 5.49.** *If $x \in \mathcal{N}(\mathcal{O})$, then $Ax \in \mathcal{N}(\mathcal{O})$; i.e., the unobservable subspace $R_{\bar{o}} = \mathcal{N}(\mathcal{O})$ is an A-invariant subspace.*

*Proof.* Let $x \in \mathcal{N}(\mathcal{O})$, so that $\mathcal{O}x = 0$. Then $CA^k x = 0$ for $0 \leq k \leq n - 1$. This statement is also true for $k > n - 1$, in view of the Cayley–Hamilton Theorem. Therefore, $\mathcal{O}Ax = 0$; i.e., $Ax \in \mathcal{N}(\mathcal{O})$.    ∎

**Theorem 5.50.** *Consider the system $x(k + 1) = Ax(k) + Bu(k)$, $y(k) = Cx(k) + Du(k)$ given in (5.80).*

*(i)  If a state $x$ is unconstructible, then it is unobservable.*

*(ii)  $R_{\overline{cn}} \subset R_{\bar{o}}$.*

*(iii) If the system is (completely state) observable, or the pair $(A, C)$ is observable, then the system is also (completely state) constructible, or the pair $(A, C)$ is constructible.*

*If $A$ is nonsingular, then relations (i) and (iii) are if and only if statements. In this case, constructibility also implies observability. Furthermore, in this case, (ii) becomes an equality; i.e., $R_{\overline{cn}} = R_{\bar{o}}$.*

*Proof.* This theorem is dual to Theorem 5.31, which relates reachability and controllability in the discrete-time case. To verify (i), assume that $x$ satisfies (5.89) and premultiply by $C$ to obtain $Cx = CA^k\hat{x}$ for every $k \geq 0$. Note that $Cx = 0$ since for $k = 0$, $x = \hat{x}$, and $C\hat{x} = 0$. Therefore, $CA^k\hat{x} = 0$ for every $k \geq 0$; i.e., $\hat{x} \in \mathcal{N}(\mathcal{O})$. In view of Lemma 5.49, $x = A^k\hat{x} \in \mathcal{N}(\mathcal{O})$, and thus, $x$ is unobservable. Since $x$ is arbitrary, we have also verified (ii). When the system is observable, $R_{\bar{o}}$ is empty, which in view of (ii), implies that $R_{\overline{cn}} = \{0\}$ or that the system is constructible. This proves (iii). Alternatively, one could also prove this directly: Assume that the system is observable but not constructible. Then there exist $x, \hat{x} \neq 0$, which satisfy (5.89). As above, this implies that $\hat{x} \in \mathcal{N}(\mathcal{O})$, which is a contradiction since the system is observable.

Consider now the case when $A$ is nonsingular and let $x$ be unobservable. Then, in view of Lemma 5.49, $\hat{x} \triangleq A^{-k}x$ is also in $\mathcal{N}(\mathcal{O})$; i.e., $C\hat{x} = 0$. Therefore, $x = A^k\hat{x}$ is unconstructible, in view of Definition 5.48. This implies also that $R_{\bar{o}} \subset R_{\overline{cn}}$, and therefore, $R_{\bar{o}} = R_{\overline{cn}}$, which proves that in the present case constructibility also implies observability. ∎

---

**Example 5.51.** Consider the system in Example 5.5, $x(k+1) = Ax(k)$, $y(k) = Cx(k)$, where $A = \begin{bmatrix} 1 & 0 \\ 1 & 1 \end{bmatrix}$ and $C = [1, 0]$. As shown, rank $\mathcal{O} = \text{rank} \begin{bmatrix} 1 & 0 \\ 1 & 0 \end{bmatrix} = 1 < 2 = n$; i.e., the system is not observable. All unobservable states are of the form $\alpha \begin{bmatrix} 0 \\ 1 \end{bmatrix}$, where $\alpha \in R$ since $\left\{ \begin{bmatrix} 0 \\ 1 \end{bmatrix} \right\}$ is a basis for $\mathcal{N}(\mathcal{O}) = R_{\bar{o}}$, the unobservable subspace.

In Example 5.6 it was shown that all the states $x$ that satisfy $CA^{-k}x = 0$ for every $k \geq 0$, i.e., all the unconstructible states, are given by $\alpha \begin{bmatrix} 0 \\ 1 \end{bmatrix}$, $\alpha \in R$. This verifies Theorem 5.50(i) and (ii) for the case when $A$ is nonsingular.

---

**Example 5.52.** Consider the system $x(k+1) = Ax(k)$, $y(k) = Cx(k)$, where $A = \begin{bmatrix} 0 & 0 \\ 1 & 0 \end{bmatrix}$ and $C = [1, 0]$. The observability matrix $\mathcal{O} = \begin{bmatrix} 1 & 0 \\ 0 & 0 \end{bmatrix}$ is of rank 1, and therefore, the system is not observable. In fact, all states of the form $\alpha \begin{bmatrix} 0 \\ 1 \end{bmatrix}$ are unobservable states since $\left\{ \begin{bmatrix} 0 \\ 1 \end{bmatrix} \right\}$ is a basis for $\mathcal{N}(\mathcal{O})$.

To check constructibility, the defining relations (5.89) must be used since $A$ is singular. $C\hat{x} = [1, \ 0]\hat{x} = 0$ implies $\hat{x} = \begin{bmatrix} 0 \\ \beta \end{bmatrix}$. Substituting into $x = A^k\hat{x}$, we obtain for $k = 0$, $x = \hat{x}$, and $x = 0$ for $k \geq 1$. Therefore, the only unconstructible state is $x = 0$, which implies that the system is constructible (although it is unobservable). This means that the initial state $x(0)$ can be uniquely determined from past measurements. In fact, from $x(k+1) = Ax(k)$ and $y(k) = Cx(k)$, we obtain $x(0) = \begin{bmatrix} x_1(0) \\ x_2(0) \end{bmatrix} = \begin{bmatrix} 0 & 0 \\ 1 & 0 \end{bmatrix} \begin{bmatrix} x_1(-1) \\ x_2(-1) \end{bmatrix} = \begin{bmatrix} 0 \\ x_1(-1) \end{bmatrix}$ and $y(-1) = Cx(-1) = [1, \ 0] \begin{bmatrix} x_1(-1) \\ x_2(-1) \end{bmatrix} = x_1(-1)$. Therefore, $x(0) = \begin{bmatrix} 0 \\ y(-1) \end{bmatrix}$.

---

When $A$ is nonsingular, the state $x_0$ at $k = 0$ can be determined from past outputs and inputs in the following manner. We consider (5.84) and note that in this case

$$\tilde{y}(k) = CA^k x_0$$

is valid for $k \leq 0$ as well. This implies that

$$\tilde{Y}_{-1,-n} = \mathcal{O}A^{-n}x_0 = \begin{bmatrix} CA^{-n} \\ \vdots \\ CA^{-1} \end{bmatrix} x_0 \qquad (5.90)$$

with $\tilde{Y}_{-1,-n} \triangleq [\tilde{y}^T(-n), \ldots, \tilde{y}^T(-1)]^T$. Equation (5.90) must be solved for $x_0$. Clearly, in the case of constructibility (and under the assumption that $A$ is nonsingular), the matrix $\mathcal{O}A^{-n}$ is of interest instead of $\mathcal{O}$ [compare this with the dual results in (5.58)]. In particular, the system is constructible if and only if $\mathrm{rank}(\mathcal{O}A^{-n}) = \mathrm{rank}\,\mathcal{O} = n$.

---

**Example 5.53.** Consider the system in Example 5.4, namely, $x(k + 1) = Ax(k), y(k) = Cx(k)$, where $A = \begin{bmatrix} 0 & 1 \\ 1 & 1 \end{bmatrix}$ and $C = [0, \ 1]$. Since $A$ is nonsingular, to check constructibility we consider $\mathcal{O}A^{-2} = \begin{bmatrix} CA^{-2} \\ CA^{-1} \end{bmatrix} = \begin{bmatrix} -1 & 1 \\ 1 & 0 \end{bmatrix}$, which has full rank. Therefore, the system is constructible (as expected), since it is observable. To determine $x(0)$, in view of (5.90), we note that $\begin{bmatrix} y(-1) \\ y(-2) \end{bmatrix} = \mathcal{O}A^{-2}x(0) = \begin{bmatrix} -1 & 1 \\ 1 & 0 \end{bmatrix} \begin{bmatrix} x_1(0) \\ x_2(0) \end{bmatrix}$, from which $\begin{bmatrix} x_1(0) \\ x_2(0) \end{bmatrix} = \begin{bmatrix} 0 & 1 \\ 1 & 1 \end{bmatrix} \begin{bmatrix} y(-1) \\ y(-2) \end{bmatrix} = \begin{bmatrix} y(-2) \\ y(-1) + y(-2) \end{bmatrix}$.

---

## 5.5 Summary and Highlights

*Reachability and Controllability*

- In continuous-time systems, reachability always implies and is implied by controllability. In discrete-time systems, reachability always implies controllability, but controllability implies reachability only when $A$ is nonsingular. See Definitions 5.8 and 5.18 and Theorems 5.20 and 5.31.
- When a discrete-time system $x(k+1) = Ax(k)+Bu(k)$ [denoted by $(A, B)$] is completely reachable (controllable-from-the-origin), the input sequence $\{u(i)\}$, $i = 0, \ldots, K-1$ that transfers the state from any $x_0(= x(0))$ to any $x_1$ in some finite time $K$ $(x_1 = x(K), K > 0)$ is determined by solving

$$x_1 = A^K x_0 + \sum_{i=0}^{K-1} A^{K-(i+1)} Bu(i) \quad \text{or}$$

$$x_1 - A^K x_0 = [B, AB, \ldots, A^{K-1}] [u^T(K-1), \ldots, u^T(0)]^T.$$

A solution for this always exists when $K = n$. See Theorem 5.27.

- $$\mathcal{C} = [B, AB, \ldots, A^{n-1}B] \ (n \times mn) \tag{5.25}$$

is the controllability matrix for both discrete- and continuous-time time-invariant systems, and it has full (row) rank when the system, denoted by $(A, B)$, is (completely) reachable (controllable-from-the-origin).
- When a continuous-time system $\dot{x} = Ax + Bu$ [denoted by $(A, B)$] is controllable, an input that transfers any state $x_0(= x(0))$ to any other state $x_1$ in some finite time $T$ $(x_1 = x(T))$ is

$$u(t) = B^T e^{A^T(T-t)} W_r^{-1}(0, T)[x_1 - e^{AT} x_0] \quad t \in [0, T], \tag{5.33}$$

where

$$W_r(0, T) = \int_0^T e^{(T-\tau)A} BB^T e^{(T-\tau)A^T} d\tau \tag{5.24}$$

is the reachability Gramian of the system.
- $(A, B)$ is reachable if and only if

$$\text{rank}[s_i I - A, B] = n \tag{5.42}$$

for $s_i$, $i = 1, \ldots, n$, all the eigenvalues of $A$.

*Observability and Constructibility*

- In continuous-time systems, observability always implies and is implied by constructibility. In discrete-time systems, observability always implies constructibility, but constructibility implies observability only when $A$ is nonsingular. See Definitions 5.34 and 5.40 and Theorems 5.41 and 5.50.

- When a discrete-time system $x(k+1) = Ax(k) + Bu(k)$, $y(k) = Cx(k) + Du(k)$ [denoted by $(A, C)$] is completely observable, any initial state $x(0) = x_0$ can be uniquely determined by observing the input and output over some finite period of time, and using the relation

$$\tilde{y}(k) = CA^k x_0 \quad k = 0, 1, \ldots, n-1, \tag{5.84}$$

where $\tilde{y}(k) = y(k) - \left[\sum_{i=0}^{k-1} CA^{k-(i+1)} Bu(i) + D(k)u(k)\right]$. To determine $x_0$, solve

$$\begin{bmatrix} \tilde{y}(0) \\ \tilde{y}(1) \\ \vdots \\ \tilde{y}(n-1) \end{bmatrix} = \begin{bmatrix} C \\ CA \\ \vdots \\ CA^{n-1} \end{bmatrix} x_0.$$

See (5.88).

- $$\mathcal{O} = \begin{bmatrix} C \\ CA \\ \vdots \\ CA^{n-1} \end{bmatrix} \quad (pn \times n) \tag{5.64}$$

is the observability matrix for both discrete- and continuous-time, time-invariant systems and it has full (column) rank when the system is completely observable.

- Consider the continuous-time system $\dot{x} = Ax + Bu$, $y = Cx + Du$. When this system [denoted by $(A, C)$] is completely observable, any initial state $x_0 = x(0)$ can be uniquely determined by observing the input and output over some finite period of time $T$ and using the relation

$$\tilde{y}(t) = Ce^{At} x_0,$$

where $\tilde{y}(t) = y(t) - \left[\int_0^t Ce^{A(t-\tau)} Bu(\tau)d\tau + Du(t)\right]$. The initial state $x_0$ may be determined from

$$x_0 = W_o^{-1}(0, T)\left[\int_0^T e^{A^T \tau} C^T \tilde{y}(\tau)d\tau\right], \tag{5.69}$$

where

$$W_o(0, T) = \int_0^T e^{A^T \tau} C^T C e^{A\tau} d\tau \tag{5.63}$$

is the observability Gramian of the system.

- $(A, C)$ is observable if and only if

$$\text{rank}\begin{bmatrix} s_i I - A \\ C \end{bmatrix} = n \tag{5.79}$$

for $s_i$, $i = 1, \ldots, n$, all the eigenvalues of $A$.

*Dual Systems*

- $(A_D = A^T, B_D = C^T, C_D = B^T, D_D = D^T)$ is the dual of $(A, B, C, D)$.
  Reachability is dual to observability. If a system is reachable (observable),
  its dual is observable (reachable).

## 5.6 Notes

The concept of controllability was first encountered as a technical condition in
certain optimal control problems and also in the so-called finite-settling-time
design problem for discrete-time systems (see Kalman [4]). In the latter, an
input must be determined that returns the state $x_0$ to the origin as quickly
as possible. Manipulating the input to assign particular values to the initial
state in (analog-computer) simulations was not an issue since the individual
capacitors could initially be charged independently. Also, observability was
not an issue in simulations due to the particular system structures that were
used (corresponding, e.g., to observer forms). The current definitions for controllability and observability and the recognition of the duality between them
were worked out by Kalman in 1959–1960 (see Kalman [7] for historical comments) and were presented by Kalman in [5]. The significance of realizations
that were both controllable and observable (see Chapter 5) was established
later in Gilbert [2], Kalman [6], and Popov [8]. For further information regarding these historical issues, consult Kailath [3] and the original sources. Note
that [3] has extensive references up to the late seventies with emphasis on the
time-invariant case and a rather complete set of original references together
with historical remarks for the period when the foundations of the state-space
system theory were set, in the late fifties and sixties.

## References

1. P.J. Antsaklis and A.N. Michel, *Linear Systems*, Birkhäuser, Boston, MA, 2006.
2. E. Gilbert, "Controllability and observability in multivariable control systems,"
   *SIAM J. Control*, Vol. 1, pp. 128–151, 1963.
3. T. Kailath, *Linear Systems*, Prentice-Hall, Englewood Cliffs, NJ, 1980.
4. R.E. Kalman, "Optimal nonlinear control of saturating systems by intermittent
   control," IRE WESCON Rec., Sec. IV, pp. 130–135, 1957.
5. R.E. Kalman, "On the general theory of control systems," in *Proc. of the First
   Intern. Congress on Automatic Control*, pp. 481–493, Butterworth, London,
   1960.
6. R.E. Kalman, "Mathematical descriptions of linear systems," *SIAM J. Control*,
   Vol. 1, pp. 152–192, 1963.
7. R.E. Kalman, *Lectures on Controllability and Observability*, C.I.M.E., Bologna,
   1968.
8. V.M. Popov, "On a new problem of stability for control systems," *Autom.
   Remote Control*, Vol. 24, No. 1, pp. 1–23, 1963.

# Exercises

**5.1.** (a) Let $\mathcal{C}_k \triangleq [B, AB, \ldots, A^{k-1}B]$, where $A \in R^{n \times n}, B \in R^{n \times m}$. Show that

$$\mathcal{R}\left(\mathcal{C}_k\right) = \mathcal{R}\left(\mathcal{C}_n\right) \text{ for } k \geq n, \quad \text{and} \quad \mathcal{R}(\mathcal{C}_k) \subset \mathcal{R}(\mathcal{C}_n) \text{ for } k < n.$$

(b) Let $\mathcal{O}_k \triangleq [C^T, (CA)^T, \ldots, (CA^{k-1})^T]^T$, where $A \in R^{n \times n}, C \in R^{p \times n}$. Show that

$$\mathcal{N}(\mathcal{O}_k) = \mathcal{N}(\mathcal{O}_n) \text{ for } k \geq n, \text{ and } \mathcal{N}(\mathcal{O}_k) \supset \mathcal{N}(\mathcal{O}_n) \text{ for } k < n.$$

**5.2.** Consider the state equation $\dot{x} = Ax + Bu$, where

$$A = \begin{bmatrix} 0 & 1 & 0 & 0 \\ 3w^2 & 0 & 0 & 2w \\ 0 & 0 & 0 & 1 \\ 0 & -2w & 0 & 0 \end{bmatrix}, \quad B = \begin{bmatrix} 0 & 0 \\ 1 & 0 \\ 0 & 0 \\ 0 & 1 \end{bmatrix},$$

which was obtained by linearizing the nonlinear equations of motion of an orbiting satellite about a steady-state solution. In the state $x = [x_1, x_2, x_3, x_4]^T$, $x_1$ is the differential radius, whereas $x_3$ is the differential angle. In the input vector $u = [u_1, u_2]^T$, $u_1$ is the radial thrust and $u_2$ is the tangential thrust.

(a) Is this system controllable from $u$? If $y = \begin{bmatrix} y_1 \\ y_2 \end{bmatrix} = \begin{bmatrix} x_1 \\ x_3 \end{bmatrix}$, is the system observable from $y$?

(b) Can the system be controlled if the radial thruster fails? What if the tangential thruster fails?

(c) Is the system observable from $y_1$ only? From $y_2$ only?

**5.3.** Consider the state equation $\begin{bmatrix} \dot{x}_1 \\ \dot{x}_2 \end{bmatrix} = \begin{bmatrix} -1/2 & 0 \\ 0 & -1 \end{bmatrix} \begin{bmatrix} x_1 \\ x_2 \end{bmatrix} + \begin{bmatrix} 1/2 \\ 1 \end{bmatrix} u.$

(a) If $x(0) = \begin{bmatrix} a \\ b \end{bmatrix}$, derive an input that will drive the state to $\begin{bmatrix} 0 \\ 0 \end{bmatrix}$ in $T$ sec.

(b) For $x(0) = \begin{bmatrix} 5 \\ -5 \end{bmatrix}$, plot $u(t), x_1(t), x_2(t)$ for $T = 1, 2$, and 5 sec. Comment on the magnitude of the input in your results.

**5.4.** Consider the state equation $x(k+1) = \begin{bmatrix} 1 & 1 & 0 \\ 0 & 1 & 0 \\ 0 & 0 & 1 \end{bmatrix} x(k) + \begin{bmatrix} 0 \\ 1 \\ 1 \end{bmatrix} u(k), y(k) = \begin{bmatrix} 1 & 1 & 0 \\ 0 & 1 & 0 \end{bmatrix} x(k).$

(a) Is $x^1 = \begin{bmatrix} 3 \\ 2 \\ 2 \end{bmatrix}$ reachable? If yes, what is the minimum number of steps

required to transfer the state from the zero state to $x^1$? What inputs do you need?

(b) Determine all states that are reachable.

(c) Determine all states that are unobservable.

(d) If $\dot{x} = Ax + Bu$ is given with $A, B$ as in (a), what is the minimum time required to transfer the state from the zero state to $x^1$? What is an appropriate $u(t)$?

**5.5.** *Output reachability (controllability)* can be defined in a manner analogous to state reachability (controllability). In particular, a system will be called output reachable if there exists an input that transfers the output from some $y_0$ to any $y_1$ in finite time.

Consider now a discrete-time time-invariant system $x(k+1) = Ax(k) + Bu(k), y(k) = Cx(k) + Du(k)$ with $A \in R^{n \times n}, B \in R^{n \times m}, C \in R^{p \times n}$ and $D \in R^{p \times m}$. Recall that

$$y(k) = CA^k x(0) + \sum_{i=0}^{k-1} CA^{k-(i+1)} Bu(i) + Du(k).$$

(a) Show that the system $\{A, B, C, D\}$ is output reachable if and only if

$$\text{rank}[D, CB, CAB, \ldots, CA^{n-1}B] = p.$$

Note that this rank condition is also the condition for output reachability for continuous-time time-invariant systems $\dot{x} = Ax + Bu, y = Cx + Du$. It should be noted that, in general, state reachability is neither necessary nor sufficient for output reachability. Notice for example that if rank $D = p$, then the system is output reachable.

(b) Let $D = 0$. Show that if $(A, B)$ is (state) reachable, then $\{A, B, C, D\}$ is output reachable if and only if rank $C = p$.

(c) Let $A = \begin{bmatrix} 1 & 0 & 0 \\ 0 & -2 & 0 \\ 0 & 0 & -1 \end{bmatrix}, B = \begin{bmatrix} 1 \\ 0 \\ 1 \end{bmatrix}, C = [1, 1, 0]$, and $D = 0$.

(i) Is the system output reachable? Is it state reachable?

(ii) Let $x(0) = 0$. Determine an appropriate input sequence to transfer the output to $y_1 = 3$ in minimum time. Repeat for $x(0) = [1, -1, 2]^T$.

**5.6.** (a) Given $\dot{x} = Ax + Bu, y = Cx + Du$, show that this system is output reachable if and only if the rows of the $p \times m$ transfer matrix $H(s)$ are linearly independent over the field of complex numbers. In view of this result, is the system $H(s) = \begin{bmatrix} \frac{1}{s+2} \\ \frac{s}{s+1} \end{bmatrix}$ output reachable?

(b) Similarly, for discrete-time systems, the system is output reachable if and only if the rows of the transfer function matrix $H(z)$ are linearly independent over the field of complex numbers. Consider now the system of Exercise 5.5 and determine whether it is output reachable.

**5.7.** Show that the circuit depicted in Figure 5.4 with input $u$ and output $y$ is neither state reachable nor observable but is output reachable.

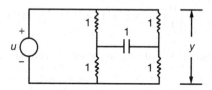

**Figure 5.4.** Circuit for Exercise 5.7

**5.8.** A system $\dot{x} = Ax + Bu, y = Cx + Du$ is called *output function controllable* if there exists an input $u(t)$, $t \in [0, \infty)$, that will cause the output $y(t)$ to follow a prescribed trajectory for $0 \le t < \infty$, assuming that the system is at rest at $t = 0$. It is easiest to derive a test for output function controllability in terms of the $p \times m$ transfer function matrix $H(s)$, and this is the approach taken in the following. We say that the $m \times p$ rational matrix $H_R(s)$ is a *right inverse of* $H(s)$ if

$$H(s)H_R(s) = I_p.$$

(a) Show that the right inverse $H_R(s)$ exists if and only if rank $H(s) = p$. *Hint:* In the sufficiency proof, select $H_R = H^T(HH^T)^{-1}$, the (right) pseudoinverse of $H$.

(b) Show that the system is output function controllable if and only if $H(s)$ has a right inverse $H_R(s)$. *Hint:* Consider $\hat{y} = H\hat{u}$. In the necessity proof, show that if rank $H < p$, then the system may not be output function controllable.

*Input function observability* is the dual to output function controllablity. Here, the *left inverse of* $H(s)$, $H_L(s)$, is of interest and is defined by

$$H_L(s)H(s) = I_m.$$

(c) Show that the left inverse $H_L(s)$ of $H(s)$ exists if and only if rank $H(s) = m$. *Hint:* This is the dual result to part (a).

(d) Let $H(s) = \left[\frac{s+1}{s}, \frac{1}{s}\right]$ and characterize all inputs $u(t)$ that will cause the system (at rest at $t = 0$) to exactly follow a step, $\hat{y}(s) = 1/s$.

Part (d) points to a variety of questions that may arise when inverses are considered, including: Is $H_R(s)$ proper? Is it unique? Is it stable? What is the minimum degree possible?

**5.9.** Consider the system $\dot{x} = Ax + Bu, y = Cx$. Show that output function controllability implies output controllability (-from-the-origin, or reachability).

**5.10.** Given $x(k+1) = \begin{bmatrix} 1 & 1 \\ 0 & 1 \end{bmatrix} x(k) + \begin{bmatrix} 1 \\ 1 \end{bmatrix} u(k), y(k) = \begin{bmatrix} 1 & 0 \\ 1 & 1 \end{bmatrix} x(k)$, and assume zero initial conditions.

(a) Is there a sequence of inputs $\{u(0), u(1), \dots\}$ that transfers the output from $y(0) = \begin{bmatrix} 0 \\ 0 \end{bmatrix}$ to $\begin{bmatrix} 0 \\ 1 \end{bmatrix}$ in finite time? If the answer is yes, determine such a sequence.

(b) Characterize all outputs that can be reached from the zero output ($y(0) = \begin{bmatrix} 0 \\ 0 \end{bmatrix}$), in one step.

**5.11.** Suppose that for system $x(k+1) = \begin{bmatrix} 1 & 1 & 0 \\ 0 & 1 & 0 \\ 0 & 0 & 1 \end{bmatrix} x(k), y(k) = \begin{bmatrix} 1 & 1 & 0 \\ 0 & 1 & 0 \end{bmatrix} x(k)$,

it is known that $y(0) = y(1) = y(2) = \begin{bmatrix} 1 \\ 0 \end{bmatrix}$. Based on this information, what can be said about the initial condition $x(0)$?

**5.12.** (a) Consider the system $\dot{x} = Ax + Bu, y = Cx + Du$, where $(A, C)$ is assumed to be observable. Express $x(t)$ as a function of $y(t), u(t)$ and their derivatives. *Hint:* Write $y(t), y^{(1)}(t), \dots, y^{(n-1)}(t)$ in terms of $x(t)$ and $u(t), u^{(1)}(t), \dots, u^{(n-1)}(t)$ ( $x(t) \in R^n$ ).

(b) Given the system $\dot{x} = Ax + Bu, y = Cx + Du$ with $(A, C)$ observable. Determine $x(0)$ in terms of $y(t), u(t)$ and their derivatives up to order $n - 1$. Note that in general this is not a practical way of determining $x(0)$, since this method requires differentiation of signals, which is very susceptible to measurement noise.

(c) Consider the system $x(k + 1) = Ax(k) + Bu(k), y(k) = Cx(k) + Du(k)$, where $(A, C)$ is observable. Express $x(k)$ as a function of $y(k), y(k + 1), \dots, y(k + n - 1)$ and $u(k), u(k + 1), \dots, y(k + n - 1)$. *Hint:* Express $y(k), \dots, y(k + n - 1)$ in terms of $x(k)$ and $u(k), u(k + 1), \dots, u(k + n - 1) [ x(k) \in R^n ]$. Note the relation to expression (5.88) in Section 5.4.

# 6

# Controllability and Observability: Special Forms

## 6.1 Introduction

In this chapter, important special forms for the state-space description of time-invariant systems are presented. These forms are obtained by means of similarity transformations and are designed to reveal those features of a system that are related to the properties of controllability and observability. In Section 6.2, special state-space forms that separate the controllable (observable) from the uncontrollable (unobservable) part of a system are presented. These forms, referred to as the standard forms for uncontrollable and unobservable systems, are very useful in establishing a number of results. In particular, these forms are used in Section 6.3 to derive alternative tests for controllability and observability and in Section 7.2 to relate state-space and input–output descriptions. In Section 6.4 the controller and observer state-space forms are introduced. These are useful in the study of state-space realizations in Chapter 8 and state feedback and state estimators in Chapter 9.

## 6.2 Standard Forms for Uncontrollable and Unobservable Systems

We consider time-invariant systems described by equations of the form

$$\dot{x} = Ax + Bu, \quad y = Cx + Du, \tag{6.1}$$

where $A \in R^{n \times n}$, $B \in R^{n \times m}$, $C \in R^{p \times n}$, and $D \in R^{p \times m}$. It was shown in the previous chapter that this system is state reachable if and only if the $n \times mn$ controllability matrix

$$\mathcal{C} \triangleq [B, AB, \dots, A^{n-1}B] \tag{6.2}$$

has full row rank $n$; i.e., $\text{rank}\,\mathcal{C} = n$. If the system is reachable (or controllable-from-the-origin), then it is also controllable (or controllable-to-the-origin), and vice versa (see Section 5.3.1).

It was also shown earlier that system (6.1) is state observable if and only if the $pn \times n$ observability matrix

$$\mathcal{O} \triangleq \begin{bmatrix} C \\ CA \\ \vdots \\ CA^{n-1} \end{bmatrix} \tag{6.3}$$

has full column rank; i.e., rank $\mathcal{O} = n$. If the system is observable, then it is also constructible, and vice versa (see Section 5.4.1).

Similar results were also derived for discrete-time time-invariant systems described by equations of the form

$$x(k+1) = Ax(k) + Bu(k), \quad y(k) = Cx(k) + Du(k). \tag{6.4}$$

Again, rank $\mathcal{C} = n$ and rank $\mathcal{O} = n$ are the necessary and sufficient conditions for state reachability and observability, respectively. Reachability always implies controllability and observability always implies constructibility, as in the continuous-time case. However, in the discrete-time case, controllability does not necessarily imply reachability and constructibility does not imply observability, unless $A$ is nonsingular (see Sections 5.3.2 and 5.4.2).

Next, we will introduce standard forms for unreachable and unobservable systems both for the continuous-time and the discrete-time time-invariant cases. These forms will be referred to as standard forms for *uncontrollable systems*, rather than unreachable systems, and standard forms for *unobservable systems*, respectively.

### 6.2.1 Standard Form for Uncontrollable Systems

If the system (6.1) [or (6.4)] is not completely reachable or controllable-from-the-origin, then it is possible to "separate" the controllable part of the system by means of an appropriate similarity transformation. This amounts to changing the basis of the state space so that all the vectors in the reachable subspace $R_r$ have a certain structure. In particular, let rank $\mathcal{C} = n_r < n$; i.e., the pair $(A, B)$ is not controllable. This implies that the subspace $R_r = \mathcal{R}(\mathcal{C})$ has dimension $n_r$. Let $\{v_1, v_2, \ldots, v_{n_r}\}$ be a basis for $R_r$. These $n_r$ vectors can be, for example, any $n_r$ linearly independent columns of $\mathcal{C}$. Define the $n \times n$ similarity transformation matrix

$$Q \triangleq [v_1, v_2, \ldots, v_{n_r}, Q_{n-n_r}], \tag{6.5}$$

where the $n \times (n - n_r)$ matrix $Q_{n-n_r}$ contains $n - n_r$ linearly independent vectors chosen so that $Q$ is nonsingular. There are many such choices. We are now in a position to prove the following result.

**Lemma 6.1.** *For $(A, B)$ uncontrollable, there exists a nonsingular matrix $Q$ such that*

$$\widehat{A} = Q^{-1}AQ = \begin{bmatrix} A_1 & A_{12} \\ 0 & A_2 \end{bmatrix} \quad and \quad \widehat{B} = Q^{-1}B = \begin{bmatrix} B_1 \\ 0 \end{bmatrix}, \qquad (6.6)$$

*where $A_1 \in R^{n_r \times n_r}, B_1 \in R^{n_r \times m}$, and the pair $(A_1, B_1)$ is controllable. The pair $(\widehat{A}, \widehat{B})$ is in the standard form for uncontrollable systems.*

*Proof.* We need to show that

$$AQ = A[v_1, \ldots, v_{n_r}, Q_{n-n_r}] = [v_1, \ldots, v_{n_r}, Q_{n-n_r}] \begin{bmatrix} A_1 & A_{12} \\ 0 & A_2 \end{bmatrix} = Q\widehat{A}.$$

Since the subspace $R_r$ is $A$-invariant (see Lemma 5.19), $Av_i \in R_r$, which can be written as a linear combination of only the $n_r$ vectors in a basis of $R_r$. Thus, $A_1$ in $\widehat{A}$ is an $n_r \times n_r$ matrix, and the $(n - n_r) \times n_r$ matrix below it in $\widehat{A}$ is a zero matrix. Similarly, we also need to show that

$$B = [v_1, \ldots, v_{n_r}, Q_{n-n_r}] \begin{bmatrix} B_1 \\ 0 \end{bmatrix} = Q\widehat{B}.$$

But this is true for similar reasons: The columns of $B$ are in the range of $\mathcal{C}$ or in $R_r$.  ∎

The $n \times nm$ controllability matrix $\widehat{\mathcal{C}}$ of $(\widehat{A}, \widehat{B})$ is

$$\widehat{\mathcal{C}} = [\widehat{B}, \widehat{A}\widehat{B}, \ldots, \widehat{A}^{n-1}\widehat{B}] = \begin{bmatrix} B_1 & A_1 B_1 & \cdots & A_1^{n-1} B_1 \\ 0 & 0 & \cdots & 0 \end{bmatrix}, \qquad (6.7)$$

which clearly has $\operatorname{rank}\widehat{\mathcal{C}} = \operatorname{rank}[B_1, A_1 B_1, \ldots, A_1^{n_r-1} B_1, \ldots, A_1^{n-1} B_1] = n_r$. Note that

$$\widehat{\mathcal{C}} = Q^{-1}\mathcal{C}. \qquad (6.8)$$

The range of $\widehat{\mathcal{C}}$ is the controllable subspace of $(\widehat{A}, \widehat{B})$. It contains vectors only of the form $[\alpha^T, 0]^T$, where $\alpha \in R^{n_r}$. Since $\dim \mathcal{R}(\widehat{\mathcal{C}}) = \operatorname{rank}\widehat{\mathcal{C}} = n_r$, every vector of the form $[\alpha^T, 0]^T$ is a controllable (state) vector. In other words, the similarity transformation has changed the basis of $R^n$ in such a manner so that all controllable vectors, expressed in terms of this new basis, have this very particular structure with zeros in the last $n - n_r$ entries.

Given system (6.1) [or (6.4)], if a new state $\hat{x}(t)$ is taken to be $\hat{x}(t) = Q^{-1}x(t)$, then

$$\dot{\hat{x}} = \widehat{A}\hat{x} + \widehat{B}u, \quad y = \widehat{C}\hat{x} + \widehat{D}u, \qquad (6.9)$$

where $\widehat{A} = Q^{-1}AQ, \widehat{B} = Q^{-1}B, \widehat{C} = CQ$, and $\widehat{D} = D$ constitutes an equivalent representation (see Section 3.4.3). For $Q$ as in Lemma 6.1, we obtain

$$\begin{bmatrix} \dot{\hat{x}}_1 \\ \dot{\hat{x}}_2 \end{bmatrix} = \begin{bmatrix} A_1 & A_{12} \\ 0 & A_2 \end{bmatrix} \begin{bmatrix} \hat{x}_1 \\ \hat{x}_2 \end{bmatrix} + \begin{bmatrix} B_1 \\ 0 \end{bmatrix} u, y = [C_1, C_2] \begin{bmatrix} \hat{x}_1 \\ \hat{x}_2 \end{bmatrix} + Du, \qquad (6.10)$$

where $\hat{x} = [\hat{x}_1^T, \hat{x}_2^T]^T$ with $\hat{x}_1 \in R^{n_r}$ and where $(A_1, B_1)$ is controllable. The matrix $\widehat{C} = [C_1, C_2]$ does not have any particular structure. This representation is called a *standard form for the uncontrollable system*. The state equation can now be written as

$$\dot{\hat{x}}_1 = A_1\hat{x}_1 + B_1u + A_{12}\hat{x}_2, \dot{\hat{x}}_2 = A_2\hat{x}_2, \qquad (6.11)$$

which shows that the input $u$ does not affect the trajectory component $\hat{x}_2(t)$ at all, and therefore, $\hat{x}_2(t)$ is determined only by the value of its initial vector. The input $u$ certainly affects $\hat{x}_1(t)$. Note also that the trajectory component $\hat{x}_1(t)$ is also influenced by $\hat{x}_2(t)$. In fact,

$$\hat{x}_1(t) = e^{A_1 t}\hat{x}_1(0) + \int_0^t e^{A_1(t-\tau)}B_1u(\tau)d\tau + \left[\int_0^t e^{A_1(t-\tau)}A_{12}e^{A_2\tau}d\tau\right]\hat{x}_2(0).$$
$$(6.12)$$

The $n_r$ eigenvalues of $A_1$ and the corresponding modes are the *controllable eigenvalues* and *controllable modes* of the pair $(A, B)$ or of system (6.1) [or of (6.4)]. The $n - n_r$ eigenvalues of $A_2$ and the corresponding modes are the *uncontrollable eigenvalues* and *uncontrollable modes*, respectively.

It is interesting to observe that in the zero-state response of the system (zero initial conditions), the uncontrollable modes are completely absent. In particular, in the solution $x(t) = e^{At}x(0) + \int_0^t e^{A(t-\tau)}Bu(\tau)d\tau$ of $\dot{x} = Ax + Bu$, given $x(0)$, notice that

$$e^{A(t-\tau)}B = [Qe^{\widehat{A}(t-\tau)}Q^{-1}][Q\widehat{B}] = Q\begin{bmatrix} e^{A_1(t-\tau)}B_1 \\ 0 \end{bmatrix},$$

where $A_1$ [from (6.6)] contains only the controllable eigenvalues. Therefore, the input $u(t)$ cannot directly influence the uncontrollable modes. Note, however, that the uncontrollable modes do appear in the zero-input response $e^{At}x(0)$. The same observations can be made for discrete-time systems (6.4) where the quantity $A^k B$ is of interest.

---

***Example 6.2.*** Given $A = \begin{bmatrix} 0 & -1 & 1 \\ 1 & -2 & 1 \\ 0 & 1 & -1 \end{bmatrix}$ and $B = \begin{bmatrix} 1 & 0 \\ 1 & 1 \\ 1 & 2 \end{bmatrix}$, we wish to reduce

system (6.1) to the standard form (6.6). Here

$$C = [B, AB, A^2B] = \begin{bmatrix} 1 & 0 & 0 & 1 & 0 & -1 \\ 1 & 1 & 0 & 0 & 0 & 0 \\ 1 & 2 & 0 & -1 & 0 & 1 \end{bmatrix}$$

and rank$C = n_r = 2 < 3 = n$. Thus, the subspace $R_r = \mathcal{R}(C)$ has dimension $n_r = 2$, and a basis $\{v_1, v_2\}$ can be found by taking two linearly independent columns of $C$, say, the first two, to obtain

$$Q = [v_1, v_2, Q_1] = \begin{bmatrix} 1 & 0 & 0 \\ 1 & 1 & 0 \\ 1 & 2 & 1 \end{bmatrix}.$$

The third column of $Q$ was selected so that $Q$ is nonsingular. Note that the first two columns of $Q$ could have been the first and fourth columns of $\mathcal{C}$ instead, or any other two linearly independent vectors obtained as a linear combination of the columns in $\mathcal{C}$. For the above choice for $Q$, we have

$$\widehat{A} = Q^{-1}AQ = \begin{bmatrix} 1 & 0 & 0 \\ -1 & 1 & 0 \\ 1 & -2 & 1 \end{bmatrix} \begin{bmatrix} 0 & -1 & 1 \\ 1 & -2 & 1 \\ 0 & 1 & -1 \end{bmatrix} \begin{bmatrix} 1 & 0 & 0 \\ 1 & 1 & 0 \\ 1 & 2 & 1 \end{bmatrix}$$

$$= \begin{bmatrix} 0 & -1 & 1 \\ 1 & -1 & 0 \\ -2 & 4 & -2 \end{bmatrix} \begin{bmatrix} 1 & 0 & 0 \\ 1 & 1 & 0 \\ 1 & 2 & 1 \end{bmatrix}$$

$$= \begin{bmatrix} 0 & 1 & 1 \\ 0 & -1 & 0 \\ 0 & 0 & -2 \end{bmatrix} = \begin{bmatrix} A_1 & A_{12} \\ 0 & A_2 \end{bmatrix},$$

$$\widehat{B} = Q^{-1}B = \begin{bmatrix} 1 & 0 & 0 \\ -1 & 1 & 0 \\ 1 & -2 & 1 \end{bmatrix} \begin{bmatrix} 1 & 0 \\ 1 & 1 \\ 1 & 2 \end{bmatrix} = \begin{bmatrix} 1 & 0 \\ 0 & 1 \\ 0 & 0 \end{bmatrix} = \begin{bmatrix} B_1 \\ 0 \end{bmatrix},$$

where $(A_1, B_1)$ is controllable. The matrix $A$ has three eigenvalues at $0, -1$, and $-2$. It is clear from $(\widehat{A}, \widehat{B})$ that the eigenvalues $0, -1$ are controllable (in $A_1$), whereas $-2$ is an uncontrollable eigenvalue (in $A_2$).

---

## 6.2.2 Standard Form for Unobservable Systems

The standard form for an unobservable system can be derived in a similar way as the standard form of uncontrollable systems. If the system (6.1) [or (6.4)] is not completely state observable, then it is possible to "separate" the unobservable part of the system by means of a similarity transformation. This amounts to changing the basis of the state space so that all the vectors in the unobservable subspace $R_{\bar{o}}$ have a certain structure.

As in the preceding discussion concerning systems or pairs $(A, B)$ that are not completely controllable, we shall select a similarity transformation $Q$ to reduce a pair $(A, C)$, which is not completely observable, to a particular form. This can be accomplished in two ways. The simplest way is to invoke duality and to work with the pair $(A_D = A^T, B_D = C^T)$, which is not controllable (refer to the discussion of dual systems in Section 5.2.3). If Lemma 6.1 is applied, then

$$\widehat{A}_D = Q_D^{-1}A_DQ_D = \begin{bmatrix} A_{D1} & A_{D12} \\ 0 & A_{D2} \end{bmatrix}, \quad \widehat{B}_D = Q_D^{-1}B_D = \begin{bmatrix} B_{D1} \\ 0 \end{bmatrix},$$

where $(A_{D1}, B_{D1})$ is controllable.

Taking the dual again, we obtain the pair $(\widehat{A}, \widehat{C})$, which has the desired properties. In particular,

$$\widehat{A} = \widehat{A}_D^T = Q_D^T A_D^T (Q_D^T)^{-1} = Q_D^T A (Q_D^T)^{-1} = \begin{bmatrix} A_{D1}^T & 0 \\ A_{D12}^T & A_{D2}^T \end{bmatrix},$$

$$\widehat{C} = \widehat{B}_D^T = B_D^T (Q_D^T)^{-1} = C(Q_D^T)^{-1} = [B_{D1}^T, 0],$$

(6.13)

where $(A_{D1}^T, B_{D1}^T)$ is completely observable by duality (see Lemma 5.7).

---

**Example 6.3.** Given $A = \begin{bmatrix} 0 & 1 & 0 \\ -1 & -2 & 1 \\ 1 & 1 & -1 \end{bmatrix}$ and $C = \begin{bmatrix} 1 & 1 & 1 \\ 0 & 1 & 2 \end{bmatrix}$, we wish to reduce system (6.1) to the standard form (6.13). To accomplish this, let $A_D = A^T$ and $B_D = C^T$. Notice that the pair $(A_D, B_D)$ is precisely the pair $(A, B)$ of Example 6.2.

---

A pair $(A, C)$ can of course also be reduced directly to the standard form for unobservable systems. This is accomplished in the following.

Consider the system (6.1) [or (6.4)] and the observability matrix $\mathcal{O}$ in (6.3). Let rank $\mathcal{O} = n_o < n$; i.e., the pair $(A, C)$ is not completely observable. This implies that the unobservable subspace $R_{\bar{o}} = \mathcal{N}(\mathcal{O})$ has dimension $n - n_o$. Let $\{v_1, \ldots, v_{n-n_o}\}$ be a basis for $R_{\bar{o}}$, and define an $n \times n$ similarity transformation matrix $Q$ as

$$Q \triangleq [Q_{n_o}, v_1, \ldots, v_{n-n_o}],$$

(6.14)

where the $n \times n_o$ matrix $Q_{n_o}$ contains $n_o$ linearly independent vectors chosen so that $Q$ is nonsingular. Clearly, there are many such choices.

**Lemma 6.4.** *For $(A, C)$ unobservable, there is a nonsingular matrix $Q$ such that*

$$\widehat{A} = Q^{-1} A Q = \begin{bmatrix} A_1 & 0 \\ A_{21} & A_2 \end{bmatrix} \quad and \quad \widehat{C} = CQ = [C_1, \, 0],$$

(6.15)

*where $A_1 \in R^{n_o \times n_o}, C_1 \in R^{p \times n_o}$, and the pair $(A_1, C_1)$ is observable. The pair $(\widehat{A}, \widehat{C})$ is in the standard form for unobservable systems.*

*Proof.* We need to show that

$$AQ = A[Q_{n_0}, v_1, \ldots, v_{n-n_o}] = [Q_{n_o}, v_1, \ldots, v_{n-n_o}] \begin{bmatrix} A_1 & 0 \\ A_{21} & A_2 \end{bmatrix} = Q\widehat{A}.$$

Since the unobservable subspace $R_{\bar{o}}$ is $A$-invariant (see Lemma 5.49), $Av_i \in R_{\bar{o}}$, which can be written as a linear combination of only the $n - n_o$ vectors in a basis of $R_{\bar{o}}$. Thus, $A_2$ in $\widehat{A}$ is an $(n - n_o) \times (n - n_o)$ matrix, and the $n_o \times (n - n_o)$ matrix above it in $\widehat{A}$ is a zero matrix. Similarly, we also need to show that

$$CQ = C[Q_{n_o}, v_1, \ldots, v_{n-n_o}] = [C_1, 0] = \widehat{C}.$$

This is true since $Cv_i = 0$. ∎

The $pn \times n$ observability matrix $\widehat{\mathcal{O}}$ of $(\widehat{A}, \widehat{C})$ is

$$\widehat{\mathcal{O}} = \begin{bmatrix} \widehat{C} \\ \widehat{C}\widehat{A} \\ \vdots \\ \widehat{C}\widehat{A}^{n-1} \end{bmatrix} = \begin{bmatrix} C_1 & 0 \\ C_1 A_1 & 0 \\ \vdots & \vdots \\ C_1 A_1^{n-1} & 0 \end{bmatrix}, \tag{6.16}$$

which clearly has

$$\operatorname{rank}\widehat{\mathcal{O}} = \operatorname{rank} \begin{bmatrix} C_1 \\ C_1 A_1 \\ \vdots \\ C_1 A_1^{n_o - 1} \\ \vdots \\ C_1 A_1^{n-1} \end{bmatrix} = n_o.$$

Note that

$$\widehat{\mathcal{O}} = \mathcal{O}Q. \tag{6.17}$$

The null space of $\widehat{\mathcal{O}}$ is the unobservable subspace of $(\widehat{A}, \widehat{C})$. It contains vectors only of the form $[0, \alpha^T]^T$, where $\alpha \in R^{n-n_o}$. Since $\dim \mathcal{N}(\widehat{\mathcal{O}}) = n - \operatorname{rank}\widehat{\mathcal{O}} = n - n_o$, every vector of the form $[0, \alpha^T]^T$ is an unobservable (state) vector. In other words, the similarity transformation has changed the basis of $R^n$ in such a manner so that all unobservable vectors expressed in terms of this new basis have this very particular structure—zeros in the first $n_o$ entries.

For $Q$ chosen as in Lemma 6.4,

$$\begin{bmatrix} \dot{\hat{x}}_1 \\ \dot{\hat{x}}_2 \end{bmatrix} = \begin{bmatrix} A_1 & 0 \\ A_{21} & A_2 \end{bmatrix} \begin{bmatrix} \hat{x}_1 \\ \hat{x}_2 \end{bmatrix} + \begin{bmatrix} B_1 \\ B_2 \end{bmatrix} u, y = [C_1, 0] \begin{bmatrix} \hat{x}_1 \\ \hat{x}_2 \end{bmatrix} + Du, \tag{6.18}$$

where $\hat{x} = [\hat{x}_1^T, \hat{x}_2^T]^T$ with $\hat{x}_1 \in R^{n_o}$ and $(A_1, C_1)$ is observable. The matrix $\widehat{B} = [B_1^T, B_2^T]^T$ does not have any particular form. This representation is called a *standard form for the unobservable system*.

The $n_o$ eigenvalues of $A_1$ and the corresponding modes are called *observable eigenvalues* and *observable modes* of the pair $(A, C)$ or of the system (6.1) [or of (6.4)]. The $n - n_o$ eigenvalues of $A_2$ and the corresponding modes are called *unobservable eigenvalues* and *unobservable modes*, respectively.

Notice that the trajectory component $\hat{x}(t)$, which is observed via the output $y$, is not influenced at all by $\hat{x}_2$, the trajectory of which is determined primarily by the eigenvalues of $A_2$.

The unobservable modes of the system are completely absent from the output. In particular, given $\dot{x} = Ax + Bu, y = Cx$ with initial state $x(0)$, we have

$$y(t) = Ce^{At}x(0) + \int_0^t Ce^{A(t-\tau)}Bu(\tau)d\tau$$

and $Ce^{At} = [\widehat{C}Q^{-1}][Qe^{\widehat{A}t}Q^{-1}] = [C_1e^{A_1t}, 0]Q^{-1}$, where $A_1$ [from (6.15)] contains only the observable eigenvalues. Therefore, the unobservable modes cannot be seen by observing the output. The same observations can be made for discrete-time systems where the quantity $CA^k$ is of interest.

---

***Example 6.5.*** Given $A = \begin{bmatrix} 0 & 1 \\ -2 & -3 \end{bmatrix}$ and $C = [1,1]$, we wish to reduce system (6.1) to the standard form (6.15). To accomplish this, we compute $\mathcal{O} = \begin{bmatrix} C \\ CA \end{bmatrix} = \begin{bmatrix} 1 & 1 \\ -2 & -2 \end{bmatrix}$, which has rank $\mathcal{O} = n_o = 1 < 2 = n$. Therefore, the unobservable subspace $R_{\bar{o}} = \mathcal{N}(\mathcal{O})$ has dimension $n - n_o = 1$. In view of (6.14),

$$Q = [Q_1, v_1] = \begin{bmatrix} 0 & 1 \\ 1 & -1 \end{bmatrix},$$

where $v_1 = [1, -1]^T$ is a basis for $R_{\bar{o}}$, and $Q_1$ was chosen so that $Q$ is nonsingular. Then

$$\widehat{A} = Q^{-1}AQ = \begin{bmatrix} 1 & 1 \\ 1 & 0 \end{bmatrix} \begin{bmatrix} 0 & 1 \\ -2 & -3 \end{bmatrix} \begin{bmatrix} 0 & 1 \\ 1 & -1 \end{bmatrix}$$

$$= \begin{bmatrix} -2 & 0 \\ 1 & -1 \end{bmatrix} = \begin{bmatrix} A_1 & 0 \\ A_{21} & A_2 \end{bmatrix},$$

$$\widehat{C} = CQ = [1, 1] \begin{bmatrix} 0 & 1 \\ 1 & -1 \end{bmatrix} = [1, 0] = [C_1, 0],$$

where $(A_1, C_1)$ is observable. The matrix $A$ has two eigenvalues at $-1, -2$. It is clear from $(\widehat{A}, \widehat{C})$ that the eigenvalue $-2$ is observable (in $A_1$), whereas $-1$ is an unobservable eigenvalue (in $A_2$).

---

### 6.2.3 Kalman's Decomposition Theorem

Lemmas 6.1 and 6.4 can be combined to obtain an equivalent representation of (6.1) where the reachable and observable parts of this system can readily be identified. We consider system (6.9) and proceed, in the following, to construct the $n \times n$ required similarity transformation matrix $Q$.

As before, we let $n_r$ denote the dimension of the controllable subspace $R_r$; i.e., $n_r = \dim R_r = \dim \mathcal{R}(\mathcal{C}) = \text{rank}\,\mathcal{C}$. The dimension of the unobservable subspace $R_{\bar{o}} = \mathcal{N}(\mathcal{O})$ is given by $n_{\bar{o}} = n - \text{rank}\,\mathcal{O} = n - n_o$. Let $n_{r\bar{o}}$ be the dimension of the subspace $R_{r\bar{o}} \triangleq R_r \cap R_{\bar{o}}$, which contains all the state vectors $x \in R^n$ that are controllable but unobservable. We choose

$$Q \triangleq [v_1, \ldots, v_{n_r-n_{r\bar{o}}+1}, \ldots, v_{n_r}, Q_N, \hat{v}_1, \ldots, \hat{v}_{n_{\bar{o}}-n_{r\bar{o}}}], \qquad (6.19)$$

where the $n_r$ vectors in $\{v_1, \ldots, v_{n_r}\}$ form a basis for $R_r$. The last $n_{r\bar{o}}$ vectors $\{v_{n_r-n_{r\bar{o}}+1}, \ldots, v_{n_r}\}$ in the basis for $R_r$ are chosen so that they form a basis for $R_{r\bar{o}} = R_r \cap R_{\bar{o}}$. The $n_{\bar{o}} - n_{r\bar{o}} = (n - n_o - n_{r\bar{o}})$ vectors $\{\hat{v}_1, \ldots, \hat{v}_{n_{\bar{o}}-n_{r\bar{o}}}\}$ are selected so that when taken together with the $n_{r\bar{o}}$ vectors $\{v_{n_r-n_{r\bar{o}}+1}, \ldots, v_{n_r}\}$ they form a basis for $R_{\bar{o}}$, the unobservable subspace. The remaining $N = n - (n_r + n_{\bar{o}} - n_{r\bar{o}})$ columns in $Q_N$ are simply selected so that $Q$ is nonsingular.

The following theorem is called the *Canonical Structure Theorem* or *Kalman's Decomposition Theorem*.

**Theorem 6.6.** *For $(A, B)$ uncontrollable and $(A, C)$ unobservable, there is a nonsingular matrix $Q$ such that*

$$\widehat{A} = Q^{-1}AQ = \begin{bmatrix} A_{11} & 0 & A_{13} & 0 \\ A_{21} & A_{22} & A_{23} & A_{24} \\ 0 & 0 & A_{33} & 0 \\ 0 & 0 & A_{43} & A_{44} \end{bmatrix}, \quad \widehat{B} = Q^{-1}B = \begin{bmatrix} B_1 \\ B_2 \\ 0 \\ 0 \end{bmatrix}, \quad (6.20)$$

$$\widehat{C} = CQ = [C_1, 0, C_3, 0],$$

*where*

*(i)* $(A_c, B_c)$ *with*

$$A_c \triangleq \begin{bmatrix} A_{11} & 0 \\ A_{21} & A_{22} \end{bmatrix} \quad and \quad B_c \triangleq \begin{bmatrix} B_1 \\ B_2 \end{bmatrix}$$

*is controllable, where $A_c \in R^{n_r \times n_r}, B_c \in R^{n_r \times m}$;*
*(ii)* $(A_o, C_o)$ *with*

$$A_o \triangleq \begin{bmatrix} A_{11} & A_{13} \\ 0 & A_{33} \end{bmatrix} \quad and \quad C_o \triangleq [C_1, C_3]$$

*is observable, where $A_o \in R^{n_o \times n_o}$ and $C_o \in R^{p \times n_o}$ and where the dimensions of the matrices $A_{ij}, B_i,$ and $C_j$ are as follows:*

$$A_{11} : (n_r - n_{r\bar{o}}) \times (n_r - n_{r\bar{o}}), \qquad A_{22} : n_{r\bar{o}} \times n_{r\bar{o}},$$
$$A_{33} : (n - (n_r + n_{\bar{o}} - n_{r\bar{o}})) \times \qquad A_{44} : (n_{\bar{o}} - n_{r\bar{o}}) \times (n_{\bar{o}} - n_{r\bar{o}}),$$
$$(n - (n_r + n_{\bar{o}} - n_{r\bar{o}})),$$
$$B_1 : (n_r - n_{r\bar{o}}) \times m, \qquad B_2 : n_{r\bar{o}} \times m,$$
$$C_1 : p \times (n_r - n_{r\bar{o}}), \qquad C_3 : p \times (n - (n_r + n_{\bar{o}} - n_{r\bar{o}}));$$

*(iii) the triple $(A_{11}, B_1, C_1)$ is such that $(A_{11}, B_1)$ is controllable and $(A_{11}, C_1)$ is observable.*

*Proof.* For details of the proof, refer to [6] and to [7], where further clarifications to [6] and an updated method of selecting Q are given. ∎

The similarity transformation (6.19) has altered the basis of the state space in such a manner that the vectors in the controllable subspace $R_r$, the vectors in the unobservable subspace $R_{\bar{o}}$, and the vectors in the subspace $R_{r\bar{o}} \cap R_{\bar{o}}$ all have specific forms. To see this, we construct the controllability matrix $\widehat{C} = [\widehat{B}, \ldots, \widehat{A}^{n-1}\widehat{B}]$ whose range is the controllable subspace and the observability matrix $\widehat{O} = [\widehat{C}^T, \ldots, (\widehat{C}\widehat{A}^{n-1})^T]^T$, whose null space is the unobservable subspace. Then, all controllable states are of the form $[x_1^T, x_2^T, 0, 0]^T$, all the unobservable ones have the structure $[0, x_2^T, 0, x_4^T]^T$, and states of the form $[0, x_2^T, 0, 0]^T$ characterize $R_{r\bar{o}}$; i.e., they are controllable but unobservable.

Similarly to the previous two lemmas, the eigenvalues of $\widehat{A}$, or of $A$, are the eigenvalues of $A_{11}, A_{22}, A_{33}$, and $A_{44}$; i.e.,

$$|\lambda I - A| = |\lambda I - \widehat{A}| = |\lambda I - A_{11}||\lambda I - A_{22}||\lambda I - A_{33}||\lambda I - A_{44}|. \quad (6.21)$$

If we consider the representation $\{\widehat{A}, \widehat{B}, \widehat{C}, \widehat{D}\}$ given in (6.20), then

$$\begin{bmatrix} \dot{\hat{x}}_1 \\ \dot{\hat{x}}_2 \\ \dot{\hat{x}}_3 \\ \dot{\hat{x}}_4 \end{bmatrix} = \begin{bmatrix} A_{11} & 0 & A_{13} & 0 \\ A_{21} & A_{22} & A_{23} & A_{24} \\ 0 & 0 & A_{33} & 0 \\ 0 & 0 & A_{43} & A_{44} \end{bmatrix} \begin{bmatrix} \hat{x}_1 \\ \hat{x}_2 \\ \hat{x}_3 \\ \hat{x}_4 \end{bmatrix} + \begin{bmatrix} B_1 \\ B_2 \\ 0 \\ 0 \end{bmatrix} u,$$

$$y = [C_1, \ 0, \ C_3, \ 0] \begin{bmatrix} \hat{x}_1 \\ \hat{x}_2 \\ \hat{x}_3 \\ \hat{x}_4 \end{bmatrix} + Du. \quad (6.22)$$

This shows that the trajectory components corresponding to $\hat{x}_3$ and $\hat{x}_4$ are not affected by the input $u$. The modes associated with the eigenvalues of $A_{33}$ and $A_{44}$ determine the trajectory components for $\hat{x}_3$ and $\hat{x}_4$ (compare this with the results in Lemma 6.1). Similarly to Lemma 6.4, the trajectory components for $\hat{x}_2$ and $\hat{x}_4$ are not influenced by $\hat{x}_1$ and $\hat{x}_3$ (observed via $y$), and they are determined by the eigenvalues of $A_{22}$ and $A_{44}$. The following is now apparent (see also Figure 6.1):

The eigenvalues of

$A_{11}$ are controllable and observable,
$A_{22}$ are controllable and unobservable,
$A_{33}$ are uncontrollable and observable,
$A_{44}$ are uncontrollable and unobservable.

---

**Example 6.7.** Given $A = \begin{bmatrix} 0 & -1 & 1 \\ 1 & -2 & 1 \\ 0 & 1 & -1 \end{bmatrix}$, $B = \begin{bmatrix} 1 & 0 \\ 1 & 1 \\ 1 & 2 \end{bmatrix}$, and $C = [0, 1, 0]$, we wish to reduce system (6.1) to the canonical structure (or Kalman decomposition) form (6.20). The appropriate transformation matrix $Q$ is given by (6.19). The matrix $\mathcal{C}$ was found in Example 6.2 and

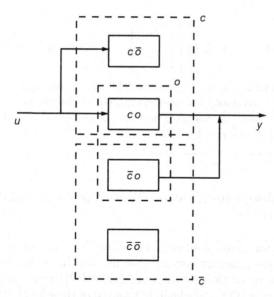

**Figure 6.1.** Canonical decomposition ($c$ and $\bar{c}$ denote controllable and uncontrollable, respectively). The connections of the $c/\bar{c}$ and $o/\bar{o}$ parts of the system to the input and output are emphasized. Note that the impulse response (transfer function) of the system, which is an input–output description only, represents the part of the system that is both controllable and observable (see Chapter 7).

$$\mathcal{O} = \begin{bmatrix} C \\ CA \\ CA^2 \end{bmatrix} = \begin{bmatrix} 0 & 1 & 0 \\ 1 & -2 & 1 \\ -2 & 4 & -2 \end{bmatrix}.$$

A basis for $R_{\bar{o}} = \mathcal{N}(\mathcal{O})$ is $\{(1, 0, -1)^T\}$. Note that $n_r = 2, n_{\bar{o}} = 1$, and $n_{r\bar{o}} = 1$. Therefore,

$$Q = [v_1, v_2, Q_N] = \begin{bmatrix} 1 & 1 & 0 \\ 1 & 0 & 0 \\ 1 & -1 & 1 \end{bmatrix}$$

is an appropriate similarity matrix (check that $\det Q \neq 0$). We compute

$$\hat{A} = Q^{-1}AQ = \begin{bmatrix} 0 & 1 & 0 \\ 1 & -1 & 0 \\ 1 & -2 & 1 \end{bmatrix} \begin{bmatrix} 0 & -1 & 1 \\ 1 & -2 & 1 \\ 0 & 1 & -1 \end{bmatrix} \begin{bmatrix} 1 & 1 & 0 \\ 1 & 0 & 0 \\ 1 & -1 & 1 \end{bmatrix}$$

$$= \begin{bmatrix} 0 & 0 & 1 \\ 0 & -1 & 0 \\ 0 & 0 & -2 \end{bmatrix} = \begin{bmatrix} A_{11} & 0 & A_{13} \\ A_{21} & A_{22} & A_{23} \\ 0 & 0 & A_{33} \end{bmatrix},$$

$$\hat{B} = Q^{-1}B = \begin{bmatrix} 0 & 1 & 0 \\ 1 & -1 & 0 \\ 1 & -2 & 1 \end{bmatrix} \begin{bmatrix} 1 & 0 \\ 1 & 1 \\ 1 & 2 \end{bmatrix} = \begin{bmatrix} 1 & 1 \\ 0 & -1 \\ 0 & 0 \end{bmatrix} = \begin{bmatrix} B_1 \\ B_2 \\ 0 \end{bmatrix},$$

and

$$\widehat{C} = CQ = [0,\ 1,\ 0] \begin{bmatrix} 1 & 1 & 0 \\ 1 & 0 & 0 \\ 1 & -1 & 1 \end{bmatrix} = [1, 0, 0] = [C_1,\ 0,\ C_3].$$

The eigenvalue 0 (in $A_{11}$) is controllable and observable, the eigenvalue $-1$ (in $A_{22}$) is controllable and unobservable and the eigenvalue $-2$ (in $A_{33}$) is uncontrollable and observable. There are no eigenvalues that are both uncontrollable and unobservable.

---

## 6.3 Eigenvalue/Eigenvector Tests for Controllability and Observability

There are tests for controllability and observability for both continuous- and discrete-time time-invariant systems that involve the eigenvalues and eigenvectors of $A$. Some of these criteria are called PBH tests, after the initials of the codiscoverers (Popov–Belevitch–Hautus) of these tests. These tests are useful in theoretical analysis, and in addition, they are also attractive as computational tools.

**Theorem 6.8.** *(i)  The pair $(A, B)$ is uncontrollable if and only if there exists a $1 \times n$ (in general) complex vector $\hat{v}_i \neq 0$ such that*

$$\hat{v}_i[\lambda_i I - A, B] = 0, \tag{6.23}$$

*where $\lambda_i$ is some complex scalar.*

*(ii) The pair $(A, C)$ is unobservable if and only if there exists an $n \times 1$ (in general) complex vector $v_i \neq 0$ such that*

$$\begin{bmatrix} \lambda_i I - A \\ C \end{bmatrix} v_i = 0, \tag{6.24}$$

*where $\lambda_i$ is some complex scalar.*

*Proof.* Only part (i) will be considered since (ii) can be proved using a similar argument or, directly, by duality arguments.

(*Sufficiency*) Assume that (6.23) is satisfied. In view of $\hat{v}_i A = \lambda_i \hat{v}_i$ and $\hat{v}_i B = 0, \hat{v}_i AB = \lambda_i \hat{v}_i B = 0$ and $\hat{v}_i A^k B = 0 \quad k = 0, 1, 2, \ldots$. Therefore, $\hat{v}_i \mathcal{C} = \hat{v}_i [B, AB, \ldots, A^{n-1}B] = 0$, which shows that $(A, B)$ is not completely controllable.

(*Necessity*) Let $(A, B)$ be uncontrollable and assume without loss of generality the standard form for $A$ and $B$ given in Lemma 6.1. We will show that there exist $\lambda_i$ and $\hat{v}_i$ so that (6.23) holds. Let $\lambda_i$ be an uncontrollable eigenvalue, and let $\hat{v}_i = [0, \alpha], \alpha^T \in C^{n-n_r}$, where $\alpha(\lambda_i I - A_2) = 0$; i.e., $\alpha$ is a left eigenvector of $A_2$ corresponding to $\lambda_i$. Then $\hat{v}_i[\lambda_i I - A, B] = [0, \alpha(\lambda_i I - A_2), 0] = 0$; i.e., (6.23) is satisfied. ∎

**Corollary 6.9.** (i) $\lambda_i$ is an uncontrollable eigenvalue of $(A, B)$ if and only if there exists a $1 \times n$ (in general) complex vector $\hat{v}_i \neq 0$ that satisfies (6.23). (ii) $\lambda_i$ is an unobservable eigenvalue of $(A, C)$ if and only if there exists an $n \times 1$ (in general) complex vector $v_i \neq 0$ that satisfies (6.24).

*Proof.* See [1, p. 273, Corollary 4.6].                                          ∎

---

***Example 6.10.*** Given are $A = \begin{bmatrix} 0 & -1 & 1 \\ 1 & -2 & 1 \\ 0 & 1 & -1 \end{bmatrix}$, $B = \begin{bmatrix} 1 & 0 \\ 1 & 1 \\ 1 & 2 \end{bmatrix}$, and $C = [0, 1, 0]$, as in Example 6.7. The matrix $A$ has three eigenvalues, $\lambda_1 = 0, \lambda_2 = -1$, and $\lambda_3 = -2$, with corresponding right eigenvectors $v_1 = [1, 1, 1]^T$, $v_2 = [1, 0, -1]^T$, $v_3 = [1, 1, -1]^T$ and with left eigenvectors $\hat{v}_1 = [1/2, 0, 1/2]$, $\hat{v}_2 = [1, -1, 0]$, and $\hat{v}_3 = [-1/2, 1, -1/2]$, respectively.

In view of Corollary 6.9, $\hat{v}_1 B = [1, 1] \neq 0$ implies that $\lambda_1 = 0$ is controllable. This is because $\hat{v}_1$ is the only nonzero vector (within a multiplication by a nonzero scalar) that satisfies $\hat{v}_1(\lambda_1 I - A) = 0$, and so $\hat{v}_1 B \neq 0$ implies that the only $1 \times 3$ vector $\alpha$ that satisfies $\alpha[\lambda_1 I - A, B] = 0$ is the zero vector, which in turn implies that $\lambda_1$ is controllable in view of (i) of Corollary 6.9. For similar reasons $C v_1 = 1 \neq 0$ implies that $\lambda_1 = 0$ is observable; see (ii) of Corollary 6.9. Similarly, $\hat{v}_2 B = [0, -1] \neq 0$ implies that $\lambda_2 = -1$ is controllable, and $C v_2 = 0$ implies that $\lambda_2 = -1$ is *unobservable*. Also, $\hat{v}_3 B = [0, 0]$ implies that $\lambda_3 = -2$ is *uncontrollable*, and $C v_3 = 1 \neq 0$ implies that $\lambda_3 = -2$ is observable. These results agree with the results derived in Example 6.7.

---

**Corollary 6.11.** (Rank Tests)

(ia) The pair $(A, B)$ is controllable if and only if

$$\text{rank}[\lambda I - A, B] = n \qquad (6.25)$$

for all complex numbers $\lambda$, or for all $n$ eigenvalues $\lambda_i$ of $A$.
(ib) $\lambda_i$ is an uncontrollable eigenvalue of $A$ if and only if

$$\text{rank}[\lambda_i I - A, B] < n. \qquad (6.26)$$

(iia) The pair $(A, C)$ is observable if and only if

$$\text{rank}\begin{bmatrix} \lambda I - A \\ C \end{bmatrix} = n \qquad (6.27)$$

for all complex numbers $\lambda$, or for all $n$ eigenvalues $\lambda_i$.
(iib) $\lambda_i$ is an unobservable eigenvalue of $A$ if and only if

$$\text{rank}\begin{bmatrix} \lambda_i I - A \\ C \end{bmatrix} < n. \qquad (6.28)$$

*Proof.* The proofs follow in a straightforward manner from Theorem 6.8. Notice that the only values of $\lambda$ that can possibly reduce the rank of $[\lambda I - A, B]$ are the eigenvalues of $A$. ∎

---

**Example 6.12.** If in Example 6.10 the eigenvalues $\lambda_1, \lambda_2, \lambda_3$ of $A$ are known, but the corresponding eigenvectors are not, consider the system matrix

$$P(s) = \begin{bmatrix} sI - A & B \\ -C & 0 \end{bmatrix} = \begin{bmatrix} s & 1 & -1 & 1 & 0 \\ -1 & s+2 & -1 & 1 & 1 \\ 0 & -1 & s+1 & 1 & 2 \\ \hline 0 & -1 & 0 & 0 & 0 \end{bmatrix}$$

and determine $\text{rank}[\lambda_i I - A, B]$ and $\text{rank}\begin{bmatrix} \lambda_i I - A \\ C \end{bmatrix}$. Notice that

$$\text{rank}\begin{bmatrix} sI - A \\ C \end{bmatrix}_{s=\lambda_2} = \text{rank}\begin{bmatrix} -1 & 1 & -1 \\ -1 & 1 & -1 \\ 0 & -1 & 0 \\ 0 & 1 & 0 \end{bmatrix} = 2 < 3 = n$$

and

$$\text{rank}[sI - A, B]_{s=\lambda_3} = \text{rank}\begin{bmatrix} -2 & 2 & -1 & 1 & 0 \\ -1 & 0 & -1 & 1 & 1 \\ 0 & -1 & -1 & 1 & 2 \end{bmatrix} = 2 < 3 = n.$$

In view of Corollary 6.11, $\lambda_2 = -1$ is unobservable and $\lambda_3 = -2$ is uncontrollable.

---

## 6.4 Controller and Observer Forms

It has been seen several times in this book that equivalent representations of systems

$$\dot{x} = Ax + Bu, \quad y = Cx + Du, \tag{6.29}$$

given by the equations

$$\dot{\hat{x}} = \widehat{A}\hat{x} + \widehat{B}u, \quad y = \widehat{C}\hat{x} + \widehat{D}u, \tag{6.30}$$

where $\hat{x} = Px$, $\widehat{A} = PAP^{-1}$, $\widehat{B} = PB$, $\widehat{C} = CP^{-1}$, and $\widehat{D} = D$ may offer advantages over the original representation when $P$ (or $Q = P^{-1}$) is chosen in an appropriate manner. This is the case when $P$ (or $Q$) is such that the new basis of the state space provides a natural setting for the properties of interest. This section shows how to select $Q$ when $(A, B)$ is controllable [or $(A, C)$ is observable] to obtain the controller and observer forms. These special forms are very useful, in realizations discussed in Chapter 8 and especially when studying state-feedback control (and state observers) discussed in Chapter 9. They are also very useful in establishing a convenient way to transition between state-space representations and another very useful class of equivalent internal representations, the polynomial matrix representations.

Controller forms are considered first. Observer forms can of course be obtained directly in a similar manner to the controller forms, or they may be obtained by duality. This is addressed in the latter part of this section.

### 6.4.1 Controller Forms

The controller form is a particular system representation where both matrices $(A, B)$ have a certain special structure. Since in this case $A$ is in the companion form, the controller form is sometimes also referred to as the *controllable companion form*. Consider the system

$$\dot{x} = Ax + Bu, \quad y = Cx + Du, \tag{6.31}$$

where $A \in R^{n \times n}$, $B \in R^{n \times m}$, $C \in R^{p \times n}$, and $D \in R^{p \times m}$ and let $(A, B)$ be controllable. Then $\mathrm{rank}\,\mathcal{C} = n$, where

$$\mathcal{C} = [B, AB, \dots, A^{n-1}B]. \tag{6.32}$$

Assume that

$$\mathrm{rank}\,B = m \leq n. \tag{6.33}$$

Under these assumptions, $\mathrm{rank}\,\mathcal{C} = n$ and $\mathrm{rank}\,B = m$. We will show how to obtain an equivalent pair $(\widehat{A}, \widehat{B})$ in controller form, first for the single-input case $(m = 1)$ and then for the multi-input case $(m > 1)$. Before this is accomplished, we discuss how to deal with two special cases that do not satisfy the above assumptions that $\mathrm{rank}\,B = m$ and that $(A, B)$ is controllable.

1. If the $m$ columns of $B$ are not linearly independent $(\mathrm{rank}\,B = r < m)$, then there exists an $m \times m$ nonsingular matrix $K$ so that $BK = [B_r, 0]$, where the $r$ columns of $B_r$ are linearly independent $(\mathrm{rank}\,B_r = r)$. Note that
$$\dot{x} = Ax + Bu = Ax + (BK)(K^{-1}u) = Ax + [B_r, 0]\begin{bmatrix} u_r \\ u_{m-r} \end{bmatrix} = Ax + B_r u_r,$$
which shows that when $\mathrm{rank}\,B = r < m$ the same input action to the system can be accomplished by only $r$ inputs, instead of $m$ inputs. The pair $(A, B_r)$, which is controllable when $(A, B)$ is controllable, can now be reduced to controller form, using the method described below.
2. When $(A, B)$ is not completely controllable, then a two-step approach can be taken. First, the controllable part is isolated (see Subsection 6.2.1) and then is reduced to the controller form, using the methods of this section. In particular, consider the system $\dot{x} = Ax + Bu$ with $A \in R^{n \times n}, B \in R^{n \times m}$, and $\mathrm{rank}\,B = m$. Let $\mathrm{rank}[B, AB, \dots, A^{n-1}B] = n_r < n$. Then

there exists a transformation $P_1$ such that $P_1 A P_1^{-1} = \begin{bmatrix} A_1 & A_{12} \\ 0 & A_2 \end{bmatrix}$ and

$P_1 B = \begin{bmatrix} B_1 \\ 0 \end{bmatrix}$, where $A_1 \in R^{n_r \times n_r}$, $B_1 \in R^{n_r \times m}$, and $(A_1, B_1)$ is controllable (Subsection 6.2.1). Since $(A_1, B_1)$ is controllable, there exists a transformation $P_2$ such that $P_2 A_1 P_2^{-1} = A_{1c}$, and $P_2 B_1 = B_{1c}$, where $A_{1c}, B_{1c}$ is in controller form, defined below. Combining, we obtain

$$PAP^{-1} = \begin{bmatrix} A_{1c} & P_2 A_{12} \\ 0 & A_2 \end{bmatrix}, \quad \text{and} \quad PB = \begin{bmatrix} B_{1c} \\ 0 \end{bmatrix} \tag{6.34}$$

[where $A_{1c} \in R^{n_r \times n_r}$, $B_{1c} \in R^{n_r \times m}$, and $(A_{1c}, B_{1c})$ is controllable], which is in controller form. Note that

$$P = \begin{bmatrix} P_2 & 0 \\ 0 & I \end{bmatrix} P_1. \tag{6.35}$$

## Single-Input Case $(m = 1)$

The representation $\{A_c, B_c, C_c, D_c\}$ in controller form is given by $A_c \triangleq \widehat{A} = PAP^{-1}$ and $B_c \triangleq \widehat{B} = PB$ with

$$A_c = \begin{bmatrix} 0 & 1 & \cdots & 0 \\ \vdots & \vdots & \ddots & \vdots \\ 0 & 0 & \cdots & 1 \\ -\alpha_0 & -\alpha_1 & \cdots & -\alpha_{n-1} \end{bmatrix}, \quad B_c = \begin{bmatrix} 0 \\ \vdots \\ 0 \\ 1 \end{bmatrix}, \tag{6.36}$$

where the coefficients $\alpha_i$ are the coefficients of the characteristic polynomial $\alpha(s)$ of $A$; that is,

$$\alpha(s) \triangleq \det(sI - A) = s^n + \alpha_{n-1} s^{n-1} + \cdots + \alpha_1 s + \alpha_0. \tag{6.37}$$

Note that $C_c \triangleq \widehat{C} = CP^{-1}$ and $D_c = D$ do not have any particular structure. The structure of $(A_c, B_c)$ is very useful (in control problems), and the representation $\{A_c, B_c, C_c, D_c\}$ shall be referred to as the *controller form* of the system. The similarity transformation matrix $P$ is obtained as follows. The controllability matrix $\mathcal{C} = [B, AB, \ldots, A^{n-1}B]$ is in this case an $n \times n$ nonsingular matrix. Let $\mathcal{C}^{-1} = \begin{bmatrix} \times \\ q \end{bmatrix}$, where $q$ is the $n$th row of $\mathcal{C}^{-1}$ and $\times$ indicates the remaining entries of $\mathcal{C}^{-1}$. Then

$$P \triangleq \begin{bmatrix} q \\ qA \\ \cdots \\ qA^{n-1} \end{bmatrix}. \tag{6.38}$$

To show that $PAP^{-1} = A_c$ and $PB = B_c$ given in (6.36), note first that $qA^{i-1}B = 0$ $i = 1, \ldots, n-1$ and $qA^{n-1}B = 1$. This can be verified from the definition of $q$, which implies that $q\,\mathcal{C} = [0, 0, \ldots, 1]$. Now

$$PC = P[B, AB, \ldots, A^{n-1}B] = \begin{bmatrix} 0 & 0 & \cdots & \cdots & 1 \\ 0 & 0 & \cdots & 1 & \times \\ & \vdots & 1 & & \vdots & \vdots \\ 1 & \times & \cdots & \times & \times \end{bmatrix} = C_c, \tag{6.39}$$

which implies that $|PC| = |P|\,|C| \neq 0$ or that $|P| \neq 0$. Therefore, $P$ qualifies as a similarity transformation matrix. In view of (6.39), $PB = [0, 0, \ldots, 1]^T = B_c$. Furthermore,

$$A_cP = \begin{bmatrix} qA \\ \vdots \\ qA^{n-1} \\ qA^n \end{bmatrix} = PA, \tag{6.40}$$

where in the last row of $A_cP$, the relation $-\sum_{i=0}^{n-1} \alpha_i A^i = A^n$ was used [which is the Cayley–Hamilton Theorem, namely, $\alpha(A) = 0$].

---

**Example 6.13.** Let $A = \begin{bmatrix} -1 & 0 & 0 \\ 0 & 1 & 0 \\ 0 & 0 & -2 \end{bmatrix}$ and $B = \begin{bmatrix} 1 \\ -1 \\ 1 \end{bmatrix}$. Since $n = 3$ and $|sI - A| = (s+1)(s-1)(s+2) = s^3 + 2s^2 - s - 2$, $\{A_c, B_c\}$ in controller form is given by

$$A_c = \begin{bmatrix} 0 & 1 & 0 \\ 0 & 0 & 1 \\ 2 & 1 & -2 \end{bmatrix} \quad \text{and} \quad B_c = \begin{bmatrix} 0 \\ 0 \\ 1 \end{bmatrix}.$$

The transformation matrix $P$ that reduces $(A, B)$ to $(A_c = PAP^{-1}, B_c = PB)$ is now derived. We have

$$C = [B, AB, A^2B] = \begin{bmatrix} 1 & -1 & 1 \\ -1 & -1 & -1 \\ 1 & -2 & 4 \end{bmatrix} \quad \text{and} \quad C^{-1} = \begin{bmatrix} 1 & -1/3 & -1/3 \\ -1/2 & -1/2 & 0 \\ -1/2 & -1/6 & 1/3 \end{bmatrix}.$$

The third (the $n$th) row of $C^{-1}$ is $q = [-1/2, -1/6, 1/3]$, and therefore,

$$P \triangleq \begin{bmatrix} q \\ qA \\ qA^2 \end{bmatrix} = \begin{bmatrix} -1/2 & -1/6 & 1/3 \\ 1/2 & -1/6 & -2/3 \\ -1/2 & -1/6 & 4/3 \end{bmatrix}.$$

It can now easily be verified that $A_c = PAP^{-1}$, or

$$A_cP = \begin{bmatrix} 1/2 & -1/6 & -2/3 \\ -1/2 & -1/6 & -2/3 \\ 1/2 & -1/6 & -8/3 \end{bmatrix} = PA,$$

and that $B_c = PB$.

---

An alternative form to (6.36) is

$$A_{c1} = \begin{bmatrix} -\alpha_{n-1} & \cdots & -\alpha_1 & -\alpha_0 \\ 1 & \cdots & 0 & 0 \\ \vdots & \ddots & \vdots & \vdots \\ 0 & \cdots & 1 & 0 \end{bmatrix}, \quad B_{c1} = \begin{bmatrix} 1 \\ 0 \\ \vdots \\ 0 \end{bmatrix}, \tag{6.41}$$

which is obtained if the similarity transformation matrix is taken to be

$$P_1 \triangleq \begin{bmatrix} qA^{n-1} \\ \vdots \\ qA \\ q \end{bmatrix}, \tag{6.42}$$

i.e., by reversing the order of the rows of $P$ in (6.38). (See Exercise 6.5 and Example 6.14.)

In the above, $A_c$ is a companion matrix of the form $\begin{bmatrix} 0 & I \\ \times & \times \end{bmatrix}$ or $\begin{bmatrix} \times & \times \\ I & 0 \end{bmatrix}$. It could also be of the form $\begin{bmatrix} 0 & \times \\ I & \times \end{bmatrix}$ or $\begin{bmatrix} \times & 0 \\ \times & I \end{bmatrix}$ with coefficients $-[\alpha_0, \ldots, \alpha_{n-1}]^T$ in the last or the first column. It is shown here, for completeness, how to determine controller forms where $A_c$ are such companion matrices. In particular, if

$$Q_2 = P_2^{-1} = [B, AB, \ldots, A^{n-1}B] = \mathcal{C}, \tag{6.43}$$

then

$$A_{c2} = Q_2^{-1}AQ_2 = \begin{bmatrix} 0 & \cdots & 0 & -\alpha_0 \\ 1 & \cdots & 0 & -\alpha_1 \\ \vdots & \ddots & \vdots & \vdots \\ 0 & \cdots & 1 & -\alpha_{n-1} \end{bmatrix}, B_{c2} = Q_2^{-1}B = \begin{bmatrix} 1 \\ 0 \\ \vdots \\ 0 \end{bmatrix}. \tag{6.44}$$

Also, if

$$Q_3 = P_3^{-1} = [A^{n-1}B, \ldots, B], \tag{6.45}$$

then

$$A_{c3} = Q_3^{-1}AQ_3 = \begin{bmatrix} -\alpha_{n-1} & 1 & \cdots & 0 \\ \vdots & \vdots & \ddots & \vdots \\ -\alpha_1 & 0 & \cdots & 1 \\ -\alpha_0 & 0 & \cdots & 0 \end{bmatrix}, B_{c3} = Q_3^{-1}B = \begin{bmatrix} 0 \\ \vdots \\ 0 \\ 1 \end{bmatrix}. \tag{6.46}$$

$(A_c, B_c)$ in (6.44) and (6.46) are also in controller canonical or controllable companion form. (See also Exercise 6.5 and Example 6.14.)

**Example 6.14.** Let $A = \begin{bmatrix} -1 & 0 & 0 \\ 0 & 1 & 0 \\ 0 & 0 & -2 \end{bmatrix}$ and $B = \begin{bmatrix} 1 \\ -1 \\ 1 \end{bmatrix}$, as in Example 6.13.

Alternative controller forms can be derived for different $P$. In particular, if

(i) $P = P_1 = \begin{bmatrix} qA^2 \\ qA \\ q \end{bmatrix} = \begin{bmatrix} -1/2 & -1/6 & 4/3 \\ 1/2 & -1/6 & -2/3 \\ -1/2 & -1/6 & 1/3 \end{bmatrix}$, as in (6.42) ($\mathcal{C}, \mathcal{C}^{-1}$, and $q$

were found in Example 6.13), then

$$A_{c1} = \begin{bmatrix} -2 & 1 & 2 \\ 1 & 0 & 0 \\ 0 & 1 & 0 \end{bmatrix}, \quad B_{c1} = \begin{bmatrix} 1 \\ 0 \\ 0 \end{bmatrix},$$

as in (6.41). Note that in the present case $A_{c1}P_1 = \begin{bmatrix} 1/2 & -1/6 & -8/3 \\ -1/2 & -1/6 & 4/3 \\ 1/2 & -1/6 & -2/3 \end{bmatrix} = $

$P_1 A$, $B_{c1} = P_1 B$.

(ii) $Q_2 = \mathcal{C} = \begin{bmatrix} 1 & -1 & 1 \\ -1 & -1 & -1 \\ 1 & -2 & 4 \end{bmatrix}$, as in (6.43). Then

$$A_{c2} = \begin{bmatrix} 0 & 0 & 2 \\ 1 & 0 & 1 \\ 0 & 1 & -2 \end{bmatrix}, \quad B_{c2} = Q_2^{-1}B = \begin{bmatrix} 1 \\ 0 \\ 0 \end{bmatrix},$$

as in (6.44).

(iii) $Q_3 = [A^2 B, AB, B] = \begin{bmatrix} 1 & -1 & 1 \\ -1 & -1 & -1 \\ 4 & -2 & 1 \end{bmatrix}$, as in (6.45). Then

$$A_{c3} = \begin{bmatrix} -2 & 1 & 0 \\ 1 & 0 & 1 \\ 2 & 0 & 0 \end{bmatrix}, \quad B_{c3} = \begin{bmatrix} 0 \\ 0 \\ 1 \end{bmatrix},$$

as in (6.46). Note that $Q_3 A_{c3} = \begin{bmatrix} -1 & 1 & -1 \\ -1 & -1 & -1 \\ -8 & 4 & -2 \end{bmatrix} = AQ_3$, $Q_3 B_{c3} = $

$\begin{bmatrix} 1 \\ -1 \\ 1 \end{bmatrix} = B$.

## Multi-Input Case ($m > 1$)

In this case, the $n \times mn$ matrix $C$ given in (6.32) is not square, and there are typically many sets of $n$ columns of $C$ that are linearly independent (rank $C = n$). Depending on which columns are chosen and in what order, different controller forms (controllable companion forms) are derived. Note that in the case when $m = 1$, four different controller forms were derived, even though there was only one set of $n$ linearly independent columns. In the present case there are many more such choices. The form that will be used most often in the following is a generalization of $(A_c, B_c)$ given in (6.36). Further discussion including derivation and alternative forms may be found in [1, Subsection 3.4D].

Let $\widehat{A} = PAP^{-1}$ and $\widehat{B} = PB$, where $P$ is constructed as follows. Consider

$$C = [B, AB, \ldots, A^{n-1}B]$$
$$= [b_1, \ldots, b_m, Ab_1, \ldots, Ab_m, \ldots, A^{n-1}b_1, \ldots, A^{n-1}b_m], \qquad (6.47)$$

where the $b_1, \ldots, b_m$ are the $m$ columns of $B$. Select, starting from the left and moving to the right, the first $n$ independent columns (rank $C = n$). Reorder these columns by taking first $b_1, Ab_1, A^2b_1$, etc., until all columns involving $b_1$ have been taken; then take $b_2, Ab_2$, etc.; and lastly, take $b_m, Ab_m$, etc., to obtain

$$\bar{C} \triangleq [b_1, Ab_1, \ldots, A^{\mu_1-1}b_1, \ldots, b_m, \ldots, A^{\mu_m-1}b_m], \qquad (6.48)$$

an $n \times n$ matrix. The integer $\mu_i$ denotes the number of columns involving $b_i$ in the set of the first $n$ linearly independent columns found in $C$ when moving from left to right.

**Definition 6.15.** *The $m$ integers $\mu_i$, $i = 1, \ldots, m$, are the* controllability indices *of the system, and $\mu \triangleq \max \mu_i$ is called the* controllability index *of the system. Note that*

$$\sum_{i=1}^{m} \mu_i = n \quad and \quad m\mu \geq n. \qquad (6.49)$$

■

An alternative but equivalent definition for $\mu$ is that $\mu$ is the minimum integer $k$ such that

$$\text{rank}[B, AB, \ldots, A^{k-1}B] = n. \qquad (6.50)$$

Notice that in (6.48) all columns of $B$ are always present since rank $B = m$. This implies that $\mu_i \geq 1$ for all $i$. Notice further that if $A^k b_i$ is present, then $A^{k-1}b_i$ must also be present.

Now define

$$\sigma_k \triangleq \sum_{i=1}^{k} \mu_i, \quad k = 1, \ldots, m; \qquad (6.51)$$

i.e., $\sigma_1 = \mu_1, \sigma_2 = \mu_1 + \mu_2, \ldots, \sigma_m = \mu_1 + \cdots + \mu_m = n$. Also, consider $\bar{C}^{-1}$ and let $q_k$, where $q_k^T \in R^n$, $k = 1, \ldots, m$, denote its $\sigma_k{}^{th}$ row; i.e.,

$$\bar{C}^{-1} = [\times, \ldots, \times, q_1^T \vdots \cdots \vdots \times, \ldots, \times, q_m^T]^T. \tag{6.52}$$

Next, define

$$P \triangleq \begin{bmatrix} q_1 \\ q_1 A \\ \vdots \\ q_1 A^{\mu_1-1} \\ \cdots \\ \vdots \\ \cdots \\ q_m \\ q_m A \\ \vdots \\ q_m A^{\mu_m-1} \end{bmatrix}. \tag{6.53}$$

It can now be shown that $PAP^{-1} = A_c$ and $PB = B_c$ with

$$A_c = [A_{ij}], \qquad i, j = 1, \ldots, m,$$

$$A_{ii} = \begin{bmatrix} 0 \\ \vdots & I_{\mu_i-1} \\ 0 \\ \times \times \cdots \times \end{bmatrix} \in R^{\mu_i \times \mu_i}, \ i = j, \quad A_{ij} = \begin{bmatrix} 0 & \cdots & 0 \\ \vdots & \vdots & \vdots \\ 0 & \cdots & 0 \\ \times \times \cdots & \times \end{bmatrix} \in R^{\mu_i \times \mu_j}, \ i \neq j,$$

and

$$B_c = \begin{bmatrix} B_1 \\ B_2 \\ \vdots \\ B_m \end{bmatrix}, \qquad B_i = \begin{bmatrix} 0 & \cdots & 0 & 0 & \cdots & 0 \\ \vdots & & \vdots & \vdots & & \vdots \\ 0 & \cdots & 0 & 1 & \times & \cdots & \times \end{bmatrix} \in R^{\mu_i \times m}, \tag{6.54}$$

where the 1 in the last row of $B_i$ occurs at the $i$th column location, $i = 1, \ldots, m$, and $\times$ denotes nonfixed entries. Note that $C_c = CP^{-1}$ does not have any particular structure. The expression (6.54) is a very useful form (in control problems) and shall be referred to as the *controller form* of the system. The derivation of this result is discussed in [1, Subsection 3.4D] .

---

***Example 6.16.*** Given are $A \in R^{n \times n}$ and $B \in R^{n \times m}$ with $(A, B)$ controllable and with rank $B = m$. Let $n = 4$ and $m = 2$. Then there must be two controllability indices $\mu_1$ and $\mu_2$ such that $n = 4 = \sum_{i=1}^{2} \mu_i = \mu_1 + \mu_2$. Under these conditions, there are three possibilities:

(i)  $\mu_1 = 2, \mu_2 = 2$,

$$A_c = \begin{bmatrix} A_{11} & A_{12} \\ A_{21} & A_{22} \end{bmatrix} = \left[\begin{array}{cc|cc} 0 & 1 & 0 & 0 \\ \times & \times & \times & \times \\ \hline 0 & 0 & 0 & 1 \\ \times & \times & \times & \times \end{array}\right], \quad B_c = \begin{bmatrix} B_1 \\ B_2 \end{bmatrix} = \begin{bmatrix} 0 & 0 \\ 1 & \times \\ \hline 0 & 0 \\ 0 & 1 \end{bmatrix}.$$

(ii)  $\mu_1 = 1, \mu_2 = 3$,

$$A_c = \left[\begin{array}{cc|cc} \times & \times & \times & \times \\ \hline 0 & 0 & 1 & 0 \\ 0 & 0 & 0 & 1 \\ \times & \times & \times & \times \end{array}\right], \quad B_c = \begin{bmatrix} 1 & \times \\ \hline 0 & 0 \\ 0 & 0 \\ 0 & 1 \end{bmatrix}.$$

(iii) $\mu_1 = 3, \mu_2 = 1$,

$$A_c = \left[\begin{array}{ccc|c} 0 & 1 & 0 & 0 \\ 0 & 0 & 1 & 0 \\ \times & \times & \times & \times \\ \hline \times & \times & \times & \times \end{array}\right], \quad B_c = \begin{bmatrix} 0 & 0 \\ 0 & 0 \\ 1 & \times \\ \hline 0 & 1 \end{bmatrix}.$$

It is possible to write $A_c, B_c$ in a systematic and perhaps more transparent way. In particular, notice that $A_c, B_c$ in (6.54) can be expressed as

$$A_c = \bar{A}_c + \bar{B}_c A_m, \quad B_c = \bar{B}_c B_m, \tag{6.55}$$

where $\bar{A}_c = $ block diag$[\bar{A}_{11}, \bar{A}_{22}, \ldots, \bar{A}_{mm}]$ with

$$\bar{A}_{ii} = \begin{bmatrix} 0 \\ \vdots & I_{\mu_i - 1} \\ 0 \\ 0 & 0\cdots 0 \end{bmatrix} \in R^{\mu_i \times \mu_i}, \quad \bar{B}_c = \text{block diag}\left(\begin{bmatrix} 0 \\ \vdots \\ 0 \\ 1 \end{bmatrix} \in R^{\mu_i \times 1}, \quad i = 1, \ldots, m\right),$$

and $A_m \in R^{m \times n}$ and $B_m \in R^{m \times m}$ are some appropriate matrices with $\sum_{i=1}^{m} \mu_i = n$. Note that the matrices $\bar{A}_c, \bar{B}_c$ are completely determined by the $m$ controllability indices $\mu_i$, $i = 1, \ldots, m$. The matrices $A_m$ and $B_m$ consist of the $\sigma_1$th, $\sigma_2$th, $\ldots, \sigma_m$th rows of $A_c$ (entries denoted by $\times$) and the same rows of $B_c$, respectively [see (6.57) and (6.58) below].

***Example 6.17.*** Let $A = \begin{bmatrix} 0 & 1 & 0 \\ 0 & 0 & 1 \\ 0 & 2 & -1 \end{bmatrix}$ and $B = \begin{bmatrix} 0 & 1 \\ 1 & 1 \\ 0 & 0 \end{bmatrix}$. To determine the controller form (6.54), consider

$$C = [B, AB, A^2B] = [b_1, b_2, Ab_1, Ab_2, A^2b_1, A^2b_2] = \begin{bmatrix} 0 & 1 & 1 & 1 & 0 & 0 \\ 1 & 1 & 0 & 0 & 2 & 2 \\ 0 & 0 & 2 & 2 & -2 & -2 \end{bmatrix},$$

where $\operatorname{rank} C = 3 = n$; i.e., $(A, B)$ is controllable. Searching from left to right, the first three columns of $C$ are selected since they are linearly independent. Then

$$\bar{C} = [b_1, Ab_1, b_2] = \begin{bmatrix} 0 & 1 & 1 \\ 1 & 0 & 1 \\ 0 & 2 & 0 \end{bmatrix}$$

and the controllability indices are $\mu_1 = 2$ and $\mu_2 = 1$. Also, $\sigma_1 = \mu_1 = 2$ and $\sigma_2 = \mu_1 + \mu_2 = 3 = n$, and

$$\bar{C}^{-1} = \begin{bmatrix} -1 & 1 & 1/2 \\ 0 & 0 & 1/2 \\ 1 & 0 & -1/2 \end{bmatrix}.$$

Notice that $q_1 = [0, 0, 1/2]$ and $q_2 = [1, 0, -1/2]$, the second and third rows of $\bar{C}^{-1}$, respectively. In view of (6.53), $P = \begin{bmatrix} q_1 \\ q_1 A \\ q_2 \end{bmatrix} = \begin{bmatrix} 0 & 0 & 1/2 \\ 0 & 1 & -1/2 \\ 1 & 0 & -1/2 \end{bmatrix}$, $P^{-1} = \begin{bmatrix} 1 & 0 & 1 \\ 1 & 1 & 0 \\ 2 & 0 & 0 \end{bmatrix}$, and $A_c = PAP^{-1} = \begin{bmatrix} A_{11} & A_{12} \\ A_{21} & A_{22} \end{bmatrix} = \begin{bmatrix} 0 & 1 & 0 \\ 2 & -1 & 0 \\ 1 & 0 & 0 \end{bmatrix}$, $B_c = PB = \begin{bmatrix} B_1 \\ B_2 \end{bmatrix} = \begin{bmatrix} 0 & 0 \\ 1 & 1 \\ 0 & 1 \end{bmatrix}$.

One can also verify (6.55) quite easily. We have

$$A_c = \begin{bmatrix} 0 & 1 & 0 \\ 2 & -1 & 0 \\ 1 & 0 & 0 \end{bmatrix} = \bar{A}_c + \bar{B}_c A_m = \begin{bmatrix} 0 & 1 & 0 \\ 0 & 0 & 0 \\ 0 & 0 & 0 \end{bmatrix} + \begin{bmatrix} 0 & 0 \\ 1 & 0 \\ 0 & 1 \end{bmatrix} \begin{bmatrix} 2 & -1 & 0 \\ 1 & 0 & 0 \end{bmatrix}$$

and

$$B_c = \begin{bmatrix} 0 & 0 \\ 1 & 1 \\ 0 & 1 \end{bmatrix} = \bar{B}_c B_m = \begin{bmatrix} 0 & 0 \\ 1 & 0 \\ 0 & 1 \end{bmatrix} \begin{bmatrix} 1 & 1 \\ 0 & 1 \end{bmatrix}.$$

It is interesting to note that in this example, the given pair $(A, B)$ could have already been in controller form if $B$ were different but $A$ were the same. For example, consider the following three cases:

1. $A = \begin{bmatrix} 0 & 1 & 0 \\ 0 & 0 & 1 \\ 0 & 2 & -1 \end{bmatrix}$, $B = \begin{bmatrix} 1 & \times \\ 0 & 0 \\ 0 & 1 \end{bmatrix}$, $\mu_1 = 1, \mu_2 = 2$,

2. $A = \begin{bmatrix} 0 & 1 & 0 \\ 0 & 0 & 1 \\ 0 & 2 & -1 \end{bmatrix}$, $B = \begin{bmatrix} 0 & 0 \\ 1 & \times \\ 0 & 1 \end{bmatrix}$, $\mu_1 = 2, \mu_1 = 1$,

3. $A = \begin{bmatrix} 0 & 1 & 0 \\ 0 & 0 & 1 \\ 0 & 2 & -1 \end{bmatrix}$, $B = \begin{bmatrix} 0 \\ 0 \\ 1 \end{bmatrix}$, $\mu_1 = 3 = n$.

Note that case 3 is the single-input case (6.36).

---

## Remarks

(i) An important result involving the controllability indices of $(A, B)$ is the following: Given $(A, B)$ controllable, then $(P(A + BGF)P^{-1}, PBG)$ will have the same controllability indices, within reordering, for any $P, F$, and $G$ ($|P| \neq 0, |G| \neq 0$) of appropriate dimensions. In other words, *the controllability indices are invariant under similarity and input transformations P and G, and state feedback F [or similarity transformation P and state feedback (F, G)]*. (For further discussion, see [1, Subsection 3.4D].)

(ii) It is not difficult to derive explicit expressions for $A_m$ and $B_m$ in (6.55). Using

$$q_i A^{k-1} b_j = 0 \quad k = 1, \ldots, \mu_j, \quad i \neq j,$$
$$q_i A^{k-1} b_i = 0 \quad k = 1, \ldots, \mu_i - 1, \text{ and } q_i A^{\mu_i - 1} b_i = 1, \quad i = j, \quad (6.56)$$

where $i = 1, \ldots, m$, and $j = 1, \ldots, m$, it can be shown that the $m$ $\sigma_1$th, $\sigma_2$th, $\ldots, \sigma_m$th rows of $A_c$ that are denoted by $A_m$ in (6.55) are given by

$$A_m = \begin{bmatrix} q_1 A^{\mu_1} \\ \vdots \\ q_m A^{\mu_m} \end{bmatrix} P^{-1}. \tag{6.57}$$

Similarly

$$B_m = \begin{bmatrix} q_1 A^{\mu_1 - 1} \\ \vdots \\ q_m A^{\mu_m - 1} \end{bmatrix} B. \tag{6.58}$$

The matrix $B_m$ is an upper triangular matrix with ones on the diagonal. (For details, see [1, Subsection 3.4D].)

---

***Example 6.18.*** We wish to reduce $A = \begin{bmatrix} 0 & 1 & 0 \\ 0 & 0 & 1 \\ 0 & 2 & -1 \end{bmatrix}$, $B = \begin{bmatrix} 1 & 1 \\ 0 & 1 \\ 0 & 0 \end{bmatrix}$ to controller form. Note that $A$ and $B$ are almost the same as in Example 6.17; however, here $\mu_1 = 1 < 2 = \mu_2$, as will be seen. We have $C = [B, AB, A^2B] = [b_1, b_2, Ab_1, Ab_2, \ldots] = \begin{bmatrix} 1 & 1 & 0 & 1 \\ 0 & 1 & 0 & 0 \\ 0 & 0 & 0 & 2 \end{bmatrix} \cdots$. Searching from left to right, the first

three linearly independent columns are $b_1, b_2, Ab_2$, and $\bar{C} = [b_1, b_2, Ab_2] =$
$\begin{bmatrix} 1 & 1 & 1 \\ 0 & 1 & 0 \\ 0 & 0 & 2 \end{bmatrix}$, from which we conclude that $\mu_1 = 1$, $\mu_2 = 2$, $\sigma_1 = 1$, and

$\sigma_2 = 3$. We compute $\bar{C}^{-1} = \begin{bmatrix} 1 & -1 & -1/2 \\ 0 & 1 & 0 \\ 0 & 0 & 1/2 \end{bmatrix}$. Note that $q_1 = [1, -1, -1/2]$

and $q_2 = [0, 0, 1/2]$, the first and third rows of $\bar{C}^{-1}$, respectively. Then

$$P = \begin{bmatrix} q_1 \\ q_2 \\ q_2 A \end{bmatrix} = \begin{bmatrix} 1 & -1 & -1/2 \\ 0 & 0 & 1/2 \\ 0 & 1 & -1/2 \end{bmatrix}, P^{-1} = \begin{bmatrix} 1 & 2 & 1 \\ 0 & 1 & 1 \\ 0 & 2 & 0 \end{bmatrix}, \text{ and}$$

$$A_c = PAP^{-1} = \begin{bmatrix} A_{11} & A_{12} \\ A_{21} & A_{22} \end{bmatrix} = \begin{bmatrix} 0 & -1 & 0 \\ 0 & 0 & 1 \\ 0 & 2 & -1 \end{bmatrix},$$

$$B_c = PB = \begin{bmatrix} B_1 \\ B_2 \end{bmatrix} = \begin{bmatrix} 1 & 0 \\ 0 & 0 \\ 0 & 1 \end{bmatrix}.$$

It is easy to verify relations (6.57) and (6.58).

---

### Structure Theorem—Controllable Version

The transfer function matrix $H(s)$ of the system $\dot{x} = Ax + Bu$, $y = Cx + Du$ is given by $H(s) = C(sI - A)^{-1}B + D$. If $(A, B)$ is in *controller form* (6.54), then $H(s)$ can alternatively be characterized by the Structure Theorem stated in Theorem 6.19 below. This result is very useful in the realization of systems, which is addressed in Chapter 8 and in the study of state feedback in Chapter 9.

Let $A = A_c = \bar{A}_c + \bar{B}_c A_m$ and $B = B_c = \bar{B}_c B_m$, as in (6.55), with $|B_m| \neq 0$, and let $C = C_c$ and $D = D_c$. Define

$$\Lambda(s) \triangleq \text{diag}[s^{\mu_1}, s^{\mu_2}, \ldots, s^{\mu_m}], \tag{6.59}$$

$$S(s) \triangleq \text{block diag}([1, s, \ldots, s^{\mu_i - 1}]^T, \quad i = 1, \ldots, m). \tag{6.60}$$

Note that $S(s)$ is an $n \times m$ polynomial matrix ($n = \sum_{i=1}^m \mu_i$), i.e., a matrix with polynomials as entries. Now define the $m \times m$ polynomial matrix $D(s)$ and the $p \times m$ polynomial matrix $N(s)$ by

$$D(s) \triangleq B_m^{-1}[\Lambda(s) - A_m S(s)], N(s) \triangleq C_c S(s) + D_c D(s). \tag{6.61}$$

The following is the controllable version of the *Structure Theorem*.

**Theorem 6.19.** $H(s) = N(s)D^{-1}(s)$, *where* $N(s)$ *and* $D(s)$ *are defined in (6.61).*

*Proof.* First, note that

$$(sI - A_c)S(s) = B_cD(s). \tag{6.62}$$

To see this, we write $B_cD(s) = \bar{B}_cB_mB_m^{-1}[\Lambda(s) - A_mS(s)] = \bar{B}_c\Lambda(s) - \bar{B}_cA_mS(s)$ and $(sI - A_c)S(s) = sS(s) - (\bar{A}_c + \bar{B}_cA_m)S(s) = (sI - \bar{A}_c)S(s) - \bar{B}_cA_mS(s) = \bar{B}_c\Lambda(s) - \bar{B}_cA_mS(s)$, which proves (6.62). Now $H(s) = C_c(sI - A_c)^{-1}B_c + D_c = C_cS(s)D^{-1}(s) + D_c = [C_cS(s) + D_cD(s)]D^{-1}(s) = ND^{-1}$. ∎

---

**Example 6.20.** Let $A_c = \begin{bmatrix} 0 & 1 & 0 \\ 2 & -1 & 0 \\ 1 & 0 & 0 \end{bmatrix}$, $B_c = \begin{bmatrix} 0 & 0 \\ 1 & 1 \\ 0 & 1 \end{bmatrix}$, as in Example 6.17. Here $\mu_1 = 2, \mu_2 = 1$ and $A_m = \begin{bmatrix} 2 & -1 & 0 \\ 1 & 0 & 0 \end{bmatrix}$, $B_m = \begin{bmatrix} 1 & 1 \\ 0 & 1 \end{bmatrix}$. Then $\Lambda(s) = \begin{bmatrix} s^2 & 0 \\ 0 & s \end{bmatrix}$, $S(s) = \begin{bmatrix} 1 & 0 \\ s & 0 \\ 0 & 1 \end{bmatrix}$ and

$$D(s) = B_m^{-1}[\Lambda(s) - A_mS(s)] = \begin{bmatrix} 1 & -1 \\ 0 & 1 \end{bmatrix} \left[ \begin{bmatrix} s^2 & 0 \\ 0 & s \end{bmatrix} - \begin{bmatrix} -s+2 & 0 \\ 1 & 0 \end{bmatrix} \right]$$

$$= \begin{bmatrix} 1 & -1 \\ 0 & 1 \end{bmatrix} \begin{bmatrix} s^2 + s - 2 & 0 \\ -1 & s \end{bmatrix} = \begin{bmatrix} s^2 + s - 1 & -s \\ -1 & s \end{bmatrix}.$$

Now $C_c = [0, 1, 1]$, and $D_c = [0, 0]$,

$$N(s) = C_cS(s) + D_cD(s) = [s, 1],$$

and

$$H(s) = [s, 1] \begin{bmatrix} s^2 + s - 1 & -s \\ -1 & s \end{bmatrix}^{-1} = [s, 1] \begin{bmatrix} s & s \\ 1 & s^2 + s - 1 \end{bmatrix} \frac{1}{s(s^2 + s - 2)}$$

$$= \frac{1}{s(s^2 + s - 2)} [s^2 + 1, 2s^2 + s - 1]$$

$$= C_c(sI - A_c)^{-1}B_c + D_c.$$

---

**Example 6.21.** Let $A_c = \begin{bmatrix} 0 & 1 & 0 \\ 0 & 0 & 1 \\ 2 & 1 & -2 \end{bmatrix}$, $B_c = \begin{bmatrix} 0 \\ 0 \\ 1 \end{bmatrix}$, $C_c = [0, 1, 0]$, and $D_c = 0$ (see Example 6.13). In the present case, we have $A_m = [2, 1, -2]$, $B_m = 1$, $\Lambda(s) = s^3$, $S(s) = [1, s, s^2]^T$, and

$$D(s) = 1 \cdot [s^3 - [2, 1, -2][1, s, s^2]^T] = s^3 + 2s^2 - s - 2, \quad N(s) = s.$$

Then

$$H(s) = N(s)D^{-1}(s) = s/(s^3 + 2s^2 - s - 2) = C_c(sI - A_c)^{-1}B_c + D_c.$$

## 6.4.2 Observer Forms

Consider the system $\dot{x} = Ax + Bu$, $y = Cx + Du$ given in (6.1) and assume that $(A, C)$ is observable; i.e., rank $\mathcal{O} = n$, where

$$\mathcal{O} = \begin{bmatrix} C \\ CA \\ \vdots \\ CA^{n-1} \end{bmatrix}. \tag{6.63}$$

Also, assume that the $p \times n$ matrix $C$ has a full row rank $p$; i.e.,

$$\operatorname{rank} C = p \le n. \tag{6.64}$$

It is of interest to determine a transformation matrix $P$ so that the equivalent system representation $\{A_o, B_o, C_o, D_o\}$ with

$$A_o = PAP^{-1}, \quad B_o = PB, \quad C_o = CP^{-1}, \quad D_o = D \tag{6.65}$$

will have $(A_o, C_o)$ in an observer form (defined below). As will become clear in the following discussion, these forms are dual to the controller forms previously discussed and can be derived by taking advantage of this fact. In particular, let $\tilde{A} \triangleq A^T$, $\tilde{B} \triangleq C^T$ $[(\tilde{A}, \tilde{B})$ is controllable], and determine a nonsingular transformation $\tilde{P}$ so that $\tilde{A}_c = \tilde{P}\tilde{A}\tilde{P}^{-1}$, $\tilde{B}_c = \tilde{P}\tilde{B}$ are in controller form given in (6.54). Then $A_o = \tilde{A}_c^T$ and $C_o = \tilde{B}_c^T$ is in observer form.

It will be demonstrated in the following discussion how to obtain observer forms directly, in a way that parallels the approach described for controller forms. This is done for the sake of completeness and to define the observability indices. The approach of using duality just given can be used in each case to verify the results.

We first note that if rank $C = r < p$, an approach analogous to the case when rank $B < m$ can be followed, as in Subsection 6.4.1. The fact that the rows of $C$ are not linearly independent means that the same information can be extracted from only $r$ outputs, and therefore, the choice for the outputs should perhaps be reconsidered. Now if $(A, C)$ is unobservable, one may use two steps to first isolate the observable part and then reduce it to the observer form, in an analogous way to the uncontrollable case previously given.

### Single-Output Case ($p = 1$)

Let

$$P^{-1} = Q \triangleq [\tilde{q}, A\tilde{q}, \ldots, A^{n-1}\tilde{q}], \tag{6.66}$$

where $\tilde{q}$ is the $n$th column in $\mathcal{O}^{-1}$. Then

$$A_0 = \begin{bmatrix} 0 \cdots 0 & -\alpha_0 \\ 1 \cdots 0 & -\alpha_1 \\ \vdots \ddots \vdots & \vdots \\ 0 \cdots 1 & -\alpha_{n-1} \end{bmatrix}, \quad C_o = [0, \ldots, 0, 1], \tag{6.67}$$

where the $\alpha_i$ denote the coefficients of the characteristic polynomial $\alpha(s) \triangleq \det(sI - A) = s^n + \alpha_{n-1}s^{n-1} + \cdots + \alpha_1 s + \alpha_0$. Here $A_o = PAP^{-1} = Q^{-1}AQ$, $C_o = CP^{-1} = CQ$, and the desired result can be established by using a proof that is completely analogous to the proof in determining the (dual) controller form presented in Subsection 6.4.1. Note that $B_o = PB$ does not have any particular structure. The representation $\{A_o, B_o, C_o, D_o\}$ will be referred to as the *observer form* of the system.

Reversing the order of columns in $P^{-1}$ given in (6.66) or selecting $P$ to be exactly $\mathcal{O}$, or to be equal to the matrix obtained after the order of the columns in $\mathcal{O}$ has been reversed, leads to alternative observer forms in a manner analogous to the controller form case.

---

**Example 6.22.** Let $A = \begin{bmatrix} -1 & 0 & 0 \\ 0 & 1 & 0 \\ 0 & 0 & -2 \end{bmatrix}$ and $C = [1, -1, 1]$. To derive the observer form (6.67), we could use duality, by defining $\tilde{A} = A^T, \tilde{B} = C^T$, and deriving the controller form of $\tilde{A}, \tilde{B}$, i.e., by following the procedure outlined above. We note that the $\tilde{A}, \tilde{B}$ are exactly the matrices given in Examples 6.13 and 6.14. As an alternative approach, the observer form is now derived directly. In particular, we have

$$\mathcal{O} = \begin{bmatrix} C \\ CA \\ CA^2 \end{bmatrix} = \begin{bmatrix} 1 & -1 & 1 \\ -1 & -1 & -2 \\ 1 & -1 & 4 \end{bmatrix}, \mathcal{O}^{-1} = \begin{bmatrix} 1 & -1/2 & -1/2 \\ -1/3 & -1/2 & -1/6 \\ -1/3 & 0 & 1/3 \end{bmatrix},$$

and in view of (6.66),

$$Q = P^{-1} = [\tilde{q}, A\tilde{q}, A^2\tilde{q}] = \begin{bmatrix} -1/2 & 1/2 & -1/2 \\ -1/6 & -1/6 & -1/6 \\ 1/3 & -2/3 & 4/3 \end{bmatrix}.$$

Note that $\tilde{q} = [-1/2, -1/6, 1/3]^T$, the last column of $\mathcal{O}^{-1}$. Then

$$A_o = Q^{-1}AQ = \begin{bmatrix} 0 & 0 & 2 \\ 1 & 0 & 1 \\ 0 & 1 & -2 \end{bmatrix}, \text{ and } C_o = CQ = [0, 0, 1],$$

where $|sI - A| = s^3 + 2s - s - 2 = s^3 + \alpha_2 s^2 + \alpha_1 s + \alpha_0$. Hence, $QA_o = \begin{bmatrix} 1/2 & -1/2 & 1/2 \\ -1/6 & -1/6 & -1/6 \\ -2/3 & 4/3 & -8/3 \end{bmatrix} = AQ.$

---

**Multi-Output Case ($p > 1$)**

Consider

$$\mathcal{O} = \begin{bmatrix} C \\ CA \\ \vdots \\ CA^{n-1} \end{bmatrix} = \begin{bmatrix} c_1 \\ \vdots \\ c_p \\ c_1 A \\ \vdots \\ c_p A \\ \vdots \\ c_1 A^{n-1} \\ \vdots \\ c_p A^{n-1} \end{bmatrix}, \tag{6.68}$$

where $c_1, \ldots, c_p$ denote the $p$ rows of $C$, and select the first $n$ linearly independent rows in $\mathcal{O}$, moving from the top to bottom (rank $\mathcal{O} = n$). Next, reorder the selected rows by first taking all rows involving $c_1$, then $c_2$, etc., to obtain

$$\bar{\mathcal{O}} \triangleq \begin{bmatrix} c_1 \\ c_1 A \\ \vdots \\ c_1 A^{\nu_1 - 1} \\ \vdots \\ c_p \\ \vdots \\ c_p A^{\nu_p - 1} \end{bmatrix}, \tag{6.69}$$

an $n \times n$ matrix. The integer $\nu_i$ denotes the number of rows involving $c_i$ in the set of the first $n$ linearly independent rows found in $\mathcal{O}$ when moving from top to bottom.

**Definition 6.23.** *The $p$ integers $\nu_i$, $i = 1, \ldots, p$, are the* observability indices *of the system, and $\nu \triangleq \max \nu_i$ is called the* observability index *of the system. Note that*

$$\sum_{i=1}^{p} \nu_i = n \quad and \quad p\nu \geq n. \tag{6.70}$$

∎

When rank $C = p$, then $\nu_i \geq 1$. Now define

$$\tilde{\sigma}_k \triangleq \sum_{i=1}^{k} \nu_i \quad k = 1, \ldots, p; \tag{6.71}$$

i.e., $\tilde{\sigma}_1 = \nu_1, \tilde{\sigma}_2 = \nu_1 + \nu_2, \ldots, \tilde{\sigma}_p = \nu_1 + \cdots + \nu_p = n$. Consider $\bar{\mathcal{O}}^{-1}$ and let $\tilde{q}_k \in R^n$, $k = 1, \ldots, p$, represent its $\tilde{\sigma}_k$th column; i.e.,

$$\bar{\mathcal{O}}^{-1} = [\times \cdots \times \tilde{q}_1 | \times \cdots \times \tilde{q}_2 | \cdots | \times \cdots \times \tilde{q}_p]. \tag{6.72}$$

Define

$$P^{-1} = Q = [\tilde{q}_1, \ldots, A^{\nu_1 - 1}\tilde{q}_1, \ldots, \tilde{q}_p, \ldots, A^{\nu_p - 1}\tilde{q}_p]. \tag{6.73}$$

Then $A_o = PAP^{-1} = Q^{-1}AQ$ and $C_o = CP^{-1} = CQ$ are given by

$$A_o = [A_{ij}], \quad i, j = 1, \ldots, p,$$

$$A_{ii} = \begin{bmatrix} 0 \cdots 0 \times \\ I_{\nu_i - 1} \; \vdots \\ \times \end{bmatrix} \in R^{\nu_i \times \nu_i}, \quad i = j, A_{ij} = \begin{bmatrix} 0 \cdots 0 \times \\ \vdots \quad \vdots \; \vdots \\ 0 \cdots 0 \times \end{bmatrix} \in R^{\nu_i \times \nu_j}, \quad i \neq j,$$

and

$$C_o = [C_1, C_2, \ldots, C_p], C_i = \begin{bmatrix} 0 \cdots 0 \; 0 \\ \vdots \quad \vdots \; \vdots \\ 0 \cdots 0 \; 0 \\ 0 \cdots 0 \; 1 \\ 0 \cdots 0 \times \\ \vdots \quad \vdots \; \vdots \\ 0 \cdots 0 \times \end{bmatrix} \in R^{p \times \nu_i}, \tag{6.74}$$

where the 1 in the last column of $C_i$ occurs at the $i$th row location ($i = 1, \ldots, p$) and $\times$ denotes nonfixed entries. Note that the matrix $B_o = PB = Q^{-1}B$ does not have any particular structure. Equation (6.74) is a very useful form (in the observer problem) and shall be referred to as the *observer form* of the system.

Analogous to (6.55), we express $A_o$ and $C_o$ as

$$A_o = \bar{A}_o + A_p\bar{C}_o, \quad C_o = C_p\bar{C}_o, \tag{6.75}$$

where $\bar{A}_o = $ block diag$[A_1, A_2, \ldots, A_p]$ with $A_i = \begin{bmatrix} 0 \cdots \; 0 \\ I_{\nu_i - 1} \; \vdots \\ 0 \end{bmatrix} \in R^{\nu_i \times \nu_i}, \bar{C}_o = $ block diag$([0, \ldots, 0, 1]^T \in R^{\nu_i}, i = 1, \ldots, p)$, and $A_p \in R^{n \times p}$, and $C_p \in R^{p \times p}$ are appropriate matrices ($\sum_{i=1}^p \nu_i = n$). Note that $\bar{A}_o, \bar{C}_o$ are completely determined by the $p$ observability indices $\nu_i$, $i = 1, \ldots, p$, and $A_p$ and $C_p$ contain this information in the $\tilde{\sigma}_1$th, $\ldots, \tilde{\sigma}_p$th columns of $A_o$ and in the same columns of $C_o$, respectively.

---

**Example 6.24.** Given $A = \begin{bmatrix} 0 & 0 & 0 \\ 1 & 0 & 2 \\ 0 & 1 & -1 \end{bmatrix}$ and $C = \begin{bmatrix} 0 & 1 & 0 \\ 1 & 1 & 0 \end{bmatrix}$, we wish to reduce these to observer form. This can be accomplished using duality, i.e., by first reducing $\tilde{A} \triangleq A^T, \tilde{B} \triangleq C^T$ to controller form. Note that $\tilde{A}, \tilde{B}$ are the matrices used in Example 6.17, and therefore, the desired answer is easily obtained. Presently, we shall follow the direct algorithm described above. We have

$$
\mathcal{O} = \begin{bmatrix} C \\ CA \\ CA^2 \end{bmatrix} = \begin{bmatrix} 0 & 1 & 0 \\ 1 & 1 & 0 \\ 1 & 0 & 2 \\ 1 & 0 & 2 \\ 0 & 2 & -2 \\ 0 & 2 & -2 \end{bmatrix}.
$$

Searching from top to bottom, the first three linearly independent rows are $c_1, c_2, c_1 A$, and

$$
\bar{\mathcal{O}} = \begin{bmatrix} c_1 \\ c_1 A \\ c_2 \end{bmatrix} = \begin{bmatrix} 0 & 1 & 0 \\ 1 & 0 & 2 \\ 1 & 1 & 0 \end{bmatrix}.
$$

Note that the observability indices are $\nu_1 = 2, \nu_2 = 1$ and $\tilde{\sigma}_1 = 2, \tilde{\sigma}_2 = 3$. We compute

$$
\bar{\mathcal{O}}^{-1} = \begin{bmatrix} -1 & 0 & 1 \\ 1 & 0 & 0 \\ 1/2 & 1/2 & -1/2 \end{bmatrix} = \begin{bmatrix} \times & 0 & 1 \\ \times & 0 & 0 \\ \times & 1/2 & -1/2 \end{bmatrix}.
$$

Then, $Q = [\tilde{q}_1, A\tilde{q}_1, \tilde{q}_2] = \begin{bmatrix} 0 & 0 & 1 \\ 0 & 1 & 0 \\ 1/2 & -1/2 & -1/2 \end{bmatrix}$ and $Q^{-1} = \begin{bmatrix} 1 & 1 & 2 \\ 0 & 1 & 0 \\ 1 & 0 & 0 \end{bmatrix}$. Therefore,

$$
A_o = Q^{-1}AQ = \begin{bmatrix} A_{11} & A_{12} \\ \hline A_{21} & A_{22} \end{bmatrix} = \begin{bmatrix} 0 & 2 & 1 \\ 1 & -1 & 0 \\ 0 & 0 & 0 \end{bmatrix}, \quad C_o = CQ = [C_1 \vdots C_2] = \begin{bmatrix} 0 & 1 & 0 \\ 0 & 1 & 1 \end{bmatrix}.
$$

We can also verify (6.47), namely

$$
A_o = \begin{bmatrix} 0 & 2 & 1 \\ 1 & -1 & 0 \\ 0 & 0 & 0 \end{bmatrix} = \bar{A}_o + A_p \bar{C}_o = \begin{bmatrix} 0 & 0 & 0 \\ 1 & 0 & 0 \\ 0 & 0 & 0 \end{bmatrix} + \begin{bmatrix} 2 & 1 \\ -1 & 0 \\ 0 & 0 \end{bmatrix} \begin{bmatrix} 0 & 1 & 0 \\ 0 & 0 & 1 \end{bmatrix}
$$

and

$$
C_o = \begin{bmatrix} 0 & 1 & 0 \\ 0 & 1 & 1 \end{bmatrix} = C_p \bar{C}_o = \begin{bmatrix} 1 & 0 \\ 1 & 1 \end{bmatrix} \begin{bmatrix} 0 & 1 & 0 \\ 0 & 0 & 1 \end{bmatrix}.
$$

## Structure Theorem—Observable Version

The transfer function matrix $H(s)$ of system $\dot{x} = Ax + Bu, y = Cx + Du$ is given by $H(s) = C(sI - A)^{-1}B + D$. If $(A, C)$ is in the *observer form*, given in (6.74), then $H(s)$ can alternatively be characterized by the Structure

Theorem stated in Theorem 6.25 below. This result will be very useful in the realization of systems, addressed in Chapter 8 and also in the study of observers in Chapter 9.

Let $A = A_o = \bar{A}_o + A_p \bar{C}_o$ and $C = C_o = C_p \bar{C}_o$ as in (6.75) with $|C_p| \neq 0$; let $B = B_o$ and $D = D_o$, and define

$$\tilde{\Lambda}(s) \triangleq \mathrm{diag}[s^{\nu_1}, s^{\nu_2}, \ldots, s^{\nu_p}], \tilde{S}(s) \triangleq \mathrm{block\ diag}([1, s, \ldots, s^{\nu_i - 1}], i = 1, \ldots, p). \tag{6.76}$$

Note that $\tilde{S}(s)$ is a $p \times n$ polynomial matrix, where $n = \sum_{i=1}^{p} \nu_i$. Now define the $p \times p$ polynomial matrix $\tilde{D}(s)$ and the $p \times m$ polynomial matrix $\tilde{N}(s)$ as

$$\tilde{D}(s) \triangleq [\tilde{\Lambda}(s) - \tilde{S}(s)A_p]C_p^{-1}, \quad \tilde{N}(s) \triangleq \tilde{S}(s)B_o + \tilde{D}(s)D_o. \tag{6.77}$$

The following result is the observable version of the *Structure Theorem*. It is the dual of Theorem 6.19 and can therefore be proved using duality arguments. The proof given is direct.

**Theorem 6.25.** $H(s) = \tilde{D}^{-1}(s)\tilde{N}(s)$, where $\tilde{N}(s), \tilde{D}(s)$ are defined in (6.77).

*Proof.* First we note that

$$\tilde{D}(s)C_o = \tilde{S}(s)(sI - A_o). \tag{6.78}$$

To see this, write $\tilde{D}(s)C_o = [\tilde{\Lambda}(s) - \tilde{S}(s)A_p]C_p^{-1}C_p\bar{C}_o = \tilde{\Lambda}(s)\bar{C}_o - \tilde{S}(s)A_p\bar{C}_o$, and also, $\tilde{S}(s)(sI - A_o) = \tilde{S}(s)s - \tilde{S}(s)(\bar{A}_o + A_p\bar{C}_o) = \tilde{S}(s)(sI - \bar{A}_o) - \tilde{S}(s)A_p\bar{C}_o = \tilde{\Lambda}(s)\bar{C}_o - \tilde{S}(s)A_p\bar{C}_o$, which proves (6.78). We now obtain $H(s) = C_o(sI - A_o)^{-1}B_o + D_o = \tilde{D}^{-1}(s)\tilde{S}(s)B_o + D_o = \tilde{D}^{-1}(s)[\tilde{S}(s)B_o + \tilde{D}(s)D_o] = \tilde{D}^{-1}(s)\tilde{N}(s)$. ∎

---

***Example 6.26.*** Consider $A_o = \begin{bmatrix} 0 & 2 & 1 \\ 1 & -1 & 0 \\ 0 & 0 & 0 \end{bmatrix}$ and $C_o = \begin{bmatrix} 0 & 1 & 0 \\ 0 & 1 & 1 \end{bmatrix}$ of Example 6.24. Here $\nu_1 = 2, \nu_2 = 1, \tilde{\Lambda}(s) = \begin{bmatrix} s^2 & 0 \\ 0 & s \end{bmatrix}$, and $\tilde{S}(s) = \begin{bmatrix} 1 & s & 0 \\ 0 & 0 & 1 \end{bmatrix}$. Then

$$\tilde{D}(s) = [\tilde{\Lambda}(s) - \tilde{S}(s)A_p]C_p^{-1} = \left[ \begin{bmatrix} s^2 & 0 \\ 0 & s \end{bmatrix} - \begin{bmatrix} 1 & s & 0 \\ 0 & 0 & 1 \end{bmatrix} \begin{bmatrix} 2 & 1 \\ -1 & 0 \\ 0 & 0 \end{bmatrix} \right] \cdot \begin{bmatrix} 1 & 0 \\ 1 & 1 \end{bmatrix}^{-1} =$$

$$\left[ \begin{bmatrix} s^2 & 0 \\ 0 & s \end{bmatrix} - \begin{bmatrix} -s+2 & 1 \\ 0 & 0 \end{bmatrix} \right] \cdot \begin{bmatrix} 1 & 0 \\ -1 & 1 \end{bmatrix} = \begin{bmatrix} s^2 + s - 2, & -1 \\ 0 & s \end{bmatrix} \cdot \begin{bmatrix} 1 & 0 \\ -1 & 1 \end{bmatrix} = \begin{bmatrix} s^2 + s - 1 & -1 \\ -s & s \end{bmatrix}.$$

Now if $B_o = [0, 1, 1]^T, D_o = 0$, and $\tilde{N}(s) = \tilde{S}(s)B_o + \tilde{D}(s)D_o = [s, 1]^T$, then $H(s) = \tilde{D}^{-1}(s)\tilde{N}(s) = \frac{1}{s(s^2+s-2)}[s^2+1, 2s^2+s-1]^T = C_o(sI - A_o)^{-1}B_o + D_o$.

---

# 6.5 Summary and Highlights

- The standard form for uncontrollable systems is

$$\widehat{A} = Q^{-1}AQ = \begin{bmatrix} A_1 & A_{12} \\ 0 & A_2 \end{bmatrix}, \quad \widehat{B} = Q^{-1}B = \begin{bmatrix} B_1 \\ 0 \end{bmatrix}, \tag{6.6}$$

  where $A_1 \in R^{n_r \times n_r}$, $B_1 \in R^{n_r \times m}$, and $(A_1, B_1)$ is controllable. $n_r < n$ is the rank of the controllability matrix $\mathcal{C} = [B, AB, \ldots, A^{n-1}B]$; i.e.,

$$\operatorname{rank} \mathcal{C} = n_r.$$

- The standard form for unobservable systems is

$$\widehat{A} = Q^{-1}AQ = \begin{bmatrix} A_1 & 0 \\ A_{21} & A_2 \end{bmatrix}, \quad \widehat{C} = CQ = \begin{bmatrix} C_1 \\ 0 \end{bmatrix}, \tag{6.15}$$

  where $A_1 \in R^{n_o \times n_o}$, $C_1 \in R^{p \times n_o}$, and $(A_1, C_1)$ is observable. $n_o < n$ is the rank of the observability matrix

$$\mathcal{O} = \begin{bmatrix} \widehat{C} \\ \widehat{C}\widehat{A} \\ \vdots \\ \widehat{C}\widehat{A}^{n-1} \end{bmatrix};$$

  i.e.,

$$\operatorname{rank} \mathcal{O} = n_o.$$

- Kalman's Decomposition Theorem.

$$\widehat{A} = Q^{-1}AQ = \begin{bmatrix} A_{11} & 0 & A_{13} & 0 \\ A_{21} & A_{22} & A_{23} & A_{24} \\ 0 & 0 & A_{33} & 0 \\ 0 & 0 & A_{43} & A_{44} \end{bmatrix}, \quad \widehat{B} = Q^{-1}B = \begin{bmatrix} B_1 \\ B_2 \\ 0 \\ 0 \end{bmatrix}, \tag{6.20}$$

$$\widehat{C} = CQ = [C_1,\ 0,\ C_3,\ 0],$$

  where $(A_{11}, B_1, C_1)$ is controllable and observable.
- $\lambda_i$ is an uncontrollable eigenvalue if and only if

$$\widehat{v}_i[\lambda_i I - A, B] = 0, \tag{6.23}$$

  where $\widehat{v}_i$ is the corresponding (left) eigenvector.
- $\lambda_i$ is an unobservable eigenvalue if and only if

$$\begin{bmatrix} \lambda_i I - A \\ C \end{bmatrix} v_i = 0, \tag{6.24}$$

  where $v_i$ is the corresponding (right) eigenvector.

*Controller Forms (for Controllable Systems)*

- $m = 1$ case.

$$A_c = \begin{bmatrix} 0 & 1 & \cdots & 0 \\ \vdots & \vdots & \ddots & \vdots \\ 0 & 0 & \cdots & 1 \\ -\alpha_0 & -\alpha_1 & \cdots & -\alpha_{n-1} \end{bmatrix}, \quad B_c = \begin{bmatrix} 0 \\ \vdots \\ 0 \\ 1 \end{bmatrix}, \quad (6.36)$$

where

$$\alpha(s) \triangleq \det(sI - A) = s^n + \alpha_{n-1}s^{n-1} + \cdots + \alpha_1 s + \alpha_0. \quad (6.37)$$

- $m > 1$ case.

$$A_c = [A_{ij}], \quad i, j = 1, \ldots, m,$$

$$A_{ii} = \begin{bmatrix} 0 & & \\ \vdots & I_{\mu_i - 1} & \\ 0 & & \\ \times & \times & \cdots & \times \end{bmatrix} \in R^{\mu_i \times \mu_i}, \ i = j, \ A_{ij} = \begin{bmatrix} 0 & \cdots & 0 \\ \vdots & \vdots & \vdots \\ 0 & \cdots & 0 \\ \times & \times & \cdots & \times \end{bmatrix} \in R^{\mu_i \times \mu_j}, \ i \neq j,$$

and

$$B_c = \begin{bmatrix} B_1 \\ B_2 \\ \vdots \\ B_m \end{bmatrix}, \quad B_i = \begin{bmatrix} 0 & \cdots & 0 & 0 & \cdots & 0 \\ \vdots & & \vdots & \vdots & & \vdots \\ 0 & \cdots & 0 & 1 & \times & \cdots & \times \end{bmatrix} \in R^{\mu_i \times m}. \quad (6.54)$$

An example for $n = 4$, $m = 2$ and $\mu_1 = 2, \mu_2 = 2$ is

$$A_c = \begin{bmatrix} A_{11} & A_{12} \\ A_{21} & A_{22} \end{bmatrix} = \begin{bmatrix} 0 & 1 & 0 & 0 \\ \times & \times & \times & \times \\ 0 & 0 & 0 & 1 \\ \times & \times & \times & \times \end{bmatrix}, \quad B_c = \begin{bmatrix} B_1 \\ B_2 \end{bmatrix} = \begin{bmatrix} 0 & 0 \\ 1 & \times \\ 0 & 0 \\ 0 & 1 \end{bmatrix}.$$

- 
$$A_c = \bar{A}_c + \bar{B}_c A_m, \quad B_c = \bar{B}_c B_m. \quad (6.55)$$

- *Structure theorem—controllable version*
  $H(s) = N(s)D^{-1}(s)$, where

$$D(s) = B_m^{-1}[\Lambda(s) - A_m S(s)], N(s) = C_c S(s) + D_c D(s). \quad (6.61)$$

Note that

$$(sI - A_c)S(s) = B_c D(s). \quad (6.62)$$

*Observer Forms (for Observable Systems)*

- $p = 1$ case.

$$A_0 = \begin{bmatrix} 0 \cdots 0 & -\alpha_0 \\ 1 \cdots 0 & -\alpha_1 \\ \vdots \ddots \vdots & \vdots \\ 0 \cdots 1 & -\alpha_{n-1} \end{bmatrix}, \quad C_o = [0, \ldots, 0, 1]. \tag{6.67}$$

- $p > 1$.

$$A_o = [A_{ij}], \quad i, j = 1, \ldots, p,$$

$$A_{ii} = \begin{bmatrix} 0 \cdots 0 & \times \\ I_{\nu_i - 1} & \vdots \\ & \times \end{bmatrix} \in R^{\nu_i \times \nu_i}, \, i = j, \; A_{ij} = \begin{bmatrix} 0 \cdots & 0 & \times \\ \vdots & \vdots & \vdots \\ 0 \cdots & 0 & \times \end{bmatrix} \in R^{\nu_i \times \nu_j}, \, i \neq j,$$

and

$$C_o = [C_1, C_2, \ldots, C_p], C_i = \begin{bmatrix} 0 \cdots 0 & 0 \\ \vdots & \vdots \vdots \\ 0 \cdots 0 & 0 \\ 0 \cdots 0 & 1 \\ 0 \cdots 0 & \times \\ \vdots & \vdots \vdots \\ 0 \cdots 0 & \times \end{bmatrix} \in R^{p \times \nu_i}, \tag{6.74}$$

If $(A_c, B_c)$ is in controller form, $(A_o = A_c^T, C_o = B_c^T)$ will be in observer form.

- 

$$A_o = \bar{A}_o + A_p \bar{C}_o, \quad C_o = C_p \bar{C}_o. \tag{6.75}$$

- *Structure theorem—observable version*
  $H(s) = \tilde{D}^{-1}(s)\tilde{N}(s)$, where

$$\tilde{D}(s) = [\tilde{\Lambda}(s) - \tilde{S}(s)A_p]C_p^{-1}, \quad \tilde{N}(s) = \tilde{S}(s)B_o + \tilde{D}(s)D_o. \tag{6.77}$$

Note that

$$\tilde{D}(s)C_o = \tilde{S}(s)(sI - A_o). \tag{6.78}$$

## 6.6 Notes

Special state-space forms for controllable and observable systems obtained by similarity transformations are discussed at length in Kailath [5]. Wolovich [13] discusses the algorithms for controller and observer forms and introduces the Structure Theorems. The controller form is based on results by Luenberger [9] (see also Popov [10]). A detailed derivation of the controller form can also be found in Rugh [12].

Original sources for the Canonical Structure Theorem include Kalman [6] and Gilbert [3].

The eigenvector and rank tests for controllability and observability are called PBH tests in Kailath [5]. Original sources for these include Popov [10], Belevich [2], and Hautus [4]. Consult also Rosenbrock [11], and for the case when $A$ can be diagonalized via a similarity transformation, see Gilbert [3]. Note that in the eigenvalue/eigenvector tests presented herein the uncontrollable (unobservable) eigenvalues are also explicitly identified, which represents a modification of the above original results.

The fact that the controllability indices appear in the work of Kronecker was recognized by Rosenbrock [11] and Kalman [8].

For an extensive introductory discussion and a formal definition of canonical forms, see Kailath [5].

# References

1. P.J. Antsaklis and A.N. Michel, *Linear Systems*, Birkhäuser, Boston, MA, 2006.
2. V. Belevich, *Classical Network Theory*, Holden-Day, San Francisco, CA, 1968.
3. E. Gilbert, "Controllability and observability in multivariable control systems," *SIAM J. Control*, Vol. 1, pp. 128–151, 1963.
4. M.L.J. Hautus, "Controllability and observability conditions of linear autonomous systems," *Proc. Koninklijke Akademie van Wetenschappen, Serie A*, Vol. 72, pp. 443–448, 1969.
5. T. Kailath, *Linear Systems*, Prentice-Hall, Englewood Cliffs, NJ, 1980.
6. R.E. Kalman, "Mathematical descriptions of linear systems," *SIAM J. Control*, Vol. 1, pp. 152–192, 1963.
7. R.E. Kalman, "On the computation of the reachable/observable canonical form," *SIAM J. Control Optimization*, Vol. 20, no. 2, pp. 258–260, 1982.
8. R.E. Kalman, "Kronecker invariants and feedback," in *Ordinary Differential Equations*, L. Weiss, ed., pp. 459–471, Academic Press, New York, NY, 1972.
9. D.G. Luenberger, "Canonical forms for linear multivariable systems," *IEEE Trans. Auto. Control*, Vol. 12, pp. 290–293, 1967.
10. V.M. Popov, "Invariant description of linear, time-invariant controllable systems," *SIAM J. Control Optimization*, Vol. 10, No. 2, pp. 252–264, 1972.
11. H.H. Rosenbrock, *State-Space and Multivariable Theory*, Wiley, New York, NY, 1970.
12. W.J. Rugh, *Linear System Theory*, Second Ed., Prentice-Hall, Englewood Cliffs, NJ, 1996.
13. W.A. Wolovich, *Linear Multivariable Systems*, Springer-Verlag, New York, NY, 1974.

# Exercises

**6.1.** Write software programs to implement the algorithms of Section 6.2. In particular:

(a) Given the pair $(A, B)$, where $A \in R^{n \times n}, B \in R^{n \times m}$ with

$$\text{rank}[B, AB, \ldots, A^{n-1}B] = n_r < n,$$

reduce this pair to the standard uncontrollable form

$$\widehat{A} = PAP^{-1} = \begin{bmatrix} A_1 & A_{12} \\ 0 & A_2 \end{bmatrix}, \widehat{B} = PB = \begin{bmatrix} B_1 \\ 0 \end{bmatrix},$$

where $(A_1, B_1)$ is controllable and $A_1 \in R^{n_r \times n_r}, B_1 \in R^{n_r \times m}$.

(b) Given the controllable pair $(A, B)$, where $A \in R^{n \times n}, B \in R^{n \times m}$ with rank $B = m$, reduce this pair to the controller form $A_c = PAP^{-1}, B_c = PB$.

**6.2.** Determine the uncontrollable modes of each pair $(A, B)$ given below by

(a) Reducing $(A, B)$, using a similarity transformation.
(b) Using eigenvalue/eigenvector criteria:

$$A = \begin{bmatrix} 1 & 0 & 0 \\ 0 & -1 & 0 \\ 0 & 0 & 2 \end{bmatrix}, \quad B = \begin{bmatrix} 1 & 0 \\ 0 & 1 \\ 0 & 0 \end{bmatrix} \quad \text{and} \quad A = \begin{bmatrix} 0 & 0 & 1 & 0 \\ 0 & 0 & 1 & 0 \\ 0 & 0 & 0 & 0 \\ 0 & 0 & 0 & -1 \end{bmatrix}, \quad B = \begin{bmatrix} 0 & 1 \\ 0 & 0 \\ 1 & 0 \\ 0 & 0 \end{bmatrix}.$$

**6.3.** Reduce the pair

$$A = \begin{bmatrix} 0 & 0 & 1 & 0 \\ 3 & 0 & -3 & 1 \\ -1 & 1 & 4 & -1 \\ 1 & 0 & -1 & 0 \end{bmatrix}, \quad B = \begin{bmatrix} 0 & 0 \\ 1 & 0 \\ 0 & 1 \\ 0 & 0 \end{bmatrix}$$

into controller form $A_c = PAP^{-1}, B_c = PB$. What is the similarity transformation matrix in this case? What are the controllability indices?

**6.4.** Consider

$$A_c = \begin{bmatrix} 0 & 1 & \cdots & 0 \\ \vdots & \vdots & \ddots & \vdots \\ 0 & 0 & \cdots & 1 \\ -\alpha_0 & -\alpha_1 & \cdots & -\alpha_{n-1} \end{bmatrix}, \quad B_c = \begin{bmatrix} 0 \\ \vdots \\ 0 \\ 1 \end{bmatrix}.$$

Show that

$$C = [B_c, A_c B_c, \ldots, A_c^{n-1} B_c] = \begin{bmatrix} 0 & 0 & 0 & \cdots & 1 \\ 0 & 0 & 0 & \cdots & c_1 \\ \vdots & \vdots & \vdots & & \vdots \\ 0 & 0 & 1 & \cdots & c_{n-3} \\ 0 & 1 & c_1 & \cdots & c_{n-2} \\ 1 & c_1 & c_2 & \cdots & c_{n-1} \end{bmatrix},$$

where $c_k = -\sum_{i=0}^{k-1} \alpha_{n-i-1} c_{k-i-1}$, $k = 1, \ldots, n-1$, with $c_0 = 1$. Also, show that

$$
C^{-1} = \begin{bmatrix} \alpha_1 & \alpha_2 & \cdots & \alpha_{n-1} & 1 \\ \alpha_2 & \alpha_3 & \cdots & 1 & 0 \\ \vdots & \vdots & & \vdots & \vdots \\ \alpha_{n-1} & 1 & \cdots & 0 & 0 \\ 1 & 0 & \cdots & 0 & 0 \end{bmatrix}.
$$

**6.5.** Show that the matrices $A_c = PAP^{-1}, B_c = PB$ are as follows:

(a) Given by (6.41) if $P$ is given by (6.42).
(b) Given by (6.44) if $Q(= P^{-1})$ is given by (6.43).
(c) Given by (6.46) if $Q(= P^{-1})$ is given by (6.45).

**6.6.** Consider the pair $(A, b)$, where $A \in R^{n \times n}, b \in R^n$. Show that if more than one linearly independent eigenvector can be associated with a single eigenvalue, then $(A, b)$ is uncontrollable. *Hint:* Use the eigenvector test. Let $\hat{v}_1, \hat{v}_2$ be linearly independent left eigenvectors associated with eigenvalue $\lambda_1 = \lambda_2 = \lambda$. Notice that if $\hat{v}_1 b = \alpha_1$ and $\hat{v}_2 b = \alpha_2$, then $(\alpha_1 \hat{v}_1 - \alpha_1 \hat{v}_2) b = 0$.

**6.7.** Show that if $(A, B)$ is controllable, where $A \in R^{n \times n}$, and $B \in R^{n \times m}$, and rank $B = m$, then rank $A \geq n - m$.

**6.8.** Given $A \in R^{n \times n}$, and $B \in R^{n \times m}$, let rank $C = n$, where $C = [B, AB, \ldots, A^{n-1}B]$. Consider $\hat{A} \in R^{n \times n}, \hat{B} \in R^{n \times m}$ with rank $\hat{C} = n$, where $\hat{C} = [\hat{B}, \hat{A}\hat{B}, \ldots, \hat{A}^{n-1}\hat{B}]$, and assume that $P \in R^{n \times n}$ with $\det P \neq 0$ exists such that

$$
P[C, A^n B] = [\hat{C}, \hat{A}^n \hat{B}].
$$

Show that $\hat{B} = PB$ and $\hat{A} = PAP^{-1}$. *Hint:* Show that $(PA - \hat{A}P)C = 0$.

**6.9.** Let $A = \bar{A}_c + \bar{B}_c A_m$ and $B = \bar{B}_c B_m$, where the $\bar{A}_c, \bar{B}_c$ are as in (6.55) with $A_m \in R^{m \times n}, B_m \in R^{m \times m}$, and $|B_m| \neq 0$. Show that $(A, B)$ is controllable with controllability indices $\mu_i$. *Hint:* Use the eigenvalue test to show that $(A, B)$ is controllable. Use state feedback to simplify $(A, B)$ (see Exercise 6.11), and show that the $\mu_i$ are the controllability indices.

**6.10.** Show that the controllability indices of the state equation $\dot{x} = Ax + BGv$, where $|G| \neq 0$ and $(A, B)$ is controllable, with $A \in R^{n \times n}, B \in R^{n \times m}$, are the same as the controllability indices of $\dot{x} = Ax + Bu$, within reordering. *Hint:* Write $\bar{C}_k = [BG, ABG, \ldots, A^{k-1}BG] = [B, AB, \ldots, A^{k-1}B] \cdot$ [block diag $G$] $= C_k \cdot$ [block diag $G$] and show that the number of linearly dependent columns in $A^k BG$ that occur while searching from left to right in $\bar{C}_n$ is the same as the corresponding number in $C_n$.

**6.11.** Consider the state equation $\dot{x} = Ax + Bu$, where $A \in R^{n \times n}, B \in R^{n \times m}$ with $(A, B)$ controllable. Let the linear state-feedback control law be $u = Fx + Gv, F \in R^{m \times n}, G \in R^{m \times m}$ with $|G| \neq 0$. Show that

(a) $(A + BF, BG)$ is controllable.
(b) The controllability indices of $(A + BF, B)$ are identical to those of $(A, B)$.
(c) The controllability indices of $(A+BF, BG)$ are equal to the controllability indices of $(A, B)$ within reordering. *Hint:* Use the eigenvalue test to show (a). To show (b), use the controller forms in Section 6.4.

**6.12.** For the system $\dot{x} = Ax + Bu, y = Cx$, consider the corresponding sampled-data system $\bar{x}(k + 1) = \bar{A}\bar{x}(k) + \bar{B}\bar{u}(k), \bar{y}(k) = \bar{C}\bar{x}(k)$, where

$$\bar{A} = e^{AT}, \bar{B} = [\int_0^T e^{A\tau} d\tau]B, \quad \text{and} \quad \bar{C} = C.$$

(a) Let the continuous-time system $\{A, B, C\}$ be controllable (observable), and assume it is a SISO system. Show that $\{\bar{A}, \bar{B}, \bar{C}\}$ is controllable (observable) if and only if the sampling period $T$ is such that

$$Im\,(\lambda_i - \lambda_j) \neq \frac{2\pi k}{T}, \text{ where } k = \pm 1, \pm 2, \ldots \text{ whenever } Re\,(\lambda_i - \lambda_j) = 0,$$

where $\{\lambda_i\}$ are the eigenvalues of $A$. *Hint:* Use the PBH test.
(b) Apply the results of (a) to the double integrator (Example 3.33 in Chapter 3), where $A = \begin{bmatrix} 0 & 1 \\ 0 & 0 \end{bmatrix}$, $B = \begin{bmatrix} 0 \\ 1 \end{bmatrix}$, and $C = [1, 0]$, and also to $A = \begin{bmatrix} 0 & 1 \\ -1 & 0 \end{bmatrix}$, $B = \begin{bmatrix} 0 \\ 1 \end{bmatrix}$, $C = [1, 0]$. Determine the values of $T$ that preserve controllability (observability).

**6.13. (Spring mass system)** Consider the spring mass given in Exercise 3.37.

(a) Is the system controllable from $[f_1, f_2]^T$? If yes, reduce $(A, B)$ to controller form.
(b) Is the system controllable from input $f_1$ only? Is it controllable from $f_2$ only? Discuss your answers.
(c) Let $y = Cx$ with $C = \begin{bmatrix} 1 & 0 & 0 & 0 \\ 0 & 1 & 0 & 0 \end{bmatrix}$. Is the system observable from $y$? If yes, reduce $(A, C)$ to observer form.

# 7

# Internal and External Descriptions: Relations and Properties

## 7.1 Introduction

In this chapter it is shown how external descriptions of a system, such as the transfer function and the impulse response, depend only on the controllable and observable parts of internal state-space descriptions (Section 7.2). Based on these results, the exact relation between internal (Lyapunov) stability and input–output stability is established in Section 7.3. In Section 7.4 the poles of the transfer function matrix, the poles of the system (eigenvalues), the zeros of the transfer function, the invariant zeros, the decoupling zeros, and their relation to uncontrollable or unobservable eigenvalues are addressed. In the final Section 7.5, polynomial matrix and matrix fractional descriptions are introduced. Polynomial matrix descriptions are generalizations of state-space internal descriptions. The matrix fractional descriptions of transfer function matrices offer a convenient way to work with transfer functions in control design and to establish the relations between internal and external descriptions of systems.

## 7.2 Relations Between State-Space and Input–Output Descriptions

In this section it is shown that the input–output description, namely the transfer function or the impulse response of a system, depends only on the part of the state-space representation that is both controllable and observable. The uncontrollable and/or unobservable parts of the system "cancel out" and play no role in the input–output system descriptions.

Consider the system

$$\dot{x} = Ax + Bu, \quad y = Cx + Du, \tag{7.1}$$

where $A \in R^{n \times n}$, $B \in R^{n \times m}$, $C \in R^{p \times n}$, $D \in R^{p \times m}$ has $p \times m$. The transfer function matrix

$$H(s) = C(sI - A)^{-1}B + D = \widehat{C}(sI - \widehat{A})^{-1}\widehat{B} + \widehat{D}, \qquad (7.2)$$

where $\{\widehat{A}, \widehat{B}, \widehat{C}, \widehat{D}\}$ is an equivalent representation given in (6.9) with $\widehat{A} = Q^{-1}AQ$, $\widehat{B} = Q^{-1}B$, $\widehat{C} = CQ$, and $\widehat{D} = D$. Consider now the Kalman Decomposition Theorem in Section 6.2.3 and the representation (6.22). We wish to investigate which of the submatrices $A_{ij}, B_i, C_j$ determine $H(s)$ and which do not. The inverse of $sI - \widehat{A}$ can be determined by repeated application of the formulas

$$\begin{bmatrix} \alpha & \beta \\ 0 & \delta \end{bmatrix}^{-1} = \begin{bmatrix} \alpha^{-1} & -\alpha^{-1}\beta\delta^{-1} \\ 0 & \delta^{-1} \end{bmatrix} \quad \text{and} \quad \begin{bmatrix} \alpha & 0 \\ \gamma & \delta \end{bmatrix}^{-1} = \begin{bmatrix} \alpha^{-1} & 0 \\ -\delta^{-1}\gamma\alpha^{-1} & \delta^{-1} \end{bmatrix},$$
$$(7.3)$$

where $\alpha, \beta, \gamma, \delta$ are matrices with $\alpha$ and $\delta$ square and nonsingular. It turns out that

$$H(s) = C_1(sI - A_{11})^{-1}B_1 + D, \qquad (7.4)$$

that is, the only part of the system that determines the external description is $\{A_{11}, B_1, C_1, D\}$, the subsystem that is both controllable and observable [see Theorem 6.6(iii)]. Analogous results exist in the time domain. Specifically, taking the inverse Laplace transform of both sides in (7.4), the impulse response of the system for $t \geq 0$ is derived as

$$H(t, 0) = C_1 e^{A_{11}t}B_1 + D\delta(t), \qquad (7.5)$$

which depends only on the controllable and observable parts of the system, as expected.

Similar results exist for discrete-time systems described by (6.4). For such systems, the transfer function matrix $H(z)$ and the pulse response $H(k, 0)$ are given by

$$H(z) = C_1(zI - A_{11})^{-1}B_1 + D \qquad (7.6)$$

and

$$H(k, 0) = \begin{cases} C_1 A_{11}^{k-1}B_1, & k > 0, \\ D, & k = 0. \end{cases} \qquad (7.7)$$

Again, these depend only on the part of the system that is both controllable and observable, as in the continuous-time case.

---

**Example 7.1.** For the system $\dot{x} = Ax + Bu$, $y = Cx$, where $A, B, C$ are as in Examples 6.7 and 6.10, we have $H(s) = C(sI - A)^{-1}B = C_1(sI - A_{11})^{-1}B_1 = (1)(1/s)[1, 1] = [1/s, \ 1/s]$. Notice that only the controllable and observable eigenvalue of $A$, $\lambda_1 = 0$ (in $A_{11}$), appears in the transfer function as a pole. All other eigenvalues ($\lambda_2 = -1, \lambda_3 = -2$) cancel out.

---

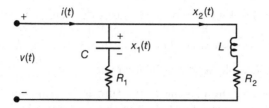

**Figure 7.1.** An $RL/RC$ circuit

---

*Example 7.2.* The circuit depicted in Figure 7.1 is described by the state-space equations

$$\begin{bmatrix} \dot{x}_1(t) \\ \dot{x}_2(t) \end{bmatrix} = \begin{bmatrix} -1/(R_1C) & 0 \\ 0 & -R_2/L \end{bmatrix} \begin{bmatrix} x_1(t) \\ x_2(t) \end{bmatrix} + \begin{bmatrix} 1/(R_1C) \\ 1/L \end{bmatrix} v(t)$$

$$i(t) = [-1/R_1, 1] \begin{bmatrix} x_1(t) \\ x_2(t) \end{bmatrix} + (1/R_1)v(t),$$

where the voltage $v(t)$ and current $i(t)$ are the input and output variables of the system, $x_1(t)$ is the voltage across the capacitor, and $x_2(t)$ is the current through the inductor. We have $\hat{i}(s) = H(s)\hat{v}(s)$ with the transfer function given by

$$H(s) = C(sI - A)^{-1}B + D = \frac{(R_1^2C - L)s + (R_1 - R_2)}{(Ls + R_2)(R_1^2Cs + R_1)} + \frac{1}{R_1}.$$

The eigenvalues of $A$ are $\lambda_1 = -1/(R_1C)$ and $\lambda_2 = -R_2/L$. Note that in general $\operatorname{rank}[\lambda_i I - A, B] = \operatorname{rank} \begin{bmatrix} \lambda_i I - A \\ C \end{bmatrix} = 2 = n$; i.e., the system is controllable and observable, unless the relation $R_1R_2C = L$ is satisfied. In this case, $\lambda_1 = \lambda_2 = -R_2/L$ and the system matrix $P(s)$ assumes the form

$$P(s) = \begin{bmatrix} sI - A, & B \\ -C, & D \end{bmatrix} = \begin{bmatrix} s + R_2/L & 0 & R_2/L \\ 0 & s + R_2/L & 1/L \\ 1/R_1 & -1 & 1/R_1 \end{bmatrix}.$$

In the following discussion, assume that $R_1R_2C = L$ is satisfied.

(i) Let $R_1 \neq R_2$ and take

$$[v_1, v_2] = \begin{bmatrix} R_2 & R_1 \\ 1 & 1 \end{bmatrix}, \quad \begin{bmatrix} \hat{v}_1 \\ \hat{v}_2 \end{bmatrix} = [v_1, v_2]^{-1} = \frac{1}{R_2 - R_1} \begin{bmatrix} 1 & -R_1 \\ -1 & R_2 \end{bmatrix}$$

to be the linearly independent right and left eigenvectors corresponding to the eigenvalues $\lambda_1 = \lambda_2 = -R_2/L$. The eigenvectors could have been any two linearly independent vectors since $\lambda_i I - A = 0$. They were chosen as above because they also have the property that $\hat{v}_2 B = 0$ and $C v_2 = 0$, which in view of Corollary 6.9, implies that $\lambda_2 = -R_2/L$ is both uncontrollable and unobservable. The eigenvalue $\lambda_1 = -R_2/L$ is both controllable and observable, as it can be seen using $Q = \begin{bmatrix} R_2 & R_1 \\ 1 & 1 \end{bmatrix}$ to reduce the representation to the canonical structure form (Kalman Decomposition Theorem). The transfer function is in this case given by

$$H(s) = \frac{(s + R_1/L)(s + R_2/L)}{R_1(s + R_2/L)(s + R_2/L)} = \frac{s + R_1/L}{R_1(s + R_2/L)};$$

that is, only the controllable and observable eigenvalue appears as a pole in $H(s)$, as expected.

(ii) Let $R_1 = R_2 = R$ and take

$$[v_1, v_2] = \begin{bmatrix} 1 & R \\ 0 & 1 \end{bmatrix}, \begin{bmatrix} \hat{v}_1 \\ \hat{v}_2 \end{bmatrix} = [v_1, v_2]^{-1} = \begin{bmatrix} 1 & -R \\ 0 & 1 \end{bmatrix}.$$

In this case $\hat{v}_1 B = 0$ and $C v_2 = 0$. Thus, one of the eigenvalues, $\lambda_1 = -R/L$, is uncontrollable (but can be shown to be observable) and the other eigenvalue, $\lambda_2 = -R/L$, is unobservable (but can be shown to be controllable). In this case, none of the eigenvalues appear in the transfer function. In fact,

$$H(s) = 1/R,$$

as can readily be verified. Thus, in this case, the network behaves as a constant resistance network.

*At this point it should be made clear that the modes that are uncontrollable and/or unobservable from certain inputs and outputs do not actually disappear; they are simply invisible from certain vantage points under certain conditions. (The voltages and currents of this network in the case of constant resistance $[H(s) = 1/R]$ are studied in Exercise 7.2.)*

---

**Example 7.3.** Consider the system $\dot{x} = Ax + Bu, y = Cx$, where $A = \begin{bmatrix} 1 & 0 & 0 \\ 0 & -2 & 0 \\ 0 & 0 & -1 \end{bmatrix}, B = \begin{bmatrix} 1 \\ 0 \\ 1 \end{bmatrix}$, and $C = [1, 1, 0]$. Using the eigenvalue/eigenvector test, it can be shown that the three eigenvalues of $A$ (resp., the three modes of $A$) are $\lambda_1 = 1$ (resp., $e^t$), which is controllable and observable; $\lambda_2 = -2$ (resp., $e^{-2t}$), which is uncontrollable and observable; and $\lambda_3 = -1$ (resp., $e^{-t}$), which is controllable and unobservable.

The response due to the initial condition $x(0)$ and the input $u(t)$ is

$$x(t) = e^{At}x(0) + \int_0^t e^{A(t-\tau)}Bu(\tau)d\tau$$

$$= \begin{bmatrix} e^t & 0 & 0 \\ 0 & e^{-2t} & 0 \\ 0 & 0 & e^{-t} \end{bmatrix} x(0) + \int_0^t \begin{bmatrix} e^{(t-\tau)} \\ 0 \\ e^{-(t-\tau)} \end{bmatrix} u(\tau)d\tau$$

and

$$y(t) = Ce^{At}x(0) + \int_0^t Ce^{A(t-\tau)}Bu(\tau)d\tau$$

$$= [e^t, e^{-2t}, 0]x(0) + \int_0^t e^{(t-\tau)}u(\tau)d\tau.$$

Notice that only controllable modes appear in $e^{At}B$ [resp., only controllable eigenvalues appear in $(sI - A)^{-1}B$], only observable modes appear in $Ce^{At}$ [resp., only observable eigenvalues appear in $C(sI - A)^{-1}$], and only modes that are both controllable and observable appear in $Ce^{At}B$ [resp., only eigenvalues that are both controllable and observable appear in $C(sI - A)^{-1}B = H(s)$]. For the discrete-time case, refer to Exercise 7.1d.

---

## 7.3 Relations Between Lyapunov and Input–Output Stability

In view of the relation between eigenvalues of $A$ and poles of $H(s)$ developed above [see also (7.20) and (7.22)] we are now in a position to provide complete insight into the relation between exponential stability or Lyapunov stability and BIBO (Bounded Input Bounded Output) stability of a system.

Consider the system $\dot{x} = Ax + Bu$, $y = Cx + Du$, and recall the following results:

(i) The system is asymptotically stable (internally stable, stable in the sense of Lyapunov) if and only if the real parts of all the eigenvalues of $A$, $Re\lambda_i(A)$ $i = 1, \ldots, n$, are negative. Recall also that asymptotic stability is equivalent to exponential stability in the case of linear time-invariant systems.

(ii) Let the transfer function be $H(s) = C(sI - A)^{-1}B + D$. The system is BIBO stable if and only if the real parts of all the poles of $H(s), Rep_i(H(s))$ $i = 1, \ldots, r$, are negative [see Section 4.7].

The relation between the eigenvalues of $A$ and the poles of $H(s)$ is

$$\{\text{eigenvalues of } A\} \supset \{\text{poles of } H(s)\} \tag{7.8}$$

with equality holding when all eigenvalues are controllable and observable [see (7.20), (7.22) and Chapter 8, Theorems 8.9 and 8.12]. Specifically, the eigenvalues of $A$ may be controllable and observable, uncontrollable and/or unobservable, and the poles of $H(s)$ are exactly the eigenvalues of $A$ that are both controllable and observable. The remaining eigenvalues of $A$, the uncontrollable and/or unobservable ones, cancel out when $H(s) = C(sI - A)^{-1}B + D$ is determined. Note also that the uncontrollable/unobservable eigenvalues that cancel correspond to input and output decoupling zeros (see Section 7.4). So the cancellations that take place in forming $H(s)$ are really pole/zero cancellations, i.e., cancellations between poles of the system (uncontrollable and unobservable eigenvalues of $A$) and zeros of the system (input and output decoupling zeros).

It is now straightforward to see that

$$\{\text{Internal stability}\} \not\Leftarrow \{\text{BIBO stability}\};$$

that is, internal stability implies, but is not necessarily implied by, BIBO stability. BIBO stability implies internal stability only when the system is completely controllable and observable ([1, p. 487, Theorem 9.4]).

---

**Example 7.4.** Consider the system $\dot{x} = Ax + Bu, y = Cx$, where $A = \begin{bmatrix} 0 & 1 \\ 2 & 1 \end{bmatrix}$, $B = \begin{bmatrix} 0 \\ 1 \end{bmatrix}$, and $C = [-2, \ 1]$. The eigenvalues of $A$ are the roots of $|sI - A| = s^2 - s - 2 = (s+1)(s-2)$ at $\{-1, 2\}$, and so the system is not internally stable (it is not stable in the sense of Lyapunov). The transfer function is

$$H(s) = C(sI - A)^{-1}B = \frac{s-2}{(s+1)(s-2)} = \frac{1}{s+1}.$$

Since there is one pole of $H(s)$ at $\{-1\}$, the system is BIBO stable, which verifies that BIBO stability does not necessarily imply internal stability. As it can be easily verified, the $-1$ eigenvalue of $A$ is controllable and observable and it is the eigenvalue that appears as a pole of $H(s)$ at $-1$. The other eigenvalue at $+2$ that is unobservable, which is also the output decoupling zero of the system, cancels in a pole/zero cancellation in $H(s)$ as expected.

---

## 7.4 Poles and Zeros

In this section the poles and zeros of a time-invariant system are defined and discussed. The poles and zeros are related to the (controllable and observable, resp., uncontrollable and unobservable) eigenvalues of $A$. These relationships shed light on the eigenvalue cancellation mechanisms encountered when input–output relations, such as transfer functions, are formed.

In the following development, the finite *poles of a transfer function matrix* $H(s)$ [or $H(z)$] are defined first (for the definition of poles at infinity, refer to the Exercise 7.9). It should be noted here that the eigenvalues of $A$ are sometimes called *poles of the system* $\{A, B, C, D\}$. To avoid confusion, we shall use the complete term *poles of $H(s)$*, when necessary. The zeros of a system are defined using internal descriptions (state-space representations).

### 7.4.1 Smith and Smith–McMillan Forms

To define the poles of $H(s)$, we shall first introduce the Smith form of a polynomial matrix $P(s)$ and the Smith–McMillan form of a rational matrix $H(s)$.

The *Smith form* $S_P(s)$ of a $p \times m$ polynomial matrix $P(s)$ (in which the entries are polynomials in $s$) is defined as

$$S_P(s) = \begin{bmatrix} \Lambda(s) & 0 \\ 0 & 0 \end{bmatrix} \tag{7.9}$$

with $\Lambda(s) \triangleq \mathrm{diag}[\epsilon_1(s), \ldots, \epsilon_r(s)]$, where $r = \mathrm{rank}\, P(s)$. The unique *monic* polynomials $\epsilon_i(s)$ (polynomials with leading coefficient equal to one) are the *invariant factors* of $P(s)$. It can be shown that $\epsilon_i(s)$ divides $\epsilon_{i+1}(s)$, $i = 1, \ldots, r - 1$. Note that $\epsilon_i(s)$ can be determined by

$$\epsilon_i(s) = D_i(s)/D_{i-1}(s), \quad i = 1, \ldots, r,$$

where $D_i(s)$ is the monic greatest common divisor of all the nonzero $i$th-order minors of $P(s)$ with $D_0(s) = 1$. The $D_i(s)$ are the *determinantal divisors* of $P(s)$. A matrix $P(s)$ can be reduced to Smith form by elementary row and column operations or by a pre- and post-multiplication by unimodular matrices, namely

$$U_L(s)P(s)U_R(s) = S_p(s). \tag{7.10}$$

*Unimodular Matrices.* Let $R[s]^{p \times m}$ denote the set of $p \times m$ matrices with entries that are polynomials in $s$ with real coefficients. A polynomial matrix $U(s) \in R[s]^{p \times p}$ is called *unimodular* (or $R[s]$-unimodular) if there exists a $\hat{U}(s) \in R[s]^{p \times p}$ such that $U(s)\hat{U}(s) = I_p$. This is the same as saying that $U^{-1}(s) = \hat{U}(s)$ exists and is a polynomial matrix. Equivalently, $U(s)$ is unimodular if $\det U(s) = \alpha \in R, \alpha \neq 0$. It can be shown that every unimodular matrix is a matrix representation of a finite number of successive elementary row and column operations. See [1, p. 526].

Consider now a $p \times m$ rational matrix $H(s)$. Let $d(s)$ be the monic least common denominator of all nonzero entries, and write

$$H(s) = \frac{1}{d(s)} N(s), \tag{7.11}$$

with $N(s)$ a polynomial matrix. Let $S_N(s) = \mathrm{diag}[n_1(s), \ldots, n_r(s), 0_{p-r, m-r}]$ be the Smith form of $N(s)$, where $r = \mathrm{rank}\, N(s) = \mathrm{rank}\, H(s)$. Divide each

$n_i(s)$ of $S_N(s)$ by $d(s)$, canceling all common factors to obtain the *Smith–McMillan* form of $H(s)$,

$$SM_H(s) = \begin{bmatrix} \tilde{\Lambda}(s) & 0 \\ 0 & 0 \end{bmatrix}, \tag{7.12}$$

with $\tilde{\Lambda}(s) \triangleq \text{diag} \left[ \frac{\epsilon_1(s)}{\psi_1(s)}, \ldots, \frac{\epsilon_r(s)}{\psi_r(s)} \right]$, where $r = \text{rank } H(s)$. Note that $\epsilon_i(s)$ divides $\epsilon_{i+1}(s)$, $i = 1, 2, \ldots, r-1$, and $\psi_{i+1}(s)$ divides $\psi_i(s)$, $i = 1, 2, \ldots, r-1$.

## 7.4.2 Poles

*Pole Polynomial of $H(s)$.* Given a $p \times m$ rational matrix $H(s)$, its *characteristic polynomial* or *pole polynomial*, $p_H(s)$, is defined as

$$p_H(s) = \psi_1(s) \cdots \psi_r(s), \tag{7.13}$$

where the $\psi_i$, $i = 1, \cdots, r$, are the denominators of the Smith–McMillan form, $SM_H(s)$, of $H(s)$. It can be shown that $p_H(s)$ is the monic least common denominator of all nonzero minors of $H(s)$.

**Definition 7.5.** *The* poles *of $H(s)$ are the roots of the pole polynomial $p_H(s)$.*
∎

Note that the monic least common denominator of all nonzero first-order minors (entries) of $H(s)$ is called the *minimal polynomial* of $H(s)$ and is denoted by $m_H(s)$. The $m_H(s)$ divides $p_H(s)$ and when the roots of $p_H(s)$ [poles of $H(s)$] are distinct, $m_H(s) = p_H(s)$, since the additional roots in $p_H(s)$ are repeated roots of $m_H(s)$.

It is important to note that when the minors of $H(s)$ [of order $1, 2, \ldots$, $\min(p, m)$] are formed by taking the determinants of all square submatrices of dimension $1 \times 1, 2 \times 2$, etc., all cancellations of common factors between numerator and denominator polynomials should be carried out.

In the scalar case, $p = m = 1$, Definition 7.5 reduces to the well-known definition of poles of a transfer function $H(s)$, since in this case there is only one minor (of order 1), $H(s)$, and the poles are the roots of the denominator polynomial of $H(s)$. Notice that in this case, it is assumed that all the possible cancellations have taken place in the transfer function of a system. Here $p_H(s) = m_H(s)$, that is, the pole or characteristic polynomial equals the minimal polynomial of $H(s)$. Thus, $p_H(s) = m_H(s)$ are equal to the (monic) denominator of $H(s)$.

---

**Example 7.6.** Let $H(s) = \begin{bmatrix} 1/[s(s+1)] & 1/s & 1 \\ 0 & 0 & 1/s^2 \end{bmatrix}$. The nonzero minors of order 1 are the nonzero entries. The least common denominator is $s^2(s+1) = m_H(s)$, the minimal polynomial of $H(s)$. The nonzero minors of order 2 are

$1/[s^3(s+1)]$ and $1/s^3$ (taking columns 1 and 3, and 2 and 3, respectively). The least common denominator of all minors (of order 1 and 2) is $s^3(s+1) = p_H(s)$, the characteristic polynomial of $H(s)$. The poles are $\{0,0,0,-1\}$. Note that $m_H(s)$ is a factor of $p_H(s)$, and the additional root at $s = 0$ in $p_H(s)$ is a repeated pole. To obtain the Smith–McMillan form of $H(s)$, write $H(s) = \frac{1}{s^2(s+1)} \begin{bmatrix} s & s(s+1) & s^2(s+1) \\ 0 & 0 & (s+1) \end{bmatrix} = \frac{1}{d(s)} N(s)$, where $d(s) = s^2(s+1) = m_H(s)$ [see (7.11)]. The Smith form of $N(s)$ is

$$S_N(s) = \begin{bmatrix} 1 & 0 & 0 \\ 0 & s(s+1) & 0 \end{bmatrix}$$

since $D_0 = 1, D_1 = 1, D_2 = s(s+1)$ [the determinantal divisors of $N(s)$], and $n_1 = D_1/D_0 = 1, n_2 = D_2/D_1 = s(s+1)$, the invariant factors of $N(s)$. Dividing by $d(s)$, we obtain the Smith–McMillan form of $H(s)$,

$$SM_H(s) = \begin{bmatrix} \epsilon_1/\psi_1 & 0 & 0 \\ 0 & \epsilon_2/\psi_2 & 0 \end{bmatrix} = \begin{bmatrix} 1/[s^2(s+1)] & 0 & 0 \\ 0 & 1/s & 0 \end{bmatrix}.$$

Note that $\psi_2$ divides $\psi_1$ and $\epsilon_1$ divides $\epsilon_2$. Now the characteristic or pole polynomial of $H(s)$ is $p_H(s) = \psi_1\psi_2 = s^3(s+1)$ and the poles are $\{0,0,0,-1\}$, as expected.

---

**Example 7.7.** Let $H(s) = \frac{1}{s+2} \begin{bmatrix} 1 & \alpha \\ 1 & 1 \end{bmatrix}$. If $\alpha \neq 1$, then the second-order minor is $|H(s)| = \frac{1-\alpha}{(s+2)^2}$. The least common denominator of this nonzero second-order minor $|H(s)|$ and of all the entries of $H(s)$ (the first-order minors) is $(s+2)^2 = p_H(s)$; i.e., the poles are at $\{-2,-2\}$. Also, $m_H(s) = s+2$.

Now if $\alpha = 1$, then there are only first-order nonzero minors ($|H(s)| = 0$). In this case $p_H(s) = m_H(s) = s+2$, which is quite different from the case when $\alpha \neq 1$. Presently, there is only one pole at $-2$.

---

As will be shown in Chapter 8 via Theorems 8.9 and 8.12, the poles of $H(s)$ are exactly the controllable and observable eigenvalues of the system (in $A_{11}$) and no factors of $|sI - A_{11}|$ in $H(s)$ cancel [see (7.52)].

In general, for the set of poles of $H(s)$ and the eigenvalues of $A$, we have

$$\{\text{Poles of } H(s)\} \subset \{\text{eigenvalues of } A\}, \tag{7.14}$$

with equality holding when all the eigenvalues of $A$ are controllable and observable eigenvalues of the system. Similar results hold for discrete-time systems and $H(z)$.

**Example 7.8.** Consider $A = \begin{bmatrix} 0 & -1 & 1 \\ 1 & -2 & 1 \\ 0 & 1 & -1 \end{bmatrix}$, $B = \begin{bmatrix} 1 & 0 \\ 1 & 1 \\ 1 & 2 \end{bmatrix}$, and $C = [0, 1, 0]$.
Then the transfer function $H(s) = [1/s, 1/s]$. $H(s)$ has only one pole, $s_1 = 0$
($p_H(s) = s$), and $\lambda_1 = 0$, is the only controllable and observable eigenvalue.
The other two eigenvalues of $A$, $\lambda_2 = -1$, $\lambda_3 = -2$, which are not both
controllable and observable, do not appear as poles of $H(s)$.

**Example 7.9.** Recall the circuit in Example 7.2 in Section 7.2. If $R_1 R_2 C \neq L$,
then {poles of $H(s)$} = {eigenvalues of $A$ at $\lambda_1 = -1/(R_1 C)$ and $\lambda_2 =$
$-R_2/L$}. In this case, both eigenvalues are controllable and observable. Now
if $R_1 R_2 C = L$ with $R_1 \neq R_2$, then $H(s)$ has only one pole, $s_1 = -R_2/L$,
since in this case only one eigenvalue $\lambda_1 = -R_2/L$ is controllable and observable. The other eigenvalue $\lambda_2$ at the same location $-R_2/L$ is uncontrollable
and unobservable. Now if $R_1 R_2 C = L$ with $R_1 = R_2 = R$, then one of the
eigenvalues becomes uncontrollable and the other (also at $-R/L$) becomes
unobservable. In this case $H(s)$ has no finite poles ($H(s) = 1/R$).

### 7.4.3 Zeros

In a scalar transfer function $H(s)$, the roots of the denominator polynomial
are the poles, and the roots of its numerator polynomial are the zeros of $H(s)$.
As was discussed, the *poles of $H(s)$* are some or all of the eigenvalues of $A$ (the
eigenvalues of $A$ are sometimes also called *poles of the system* $\{A, B, C, D\}$). In
particular, the uncontrollable and/or unobservable eigenvalues of $A$ can never
be poles of $H(s)$. In Chapter 8 (Theorems 8.9 and 8.12), it is shown that only
those eigenvalues of $A$ that are both controllable and observable appear as
poles of the transfer function $H(s)$. Along similar lines, the *zeros of $H(s)$* (to
be defined later) are some or all of the characteristic values of another matrix,
the system matrix $P(s)$. These characteristic values are called the *zeros of the
system* $\{A, B, C, D\}$.

The *zeros of a system* for both the continuous- and the discrete-time cases
are defined and discussed next. We consider now only finite zeros. For the case
of zeros at infinity, refer to the exercises.

Let the *system matrix* (also called *Rosenbrock's system matrix*) of
$\{A, B, C, D\}$ be

$$P(s) \triangleq \begin{bmatrix} sI - A & B \\ -C & D \end{bmatrix}. \tag{7.15}$$

Note that in view of the system equations $\dot{x} = Ax + Bu, y = Cx + Du$, we
have

$$P(s) \begin{bmatrix} -\hat{x}(s) \\ \hat{u}(s) \end{bmatrix} = \begin{bmatrix} 0 \\ \hat{y}(s) \end{bmatrix},$$

where $\hat{x}(s)$ denotes the Laplace transform of $x(t)$.

*Zero Polynomial of* $(A, B, C, D)$. Let $r = \text{rank} P(s)$ [note that $n \leq r \leq \min(p + n, m + n)$], and consider all those $r$th order nonzero minors of $P(s)$ that are formed by taking the first $n$ rows and $n$ columns of $P(s)$, i.e., all rows and columns of $sI - A$, and then adding appropriate $r - n$ rows (of $[-C, D]$ ) and columns (of $[B^T, D^T]^T$ ). The *zero polynomial of the system* $\{A, B, C, D\}$, $z_p(s)$, is defined as the monic greatest common divisor of all these minors.

**Definition 7.10.** *The* zeros of the system $\{A, B, C, D\}$ *or the* system zeros *are the roots of the zero polynomial of the system,* $z_P(s)$. ∎

In addition, we define the *invariant zeros of the system* as the roots of the invariant polynomials of $P(s)$.

In particular, consider the $(p + n) \times (m + n)$ system matrix $P(s)$ and let

$$S_P(s) = \begin{bmatrix} \Lambda(s) & 0 \\ 0 & 0 \end{bmatrix}, \quad \Lambda(s) = \text{diag}[\epsilon_1(s), \ldots, \epsilon_r(s), 0] \quad (7.16)$$

be its Smith form. The *invariant zero polynomial of the system* $\{A, B, C, D\}$ is defined as

$$z_P^I(s) = \epsilon_1(s)\epsilon_2(s) \cdots \epsilon_r(s), \quad (7.17)$$

and its roots are the *invariant zeros of the system*. It can be shown that the monic greatest common divisor of all the highest order nonzero minors of $P(s)$ equals $z_P^I(s)$.

In general,

$$\{\text{zeros of the system}\} \supset \{\text{invariant zeros of the system}\}.$$

When $p = m$ with $\det P(s) \neq 0$, then the zeros of the system coincide with the invariant zeros.

Now consider the $n \times (m + n)$ matrix $[sI - A, B]$ and determine its $n$ invariant factors $\epsilon_i(s)$ and its Smith form. The product of its invariant factors is a polynomial, the roots of which are the *input-decoupling zeros of the system* $\{A, B, C, D\}$. Note that this polynomial equals the monic greatest common divisor of all the highest order nonzero minors (of order $n$) of $[sI - A, B]$. Similarly, consider the $(p + n) \times n$ matrix $\begin{bmatrix} sI - A \\ -C \end{bmatrix}$ and its invariant polynomials, the roots of which define the *output-decoupling zeros of the system* $\{A, B, C, D\}$.

Using the above definitions, it is not difficult to show that the input-decoupling zeros of the system are eigenvalues of $A$ and also zeros of the system $\{A, B, C, D\}$. In addition note that if $\lambda_i$ is such an input-decoupling zero, then $\text{rank}[\lambda_i I - A, B] < n$, and therefore, there exists a $1 \times n$ vector $\hat{v}_i \neq 0$ such that $\hat{v}_i[\lambda_i I - A, B] = 0$. This, however, implies that $\lambda_i$ is an uncontrollable eigenvalue of $A$ (and $\hat{v}_i$ is the corresponding left eigenvector), in view of Section 6.3. Conversely, it can be shown that an uncontrollable eigenvalue is an input-decoupling zero. Therefore, *the input-decoupling zeros of the system* $\{A, B, C, D\}$ *are the uncontrollable eigenvalues of* $A$. Similarly, it can be shown that the *output-decoupling zeros of the system* $\{A, B, C, D\}$ *are the unobservable eigenvalues of* $A$. They are also zeros of the system, as can easily be seen from the definitions.

There are eigenvalues of $A$ that are both uncontrollable and unobservable. These can be determined using the left and right corresponding eigenvector test or by the Canonical Structure Theorem (Kalman Decomposition Theorem) (see Sections 6.2 and 6.3). These uncontrollable and unobservable eigenvalues of $A$ are zeros of the system that are both input- and output-decoupling zeros and are called *input–output decoupling zeros*. These input–output decoupling zeros can also be defined directly from $P(s)$ given in (7.15); however, care should be taken in the case of repeated zeros.

If the zeros of a system are determined and the zeros that are input- and/or output-decoupling zeros are removed, then the zeros that remain are the *zeros of* $H(s)$ and can be found directly from the transfer function $H(s)$.

*Zero Polynomial of* $H(s)$. In particular, if the Smith–McMillan form of $H(s)$ is given by (7.12), then

$$z_H(s) = \epsilon_1(s)\epsilon_2(s)\cdots\epsilon_r(s) \tag{7.18}$$

is the *zero polynomial of* $H(s)$ and its roots are the *zeros of* $H(s)$. These are also called the *transmission zeros of the system*.

**Definition 7.11.** *The zeros of* $H(s)$ *or the* transmission zeros of the system *are the roots of the zero polynomial of* $H(s), z_H(s)$.    ∎

When $P(s)$ is square and nonsingular, the relationship between the zeros of the system and the zeros of $H(s)$ can easily be determined. Consider the identity

$$P(s) = \begin{bmatrix} sI - A & B \\ -C & D \end{bmatrix} = \begin{bmatrix} sI - A & 0 \\ -C & I \end{bmatrix} \begin{bmatrix} I & (sI - A)^{-1}B \\ 0 & H(s) \end{bmatrix}$$

and note that $|P(s)| = |sI - A|\, |H(s)|$. In this case, the invariant zeros of the system [the roots of $|P(s)|$], which are equal here to the zeros of the system, are the zeros of $H(s)$ [the roots of $|H(s)|$] *and* those eigenvalues of $A$ that are not both controllable and observable [the ones that do not cancel in $|sI - A||H(s)|$].

Note that the zero polynomial of $H(s), z_H(s)$, equals the monic greatest common divisor of the numerators of all the highest order nonzero minors in $H(s)$ after all their denominators have been set equal to $p_H(s)$, the characteristic polynomial of $H(s)$. In the scalar case ($p = m = 1$), our definition of the zeros of $H(s)$ reduces to the well-known definition of zeros, namely, the roots of the numerator polynomial of $H(s)$.

**Example 7.12.** Consider $H(s)$ of Example 7.6. From the Smith–McMillan form of $H(s)$, we obtain the zero polynomial $z_H(s) = 1$, and $H(s)$ has no (finite) zeros. Alternatively, the highest order nonzero minors are $1/[s^3(s + 1)]$ and $1/s^3 = (s + 1)/[s^3(s + 1)]$ and the greatest common divisor of the numerators is $z_H(s) = 1$.

**Example 7.13.** We wish to determine the zeros of $H(s) = \begin{bmatrix} \frac{s}{s+1} & 0 \\ \frac{1}{s+1} & \frac{s+1}{s^2} \end{bmatrix}$. The first-order minors are the entries of $H(s)$, namely $\frac{s}{s+1}, \frac{1}{s+1}, \frac{s+1}{s^2}$, and there is only one second-order minor $\frac{s}{s+1} \cdot \frac{s+1}{s^2} = \frac{1}{s}$. Then $p_H(s) = s^2(s+1)$, the least common denominator, is the characteristic polynomial. Next, write the highest (second-) order minor as $\frac{1}{s} = \frac{s(s+1)}{s^2(s+1)} = \frac{s(s+1)}{p_H(s)}$ and note that $s(s+1)$ is the zero polynomial of $H(s)$, $z_H(s)$, and the zeros of $H(s)$ are $\{0, -1\}$. It is worth noting that the poles and zeros of $H(s)$ are at the same locations. This may happen only when $H(s)$ is a matrix.

If the Smith–McMillan form of $H(s)$ is to be used, write $H(s) = \frac{1}{s^2(s+1)} \begin{bmatrix} s^3 & 0 \\ s^2 & (s+1)^2 \end{bmatrix} = \frac{1}{d(s)} N(s)$. The Smith form of $N(s)$ is now $\begin{bmatrix} 1 & 0 \\ 0 & s^3(s+1)^2 \end{bmatrix}$ since $D_0 = 1, D_1 = 1, D_2 = s^3(s+1)^2$ with invariant factors of $N(s)$ given by $n_1 = D_1/D_0 = 1$ and $n_2 = D_2/D_1 = s^3(s+1)^2$. Therefore, the Smith–McMillan form (7.12) of $H(s)$ is

$$SM_H(s) = \begin{bmatrix} \frac{1}{s^2(s+1)} & 0 \\ 0 & \frac{s(s+1)}{1} \end{bmatrix} = \begin{bmatrix} \epsilon_1/\psi_1 & 0 \\ 0 & \epsilon_2/\psi_2 \end{bmatrix}.$$

The zero polynomial is then $z_H(s) = \epsilon_1 \epsilon_2 = s(s+1)$, and the zeros of $H(s)$ are $\{0, -1\}$, as expected. Also, the pole polynomial is $p_H(s) = \psi_1 \psi_2 = s^2(s+1)$, and the poles are $\{0, 0, -1\}$.

**Example 7.14.** We wish to determine the zeros of $H(s) = \begin{bmatrix} \frac{s}{s+1} & 0 \\ \frac{1}{s+1} & \frac{s+1}{s^2} \\ 0 & \frac{1}{s} \end{bmatrix}$. The second-order minors are $\frac{1}{s}, \frac{1}{s+1}, \frac{1}{s(s+1)}$, and the characteristic polynomial is $p_H(s) = s^2(s+1)$. Rewriting the highest (second-) order minors as $s(s+1)/p_H(s), s^2/p_H(s)$, and $s/p_H(s)$, the greatest common divisor of the numerators is $s$; i.e., the zero polynomial of $H(s)$ is $z_H(s) = s$. Thus, there is only one zero of $H(s)$ located at 0. Alternatively, note that the Smith–McMillan form is

$$SM_H(s) = \begin{bmatrix} 1/[s^2(s+1)] & 0 \\ 0 & s/1 \\ 0 & 0 \end{bmatrix}.$$

### 7.4.4 Relations Between Poles, Zeros, and Eigenvalues of $A$

Consider the system $\dot{x} = Ax + Bu, y = Cx + Du$ and its transfer function matrix $H(s) = C(sI - A)^{-1}B + D$. Summarizing the above discussion, the following relations can be shown to be true.

1. We have the set relationship

$$\{\text{zeros of the system}\} = \{\text{zeros of } H(s)\}$$
$$\cup \{\text{input-decoupling zeros}\} \cup \{\text{output-decoupling zeros}\}$$
$$- \{\text{input-output decoupling zeros}\}. \tag{7.19}$$

Note that the invariant zeros of the system contain all the zeros of $H(s)$ (transmission zeros), but not all the decoupling zeros (see Example 7.15). When $P(s)$ is square and nonsingular, the zeros of the system are exactly the invariant zeros of the system. Also, in the case when $\{A, B, C, D\}$ is controllable and observable, the zeros of the system, the invariant zeros, and the transmission zeros [zeros of $H(s)$] all coincide.

2. We have the set relationship

$$\{\text{eigenvalues of } A \text{ (or poles of the system)}\} = \{\text{poles of } H(s)\}$$
$$\cup \{\text{uncontrollable eigenvalues of } A\} \cup \{\text{unobservable eigenvalues of } A\}$$
$$- \{\text{both uncontrollable and unobservable eigenvalues of } A\}. \tag{7.20}$$

3. We have the set relationships

$$\{\text{input-decoupling zeros}\} = \{\text{uncontrollable eigenvalues of } A \},$$
$$\{\text{output-decoupling zeros}\} = \{\text{unobservable eigenvalue of } A\},$$

and

$$\{\text{input-output decoupling zeros}\} =$$
$$\{\text{eigenvalues of } A \text{ that are both uncontrollable and unobservable}\}. \tag{7.21}$$

4. When the system $\{A, B, C, D\}$ is controllable and observable, then

$$\{\text{zeros of the system}\} = \{\text{zeros of} H(s)\}$$
$$\text{and } \{\text{eigenvalues of } A \text{ (or poles of the system)}\} = \{\text{poles of } H(s)\}. \tag{7.22}$$

Note that *the eigenvalues of $A$ (the poles of the system)* can be defined as the roots of the invariant factors of $sI - A$ in $P(s)$ given in (7.15).

**Example 7.15.** Consider the system $\{A, B, C\}$ of Example 7.8. Let

$$P(s) = \begin{bmatrix} sI - A & B \\ -C & D \end{bmatrix} = \begin{bmatrix} s & 1 & -1 & 1 & 0 \\ -1 & s+2 & -1 & 1 & 1 \\ 0 & -1 & s+1 & 1 & 2 \\ \hline 0 & -1 & 0 & 0 & 0 \end{bmatrix}.$$

There are two fourth-order minors that include all columns of $sI - A$ obtained by taking columns 1, 2, 3, 4 and columns 1, 2, 3, 5 of $P(s)$; they are $(s + 1)(s + 2)$ and $(s + 1)(s + 2)$. The zero polynomial of the system is $z_P = (s+1)(s+2)$, and the zeros of the system are $\{-1, -2\}$. To determine the input-decoupling zeros, consider all the third-order minors of $[sI - A, B]$. The greatest common divisor is $s + 2$, which implies that the input-decoupling zeros are $\{-2\}$. Similarly, consider $\begin{bmatrix} sI - A \\ -C \end{bmatrix}$ and show that $s + 1$ is the greatest common divisor of all the third-order minors and that the output-decoupling zeros are $\{-1\}$. The transfer function for this example was found in Example 7.8 to be $H(s) = [1/s, 1/s]$. The zero polynomial of $H(s)$ is $z_H(s) = 1$, and there are no zeros of $H(s)$. Notice that there are no input–output decoupling zeros. It is now clear that relation (7.19) holds.

The controllable (resp., uncontrollable) and the observable (resp., unobservable) eigenvalues of $A$ (poles of the system) have been found in Example 6.10. Compare these results to show that (7.21) holds. The poles of $H(s)$ are $\{0\}$. Verify that (7.20) holds.

One could work with the Smith form of the matrices of interest and the Smith–McMillan form of $H(s)$. In particular, it can be shown that the

Smith form of $P(s)$ is $\begin{bmatrix} 1 & 0 & 0 & 0 & 0 \\ 0 & 1 & 0 & 0 & 0 \\ 0 & 0 & 1 & 0 & 0 \\ 0 & 0 & 0 & (s+2) & 0 \end{bmatrix}$, of $[sI - A, B]$ is $\begin{bmatrix} 1 & 0 & 0 & 0 & 0 \\ 0 & 1 & 0 & 0 & 0 \\ 0 & 0 & s+2 & 0 & 0 \end{bmatrix}$,

of $\begin{bmatrix} sI - A \\ -C \end{bmatrix}$ is $\begin{bmatrix} 1 & 0 & 0 \\ 0 & 1 & 0 \\ 0 & 0 & s+1 \\ 0 & 0 & 0 \end{bmatrix}$, and of $[sI - A]$ is $\begin{bmatrix} 1 & 0 & 0 \\ 0 & 1 & 0 \\ 0 & 0 & s(s+1)(s+2) \end{bmatrix}$. Also,

it can be shown that the Smith–McMillan form of $H(s)$ is

$$SM_H(s) = [1/s, 0].$$

It is straightforward to verify the above results. Note that in the present case the invariant zero polynomial is $z_P^I(s) = s + 2$ and there is only one invariant zero at $-2$.

**Example 7.16.** Consider the circuit of Example 7.9 and of Example 7.2 and the system matrix $P(s)$ for the case when $R_1 R_2 C = L$ given by

$$P(s) = \begin{bmatrix} sI - A & B \\ -C & D \end{bmatrix} = \begin{bmatrix} s + R_2/L & 0 & R_2/L \\ 0 & s + R_2/L & 1/L \\ 1/R_1 & -1 & 1/R_1 \end{bmatrix}.$$

(i) First, let $R_1 \neq R_2$. To determine the zeros of the system, consider $|P(s)| = (1/R_1)(s+R_1/L)(s+R_2/L)$, which implies that the zeros of the system are $\{-R_1/L, -R_2/L\}$. Consider now all second-order (nonzero) minors of $[sI - A, B]$, namely, $(s+R_2/L)^2, (1/L)(s+R_2/L)$ and $-(R_2/L)(s+R_2/L)$, from which we see that $\{-R_2/L\}$ is the input-decoupling zero. Similarly, we also see that $\{-R_2/L\}$ is the output-decoupling zero. Therefore, $\{-R_2/L\}$ is the input–output decoupling zero. Compare this with the results in Example 7.9 to verify (7.22).

(ii) When $R_1 = R_2 = R$, then $|P(s)| = (1/R)(s + R/L)^2$, which implies that the zeros of the system are at $\{-R/L, -R/L\}$. Proceeding as in (i), it can readily be shown that $\{-R/L\}$ is the input-decoupling zero and $\{-R/L\}$ is the output-decoupling zero. To determine which are the input–output decoupling zeros, one needs additional information to the zero location. This information can be provided by the left and right eigenvectors of the two zeros at $-R/L$ to determine that there is no input–output decoupling zero in this case (see Example 7.2).

In both cases (i) and (ii), $H(s)$ has been derived in Example 7.2. Verify relation (7.19).

Finally, note that there are characteristic vectors or zero directions, associated with each invariant and decoupling zero of the system $\{A, B, C, D\}$, just as there are characteristic vectors or eigenvectors, associated with each eigenvalue of $A$ (pole of the system) (see [1, p. 306, Section 3.5]). For pole-zero cancellations to take place in the case of multi-input or output systems when the transfer function matrix is formed, not only the pole, zero locations must be the same but also their characteristic directions must be aligned.

## 7.5 Polynomial Matrix and Matrix Fractional Descriptions of Systems

In this section, representations of linear time-invariant systems based on polynomial matrices, called *Polynomial Matrix Description (PMD)* [or *Differential (Difference) Operator Representation (DOR)*] are introduced. Such representations arise naturally when differential (or difference) equations are used to describe the behavior of systems, and the differential (or difference) operator is introduced to represent the operation of differentiation (or of time-shift). Polynomial matrices in place of polynomials are involved since this approach is typically used to describe multi-input, multi-output systems. Note that state-space system descriptions only involve first-order differential (or difference) equations, and as such, PMDs include the state-space descriptions as special cases.

A rational function matrix can be written as a ratio or fraction of two polynomial matrices or of two rational matrices. If the transfer function matrix of a system is expressed as a fraction of two polynomial or rational matrices, this leads to a *Matrix Fraction(al) Description (MFD)* of the system. The MFDs that involve polynomial matrices, called polynomial MFDs, can be viewed as representations of internal realizations of the transfer function matrix; that is, they can be viewed as system PMDs of special form. *These polynomial fractional descriptions (PMFDs) help establish the relationship between internal and external system representations in a clear and transparent manner.* This can be used to advantage, for example, in the study of feedback control problems, leading to clearer understanding of the phenomena that occur when systems are interconnected in feedback configurations. The MFDs that involve ratios of rational matrices, in particular ratios of proper and stable rational matrices, offer convenient characterizations of transfer functions in feedback control problems.

MFDs that involve ratios of polynomial matrices and ratios of proper and stable rational matrices are essential in parameterizing all stabilizing feedback controllers. Appropriate selection of the parameters guarantees that a closed-loop system is not only stable, but it will also satisfy additional control criteria. This is precisely the approach taken in optimal control methods, such as $H^\infty$-optimal control. Parameterizations of all stabilizing feedback controllers are studied in Chapter 10. We note that extensions of MFDs are also useful in linear, time-varying systems and in nonlinear systems. These extensions are not addressed here.

In addition to the importance of MFDs in characterizing all stabilizing controllers, and in $H^\infty$-optimal control, PMFDs and PMDs have been used in other control design methodologies as well (e.g., self-tuning control). The use of PMFDs in feedback control leads in a natural way to the polynomial Diophantine matrix equation, which is central in control design when PMDs are used and which directly leads to the characterization of all stabilizing controllers. Finally, PMDs are generalizations of state-space descriptions, and the use of PMDs to characterize the behavior of systems offers additional insight and flexibility. Detailed treatment of all these issues may be found in [1, Chapter 7]. The development of the material in this section is concerned only with continuous-time systems; however, completely analogous results are valid for discrete-time systems and can easily be obtained by obvious modifications. In this section we emphasize PMFD and discuss controllability, observability, and stability.

*An Important Comment on Notation.* We will be dealing with matrices with entries polynomials in $s$ or $q$, denoted by, e.g., $D(s)$ or $D(q)$, where $s$ is the Laplace variable and $q \triangleq d/dt$, the differential operator. For simplicity of notation we frequently omit the argument $s$ or $q$ and we write $D$ to denote the polynomial matrix on hand. When ambiguity may arise, or when it is important to stress the fact that the matrix in question is a polynomial matrix, the argument will be included.

### 7.5.1 A Brief Introduction to Polynomial and Fractional Descriptions

Below, the Polynomial Matrix Description (PMD) and the Matrix Fractional Description (MFD) of a linear, time-invariant system are introduced via a simple illustrating example.

---

**Example 7.17.** In the ordinary differential equation representation of a system given by

$$\ddot{y}_1(t) + y_1(t) + y_2(t) = \dot{u}_2(t) + u_1(t),$$
$$\dot{y}_1(t) + \dot{y}_2(t) + 2y_2(t) = \dot{u}_2(t), \tag{7.23}$$

$y_1(t), y_2(t)$ and $u_1(t), u_2(t)$ denote, respectively, outputs and inputs of interest. We assume that appropriate initial conditions for the $u_i(t), y_i(t)$ and their derivatives at $t = 0$ are given.

By changing variables, one can express (7.23) by an equivalent set of first-order ordinary differential equations, in the sense that this set of equations will generate all solutions of (7.23), using appropriate initial conditions and the same inputs. To this end, let

$$x_1 = \dot{y}_1 - u_2, \quad x_2 = y_1, \quad x_3 = y_1 + y_2 - u_2. \tag{7.24}$$

Then (7.23) can be written as

$$\dot{x} = Ax + Bu, \quad y = Cx + Du, \tag{7.25}$$

where $x(t) = \begin{bmatrix} x_1(t) \\ x_2(t) \\ x_3(t) \end{bmatrix}$, $u(t) = \begin{bmatrix} u_1(t) \\ u_2(t) \end{bmatrix}$, $y(t) = \begin{bmatrix} y_1(t) \\ y_2(t) \end{bmatrix}$, and

$$A = \begin{bmatrix} 0 & 0 & -1 \\ 1 & 0 & 0 \\ 0 & 2 & -2 \end{bmatrix}, \quad B = \begin{bmatrix} 1 & -1 \\ 0 & 1 \\ 0 & -2 \end{bmatrix}, \quad C = \begin{bmatrix} 0 & 1 & 0 \\ 0 & -1 & 1 \end{bmatrix}, \quad D = \begin{bmatrix} 0 & 0 \\ 0 & 1 \end{bmatrix}$$

with initial conditions $x(0)$ calculated by using (7.24).

More directly, however, system (7.23) can be represented by

$$P(q)z(t) = Q(q)u(t), \quad y(t) = R(q)z(t) + W(q)u(t), \tag{7.26}$$

where $z(t) = \begin{bmatrix} z_1(t) \\ z_2(t) \end{bmatrix}$, $u(t) = \begin{bmatrix} u_1(t) \\ u_2(t) \end{bmatrix}$, $y(t) = \begin{bmatrix} y_1(t) \\ y_2(t) \end{bmatrix}$, and

$$P(q) = \begin{bmatrix} q^2 + 1 & 1 \\ q & q+2 \end{bmatrix}, \quad Q(q) = \begin{bmatrix} 1 & q \\ 0 & q \end{bmatrix}, \quad R(q) = \begin{bmatrix} 1 & 0 \\ 0 & 1 \end{bmatrix}, \quad W(q) = \begin{bmatrix} 0 & 0 \\ 0 & 0 \end{bmatrix}$$

with $q \triangleq \frac{d}{dt}$, the differential operator. The variables $z_1(t), z_2(t)$ are called *partial state variables*, $z(t)$ denotes the *partial state* of the system description (7.26), and $u(t)$ and $y(t)$ denote the input and output vectors, respectively.

---

*Polynomial Matrix Descriptions (PMDs)*

Representation (7.26), also denoted as $\{P(q), Q(q), R(q), W(q)\}$, is an example of a *Polynomial Matrix Description (PMD)* of a system. Note that the state-space description (7.25) is a special case of (7.26). To see this, write (7.25) as

$$(qI - A)x(t) = Bu(t), \quad y(t) = Cx(t) + Du(t). \tag{7.27}$$

Clearly, description $\{qI - A, B, C, D\}$ is a special case of the general Polynomial Matrix Description $\{P(q), Q(q), R(q), W(q)\}$ with

$$P(q) = qI - A, Q(q) = B, R(q) = C, W(q) = D. \tag{7.28}$$

The above example points to the fact that a PMD of a system can be derived in a natural way from differential (or difference) equations that involve variables that are directly connected to physical quantities. By this approach, it is frequently possible to study the behavior of physical variables directly without having to transform the system to a state-space description. The latter may involve (state) variables that are quite removed from the physical phenomena they represent, thus losing physical insight when studying a given problem. The price to pay for this additional insight is that one has to deal with differential (or difference) equations of order greater than one. This typically adds computational burdens. We note that certain special forms of PMDs, namely the polynomial Matrix Fractional Descriptions, are easier to deal with than general forms. However, a change of variables may again be necessary to obtain such forms.

Consider a general PMD of a system given by

$$P(q)z(t) = Q(q)u(t), \quad y(t) = R(q)z(t) + W(q)u(t), \tag{7.29}$$

with $P(q) \in R[q]^{l \times l}, Q(q) \in R[q]^{l \times m}$, and $R(q) \in R[q]^{p \times l}, W(q) \in R[q]^{p \times m}$, where $R[q]^{l \times l}$ denotes the set of $l \times l$ matrices with entries that are real polynomials in $q$. The transfer function matrix $H(s)$ of (7.29) can be determined by taking the Laplace transform of both sides of the equation assuming zero initial conditions ($z(0) = \dot{z}(0) = \cdots = 0, u(0) = \dot{u}(0) = \cdots = 0$). Then

$$H(s) = R(s)P^{-1}(s)Q(s) + W(s). \tag{7.30}$$

For the special case of state-space representations, $H(s)$ in (7.30) assumes the well-known expression $H(s) = C(sI - A)^{-1}B + D$. For the study of the relationship between external and internal descriptions, (7.30) is not particularly convenient. There are, however, special cases of (7.30) that are very convenient to use in this regard. In particular, it can be shown [1, Section 7.3] that if the system is controllable, then there exists a representation equivalent to (7.29), which is of the form

$$D_c(q)z_c(t) = u(t), \quad y(t) = N_c(q)z_c(t), \tag{7.31}$$

where $D_c(q) \in R[q]^{m \times m}$ and $N_c(q) \in R[q]^{p \times m}$. Representation (7.31) is obtained by letting $Q(q) = I_m$ and $W(q) = 0$ in (7.29) and using $D$ and $N$ instead of $P$ and $R$. Equation (7.30) now becomes

$$H(s) = N_c(s)D_c(s)^{-1}, \tag{7.32}$$

where $N_c(s)$ and $D_c(s)$ represent the matrix numerator and matrix demoninator of the transfer function, respectively. Similarly, if the system is observable, there exists a representation equivalent to (7.29), which is of the form

$$D_o(q)z_o(t) = N_o(q)u(t), \quad y(t) = z_o(t), \tag{7.33}$$

where $D_o(q) \in R[q]^{p \times p}$ and $N_o(q) \in R[q]^{p \times m}$. Representation (7.33) is obtained by letting in (7.29) $R(q) = I_p$ and $W(q) = 0$ with $P(q) = D_o(q)$ and $Q(q) = N_o(q)$. Here,

$$H(s) = D_o^{-1}(s)N_o(s). \tag{7.34}$$

Note that (7.32) and (7.34) are generalizations to the MIMO case of the SISO system expression $H(s) = n(s)/d(s)$. As $H(s) = n(s)/d(s)$ can be derived directly from the differential equation $d(q)y(t) = n(q)u(t)$, by taking the Laplace transform and assuming that the initial conditions are zero, (7.34) can be derived directly from (7.33).

Returning now to (7.25) in Example 7.17, notice that the system is observable (state observable from the output $y$). Therefore, the system in this case can be represented by a description of the form $\{D_o, N_o, I_2, 0\}$. In fact, (7.26) is such a description, where $D_o$ and $N_o$ are equal to $P$ and $Q$, respectively, i.e.,

$D_o(q) = \begin{bmatrix} q^2 + 1 & 1 \\ q & q + 2 \end{bmatrix}$, and $N_o(q) = \begin{bmatrix} 1 & q \\ 0 & q \end{bmatrix}$. The transfer function matrix is given by

$$H(s) = C(sI - A)^{-1}B + D = \begin{bmatrix} 0 & 1 & 0 \\ 0 & -1 & 1 \end{bmatrix} \begin{bmatrix} s & 0 & 1 \\ -1 & s & 0 \\ 0 & -2 & s+2 \end{bmatrix}^{-1} \begin{bmatrix} 1 & -1 \\ 0 & 1 \\ 0 & -2 \end{bmatrix} + \begin{bmatrix} 0 & 0 \\ 0 & 1 \end{bmatrix}$$

$$= D_o^{-1}(s)N_o(s) = \begin{bmatrix} s^2 + 1, & 1 \\ s, & s+2 \end{bmatrix}^{-1} \begin{bmatrix} 1 & s \\ 0 & s \end{bmatrix}$$

$$= \tfrac{1}{s^3 + 2s^2 + 2} \begin{bmatrix} s+2 & -1 \\ -s & s^2 + 1 \end{bmatrix} \begin{bmatrix} 1 & s \\ 0 & s \end{bmatrix} = \tfrac{1}{s^3 + 2s^2 + 2} \begin{bmatrix} s+2 & s(s+1) \\ -s & s(s^2 - s + 1) \end{bmatrix}.$$

*Matrix Fractional Descriptions (MFDs) of System Transfer Matrices*

A given $p \times m$ proper, rational transfer function matrix $H(s)$ of a system can be represented as

$$H(s) = N_R(s)D_R^{-1}(s) = D_L^{-1}(s)N_L(s), \tag{7.35}$$

where $N_R(s) \in R[s]^{p \times m}$, $D_R(s) \in R[s]^{m \times m}$ and $N_L(s) \in R[s]^{p \times m}$, $D_L(s) \in R[s]^{p \times p}$. The pairs $\{N_R(s), D_R(s)\}$ and $\{D_L(s), N_L(s)\}$ are called *Polynomial Matrix Fractional Descriptions (PMFDs)* of the system transfer matrix with $\{N_R(s), D_R(s)\}$ termed a *right Fractional Description* and $\{D_L(s), N_L(s)\}$ a *left Fractional Description*. Notice that in view of (7.32), the right Polynomial Matrix Fractional Description (rPMFD) corresponds to the controllable Polynomial Matrix Description (PMD) given in (7.31). That is, $\{D_R, I_m, N_R, 0\}$, or

$$D_R(q)z_R(t) = u(t), \quad y(t) = N_R(q)z_R(t), \tag{7.36}$$

is a controllable PMD of the system with transfer function $H(s)$. The subscript $c$ was used in (7.31) and (7.32) to emphasize the fact that $N_c, D_c$ originated from an internal description that was controllable. In (7.35) and (7.36), the subscript $R$ is used to emphasize that $\{N_R, D_R\}$ is a right fraction representation of the external description $H(s)$.

Similarly, in view of (7.34), the left Polynomial Matrix Fractional Description (lPMFD) corresponds to the observable Polynomial Matrix Description (PMD) given in (7.33). That is, $\{D_L, N_L, I_p, 0\}$, or

$$D_L(q)z_L(t) = N_L(q)u(t), \quad y(t) = z_L(t), \tag{7.37}$$

is an observable PMD of the system with transfer function $H(s)$. Comments analogous to the ones made above concerning controllable and right fractional descriptions (subscripts $c$ and $R$) can also be made here concerning the subscripts $o$ and $L$.

An *MFD* of a transfer function may not consist necessarily of ratios of polynomial matrices. In particular, given a $p \times m$ proper transfer function matrix $H(s)$, one can write

$$H(s) = \widehat{N}_R(s)\widehat{D}_R^{-1}(s) = \widehat{D}_L^{-1}(s)\widehat{N}_L(s), \tag{7.38}$$

where $\widehat{N}_R, \widehat{D}_R, \widehat{D}_L, \widehat{N}_L$ are proper and stable rational matrices. To illustrate, in the example considered above, $H(s)$ can be written as

$$H(s) = \frac{1}{s^3 + 2s^2 + 2} \begin{bmatrix} s+2 & s(s+1) \\ -s & s(s^2 - s + 1) \end{bmatrix}$$

$$= \left[ \begin{bmatrix} (s+1)^2 & 0 \\ 0 & s+2 \end{bmatrix}^{-1} \begin{bmatrix} s^2 + 1 & 1 \\ s & s+2 \end{bmatrix} \right]^{-1} \left[ \begin{bmatrix} (s+1)^2 & 0 \\ 0 & s+2 \end{bmatrix}^{-1} \begin{bmatrix} 1 & s \\ 0 & s \end{bmatrix} \right]$$

$$= \begin{bmatrix} \frac{s^2+1}{(s+1)^2} & \frac{1}{(s+1)^2} \\ \frac{s}{s+2} & 1 \end{bmatrix}^{-1} \begin{bmatrix} \frac{1}{(s+1)^2} & \frac{s}{(s+1)^2} \\ 0 & \frac{s}{s+2} \end{bmatrix} = \widehat{D}_L^{-1}(s)\widehat{N}_L(s).$$

Note that $\widehat{D}_L(s)$ and $\widehat{N}_L(s)$ are proper and stable rational matrices.

Such representations of proper transfer functions offer certain advantages when designing feedback control systems. They are discussed further in [1, Section 7.4D].

## 7.5.2 Coprimeness and Common Divisors

Coprimeness of polynomial matrices is one of the most important concepts in the polynomial matrix representation of systems since it is directly related to controllability and observability.

A polynomial $g(s)$ is a *common divisor* (cd) of polynomials $p_1(s), p_2(s)$ if and only if there exist polynomials $\tilde{p}_1(s), \tilde{p}_2(s)$ such that

$$p_1(s) = \tilde{p}_1(s)g(s), \quad p_2(s) = \tilde{p}_2(s)g(s). \tag{7.39}$$

The highest degree cd of $p_1(s), p_2(s), g^*(s)$, is a *greatest common divisor* (gcd) of $p_1(s), p_2(s)$. It is unique within multiplication by a nonzero real number. Alternatively, $g^*(s)$ is a gcd of $p_1(s), p_2(s)$ if and only if *any* cd $g(s)$ of $p_1(s), p_2(s)$ is a divisor of $g^*(s)$ as well; that is,

$$g^*(s) = m(s)g(s) \tag{7.40}$$

with $m(s)$ a polynomial. The polynomials $p_1(s), p_2(s)$ are *coprime* (cp) if and only if a gcd $g^*(s)$ is a nonzero real.

The above can be extended to matrices. In this case, both right divisors and left divisors must be defined, since in general, two polynomial matrices do not commute. Note that one may talk about right or left divisors of polynomial matrices only when the matrices have the same number of columns or rows, respectively.

An $m \times m$ matrix $G_R(s)$ is a *common right divisor* (crd) of the $p_1 \times m$ polynomial matrix $P_1(s)$ and the $p_2 \times m$ polynomial matrix $P_2(s)$, if there exist polynomial matrices $P_{1R}(s), P_{2R}(s)$ so that

$$P_1(s) = P_{1R}(s)G_R(s), \quad P_2(s) = P_{2R}(s)G_R(s). \tag{7.41}$$

Similarly, a $p \times p$ polynomial matrix $G_L(s)$ is a *common left divisor* (cld) of the $p \times m_1$ polynomial matrix $\widehat{P}_1(s)$ and the $p \times m_2$ matrix $\widehat{P}_2(s)$, if there exist polynomial matrices $\widehat{P}_{1L}(s), \widehat{P}_{2L}(s)$ so that

$$\widehat{P}_1(s) = G_L(s)\widehat{P}_{1L}(s), \quad \widehat{P}_2(s) = G_L(s)\widehat{P}_{2L}(s). \tag{7.42}$$

Also $G_R^*(s)$ is a *greatest common right divisor* (gcrd) of $P_1(s)$ and $P_2(s)$ if and only if any crd $G_R(s)$ is an rd of $G_R^*(s)$. Similarly, $G_L^*(s)$ is a *greatest common left divisor* (gcld) of $\widehat{P}_1(s)$ and $\widehat{P}_2(s)$ if and only if any cld $G_L(s)$ is a ld of $G_L^*(s)$. That is,

$$G_R^*(s) = M(s)G_R(s), \quad G_L^*(s) = G_L(s)N(s), \tag{7.43}$$

with $M(s)$ and $N(s)$ polynomial matrices and $G_R(s)$ and $G_L(s)$ any crd and cld of $P_1(s), P_2(s)$, respectively.

Alternatively, it can be shown that any crd $G_R^*(s)$ of $P_1(s)$ and $P_2(s)$ [or a cld $G_L^*(s)$ of $\widehat{P}_1(s)$ and $\widehat{P}_2(s)$] with determinant of the highest degree possible is a gcrd (gcld) of the matrices. It is unique within a pre-multiplication (post-multiplication) by a unimodular matrix. Here it is assumed that $G_R(s)$ is nonsingular. Note that if $\operatorname{rank} \begin{bmatrix} P_1(s) \\ P_2(s) \end{bmatrix} = m$ (a $(p_1+p_2) \times m$ matrix), which is a typical case in polynomial matrix system descriptions, then $\operatorname{rank} G_R(s) = m$; that is, $G_R(s)$ is nonsingular.

The polynomial matrices $P_1(s)$ and $P_2(s)$ are *right coprime* (rc) if and only if a gcrd $G_R^*(s)$ is a unimodular matrix. Similarly, $\widehat{P}_1(s)$ and $\widehat{P}_2(s)$ are *left coprime* (lc) if and only if a gcld $G_2^*(s)$ is a unimodular matrix.

---

**Example 7.18.** Let $P_1 = \begin{bmatrix} s(s+2) & 0 \\ 0 & (s+1)^2 \end{bmatrix}$, $P_2 = \begin{bmatrix} (s+1)(s+2) & s+1 \\ 0 & s(s+1) \end{bmatrix}$. Two distinct common right divisors are $G_{R_1} = \begin{bmatrix} 1 & 0 \\ 0 & s+1 \end{bmatrix}$ and $G_{R_2} = \begin{bmatrix} s+2 & 0 \\ 0 & 1 \end{bmatrix}$

since $\begin{bmatrix} P_1 \\ P_2 \end{bmatrix} = \begin{bmatrix} s(s+2) & 0 \\ 0 & s+1 \\ \hline (s+1)(s+2) & 1 \\ 0 & s \end{bmatrix} G_{R_1} = \begin{bmatrix} s & 0 \\ 0 & (s+1)^2 \\ \hline s+2 & s+1 \\ 0 & s(s+1) \end{bmatrix} G_{R_2}$. A great-

est common right divisor (gcrd) is $G_R^* = \begin{bmatrix} s+2 & 0 \\ 0 & s+1 \end{bmatrix} = \begin{bmatrix} s+2 & 0 \\ 0 & 1 \end{bmatrix} G_{R_1} =$

$\begin{bmatrix} 1 & 0 \\ 0 & s+1 \end{bmatrix} G_{R_2}$. Now, $\begin{bmatrix} P_1 \\ P_2 \end{bmatrix} G_R^{*-1} = \begin{bmatrix} P_{1R}^* \\ P_{2R}^* \end{bmatrix} = \begin{bmatrix} s & 0 \\ 0 & s+1 \\ \hline s+1 & 1 \\ 0 & s \end{bmatrix}$ where $P_{1R}^*$ and

$P_{2R}^*$ are right coprime (rc). Note that a greatest common left divisor (gcld) of $P_1$ and $P_2$ is $G_L^* = \begin{bmatrix} 1 & 0 \\ 0 & s+1 \end{bmatrix}$. Both $G_R^*$ and $G_L^*$ can be determined using an algorithm to derive the Hermite form of $\begin{bmatrix} P_1 \\ P_2 \end{bmatrix}$; see [1, p. 532].

---

*Remarks*

It can be shown that two square $p \times p$ nonsingular polynomial matrices with determinants that are prime polynomials are both right and left coprime. The converse of this is not true; that is, two right coprime polynomial matrices do not necessarily have prime determinant polynomials. A case in point is Example 7.18, where $P_{1R}^*$ and $P_{2R}^*$ are right coprime; however, $\det P_{1R}^* = \det P_{2R}^* = s(s+1)$.

Left and right coprimeness of two polynomial matrices (provided that the matrices are compatible) are quite distinct properties. For example, two matrices can be left coprime but not right coprime, and vice versa (refer to Example 7.19).

---

**Example 7.19.** $P_1 = \begin{bmatrix} s(s+2) & 0 \\ 0 & s+1 \end{bmatrix}$ and $P_2 = \begin{bmatrix} (s+1)(s+2) & 1 \\ 0 & s \end{bmatrix}$ are left

coprime but not right coprime since a gcrd is $G_R^* = \begin{bmatrix} s+2 & 0 \\ 0 & 1 \end{bmatrix}$ with $\det G_R^* = (s+2)$.

---

Finally, we note that all of the above definitions apply also to more than two polynomial matrices. To see this, replace in all definitions $P_1, P_2$ by $P_1, P_2, \ldots, P_k$. This is not surprising in view of the fact that the $p_1 \times m$ matrix $P_1(s)$ and the $p_2 \times m$ matrix $P_2(s)$ consist of $p_1$ and $p_2$ rows, respectively, each of which can be viewed as a $1 \times m$ polynomial matrix; that is, instead of, e.g., the coprimeness of $P_1$ and $P_2$, one could speak of the coprimeness of the $(p_1 + p_2)$ rows of $P_1$ and $P_2$.

*How to Determine a Greatest Common Right Divisor*

**Lemma 7.20.** *Let $P_1(s) \in R[s]^{p_1 \times m}$ and $P_2(s) \in R[s]^{p_2 \times m}$ with $p_1 + p_2 \geq m$. Let the unimodular matrix $U(s)$ be such that*

$$U(s) \begin{bmatrix} P_1(s) \\ P_2(s) \end{bmatrix} = \begin{bmatrix} G_R^*(s) \\ 0 \end{bmatrix}. \tag{7.44}$$

*Then $G_R^*(s)$ is a greatest common right divisor (gcrd) of $P_1(s), P_2(s)$.*

*Proof.* Let

$$U = \begin{bmatrix} \bar{X} & \bar{Y} \\ -\tilde{P}_2 & \tilde{P}_1 \end{bmatrix}, \tag{7.45}$$

with $\bar{X} \in R[s]^{m \times p_1}, \bar{Y} \in R[s]^{m \times p_2}, \tilde{P}_2 \in R[s]^{q \times p_1}$, and $\tilde{P}_1 \in R[s]^{q \times p_2}$, where $q \triangleq (p_1 + p_2) - m$. Note that $\bar{X}, \bar{Y}$ and $\tilde{P}_2, \tilde{P}_1$ are left coprime (lc) pairs. If they were not, then $\det U \neq \alpha$, a nonzero real number. Similarly, $\bar{X}, \tilde{P}_2$ and $\bar{Y}, \tilde{P}_1$ are right coprime (rc) pairs. Let

$$U^{-1} = \begin{bmatrix} \bar{P}_1 & -\tilde{Y} \\ \bar{P}_2 & \tilde{X} \end{bmatrix}, \tag{7.46}$$

where $\bar{P}_1 \in R[s]^{p_1 \times m}, \bar{P}_2 \in R[s]^{p_2 \times m}$ are rc and $\tilde{X} \in R[s]^{p_2 \times q}, \tilde{Y} \in R[s]^{p_1 \times q}$ are rc. Equation (7.44) implies that

$$\begin{bmatrix} P_1 \\ P_2 \end{bmatrix} = U^{-1} \begin{bmatrix} G_R^* \\ 0 \end{bmatrix} = \begin{bmatrix} \bar{P}_1 \\ \bar{P}_2 \end{bmatrix} G_R^*; \tag{7.47}$$

i.e., $G_R^*$ is a common right divisor of $P_1, P_2$. Equation (7.44) implies also that

$$\bar{X}P_1 + \bar{Y}P_2 = G_R^*. \tag{7.48}$$

This relationship shows that any crd $G_R$ of $P_1, P_2$ will also be a right divisor of $G_R^*$. This can be seen directly by expressing (7.48) as $MG_R = G_R^*$, where $M$ is a polynomial matrix. Thus, $G_R^*$ is a crd of $P_1, P_2$ with the property that any crd $G_R$ of $P_1, P_2$ is a rd of $G_R^*$. This implies that $G_R^*$ is a gcrd of $P_1, P_2$. ∎

---

**Example 7.21.** Let $P_1 = \begin{bmatrix} s(s+2) & 0 \\ 0 & (s+1)^2 \end{bmatrix}$, $P_2 = \begin{bmatrix} (s+1)(s+2) & s+1 \\ 0 & s(s+1) \end{bmatrix}$.

Then

$$U \begin{bmatrix} P_1 \\ P_2 \end{bmatrix} = \begin{bmatrix} \bar{X} & \bar{Y} \\ -\tilde{P}_2 & \tilde{P}_1 \end{bmatrix} \begin{bmatrix} P_1 \\ P_2 \end{bmatrix} = \begin{bmatrix} -(s+2) & -1 & s+1 & 0 \\ s+1 & 1 & -s & 0 \\ -(s+1)^2 & -s & s(s+1) & 0 \\ -(s+1) & 0 & s & -1 \end{bmatrix} \begin{bmatrix} P_1 \\ P_2 \end{bmatrix}$$

$$= \begin{bmatrix} s+2 & 0 \\ 0 & s+1 \\ 0 & 0 \\ 0 & 0 \end{bmatrix} = \begin{bmatrix} G_R^* \\ 0 \end{bmatrix}.$$

In view of Lemma 7.20, $G_{R*} = \begin{bmatrix} s+2 & 0 \\ 0 & s+1 \end{bmatrix}$ is a gcrd (see also Example 7.18).

---

Note that in order to derive (7.44) and thus determine a gcrd $G_R^*$ of $P_1$ and $P_2$, one could use the algorithm to obtain the Hermite form [1, p. 532]. Finally, note also that if the Smith form of $\begin{bmatrix} P_1 \\ P_2 \end{bmatrix}$ is known, i.e., $U_L \begin{bmatrix} P_1 \\ P_2 \end{bmatrix} U_R = S_P = \begin{bmatrix} \text{diag}[\epsilon_i] & 0 \\ 0 & 0 \end{bmatrix}$, then $(\text{diag}[\epsilon_i], 0)U_R^{-1}$ is a gcrd of $P_1$ and $P_2$ in view of Lemma 7.20. When rank $\begin{bmatrix} P_1 \\ P_2 \end{bmatrix} = m$, which is the case of interest in systems, then a gcrd of $P_1$ and $P_2$ is $\text{diag}[\epsilon_i]U_R^{-1}$.

*Criteria for Coprimeness*

There are several ways of testing the coprimeness of two polynomial matrices as shown in the following Theorem.

**Theorem 7.22.** Let $P_1 \in R[s]^{p_1 \times m}$ and $P_2 \in R[s]^{p_2 \times m}$ with $p_1 + p_2 \geq m$. The following statements are equivalent:

(a) $P_1$ and $P_2$ are right coprime.
(b) A gcrd of $P_1$ and $P_2$ is unimodular.

(c) *There exist polynomial matrices* $X \in R[s]^{m \times p_1}$ *and* $Y \in R[s]^{m \times p_2}$ *such that*

$$XP_1 + YP_2 = I_m. \tag{7.49}$$

(d) *The Smith form of* $\begin{bmatrix} P_1 \\ P_2 \end{bmatrix}$ *is* $\begin{bmatrix} I \\ 0 \end{bmatrix}$.

(e) rank $\begin{bmatrix} P_1(s_i) \\ P_2(s_i) \end{bmatrix} = m$ *for any complex number* $s_i$.

(f) $\begin{bmatrix} P_1 \\ P_2 \end{bmatrix}$ *constitutes* $m$ *columns of a unimodular matrix.*

*Proof.* See [1, p. 538, Section 7.2D, Theorem 2.4]. ■

---

**Example 7.23.** (a) The polynomial matrices $P_1 = \begin{bmatrix} s & 0 \\ 0 & s+1 \end{bmatrix}, P_2 = \begin{bmatrix} s+1 & 1 \\ 0 & s \end{bmatrix}$ are right coprime in view of the following relations. To use condition (b) of the above theorem, let $U \begin{bmatrix} P_1 \\ P_2 \end{bmatrix} = \begin{bmatrix} -(s+2) & -1 & s+1 & 0 \\ s+1 & 1 & -s & 0 \\ -(s+1)^2 & -s & s(s+1) & 0 \\ -(s+1) & 0 & s & -1 \end{bmatrix} \begin{bmatrix} P_1 \\ P_2 \end{bmatrix} =$

$\begin{bmatrix} 1 & 0 \\ 0 & 1 \\ 0 & 0 \\ 0 & 0 \end{bmatrix} = \begin{bmatrix} G_R^* \\ 0 \end{bmatrix}$. Then $G_R^* = I_2$, which is unimodular. Applying condition (c)

$$XP_1 + YP_2 = \begin{bmatrix} -(s+2) & -1 \\ s+1 & 1 \end{bmatrix} P_1 + \begin{bmatrix} s+1 & 0 \\ -s & 0 \end{bmatrix} P_2 = I_2.$$

To use (d), note that the invariant polynomials of $\begin{bmatrix} P_1 \\ P_2 \end{bmatrix}$ are $\epsilon_1 = \epsilon_2 = 1$;

and the Smith form is then $\begin{bmatrix} I_2 \\ 0 \end{bmatrix}$. To use condition (e), note that the only

complex values $s_i$ that may reduce the rank of $\begin{bmatrix} P_1(s_i) \\ P_2(s_i) \end{bmatrix}$ are those for which

det $P_1(s_i)$ or det $P_2(s_i) = 0$; i.e., $s_1 = 0$ and $s_2 = -1$. For these values we have

$$\text{rank} \begin{bmatrix} P_1(s_1) \\ P_2(s_1) \end{bmatrix} = \text{rank} \begin{bmatrix} 0 & 0 \\ 0 & 1 \\ 1 & 1 \\ 0 & 0 \end{bmatrix} = 2 \text{ and rank} \begin{bmatrix} P_1(s_2) \\ P_2(s_2) \end{bmatrix} = \text{rank} \begin{bmatrix} -1 & 0 \\ 0 & 0 \\ 0 & 1 \\ 0 & -1 \end{bmatrix} = 2;$$

i.e., both are of full rank.

---

The following Theorem 7.24 is the corresponding to Theorem 7.22 result for (left coprime) proper and stable matrices. Note that $\widehat{U}$ proper and stable is a unimodular matrix if $\widehat{U}^{-1}$ is also a proper and stable matrix.

**Theorem 7.24.** *Let* $\widehat{P}_1 \in R[s]^{p \times m_1}$ *and* $\widehat{P}_2 \in R[s]^{p \times m_2}$ *with* $m_1 + m_2 \geq p$. *The following statements are equivalent:*

*(a)* $\widehat{P}_1$ *and* $\widehat{P}_2$ *are left coprime.*
*(b)* *A gcld of* $\widehat{P}_1$ *and* $\widehat{P}_2$ *is unimodular.*
*(c)* *There exist polynomial matrices* $\widehat{X} \in R[s]^{m_1 \times p}$ *and* $\widehat{Y} \in R[s]^{m_2 \times p}$ *such that*

$$\widehat{P}_1\widehat{X} + \widehat{P}_2\widehat{Y} = I_p. \tag{7.50}$$

*(d)* *The Smith form of* $[\widehat{P}_1, \widehat{P}_2]$ *is* $[I, 0]$.
*(e)* $\mathrm{rank}[\widehat{P}_1(s_i), \widehat{P}_2(s_i)] = p$ *for any complex number* $s_i$.
*(f)* $[\widehat{P}_1, \widehat{P}_2]$ *are p rows of a unimodular matrix.*

*Proof.* The proof is completely analogous to the proof of Theorem 7.22 and is omitted.  ∎

### 7.5.3 Controllability, Observability, and Stability

Consider now the Polynomial Matrix Description

$$P(q)z(t) = Q(q)u(t), \quad y(t) = R(q)z(t) + W(q)u(t), \tag{7.51}$$

where $P(q) \in R[q]^{l \times l}$, $Q(q) \in R[q]^{l \times m}$, $R(q) \in R[q]^{p \times l}$, and $W(q) \in R[q]^{p \times m}$.

Assume that the PMD given in (7.51) is equivalent to some state-space representation

$$\dot{x}(t) = Ax(t) + Bu(t), \quad y(t) = Cx(t) + Du(t), \tag{7.52}$$

where $A \in R^{n \times n}$, $B \in R^{n \times m}$, $C \in R^{p \times n}$, and $D \in R^{p \times m}$ [1, p. 553, Section 7.3A].

*Controllability*

**Definition 7.25.** *The representation* $\{P, Q, R, W\}$ *given in (7.51) is said to be* controllable *if its equivalent state-space representation* $\{A, B, C, D\}$ *given in (7.52) is state controllable.*

**Theorem 7.26.** *The following statements are equivalent:*

*(a)* $\{P, Q, R, W\}$ *is controllable.*
*(b)* *The Smith form of* $[P, Q]$ *is* $[I, 0]$.
*(c)* $\mathrm{rank}[P(s_i), Q(s_i)] = l$ *for any complex number* $s_i$.
*(d)* *P, Q are left coprime.*

*Proof.* See [1, p. 561, Theorem 3.4].  ∎

The right Polynomial Matrix Fractional Description, $\{D_R, I_m, N_R\}$, is controllable since $D_R$ and $I$ are left coprime.

*Observability*

Observability can be introduced in a completely analogous manner to controllability. This leads to the following concept and result.

**Definition 7.27.** *The representation $\{P, Q, R, W\}$ given in (7.51) is said to be* observable *if its equivalent state-space representation $\{A, B, C, D\}$ given in (7.52) is state observable.* ∎

**Theorem 7.28.** *The following statements are equivalent:*

*(a) $\{P, Q, R, W\}$ is observable.*

*(b) The Smith form of $\begin{bmatrix} P \\ R \end{bmatrix}$ is $\begin{bmatrix} I \\ 0 \end{bmatrix}$.*

*(c)* rank $\begin{bmatrix} P(s_i) \\ R(s_i) \end{bmatrix} = l$ *for any complex number $s_i$.*

*(d) $P, R$ are right coprime.*

*Proof.* It is analogous to the proof of Theorem 7.26. ∎

The left Polynomial Matrix Fractional Description (PMFD), $\{D_L, N_L, I_p\}$, is observable since $D_L$ and $I_p$ are right coprime.

*Stability*

**Definition 7.29.** *The representation $\{P, Q, R, W\}$ given in (7.51) is said to be* asymptotically stable *if for its equivalent state-space representation $\{A, B, C, D\}$ given in (7.52) the equilibrium $x = 0$ of the free system $\dot{x} = Ax$ is asymptotically stable.*

**Theorem 7.30.** *The representation $\{P, Q, R, W\}$ is asymptotically stable if and only if $Re\lambda_i < 0$, $i = 1, \ldots, n$, where $\lambda_i$, $i = 1, \ldots, n$ are the roots of $\det P(s)$; the $\lambda_i$ are the eigenvalues or poles of the system.*

*Proof.* See [1, p. 563, Theorem 3.6]. ∎

### 7.5.4 Poles and Zeros

*Poles and zeros* can be defined in a completely analogous way for system (7.51) as was done in Section 7.4 for state-space representations.

It is straightforward to show that

$$\{\text{poles of } H(s) \} \subset \{\text{roots of } \det P(s) \}. \tag{7.53}$$

The roots of $\det P$ are the *eigenvalues or the poles of the system* $\{P, Q, R, W\}$ and are equal to the eigenvalues of $A$ in any equivalent state-space representation $\{A, B, C, D\}$. Relation (7.53) becomes an equality when the system is controllable and observable, since in this case the poles of the transfer function matrix $H$ are exactly those eigenvalues of the system that are both controllable and observable.

Consider the *system matrix* or *Rosenbrock Matrix* of the representation $\{P, Q, R, W\}$,

$$S(s) = \begin{bmatrix} P(s) & Q(s) \\ -R(s) & W(s) \end{bmatrix}. \tag{7.54}$$

The *invariant zeros of the system* are the roots of the invariant zero polynomial, which is the product of all the invariant factors of $S(s)$. The *input-decoupling*, *output-decoupling*, and the *input–output decoupling* zeros of $\{P, Q, R, W\}$ can be defined in a manner completely analogous to the state-space case. For example, the roots of the product of all invariant factors of $[P(s), Q(s)]$ are the input-decoupling zeros of the system; they are also the uncontrollable eigenvalues of the system. Note that the input-decoupling zeros are the roots of $\det G_L(s)$, where $G_L(s)$ is a gcld of all the columns of $[P(s), Q(s)] = G_L(s)[\bar{P}(s), \bar{Q}(s)]$. Similar results hold for the output-decoupling zeros.

The *zeros of $H(s)$*, also called the *transmission zeros of the system*, are defined as the roots of *the zero polynomial of $H(s)$*,

$$z_H(s) = \epsilon_1(s) \ldots \epsilon_r(s), \tag{7.55}$$

where the $\epsilon_i$ are the numerator polynomials in the Smith–McMillan form of $H(s)$. When $\{P, Q, R, W\}$ is controllable and observable, the zeros of the system, the invariant zeros, and the transmission zeros coincide.

Consider the representation $D_R z_R = u, y = N_R z_R$ with $D_R \in R[s]^{m \times m}$ and $N_R \in R[s]^{p \times m}$ and notice that in this case the Rosenbrock matrix (7.54) can be reduced via elementary column operations to the form

$$\begin{bmatrix} D_R & I \\ -N_R & 0 \end{bmatrix} \begin{bmatrix} I & 0 \\ -D_R & I \end{bmatrix} \begin{bmatrix} 0 & I \\ I & 0 \end{bmatrix} = \begin{bmatrix} 0 & I \\ -N_R & 0 \end{bmatrix} \begin{bmatrix} 0 & I \\ I & 0 \end{bmatrix} = \begin{bmatrix} I & 0 \\ 0 & -N_R \end{bmatrix}.$$

In view of the fact that the invariant factors of $S$ do not change under elementary matrix operations, the nonunity invariant factors of $S$ are the nonunity invariant factors of $N_R$. Therefore, the *invariant zero polynomial of the system* equals the product of all invariant factors of $N_R$ and its roots are the *invariant zeros of the system*. Note that when $\mathrm{rank}\, N_R = p \le m$, the invariant zeros of the system are the roots of $\det G_L$, where $G_L$ is the gcld of all the columns of $N_R$; i.e., $N_R = G_L \bar{N}_R$. When $N_R, D_R$ are right coprime, the system is controllable and observable. In this case it can be shown that the *zeros of $H$* $(= N_R D_R^{-1})$, also called the *transmission zeros of the system*, are equal to the *invariant zeros* (and to the *system zeros* of $\{D_R, I, N_R\}$) and can be determined from $N_R$. In fact, the zero polynomial of the system, $z_s(s)$, equals $z_H(s)$, the zero polynomial of $H$, which equals $\epsilon_1(s) \ldots \epsilon_r(s)$, the product of the invariant factor of $N_R$; i.e.,

$$z_s(s) = z_H(s) = \epsilon_1(s) \ldots \epsilon_r(s). \tag{7.56}$$

The pole polynomial of $H(s)$ is

$$p_H(s) = k \quad \det D_R(s), \tag{7.57}$$

where $k \in R$.

When $H(s)$ is square and nonsingular and $H(s) = N_R(s)D_R^{-1}(s) = D_L^{-1}(s)N_L(s)$ rc and lc, respectively, the poles of $H(s)$ are the roots of $\det D_R(s)$ or of $\det D_L(s)$ and the zeros of $H(s)$ are the roots of $N_R(s)$ or of $N_L(s)$. An important well-known special case is the case of a scalar $H(s) = n(s)/d(s)$, where the poles of $H(s)$ are the roots of $d(s)$ and the zeros of $H(s)$ are the roots of $n(s)$.

## 7.6 Summary and Highlights

- The transfer function

$$H(s) = C_1(sI - A_{11})^{-1}B_1 + D \tag{7.4}$$

and the impulse response

$$H(t, 0) = C_1 e^{A_{11}t} B_1 + D\delta(t) \tag{7.5}$$

depend only on the controllable and observable parts of the system, $(A_{11}, B_1, C_1)$. Similar results hold for the discrete-time case in (7.6) and (7.7).

- Since

$$\{\text{eigenvalues of } A\} \supset \{\text{poles of } H(s)\}, \tag{7.9}$$

internal stability always implies BIBO stability but not necessarily vice versa. Recall that the system is stable in the sense of Lyapunov (or internally stable) if and only if all eigenvalues of $A$ have negative real parts; the system is BIBO stable if and only if all poles of $H(s)$ have negative real parts. BIBO stability implies internal stability only when the eigenvalues of $A$ are exactly the poles of $H(s)$, which is the case when the system is both controllable and observable.

- When $H(s) = C(sI - A)^{-1}B + D$ and $(A, B)$ is controllable and $(A, C)$ is observable, then

$$\{\text{eigenvalues of } A \text{ (poles of the system)}\} = \{\text{poles of } H(s)\}, \tag{7.58}$$
$$\{\text{zeros of the system}\} = \{\text{zeros of } H(s)\}. \tag{7.22}$$

When the system is not controllable and observable

$$\{\text{eigenvalues of } A \text{ (poles of the system)}\} =$$
$$\{\text{poles of } H(s)\} \cup \{\text{uncontrollable and/or unobservable eigenvalues}\}. \tag{7.20}$$

- If the system $\{A, B, C, D\}$ is not both controllable and observable, then the uncontrollable and/or unobservable eigenvalues cancel out when the transfer functions $H(s) = C(sI - A)^{-1}B + D$ is determined.

*Poles and Zeros*

- The Smith–McMillan form of a transfer function matrix $H(s)$ is

$$SM_H(s) = \begin{bmatrix} \tilde{\Lambda}(s) & 0 \\ 0 & 0 \end{bmatrix}, \tag{7.12}$$

  with $\tilde{\Lambda}(s) \triangleq \operatorname{diag}\left[\frac{\epsilon_1(s)}{\psi_1(s)}, \ldots, \frac{\epsilon_r(s)}{\psi_r(s)}\right]$ are the invariant factors of $N(s)$ in $H(s) = \frac{1}{d(s)}N(s)$ and $r = \operatorname{rank} H(s)$.
- The characteristic or pole polynomial of $H(s)$ is

$$p_H(s) = \psi_1(s) \cdots \psi_r(s). \tag{7.13}$$

  $p_H$ is also the monic least common denominator of all nonzero minors of $H(s)$. The roots of $p_H(s)$ are the poles of $H(s)$.
- The zero polynomial of $H(s)$ is

$$z_H(s) = \epsilon_1(s)\epsilon_2(s) \cdots \epsilon_r(s). \tag{7.18}$$

  The roots of $z_H(s)$ are the zeros of $H(s)$ (or the transmission zeros of the system). When $H(s) = N(s)D(s)^{-1}$ a right coprime polynomial factorization, the zeros of $H(s)$ are the invariant zeros of $N(s)$. When $N(s)$ is square, the zeros are the roots of $|N(s)|$.

*Polynomial Matrix Descriptions*

- PMDs are given by

$$P(q)z(t) = Q(q)u(t), \quad y(t) = R(q)z(t) + W(q)u(t), \tag{7.29}$$

  where $q \triangleq d/dt$ the differential operator ($qz = \dot{z}$). PMDs are, in general, equivalent to state-space representations of the form

$$\dot{x} = Ax + Bu, \quad y = Cx + D(q)u,$$

  and so they are more general than the $\{A, B, C, D\}$ descriptions.
- The transfer function matrix is

$$H(s) = R(s)P^{-1}(s)Q(s) + W(s). \tag{7.30}$$

- The system is *controllable* if and only if $(P, Q)$ are *left coprime* (lc). It is *observable* if and only if $(P, R)$ are *right coprime* (rc). (See Theorems 7.26 and 7.28.) The system is *asymptotically stable* if all the eigenvalues of the system, the roots of $|P(q)|$, have negative real parts. (See Theorem 7.30.)

- Polynomial Matrix Fractional Descriptions (PMFDs) are given by

$$H(s) = N_R(s)D_R^{-1}(s) = D_L^{-1}(s)N_L(s), \qquad (7.35)$$

where $(N_R, D_R)$ are rc and $(D_L, N_L)$ are lc. They correspond to the PMD

$$D_R(q)z_R(t) = u(t), \quad y(t) = N_R(q)z_R(t) \qquad (7.36)$$

and

$$D_L(q)z_L(t) = N_L(q)u(t), \quad y(t) = z_L(t), \qquad (7.37)$$

which are both controllable and observable representations.
- Proper and stable Matrix Fractional Descriptions (MFDs) are given by

$$H(s) = \widehat{N}_R(s)\widehat{D}_R(s)^{-1} = \widehat{D}_L^{-1}(s)\widehat{N}_L(s), \qquad (7.38)$$

where $\widehat{N}_R, \widehat{D}_R, \widehat{D}_L, \widehat{N}_L$ are proper and stable matrices with $(\widehat{N}_R, \widehat{D}_R)$ rc and $(\widehat{D}_L, \widehat{N}_L)$ lc.

## 7.7 Notes

The role of controllability and observability in the relation between propertiers of internal and external descriptions are found in Gilbert [2], Kalman [4], and Popov [5]. For further information regarding these historical issues, consult Kailath [3] and the original sources.

Multivariable zeros have an interesting history. For a review, see Schrader and Sain [7] and the references therein. Refer also to Vardulakis [8]. Polynomial matrix descriptions were used by Rosenbrock [6] and Wolovich [9]. See [1, Sections 7.6 and 7.7] for extensive notes and references.

## References

1. P.J. Antsaklis and A.N. Michel, *Linear Systems*, Birkhäuser, Boston, MA, 2006.
2. E. Gilbert, "Controllability and observability in multivariable control systems," *SIAM J. Control*, Vol. 1, pp. 128–151, 1963.
3. T. Kailath, *Linear Systems*, Prentice-Hall, Englewood, NJ, 1980.
4. R.E. Kalman, "Mathematical descriptions of linear systems," *SIAM J. Control*, Vol. 1, pp. 152–192, 1963.
5. V.M. Popov, "On a new problem of stability for control systems," *Autom. Remote Control*, pp. 1–23, Vol. 24, No. 1, 1963.
6. H.H. Rosenbrock, *State-Space and Multivariable Theory*, Wiley, New York, NY, 1970.
7. C.B. Schrader and M.K. Sain, "Research on system zeros: a survey," *Int. Journal of Control*, Vol. 50, No. 4, pp. 1407–1433, 1989.
8. A.I.G. Vardulakis, *Linear Multivariable Control. Algebraic Analysis and Synthesis Methods*, Wiley, New York, NY, 1991.
9. W.A. Wolovich, *Linear Multivariable Systems*, Springer-Verlag, New York, NY, 1974.

# Exercises

**7.1.** Consider the system $\dot{x} = Ax + Bu, y = Cx + Du$.

(a) Show that only controllable modes appear in $e^{At}B$ and therefore in the zero-state response of the state.

(b) Show that only observable modes appear in $Ce^{At}$ and therefore in the zero-input response of the system.

(c) Show that only modes that are both controllable and observable appear in $Ce^{At}B$ and therefore in the impulse response and the transfer function matrix of the system. Consider next the system $x(k+1) = Ax(k) + Bu(k)$, $y(k) = Cx(k) + Du(k)$.

(d) Show that only controllable modes appear in $A^k B$, only observable modes in $CA^k$, and only modes that are both controllable and observable appear in $CA^k B$ [that is, in $H(z)$].

(e) Let $A = \begin{bmatrix} 1 & 0 & 0 \\ 0 & -2 & 0 \\ 0 & 0 & -1 \end{bmatrix}$, $B = \begin{bmatrix} 1 \\ 0 \\ 1 \end{bmatrix}$, $C = [1, 1, 0]$, and $D = 0$. Verify the results obtained in (d).

**7.2.** In the circuit of Example 7.2, let $R_1 R_2 C = L$ and $R_1 = R_2 = R$. Determine $x(t) = [x_1(t), x_2(t)]^T$ and $i(t)$ for unit step input voltage, $v(t)$, and initial conditions $x(0) = [a, b]^T$. Comment on your results.

**7.3.** (a) Consider the state equation $\dot{x} = Ax + Bu, x(0) = x_0$, where $A = \begin{bmatrix} 0 & -1 & 1 \\ 1 & -2 & 1 \\ 0 & 1 & -1 \end{bmatrix}$ and $B = \begin{bmatrix} 1 & 0 \\ 1 & 1 \\ 1 & 2 \end{bmatrix}$. Determine $x(t)$ as a function of $u(t)$ and $x_0$, and verify that the uncontrollable modes do not appear in the zero-state response but do appear in the zero-input response.

(b) Consider the state equation $x(k + 1) = Ax(k) + Bu(k)$ and $x(0) = x_0$, where $A$ and $B$ are as in (a). Demonstrate for this case results corresponding to (a).

In (a) and (b), determine $x(t)$ and $x(k)$ for unit step inputs and $x(0) = [1, 1, 1]^T$.

**7.4.** (a) Consider the system $\dot{x} = Ax + Bu, y = Cx$ with $x(0) = x_0$, where $A = \begin{bmatrix} 0 & 1 \\ -2 & -3 \end{bmatrix}$, $B = \begin{bmatrix} 0 \\ 1 \end{bmatrix}$, and $C = [1, 1]$. Determine $y(t)$ as a function of $u(t)$ and $x_0$, and verify that the unobservable modes do not appear in the output.

(b) Consider the system $x(k+1) = Ax(k) + Bu(k), y(k) = Cx(k)$ with $x(0) = x_0$, where $A, B$, and $C$ are as in (a). Demonstrate for this case results that correspond to (a).

In (a) and (b), determine and plot $y(t)$ and $y(k)$ for unit step inputs and $x(0) = 0$.

**7.5.** Consider the system $x(k + 1) = Ax(k) + Bu(k), y(k) = Cx(k)$, where

$$A = \begin{bmatrix} 1 & 0 & 0 \\ 0 & -1/2 & 0 \\ 0 & 0 & -1/2 \end{bmatrix}, \qquad B = \begin{bmatrix} 1 \\ 0 \\ 1 \end{bmatrix}, \qquad C = [1, 1, 0].$$

Determine the eigenvalues that are uncontrollable and/or unobservable. Determine $x(k), y(k)$ for $k \geq 0$, given $x(0)$ and $u(k)$, $k \geq 0$, and show that only controllable eigenvalues (resp., modes) appear in $A^k B$, only observable ones appear in $CA^k$, and only eigenvalues (resp., modes) that are both controllable and observable appear in $CA^k B$ [in $H(z)$ ].

**7.6.** Given is the system $\dot{x} = \begin{bmatrix} -1 & 0 & 0 \\ 0 & -1 & 0 \\ 0 & 0 & 2 \end{bmatrix} x + \begin{bmatrix} 1 & 0 \\ 0 & 1 \\ 0 & 0 \end{bmatrix} u, \ y = \begin{bmatrix} 1 & 1 & 0 \\ 1 & 0 & 0 \end{bmatrix} x.$

(a) Determine the uncontrollable and the unobservable eigenvalues (if any).
(b) What is the impulse response of this system? What is its transfer function matrix?
(c) Is the system asymptotically stable?

**7.7.** Given is the transfer function matrix $H(s) = \begin{bmatrix} \frac{s-1}{s} & 0 & \frac{s-2}{s+2} \\ 0 & \frac{s+1}{s} & 0 \end{bmatrix}.$

(a) Determine the Smith–McMillan form of $H(s)$ and its characteristic (pole) polynomial and minimal polynomial. What are the poles of $H(s)$?
(b) Determine the zero polynomial of $H(s)$. What are the zeros of $H(s)$?

**7.8.** Let $H(s) = \begin{bmatrix} \frac{s^2+1}{s^2} \\ \frac{s+1}{s^3} \end{bmatrix}.$

(a) Determine the Smith–McMillan form of $H(s)$ and its characteristic (pole) polynomial and minimal polynomial. What are the poles of $H(s)$?
(b) Determine the zero polynomial of $H(s)$. What are the zeros of $H(s)$?

**7.9.** A rational function matrix $R(s)$ may have, in addition to finite poles and zeros, *poles and zeros at infinity* ($s = \infty$). To study the poles and zeros at infinity, the bilinear transformation

$$s = \frac{b_1 w + b_0}{a_1 w + a_0}$$

with $a_1 \neq 0, b_1 a_0 - b_0 a_1 \neq 0$ may be used, where $b_1/a_1$ is not a finite pole or zero of $R(s)$. This transformation maps the point $s = b_1/a_1$ to $w = \infty$ and the point of interest, $s = \infty$, to $w = -a_0/a_1$. The rational matrix $\widehat{R}(w)$ is now obtained as

$$\widehat{R}(w) = R\left(\frac{b_1 w + b_0}{a_1 w + a_0}\right),$$

and the finite poles and zeros of $\widehat{R}(w)$ are determined. The poles and zeros at $w = -a_0/a_1$ are the poles and zeros of $R(s)$ at $s = \infty$. Note that frequently a good choice for the bilinear transformation is $s = 1/w$; that is, $b_1 = 0, b_0 = 1$ and $a_1 = 1, a_0 = 0$.

(a) Determine the poles and zeros at infinity of

$$R_1(s) = \frac{1}{s+1}, \qquad R_2(s) = s, \qquad R_3(s) = \begin{bmatrix} 1 & 0 \\ s+1 & 1 \end{bmatrix}.$$

Note that a rational matrix may have both poles and zeros at infinity.

(b) Show that if $R(s)$ has a pole at $s = \infty$, then it is not a proper rational function ($\lim\limits_{s \to \infty} R(s) \to \infty$).

**7.10.** Consider the polynomial matrices $P(s) = \begin{bmatrix} s^2 + s & -s \\ -s^2 - 1 & s^2 \end{bmatrix}$, $R(s) = \begin{bmatrix} s & 0 \\ -s - 1 & 1 \end{bmatrix}$.

(a) Are they right coprime (rc)? If they are not, find a greatest common right divisor (gcrd).

(b) Are they left coprime (lc)? If they are not, find a greatest common left divisor (gcld).

**7.11.** (a) Show that two square and nonsingular polynomial matrices, the determinants of which are coprime polynomials, are both right and left coprime. *Hint:* Assume they are not, say, right coprime and then use the determinants of their gcrd to arrive at a contradiction.

(b) Show that the opposite is not true; i.e., two right (left) coprime polynomial matrices do not necessarily have determinants which are coprime polynomials.

**7.12.** Let $P(s)$ be a polynomial matrix of full column rank, and let $y(s)$ be a given polynomial vector. Show that the equation $P(s)x(s) = y(s)$ will have a polynomial solution $x(s)$ for any $y(s)$ if and only if the columns of $P(s)$ are lc, or equivalently, if and only if $P(\lambda)$ has full column rank for any complex number $\lambda$.

**7.13.** Consider $P(q)z(t) = Q(q)u(t)$ and $y(t) = R(q)z(t) + W(q)u(t)$, where

$$P(q) = \begin{bmatrix} q^3 - q & q^2 - 1 \\ -q - 2 & 0 \end{bmatrix}, \qquad Q(q) = \begin{bmatrix} q - 1 & -2q + 2 \\ 1 & 3q \end{bmatrix},$$

$$R(q) = \begin{bmatrix} 2q^2 + q + 2 & 2q \\ -q - 2 & 0 \end{bmatrix}, \qquad W(q) = \begin{bmatrix} -1 & 3q + 4 \\ -1 & -3q \end{bmatrix},$$

with $q \triangleq \frac{d}{dt}$.

(a) Is this system representation controllable? Is it observable?

(b) Find the transfer function matrix $H(s)(\hat{y}(s) = H(s)\hat{u}(s))$.

(c) Determine an equivalent state-space representation $\dot{x} = Ax + Bu$, $y = Cx + Du$, and repeat (a) and (b) for this representation.

**7.14.** Use system theoretic arguments to show that two polynomials $d(s) = s^n + d_{n-1}s^{n-1} + \cdots + d_1 s + d_0$ and $n(s) = n_{n-1}s^{n-1} + n_{n-2}s^{n-2} + \cdots + n_1 s + n_0$ are coprime if and only if

$$\text{rank} \begin{bmatrix} C_c \\ C_c A_c \\ \vdots \\ C_c A_c^{n-1} \end{bmatrix} = n,$$

where $A_c = \begin{bmatrix} 0 & 1 & 0 & \cdots & 0 \\ 0 & 0 & 1 & \cdots & 0 \\ \vdots & \vdots & \vdots & & \vdots \\ 0 & 0 & 0 & \cdots & 1 \\ -d_0 & -d_1 & -d_2 & & -d_{n-1} \end{bmatrix}$ and $C_c = [n_0, n_1, \ldots, n_{n-1}]$.

**7.15.** Consider the system $Dz = u$, $y = Nz$, where $D = \begin{bmatrix} s^2 & 0 \\ 0 & s^3 \end{bmatrix}$ and $N = [s^2 - 1, s + 1]$.

(a) Is the system controllable? Is it observable? Determine all uncontrollable and/or unobservable eigenvalues, if any.

(b) Determine the invariant and the transmission zeros of the system.

# 8

---

# Realization Theory and Algorithms

## 8.1 Introduction

In this chapter the following problem is being addressed: Given an external description of a linear system, specifically, its transfer function or its impulse response, determine an internal, state-space description for the system that generates the given transfer function. This is the problem of *system realization*. The name reflects the fact that if a (continuous-time) state-space description is known, an operational amplifier circuit can be built in a straightforward manner to realize (actually simulate) the system response.

There are many ways, an infinite number in fact, of realizing a given transfer function. Presently, we are interested in realizations that contain the least possible number of energy or memory storage elements, i.e., in realizations of least order (in terms of differential or difference equations). To accomplish this, the concepts of controllability and observability play a central role. Indeed, it turns out that realizations of transfer functions of least order are both controllable and observable. In Section 8.2, the problem of state-space realizations of input–output descriptions is defined and the existence of such realizations is addressed. The minimality of realizations of $H(s)$ is studied in Section 8.3, culminating in two results, Theorem 8.9 and Theorem 8.10, where it is first shown that a realization is minimal if and only if it is controllable and observable, and next, that if a realization is minimal, all other minimal realizations of a given $H(s)$ can be found via similarity transformations. It is also shown how to determine the order of minimal realizations directly from $H(s)$. Several realization algorithms are presented in Section 8.4, and the role of duality is emphasized in Subsection 8.4.1.

## 8.2 State-Space Realizations of External Descriptions

In this section, state-space realizations of impulse responses and of transfer functions for time-invariant systems are introduced. Continuous-time systems are discussed first in Subsection 8.2.1, followed by discrete-time systems in Subsection 8.2.2.

### 8.2.1 Continuous-Time Systems

Before formally defining the problem of system realization, we first review some of the relations that were derived in Chapter 3.

We consider a time-invariant system described by equations of the form

$$\dot{x} = Ax + Bu, \quad y = Cx + Du, \tag{8.1}$$

where $A \in R^{n \times n}, B \in R^{n \times m}, C \in R^{p \times n}$, and $D \in R^{p \times m}$. The response of this system is given by

$$y(t) = Ce^{At}x_0 + \int_0^t H(t, \tau)u(\tau)d\tau, \tag{8.2}$$

where, without loss of generality, the initial time $t_0$ was taken to be zero. The impulse response is now given by the expression

$$H(t, \tau) = \begin{cases} Ce^{A(t-\tau)}B + D\delta(t - \tau), & \text{for } t \geq \tau, \\ 0, & \text{for } t < \tau. \end{cases} \tag{8.3}$$

Recall that the time invariance of system (8.1) implies that $H(t, \tau) = H(t - \tau, 0)$, and therefore, $\tau$, which is the time at which a unit impulse input is applied to the system, can be taken to equal zero ($\tau = 0$), without loss of generality, to yield $H(t, 0)$. The transfer function matrix of the system is the (one-sided) Laplace transform of $H(t, 0)$, namely,

$$H(s) = \mathcal{L}[H(t, 0)] = C(sI - A)^{-1}B + D. \tag{8.4}$$

In the time-invariant case, a realization is commonly defined in terms of the transfer function matrix. We let $\{A, B, C, D\}$ denote the system description given in (8.1), and we let $H(s)$ be a $p \times m$ matrix with entries that are functions of $s$.

**Definition 8.1.** *A realization of $H(s)$ is any set $\{A, B, C, D\}$, the transfer function matrix of which is $H(s)$; i.e., $\{A, B, C, D\}$ is a realization of $H(s)$ if (8.4) is satisfied. (See Figure 8.1.)* ∎

As will be shown in the next section, given $H(s)$, a condition for a realization $\{A, B, C, D\}$ of $H(s)$ to exist is that all entries in $H(s)$ are proper, rational functions. Alternative conditions under which a given set $\{A, B, C, D\}$ is a realization of some $H(s)$ can easily be derived. To this end, we expand $H(s)$ in a Laurent series to obtain

$$H(s) = H_0 + H_1 s^{-1} + H_2 s^{-2} + \cdots. \tag{8.5}$$

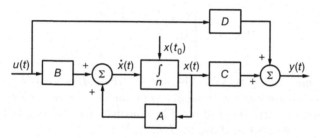

**Figure 8.1.** Block diagram realization of $\{A, B, C, D\}$

**Definition 8.2.** *The terms $H_i$, $i = 0, 1, 2, \ldots$, in (8.5) are the* Markov parameters *of the system.*    ∎

The Markov parameters can be determined by the formulas

$$H_0 = \lim_{s \to \infty} H(s), \quad H_1 = \lim_{s \to \infty} s(H(s) - H_0), \quad H_2 = \lim_{s \to \infty} s^2(H(s) - H_0 - H_1 s^{-1}),$$

and so forth. Recall that relations involving the Markov parameters were used in Exercise 3.34 of Chapter 3.

**Theorem 8.3.** *The set $\{A, B, C, D\}$ is a realization of $H(s)$ if and only if*

$$H_0 = D \text{ and } H_i = CA^{i-1}B, \quad i = 1, 2, \ldots. \tag{8.6}$$

*Proof.* $H(s) = D + C(sI - A)^{-1}B = D + Cs^{-1}(I - s^{-1}A)^{-1}B = D + Cs^{-1}[\sum_{i=0}^{\infty}(s^{-1}A)^i]B = D + \sum_{i=1}^{\infty}[CA^{i-1}B]s^{-i}$, from which (8.6) is derived in view of (8.5).    ∎

### 8.2.2 Discrete-Time Systems

The realization theory in the discrete-time case essentially parallels the continuous-time case. There are of course certain notable differences because in the present case the realizations are difference equations instead of differential equations. We point to these differences in the subsequent sections.

Some of the relations derived in Section 3.4 will be recalled next. We consider systems described by equations of the form

$$x(k+1) = Ax(k) + Bu(k), \quad y(k) = Cx(k) + Du(k), \tag{8.7}$$

where $A \in R^{n \times n}$, $B \in R^{n \times m}$, $C \in R^{p \times n}$, and $D \in R^{p \times m}$. The response of this system is given by

$$y(k) = CA^k x_0 + \sum_{i=0}^{k-1} H(k, i)u(i), \quad k > 0, \tag{8.8}$$

where, without loss of generality, $k_0$ was taken to be zero. The unit pulse (discrete impulse) response is now given by

$$H(k,i) = \begin{cases} CA^{k-(i+1)}B, & k > i, \\ D, & k = i, \\ 0, & k < i. \end{cases} \qquad (8.9)$$

Recall that since the system (8.7) is time-invariant, $H(k,i) = H(k-i,0)$ and $i$, the time the pulse input is applied, can be taken to be zero, to yield $H(k,0)$ as the external system description. The transfer function matrix for (8.7) is now the (one-sided) $z$-transform of $H(k,0)$. We have

$$H(z) = \mathcal{Z}\{H(k,0)\} = C(zI - A)^{-1}B + D. \qquad (8.10)$$

Now let $\{A, B, C, D\}$ denote the system description (8.7) and let $H(z)$ be a $p \times m$ matrix with functions of $z$ as entries.

**Definition 8.4.** *A realization of $H(z)$ is any set $\{A, B, C, D\}$, the transfer function matrix of which is $H(z)$; i.e., it satisfies (8.10).* ■

A result that is analogous to Theorem 8.3 is also valid in the discrete-time case [with $H(s)$ replaced by $H(z)$].

## 8.3 Existence and Minimality of Realizations

The existence of realizations is examined first. Given a $p \times m$ matrix $H(s)$, conditions for $H(s)$ to be the transfer function matrix of a system described by equations of the form $\dot{x} = Ax + Bu$, $y = Cx + Du$ are given in Theorem 8.5. It is shown that such realizations exist if and only if $H(s)$ is a matrix of rational functions with the property that $\lim_{s \to \infty} H(s)$ is finite. The corresponding results for discrete-time systems are also presented.

Realizations of least order, also called minimal or irreducible realizations, are of interest to us since they realize a system, using the least number of dynamical elements (minimum number of elements with memory). The principal results are given in Theorems 8.9 and 8.10, where it is shown that minimal realizations are controllable (-from-the-origin) and observable and that all minimal realizations of $H(s)$ are equivalent representations. The order of any minimal realization can be determined directly without first determining a minimal realization, and this can be accomplished by using the characteristic polynomial and the degree of $H(s)$ (Theorem 8.12) or from the rank of a Hankel matrix (Theorem 8.16). All the results on minimality of realizations apply to the discrete-time case as well with no substantial changes. This is discussed at the end of the section.

### 8.3.1 Existence of Realizations

*Continuous-Time Systems.* Given a $p \times m$ matrix $H(s)$, the following result establishes necessary and sufficient conditions for the existence of time-invariant realizations.

**Theorem 8.5.** $H(s)$ *is realizable as the transfer function matrix of a time-invariant system described by (8.1) if and only if $H(s)$ is a matrix of rational functions and satisfies*

$$\lim_{s \to \infty} H(s) < \infty, \tag{8.11}$$

*i.e., if and only if $H(s)$ is a* proper rational matrix.

*Proof.* (*Necessity*) If the system $\dot{x} = Ax + Bu, y = Cx + Du$ is a realization of $H(s)$, then $C(sI - A)^{-1}B + D = H(s)$, which shows that $H(s)$ must be a rational matrix. Furthermore,

$$\lim_{s \to \infty} H(s) = D, \tag{8.12}$$

which is a real finite matrix.

(*Sufficiency*) If $H(s)$ is a proper rational matrix, then any of the algorithms discussed in the next section can be applied to derive a realization. ∎

*Discrete-Time Systems.* Given a $p \times m$ matrix $H(z)$, the next theorem establishes necessary and sufficient conditions for time-invariant realizations. This result corresponds to Theorem 8.5 for the continuous-time case. Notice that the conditions in these results are identical.

**Theorem 8.6.** $H(z)$ *is realizable as the transfer function matrix of a time-invariant system described by (8.7) if and only if $H(z)$ is a matrix of rational functions and satisfies the condition that*

$$\lim_{z \to \infty} H(z) < \infty. \tag{8.13}$$

*Proof.* Similar to the proof of Theorem 8.5. ∎

## 8.3.2 Minimality of Realizations

Realizations of a transfer function matrix $H(s)$ can be expected to generate only the *zero-state response* of a system, since the external description $H(s)$ has, by definition, no information about the initial conditions and the zero-input response of the system.

A second important point is the fact that *if a realization of a given $H(s)$ exists, then there exists an infinite number of realizations.* If (8.1) is a realization of the $p \times m$ matrix $H(s)$, then realizations of the same order $n$, i.e., of the same dimension $n$ of the state vector, can readily be generated by an equivalence transformation. There are, of course, other ways of generating alternative realizations. In particular, if (8.1) is a realization of $H(s)$, then, for example, the system

$$\begin{aligned} \dot{x} &= Ax + Bu, \quad y = Cx + Du, \\ \dot{z} &= Fz + Gu \end{aligned} \tag{8.14}$$

is also a realization. This was accomplished by adding to (8.1) a state equation $\dot{z} = Fz + Gu$ that does not affect the system output. The dimension of $F$, $\dim F$, and consequently the order of the realization, $n + \dim F$, can be larger than any given finite number. In other words, *there may be no upper bound to the order of the realizations of a given $H(s)$*. There exists, however, a lower bound, and a realization of such lowest order is called a least-order minimal or irreducible realization.

**Definition 8.7.** *A realization*

$$\dot{x} = Ax + Bu, \quad y = Cx + Du \tag{8.15}$$

*of the transfer function matrix $H(s)$ of least order $n$ ($A \in R^{n \times n}$) is called a least-order, or a minimal, or an irreducible realization of $H(s)$.* ∎

Theorems 8.9 and 8.10 below completely solve the minimal realization problem. The first of these results shows that a realization is minimal if and only if it is controllable (-from-the-origin or reachable) and observable, whereas the second result shows that if a minimal realization has been found, then all other minimal realizations can be obtained from the determined realization, using equivalence of representations.

Controllability (-from-the-origin, or reachability) and observability play an important role in the minimality of realizations. Indeed, it was shown in Section 7.2 that only that part of a system that is both controllable and observable appears in $H(s)$. In other words, $H(s)$ *contains no information about the uncontrollable and/or unobservable parts of the system.* To illustrate this, consider the following specific case.

---

***Example 8.8.*** Let $H(s) = 1/(s+1)$. Four different realizations of $H(s)$ are given by

(i) $\{A = \begin{bmatrix} 0 & 1 \\ 1 & 0 \end{bmatrix}$, $B = \begin{bmatrix} 0 \\ 1 \end{bmatrix}$, $C = [-1, 1]$, $D = 0\}$,

(ii) $\{A = \begin{bmatrix} 0 & 1 \\ 1 & 0 \end{bmatrix}$, $B = \begin{bmatrix} -1 \\ 1 \end{bmatrix}$, $C = [0, 1]$, $D = 0\}$,

(iii) $\{A = \begin{bmatrix} 1 & 0 \\ 0 & -1 \end{bmatrix}$, $B = \begin{bmatrix} 0 \\ 1 \end{bmatrix}$, $C = [0, 1]$, $D = 0\}$,

(iv) $\{A = -1, B = 1, C = 1, D = 0\}$.

The eigenvalue $+1$ in (i) is unobservable, in (ii) is uncontrollable, and in (iii) is both uncontrollable and unobservable and does not appear in $H(s)$ at all. Realization (iv), which is of order 1, is a minimal realization. It is controllable and observable.

---

**Theorem 8.9.** *An $n$-dimensional realization $\{A, B, C, D\}$ of $H(s)$ is minimal (irreducible, of least order) if and only if it is both controllable and observable.*

*Proof.* (*Necessity*) Assume that $\{A, B, C, D\}$ is a minimal realization but is not both controllable and observable. Using Kalman's Canonical Decomposition (see Subsection 6.2.3), one may find another realization of lower dimension that is both controllable and observable. This contradicts the assumption that $\{A, B, C, D\}$ is a minimal realization. Therefore, it must be both controllable and observable.

(*Sufficiency*) Assume that the realization $\{A, B, C, D\}$ is controllable and observable, but there exists another realization, say, $\{\bar{A}, \bar{B}, \bar{C}, \bar{D}\}$ of order $\bar{n} < n$. Since they are both realizations of $H(s)$, or of the impulse response $H(t, 0)$, then

$$Ce^{At}B + D\delta(t) = \bar{C}e^{\bar{A}t}\bar{B} + \bar{D}\delta(t) \tag{8.16}$$

for all $t \geq 0$. Clearly, $D = \bar{D} = \lim_{s\to\infty} H(s)$. Using the power series expansion of the exponential and equating coefficients of the same power of $t$, we obtain

$$CA^k B = \bar{C}\bar{A}^k\bar{B}, \quad k = 0, 1, 2, \ldots; \tag{8.17}$$

i.e., the Markov parameters of the two representations are the same (see Theorem 8.3). Let

$$\mathcal{C}_n \triangleq [B, AB, \ldots, A^{n-1}B] \in R^{n \times mn}$$

and

$$\mathcal{O}_n \triangleq \begin{bmatrix} C \\ CA \\ \vdots \\ CA^{n-1} \end{bmatrix} \in R^{pn \times n}. \tag{8.18}$$

Then the $pn \times mn$ matrix product $\mathcal{O}_n\mathcal{C}_n$ assumes the form

$$\mathcal{O}_n\mathcal{C}_n = \begin{bmatrix} CB & CAB & \cdots & CA^{n-1}B \\ CAB & CA^2B & \cdots & CA^nB \\ \vdots & \vdots & & \vdots \\ CA^{n-1}B & CA^nB & \cdots & CA^{2n-2}B \end{bmatrix}$$

$$= \begin{bmatrix} \bar{C}\bar{B} & \bar{C}\bar{A}\bar{B} & \cdots & \bar{C}\bar{A}^{n-1}\bar{B} \\ \bar{C}\bar{A}\bar{B} & \bar{C}\bar{A}^2\bar{B} & \cdots & \bar{C}\bar{A}^n\bar{B} \\ \vdots & \vdots & & \vdots \\ \bar{C}\bar{A}^{n-1}\bar{B} & \bar{C}\bar{A}^n\bar{B} & \cdots & \bar{C}\bar{A}^{2n-2}\bar{B} \end{bmatrix} = \bar{\mathcal{O}}_n\bar{\mathcal{C}}_n. \tag{8.19}$$

In view of *Sylvester's Rank Inequality*, which relates the rank of the product of two matrices to the rank of its factors, we have

$$\text{rank}\,\mathcal{O}_n + \text{rank}\,\mathcal{C}_n - n \leq \text{rank}(\bar{\mathcal{O}}_n\bar{\mathcal{C}}_n) \leq \min(\text{rank}\,\mathcal{O}_n, \text{rank}\,\mathcal{C}_n) \tag{8.20}$$

and we obtain that $\text{rank}\,\mathcal{O}_n = \text{rank}\,\mathcal{C}_n = n$, $\text{rank}(\bar{\mathcal{O}}_n\bar{\mathcal{C}}_n) = n$. This result, however, contradicts our assumptions, since $n = \text{rank}(\bar{\mathcal{O}}_n\bar{\mathcal{C}}_n) \leq \min(\text{rank}\,\bar{\mathcal{O}}_n, \text{rank}\,\bar{\mathcal{C}}_n) \leq \bar{n}$ because $\bar{n}$ is the order of $\{\bar{A}, \bar{B}, \bar{C}, \bar{D}\}$. Therefore $n \leq \bar{n}$. Hence, $\bar{n}$ cannot be less than $n$ and they can only be equal. Thus, $n = \bar{n}$ and $\{A, B, C, D\}$ is indeed a minimal realization. ∎

Theorem 8.9 suggests the following procedure to realize $H(s)$. First, we obtain a controllable (observable) realization of $H(s)$. Next, using a similarity transformation, we obtain an observable standard form to separate the observable from the unobservable parts (controllable from the uncontrollable parts), using the approach of Subsection 6.2.1. Finally, we take the observable (controllable) part that will also be controllable (observable) as the minimal realization. We shall use this procedure in the next section.

Is the minimal realization unique? The answer to this question is of course "no" since we know that equivalent representations, which are of the same order, give the same transfer function matrix. The following theorem shows how to obtain *all* minimal realizations of $H(s)$.

**Theorem 8.10.** *Let* $\{A, B, C, D\}$ *and* $\{\bar{A}, \bar{B}, \bar{C}, \bar{D}\}$ *be realizations of* $H(s)$. *If* $\{A, B, C, D\}$ *is a minimal realization, then* $\{\bar{A}, \bar{B}, \bar{C}, \bar{D}\}$ *is also a minimal realization if and only if the two realizations are equivalent, i.e., if and only if* $\bar{D} = D$ *and there exists a nonsingular matrix* $P$ *such that*

$$\bar{A} = PAP^{-1}, \quad \bar{B} = PB, \text{ and } \bar{C} = CP^{-1}. \tag{8.21}$$

*Furthermore, if* $P$ *exists, it is given by*

$$P = C\bar{C}^T(\bar{C}\bar{C}^T)^{-1} \text{ or } P = (\bar{\mathcal{O}}^T\bar{\mathcal{O}})^{-1}\bar{\mathcal{O}}^T\mathcal{O}. \tag{8.22}$$

*Proof. (Sufficiency)* Let the realizations be equivalent. Since $\{A, B, C, D\}$ is minimal, it is controllable and observable and its equivalent representation $\{\bar{A}, \bar{B}, \bar{C}, \bar{D}\}$ is also controllable and observable and, therefore, minimal. Alternatively, since equivalence preserves the dimension of $A$, the equivalent realization $\{\bar{A}, \bar{B}, \bar{C}, \bar{D}\}$ is also minimal.

*(Necessity)* Suppose $\{\bar{A}, \bar{B}, \bar{C}, \bar{D}\}$ is also minimal. We shall show that it is equivalent to $\{A, B, C, D\}$. Since they are both realizations of $H(s)$, they satisfy $D = \bar{D}$ and

$$CA^kB = \bar{C}\bar{A}^k\bar{B}, \bar{k} = 0, 1, 2\ldots, \tag{8.23}$$

as was shown in the proof of Theorem 8.9. Here, both realizations are minimal, and therefore, they are both of the same order $n$ and are both controllable and observable.

Define $\mathcal{C} = \mathcal{C}_n$ and $\mathcal{O} = \mathcal{O}_n$, as in (8.18). Then, in view of (8.19), $\mathcal{O}\mathcal{C} = \bar{\mathcal{O}}\bar{\mathcal{C}}$ and premultiplying by $\bar{\mathcal{O}}^T$, we obtain $\bar{\mathcal{O}}^T\mathcal{O}\mathcal{C} = \bar{\mathcal{O}}^T\bar{\mathcal{O}}\bar{\mathcal{C}}$. Using Sylvester's Inequality, we obtain rank $\bar{\mathcal{O}}^T\bar{\mathcal{O}} = n$, and therefore,

$$\bar{\mathcal{C}} = [(\bar{\mathcal{O}}^T\bar{\mathcal{O}})^{-1}\bar{\mathcal{O}}^T\mathcal{O}]\mathcal{C} = P\mathcal{C}, \tag{8.24}$$

where $P \triangleq (\bar{\mathcal{O}}^T\bar{\mathcal{O}})^{-1}\bar{\mathcal{O}}^T\mathcal{O} \in R^{n\times n}$. Note that rank $P = n$ since rank $\bar{\mathcal{O}}^T\mathcal{O}$ is also equal to $n$ as can be seen from rank $\bar{\mathcal{O}}^T\mathcal{O}\mathcal{C} = n$ and from Sylvester's Inequality. Therefore, $P$ qualifies as a similarity transformation. Similarly, $\mathcal{O}\mathcal{C} = \bar{\mathcal{O}}\bar{\mathcal{C}}$ implies that $\mathcal{O}\mathcal{C}\mathcal{C}^T = \bar{\mathcal{O}}\bar{\mathcal{C}}\mathcal{C}^T$, and

$$\mathcal{O} = \bar{\mathcal{O}}[\bar{\mathcal{C}}\mathcal{C}^T(\mathcal{C}\mathcal{C}^T)^{-1}] = \bar{\mathcal{O}}\bar{P}, \tag{8.25}$$

where $\bar{P} \triangleq \bar{\mathcal{C}}\mathcal{C}^T(\mathcal{C}\mathcal{C}^T)^{-1} \in R^{n \times n}$ with rank $\bar{P} = n$. Note that $P = (\bar{\mathcal{O}}^T\bar{\mathcal{O}})^{-1}\bar{\mathcal{O}}^T(\bar{\mathcal{O}}\bar{P}) = \bar{P}$. To show that $P$ is the equivalence transformation given in (8.21), we note that $\mathcal{O}AC = \bar{\mathcal{O}}\bar{A}\bar{\mathcal{C}}$ from (8.19). Premultiplying by $\bar{\mathcal{O}}^T$ and postmultiplying by $\mathcal{C}^T$, we obtain $PA = \bar{A}P$, in view of (8.24) and (8.25). To show that $PB = \bar{B}$ and $C = \bar{C}P$, we simply use the relations $PC = \bar{C}$ and $\mathcal{O} = \bar{\mathcal{O}}P$, respectively.  ∎

### 8.3.3 The Order of Minimal Realizations

One could ask the question whether the order of a minimal realization of $H(s)$ can be determined directly, without having to actually derive a minimal realization. The answer to this question is yes, and in the following we will show how this can be accomplished.

#### Determination via the Characteristic or Pole Polynomial of $H(s)$.

The *characteristic polynomial (or pole polynomial)*, $p_H(s)$, *of a transfer function* matrix $H(s)$ was defined in Section 7.4 using the Smith–McMillan form of $H(s)$. The polynomial $p_H(s)$ is equal to the monic least common denominator of all nonzero minors of $H(s)$. *The minimal polynomial of a transfer function* matrix $H(s)$, $m_H(s)$, was defined as the monic least common denominator of all nonzero first-order minors (entries) of $H(s)$.

**Definition 8.11.** *The* McMillan degree *of $H(s)$ is the degree of $p_H(s)$.*  ∎

The number of poles in $H(s)$, which are defined as the zeros of $p_H(s)$, is equal to the McMillan degree of $H(s)$. The degree of $H(s)$ is in fact the order of any minimal realization of $H(s)$, as the following result shows.

**Theorem 8.12.** *Let $\{A, B, C, D\}$ be a minimal realization of $H(s)$. Then the characteristic polynomial of $H(s), p_H(s)$, is equal to the characteristic polynomial of $A, \alpha(s) \triangleq |sI - A|$; i.e., $p_H(s) = \alpha(s)$. Therefore, the McMillan degree of $H(s)$ equals the order of any minimal realization.*

*Proof.* See [1, p. 397, Chapter 5, Theorem 3.11].  ∎

It can also be shown that the minimal polynomial of $H(s), m_H(s)$, is equal to the minimal polynomial of $A, \alpha_m(s)$, where $\{A, B, C, D\}$ is any controllable and observable realization of $H(s)$. This is illustrated in the following example.

---

**Example 8.13.** Let $H(s) = \begin{bmatrix} 1/s & 2/s \\ 0 & -1/s \end{bmatrix}$. The first-order minors, the entries of $H(s)$, have denominators $s, s$, and $s$, and therefore, $m_H(s) = s$. The only second-order minor is $-1/s^2$ and $p_H(s) = s^2$ with $\deg p_H(s) = 2$. Therefore, the order of a minimal realization is 2. Such a realization is given by $\dot{x} = Ax +$ $Bu$ and $y = Cx$ with $A = \begin{bmatrix} 0 & 0 \\ 0 & 0 \end{bmatrix}$, $B = \begin{bmatrix} 1 & 2 \\ 0 & -1 \end{bmatrix}$, $C = \begin{bmatrix} 1 & 0 \\ 0 & 1 \end{bmatrix}$. We verify first that this system is a realization of $H(s)$ and then that it is controllable and observable and, therefore, minimal. Notice that the characteristic polynomial of $A$ is $\alpha(s) = s^2 = p_H(s)$ and that its minimal polynomial is $\alpha_m(s) = s = m_H(s)$.

---

In the case when $H(s)$ is a scalar, the roots of $m_H = p_H$ are the eigenvalues of any minimal realization of $H(s)$.

**Corollary 8.14.** *Let $H(s) = n(s)/d(s)$ be a scalar proper rational function. If $\{A, B, C, D\}$ is a minimal realization of $H(s)$, then*

$$kd(s) = \alpha(s) = \alpha_m(s), \tag{8.26}$$

*where $\alpha(s) = \det(sI - A)$ and $\alpha_m(s)$ are the characteristic and minimal polynomials of $A$, respectively, and $k$ is a real scalar so that $kd(s)$ is a monic polynomial.*

*Proof.* The characteristic and minimal polynomials of $H(s), p_H(s)$, and $m_H(s)$ are by definition equal to $d(s)$ in the scalar case. Applying Theorem 8.12 proves the result. ∎

### Determination via the Hankel Matrix

There is an alternative way of determining the order of a minimal realization of $H(s)$. This is accomplished via the Hankel matrix, associated with $H(s)$.

Given $H(s)$, we express $H(s)$ as a Laurent series expansion to obtain

$$H(s) = H_0 + \widehat{H}(s) = H_0 + H_1 s^{-1} + H_2 s^{-2} + H_3 s^{-3} + \dots, \tag{8.27}$$

where $\widehat{H}(s)$ is strictly proper and the real $p \times m$ matrices $H_0, H_1, \dots$ are the Markov parameters of the system. They can be determined by the formulas

$$H_0 = \lim_{s \to \infty} H(s), \quad H_1 = \lim_{s \to \infty} s(H(s) - H_0), \quad H_2 = \lim_{s \to \infty} s^2(H(s) - H_0 - H_1 s^{-1}),$$

and so forth.

**Definition 8.15.** *The* Hankel matrix $M_H(i,j)$ *of order $(i,j)$ corresponding to the (Markov parameter) sequence $H_1, H_2, \dots$ is defined as the $ip \times jm$ matrix given by*

$$M_H(i,j) \triangleq \begin{bmatrix} H_1 & H_2 & \cdots & H_j \\ H_2 & H_3 & \cdots & H_{j+1} \\ \vdots & \vdots & & \vdots \\ H_i & H_{i+1} & \cdots & H_{i+j-1} \end{bmatrix}. \tag{8.28}$$

∎

**Theorem 8.16.** *The order of a minimal realization of $H(s)$ is the rank of $M_H(r,r)$ where $r$ is the degree of the least common denominator of the entries of $H(s)$; i.e., $r = \deg m_H(s)$.*

*Proof.* See [1, p. 399, Chapter 5, Theorem 3.13]. ∎

---

**Example 8.17.** Let $H(s) = \begin{bmatrix} \frac{1}{s+1} & \frac{2}{s+1} \\ \frac{-1}{(s+1)(s+2)} & \frac{1}{s+2} \end{bmatrix}$. Here the minimal polynomial is $m_H(s) = (s+1)(s+2)$, and therefore, $r = \deg m_H(s) = 2$. The Hankel matrix $M_H(r,r)$ is then

$$M_H(r,r) = M_H(2,2) = \begin{bmatrix} H_1 & H_2 \\ H_2 & H_3 \end{bmatrix},$$

an $rp \times rm = 4 \times 4$ matrix, and $H_1 = \lim_{s\to\infty} sH(s) = \lim_{s\to\infty} \begin{bmatrix} \frac{s}{s+1} & \frac{2s}{s+1} \\ \frac{-s}{(s+1)(s+2)} & \frac{s}{s+2} \end{bmatrix} = \begin{bmatrix} 1 & 2 \\ 0 & 1 \end{bmatrix}$ and $H_2 = \lim_{s\to\infty} s^2(H(s) - H_1 s^{-1}) = \lim_{s\to\infty} \begin{bmatrix} \frac{s^2}{s+1} - s & \frac{2s^2}{s+1} - 2s \\ \frac{-s^2}{(s+1)(s+2)} & \frac{s^2}{s+2} - s \end{bmatrix} = \lim_{s\to\infty} \begin{bmatrix} \frac{-s}{s+1} & \frac{-2s}{s+1} \\ \frac{-s^2}{(s+1)(s+2)} & \frac{-2s}{s+2} \end{bmatrix} = \begin{bmatrix} -1 & -2 \\ -1 & -2 \end{bmatrix}$. Similarly, $H_3 = \begin{bmatrix} 1 & 2 \\ 3 & 4 \end{bmatrix}$. Now

$$\text{rank}\, M_H(2,2) = \text{rank} \begin{bmatrix} 1 & 2 & -1 & -2 \\ 0 & 1 & -1 & -2 \\ -1 & -2 & 1 & 2 \\ -1 & -2 & 3 & 4 \end{bmatrix} = 3,$$

which is the order of any minimal realization, in view of Theorem 8.16. The reader should verify this result, using Theorem 8.12.

---

**Example 8.18.** Consider the transfer function matrix $H(s) = \begin{bmatrix} 1/s & 2/s \\ 0 & -1/s \end{bmatrix}$, as in Example 8.13. Here $r = \deg m_H(s) = \deg s = 1$. Now, the Hankel matrix $M_H(r,r) = M_H(1,1) = H_1 = \lim_{s\to\infty} sH(s) = \begin{bmatrix} 1 & 2 \\ 0 & -1 \end{bmatrix}$. Its rank is 2, which is the order of a minimal realization of $H(s)$. This agrees with the results in Example 8.13.

---

### 8.3.4 Minimality of Realizations: Discrete-Time Systems

The fact that the results on minimality of realizations in the discrete-time case are essentially identical to the corresponding results for the continuous-time case is not surprising since we are concentrating here on the time-invariant cases for which the transfer function matrices have the same forms: $H(s) = C(sI - A)^{-1}B + D$ and $H(z) = C(zI - A)^{-1}B + D$. Accordingly, the results on how to generate 4-tuples $\{A, B, C, D\}$ to satisfy these relations are, of course, also the same.

## 8.4 Realization Algorithms

In this section, algorithms for generating time-invariant state-space realizations of external system descriptions are introduced. A brief outline of the contents of this section follows.

Realizations of $H(s)$ can often be derived in an easier manner if duality is used, and this is demonstrated first in this section. Realizations of minimal order are both controllable and observable, as was shown in the previous section. To derive a minimal realization of $H(s)$, one typically derives a realization that is controllable (observable) and then extracts the part that is also observable (controllable). This involves in general a two-step procedure. However, in certain cases, a minimal realization can be derived in one step, as for example, when $H(s)$ is a scalar transfer function. Algorithms for realizations in a controller/observer form are discussed first. In the interest of clarity, the SISO case is presented separately, thus providing an introduction to the general MIMO case. Realization algorithms, where A is diagonal, are introduced next. Finally, balanced realizations are addressed.

It is not difficult to see that the above algorithms can also be used to derive realizations described by equations of the form $x(k + 1) = Ax(k) + Bu(k), y(k) = Cx(k) + Du(k)$ of transfer function matrices $H(z)$ for discrete-time time-invariant systems. Accordingly, the discrete-time case will not be treated separately in this section. Additional details, algorithms, and proofs may be found in [1, Section 5.4].

### 8.4.1 Realizations Using Duality

If the system described by the equations $\dot{x} = Ax + Bu$, $y = Cx + Du$ is a realization of $H(s)$, then

$$H(s) = C(sI - A)^{-1}B + D. \tag{8.29}$$

If $\widetilde{H}(s) \triangleq H^T(s)$, then $\dot{\tilde{x}} = \widetilde{A}\tilde{x} + \widetilde{B}\tilde{u}$ and $\tilde{y} = \widetilde{C}\tilde{x} + \widetilde{D}\tilde{u}$, where $\widetilde{A} = A^T, \widetilde{B} = C^T, \widetilde{C} = B^T$, and $\widetilde{D} = D^T$, is a realization of $\widetilde{H}(s)$ since in view of (8.29),

$$\begin{aligned} \widetilde{H}(s) &= H^T(s) \\ &= B^T(sI - A^T)^{-1}C^T + D^T \\ &= \widetilde{C}(sI - \widetilde{A})^{-1}\widetilde{B} + \widetilde{D}. \end{aligned} \tag{8.30}$$

The representation $\{\tilde{A}, \tilde{B}, \tilde{C}, \tilde{D}\}$ is *the dual representation* to $\{A, B, C, D\}$, and if $\{A, B, C, D\}$ is controllable (observable), then $\{\tilde{A}, \tilde{B}, \tilde{C}, \tilde{D}\}$ is observable (controllable) (see Section 5.2.3). In other words, if a controllable (observable) realization $\{A, B, C, D\}$ of the $p \times m$ transfer function matrix $H(s)$ is known, then an observable (controllable) realization of the $m \times p$ transfer function matrix $\tilde{H}(s) = H^T(s)$ can be derived immediately: It is the dual representation, namely, $\{\tilde{A}, \tilde{B}, \tilde{C}, \tilde{D}\} = \{A^T, C^T, B^T, D^T\}$. This fact is used to advantage in deriving realizations in the MIMO case, since obtaining first a realization of $H^T(s)$ instead of $H(s)$ and then using duality leads sometimes to simpler, lower order, realizations.

Duality is very useful in realizations of symmetric transfer functions, which have the property that $H(s) = H^T(s)$, as, e.g., in the case of SISO systems where $H(s)$ is a scalar. Under these conditions, if $\{A, B, C, D\}$ is a controllable (observable) realization of $H(s)$, then $\{A^T, C^T, B^T, D^T\}$ is an observable (controllable) realization of the same $H(s)$. Note that in this case,

$$H(s) = C(sI - A)^{-1}B + D = H^T(s) = B^T(sI - A^T)^{-1}C^T + D^T.$$

In realization algorithms of MIMO systems, a realization that is either controllable or observable is typically obtained first. Next, this realization is reduced to a minimal one by extracting the part of the system that is both controllable and observable, using the methods of Subsection 6.2.1. Dual representations may simplify this process considerably. In the following discussion, we summarize the process of deriving minimal realizations for the reader's convenience.

Given a proper rational $p \times m$ transfer function matrix $H(s)$, with $\lim_{s \to \infty} H(s) < \infty$, we consider the strictly proper part $\hat{H}(s) = H(s) - \lim_{s \to \infty} H(s) = H(s) - D$ [noting that working with $\hat{H}(s)$ instead of $H(s)$ is optional].

1. If a realization algorithm leading to a controllable realization is used, then the following steps are taken:

$$\hat{H}(s) \to (\tilde{H}(s) = \hat{H}^T(s)) \to \{\tilde{A}, \tilde{B}, \tilde{C}\} \to \{A = \tilde{A}^T, B = \tilde{C}^T, C = \tilde{B}^T\},$$
$$(8.31a)$$

where $\{\tilde{A}, \tilde{B}, \tilde{C}\}$ is a controllable realization of $\tilde{H}(s)$ and $\{A, B, C\}$ is an observable realization of $\hat{H}(s)$.

2. To obtain a minimal realization,

$$\{A, B, C\} \to \left\{ \begin{bmatrix} A_1 & A_{12} \\ 0 & A_2 \end{bmatrix}, \begin{bmatrix} B_1 \\ 0 \end{bmatrix}, [C_1, C_2] \right\},
\qquad (8.31b)$$

where $\{A, B, C\}$ is an observable realization of $\hat{H}(s)$ obtained from step (1), and $(A_1, B_1)$ is controllable (derived by using the method of Subsection 6.2.1), then $\{A_1, B_1, C_1\}$ is a controllable and observable, and therefore, a minimal realization of $\hat{H}(s)$, and furthermore, $\{A_1, B_1, C_1, D\}$, is a minimal realization of $H(s)$.

## 8.4.2 Realizations in Controller/Observer Form

We shall first consider realizations of scalar transfer functions $H(s)$.

### Single-Input/Single-Output (SISO) Systems ($p = m = 1$)

Let

$$H(s) = \frac{n(s)}{d(s)} = \frac{b_n s^n + \cdots + b_1 s + b_0}{s^n + a_{n-1} s^{n-1} + \cdots + a_1 s + a_0}, \quad (8.32)$$

where $n(s)$ and $d(s)$ are prime polynomials. This is the general form of a proper transfer function of (McMillan) degree $n$. Note that if the leading coefficient in the numerator $n(s)$ is zero, i.e., $b_n = 0$, then $H(s)$ is strictly proper. Also, recall that

$$y^{(n)} + a_{n-1} y^{(n-1)} + \cdots + a_1 y^{(1)} + a_0 y = b_n u^{(n)} + \cdots + b_1 u^{(1)} + b_0 u \quad (8.33a)$$

or

$$
\begin{aligned}
d(q)y(t) &= (q^n + a_{n-1}q^{n-1} + \cdots + a_1 q + a_0)y(t) \\
&= (b_n q^n + \ldots b_1 q + b_0)u(t) = n(q)u(t),
\end{aligned}
\quad (8.33b)
$$

where $q \triangleq d/dt$, the differential operator. This is the corresponding $n$th-order differential equation that directly gives rise to the map $\hat{y}(s) = H(s)\hat{u}(s)$ if the Laplace transform of both sides is taken, assuming that all variables and their derivatives are zero at $t = 0$.

*Controller Form Realizations*

Given $n(s)$ and $d(s)$, we proceed as follows to derive a realization in controller form.

1. Determine $C_c^T \in R^n$ and $D_c \in R$ so that

$$n(s) = C_c S(s) + D_c d(s), \quad (8.34)$$

where $S(s) \triangleq [1, s, \ldots, s^{n-1}]^T$ is an $n \times 1$ vector of polynomials. Equation (8.34) implies that

$$D_c = \lim_{s \to \infty} H(s) = b_n. \quad (8.35)$$

Then $n(s) - b_n d(s)$ is in general a polynomial of degree $n - 1$, which shows that a real vector $C_c$ that satisfies (8.34) always exists.

If $b_n = 0$, i.e., if $H(s)$ is strictly proper, then from (8.34) we obtain $C_c = [b_0, \ldots, b_{n-1}]$; i.e., $C_c$ consists of the coefficients of the $n - 1$ degree numerator.

If $b_n \neq 0$, then (8.34) implies that the entries of $C_c$ are a combination of the coefficients $b_i$ and $a_i$. In particular,

$$C_c = [b_0 - b_n a_0, b_1 - b_n a_1, \ldots, b_{n-1} - b_n a_{n-1}]. \quad (8.36)$$

2. A realization of $H(s)$ in controller form is given by the equations

$$\dot{x}_c = A_c x_c + B_c u = \begin{bmatrix} 0 & 1 & \cdots & 0 \\ \vdots & \vdots & \ddots & \vdots \\ 0 & 0 & \cdots & 1 \\ -a_0 & -a_1 & \cdots & -a_{n-1} \end{bmatrix} x_c + \begin{bmatrix} 0 \\ 0 \\ \vdots \\ 1 \end{bmatrix} u,$$

$$y = C_c x_c + D_c u. \tag{8.37}$$

The $n$ states of the realization in (8.37) are related by $x_{i+1} = \dot{x}_i$, $i = 1, \ldots, n-1$, or $x_{i+1} = x_1^{(i)}$, $i = 1, \ldots, n-1$, and $\dot{x}_n = -a_0 x_1 - \sum_{i=1}^{n-1} a_i x_{i+1} + u = -a_0 x_1 - \sum_{i=1}^{n-1} a_i x_1^{(i)} + u$. It can now be shown that $x_1$ satisfies the relationship

$$d(q)x_1(t) = u(t), \quad y(t) = n(q)x_1(t), \tag{8.38}$$

where $q \triangleq d/dt$, the differential operator. Note that $d(q)x_1(t) = u(t)$ because $\dot{x}_n = -\sum_{i=0}^{n-1} a_i x_1^{(i)} + x_1^{(n)} + u = -d(q)x_1 + u + x_1^{(n)}$, which in view of $\dot{x}_n = x_1^{(n)}$, derived from $x_n = x_1^{(n-1)}$, implies that $-d(q)x_i + u = 0$. The relation $y(t) = n(q)x_1(t)$ can easily be verified by multiplying both sides of $n(q) = C_c S(q) + D_c d(q)$ given in (8.34) by $x_1$.

**Lemma 8.19.** *The representation (8.37) is a minimal realization of $H(s)$ given in (8.32).*

*Proof.* We must first show that (8.37) is indeed a realization, i.e., that it satisfies (8.29). This is of course true in view of the Structure Theorem in Subsection 6.4.1. Presently, this will be shown directly, using (8.38).

Relation $d(q)x_1(t) = u(t)$ implies that $\hat{x}_1(s) = (d(s))^{-1}\hat{u}(s)$. This yields for the state that $\hat{x}(s) = [\hat{x}_1(s), \ldots, \hat{x}_n(s)]^T = [1, s, \ldots, s^{n-1}]^T \hat{x}_1(s) = S(s)(d(s))^{-1}\hat{u}(s)$. However, we also have $\hat{x}(s) = (sI - A_c)^{-1} B_c \hat{u}(s)$. Therefore,

$$(sI - A_c)S(s) = B_c d(s). \tag{8.39}$$

Now $C_c(sI - A_c)^{-1}B_c + D_c = C_c S(s)(d(s))^{-1} + D_c = (C_c S(s) + D_c d(s))(d(s))^{-1} = \frac{n(s)}{d(s)} = H(s)$; i.e., (8.37) is indeed a realization.

System (8.37) is of order $n$ and is therefore, a minimal, controllable, and observable realization. This is because the degree of $H(s)$ is $n$, which in view of Theorem 8.12, is the order of any minimal realization. Controllability and observability can also be established directly by forming the controllability and observability matrices. The reader is encouraged to pursue this approach. ∎

According to Definition 8.11, the McMillan degree of a rational scalar transfer function $H(s) = n(s)/d(s)$ is the degree of $d(s)$ only when $n(s)$ and $d(s)$ are prime polynomials; if they are not, all cancellations must first take place before the degree can be determined. If $n(s)$ and $d(s)$ are not prime, then the above algorithm will yield a realization that is not observable. Notice that realization (8.37) is always controllable, since it is in controller form. This can also be seen directly from the expression

$$[B_c, A_c B_c, \ldots, A_c^{n-1} B_c] = \begin{bmatrix} 0 & 0 & \cdots & 1 \\ \vdots & \vdots & & \vdots \\ 0 & 1 & \cdots & \times \\ 1 & \times & \cdots & \times \end{bmatrix}, \tag{8.40}$$

which is of full rank. The realization (8.37) is observable if and only if the polynomials $d(s)$ and $n(s)$ are prime.

In Figure 8.2 a block realization diagram of the form (8.37) for a second-order transfer function is shown. Note that the states $x_1(t)$ and $x_2(t)$ are taken to be the voltages at the outputs of the integrators.

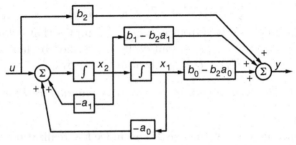

**Figure 8.2.** Block realization of $H(s)$ in controller form of the system $\begin{bmatrix} \dot{x}_1 \\ \dot{x}_2 \end{bmatrix} = \begin{bmatrix} 0 & 1 \\ -a_0 & -a_1 \end{bmatrix} \begin{bmatrix} x_1 \\ x_2 \end{bmatrix} + \begin{bmatrix} 0 \\ 1 \end{bmatrix} u$, $y = [b_0 - b_2 a_0, b_1 - b_2 a_1] \begin{bmatrix} x_1 \\ x_2 \end{bmatrix} + b_2 u$; $H(s) = \frac{b_2 s^2 + b_1 s + b_0}{s^2 + a_1 s + a_0}$

*Observer Form Realizations*

Given the transfer function (8.32), the $n$th-order realization in observer form is given by

$$\dot{x}_o = A_o x_o + B_o u$$

$$= \begin{bmatrix} 0 & \cdots & 0 & -a_0 \\ 1 & & 0 & -a_1 \\ \vdots & \ddots & \vdots & \vdots \\ 0 & \cdots & 1 & -a_{n-1} \end{bmatrix} x_o + \begin{bmatrix} b_0 - b_n a_0 \\ b_1 - b_n a_1 \\ \vdots \\ b_{n-1} - b_n a_{n-1} \end{bmatrix} u,$$

$$y = C_o x_o + D_o u = [0, 0, \ldots, 0, 1] x_o + b_n u. \tag{8.41}$$

This realization was derived by taking the dual of realization (8.37). Notice that $A_o = A_c^T, B_o = C_c^T, C_o = B_c^T$, and $D_o = D_c^T$.

**Lemma 8.20.** *The representation (8.41) is a minimal realization of $H(s)$ given in (8.32).*

*Proof.* Note that the observer form realization $\{A_o, B_o, C_o, D_o\}$ described by (8.41) is the dual of the controller form realization $\{A_c, B_c, C_c, D_c\}$ described by (8.37), used in Lemma 8.19. ∎

The realization (8.41) can also be derived directly from $H(s)$, using defining relations similar to (8.34). In particular, $B_o$ and $D_o$ can be determined from the expression [see Subsection 6.4.2]

$$n(s) = \widetilde{S}(s)B_o + d(s)D_o, \tag{8.42}$$

where $\widetilde{S}(s) = [1, s, \ldots, s^{n-1}]$.

It can be shown (by taking transposes) that the corresponding relation to (8.39) is now given by

$$\widetilde{S}(s)(sI - A_o) = d(s)C_o \tag{8.43}$$

and that

$$d(q)z(t) = n(q)u(t), \quad y(t) = z(t) \tag{8.44}$$

corresponds to (8.38).

Figure 8.3 depicts a block realization diagram of the form (8.41) for a second-order transfer function.

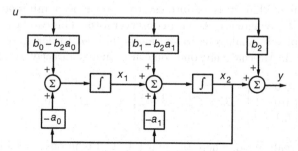

**Figure 8.3.** Block realization of $H(s)$ in observer form of the system $\begin{bmatrix} \dot{x}_1 \\ \dot{x}_2 \end{bmatrix} = \begin{bmatrix} 0 & -a_0 \\ 1 & -a_1 \end{bmatrix} \begin{bmatrix} x_1 \\ x_2 \end{bmatrix} + \begin{bmatrix} b_0 - b_2 a_0 \\ b_1 - b_2 a_1 \end{bmatrix} u$, $y = [0, \ 1]\begin{bmatrix} x_1 \\ x_2 \end{bmatrix} + b_2 u$; $H(s) = \frac{b_2 s^2 + b_1 s + b_0}{s^2 + a_1 s + a_0}$.

---

**Example 8.21.** We wish to derive a minimal realization for the transfer function $H(s) = \frac{s^3 + s - 1}{s^3 + 2s^2 - s - 2}$. Consider a realization $\{A_c, B_c, C_c, D_c\}$, where $(A_c, B_c)$ is in controller form. In view of (8.34) to (8.37), $D_c = \lim_{s \to \infty} H(s) = 1$ and $n(s) = s^3 + s - 1 = C_c S(s) + D_c d(s)$, from which we have

$C_c S(s) = (s^3 + s - 1) - (s^3 + 2s^2 - s - 2) = -2s^2 + 2s + 1 = [1, 2, -2][1, s, s^2]^T$.

Therefore, a realization of $H(s)$ is $\dot{x}_c = A_c x_c + B_c u$, $y = C_c x_c + D_c u$, where

$$A_c = \begin{bmatrix} 0 & 1 & 0 \\ 0 & 0 & 1 \\ 2 & 1 & -2 \end{bmatrix}, \quad B_c = \begin{bmatrix} 0 \\ 0 \\ 1 \end{bmatrix}, \quad C_c = [1, 2, -2], \quad D_c = 1.$$

This is a minimal realization. Instead of solving $n(s) = C_c S(s) + D_c d(s)$ for $C_c$ as was done above, it is possible to derive $C_c$ by inspection after $H(s)$ is written as

$$H(s) = \widehat{H}(s) + \lim_{s \to \infty} H(s) = \widehat{H}(s) + D_c, \tag{8.45}$$

where $\widehat{H}(s)$ is now strictly proper. Notice that if $H(s)$ is given by (8.32), then $D_c = b_n$ and

$$\widehat{H}(s) = \frac{c_{n-1} s^{n-1} + \cdots + c_1 s + c_0}{s^n + a_{n-1} s^{n-1} + \cdots + a_1 s + a_0}, \tag{8.46}$$

where in fact, $c_i = b_i - b_n a_i$, $i = 0, \ldots, n-1$. The realization $\{A_c, B_c, C_c\}$ of $\widehat{H}(s)$ has $(A_c, B_c)$ precisely the same as before; however, $C_c$ can now be written directly as

$$C_c = [c_0, c_1, \ldots, c_{n-1}]; \tag{8.47}$$

i.e., given $H(s)$ there are three ways of determining $C_c$: (i) using formula (8.36), (ii) solving $C_c S(s) = n(s) - D_c d(s)$ as in (8.34), and (iii) calculating $\widehat{H}(s) = H(s) - \lim_{s \to \infty} H(s)$. The reader should verify that for this example, (i) and (iii) yield the same $C_c = [1, 2, -2]$ as in method (ii).

Suppose now that it is of interest to determine a minimal realization $\{A_o, B_o, C_o, D_o\}$, where $(A_o, C_o)$ is in observer form. This can be accomplished in ways completely analogous to the methods used to derive realizations in controller form. Alternatively, one could use duality directly and show that

$$A_o = A_c^T = \begin{bmatrix} 0 & 0 & 2 \\ 1 & 0 & 1 \\ 0 & 1 & -2 \end{bmatrix}, B_o = C_c^T = \begin{bmatrix} 1 \\ 2 \\ -2 \end{bmatrix}, C_o = B_c^T = [0, 0, 1], D_o = D_c^T = 1$$

is a minimal realization, where the pair $(A_o, C_o)$ is in observer form.

---

**Example 8.22.** Consider now the transfer function $H(s) = \frac{s^3 - 1}{s^3 + 2s^2 - s - 2}$, where the numerator is $n(s) = s^3 - 1$ instead of $s^3 + s - 1$, as in Example 8.21. We wish to derive a minimal realization of $H(s)$. Using the same procedure as in the previous example, it is not difficult to derive the realization

$$A_c = \begin{bmatrix} 0 & 1 & 0 \\ 0 & 0 & 1 \\ 2 & 1 & -2 \end{bmatrix}, \quad B_c = \begin{bmatrix} 0 \\ 0 \\ 1 \end{bmatrix}, \quad C_c = [1, 1, -2], \quad D_c = 1.$$

This realization is controllable, since $(A_c, B_c)$ is in controller form (see Exercise 8.4); however, it is not observable, since $\text{rank}\,\mathcal{O} = 2 < 3 = n$, where $\mathcal{O}$ denotes the observability matrix given by

$$\mathcal{O} = \begin{bmatrix} C_c \\ C_c A_c \\ C_c A_c^2 \end{bmatrix} = \begin{bmatrix} 1 & 1 & -2 \\ -4 & -1 & 5 \\ 10 & 1 & -11 \end{bmatrix}.$$

Therefore, the above matrix is not a minimal realization. This has occurred because the numerator and denominator of $H(s)$ are not prime polynomials; i.e., $s - 1$ is a common factor. Thus, strictly speaking, the $H(s)$ given above is not a transfer function, since it is assumed that in a transfer function all cancellations of common factors have taken place. (See also the discussion following Lemma 8.19.) Correspondingly, if the algorithm for deriving an observer form would be applied to the present case, the realization $\{A_o, B_o, C_o, D_o\}$ would be an observable realization, but not a controllable one, and would therefore not be a minimal realization.

To obtain a minimal realization of the above transfer function $H(s)$, one could either extract the part of the controllable realization $\{A_c, B_c, C_c, D_c\}$ that is also observable or simply cancel the factor $s - 1$ in $H(s)$ and apply the algorithm again. The former approach of reducing a controllable realization will be illustrated when discussing the MIMO case. The latter approach is perhaps the easiest one to apply in this case. We have

$$H(s) = \frac{s^3 - 1}{s^3 + 2s^2 - s - 2} = \frac{s^2 + s + 1}{s^2 + 3s + 2} = \frac{-2s - 1}{s^2 + 3s + 2} + 1,$$

and a minimal realization of this is then determined as

$$A_c = \begin{bmatrix} 0 & 1 \\ -2 & -3 \end{bmatrix}, \quad B_c = \begin{bmatrix} 0 \\ 1 \end{bmatrix}, \quad C_c = [-1, \ -2], \quad D_c = 1.$$

---

## Multi-Input/Multi-Output (MIMO) Systems ($pm > 1$)

Let a $(p \times m)$ proper rational matrix $H(s)$ be given with $\lim_{s \to \infty} H(s) < \infty$. We now present alogrithms to obtain realizations $\{A_c, B_c, C_c, D_c\}$ of $H(s)$ in controller form and realizations $\{A_o, B_o, C_o, D_o\}$ of $H(s)$ in observer form. Minimal realizations can then be obtained by separating the observable (controllable) part of the controllable (observable) realization.

*Controller Form Realizations*

Consider a transfer function matrix $H(s) = [n_{ij}(s)/d_{ij}(s)]$, $i = 1, \ldots, p$, $j = 1, \ldots, m$, and let $\ell_j(s)$ denote the (monic) least common denominator of all entries in the $j$th column of $H(s)$. The $\ell_j(s)$ is the least degree polynomial divisible by all $d_{ij}(s)$, $i = 1, \ldots, p$. Then $H(s)$ can be written as

$$H(s) = N(s)D^{-1}(s), \tag{8.48}$$

a ratio of two polynomial matrices, where $N(s) \triangleq [\bar{n}_{ij}(s)]$ and $D(s) \triangleq \mathrm{diag}[\ell_1(s), \ldots, \ell_m(s)]$. Note that $\bar{n}_{ij}(s)/\ell_j(s) = n_{ij}(s)/d_{ij}(s)$ for $i = 1, \ldots, p$, and all $j = 1, \ldots, m$. Let $d_j \triangleq \deg \ell_j(s)$, and assume that $d_j \geq 1$. Define

$$\Lambda(s) \triangleq \mathrm{diag}(s^{d_1}, \ldots, s^{d_m})$$

and

$$S(s) \triangleq \text{block diag}\left( \begin{bmatrix} 1 \\ s \\ \vdots \\ s^{d_j - 1} \end{bmatrix} \; j = 1, \ldots, m \right), \tag{8.49}$$

and note that $S(s)$ is an $n \left( \triangleq \sum_{j=1}^m d_j \right) \times m$ polynomial matrix. Write

$$D(s) = D_h \Lambda(s) + D_\ell S(s), \tag{8.50}$$

and note that $D_h$ is the highest column degree coefficient matrix of $D(s)$. Here $D(s)$ is diagonal with monic polynomial entries, and therefore, $D_h = I_m$. If, for example, $D(s) = \begin{bmatrix} 3s^2 + 1 & 2s \\ 2s & s \end{bmatrix}$, then the highest column degree coefficient matrix $D_h = \begin{bmatrix} 3 & 2 \\ 0 & 1 \end{bmatrix}$, and $D_\ell S(s)$ given in (8.50) accounts for the remaining lower column degree terms in $D(s)$, with $D_\ell$ being a matrix of coefficients.

Observe that $|D_h| \neq 0$, and define the $m \times m$ and $m \times n$ matrices

$$B_m = D_h^{-1}, \quad A_m = -D_h^{-1} D_\ell, \tag{8.51}$$

respectively. Also, determine $C_c$ and $D_c$ such that

$$N(s) = C_c S(s) + D_c D(s), \tag{8.52}$$

and note that

$$D_c = \lim_{s \to \infty} H(s). \tag{8.53}$$

We have $H(s) = N(s)D^{-1}(s) = C_c S(s)D^{-1}(s) + D_c$ with $C_c S(s)D^{-1}(s)$ being strictly proper (show this). Therefore, only $C_c$ needs to be determined from (8.52).

A controllable realization of $H(s)$ in controller form is now given by the equations

$$\dot{x}_c = A_c x_c + B_c u, \quad y = C_c x_c + D_c u.$$

Here $C_c$ and $D_c$ were defined in (8.52) and (8.53), respectively,

$$A_c = \bar{A}_c + \bar{B}_c A_m, \quad B_c = \bar{B}_c B_m, \tag{8.54}$$

where $\bar{A}_c = \text{block diag}[A_1, A_2 \dots, A_m]$ with

$$A_j = \begin{bmatrix} 0 & \\ \vdots & I_{d_j-1} \\ 0\,0\cdots0 & \end{bmatrix} \in R^{d_j \times d_j},$$

$$\bar{B}_c = \text{block diag}\left( \begin{bmatrix} 0 \\ \vdots \\ 0 \\ 1 \end{bmatrix} \in R^{d_j}, \; j = 1,\dots,m \right),$$

and $A_m, B_m$ were defined in (8.51). Note that if $d_j = \mu_j$, $j = 1,\dots,m$, the controllability indices, then (8.54) is precisely the relation (6.56) of Section 6.4.

**Lemma 8.23.** *The system $\{A_c, B_c, C_c, D_c\}$ is an $n(= \sum_{j=1}^m d_j)$-th-order controllable realization of $H(s)$ with $(A_c, B_c)$ in controller form.*

*Proof.* First, to show that $\{A_c, B_c, C_c, D_c\}$ is a realization of $H(s)$, we note that in view of the Structure Theorem given in Subsection 6.4.1, we have $C_c(sI - A_c)^{-1}B_c + D_c = \bar{N}(s)\bar{D}(s)^{-1}$, where

$$\bar{D}(s) \triangleq B_m^{-1}[\Lambda(s) - A_m S(s)], \quad \bar{N}(s) \triangleq C_c S(s) + D_c D(s).$$

However, $\bar{D}(s) = D(s)$ and $\bar{N}(s) = N(s)$, in view of (8.50) to (8.52). Therefore, $C_c(sI - A_c)^{-1}B_c + D_c = N(s)D^{-1}(s) = H(s)$, in view of (8.48).

It is now shown that $(A_c, B_c)$ is controllable. We write

$$[sI - A_c, B_c] = [sI - \bar{A}_c - \bar{B}_c A_m, \bar{B}_c B_m]$$

$$= [sI - \bar{A}_c, \bar{B}_c] \begin{bmatrix} I & 0 \\ -A_m & B_m \end{bmatrix} \tag{8.55}$$

and notice that $\text{rank}[s_j I - \bar{A}_c, \bar{B}_c] = n$ for any complex $s_j$. This is so because of the special form of $\bar{A}_c, \bar{B}_c$. (This is, in fact, the Brunovski canonical form.) Now since $|B_m| \neq 0$, Sylvester's Rank Inequality implies that $\text{rank}[s_j I - A_c, B_c] = n$ for any complex $s_j$, which in view of Section 6.3 implies that $(A_c, B_c)$ is controllable. In addition, since $B_m = I_m$, it follows that $(A_c, B_c)$ is of the form (6.55) of Section 6.4. With $d_j = \mu_i$, the pair $(A_c, B_c)$ is in controller form. ∎

An alternative way of determining $C_c$ is to first write $H(s)$ in the form

$$H(s) = \widehat{H}(s) + \lim_{s \to \infty} H(s) = \widehat{H}(s) + D_c, \tag{8.56}$$

where $\widehat{H}(s) \triangleq H(s) - D_c$ is strictly proper. Now applying the above algorithm to $\widehat{H}(s)$, one obtains $\widehat{H}(s) = \widehat{N}(s)D^{-1}(s)$, where $D(s)$ is precisely equal to the expression given in (8.50). We note, however, that $\widehat{N}(s)$ is different. In fact, $\widehat{N}(s) = N(s) - D_c D(s)$. In view of (8.52) the matrix $C_c$ is now found to be of the form

$$\widehat{N}(s) = C_c S(s). \tag{8.57}$$

Note that this is a generalization of the scalar case discussed in Example 8.21 [see (8.45) to (8.47)].

In the above algorithm the assumption that $d_j \geq 1$ for all $j = 1, \ldots, m$, was made. If for some $j$, $d_j = 0$, this would mean that the $j$th column of $H(s)$ will be a real $m \times 1$ vector that will be equal to the $j$th column of $D_c$ [recall that $D_c = \lim_{s \to \infty} H(s)$]. The strictly proper $\widehat{H}(s)$ in (8.56) will then have its $j$th column equal to zero, and this zero column can be generated by a realization where the $j$th column of $B_c$ is set to zero. Therefore, the zero column (the $j$th column) of $\widehat{H}(s)$ is ignored in this case and the algorithm is applied to obtain a controllable realization. A zero column is then added to $B_c$. (See Example 8.26 below.)

*Observer Form Realizations*

These realizations are dual to the controller form realizations and can be obtained by duality arguments. In the following discussion, observer form realizations are obtained directly for completeness of exposition.

We consider the transfer function matrix $H(s) = [n_{ij}(s)/d_{ij}(s)]$, $i = 1, \ldots, p$, $j = 1, \ldots, m$, and let $\tilde{\ell}_i(s)$ be the (monic) least common denominator of all entries in the $i$th row of $H(s)$. Then $H(s)$ can be written as

$$H(s) = \tilde{D}^{-1}(s)\tilde{N}(s), \tag{8.58}$$

where $\tilde{N}(s) \triangleq [\bar{n}_{ij}(s)]$ and $\tilde{D}(s) \triangleq \text{diag}[\tilde{\ell}_1(s), \ldots, \tilde{\ell}_p(s)]$. Note that $\bar{n}_{ij}(s)/\tilde{\ell}_i(s) = n_{ij}(s)/d_{ij}(s)$ for $j = 1, \ldots, m$, and all $i = 1, \ldots, p$.

Let $\tilde{d}_i \triangleq \deg \ell_i(s)$, assume that $\tilde{d}_i \geq 1$, define

$$\tilde{\Lambda}(s) \triangleq \text{diag}(s^{\tilde{d}_1}, \ldots, s^{\tilde{d}_p}), \tilde{S}(s) \triangleq \text{block diag}([1, s, \ldots, s^{\tilde{d}_i - 1}], i = 1, \ldots, p), \tag{8.59}$$

and note that $\tilde{S}(s)$ is a $p \times n (\triangleq \sum_{i=1}^{p} \tilde{d}_i)$ polynomial matrix. Now, write

$$\tilde{D}(s) = \tilde{\Lambda}(s)\tilde{D}_h + \tilde{S}(s)\tilde{D}_\ell \tag{8.60}$$

and note that $\tilde{D}_h$ is the highest row degree coefficient matrix of $\tilde{D}(s)$. Note that $\tilde{D}(s)$ is diagonal, with entries monic polynomials, so that $\tilde{D}_h = I_p$, the $p \times p$ identity matrix. If for example, $\tilde{D}(s) = \begin{bmatrix} 3s^2 + 1 & 2s \\ 2s & s \end{bmatrix}$, then the highest row degree coefficient matrix is $\tilde{D}_h = \begin{bmatrix} 3 & 0 \\ 2 & 1 \end{bmatrix}$ and $\tilde{S}(s)\tilde{D}_\ell$ in (8.60) accounts for the remaining lower row degree terms of $\tilde{D}(s)$, with $\tilde{D}_\ell$ a matrix of coefficients.

Observe that $|\widetilde{D}_h| \neq 0$, in fact $\widetilde{D}_h = I_p$. Define the $p \times p$ and $n \times p$ matrices

$$C_p = \widetilde{D}_h^{-1} \quad \text{and} \quad A_p = -\widetilde{D}_\ell \widetilde{D}_h^{-1}, \tag{8.61}$$

respectively. Also, determine $B_o$ and $D_o$ such that

$$\widetilde{N}(s) = \widetilde{S}(s)B_o + \widetilde{D}(s)D_o. \tag{8.62}$$

Note that

$$D_o = \lim_{s \to \infty} H(s), \tag{8.63}$$

and therefore, only $B_o$ needs to be determined from (8.62).

An observable realization of $H(s)$ in observer form is now given by

$$\dot{x}_o = A_o x_o + B_o u, \quad y = C_o x_o + D_o u,$$

where $B_o$ and $D_o$ were defined in (8.62) and (8.63), respectively, and

$$A_o = \bar{A}_o + A_p \bar{C}_o, \quad C_o = C_p \bar{C}_o, \tag{8.64}$$

where $\bar{A}_o = \text{block diag}[A_1, A_2, \dots, A_p]$ with

$$A_i = \begin{bmatrix} 0 \cdots 0 & 0 \\ & & 0 \\ I_{\tilde{d}_i - 1} & & \vdots \\ & & 0 \end{bmatrix} \in R^{\tilde{d}_i \times \tilde{d}_i},$$

$\bar{C}_o = \text{block diag}([0, \dots, 0, 1] \in R^{1 \times \tilde{d}_i} \ i = 1, \dots, p)$, and $A_p, C_p$ is defined in (8.61). Note that (8.64) is exactly relation (6.76) of Section 6.4 if $\tilde{d}_i = \nu_i$, $i = 1, \dots, p$, the observability indices.

**Lemma 8.24.** *The system $\{A_o, B_o, C_o, D_o\}$ is an $n(\triangleq \sum_{i=1}^p \tilde{d}_i)$-th-order observable realization of $H(s)$ with $(A_o, C_o)$ in observer form.*

*Proof.* This is the dual result to Lemma 8.23. The proof is completely analogous and is omitted. ∎

We conclude by noting that results dual to the results discussed after Lemma 8.23 are also valid here, i.e., results involving (i) a strictly proper $\widehat{H}(s)$, (ii) an $H(s)$ with $\tilde{d}_i = 0$ for some row $i$, and (iii) $H(s) = \widetilde{D}^{-1}(s)\widetilde{N}(s)$, where $\widetilde{D}(s), \widetilde{N}(s)$ are not necessarily determined using (8.58) (refer to the following examples).

---

***Example 8.25.*** Let $H(s) = \left[\frac{s^2+1}{s^2}, \frac{s+1}{s^3}\right]$. We wish to derive a minimal realization for $H(s)$. To this end we consider realizations $\{A_c, B_c, C_c, D_c\}$, where $(A_c, B_c)$ is in controller form. Here $\ell_1(s) = s^2, \ell_2(s) = s^3$ and $H(s)$ can therefore be written in the form (8.48) as

$$H(s) = N(s)D^{-1}(s) = [s^2 + 1, s + 1] \begin{bmatrix} s^2 & 0 \\ 0 & s^3 \end{bmatrix}^{-1}.$$

Here $d_1 = 2$, $d_2 = 3$ and $\Lambda(s) = \begin{bmatrix} s^2 & 0 \\ 0 & s^3 \end{bmatrix} 4$, $S(s) = \begin{bmatrix} 1\ s\ 0\ 0\ 0 \\ 0\ 0\ 1\ s\ s^2 \end{bmatrix}^T$. Note that $n = d_1 + d_2 = 5$, and therefore, the realization will be of order 5. Write $D(s) = D_h \Lambda(s) + D_\ell S(s)$, and note that $D_h = I_2$, $D_\ell = \begin{bmatrix} 0\ 0\ 0\ 0\ 0 \\ 0\ 0\ 0\ 0\ 0 \end{bmatrix}$. Therefore, in view of (8.51),

$$B_m = \begin{bmatrix} 1 & 0 \\ 0 & 1 \end{bmatrix} \quad \text{and} \quad A_m = -D_\ell = \begin{bmatrix} 0\ 0\ 0\ 0\ 0 \\ 0\ 0\ 0\ 0\ 0 \end{bmatrix}.$$

Here $D_c = \lim_{s \to \infty} H(s) = [1, 0]$ and (8.52) implies that $C_c S(s) = N(s) - D_c D(s) = [s^2 + 1, s + 1] - [s^2, 0] = [1, s + 1]$, from which we have $C_c = [1, 0, 1, 1, 0]$. A controllable realization in controller form is therefore given by $\dot{x} = A_c x_c + B_c u$ and $y = C_c x_c + D_c u$, where

$$A_c = \begin{bmatrix} 0 & 1 & 0 & 0 & 0 \\ 0 & 0 & 0 & 0 & 0 \\ 0 & 0 & 0 & 1 & 0 \\ 0 & 0 & 0 & 0 & 1 \\ 0 & 0 & 0 & 0 & 0 \end{bmatrix}, \quad B_c = \begin{bmatrix} 0 & 0 \\ 1 & 0 \\ 0 & 0 \\ 0 & 0 \\ 0 & 1 \end{bmatrix},$$

$$C_c = [1,\ 0,\ 1,\ 1,\ 0], \quad \text{and} \quad D_c = [1, 0].$$

Note that the characteristic (pole) polynomial of $H(s)$ is $s^3$ and that the McMillan degree of $H(s)$ is 3. The order of any minimal realization of $H(s)$ is therefore 3. This implies that the controllable fifth-order realization derived above cannot be observable [verify that $(A_c, C_c)$ is not observable]. To derive a minimal realization, the observable part of the system $\{A_c, B_c, C_c, D_c\}$ needs to be extracted, using the method described in Section 6.2. In particular, a transformation matrix $P$ needs to be determined so that

$$\hat{A} = P A_c P^{-1} = \begin{bmatrix} A_1 & 0 \\ A_{21} & A_2 \end{bmatrix} \quad \text{and} \quad \hat{C} = C_c P^{-1} = [C_1,\ 0],$$

where $(A_1, C_1)$ is observable. If $\hat{B} = P B_c = \begin{bmatrix} B_1 \\ B_2 \end{bmatrix}$, then $\{A_1, B_1, C_1, D_1\}$ is a minimal realization of $H(s)$. To reduce $(A_c, C_c)$ to such a standard form for unobservable systems, we let $A_D = A^T$, $B_D = C_c^T$, and $C_D = B_c^T$ and we reduce $(A_D, B_D)$ to a standard form for uncontrollable systems. Here the controllability matrix is

$$\mathcal{C}_D = \begin{bmatrix} 1 & 0 & 0 & 0 & 0 \\ 0 & 1 & 0 & 0 & 0 \\ 1 & 0 & 0 & 0 & 0 \\ 1 & 1 & 0 & 0 & 0 \\ 0 & 1 & 1 & 0 & 0 \end{bmatrix}.$$

Note that $\operatorname{rank} \mathcal{C}_D = 3$. Now if the first three columns of $Q_D = P_D^{-1}$ are taken to be the first three linearly independent columns of $\mathcal{C}_D$, whereas the rest are chosen so that $|Q_D| \neq 0$, then

$$Q_D = \begin{bmatrix} 1 & 0 & 0 & 0 & 1 \\ 0 & 1 & 0 & 0 & 0 \\ 1 & 0 & 0 & 1 & 0 \\ 1 & 1 & 0 & 0 & 0 \\ 0 & 1 & 1 & 0 & 0 \end{bmatrix} \quad \text{and} \quad Q_D^{-1} = \begin{bmatrix} 0 & -1 & 0 & 1 & 0 \\ 0 & 1 & 0 & 0 & 0 \\ 0 & -1 & 0 & 0 & 1 \\ 0 & 1 & 1 & -1 & 0 \\ 1 & 1 & 0 & -1 & 0 \end{bmatrix}.$$

This implies that

$$\hat{A}_D = Q_D^{-1} A_D Q_D = \begin{bmatrix} A_{D1} & A_{D12} \\ 0 & A_{D2} \end{bmatrix} = \begin{bmatrix} 0 & 0 & 0 & 1 & -1 \\ 1 & 0 & 0 & 0 & 1 \\ 0 & 1 & 0 & 0 & -1 \\ 0 & 0 & 0 & -1 & 1 \\ 0 & 0 & 0 & -1 & 1 \end{bmatrix},$$

$$\hat{B}_D = Q_D^{-1} B_D = \begin{bmatrix} B_{D1} \\ B_{D2} \end{bmatrix} = \begin{bmatrix} 1 \\ 0 \\ 0 \\ 0 \\ 0 \end{bmatrix}, \quad \text{and} \quad \hat{C}_D = C_D Q_D = \begin{bmatrix} 0 & 1 & 0 & 0 & 0 \\ 0 & 1 & 1 & 0 & 0 \end{bmatrix}.$$

Then

$$\hat{A} = \begin{bmatrix} A_1 & 0 \\ A_{21} & A_2 \end{bmatrix} = \hat{A}_D^T = \begin{bmatrix} 0 & 1 & 0 & 0 & 0 \\ 0 & 0 & 1 & 0 & 0 \\ 0 & 0 & 0 & 0 & 0 \\ 1 & 0 & 0 & -1 & -1 \\ -1 & 1 & -1 & 1 & 1 \end{bmatrix},$$

$$\hat{B} = \hat{C}_D^T = \begin{bmatrix} 0 & 0 \\ 1 & 1 \\ 0 & 1 \\ 0 & 0 \\ 0 & 0 \end{bmatrix}, \quad \text{and} \quad \hat{C} = \hat{B}_D^T = [C_1, 0] = [1, 0, 0, \vdots\ 0, 0].$$

Clearly, $\hat{A} = \hat{A}_D^T, \hat{C} = \hat{B}_D^T$ is in standard form. Therefore, a controllable and observable realization, which is a minimal realization, is given by $\dot{x}_{co} = A_{co} x_{co} + B_{co} u$ and $y = C_{co} x_{co} + D_{co} u$, where

$$A_{co} = \begin{bmatrix} 0 & 1 & 0 \\ 0 & 0 & 1 \\ 0 & 0 & 0 \end{bmatrix}, \quad B_{co} = \begin{bmatrix} 0 & 0 \\ 1 & 1 \\ 0 & 1 \end{bmatrix}, \quad C_{co} = [1, 0, 0], \quad D_{co} = [1, 0].$$

A minimal realization could also have been derived directly in the present case if a realization $\{A_o, B_o, C_o, D_o\}$ of $H(s)$, where $(A_o, B_o)$ is in observer form, had been considered first, as is shown next. Notice that the McMillan degree of $H(s)$ is 3, and therefore, any realization of order higher than 3 will not be minimal. Here, however, the degree of the least common denominator of the (only) row is 3, and therefore, it is known in advance that the realization in observer form, which is of order 3, will be minimal.

A realization $\{A_o, B_o, C_o, D_o\}$ of $H(s)$ in observer form can also be derived by considering $H^T(s)$ and deriving a realization in controller form. Presently, $\{A_o, B_o, C_o, D_o\}$ is derived directly. In particular, we write $H(s) = \widetilde{D}^{-1}(s)\widetilde{N}(s) = (s^3)^{-1}[s(s^2+1), s+1]$. Then $\tilde{d}_1 = 3$ $[= \deg \tilde{\ell}_1(s) = \deg s^3]$ and $\widetilde{\Lambda}(s) = s^3$, $\widetilde{S}(s) = [1, s, s^2]$. Then $\widetilde{D}(s) = s^3 = \widetilde{\Lambda}(s)\widetilde{D}_h + \widetilde{S}(s)\widetilde{D}_\ell$ implies that $\widetilde{D}_h = 1$ and $\widetilde{D}_\ell = [0, 0, 0]^T$. In view of (8.61), we have

$$C_p = 1, \quad A_p = [0, 0, 0]^T,$$

$D_o = \lim_{s \to \infty} H(s) = [1, 0]$, and (8.62) implies that $\widetilde{S}(s)B_o = \widetilde{N}(s) - \widetilde{D}(s)D_o$
$= [s(s^2+1), s+1] - [s^3, 0] = [s, s+1]$, from which we have $B_o = \begin{bmatrix} 0 & 1 & 0 \\ 1 & 1 & 0 \end{bmatrix}^T$. An observable realization of $H(s)$ is the system $\dot{x} = A_o x_o + B_o u, y = C_o x_o + D_o u$, where

$$A_o = \begin{bmatrix} 0 & 0 & 0 \\ 1 & 0 & 0 \\ 0 & 1 & 0 \end{bmatrix}, \quad B_o = \begin{bmatrix} 0 & 1 \\ 1 & 1 \\ 0 & 0 \end{bmatrix}, \quad C_o = [0, 0, 1], \quad D_o = [1, 0],$$

with $(A_o, C_o)$ in observer form (see Lemma 8.24). This realization is minimal since it is of order 3, which is the McMillan degree of $H(s)$. (The reader should verify this.) Note how much easier it was to derive a minimal realization, using the second approach.

---

**Example 8.26.** Let $H(s) = \begin{bmatrix} \frac{2}{s+1} & 1 \\ \frac{1}{s} & 0 \end{bmatrix}$. We wish to derive a minimal realization. Here $\ell_1(s) = s(s+1)$ with $d_1 = 2$ and $\ell_2(s) = 1$ with $d_2 = 0$. In view of the discussion following Lemma 8.23, we let $D_c = \lim_{s \to \infty} H(s) = \begin{bmatrix} 0 & 1 \\ 0 & 0 \end{bmatrix}$ and $\widehat{H}(s) = \begin{bmatrix} \frac{2}{s+1} & 0 \\ \frac{1}{s} & 0 \end{bmatrix}$. We now consider the transfer function $\widehat{\widehat{H}}(s) = \begin{bmatrix} \frac{2}{s+1} \\ \frac{1}{s} \end{bmatrix}$ and determine a minimal realization.

Note that the McMillan degree of $\widehat{\widehat{H}}(s)$ is 2, and therefore, any realization of order 2 will be minimal. Minimal realizations are now derived using two alternative approaches:

1. *Via a controller form realization.* Here $\ell_1(s) = s(s+1), d_1 = 2$, and
$$\widehat{\widehat{H}}(s) = \begin{bmatrix} 2s \\ s+1 \end{bmatrix}[s(s+1)]^{-1} = N(s)D^{-1}(s). \text{ Then } \Lambda(s) = s^2 \text{ and } S(s) =$$
$[1, \quad s]^T, D(s) = s(s+1) = 1s^2 + [0, 1][1, s]^T = D_h\Lambda(s) + D_\ell S(s).$

Therefore, $B_m = 1$ and $A_m = -[0, 1]$. Also, $C_c = \begin{bmatrix} 0 & 2 \\ 1 & 1 \end{bmatrix}$, which follows

from $N(s) = \begin{bmatrix} 2s \\ s+1 \end{bmatrix} = \begin{bmatrix} 0 & 2 \\ 1 & 1 \end{bmatrix} \begin{bmatrix} 1 \\ s \end{bmatrix} = C_c S(s)$. Then a minimal realization

for $H(s)$ is $A_c = \begin{bmatrix} 0 & 1 \\ 0 & -1 \end{bmatrix}$, $B_c = \begin{bmatrix} 0 \\ 1 \end{bmatrix}$, $C_c = \begin{bmatrix} 0 & 2 \\ 1 & 1 \end{bmatrix}$. Adding a zero column
to $B_c$, a minimal realization of $H(s)$ is now derived as

$$A = \begin{bmatrix} 0 & 1 \\ 0 & -1 \end{bmatrix}, \quad B = \begin{bmatrix} 0 & 0 \\ 1 & 0 \end{bmatrix}, \quad C = \begin{bmatrix} 0 & 2 \\ 1 & 1 \end{bmatrix}, \quad D = \begin{bmatrix} 0 & 1 \\ 0 & 0 \end{bmatrix}.$$

We ask the reader to verify that by adding a zero column to $B_c$, controllability is preserved.

2. *Via an observer form realization.* We consider $\widehat{H}^T(s) = [2/(s+1), 1/s]$
and derive a realization in controller form. In particular, $\ell_1 = s + 1, \ell_2 = s, \widehat{H}^T(s) = [2,1] \begin{bmatrix} s+1 & 0 \\ 0 & s \end{bmatrix}^{-1}$, $d_1 = d_2 = 1$, $\Lambda(s) = \begin{bmatrix} s & 0 \\ 0 & s \end{bmatrix}$, and

$S(s) = \begin{bmatrix} 1 & 0 \\ 0 & 1 \end{bmatrix}$. Then $D(s) = \begin{bmatrix} s+1 & 0 \\ 0 & s \end{bmatrix} = \begin{bmatrix} 1 & 0 \\ 0 & 1 \end{bmatrix} \begin{bmatrix} s & 0 \\ 0 & s \end{bmatrix} + \begin{bmatrix} 1 & 0 \\ 0 & 0 \end{bmatrix} \begin{bmatrix} 1 & 0 \\ 0 & 1 \end{bmatrix} =$

$D_h \Lambda(s) + D_\ell S(s)$ and $B_m = \begin{bmatrix} 1 & 0 \\ 0 & 1 \end{bmatrix}$, $A_m = \begin{bmatrix} -1 & 0 \\ 0 & 0 \end{bmatrix}$. Also, $C_c = [2, 1]$, from

which we obtain $N(s) = [2,1] = [2, 1] \begin{bmatrix} 1 & 0 \\ 0 & 1 \end{bmatrix} = C_c S(s)$. Therefore, a min-

imal realization $\{A, B, C\}$ of $\widehat{H}^T(s)$ is $\left\{ \begin{bmatrix} -1 & 0 \\ 0 & 0 \end{bmatrix}, \begin{bmatrix} 1 & 0 \\ 0 & 1 \end{bmatrix}, [2,1] \right\}$. The dual

of this is a minimal realization of $\widehat{H}(s)$, namely, $A_o = \begin{bmatrix} -1 & 0 \\ 0 & 0 \end{bmatrix}$, $B_o = \begin{bmatrix} 2 \\ 1 \end{bmatrix}$,

and $C_o = \begin{bmatrix} 1 & 0 \\ 0 & 1 \end{bmatrix}$. Therefore, a minimal realization of $H(s)$ is

$$A = \begin{bmatrix} -1 & 0 \\ 0 & 0 \end{bmatrix}, \quad B = \begin{bmatrix} 2 & 0 \\ 1 & 0 \end{bmatrix}, \quad C = \begin{bmatrix} 1 & 0 \\ 0 & 1 \end{bmatrix}, \quad D = \begin{bmatrix} 0 & 1 \\ 0 & 0 \end{bmatrix}.$$

### 8.4.3 Realizations with Matrix $A$ Diagonal

When the roots of the minimal polynomial $m_H(s)$ of $H(s)$ are distinct, there
is a realization algorithm that provides a minimal realization of $H(s)$ with $A$
diagonal. Let

$$m_H(s) = s^r + d_{r-1}s^{r-1} + \cdots + d_1 s + d_0 \tag{8.65}$$

be the (monic) least common denominator of all nonzero entries of the $p \times m$
matrix $H(s)$, which in view of Section 7.4, is the minimal polynomial of $H(s)$.
We assume that its $r$ roots $\lambda_i$ are distinct, and we write

$$m_H(s) = \prod_{i=1}^{r}(s - \lambda_i). \tag{8.66}$$

Note that the pole polynomial of $H(s), p_H(s)$, will have repeated roots (poles) if $p_H(s) \neq m_H(s)$. We now consider the strictly proper matrix $\widehat{H}(s) \triangleq H(s) - \lim_{s \to \infty} H(s) = H(s) - D$, and we expand it into partial fractions to obtain

$$\widehat{H}(s) = \widehat{N}(s)/m_H(s) = \sum_{i=1}^{r} \frac{1}{s - \lambda_i} R_i. \tag{8.67}$$

The $p \times m$ residue matrices $R_i$ can be found from the relation

$$R_i = \lim_{s \to \lambda_i} (s - \lambda_i)\widehat{H}(s). \tag{8.68}$$

We write

$$R_i = C_i B_i, \quad i = 1, \dots, r, \tag{8.69}$$

where $C_i$ is a $p \times \rho_i$ and $B_i$ is a $\rho_i \times m$ matrix with $\rho_i \triangleq \operatorname{rank} R_i \leq \min(p, m)$. Note that the above expression is always possible. Indeed, there is a systematic procedure of generating it, namely, by obtaining an LU decomposition of $R_i$. Then

$$A = \begin{bmatrix} \lambda_1 I_{\rho_1} & & & \\ & \lambda_2 I_{\rho_2} & & \\ & & \ddots & \\ & & & \lambda_r I_{\rho_r} \end{bmatrix}, \quad B = \begin{bmatrix} B_1 \\ B_2 \\ \vdots \\ B_r \end{bmatrix}, \tag{8.70}$$

$$C = [C_1, C_2, \dots, C_r], \quad D = \lim_{s \to \infty} H(s)$$

is a minimal realization of order $n \triangleq \sum_{i=1}^{r} \rho_i$.

**Lemma 8.27.** *Representation (8.70) is a minimal realization of $H(s)$.*

*Proof.* It can be verified directly that $C(sI - A)^{-1}B + D = H(s)$, i.e., that (8.70) is a realization of $H(s)$. To verify controllability, we write

$$\mathcal{C} = [B, AB, \dots, A^{n-1}B] = \begin{bmatrix} B_1 & & & \\ & B_2 & & \\ & & \ddots & \\ & & & B_r \end{bmatrix} \begin{bmatrix} I_m, \lambda_1 I_m, \dots, \lambda_1^{n-1} I_m \\ \vdots \quad \vdots \qquad\qquad \vdots \\ I_m, \lambda_r I_m, \dots, \lambda_r^{n-1} I_m \end{bmatrix}.$$

The second matrix in the product is a block Vandermonde matrix of dimensions $mr \times mn$. It can be shown that this matrix has full rank $mr$ since all $\lambda_i$ are assumed to be distinct. Also note that the $(n = \Sigma \rho_i) \times mr$ matrix block diag$[B_i]$ has rank equal to $\sum_{i=1}^{r} \operatorname{rank} B_i = \sum_{i=1}^{r} \rho_i = n \leq mr$. Now, in view of Sylvester's Rank Inequality, as applied to the above matrix product, we have $n + mr - mr \leq \operatorname{rank}\mathcal{C} \leq \min(n, mr)$, from which $\operatorname{rank}\mathcal{C} = n$. Therefore, $\{A, B, C, D\}$ is controllable. Observability is shown in a similar way. Therefore, representation (8.70) is minimal. ∎

**Example 8.28.** Let $H(s) = \begin{bmatrix} \frac{1}{s} & 0 \\ \frac{2}{s+1} & \frac{1}{s(s+1)} \end{bmatrix}$. Here $m_H(s) = s(s+1)$ with roots $\lambda_1 = 0, \lambda_2 = -1$ distinct. We write $H(s) = \frac{1}{s}R_1 + \frac{1}{s+1}R_2$, where $R_1 = \lim_{s\to 0} sH(s) = \lim_{s\to 0} \begin{bmatrix} 1 & 0 \\ \frac{2s}{s+1} & \frac{1}{s+1} \end{bmatrix} = \begin{bmatrix} 1 & 0 \\ 0 & 1 \end{bmatrix}$, $R_2 = \lim_{s\to -1}(s+1)H(s) = \lim_{s\to -1} \begin{bmatrix} \frac{s+1}{s} & 0 \\ 2 & \frac{1}{s} \end{bmatrix} = \begin{bmatrix} 0 & 0 \\ 2 & -1 \end{bmatrix}$, $\rho_1 = \mathrm{rank}\, R_1 = 2$, and $\rho_2 = \mathrm{rank}\, R_2 = 1$; i.e., the order of a minimal realization is $n = \rho_1 + \rho_2 = 3$. We now write

$$R_1 = \begin{bmatrix} 1 & 0 \\ 0 & 1 \end{bmatrix} = \begin{bmatrix} 1 & 0 \\ 0 & 1 \end{bmatrix}\begin{bmatrix} 1 & 0 \\ 0 & 1 \end{bmatrix} = C_1 B_1,$$

$$R_2 = \begin{bmatrix} 0 & 0 \\ 2 & -1 \end{bmatrix} = \begin{bmatrix} 0 \\ 1 \end{bmatrix}[2 \ -1] = C_2\, B_2.$$

Then

$$A = \begin{bmatrix} \lambda_1 I_2 & 0 \\ 0 & \lambda_2 \end{bmatrix} = \begin{bmatrix} 0 & 0 & 0 \\ 0 & 0 & 0 \\ 0 & 0 & -1 \end{bmatrix}, \quad B = \begin{bmatrix} B_1 \\ B_2 \end{bmatrix} = \begin{bmatrix} 1 & 0 \\ 0 & 1 \\ 2 & -1 \end{bmatrix},$$

$$C = [C_1,\ C_2] = \begin{bmatrix} 1 & 0 & 0 \\ 0 & 1 & 1 \end{bmatrix}$$

is a minimal realization with $A$ diagonal (show this). Note that the characteristic polynomial of $H(s)$ is $p_H(s) = s^2(s+1)$, and therefore, the McMillan degree, which is equal to the order of any minimal realization, is 3, as expected.

### 8.4.4 Realizations Using Singular-Value Decomposition

*Internally Balanced Realizations.* Given a proper $p \times m$ matrix $H(s)$, we let $r$ denote the degree of its minimal polynomial $m_H(s)$, we write

$$H(s) = H_0 + H_1 s^{-1} + H_2 s^{-2} + \dots$$

to obtain the Markov parameters $H_i$, and we define

$$T \triangleq M_H(r,r) = \begin{bmatrix} H_1 \cdots H_r \\ \vdots \\ H_r \cdots H_{2r-1} \end{bmatrix}, \hat{T} \triangleq \begin{bmatrix} H_2 \cdots H_{r+1} \\ \vdots \\ H_{r+1} \cdots H_{2r} \end{bmatrix}, \tag{8.71}$$

where $M_H(r,r)$ is the Hankel matrix (see Definition 8.15) and $T, \hat{T}$ are real matrices of dimension $rp \times rm$.

Using *singular-value decomposition* (see Section A.9), we write

$$T = K \begin{bmatrix} \Sigma & 0 \\ 0 & 0 \end{bmatrix} L, \tag{8.72}$$

where $\sum = \mathrm{diag}[\lambda_1, \ldots, \lambda_n] \in R^{n \times n}$ with $n = \mathrm{rank}\, T = \mathrm{rank}\, M_H(r, r)$, which in view of Theorem 8.16 is the order of a minimal realization of $H(s)$. The $\lambda_i$ with $\lambda_1 \geq \lambda_2 \geq \cdots \geq \lambda_n > 0$ are the singular values of $T$, i.e., the nonzero eigenvalues of $T^T T$. Furthermore, $KK^T = K^T K = I_{pr}$ and $LL^T = L^T L = I_{mr}$. We write

$$T = K_1 \Sigma L_1 = (K_1 \Sigma^{1/2})(\Sigma^{1/2} L_1) = VU, \tag{8.73}$$

where $K_1$ denotes the first $n$ columns of $K$, $L_1$ denotes the first $n$ rows of $L$, $K_1^T K_1 = I_n$, and $L_1 L_1^T = I_n$. Also, $V \in R^{rp \times n}$ and $U \in R^{n \times rm}$.

We let $V^+$ and $U^+$ denote pseudoinverses of $V$ and $U$, respectively (see the appendix); i.e.,

$$V^+ = \Sigma^{-1/2} K_1^T \quad \text{and} \quad U^+ = L_1^T \Sigma^{-1/2}, \tag{8.74}$$

where $V^+ V = I_n$ and $UU^+ = I_n$. Now define

$$A = V^+ \widehat{T} U^+, \quad B = U I_{m,mr}^T, \quad C = I_{p,pr} V, \quad D = H_0, \tag{8.75}$$

where $I_{k,\ell} \triangleq [I_k, 0_{\ell-k}]$, $k > \ell$; i.e., $I_{k,\ell}$ is a $k \times \ell$ matrix with its first $k$ columns determining an identity matrix and the remaining $\ell - k$ columns being equal to zero. Thus, $B$ is defined as the first $m$ columns of $U$, and $C$ is defined as the first $p$ rows of $V$. Note that $A \in R^{n \times n}$, $B \in R^{n \times m}$, $C \in R^{p \times n}$, and $D \in R^{p \times m}$.

**Lemma 8.29.** *The representation (8.75) is a minimal realization of $H(s)$.*

*Proof.* It can be shown that $CA^{i-1}B = H_i$, $i = 1, 2, \ldots$. Thus, $\{A, B, C, D\}$ is a realization. We note that $V$ and $U$ are the observability and controllability matrices, respectively, and that both are of full rank $n$. Therefore, the realization is minimal. Furthermore, we notice that $V^T V = UU^T = \Sigma$. Realizations of this type are called *internally balanced realizations*. ∎

The term *internally balanced* emphasizes the fact that realizations of this type are "as much controllable as they are observable," since their controllability and observability Gramians are equal and diagonal. Using such representations, it is possible to construct reasonable reduced-order models of systems by deleting that part of the state space that is "least controllable" and therefore "least observable" in accordance with some criterion. In fact, the realization procedure described can be used to obtain a *reduced-order model* for a given system. Specifically, if the system is to be approximated by a $q$-dimensional model with $q < n$, then the reduced-order model can be obtained from

$$T = K_q \, \mathrm{diag}[\lambda_1, \ldots, \lambda_q] L_q, \tag{8.76}$$

where $K_q$ denotes the first $q$ columns of $K$ in (8.72) and $L_q$ denotes the first $q$ rows of $L$.

## 8.5 Polynomial Matrix Realizations

It is rather straightforward to derive a realization of $H$ in PMD form [see Section 7.5]. In fact, realizations in right (left) Polynomial Matrix Fractional Description (PMFD) form were derived as a step toward determining a state-space realization in controller (observer) form (see Subsection 8.4.2). However, these realizations, of the form $\{D_R, I_m, N_R\}$ and $\{D_L, N_L, I_p\}$, are typically not of minimal order; i.e., they are not controllable and observable. This implies that the controllable realization $\{D_R, I_m, N_R\}$, for example, is not observable; i.e., $D_R, N_R$ are not right coprime. Similarly, the observable realization $\{D_L, N_L, I_p\}$ is not controllable; i.e., $D_L, N_L$ are not left coprime. To obtain a minimal realization, a greatest common right divisor (gcrd) must be extracted from $D_R, N_R$, and similarly, a gcld must be extracted from $D_L, N_L$. This leads to the following realization algorithm, which results in a minimal realization $\{D, I_m, N\}$ of $H$. A minimal realization of the form $\{D, N, I_p\}$ is obtained in an analogous (dual) manner.

Consider $H(s) = [n_{ij}(s)/d_{ij}(s)]$, $i = 1, \ldots, p$, $j = 1, \ldots, m$, and let $l_j(s)$ be the (monic) least common denominator of all entries in the $j$th column of $H(s)$. Note that $l_j(s)$ is the (monic) least degree polynomial divisible by all $d_{ij}(s)$, $i = 1, \ldots, p$. Then $H(s)$ can be written as

$$H(s) = N_R(s)D_R^{-1}(s), \tag{8.77}$$

where $N_R(s) \triangleq = [\bar{n}_{ij}(s)]$ and $D_R(s) \triangleq = \mathrm{diag}(l_1(s), \ldots, l_m(s))$. Note that $\bar{n}_{ij}/l_j(s) = n_{ij}(s)/d_{ij}(s)$ for $i = 1, \ldots, p$ and all $j = 1, \ldots, m$. Now

$$D_R(q)z_R(t) = u(t), \quad y(t) = N_R(q)z_R(t) \tag{8.78}$$

is a controllable realization of $H(s)$. If $D_R, N_R$ are right coprime, it is observable as well and therefore minimal. If $D_R$ and $N_R$ are not right coprime, let $G_R$ be a greatest common right divisor (gcrd) and let $D = D_R G_R^{-1}$ and $N = N_R G_R^{-1}$. Then

$$D(q)z(t) = u(t), \quad y(t) = N(q)z(t) \tag{8.79}$$

is a controllable and observable, and therefore, minimal realization of $H(s)$ since $D, I$ and $D, N$ are left and right coprime polynomial matrix pairs, respectively. Note that $ND^{-1} = (N_R G_R^{-1})(D_R G_R^{-1})^{-1} = (N_R G_R^{-1})(G_R D_R^{-1}) = N_R D_R^{-1} = H$.

There is a dual algorithm that extracts a left PMFD resulting in

$$H(s) = D_L^{-1}(s)N_L(s), \tag{8.80}$$

which corresponds to an observable realization of $H(s)$, given by

$$D_L(q)z_L(t) = N_L(q)u(t), y(t) = z_L(t). \tag{8.81}$$

The details of this procedure are completely analogous to the above procedure that led to (8.77). If $D_L, N_L$ are not left coprime, let $G_L$ be a greatest common left divisor and let $\tilde{D} = G_L^{-1}D_L$ and $\tilde{N} = G_L^{-1}N_L$. Then a controllable and observable and, therefore, minimal realization of $H(s)$ is given by

$$\tilde{D}(q)\tilde{z}(t) = \tilde{N}(q)u(t), \quad y(t) = \tilde{z}(t). \tag{8.82}$$

The following example illustrates the above realization algorithms.

---

**Example 8.30.** Let us derive a minimal realization for $H(s) = \left[\frac{s^2+1}{s^2}, \frac{s+1}{s^3}\right]$. Note that this is the same $H(s)$ as in Example 8.25 of Section 8.4.2, where minimal state-space realizations were derived. The reader is encouraged to compare those results with the realizations derived below. We shall begin with a controllable realization. In view of (8.77) $l_1 = s^2, l_2 = s^3$, and $H =$

$$N_R D_R^{-1} = [s^2 + 1, s + 1]\begin{bmatrix} s^2 & 0 \\ 0 & s^3 \end{bmatrix}^{-1}. \text{ Therefore, } D_R z_R = u \text{ and } y = N_R z_R$$

constitute a controllable realization. This realization is not observable since

$$\text{rank}\begin{bmatrix} D_R(s) \\ N_R(s) \end{bmatrix}_{s=0} = \text{rank}\begin{bmatrix} 0 & 0 \\ 0 & 0 \\ 1 & 1 \end{bmatrix} = 1 < m = 2; \text{ i.e., } D_R \text{ and } N_R \text{ are not right}$$

coprime.

Another way of determining that $D_R$ and $N_R$ are not right coprime would have been to observe that $\deg \det D(s) = 5 = $ order of the realization $\{D_R, I, N_R\}$. Now the McMillan degree of $H$, which is easily derived in the present case, is three. Therefore, the order of any minimal realization for this example is three. Since $\{D_R, I, N_R\}$ is of order five and is controllable, it cannot be observable; i.e., $D_R$ and $N_R$ cannot be right coprime.

We shall now extract a gcrd from $D_R$ and $N_R$ (using the procedure described in [1, Section 7.2D]). We have

$$\begin{bmatrix} D_R \\ N_R \end{bmatrix} = \begin{bmatrix} s^2 & 0 \\ 0 & s^3 \\ s^2+1 & s+1 \end{bmatrix} \longrightarrow \begin{bmatrix} s^2 & 0 \\ 0 & s^3 \\ 1 & s+1 \end{bmatrix} \longrightarrow \begin{bmatrix} 1 & s+1 \\ s^2 & 0 \\ 0 & s^3 \end{bmatrix} \longrightarrow$$

$$\begin{bmatrix} 1 & s+1 \\ 0 & -s^3-s^2 \\ 0 & s^3 \end{bmatrix} \longrightarrow \begin{bmatrix} 1 & s+1 \\ 0 & s^2 \\ 0 & s^3 \end{bmatrix} \longrightarrow \begin{bmatrix} 1 & s+1 \\ 0 & s^2 \\ 0 & 0 \end{bmatrix}.$$

Therefore, $G_R = \begin{bmatrix} 1 & s+1 \\ 0 & s^2 \end{bmatrix}$ is a gcrd. We now determine $D = D_R G_R^{-1}$ and

$N = N_R G_R^{-1}$, using $D_R = \begin{bmatrix} s^2 & 0 \\ 0 & s^3 \end{bmatrix} = \begin{bmatrix} s^2 & -(s+1) \\ 0 & s \end{bmatrix}\begin{bmatrix} 1 & s+1 \\ 0 & s^2 \end{bmatrix} = DG_R$ and

$N_R = [s^2+1, s+1] = [s^2+1, -(s+1)]\begin{bmatrix} 1 & s+1 \\ 0 & s^2 \end{bmatrix} = NG_R$, and we verify that

they are right coprime. Then

$$\{D_R, I, N_R\} = \left\{\begin{bmatrix} q^2 & -(q+1) \\ 0 & q \end{bmatrix}, \begin{bmatrix} 1 & 0 \\ 0 & 1 \end{bmatrix}, [q^2 + 1, -(q+1)]\right\}$$

is a minimal realization of $H(s)$.

Alternatively, given $H$, we shall first derive an observable realization. In view of (8.80),

$$H = D_L^{-1} N_L = (s^3)^{-1}[s(s^2 + 1), s + 1].$$

Here $D_L(q)$ and $N_L(q)$ are left coprime, and therefore, $\widetilde{D}(q)\tilde{z}(t) = \widetilde{N}(q)u(t)$ and $y(t) = \tilde{z}(t)$ with $\widetilde{D}(q) = D_L(q)$ and $\widetilde{N}(q) = N_L(q)$ is controllable and observable and is a minimal realization. Note that the order of this realization is $\deg \det D_L(s) = 3$, which equals the McMillan degree of $H(s)$.

---

## 8.6 Summary and Highlights

*Realizations*

- $\dot{x} = Ax + Bu$, $y = Cx + Du$ is a realization of $H(s)$ ($\hat{y} = H(s)\hat{u}$) if $\hat{y} = [C(sI - As^{-1}B + D]\hat{u}$.
- Realizations of $H(s)$ exist if and only if $H(s)$ is a proper rational matrix. $\lim_{s\to\infty} H(s) = D < \infty$. (See Theorem 8.5.)
- The Markov parameters $H_i$ of the system in

$$H(s) = H_0 + H_1 s^{-1} + H_2 s^{-2} + \ldots$$

satisfy

$$H_0 = D \text{ and } H_i = CA^{i-1}B, \quad i = 1, 2, \ldots.$$

(See Theorem 8.3.)
- A realization $\{A, B, C, D\}$ of $H(s)$ is minimal if and only if it is both controllable and observable. (See Theorem 8.9.)
- Two realizations of $H(s)$ that are minimal must be equivalent representations. (See Theorem 8.10.)
- The order of a minimal realization of $H(s)$ equals its McMillan degree, the order of its characteritic or pole polynomial $p_H(s)$. (See Theorem 8.12.) The order of a minimal realization of $H(s)$ is also given by the rank of the Hankel matrix $M_H(r, r)$. (See Theorem 8.16.)
- Duality can be very useful in obtaining realizations. (See Subsection 8.4.1.)
- Realization algorithms are presented to obtain realizations in controller/observer form [Subsection 8.4.2], realizations with $A$ diagonal [Subsection 8.4.3], and balanced realizations via singular-value decomposition [Subsection 8.4.4].

## 8.7 Notes

A clear understanding of the relationship between external and internal descriptions of systems is one of the principal contributions of systems theory. This topic was developed in the early sixties with original contributions by Gilbert [3] and Kalman [5]. The role of controllability and observability in minimal realizations is due to Kalman [5]. See also Kalman, Falb, and Arbib [6]. The first realization method for MIMO systems is attributed to Gilbert [3]. It was developed for systems where the matrix $A$ can be taken to be diagonal. This method is presented in this chapter. For extensive historical comments concerning this topic, see Kailath [4]. Additional information concerning realizations for the time-varying case can be found, for example, in Brockett [2], Silverman [10], Kamen [7], Rugh [9], and the literature cited in these references. Balanced realizations were introduced in Moore [8].

## References

1. P.J. Antsaklis and A.N. Michel, *Linear Systems*, Birkhäuser, Boston, MA, 2006.
2. R.W. Brockett, *Finite Dimensional Linear Systems*, Wiley, New York, NY, 1970.
3. E. Gilbert, "Controllability and observability in multivariable control systems," *SIAM J. Control*, Vol. 1, pp. 128–151, 1963.
4. T. Kailath, *Linear Systems*, Prentice Hall, Englewood Cliffs, NJ, 1980.
5. R.E. Kalman, "Mathematical description of linear systems," *SIAM J. Control*, Vol. 1, pp. 152–192, 1963.
6. R.E. Kalman, P.L. Falb, and M.A. Arbib, *Topics in Mathematical System Theory*, McGraw-Hill, New York, NY, 1969.
7. E.W. Kamen, "New results in realization theory for linear time-varying analytic systems," *IEEE Trans. Auto. Control*, Vol. AC-24, pp. 866–877, 1979.
8. B.C. Moore, "Principal component analysis in linear systems: controllability, observability and model reduction," *IEEE Trans. Auto. Control*, Vol. AC-26, pp. 17–32, 1981.
9. W.J. Rugh, *Linear System Theory, Second Edition*, Prentice-Hall, Englewood Cliffs, NJ, 1996.
10. L.M. Silverman, "Realization of linear dynamical systems," *IEEE Trans. Auto. Control*, Vol. AC-16, pp. 554–567, 1971.

## Exercises

**8.1.** Consider a scalar proper rational transfer function $H(s) = n(s)/d(s)$, and let $\dot{x} = A_c x_c + B_c u$, $y = C_c x_c + D_c u$ be a realization of $H(s)$ in controller form.

(a) Show that the realization $\{A_c, B_c, C_c, D_c\}$ is always controllable.

(b) Show that $\{A_c, B_c, C_c, D_c\}$ is observable if and only if $n(s)$ and $d(s)$ do not have any factors in common; i.e., they are prime polynomials.
(c) State the dual results to (a) and (b) involving a realization in observer form.

**8.2.** Let $H(s) = \frac{n(s)}{d(s)} = \frac{s^2-s+1}{s^5-s^4+s^3-s^2+s-1}$. Determine a realization in controller form. Is your realization minimal? Explain your answer. *Hint*: Use the results of Exercise 8.1.

**8.3.** For the transfer function $H(s) = \frac{s+1}{s^2+2}$, find

(a) an uncontrollable realization,
(b) an unobservable realization,
(c) an uncontrollable and unobservable realization,
(d) a minimal realization.

**8.4.** Consider the transfer function matrix $H(s) = \begin{bmatrix} \frac{s-1}{s+1} & \frac{1}{s^2-1} \\ 1 & 0 \end{bmatrix}$.

(a) Determine the pole polynomial and the McMillan degree of $H(s)$, using both the Smith–McMillan form and the Hankel matrix.
(b) Determine an observable realization of $H(s)$.
(c) Determine a minimal realization of $H(s)$. *Hint*: Obtain realizations for $\begin{bmatrix} \frac{s-1}{s+1}, & \frac{1}{s^2-1} \end{bmatrix}$.

**8.5.** Consider the transfer function matrix $H(s) = \begin{bmatrix} \frac{(s+1)(-s+5)}{(s-1)(s^2-9)}, & \frac{s}{s-1} \end{bmatrix}^T$, and determine for $H(s)$ a minimal realization in controller form.

**8.6.** Consider the transfer function $H(s) = \begin{bmatrix} \frac{1}{s} & \frac{s+3}{s+1} \\ \frac{1}{s+3} & \frac{s}{s+1} \end{bmatrix}$.

(a) Determine the pole polynomial of $H(s)$ and the McMillan degree of $H(s)$.
(b) Determine a minimal realization $\{A, B, C, D\}$ of $H(s)$, where $A$ is a diagonal matrix.

**8.7.** Given is the system depicted in the block diagram of Figure 8.4, where $H(s) = \frac{s^2+1}{(s+1)(s+2)(s+3)}$. Determine a minimal state-space representation for the closed-loop system, using two approaches. In particular:

(a) First, determine a state-space realization for $H(s)$, and then, determine a minimal state-space representation for the closed-loop system;
(b) first, find the closed-loop transfer function, and then, determine a minimal state-space representation for the closed-loop system.

Compare the two approaches.

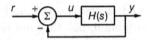

**Figure 8.4.** Block diagram of the system in Exercise 8.7

**8.8.** Consider the system depicted in the block diagram of Figure 8.5, where $H(s) = \frac{s+1}{s(s+3)}$ and $G(s) = \frac{k}{s+a}$ with $k, a \in R$. Presently, $H(s)$ could be viewed as the system to be controlled and $G(s)$ could be regarded as a feedback controller.

(a) Obtain a state-space representation of the closed-loop system by
   (i) first, determining realizations for $H(s)$ and $G(s)$ and then combining them;
   (ii) first, determining $H_c(s)$, the closed-loop transfer function.
(b) Are there any choices for the parameters $k$ and $a$ for which your closed-loop state-space representation is uncontrollable *and* unobservable? If your answer is affirmative, state why.

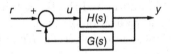

**Figure 8.5.** Block diagram of the system in Exercise 8.8

**8.9.** Consider the controllable and observable system given by $\dot{x} = Ax + Bu$, $y = Cx + Du$, and its equivalent representation $\dot{\hat{x}} = \hat{A}\hat{x} + \hat{B}u$, $y = \hat{C}\hat{x} + \hat{D}u$, where $\hat{A} = PAP^{-1}, \hat{B} = PB, \hat{C} = CP^{-1}$, and $\hat{D} = D$. Let $W_r$ and $W_o$ denote the reachability and observability Gramians, respectively.

(a) Show that $\widehat{W}_r = PW_rP^*$ and $\widehat{W}_o = (P^{-1})^*W_oP^{-1}$, where $P^*$ denotes the complex conjugate transpose of $P$. Note that $P^* = P^T$ when only real coefficients in the system equations are involved.
   Using singular-value decomposition (refer to Section A.9), write

$$W_r = U_r\Sigma_rV_r^* \quad \text{and} \quad W_o = U_o\Sigma_oV_o^*,$$

where $U^*U = I$, $VV^* = I$, and $\Sigma = \text{diag}(\sigma_1, \sigma_2, \ldots, \sigma_n)$ with $\sigma_i$ the singular values of $W$. Define

$$H = (\Sigma_o^{1/2})^*U_o^*U_r(\Sigma_r^{1/2}),$$

and using singular-value decomposition, write

$$H = U_H\Sigma_HV_H,$$

where $U_H^*U_H = I$, $V_HV_H^* = I$. Prove the following:

(b) If $P = P_{\text{in}} \triangleq V_H (\sum_r^{1/2})^{-1} V_r^*$, then $\widehat{W}_r = I$, $\widehat{W}_o = \Sigma_H^2$.

(c) If $P = P_{\text{out}} \triangleq U_H^* (\sum_o^{1/2})^* V_o^*$, then $\widehat{W}_r = \Sigma_H^2$, $\widehat{W}_o = I$.

(d) If $P = P_{ib} = P_{\text{in}} \sum_H^{1/2} = \sum_H^{1/2} P_{\text{out}}$, then $\widehat{W}_r = \widehat{W}_o = \Sigma_H$. Note that the equivalent representations $\{\widehat{A}, \widehat{B}, \widehat{C}, \widehat{D}\}$ in (b), (c), and (d) are called, respectively, *input-normal, output-normal,* and *internally balanced representations.*

**8.10.** Consider a system described by

$$\begin{bmatrix} \hat{y}_1(s) \\ \hat{y}_2(s) \end{bmatrix} = \begin{bmatrix} \frac{1}{(s+1)^2} & \frac{2}{s^2} \\ 0 & \frac{s+1}{s} \end{bmatrix} \begin{bmatrix} \hat{u}_1(s) \\ \hat{u}_2(s) \end{bmatrix}.$$

(a) What is the order of a controllable and observable realization of this system?

(b) If we consider such a realization, is the resulting system controllable from the input $u_2$? Is it observable from the output $y_1$? Explain your answers.

**8.11.** Consider the system described by $H(s) = \frac{1}{s-(1+\epsilon)}(\hat{y}(s) = H(s)\hat{u}(s))$ and $C(s) = \frac{s-1}{s+2}(\hat{u}(s) = C(s)\hat{r}(s))$ connected in series ($\epsilon \in R$).

(a) Derive minimal state-space realizations for $H(s)$ and $C(s)$, and determine a (second order) state-space description for the system $\hat{y}(s) = H(s)C(s)\hat{r}(s)$.

(b) Let $\epsilon = 0$, and discuss the implications regarding the overall transfer function and your state-space representations in (a). Is the overall system now controllable, observable, asymptotically stable? Are the poles of the overall transfer function stable? [That is, is the overall system BIBO stable? (See Chapter 4.)] Plot the states and the output for some nonzero initial condition and a unit step input, and comment on your results.

(c) In practice, if $H(s)$ is a given system to be controlled and $C(s)$ is a controller, it is unlikely that $\epsilon$ will be exactly equal to zero and therefore the situation in (a), rather than (b), will arise. In view of this, comment on whether open-loop stabilization can be used in practice. Carefully explain your reasoning.

**8.12.** Consider the transfer function $H(s) = \begin{bmatrix} 1 & \frac{1}{s} & \frac{s-1}{s} \\ 0 & \frac{s+1}{s^2} & 0 \end{bmatrix}$. Determine a minimal realization in

(a) Polynomial Matrix Fractional Description (PMFD) form,

(b) State-Space Description (SSD) form.

# 9

## State Feedback and State Observers

## 9.1 Introduction

Feedback is a fundamental mechanism arising in nature and is present in many natural processes. Feedback is also common in manufactured systems and is essential in automatic control of dynamic processes with uncertainties in their model descriptions and their interactions with the environment. When feedback is used, the actual values of system variables are sensed, fed back, and used to control the system. Hence, a control law decision process is based not only on predictions about the system behavior derived from a process model, but also on information about the actual behavior. A common example of an automatic feedback control system is the cruise control system in an automobile, which maintains the speed of the automobile at a certain desired value within acceptable tolerances. In this chapter, feedback is introduced and the problem of pole or eigenvalue assignment by means of state feedback is discussed at length in Section 9.2. It is possible to arbitrarily assign all closed-loop eigenvalues by linear static state feedback if and only if the system is completely controllable.

In the study of state feedback, it is assumed that it is possible to measure the values of the states using appropriate sensors. Frequently, however, it may be either impossible or impractical to obtain measurements for all states. It is therefore desirable to be able to estimate the states from measurements of input and output variables that are typically available. In addition to feedback control problems, there are many other problems where knowledge of the state vector is desirable, since such knowledge contains useful information about the system. This is the case, for example, in navigation systems. State estimation is related to observability in an analogous way that state feedback control is related to controllability. The duality between controllability and observability makes it possible to easily solve the estimation problem once the control problem has been solved, and vice versa. In this chapter, asymptotic state estimators, also called state observers, are discussed at length in Section 9.3.

Finally, state feedback static controllers and state dynamic observers are combined to form dynamic output feedback controllers. Such controllers are studied in Section 9.4, using both state-space and transfer function matrix descriptions. In the following discussion, state feedback and state estimation are introduced for continuous- and discrete-time time-invariant systems.

## 9.2 Linear State Feedback

### 9.2.1 Continuous-Time Systems

We consider linear, time-invariant, continuous-time systems described by equations of the form

$$\dot{x} = Ax + Bu, \quad y = Cx + Du, \tag{9.1}$$

where $A \in R^{n \times n}, B \in R^{n \times m}, C \in R^{p \times n}$, and $D \in R^{p \times m}$.

**Definition 9.1.** *The* linear, time-invariant, state feedback control law *is defined by*

$$u = Fx + r, \tag{9.2}$$

*where $F \in R^{m \times n}$ is a gain matrix and $r(t) \in R^m$ is an external input vector.* ∎

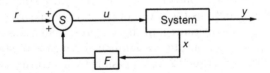

**Figure 9.1.** Linear state feedback configuration

Note that $r(t)$ is an *external input*, also called a *command or reference input* (see Figure 9.1). It is used to provide an input to the compensated closed-loop system and is omitted when such input is not necessary in a given discussion $[r(t) = 0]$. This is the case, e.g., when the Lyapunov stability of a system is studied. Note that the vector $r(t)$ in (9.2) has the same dimension as $u(t)$. If a different number of inputs is desired, then an input transformation map may be used to accomplish this.

The *compensated closed-loop system* of Figure 9.1 is described by the equations

$$\dot{x} = (A + BF)x + Br,$$
$$y = (C + DF)x + Dr, \tag{9.3}$$

which were determined by substituting $u = Fx + r$ into the description of the *uncompensated open-loop system* (9.1).

The *state feedback gain matrix* $F$ affects the closed-loop system behavior. This is accomplished by altering the matrices $A$ and $C$ of (9.1). In fact, the main influence of $F$ is exercised through the matrix $A$, resulting in the matrix $A + BF$ of the closed-loop system. The matrix $F$ affects the eigenvalues of $A + BF$ and, therefore, the modes of the closed-loop system. The effects of $F$ can also be thought of as restricting the choices for $u$ ($= Fx$ for $r = 0$) so that for apppropriate $F$, certain properties, such as asymptotic Lyapunov stability, of the equilibrium $x = 0$ are obtained.

## Open- Versus Closed-Loop Control

The linear state feedback control law (9.2) can be expressed in terms of the initial state $x(0) = x_0$. In particular, working with Laplace transforms, we obtain $\hat{u} = F\hat{x} + \hat{r} = F[(sI - A)^{-1}x_0 + (sI - A)^{-1}B\hat{u}] + \hat{r}$, in view of $s\hat{x} - x_0 = A\hat{x} + B\hat{u}$, derived from $\dot{x} = Ax + Bu$. Collecting terms, we have $[I - F(sI - A)^{-1}B]\hat{u} = F(sI - A)^{-1}x_0 + \hat{r}$. This yields

$$\hat{u} = F[sI - (A + BF)]^{-1}x_0 + [I - F(sI - A)^{-1}B]^{-1}\hat{r}, \qquad (9.4)$$

where the matrix identities $[I - F(sI - A)^{-1}B]^{-1}F(sI - A)^{-1} \equiv F(sI - A)^{-1}[I - BF(sI - A)^{-1}]^{-1} \equiv F[sI - (A + BF)]^{-1}$ have been used.

Expression (9.4) is an *open-loop (feedforward) control law*, expressed in the Laplace transform domain. It is phrased in terms of the initial conditions $x(0) = x_0$, and if it is applied to the open-loop system (9.1), it generates exactly the same control action $u(t)$ for $t \geq 0$ as the state feedback $u = Fx + r$ in (9.2). It can readily be verified that the descriptions of the compensated system are exactly the same when either control expressions, (9.2) or (9.4), are used. In practice, however, these two control laws hardly behave the same, as explained in the following.

First, notice that in the open-loop scheme (9.4), the initial conditions $x_0$ are assumed to be known exactly. It is also assumed that the plant parameters in $A$ and $B$ are known exactly. If there are *uncertainties* in the data, this control law may fail miserably, even when the differences are small, since it is based on incorrect information without any way of knowing that these data are not valid. In contrast to the above, the feedback law (9.2) does not require knowledge of $x_0$. Moreover, it receives feedback information from $x(t)$ and adjusts $u(t)$ to reflect the current system parameters, and consequently, it is more robust to parameter variations. Of course the feedback control law (9.2) will also fail when the parameter variations are too large. In fact, the area of *robust control* relates feedback control law designs to bounds on the uncertainties (due to possible changes) and aims to derive the best design possible under the circumstances.

The point we wish to emphasize here is that although open- and closed-loop control laws may appear to produce identical effects, typically they do not, the reason being that the mathematical system models used are not sufficiently accurate, by necessity or design. Feedback control and closed-loop control are preferred to accommodate ever-present modeling uncertainties in the plant and the environment.

At this point, a few observations are in order. First, we note that feeding back the state in synthesizing a control law is a very powerful mechanism, since the state contains all the information about the history of a system that is needed to uniquely determine the future system behavior, given the input. We observe that the state feedback control law considered presently is linear, resulting in a closed-loop system that is also linear. Nonlinear state feedback control laws are of course also possible. Notice that when a time-invariant system is considered, the state feedback is typically static, unless there is no choice (as in certain optimal control problems), resulting in a closed-loop system that is also time-invariant. These comments justify to a certain extent the choice of linear, time-invariant, state feedback control to compensate linear time-invariant systems.

The problem of stabilizing a system by using state feedback is considered next.

## Stabilization

The problem we wish to consider now is to determine a state feedback control law (9.2) having the property that the resulting compensated closed-loop system has an equilibrium $x = 0$ that is asymptotically stable (in the sense of Lyapunov) when $r = 0$. (For a discussion of asymptotic stability, refer to Subsection 3.3.3 and to Chapter 4.) In particular, we wish to determine a matrix $F \in R^{m \times n}$ so that the system

$$\dot{x} = (A + BF)x, \tag{9.5}$$

where $A \in R^{n \times n}$ and $B \in R^{n \times m}$ has equilibrium $x = 0$ that is asymptotically stable. Note that (9.5) was obtained from (9.3) by letting $r = 0$.

One method of deriving such stabilizing $F$ is by formulating the problem as an optimal control problem, e.g., as the Linear Quadratic Regulator (LQR) problem. This is discussed at the end of this section.

Alternatively, in view of Subsection 3.3.3, the equilibrium $x = 0$ of (9.5) is asymptotically stable if and only if the eigenvalues $\lambda_i$ of $A + BF$ satisfy $Re\lambda_i < 0$, $i = 1, \cdots, n$. Therefore, the stabilization problem for the time-invariant case reduces to the problem of selecting $F$ in such a manner that the eigenvalues of $A + BF$ are shifted into desired locations. This will be studied in the following subsection. Note that stabilization is only one of the control objectives, although a most important one, that can be achieved by shifting eigenvalues. Control system design via eigenvalue (pole) assignment is a topic that is addressed in detail in a number of control books.

### 9.2.2 Eigenvalue Assignment

Consider again the closed-loop system $\dot{x} = (A + BF)x$ given in (9.5). We shall show that if $(A, B)$ is fully controllable (-from-the-origin, or reachable), all eigenvalues of $A + BF$ can be arbitrarily assigned by appropriately selecting $F$. In other words, "the eigenvalues of the original system can arbitrarily be changed in this case." This last statement, commonly used in the literature, is rather confusing: The eigenvalues of a given system $\dot{x} = Ax + Bu$ are *not* physically changed by the use of feedback. They are the same as they used to be before the introduction of feedback. Instead, the feedback law $u = Fx + r, r = 0$, generates an input $u(t)$ that, when fed back to the system, makes it behave *as if* the eigenvalues of the system were at different locations [i.e., the input $u(t)$ makes it behave as a different system, the behavior of which is, we hope, more desirable than the behavior of the original system].

**Theorem 9.2.** *Given $A \in R^{n \times n}$ and $B \in R^{n \times m}$, there exists $F \in R^{m \times n}$ such that the $n$ eigenvalues of $A + BF$ can be assigned to arbitrary, real, or complex conjugate locations if and only if $(A, B)$ is controllable (-from-the-origin, or reachable).*

*Proof.* (*Necessity*): Suppose that the eigenvalues of $A + BF$ have been arbitrarily assigned, and assume that $(A, B)$ in (9.1) is not fully controllable. We shall show that this leads to a contradiction. Since $(A, B)$ is not fully controllable, in view of the results in Section 6.2, there exists a similarity transformation that will separate the controllable part from the uncontrollable part in (9.5). In particular, there exists a nonsingular matrix $Q$ such that

$$Q^{-1}(A + BF)Q = Q^{-1}AQ + (Q^{-1}B)(FQ) = \begin{bmatrix} A_1 & A_{12} \\ 0 & A_2 \end{bmatrix} + \begin{bmatrix} B_1 \\ 0 \end{bmatrix} [F_1, F_2]$$

$$= \begin{bmatrix} A_1 + B_1F_1 & A_{12} + B_1F_2 \\ 0 & A_2 \end{bmatrix}, \tag{9.6}$$

where $[F_1, F_2] \triangleq FQ$ and $(A_1, B_1)$ is controllable. The eigenvalues of $A + BF$ are the same as the eigenvalues of $Q^{-1}(A + BF)Q$, which implies that $A + BF$ has certain fixed eigenvalues, the eigenvalues of $A_2$, that cannot be shifted via $F$. These are the uncontrollable eigenvalues of the system. Therefore, the eigenvalues of $A + BF$ have not been arbitrarily assigned, which is a contradiction. Thus, $(A, B)$ is fully controllable.

(*Sufficiency*): Let $(A, B)$ be fully controllable. Then by using any of the eigenvalue assignment algorithms presented later in this section, all the eigenvalues of $A + BF$ can be arbitrarily assigned. ∎

**Lemma 9.3.** *The uncontrollable eigenvalues of $(A, B)$ cannot be shifted via state feedback.*

*Proof.* See the necessity part of the proof of Theorem 9.2. Note that the uncontrollable eigenvalues are the eigenvalues of $A_2$. ∎

**Example 9.4.** Consider the uncontrollable pair $(A, B)$, where $A = \begin{bmatrix} 0 & -2 \\ 1 & -3 \end{bmatrix}$, $B = \begin{bmatrix} 1 \\ 1 \end{bmatrix}$. This pair can be transformed to a standard form for uncontrollable systems, namely, $\hat{A} = \begin{bmatrix} -2 & 1 \\ 0 & -1 \end{bmatrix}$, $\hat{B} = \begin{bmatrix} 1 \\ 0 \end{bmatrix}$, from which it can easily be seen that $-1$ is the uncontrollable eigenvalue, whereas $-2$ is the controllable eigenvalue.

Now if $F = [f_1, f_2]$, then $\det(sI - (A + BF)) = \det \begin{bmatrix} s - f_1, & 2 - f_2 \\ -1 - f_1 & s + 3 - f_2 \end{bmatrix} = s^2 + s(-f_1 - f_2 + 3) + (-f_1 - f_2 + 2) = (s + 1)(s + (-f_1 - f_2 + 2))$. Clearly, the uncontrollable eigenvalue $-1$ cannot be shifted via state feedback. The controllable eigenvalue $-2$ can be shifted arbitrarily to $(f_1 + f_2 - 2)$ by $F = [f_1, f_2]$.

It is now quite clear that a given system (9.1) can be made asymptotically stable via the state feedback control law (9.2) only when all the uncontrollable eigenvalues of $(A, B)$ are already in the open left part of the $s$-plane. This is so because state feedback can alter only the controllable eigenvalues.

**Definition 9.5.** *The pair* $(A, B)$ *is called* stabilizable *if all its uncontrollable eigenvalues are stable.* ∎

Before presenting methods to select $F$ for eigenvalue assignment, it is of interest to examine how the linear feedback control law $u = Fx + r$ given in (9.2) affects controllability and observability. We write

$$\begin{bmatrix} sI - (A + BF) & B \\ -(C + DF) & D \end{bmatrix} = \begin{bmatrix} sI - A & B \\ -C & D \end{bmatrix} \begin{bmatrix} I & 0 \\ -F & I \end{bmatrix} \tag{9.7}$$

and note that

$$\text{rank}[\lambda I - (A + BF), B] = \text{rank}[\lambda I - A, B]$$

for all complex $\lambda$. Thus, if $(A, B)$ is controllable, then so is $(A + BF, B)$ for any $F$. Furthermore, notice that in view of

$$\mathcal{C}_F = [B, (A + BF)B, (A + BF)^2 B, \ldots, (A + BF)^{n-1} B]$$

$$= [B, AB, A^2 B, \ldots, A^{n-1} B] \begin{bmatrix} I & FB & F(A + BF)B & \cdot \\ 0 & I & FB & \cdot \\ & & I & \cdot \\ & & & \ddots \\ & & & & I \end{bmatrix}, \tag{9.8}$$

$\mathcal{R}(\mathcal{C}_F) = \mathcal{R}([B, AB, \ldots, A^{n-1}B]) = \mathcal{R}(\mathcal{C})$. This shows that $F$ does not alter the controllability subspace of the system. This in turn proves the following lemma.

**Lemma 9.6.** *The controllability subspaces of* $\dot{x} = Ax + Bu$ *and* $\dot{x} = (A + BF)x + Br$ *are the same for any* $F$. ∎

Although the controllability of the system is not altered by linear state feedback $u = Fx + r$, this is not true for the observability property. Note that the observability of the closed-loop system (9.3) depends on the matrices $(A + BF)$ and $(C + DF)$, and it is possible to select $F$ to make certain eigenvalues unobservable from the output. In fact this mechanism is quite common and is used in several control design methods. It is also possible to make observable certain eigenvalues of the open-loop system that were unobservable.

Several methods are now presented to select $F$ to arbitrarily assign the closed-loop eigenvalues.

## Methods for Eigenvalue Assignment by State Feedback

In view of Theorem 9.2, the *eigenvalue assignment problem* can now be stated as follows. Given a controllable pair $(A, B)$, determine $F$ to assign the $n$ eigenvalues of $A + BF$ to arbitrary real and/or complex conjugate locations. This problem is also known as the *pole assignment problem*, where by the term "pole" is meant a "pole of the system" (or an eigenvalue of the "A" matrix). This is to be distinguished from the "poles of the transfer function."

Note that all matrices $A, B$, and $F$ are real, so the coefficients of the polynomial $\det[sI - (A + BF)]$ are also real. This imposes the restriction that the complex roots of this polynomial must appear in conjugate pairs. Also, note that if $(A, B)$ is not fully controllable, then (9.6) can be used together with the methods described a little later, to assign all the controllable eigenvalues; the uncontrollable ones will remain fixed.

It is assumed in the following discussion that $B$ has full column rank; i.e.,

$$\text{rank}\, B = m. \tag{9.9}$$

This means that the system $\dot{x} = Ax + Bu$ has $m$ independent inputs. If $\text{rank}\, B = r < m$, this would imply that one could achieve the same result by manipulating only $r$ inputs (instead of $m > r$). To assign eigenvalues in this case, one can proceed by writing

$$A + BF = A + (BM)(M^{-1}F) = A + [B_1, 0] \begin{bmatrix} F_1 \\ F_2 \end{bmatrix} = A + B_1 F_1, \tag{9.10}$$

where $M$ is chosen so that $BM = [B_1, 0]$ with $B_1 \in R^{n \times r}$ and $\text{rank}\, B_1 = r$. Then $F_1 \in R^{r \times n}$ can be determined to assign the eigenvalues of $A + B_1 F_1$, using any one of the methods presented next. Note that $(A, B)$ is controllable implies that $(A, B_1)$ is controllable. The state feedback matrix $F$ is given in this case by

$$F = M \begin{bmatrix} F_1 \\ F_2 \end{bmatrix}, \qquad (9.11)$$

where $F_2 \in R^{(m-r) \times n}$ is arbitrary.

1. Direct Method

Let $F = [f_{ij}]$, $i = 1, \ldots, m$, $j = 1, \ldots, n$, and express the coefficients of the characteristic polynomial of $A + BF$ in terms of $f_{ij}$; i.e.,

$$\det(sI - (A + BF)) = s^n + g_{n-1}(f_{ij})s^{n-1} + \cdots + g_0(f_{ij}).$$

Now if the roots of the polynomial

$$\alpha_d(s) = s^n + d_{n-1}s^{n-1} + \cdots + d_1 s + d_0$$

are the $n$ desired eigenvalues, then the $f_{ij}$, $i = 1, \ldots, m$, $j = 1, \ldots, n$, must be determined so that

$$g_k(f_{ij}) = d_k, \quad k = 0, 1, \ldots, n - 1. \qquad (9.12)$$

In general, (9.12) constitutes a nonlinear system of algebraic equations; however, it is linear in the single-input case, $m = 1$. The main difficulty in this method is not so much in deriving a numerical solution for the nonlinear system of equation, but in carrying out the symbolic manipulations needed to determine the coefficients $g_k$ in terms of the $f_{ij}$ in (9.12). This difficulty usually restricts this method to the simplest cases, with $n = 2$ or 3 and $m = 1$ or 2 being typical.

---

**Example 9.7.** For $A = \begin{bmatrix} 1/2 & 1 \\ 1 & 2 \end{bmatrix}$, $B = \begin{bmatrix} 1 \\ 1 \end{bmatrix}$, we have $\det(sI - A) = s(s - 5/2)$, and therefore, the eigenvalues of $A$ are 0 and $5/2$. We wish to determine $F$ so that the eigenvalues of $A + BF$ are at $-1 \pm j$.

If $F = [f_1, f_2]$, then $\det(sI - (A + BF)) = \det\left(\begin{bmatrix} s - 1/2 & -1 \\ -1 & s - 2 \end{bmatrix} - \begin{bmatrix} 1 \\ 1 \end{bmatrix}[f_1, f_2]\right) =$

$\det \begin{bmatrix} s - 1/2 - f_1, & -1 - f_2 \\ -1 - f_1, & s - 2 - f_2 \end{bmatrix} = s^2 + s(-\frac{5}{2} - f_1 - f_2) + f_1 - \frac{1}{2}f_2$. The desired eigenvalues are the roots of the polynomial

$$\alpha_d(s) = (s - (-1 + j))(s - (-1 - j)) = s^2 + 2s + 2.$$

Equating coefficients, one obtains $-\frac{5}{2} - f_1 - f_2 = 2$, $f_1 - \frac{1}{2}f_2 = 2$, a linear system of equations. Note that it is linear because $m = 1$. In general one must solve a set of nonlinear algebraic equations. We have

$$F = [f_1, f_2] = [-1/6, -13/3]$$

as the appropriate state feedback matrix.

---

## 2. The Use of Controller Forms

Given that the pair $(A, B)$ is controllable, there exists an equivalence transformation matrix $P$ so that the pair $(A_c = PAP^{-1}, B_c = PB)$ is in controller form (see Section 6.4). The matrices $A + BF$ and $P(A + BF)P^{-1} = PAP^{-1} + PBFP^{-1} = A_c + B_cF_c$ have the same eigenvalues, and the problem is to determine $F_c$ so that $A_c + B_cF_c$ has desired eigenvalues. This problem is easier to solve than the original one because of the special structures of $A_c$ and $B_c$. Once $F_c$ has been determined, then the original feedback matrix $F$ is given by

$$F = F_cP. \tag{9.13}$$

We shall now assume that $(A, B)$ has already been reduced to $(A_c, B_c)$ and describe methods of deriving $F_c$ for eigenvalue assignment.

*Single-Input Case* $(m = 1)$. We let

$$F_c = [f_0, \ldots, f_{n-1}]. \tag{9.14}$$

In view of Section 6.4, since $A_c, B_c$ are in controller form, we have

$$
A_{cF} \triangleq A_c + B_cF_c
$$

$$
= \begin{bmatrix}
0 & 1 & \cdots & 0 \\
\vdots & \vdots & & \vdots \\
0 & 0 & \cdots & 1 \\
-\alpha_0 & -\alpha_1 & \cdots & -\alpha_{n-1}
\end{bmatrix}
+ \begin{bmatrix} 0 \\ \vdots \\ 0 \\ 1 \end{bmatrix} [f_0, \ldots, f_{n-1}]
$$

$$
= \begin{bmatrix}
0 & 1 & \cdots & 0 \\
\vdots & \vdots & \ddots & \vdots \\
0 & 0 & \cdots & 1 \\
-(\alpha_0 - f_0) & -(\alpha_1 - f_1) & \cdots & -(\alpha_{n-1} - f_{n-1})
\end{bmatrix}, \tag{9.15}
$$

where $\alpha_i$, $i = 0, \ldots, n-1$, are the coefficients of the characteristic polynomial of $A_c$; i.e.,

$$
\det(sI - A_c) = s^n + \alpha_{n-1}s^{n-1} + \cdots + \alpha_1 s + \alpha_0. \tag{9.16}
$$

Notice that $A_{cF}$ is also in companion form, and its characteristic polynomial can be written directly as

$$
\det(sI - A_{cF}) = s^n + (\alpha_{n-1} - f_{n-1})s^{n-1} + \cdots + (\alpha_0 - f_0). \tag{9.17}
$$

If the desired eigenvalues are the roots of the polynomial

$$
\alpha_d(s) = s^n + d_{n-1}s^{n-1} + \cdots + d_0, \tag{9.18}
$$

then by equating coefficients, $f_i$, $i = 0, 1, \ldots, n-1$, must satisfy the relations $d_i = \alpha_i - f_i$, $i = 0, 1, \ldots, n-1$, from which we obtain

$$f_i = \alpha_i - d_i, \quad i = 0, \ldots, n - 1. \tag{9.19}$$

Alternatively, note that there exists a matrix $A_d$ in companion form, the characteristic polynomial of which is (9.18). An alternative way of deriving (9.19) is then to set $A_{cF} = A_c + B_c F_c = A_d$, from which we obtain

$$F_c = B_m^{-1}[A_{dm} - A_m], \tag{9.20}$$

where $B_m = 1$, $A_{dm} = [-d_0, \ldots, -d_{n-1}]$ and $A_m = [-\alpha_0, \ldots, -\alpha_{n-1}]$. Therefore, $B_m$, $A_{dm}$, and $A_m$ are the $n$th rows of $B_c$, $A_d$, and $A_c$, respectively (see Section 6.4). Relationship (9.20), which is an alternative formula to (9.19), has the advantage that it is in a form that can be generalized to the multi-input case studied below.

---

**Example 9.8.** Consider the matrices $A = \begin{bmatrix} 1/2 & 1 \\ 1 & 2 \end{bmatrix}$, $B = \begin{bmatrix} 1 \\ 1 \end{bmatrix}$ of Example 9.7. Determine $F$ so that the eigenvalues of $A + BF$ are $-1 \pm j$, i.e., so that they are the roots of the polynomial $\alpha_d(s) = s^2 + 2s + 2$.

To reduce $(A, B)$ into the controller form, let

$$C = [B, AB] = \begin{bmatrix} 1 & 3/2 \\ 1 & 3 \end{bmatrix} \quad \text{and} \quad C^{-1} = \frac{2}{3} \begin{bmatrix} 3 & -3/2 \\ -1 & 1 \end{bmatrix},$$

from which $P = \begin{bmatrix} q \\ qA \end{bmatrix} = \frac{1}{3} \begin{bmatrix} -2 & 2 \\ 1 & 2 \end{bmatrix}$ [see (6.38) in Section 6.4]. Then $P^{-1} = \begin{bmatrix} -1 & 1 \\ 1/2 & 1 \end{bmatrix}$ and

$$A_c = PAP^{-1} = \begin{bmatrix} 0 & 1 \\ 0 & 5/2 \end{bmatrix}, \quad B_c = \begin{bmatrix} 0 \\ 1 \end{bmatrix}.$$

Thus, $A_m = [0, 5/2]$ and $B_m = 1$. Now $A_d = \begin{bmatrix} 0 & 1 \\ -2 & -2 \end{bmatrix}$ and $A_{dm} = [-2, -2]$ since the characteristic polynomial of $A_d$ is $s^2 + 2s + 2 = \alpha_d(s)$. Applying (9.20), we obtain that

$$F_c = B_m^{-1}[A_{dm} - A_m] = [-2, -9/2]$$

and $F = F_c P = [-2, -9/2] \begin{bmatrix} -2/3 & 2/3 \\ 1/3 & 2/3 \end{bmatrix} = [-1/6, -13/3]$ assigns the eigenvalues of the closed-loop system at $-1 \pm j$. This is the same result as the one obtained by the direct method given in Example 9.7. If $\alpha_d(s) = s^2 + d_1 s + d_0$, then $A_{dm} = [-d_0, -d_1]$, $F_c = B_m^{-1}[A_{dm} - A_m] = [-d_0, -d_1 - 5/2]$, and

$$F = F_c P = \frac{1}{3}[2d_0 - d_1 - \frac{5}{2}, -2d_0 - 2d_1 - 5].$$

In general the larger the difference between the coefficients of $\alpha_d(s)$ and $\alpha(s)$, $(A_{dm} - A_m)$, the larger the gains in $F$. This is as expected, since larger changes require in general larger control action.

---

Note that (9.20) can also be derived using (6.55) of Section 6.4. To see this, write

$$A_{cF} = A_c + B_c F_c = (\bar{A}_c + \bar{B}_c A_m) + (\bar{B}_c B_m) F_c = \bar{A}_c + \bar{B}_c (A_m + B_m F_c).$$

Selecting $A_d = \bar{A}_c + \bar{B}_c A_{d_m}$ and requiring $A_{cF} = A_d$ implies

$$\bar{B}_c [A_m + B_m F_c] = \bar{B}_c A_{d_m},$$

from which $A_m + B_m F_c = A_{d_m}$, which in turn implies (9.20).

After $F_c$ has been found, to determine $F$ so that $A + BF$ has desired eigenvalues, one should use $F = F_c P$ given in (9.13). Note that $P$, which reduces $(A, B)$ to the controller form, has a specific form in this ($m = 1$) case [see (6.38) of Section 6.4]. Combining these results, it is possible to derive a formula for the eigenvalue assigning $F$ in terms of the original pair $(A, B)$ and the coefficients of the desired polynomial $\alpha_d(s)$. In particular, the $1 \times n$ matrix $F$ that assigns the $n$ eigenvalues of $A + BF$ at the roots of $\alpha_d(s)$ is unique and is given by

$$F = -e_n^T C^{-1} \alpha_d(A), \tag{9.21}$$

where $e_n = [0, \ldots, 0, 1]^T \in R^n$ and $C = [B, AB, \ldots, A^{n-1}B]$ is the controllability matrix. Relation (9.21) is known as *Ackermann's formula*; for details, see [1, p. 334].

---

**Example 9.9.** To the system of Example 9.8, we apply (9.21) and obtain

$$F = -e_2^T C^{-1} \alpha_d(A)$$

$$= -[0, 1] \begin{bmatrix} 2 & -1 \\ -2/3 & 2/3 \end{bmatrix} \left( \begin{bmatrix} 1/2 & 1 \\ 1 & 2 \end{bmatrix}^2 + 2 \begin{bmatrix} 1/2 & 1 \\ 1 & 2 \end{bmatrix} + 2 \begin{bmatrix} 1 & 0 \\ 0 & 1 \end{bmatrix} \right)$$

$$= -[-2/3, 2/3] \begin{bmatrix} 17/4 & 9/2 \\ 9/2 & 11 \end{bmatrix} = [-1/6, -13/3],$$

which is identical to the $F$ found in Example 9.8.

---

*Multi-Input Case ($m > 1$).* We proceed in a way completely analogous to the single-input case. Assume that $A_c$ and $B_c$ are in the controller form, (6.54). Notice that $A_{cF} \triangleq A_c + B_c F_c$ is also in (controller) companion form with an identical block structure as $A_c$ for any $F_c$. In fact, the pair $(A_{cF}, B_c)$ has the same controllability indices $\mu_i, i = 1, \ldots, m$, as $(A_c, B_c)$. This can be seen directly, since

$$A_c + B_c F_c = (\bar{A}_c + \bar{B}_c A_m) + (\bar{B}_c B_m) F_c = \bar{A}_c + \bar{B}_c (A_m + B_m F_c), \tag{9.22}$$

where $\bar{A}_c$ and $\bar{B}_c$ are defined in (6.55). We can now select an $n \times n$ matrix $A_d$ with desired characteristic polynomial

$$\det(sI - A_d) = \alpha_d(s) = s^n + d_{n-1}s^{n-1} + \cdots + d_0, \qquad (9.23)$$

and in companion form, having the same block structure as $A_{cF}$ or $A_c$; that is, $A_d = \bar{A}_c + \bar{B}_c A_{dm}$. Now if $A_{cF} = A_d$, then in view of (9.22), $\bar{B}_c(A_m + B_m F_c) = \bar{B}_c A_{dm}$. From this, it follows that

$$F_c = B_m^{-1}[A_{dm} - A_m], \qquad (9.24)$$

where $B_m$, $A_{dm}$, and $A_m$ are the $m$ $\sigma_j$th rows of $B_c$, $A_d$, and $A_c$, respectively, and $\sigma_j = \sum_{i=1}^{j} \mu_i$, $j = 1, \ldots, m$. Note that this is a generalization of (9.20) of the single-input case.

We shall now show how to select an $n \times n$ matrix $A_d$ in multivariable companion form to have the desired characteristic polynomial.

One choice is

$$A_d = \begin{bmatrix} 0 & 1 & \cdots & 0 \\ \vdots & \vdots & \ddots & \vdots \\ 0 & 0 & \cdots & 1 \\ -d_0 & -d_1 & \cdots & -d_{n-1} \end{bmatrix},$$

the characteristic polynomial of which is $\alpha_d(s)$. In this case the $m \times n$ matrix $A_{dm}$ is given by

$$A_{dm} = \begin{bmatrix} 0 & \cdots & 0 & 1 \cdots 0 & \cdots & 0 \\ \vdots & & \vdots & \vdots \; \vdots & & \vdots \\ 0 & \cdots & 0 & 0 \cdots 1 & \cdots & 0 \\ -d_0 & & \cdots & & & -d_{n-1} \end{bmatrix},$$

where the $i$th row, $i = 1, \ldots, m - 1$, is zero everywhere except at the $\sigma_i + 1$ column location, where it is one.

Another choice is to select $A_d = [A_{ij}]$, $i, j = 1, \ldots, m$, with $A_{ij} = 0$ for $i \neq j$, i.e.,

$$A_d = \begin{bmatrix} A_{11} & 0 & \cdots & 0 \\ 0 & A_{22} & \cdots & 0 \\ \vdots & \vdots & \ddots & \vdots \\ 0 & 0 & \cdots & A_{mm} \end{bmatrix},$$

noting that $\det(sI - A_d) = \det(sI - A_{11}) \ldots \det(sI - A_{mm})$. Then

$$A_{ii} = \begin{bmatrix} 0 & 1 \cdots & 0 \\ \vdots & \ddots & \vdots \\ 0 & & 1 \\ \times & \cdots & \times \end{bmatrix},$$

where the last row is selected so that $\det(sI - A_{ii})$ has desired roots. The disadvantage of this selection is that it may impose unnecessary restrictions on the number of real eigenvalues assigned. For example, if $n = 4, m = 2$ and the dimensions of $A_{11}$ and $A_{22}$, which are equal to the controllability indices, are $d_1 = 3$ and $d_2 = 1$, then two eigenvalues must be real.

There are of course other selections for $A_d$, and the reader is encouraged to come up with additional choices. A point that should be quite clear by now is that $F_c$ (or $F$) is not unique in the present case, since different $F_c$ can be derived for different $A_{dm}$, all assigning the eigenvalues at the same desired locations. In the single-input case, $F_c$ is unique, as was shown. Therefore, the following result has been established.

**Lemma 9.10.** *Let $(A, B)$ be controllable, and suppose that $n$ desired real complex conjugate eigenvalues for $A + BF$ have been selected. The state feedback matrix $F$ that assigns all eigenvalues of $A + BF$ to desired locations is not unique in the multi-input case $(m > 1)$. It is unique in the single-input case $m = 1$.* ∎

---

**Example 9.11.** Consider the controllable pair $(A, B)$, where $A = \begin{bmatrix} 0 & 1 & 0 \\ 0 & 0 & 1 \\ 0 & 2 & -1 \end{bmatrix}$

and $B = \begin{bmatrix} 0 & 1 \\ 1 & 1 \\ 0 & 0 \end{bmatrix}$. It was shown in Example 6.17, that this pair can be reduced to its controller form

$$A_c = PAP^{-1} = \begin{bmatrix} 0 & 1 & 0 \\ 2 & -1 & 0 \\ 1 & 0 & 0 \end{bmatrix}, \quad B_c = PB = \begin{bmatrix} 0 & 0 \\ 1 & 1 \\ 0 & 1 \end{bmatrix},$$

where $P = \begin{bmatrix} 0 & 0 & 1/2 \\ 0 & 1 & -1/2 \\ 1 & 0 & -1/2 \end{bmatrix}$. Suppose we desire to assign the eigenvalues of $A + BF$ to the locations $\{-2, -1 \pm j\}$, i.e., at the roots of the polynomial $\alpha_d(s) = (s + 2)(s^2 + 2s + 2) = s^3 + 4s^2 + 6s + 4$. A choice for $A_d$ is

$$A_{d1} = \begin{bmatrix} 0 & 1 & 0 \\ 0 & 0 & 1 \\ -4 & -6 & -4 \end{bmatrix}, \quad \text{leading to } A_{dm_1} = \begin{bmatrix} 0 & 0 & 1 \\ -4 & -6 & -4 \end{bmatrix},$$

and

$$F_{c1} = B_m^{-1}[A_{dm_1} - A_m] = \begin{bmatrix} 1 & 1 \\ 0 & 1 \end{bmatrix}^{-1} \left[ \begin{bmatrix} 0 & 0 & 1 \\ -4 & -6 & -4 \end{bmatrix} - \begin{bmatrix} 2 & -1 & 0 \\ 1 & 0 & 0 \end{bmatrix} \right]$$

$$= \begin{bmatrix} 1 & -1 \\ 0 & 1 \end{bmatrix} \begin{bmatrix} -2 & 1 & 1 \\ -5 & -6 & -4 \end{bmatrix} = \begin{bmatrix} 3 & 7 & 5 \\ -5 & -6 & -4 \end{bmatrix}.$$

Alternatively,

$$A_{d_2} = \begin{bmatrix} 0 & 1 & 0 \\ -2 & -2 & 0 \\ 0 & 0 & -2 \end{bmatrix}, \text{ from which } A_{dm2} = \begin{bmatrix} -2 & -2 & 0 \\ 0 & 0 & -2 \end{bmatrix}$$

and

$$F_{c2} = B_m^{-1}[A_{dm2} - A_m] = \begin{bmatrix} 1 & -1 \\ 0 & 1 \end{bmatrix} \begin{bmatrix} -4 & -1 & 0 \\ -1 & 0 & -2 \end{bmatrix}$$

$$= \begin{bmatrix} -3 & -1 & 2 \\ -1 & 0 & -2 \end{bmatrix}.$$

Both $F_1 = F_{c1}P = \begin{bmatrix} 5 & 7 & -9/2 \\ -4 & -6 & 5/2 \end{bmatrix}$ and $F_2 = F_{c2}P = \begin{bmatrix} 2 & -1 & -2 \\ -2 & 0 & 1/2 \end{bmatrix}$ assign
the eigenvalues of $A + BF$ to the locations $\{-2, -1 \pm j\}$.

The reader should plot the states of the equation $\dot{x} = (A + BF)x$ for
$F = F_1$ and $F = F_2$ when $x(0) = [1, \ 1, \ 1]^T$ and should comment on the
differences between the trajectories.

---

Relation (9.24) gives *all feedback matrices*, $F_c$ (or $F = F_cP$), that assign
the $n$ eigenvalues of $A_c + B_cF_c$ (or $A + BF$) to desired locations. The freedom
in selecting such $F_c$ is expressed in terms of the different $A_d$, all in companion
form, with $A_d = [A_{ij}]$ and $A_{ij}$ of dimensions $\mu_i \times \mu_j$, which have the same
characteristic polynomial. Deciding which one of all the possible matrices
$A_d$ to select, so that in addition to eigenvalue assignment other objectives
can be achieved, is not apparent. This flexibility in selecting $F$ can also be
expressed in terms of other parameters, where both eigenvalue and eigenvector
assignment are discussed, as will now be shown.

*3. Assigning Eigenvalues and Eigenvectors*

Suppose now that $F$ was selected so that $A + BF$ has a desired eigenvalue $s_j$
with corresponding eigenvector $v_j$. Then $[s_jI - (A + BF)]v_j = 0$, which can
be written as

$$[s_jI - A, B] \begin{bmatrix} v_j \\ -Fv_j \end{bmatrix} = 0. \tag{9.25}$$

To determine an $F$ that assigns $s_j$ as a closed-loop eigenvalue, one could first
determine a basis for the right kernel (null space) of $[s_jI - A, B]$, i.e., one
could determine a basis $\begin{bmatrix} M_j \\ -D_j \end{bmatrix}$ such that

$$[s_jI - A, B] \begin{bmatrix} M_j \\ -D_j \end{bmatrix} = 0. \tag{9.26}$$

Note that the dimension of this basis is $(n + m) - \text{rank}[s_jI - A, B] = (n +
m) - n = m$, where $\text{rank}[s_jI - A, B] = n$ since the pair $(A, B)$ is controllable.
Since it is a basis, there exists a nonzero $m \times 1$ vector $a_j$ so that

$$\begin{bmatrix} M_j \\ -D_j \end{bmatrix} a_j = \begin{bmatrix} v_j \\ -Fv_j \end{bmatrix}. \tag{9.27}$$

Combining the relations $-D_j a_j = -Fv_j$ and $M_j a_j = v_j$, one obtains

$$FM_j a_j = D_j a_j. \tag{9.28}$$

This is the relation that $F$ must satisfy for $s_j$ to be a closed-loop eigenvalue. The nonzero $m \times 1$ vector $a_j$ can be chosen arbitrarily. Note that $M_j a_j = v_j$ is the eigenvector corresponding to $s_j$. Note also that $a_j$ represents the flexibility one has in selecting the corresponding eigenvector, in addition to assigning an eigenvalue. The $n \times 1$ eigenvector $v_j$ cannot be arbitrarily assigned; rather, the $m \times 1$ vector $a_j$ can be (almost) arbitrarily selected. These mild conditions on $a_j$ are discussed below.

**Theorem 9.12.** *The pair $(s_j, v_j)$ is an (eigenvalue, eigenvector)-pair of $A + BF$ if and only if $F$ satisfies (9.28) for some nonzero vector $a_j$ such that $v_j = M_j a_j$ with $\begin{bmatrix} M_j \\ -D_j \end{bmatrix}$ a basis of the null space of $[s_j I - A, B]$ as in (9.26).*

*Proof.* Necessity has been shown. To prove sufficiency, postmultiply $s_j I - (A + BF)$ by $M_j a_j$ and use (9.28) to obtain $(s_j I - A)M_j a_j - BD_j a_j = 0$ in view of (9.26). Thus,

$$[s_j I - (A + BF)]M_j a_j = 0,$$

which implies that $s_j$ is an eigenvalue of $A + BF$ and $M_j a_j = v_j$ is the corresponding eigenvector. ∎

If relation (9.28) is written for $n$ desired eigenvalues $s_j$, where the $a_j$ *are selected so that the corresponding eigenvectors $v_j = M_j a_j$ are linearly independent*, then

$$FV = W, \tag{9.29}$$

where $V \triangleq [M_1 a_1, \ldots, M_n a_n]$ and $W \triangleq [D_1 a_1, \ldots, D_n a_n]$ uniquely specify $F$ as the solution to these $n$ linearly independent equations. When $s_j$ are distinct, the $n$ vectors $M_j a_j$, $j = 1, \ldots, n$, are linearly independent for almost any nonzero $a_j$, and so $V$ has full rank. When $s_j$ have repeated values, it may still be possible under certain conditions to select $a_j$ so that $M_j a_j$ are linearly independent; however, in general, for multiple eigenvalues, (9.29) needs to be modified, and the details for this can be found in the literature. Also note that if $s_{j+1} = s_j^*$, the complex conjugate of $s_j$, then the corresponding eigenvector $v_{i+1} = v_j^* = M_j^* a_j^*$.

Relation (9.29) clearly shows that the $F$ that assigns all $n$ closed-loop eigenvalues is not unique (see also Lemma 9.10). All such $F$ are parameterized by the vectors $a_j$ that in turn characterize the corresponding eigenvectors. If the corresponding eigenvectors have been decided upon—of course within the set of possible eigenvectors $v_j = M_j a_j$—then $F$ is uniquely specified. Note that in the single-input case, (9.28) becomes $FM_j = D_j$, where $v_j = M_j$. In this case, $F$ is unique.

---

**Example 9.13.** Consider the controllable pair $(A, B)$ of Example 9.11 given by

$$A = \begin{bmatrix} 0 & 1 & 0 \\ 0 & 0 & 1 \\ 0 & 2 & -1 \end{bmatrix}, \quad B = \begin{bmatrix} 0 & 1 \\ 1 & 1 \\ 0 & 0 \end{bmatrix}.$$

Again, it is desired to assign the eigenvalues of $A + BF$ at $-2, -1 \pm j$. Let $s_1 = -2, s_2 = -1 + j$, and $s_3 = -1 - j$. Then, in view of (9.26),

$$\begin{bmatrix} M_1 \\ -D_1 \end{bmatrix} = \begin{bmatrix} 1 & 1 \\ -1 & 0 \\ 2 & 0 \\ -1 & -2 \\ 1 & 2 \end{bmatrix}, \quad \begin{bmatrix} M_2 \\ -D_2 \end{bmatrix} = \begin{bmatrix} 1 & 1 \\ j & 0 \\ 2 & 0 \\ 2+j & -1+j \\ 1 & 1-j \end{bmatrix},$$

and $\begin{bmatrix} M_3 \\ -D_3 \end{bmatrix} = \begin{bmatrix} M_2^* \\ -D_2^* \end{bmatrix}$, the complex conjugate, since $s_3 = s_2^*$.

Each eigenvector $v_i = M_i a_i$, $i = 1, 2, 3$, is a linear combination of the columns of $M_i$. Note that $v_3 = v_2^*$. If we select the eigenvectors to be

$$V = [v_1, v_2, v_3] = \begin{bmatrix} 1 & 1 & 1 \\ 0 & j & -j \\ 0 & 2 & 2 \end{bmatrix},$$

i.e., $a_1 = \begin{bmatrix} 0 \\ 1 \end{bmatrix}, a_2 = \begin{bmatrix} 1 \\ 0 \end{bmatrix}$, and $a_3 = \begin{bmatrix} 1 \\ 0 \end{bmatrix}$, then (9.29) implies that

$$F \begin{bmatrix} 1 & 1 & 1 \\ 0 & j & -j \\ 0 & 2 & 2 \end{bmatrix} = \begin{bmatrix} 2 & -2-j & -2+j \\ -2 & -1 & -1 \end{bmatrix},$$

from which we have

$$F = \frac{1}{4j} \begin{bmatrix} 2 & -2-j & -2+j \\ -2 & -1 & -1 \end{bmatrix} \begin{bmatrix} 4j & 0 & -2j \\ 0 & 2 & j \\ 0 & -2 & j \end{bmatrix}$$

$$= \begin{bmatrix} 2 & -1 & -2 \\ -2 & 0 & 1/2 \end{bmatrix}.$$

As it can be verified, this matrix $F$ is such that $A + BF$ has the desired eigenvalues and eigenvectors.

---

*Remarks*

At this point, several comments are in order.

1. In Example 9.13, if the eigenvectors were chosen to be the eigenvectors of $A + BF_1$ (instead of $A + BF_2$) of Example 9.11, then from $FV = W$, it follows that $F$ would have been $F_1$ (instead of $F_2$).

2. When $s_i = s_{i+1}^*$, then the corresponding eigenvectors are also complex conjugates; i.e., $v_i = v_{i+1}^*$. In this case we obtain from (9.29) that

$$FV = F[\ldots, v_{iR} + jv_{iI}, v_{iR} - jv_{iI}, \ldots]$$
$$= [\ldots, w_{iR} + jw_{iI}, w_{iR} - jw_{iI}, \ldots] = W.$$

Although these calculations could be performed over the complex numbers (as was done in the example), this is not necessary, since postmultiplication of $FV = W$ by

$$\begin{bmatrix} I & & & \\ & \frac{1}{2} & -j\frac{1}{2} & \\ & \frac{1}{2} & +j\frac{1}{2} & \\ & & & I \end{bmatrix}$$

shows that the above equation $FV = W$ is equivalent to

$$F[\ldots, v_{iR}, v_{iI}, \ldots] = [\ldots, w_{iR}, w_{iI}, \ldots],$$

which involves only reals.

3. The bases $\begin{bmatrix} M_j \\ -D_j \end{bmatrix}$, $j = 1, \ldots, n$, in (9.26) can be determined in an alternative way and the calculations can be simplified if the controller form of the pair $(A, B)$ is known. In particular, note that $[sI - A, B] \begin{bmatrix} P^{-1}S(s) \\ D(s) \end{bmatrix} = 0$, where the $n \times m$ matrix $S(s)$ is given by $S(s) = $ block $\text{diag}[1, s, \ldots, s^{\mu_i - 1}]$ and the $\mu_i$, $i = 1, \ldots, m$, are the controllability indices of $(A, B)$. Also, the $m \times m$ matrix $D(s)$ is given by $D(s) = B_m^{-1}[\text{diag}[s^{\mu_1}, \cdots s^{\mu_m}] - A_m S(s)]$. Note that $S(s)$ and $D(s)$ were defined in the Structure Theorem (controllable version) in Section 6.4. It was shown there that $(sI - A_c)S(s) = B_c D(s)$, from which it follows that $(sI - A)P^{-1}S(s) = BD(s)$, where $P$ is a similarity transformation matrix that reduces $(A, B)$ to the controller form $(A_c = PAP^{-1}, B_c = PB)$. Since $P^{-1}S(s)$ and $D(s)$ are right coprime polynomial matrices (see Section 7.5), we have rank $\begin{bmatrix} P^{-1}S(s_j) \\ D(s_j) \end{bmatrix} = m$ for any $s_j$, and therefore, $\begin{bmatrix} P^{-1}S(s_j) \\ D(s_j) \end{bmatrix}$ qualifies as a basis for the null space of the matrix $[s_j I - A, B]$ ($P = I$ when $A, B$ are in controller form; i.e., $A = A_c$ and $B = B_c$.)

*Example 9.13 continued.* Continuing the above example, the controller form of $(A, B)$ was found in Example 9.11 using

$$P^{-1} = \begin{bmatrix} 1 & 0 & 1 \\ 1 & 1 & 0 \\ 2 & 0 & 0 \end{bmatrix}.$$

Here $S(s) = \begin{bmatrix} 1 & 0 \\ s & 0 \\ 0 & 1 \end{bmatrix}$, $D(s) = \begin{bmatrix} s^2 + s - 1 & -s \\ -1 & s \end{bmatrix}$,

and $\begin{bmatrix} M(s) \\ -D(s) \end{bmatrix} = \begin{bmatrix} P^{-1}S(s) \\ -D(s) \end{bmatrix} = \begin{bmatrix} 1 & 1 \\ s+1 & 0 \\ 2 & 0 \\ \hline -(s^2 + s - 1) & s \\ 1 & -s \end{bmatrix}.$

Then

$$\begin{bmatrix} M_1 \\ -D_1 \end{bmatrix} = \begin{bmatrix} M(-2) \\ -D(-2) \end{bmatrix} = \begin{bmatrix} 1 & 1 \\ -1 & 0 \\ 2 & 0 \\ \hline -1 & -2 \\ 1 & 2 \end{bmatrix},$$

$$\begin{bmatrix} M_2 \\ -D_2 \end{bmatrix} = \begin{bmatrix} M(-1+j) \\ -D(-1+j) \end{bmatrix} = \begin{bmatrix} 1 & 1 \\ j & 0 \\ 2 & 0 \\ \hline 2+j & -1+j \\ 1 & 1-j \end{bmatrix},$$

and $\begin{bmatrix} M_3 \\ -D_3 \end{bmatrix} = \begin{bmatrix} M_2^* \\ -D_2^* \end{bmatrix}$, which are precisely the bases used above.

*Remarks (cont.)*

4. If in Example 9.13 the only requirement were that $(s_1, v_1) = (-2, (1, 0, 0)^T)$, then $F(1, 0, 0)^T = (2, -2)^T$; i.e., any $F = \begin{bmatrix} 2 & f_{12} & f_{13} \\ 2 & f_{22} & f_{23} \end{bmatrix}$ will assign the desired values to an eigenvalue of $A + BF$ and its corresponding eigenvector.

5. All possible eigenvectors $v_1$ and $v_2(v_3 = v_2^*)$ in Example 9.13 are given by

$$v_1 = M_1 a_1 = \begin{bmatrix} 1 & 1 \\ -1 & 0 \\ 2 & 0 \end{bmatrix} \begin{bmatrix} a_{11} \\ a_{12} \end{bmatrix} \text{ and } v_2 = M_2 a_2 = \begin{bmatrix} 1 & 1 \\ j & 0 \\ 2 & 0 \end{bmatrix} \begin{bmatrix} a_{21} + j a_{31} \\ a_{22} + j a_{32} \end{bmatrix},$$

where the $a_{ij}$ are such that the set $\{v_1, v_2, v_3\}$ is linearly independent (i.e., $V = [v_1, v_2, v_3]$ is nonsingular) but otherwise arbitrary. Note that in this case ($s_j$ distinct), almost any arbitrary choice for $a_{ij}$ will satisfy the above requirement; see [1, Appendix A.4].

### 9.2.3 The Linear Quadratic Regulator (LQR): Continuous-Time Case

A linear state feedback control law that is optimal in some appropriate sense can be determined as a solution to the so-called Linear Quadratic Regulator (LQR) problem (also called the $H_2$ optimal control problem). The LQR problem has been studied extensively, and the interested reader should consult the extensive literature on *optimal control* for additional information on the subject. In the following discussion, we give a brief outline of certain central results of this topic to emphasize the fact that the state feedback gain $F$ can be determined to satisfy, in an optimal fashion, requirements other than eigenvalue assignment, discussed above. The LQR problem has been studied for the time-varying and time-invariant cases. Presently, we will concentrate on the time-invariant optimal regulator problem.

Consider the time-invariant linear system given by

$$\dot{x} = Ax + Bu, \quad z = Mx, \tag{9.30}$$

where the vector $z(t)$ represents the variables to be regulated—to be driven to zero.

We wish to determine $u(t)$, $t \geq 0$, which minimizes the quadratic cost

$$J(u) = \int_0^\infty [z^T(t)Qz(t) + u^T(t)Ru(t)]dt \tag{9.31}$$

for any initial state $x(0)$. The weighting matrices $Q, R$ are real, symmetric, and positive definite; i.e., $Q = Q^T, R = R^T$, and $Q > 0, R > 0$. This is the most common version of the LQR problem. The term $z^T Qz = x^T(M^T QM)x$ is nonnegative, and it minimizes its integral forces $z(t)$ to approach zero as $t$ goes to infinity. The matrix $M^T QM$ is in general positive semidefinite, which allows some states to be treated as "do not care" states. The term $u^T Ru$ with $R > 0$ is always positive for $u \neq 0$, and it minimizes its integral forces $u(t)$ to remain small. The relative "size" of $Q$ and $R$ enforces tradeoffs between the size of the control action and the speed of response.

Assume that $(A, B, Q^{1/2}M)$ is controllable (-from-the-origin) and observable. It turns out that the solution $u^*(t)$ to this optimal control problem can be expressed in state feedback form, which is independent of the initial condition $x(0)$. In particular, the optimal control $u^*$ is given by

$$u^*(t) = F^*x(t) = -R^{-1}B^T P_c^* x(t), \tag{9.32}$$

where $P_c^*$ denotes the symmetric positive definite solution of the *algebraic Riccati equation*

$$A^T P_c + P_c A - P_c B R^{-1} B^T P_c + M^T Q M = 0. \tag{9.33}$$

This equation may have more than one solution but only one that is positive definite (see Example 9.14). It can be shown that $u^*(t) = F^* x(t)$ is a stabilizing feedback control law and that the minimum cost is given by $J_{\min} = J(u^*) = x^T(0) P_c^* x(0)$.

The assumptions that $(A, B, Q^{1/2}M)$ are controllable and observable may be relaxed somewhat. If $(A, B, Q^{1/2}M)$ is stabilizable and detectable, then the uncontrollable and unobservable eigenvalues, respectively, are stable, and $P_c^*$ is the unique, symmetric, but now positive-semidefinite solution of the algebraic Riccati equation. The matrix $F^*$ is still a stabilizing gain, but it is understood that the uncontrollable and unobservable (but stable) eigenvalues will not be affected by $F^*$.

Note that if the time interval of interest in the evaluation of the cost goes from 0 to $t_1 < \infty$, instead of 0 to $\infty$, that is, if

$$J(u) = \int_0^{t_1} [z^T(t) Q z(t) + u^T(t) R u(t)] dt, \tag{9.34}$$

then the optimal control law is time-varying and is given by

$$u^*(t) = -R^{-1} B^T P^*(t) x(t), \quad 0 \le t \le t_1, \tag{9.35}$$

where $P^*(t)$ is the unique, symmetric, and positive-semidefinite solution of the Riccati equation, which is a matrix differential equation of the form

$$-\frac{d}{dt} P(t) = A^T P(t) + P(t) A - P(t) B R^{-1} B^T P(t) + M^T Q M, \tag{9.36}$$

where $P(t_1) = 0$. It is interesting to note that if $(A, B, Q^{1/2}M)$ is stabilizable and detectable (or controllable and observable), then the solution to this problem as $t_1 \to \infty$ approaches the steady-state value $P_c^*$ given by the algebraic Riccati equation; that is, when $t_1 \to \infty$ the optimal control policy is the time-invariant control law (9.32), which is much easier to implement than time-varying control policies.

---

**Example 9.14.** Consider the system described by the equations $\dot{x} = Ax + Bu$, $y = Cx$, where $A = \begin{bmatrix} 0 & 1 \\ 0 & 0 \end{bmatrix}$, $B = \begin{bmatrix} 0 \\ 1 \end{bmatrix}$, $C = [1, \ 0]$. Then $(A, B, C)$ is controllable and observable and $C(sI - A)^{-1}B = 1/s^2$. We wish to determine the optimal control $u^*(t), t \ge 0$, which minimizes the performance index

$$J = \int_0^\infty (y^2(t) + \rho u^2(t)) dt,$$

where $\rho$ is positive and real. Then $R = \rho > 0, z(t) = y(t)$, $M = C$, and $Q = 1 > 0$. In the present case the algebraic Riccati equation (9.33) assumes the form

$$A^T P_c + P_c A - P_c B R^{-1} B^T P_c + M^T Q M$$

$$= \begin{bmatrix} 0 & 0 \\ 1 & 0 \end{bmatrix} P_c + P_c \begin{bmatrix} 0 & 1 \\ 0 & 0 \end{bmatrix} - \frac{1}{\rho} P_c \begin{bmatrix} 0 \\ 1 \end{bmatrix} [0 \ 1] P_c + \begin{bmatrix} 1 \\ 0 \end{bmatrix} [1 \ 0]$$

$$= \begin{bmatrix} 0 & 0 \\ 1 & 0 \end{bmatrix} \begin{bmatrix} p_1 & p_2 \\ p_2 & p_3 \end{bmatrix} + \begin{bmatrix} p_1 & p_2 \\ p_2 & p_3 \end{bmatrix} \begin{bmatrix} 0 & 1 \\ 0 & 0 \end{bmatrix} - \frac{1}{\rho} \begin{bmatrix} p_1 & p_2 \\ p_2 & p_3 \end{bmatrix} \begin{bmatrix} 0 & 0 \\ 0 & 1 \end{bmatrix} \begin{bmatrix} p_1 & p_2 \\ p_2 & p_3 \end{bmatrix}$$

$$+ \begin{bmatrix} 1 & 0 \\ 0 & 0 \end{bmatrix} = \begin{bmatrix} 0 & 0 \\ 0 & 0 \end{bmatrix},$$

where $P_c = \begin{bmatrix} p_1 & p_2 \\ p_2 & p_3 \end{bmatrix} = P_c^T$. This implies that

$$-\frac{1}{\rho} p_2^2 + 1 = 0, \quad p_1 - \frac{1}{\rho} p_2 p_3 = 0, \quad 2 p_2 - \frac{1}{\rho} p_3^2 = 0.$$

Now $P_c$ is positive definite if and only if $p_1 > 0$ and $p_1 p_3 - p_2^2 > 0$. The first equation above implies that $p_2 = \pm\sqrt{\rho}$. However, the third equation, which yields $p_3^2 = 2\rho p_2$, implies that $p_2 = +\sqrt{\rho}$. Then $p_3^2 = 2\rho\sqrt{\rho}$ and $p_3 = \pm\sqrt{2\rho\sqrt{\rho}}$. The second equation yields $p_1 = \frac{1}{\rho} p_2 p_3$ and implies that only $p_3 = +\sqrt{2\rho\sqrt{\rho}}$ is acceptable, since we must have $p_1 > 0$ for $P_c$ to be positive definite. Note that $p_1 > 0$ and $p_3 - p_2^2 = 2\rho - \rho = \rho > 0$, which shows that

$$P_c^* = \begin{bmatrix} \sqrt{2\sqrt{\rho}} & \sqrt{\rho} \\ \sqrt{\rho} & \sqrt{2\rho\sqrt{\rho}} \end{bmatrix}$$

is the positive definite solution of the algebraic Riccati equation. The optimal control law is now given by

$$u^*(t) = F^* x(t) = -R^{-1} B^T P_c^* x(t) = -\frac{1}{\rho} [0, \ 1] P_c^* x(t).$$

The eigenvalues of the compensated system, i.e., the eigenvalues of $A + BF^*$, can now be determined for different $\rho$. Also, the corresponding $u^*(t)$ and $y(t)$ for given $x(0)$ can be plotted. As $\rho$ increases, the control energy expended to drive the output to zero is forced to decrease. The reader is asked to verify this by plotting $u^*(t)$ and $y(t)$ for different values of $\rho$ when $x(0) = [1, \ 1]^T$. Also, the reader is asked to plot the eigenvalues of $A + BF^*$ as a function of $\rho$ and to comment on the results.

---

It should be pointed out that the locations of the closed-loop eigenvalues, as the weights $Q$ and $R$ vary, have been studied extensively. Briefly, for the single-input case and for $Q = qI$ and $R = r$ in (9.31), it can be shown that the optimal closed-loop eigenvalues are the stable zeros of $1 + (q/r)H^T(-s)H(s)$, where $H(s) = M(sI - A)^{-1}B$. As $q/r$ varies from zero (no state weighting) to infinity (no control weighting), the optimal closed-loop eigenvalues move from the stable poles of $H^T(-s)H(s)$ to the stable zeros of $H^T(-s)H(s)$. Note that the stable poles of $H^T(-s)H(s)$ are the stable poles of $H(s)$ and the stable reflections of its unstable poles with respect to the imaginary axis in the complex plane, whereas its stable zeros are the stable zeros of $H(s)$ and the stable reflections of its unstable zeros.

The solution of the LQR problem relies on solving the Riccati equation. A number of numerically stable algorithms exist for solving the algebraic Riccati equation. The reader is encouraged to consult the literature for computer software packages that implement these methods. A rather straightforward method for determining $P_c^*$ is to use the *Hamiltonian matrix* given by

$$H \triangleq \begin{bmatrix} A & -BR^{-1}B^T \\ -M^TQM & -A^T \end{bmatrix}. \tag{9.37}$$

Let $[V_1^T, V_2^T]^T$ denote the $n$ eigenvectors of $H$ that correspond to the $n$ stable $[Re\,(\lambda) < 0]$ eigenvalues. Note that of the $2n$ eigenvalues of $H$, $n$ are stable and are the mirror images reflected on the imaginary axis of its $n$ unstable eigenvalues. When $(A, B, Q^{1/2}M)$ is controllable and observable, then $H$ has no eigenvalues on the imaginary axis $[Re(\lambda) = 0]$. In this case the $n$ stable eigenvalues of $H$ are in fact the closed-loop eigenvalues of the optimally controlled system, and the solution to the algebraic Riccati equation is then given by

$$P_c^* = V_2V_1^{-1}. \tag{9.38}$$

Note that in this case the matrix $V_1$ consists of the $n$ eigenvectors of $A+BF^*$, since for $\lambda_1$ a stable eigenvalue of $H$, and $v_1$ the corresponding (first) column of $V_1$, we have

$$[\lambda_1 I - (A + BF^*)]v_1 = [\lambda_1 I - A + BR^{-1}B^TV_2V_1^{-1}]v_1$$

$$= \left[ [\lambda_1 I, 0] \begin{bmatrix} V_1 \\ V_2 \end{bmatrix} - [A, -BR^{-1}B^T] \begin{bmatrix} V_1 \\ V_2 \end{bmatrix} \right] V_1^{-1}v_1$$

$$= \begin{bmatrix} 0 & \times & \cdots & \times \\ \vdots & \vdots & & \vdots \\ 0 & \times & \cdots & \times \end{bmatrix} V_1^{-1}v_1 = \begin{bmatrix} 0 & \times & \cdots & \times \\ \vdots & \vdots & & \vdots \\ 0 & \times & \cdots & \times \end{bmatrix} \begin{bmatrix} 1 \\ 0 \\ \vdots \\ 0 \end{bmatrix} = \begin{bmatrix} 0 \\ \vdots \\ 0 \end{bmatrix},$$

where the fact that $\begin{bmatrix} V_1 \\ V_2 \end{bmatrix}$ are eigenvectors of $H$ was used. It is worth reflecting for a moment on the relationship between (9.38) and (9.29). The optimal control $F$ derived by (9.38) is in the class of $F$ derived by (9.29).

### 9.2.4 Input–Output Relations

It is useful to derive the input–output relations for a closed-loop system that is compensated by linear state feedback. Given the uncompensated or open-loop system $\dot{x} = Ax + Bu$, $y = Cx + Du$, with initial conditions $x(0) = x_0$, we have

$$\hat{y}(s) = C(sI - A)^{-1}x_0 + H(s)\hat{u}(s), \tag{9.39}$$

where the open-loop transfer function $H(s) = C(sI - A)^{-1}B + D$. Under the feedback control law $u = Fx + r$, the compensated closed-loop system is described by the equations $\dot{x} = (A + BF)x + Br$, $y = (C + DF)x + Dr$, from which we obtain

$$\hat{y}(s) = (C + DF)[sI - (A + BF)]^{-1}x_0 + H_F(s)\hat{r}(s), \tag{9.40}$$

where the closed-loop transfer function $H_F(s)$ is given by

$$H_F(s) = (C + DF)[sI - (A + BF)]^{-1}B + D.$$

Alternative expressions for $H_F(s)$ can be derived rather easily by substituting (9.4), namely,

$$\hat{u}(s) = F[sI - (A + BF)]^{-1}x_0 + [I - F(sI - A)^{-1}B]^{-1}\hat{r}(s),$$

into (9.39). This corresponds to working with an open-loop control law that nominally produces the same results when applied to the system [see the discussion on open- and closed-loop control that follows (9.4)]. Substituting, we obtain

$$\hat{y}(s) = [C(sI - A)^{-1} + H(s)F[sI - (A + BF)]^{-1}]x_0$$
$$+ H(s)[I - F(sI - A)^{-1}B]^{-1}\hat{r}(s). \tag{9.41}$$

Comparing with (9.40), we see that $(C + DF)[sI - (A + BF)]^{-1} = C(sI - A)^{-1} + H(s)F[sI - (A + BF)]^{-1}$, and that

$$H_F(s) = (C + DF)[sI - (A + BF)]^{-1}B + D$$
$$= [C(sI - A)^{-1}B + D][I - F(sI - A)^{-1}B]^{-1}$$
$$= H(s)[I - F(sI - A)^{-1}B]^{-1}. \tag{9.42}$$

The last relation points out the fact that $\hat{y}(s) = H_F(s)\hat{r}(s)$ can be obtained from $\hat{y}(s) = H(s)\hat{u}(s)$ using the open-loop control $\hat{u}(s) = [I - F(sI - A)^{-1}B]^{-1}\hat{r}(s)$.

## Using Matrix Fractional Descriptions

Relation (9.42) can easily be derived in an alternative manner, using fractional matrix descriptions for the transfer function, introduced in Section 7.5. In particular, the transfer function $H(s)$ of the open-loop system $\{A, B, C, D\}$ is given by

$$H(s) = N(s)D^{-1}(s),$$

where $N(s) = CS(s) + DD(s)$, with $S(s)$ and $D(s)$ satisfying $(sI - A)S(s) = BD(s)$ (refer to the proof of the controllable version of the Structure Theorem given in Section 6.4). Notice that it has been assumed, without loss of generality, that the pair $(A, B)$ is in controller form.

Similarly, the transfer function $H_F(s)$ of the compensated system $\{A + BF, B, C + DF, D\}$ is given by

$$H_F(s) = N_F(s)D_F^{-1}(s),$$

where $N_F(s) = (C + DF)S(s) + DD_F(s)$, with $S(s)$ and $D_F(s)$ satisfying $[sI - (A + BF)]S(s) = BD_F(s)$. This relation implies that $(sI - A)S(s) = B[D_F(s) + FS(s)]$, from which we obtain $D_F(s) + FS(s) = D(s)$. Then $N_F(s) = CS(s) + D[FS(s) + D_F(s)] = CS(s) + DD(s) = N(s)$; that is,

$$H_F(s) = N(s)D_F^{-1}(s), \tag{9.43}$$

where $D_F(s) = D(s) - FS(s)$.

Note that $I - F(sI - A)^{-1}B$ in (9.42) is the transfer function of the system $\{A, B, -F, I\}$ and can be expressed as $D_F(s)D^{-1}(s)$, where $D_F(s) = -FS(s) + ID(s)$. Let $M(s) = (D_F(s)D^{-1}(s))^{-1}$. Then (9.43) assumes the form

$$H_F(s) = N(s)D_F^{-1}(s) = (N(s)D^{-1}(s))(D(s)D_F^{-1}(s)) = H(s)M(s). \tag{9.44}$$

Relation $H_F(s) = N(s)D_F^{-1}(s)$ also shows that the zeros of $H(s)$ [in $N(s)$; see also Subsection 7.5.4] are invariant under linear state feedback; they can be changed only via cancellations with poles. Also observe that $M(s) = D(s)D_F^{-1}(s)$ is the transfer function of the system $\{A + BF, B, F, I\}$. This implies that $H_F(s)$ in (9.42) can also be written as

$$H_F(s) = H(s)[F(sI - (A + BF))^{-1}B + I], \tag{9.45}$$

which is a result that could also be shown directly using matrix identities.

---

**Example 9.15.** Consider the system $\dot{x} = Ax + Bu, y = Cx$, where

$$A = A_c = \begin{bmatrix} 0 & 1 & 0 \\ 2 & -1 & 0 \\ 1 & 0 & 0 \end{bmatrix} \quad \text{and} \quad B = B_c = \begin{bmatrix} 0 & 0 \\ 1 & 1 \\ 0 & 1 \end{bmatrix}$$

as in Example 9.11, and let $C = C_c = [1, 1, 0]$. $H_F(s)$ will now be determined. In view of the Structure Theorem developed in Section 6.4, the transfer function is given by $H(s) = N(s)D^{-1}(s)$, where

$$N(s) = C_c S(s) = [1, 1, 0] \begin{bmatrix} 1 & 0 \\ s & 0 \\ 0 & 1 \end{bmatrix} = [s + 1, 0]$$

and

$$D(s) = B_m^{-1}[A(s) - A_m S(s)] = \begin{bmatrix} 1 & 1 \\ 0 & 1 \end{bmatrix}^{-1} \left[ \begin{bmatrix} s^2 & 0 \\ 0 & s \end{bmatrix} - \begin{bmatrix} 2 & -1 & 0 \\ 1 & 0 & 0 \end{bmatrix} \begin{bmatrix} 1 & 0 \\ s & 0 \\ 0 & 1 \end{bmatrix} \right]$$

$$= \begin{bmatrix} 1 & -1 \\ 0 & 1 \end{bmatrix} \begin{bmatrix} s^2 + s - 2 & 0 \\ -1 & s \end{bmatrix} = \begin{bmatrix} s^2 + s - 1 & -s \\ -1 & s \end{bmatrix}.$$

Then

$$H(s) = N(s)D^{-1}(s) = [s+1, 0] \begin{bmatrix} s^2 + s - 1 & -s \\ -1 & s \end{bmatrix}^{-1}$$

$$= [s+1, 0] \begin{bmatrix} s & s \\ 1 & s^2 + s - 1 \end{bmatrix} \frac{1}{s^3 + s^2 - 2s}$$

$$= \frac{1}{s(s^2 + s - 2)}[s(s+1), s(s+1)] = \frac{s+1}{s^2 + s - 2}[1, 1].$$

If $F_c = \begin{bmatrix} 3 & 7 & 5 \\ -5 & -6 & -4 \end{bmatrix}$ (which is $F_{c1}$ of Example 9.11), then

$$D_F(s) = D(s) - F_c S(s) = \begin{bmatrix} s^2 + s - 1 & -s \\ -1 & s \end{bmatrix} - \begin{bmatrix} 3 & 7 & 5 \\ -5 & -6 & -4 \end{bmatrix} \begin{bmatrix} 1 & 0 \\ s & 0 \\ 0 & 1 \end{bmatrix}$$

$$= \begin{bmatrix} s^2 - 6s - 4 & -s - 5 \\ 6s + 4 & s + 4 \end{bmatrix}.$$

Note that $\det D_F(s) = s^3 + 4s^2 + 6s + 4 = (s+2)(s^2 + 2s + 2)$ with roots $-2, -1 \pm j$, as expected. Now

$$H_F(s) = N(s)D_F^{-1}(s) = [s+1, 0] \begin{bmatrix} s + 4 & s + 5 \\ -6s - 4 & s^2 - 6s - 4 \end{bmatrix} \frac{1}{(s+2)(s^2 + 2s + 2)}$$

$$= \frac{s+1}{(s+2)(s^2 + 2s + 2)}[s+4, s+5].$$

Note that the zeros of $H(s)$ and $H_F(s)$ are identical, located at $-1$. Then $H_F(s) = H(s)M(s)$, where

$$M(s) = D(s)D_F^{-1}(s) = \begin{bmatrix} s^2 + s - 1 & -s \\ -1 & s \end{bmatrix} \begin{bmatrix} s + 4 & s + 5 \\ -6s - 4 & s^2 - 6s - 4 \end{bmatrix} \frac{1}{s^3 + 4s^2 + 6s + 4}$$

$$= \begin{bmatrix} s^3 + 11s^2 + 7s - 4 & 12s^2 + 8s - 5 \\ -6s^2 - 5s - 4 & s^3 - 6s^2 - 5s - 5 \end{bmatrix} \frac{1}{s^3 + 4s^2 + 6s + 4}$$

$$= [I - F_c(sI - A_c)^{-1}B_c]^{-1}.$$

Note that the open-loop uncompensated system is unobservable, with 0 being the unobservable eigenvalue, whereas the closed-loop system is observable; i.e., the control law changed the observability of the system.

## 9.2.5 Discrete-Time Systems

Linear state feedback control for discrete-time systems is defined in a way that is analogous to the continuous-time case. The definitions are included here for purposes of completeness.

We consider a linear, time-invariant, discrete-time system described by equations of the form

$$x(k+1) = Ax(k) + Bu(k), y(k) = Cx(k) + Du(k), \qquad (9.46)$$

where $A \in R^{n \times n}$, $B \in R^{n \times m}$, $C \in R^{p \times n}$, $D \in R^{p \times m}$, and $k \geq k_0$, with $k \geq k_0 = 0$ being typical.

**Definition 9.16.** *The linear (discrete-time, time-invariant) state feedback control law is defined by*

$$u(k) = Fx(k) + r(k), \qquad (9.47)$$

*where $F \in R^{m \times n}$ is a gain matrix and $r(k) \in R^m$ is the external input vector.* ∎

The compensated closed-loop system is now given by

$$
\begin{aligned}
x(k+1) &= (A + BF)x(k) + Br(k), \\
y(k) &= (C + DF)x(k) + Dr(k).
\end{aligned}
\qquad (9.48)
$$

In view of Section 3.5, the system $x(k+1) = (A + BF)x(k)$ is asymptotically stable if and only if the eigenvalues of $A + BF$ satisfy $|\lambda_i| < 1$, i.e., if they lie strictly within the unit disk of the complex plane. The stabilization problem for the time-invariant case therefore becomes a problem of shifting the eigenvalues of $A + BF$, which is precisely the problem studied before for the continuous-time case. Theorem 9.2 and Lemmas 9.3 and 9.6 apply without change, and the methods developed before for eigenvalue assignment can be used here as well. The only difference in this case is the location of the desired eigenvalues: They are assigned to be within the unit circle to achieve stability. We will not repeat here the details for these results.

Input–output relations for discrete-time systems, which are in the spirit of the results developed in the preceding subsection for continuous-time systems, can be derived in a similar fashion, this time making use of the $z$-transform of $x(k+1) = Ax(k) + Bu(k), x(0) = x_0$ to obtain

$$\hat{x}(z) = z(zI - A)^{-1}x_0 + (zI - A)^{-1}B\hat{u}(z). \qquad (9.49)$$

[Compare expression (9.49) with $\hat{x}(s) = (sI - A)^{-1}x_0 + (sI - A)^{-1}B\hat{u}(s).$]

## 9.2.6 The Linear Quadratic Regulator (LQR): Discrete-Time Case

The formulation of the LQR problem in the discrete-time case is analogous to the continuous-time LQR problem. Consider the time-invariant linear system

$$x(k + 1) = Ax(k) + Bu(k), z(k) = Mx(k), \qquad (9.50)$$

where the vector $z(t)$ represents the variables to be regulated. The LQR problem is to determine a control sequence $\{u^*(k)\}$, $k \geq 0$, which minimizes the cost function

$$J(u) = \sum_{k=0}^{\infty} [z^T(k)Qz(k) + u^T(k)Ru(k)] \qquad (9.51)$$

for any initial state $x(0)$, where the weighting matrices $Q$ and $R$ are real symmetric and positive definite.

Assume that $(A, B, Q^{1/2}M)$ is reachable and observable. Then the solution to the $LQR$ problem is given by the linear state feedback control law

$$u^*(k) = F^*x(k) = -[R + B^T P_c^* B]^{-1} B^T P_c^* Ax(k), \qquad (9.52)$$

where $P_c^*$ is the unique, symmetric, and positive definite solution of the (discrete-time) algebraic Riccati equation, given by

$$P_c = A^T[P_c - P_c B[R + B^T P_c B]^{-1} B^T P_c]A + M^T QM. \qquad (9.53)$$

The minimum value of $J$ is $J(u^*) = J_{\min} = x^T(0)P_c^* x(0)$.

As in the continuous-time case, it can be shown that the solution $P_c^*$ can be determined from the eigenvectors of the *Hamiltonian matrix*, which in this case is

$$H = \begin{bmatrix} A + BR^{-1}B^T A^{-T} M^T QM & -BR^{-1}B^T A^{-T} \\ -A^{-T} M^T QM & A^{-T} \end{bmatrix}, \qquad (9.54)$$

where it is assumed that $A^{-1}$ exists. Variations of the above method that relax this assumption exist and can be found in the literature. Let $[V_1^T, V_2^T]^T$ be $n$ eigenvectors corresponding to the $n$ stable ($|\lambda| < 1$) eigenvalues of $H$. Note that out of the $2n$ eigenvalues of $H$, $n$ of them are stable (i.e., within the unit circle) and are the reciprocals of the remaining $n$ unstable eigenvalues (located outside the unit circle). When $(A, B, Q^{1/2}M)$ is controllable (-from-the-origin) and observable, then $H$ has no eigenvalues on the unit circle ($|\lambda| = 1$). In fact the $n$ stable eigenvalues of $H$ are in this case the closed-loop eigenvalues of the optimally controlled system.

The solution to the algebraic Riccati equation is given by

$$P_c^* = V_2 V_1^{-1}. \qquad (9.55)$$

As in the continuous-time case, we note that $V_1$ consists of the $n$ eigenvectors of $A + BF^*$.

**Example 9.17.** We consider the system $x(k+1) = Ax(k) + Bu(k)$, $y(k) = Cx(k)$, where $A = \begin{bmatrix} 0 & 1 \\ 0 & 0 \end{bmatrix}$, $B = \begin{bmatrix} 0 \\ 1 \end{bmatrix}$, $C = [1, 0]$ and we wish to determine the optimal control sequence $\{u^*(k)\}$, $k \geq 0$, that minimizes the performance index

$$J(u) = \sum_{k=0}^{\infty} (y^2(k) + \rho u^2(k)),$$

where $\rho > 0$. In (9.51), $z(k) = y(k)$, $M = C$, $Q = 1$, and $R = \rho$. The reader is asked to determine $u^*(k)$ given in (9.52) by solving the discrete-time algebraic Riccati equation (9.53) in a manner analogous to the solution in Example 9.14 (for the continuous-time algebraic Riccati equation).

## 9.3 Linear State Observers

Since the states of a system contain a great deal of useful information, there are many applications where knowledge of the state vector over some time interval is desirable. It may be possible to measure states of a system by appropriately positioned sensors. This was in fact assumed in the previous section, where the state values were multiplied by appropriate gains and then fed back to the system in the state feedback control law. Frequently, however, it may be either impossible or simply impractical to obtain measurements for all states. In particular, some states may not be available for measurement at all (as in the case, for example, with temperatures and pressures in inaccessible parts of a jet engine). There are also cases where it may be impractical to obtain state measurements from otherwise available states because of economic reasons (e.g., some sensors may be too expensive) or because of technical reasons (e.g., the environment may be too noisy for any useful measurements). Thus, there is a need to be able to estimate the values of the state of a system from available measurements, typically outputs and inputs (see Figure 9.2). Given the system parameters $A$, $B$, $C$, $D$ and the values of the inputs and outputs over a time interval, it is possible to estimate the state when the system is observable. This problem, a problem in state estimation, is discussed in this section. In particular, we will address the so-called full-order and reduced-order asymptotic estimators, which are also called full-order and reduced-order observers.

### 9.3.1 Full-Order Observers: Continuous-Time Systems

We consider systems described by equations of the form

$$\dot{x} = Ax + Bu, \quad y = Cx + Du, \tag{9.56}$$

where $A \in R^{n \times n}$, $B \in R^{n \times m}$, $C \in R^{p \times n}$, and $D \in R^{p \times m}$.

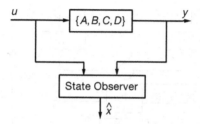

**Figure 9.2.** Linear state observer configuration

## Full-State Observers: The Identity Observer

An estimator of the full state $x(t)$ can be constructed in the following manner. We consider the system

$$\dot{\hat{x}} = A\hat{x} + Bu + K(y - \hat{y}), \tag{9.57}$$

where $\hat{y} \triangleq C\hat{x} + Du$. Note that (9.57) can be written as

$$\dot{\hat{x}} = (A - KC)\hat{x} + [B - KD, K] \begin{bmatrix} u \\ y \end{bmatrix}, \tag{9.58}$$

which clearly reveals the role of $u$ and $y$ (see Figure 9.3). The error between the actual state $x(t)$ and the estimated state $\hat{x}(t)$, $e(t) = x(t) - \hat{x}(t)$, is governed by the differential equation

$$\dot{e}(t) = \dot{x}(t) - \dot{\hat{x}}(t) = [Ax + Bu] - [A\hat{x} + Bu + KC(x - \hat{x})]$$

or

$$\dot{e}(t) = [A - KC]e(t). \tag{9.59}$$

Solving (9.59), we obtain

$$e(t) = \exp[(A - KC)t]e(0). \tag{9.60}$$

Now if the eigenvalues of $A - KC$ are in the left half-plane, then $e(t) \to 0$ as $t \to \infty$, independently of the initial condition $e(0) = x(0) - \hat{x}(0)$. This *asymptotic state estimator* is known as the *Luenberger observer*.

**Lemma 9.18.** *There exists $K \in R^{n \times p}$ so that the eigenvalues of $A - KC$ are assigned to arbitrary real or complex conjugate locations if and only if $(A, C)$ is observable.*

*Proof.* The eigenvalues of $(A - KC)^T = A^T - C^T K^T$ are arbitarily assigned via $K^T$ if and only if the pair $(A^T, C^T)$ is controllable (see Theorem 9.2 of the previous section) or, equivalently, if and only if the pair $(A, C)$ is observable.  ∎

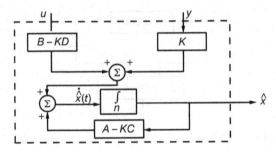

**Figure 9.3.** Full-state identity observer configuration

*Discussion*

If $(A, C)$ is not observable, but the unobservable eigenvalues are stable, i.e., $(A, C)$ is detectable, then the error $e(t)$ will still tend to zero asymptotically. However, the unobservable eigenvalues will appear in this case as eigenvalues of $A - KC$, and they may affect the speed of the response of the estimator in an undesirable way. For example, if the unobservable eigenvalues are stable but are located close to the imaginary axis, then their corresponding modes will tend to dominate the response, most likely resulting in a state estimator that converges too slowly to the actual value of the state.

Where should the eigenvalues of $A - KC$ be located? This problem is dual to the problem of closed-loop eigenvalue placement via state feedback and is equally difficult to resolve. On the one hand, the observer must estimate the state sufficiently fast, which implies that the eigenvalues should be placed sufficiently far from the imaginary axis so that the error $e(t)$ will tend to zero sufficiently fast. On the other hand, this requirement may result in a high gain $K$, which tends to amplify existing noise, thus reducing the accuracy of the estimate. Note that in this case, noise is the only limiting factor of how fast an estimator may be, since the gain $K$ is realized by an algorithm and is typically implemented by means of a digital computer. Therefore, gains of any size can easily be introduced. Compare this situation with the limiting factors in the control case, which is imposed by the magnitude of the required control action (and the limits of the corresponding actuator). Typically, the faster the compensated system, the larger the required control magnitude.

One may of course balance the tradeoffs between speed of response of the estimator and effects of noise by formulating an *optimal estimation problem* to derive the best $K$. To this end, one commonly assumes certain probabilistic properties for the process. Typically, the measurement noise and the initial condition of the plant are assumed to be Gaussian random variables, and one tries to minimize a quadratic performance index. This problem is typically referred to as the *Linear Quadratic Gaussian (LQG)* estimation problem. This optimal estimation or filtering problem can be seen to be the dual of the quadratic optimal control problem of the previous section, a fact that will be exploited in deriving its solution. Note that the well-known *Kalman filter* is such an estimator. In the following discussion, we shall briefly discuss the optimal estimation problem. First, however, we shall address the following related issues.

1. Is it possible to take $K = 0$ in the estimator (9.57)? Such a choice would eliminate the information contained in the term $y - \hat{y}$ from the estimator, which would now be of the form

$$\dot{\hat{x}} = A\hat{x} + Bu. \tag{9.61}$$

In this case, the estimator would operate without receiving any information on how accurate the estimate $\hat{x}$ actually is. The error $e(t) = x(t) - \hat{x}(t)$ would go to zero only when $A$ is stable. There is no mechanism to affect the speed by which $\hat{x}(t)$ would approach $x(t)$ in this case, and this is undesirable. One could perhaps determine $x(0)$, using the methods in Section 5.4, assuming that the system is observable. Then, by setting $\hat{x}(0) = x(0)$, presumably $\hat{x}(t) = x(t)$ for all $t \geq 0$, in view of (9.61). This of course is not practical for several reasons. First, the calculated $\hat{x}(0)$ is never exactly equal to the actual $x(0)$, which implies that $e(0)$ would be nonzero. Therefore, the method would rely again on $A$ being stable, as before, with the advantage here that $e(0)$ would be small in some sense and so $e(t) \to 0$ faster. Second, this scheme assumes that sufficient data have been collected in advance to determine (an approximation to) $x(0)$ and to initialize the estimator, which may not be possible. Third, it is assumed that this initialization process is repeated whenever the estimator is restarted, which may be impractical.

2. If derivatives of the inputs and outputs are available, then the state $x(t)$ may be determined directly (see Exercise 5.12 in Chapter 5). The estimate $\hat{x}(t)$ is in this case produced instantaneously from the values of the inputs and outputs and their derivatives. Under these circumstances, $\hat{x}(t)$ is the output of a static state estimator, as opposed to the above dynamic state estimator, which leads to a state estimate $\hat{x}(t)$ that only approaches the actual state $x(t)$ asymptotically as $t \to \infty$ [$e(t) = x(t) - \hat{x}(t) \to 0$ as $t \to \infty$]. Unfortunately, this approach is in general not viable since noise present in the measurements of $u(t)$ and $y(t)$ makes accurate calculations of the derivatives problematic, and since errors in $u(t), y(t)$ and their derivatives are not smoothed by the algebraic equations of the static estimator (as opposed to the smoothing effects introduced by integration in dynamic systems). It follows that in this case the state estimates may be erroneous.

---

**Example 9.19.** Consider the observable pair

$$A = \begin{bmatrix} 0 & 1 & 0 \\ 0 & 0 & 1 \\ 0 & 2 & -1 \end{bmatrix}, \quad C = [1, 0, 0].$$

We wish to assign the eigenvalues of $A - KC$ in a manner that enables us to design a full-order/full-state asymptotic observer. Let the desired characteristic polynomial be $\alpha_d(s) = s^3 + d_2 s^2 + d_1 s + d_0$, and consider

$$A_D = A^T = \begin{bmatrix} 0 & 0 & 0 \\ 1 & 0 & 2 \\ 0 & 1 & -1 \end{bmatrix} \quad \text{and} \quad B_D = C^T = \begin{bmatrix} 1 \\ 0 \\ 0 \end{bmatrix}.$$

To reduce $(A_D, B_D)$ to controller form, we consider

$$\mathcal{C} = [B_D, A_D B_D, A_D^2 B_D] = \begin{bmatrix} 1 & 0 & 0 \\ 0 & 1 & 0 \\ 0 & 0 & 1 \end{bmatrix} = \mathcal{C}^{-1}.$$

Then $P = \begin{bmatrix} q \\ q A_D \\ q A_D^2 \end{bmatrix} = \begin{bmatrix} 0 & 0 & 1 \\ 0 & 1 & -1 \\ 1 & -1 & 3 \end{bmatrix}$ and $P^{-1} = \begin{bmatrix} -2 & 1 & 1 \\ 1 & 1 & 0 \\ 1 & 0 & 0 \end{bmatrix}$,

from which we obtain

$$A_{D_c} = P A_D P^{-1} = \begin{bmatrix} 0 & 1 & 0 \\ 0 & 0 & 1 \\ 0 & 2 & -1 \end{bmatrix} \quad \text{and} \quad B_{D_c} = P B_D = \begin{bmatrix} 0 \\ 0 \\ 1 \end{bmatrix}.$$

The state feedback is then given by $F_{D_c} = B_m^{-1}[A_{d_m} - A_m] = [-d_0, -d_1 - 2, -d_2 + 1]$ and $F_D = F_{D_c} P = [-d_2 + 1, d_2 - d_1 - 3, d_1 - d_0 - 3d_2 + 5]$. Then

$$K = -F_D^T = [d_2 - 1, d_1 - d_2 + 3, d_0 - d_1 + 3d_2 - 5]^T$$

assigns the eigenvalues of $A - KC$ at the roots of $\alpha_d(s) = s^3 + d_2 s^2 + d_1 s + d_0$. Note that the same result could also have been derived using the direct method for eigenvalue assignment, using $|sI - (A - (k_0, k_1, k_2)^T C)| = \alpha_d(s)$. Also, the result could have been derived using the *observable version of Ackermann's formula*, namely,

$$K = -F_D^T = \alpha_d(A) \mathcal{O}^{-1} e_n,$$

where $F_D = -e_n^T \mathcal{C}_D^{-1} \alpha_d(A_D)$ from (9.21). Note that the given system has eigenvalues at $0, 1, -2$ and is therefore unstable. The observer derived in this case will be used in the next section (Example 9.25) in combination with state feedback to stabilize the system $\dot{x} = Ax + Bu, y = Cx$, where

$$A = \begin{bmatrix} 0 & 1 & 0 \\ 0 & 0 & 1 \\ 0 & 2 & -1 \end{bmatrix}, \quad B = \begin{bmatrix} 0 & 1 \\ 1 & 1 \\ 0 & 0 \end{bmatrix}, \quad \text{and} \quad C = [1, 0, 0]$$

(see Example 9.11), using only output measurements.

***Example 9.20.*** Consider the system $\dot{x} = Ax$, $y = Cx$, where $A = \begin{bmatrix} 0 & -2 \\ 1 & -2 \end{bmatrix}$ and $C = [0, 1]$, and where $(A, C)$ is in observer form. It is easy to show that $K = [d_0 - 2, d_1 - 2]^T$ assigns the eigenvalues of $A - KC$ at the roots of $s^2 + d_1 s + d_0$. To verify this, note that

$$\det(sI - (A - KC)) = \det\left( \begin{bmatrix} s & 0 \\ 0 & s \end{bmatrix} - \begin{bmatrix} 0 & -d_0 \\ 1 & -d_1 \end{bmatrix} \right) = s^2 + d_1 + d_0.$$

The error $e(t) = x(t) - \hat{x}(t)$ is governed by the equation $\dot{e}(t) = (A - KC)e(t)$ given in (9.59). Noting that the eigenvalues of $A$ are $-1 \pm j$, select different sets of eigenvalues for the observer and plot the states $x(t), \hat{x}(t)$ and the error $e(t)$ for $x(0) = [2, 2]^T$ and $\hat{x}(0) = [0, 0]^T$. The further away the eigenvalues of the observer are selected from the imaginary axis (with negative real parts), the larger the gains in $K$ will become and the faster $\hat{x}(t) \to x(t)$.

## Partial or Linear Functional State Observers

The state estimator studied above is a full-state estimator or observer; i.e., $\hat{x}(t)$ is an estimate of the full-state vector $x(t)$. There are cases where only part of the state vector, or a linear combination of the states, is of interest. In control problems, for example, $F\hat{x}(t)$ is used and fed back, instead of $Fx(t)$, where $F$ is an $m \times n$ state feedback gain matrix. An interesting question that arises at this point is as follows: Is it possible to estimate directly a linear combination of the state, say, $Tx$, where $T \in R^{\tilde{n} \times n}$, $\tilde{n} \le n$? For details of this problem see materials starting with [1, p. 354].

### 9.3.2 Reduced-Order Observers: Continuous-Time Systems

Suppose that $p$ states, out of the $n$ state, can be measured directly. This information can then be used to reduce the order of the full-state estimator from $n$ to $n - p$. Similar results are true for the estimator of a linear function of the state, but this problem will not be addressed here. To determine a full-state estimator of order $n - p$, first consider the case when $C = [I_p, 0]$. In particular, let

$$\begin{bmatrix} \dot{x}_1 \\ \dot{x}_2 \end{bmatrix} = \begin{bmatrix} A_{11} & A_{12} \\ A_{21} & A_{22} \end{bmatrix} \begin{bmatrix} x_1 \\ x_2 \end{bmatrix} + \begin{bmatrix} B_1 \\ B_2 \end{bmatrix} u$$

$$z = [I_p, 0] \begin{bmatrix} x_1 \\ x_2 \end{bmatrix}, \tag{9.62}$$

where $z = x_1$ represents the $p$ measured states. Therefore, only $x_2(t) \in R^{(n-p) \times 1}$ is to be estimated. The system whose state is to be estimated is now given by

$$\dot{x}_2 = A_{22}x_2 + [A_{21}, B_2]\begin{bmatrix} x_1 \\ u \end{bmatrix}$$

$$= A_{22}x_2 + \tilde{B}\tilde{u}, \tag{9.63}$$

where $\tilde{B} \triangleq [A_{21}, B_2]$ and $\tilde{u} \triangleq \begin{bmatrix} x_1 \\ u \end{bmatrix} = \begin{bmatrix} z \\ u \end{bmatrix}$ is a known signal. Also,

$$\tilde{y} \triangleq \dot{x}_1 - A_{11}x_1 - B_1u = A_{12}x_2, \tag{9.64}$$

where $\tilde{y}$ is known. An estimator for $x_2$ can now be constructed. In particular, in view of (9.57), we have that the system

$$\dot{\hat{x}}_2 = A_{22}\hat{x}_2 + \tilde{B}\tilde{u} + \tilde{K}(\tilde{y} - A_{12}\hat{x}_2)$$
$$= (A_{22} - \tilde{K}A_{12})\hat{x}_2 + (A_{21}z + B_2u) + \tilde{K}(\dot{z} - A_{11}z - B_1u) \tag{9.65}$$

is an asymptotic state estimator for $x_2$. Note that the error $e$ satisfies the equation

$$\dot{e} = \dot{x}_2 - \dot{\hat{x}}_2 = (A_{22} - \tilde{K}A_{12})e, \tag{9.66}$$

and if $(A_{22}, A_{12})$ is observable, then the eigenvalues of $A_{22} - \tilde{K}A_{12}$ can be arbitrarily assigned making use of $\tilde{K}$. It can be shown that if the pair $(A = [A_{ij}], C = [I_p, 0])$ is observable, then $(A_{22}, A_{12})$ is also observable (prove this using the eigenvalue observability test of Section 6.3). System (9.65) is an estimator of order $n - p$, and therefore, the estimate of the entire state $x$ is $\begin{bmatrix} z \\ \hat{x}_2 \end{bmatrix}$. To avoid using $\dot{z} = \dot{x}_1$ in $\tilde{y}$ given by (9.64), one could use $\hat{x}_2 = w + \tilde{K}z$ and obtain from (9.65) an estimator in terms of $w, z$, and $u$. In particular,

$$\dot{w} = (A_{22} - \tilde{K}A_{12})w + [(A_{22} - \tilde{K}A_{12})\tilde{K} + A_{21} - \tilde{K}A_{11}]z + [B_2 - \tilde{K}B_1]u. \tag{9.67}$$

Then $w$ is an estimate of $\hat{x}_2 - \tilde{K}z$ and of course $w + \tilde{K}z$ is an estimate for $\hat{x}_2$.

In the above derivation, it was assumed for simplicity that a part of the state $x_1$, is measured directly; i.e., $C = [I_p, 0]$. One could also derive a reduced-order estimator for the system

$$\dot{x} = Ax + Bu, \quad y = Cx.$$

To see this, let $\text{rank } C = p$ and define a similarity transformation matrix $P = \begin{bmatrix} C \\ \hat{C} \end{bmatrix}$, where $\hat{C}$ is such that $P$ is nonsingular. Then

$$\dot{\bar{x}} = \bar{A}\bar{x} + \bar{B}u, \quad y = \bar{C}\bar{x} = [I_p, 0]\bar{x}, \tag{9.68}$$

where $\bar{x} = Px, \bar{A} = PAP^{-1}, \bar{B} = PB$, and $\bar{C} = CP^{-1} = [I_p, 0]$. The transformed system is now in an appropriate form for an estimator of order $n - p$ to be derived, using the procedure discussed above. The estimate of $\bar{x}$ is

$\begin{bmatrix} y \\ \hat{\bar{x}}_2 \end{bmatrix}$, and the estimate of the original state $x$ is $P^{-1} \begin{bmatrix} y \\ \hat{\bar{x}}_2 \end{bmatrix}$. In particular, $\bar{x}_2 = w + \widetilde{K}y$, where $w$ satisfies (9.67) with $z = y$, $[A_{ij}] = \bar{A} = PAP^{-1}$, and $\begin{bmatrix} B_1 \\ B_2 \end{bmatrix} = \bar{B} = PB$. The interested reader should verify this result.

---

**Example 9.21.** Consider the system $\dot{x} = Ax + Bu$, $y = Cx$, where $A = \begin{bmatrix} 0 & -2 \\ 1 & -2 \end{bmatrix}$, $B = \begin{bmatrix} 0 \\ 1 \end{bmatrix}$, and $C = [0, 1]$. We wish to design a reduced $n - p = n - 1 = 2 - 1 = 1$, a first-order asymptotic state estimator.

The similarity transformation matrix $P = \begin{bmatrix} C \\ \hat{C} \end{bmatrix} = \begin{bmatrix} 0 & 1 \\ 1 & 0 \end{bmatrix}$ leads to (9.68), where $\bar{x} = Px$ and $\bar{A} = PAP^{-1} = \begin{bmatrix} 0 & 1 \\ 1 & 0 \end{bmatrix} \begin{bmatrix} 0 & -2 \\ 1 & -2 \end{bmatrix} \begin{bmatrix} 0 & 1 \\ 1 & 0 \end{bmatrix} = \begin{bmatrix} -2 & 1 \\ -2 & 0 \end{bmatrix}$, $\bar{B} = PB = \begin{bmatrix} 1 \\ 0 \end{bmatrix}$, and $\bar{C} = CP^{-1} = [1, 0]$. The system $\{\bar{A}, \bar{B}, \bar{C}\}$ is now in an appropriate form for use of (9.67). We have $\bar{A} = \begin{bmatrix} A_{11} & A_{12} \\ A_{21} & A_{22} \end{bmatrix} = \begin{bmatrix} -2 & 1 \\ -2 & 0 \end{bmatrix}$, $\bar{B} = \begin{bmatrix} B_1 \\ B_2 \end{bmatrix} = \begin{bmatrix} 1 \\ 0 \end{bmatrix}$, and (9.67) assumes the form

$$\dot{w} = (-\widetilde{K})w + [-\widetilde{K}^2 + (-2) - \widetilde{K}(-2)]y + (-\widetilde{K})u,$$

which is a system observer of order 1.

For $\widetilde{K} = -10$ we have $\dot{w} = 10w - 122y + 10u$, and $w + \widetilde{K}y = w - 10y$ is an estimate for $\hat{\bar{x}}_2$. Therefore, $\begin{bmatrix} y \\ w - 10y \end{bmatrix}$ is an estimate of $\bar{x}$, and

$$P^{-1} \begin{bmatrix} y \\ w - 10y \end{bmatrix} = \begin{bmatrix} 0 & 1 \\ 1 & 0 \end{bmatrix} \begin{bmatrix} y \\ w - 10y \end{bmatrix} = \begin{bmatrix} w - 10y \\ y \end{bmatrix}$$

is an estimate of $x(t)$ for the original system.

---

### 9.3.3 Optimal State Estimation: Continuous-Time Systems

The gain $K$ in the estimator (9.57) above can be determined so that it is optimal in an appropriate sense. This is discussed briefly below. The interested reader should consult the extensive literature on *filtering theory* for additional information, in particular, the literature on the *Kalman–Bucy filter*.

In addressing optimal state estimation, noise with certain statistical properties is introduced in the model and an appropriate cost functional is set up that is then minimized. In the following discussion, we shall introduce some of the key equations of the *Kalman–Bucy filter* and we will point out the duality between the optimal control and estimation problems. We concentrate on the time-invariant case, although, as in the LQR control problem discussed earlier, more general results for the time-varying case do exist.

We consider the linear time-invariant system

$$\dot{x} = Ax + Bu + \Gamma w, y = Cx + v, \tag{9.69}$$

where $w$ and $v$ represent process and measurement noise terms. Both $w$ and $v$ are assumed to be white, zero-mean Gaussian stochastic processes; i.e., they are uncorrelated in time and have expected values $E[w] = 0$ and $E[v] = 0$. Let

$$E[ww^T] = W, \quad E[vv^T] = V \tag{9.70}$$

denote their covariances, where $W$ and $V$ are real, symmetric, and positive definite matrices, i.e., $W = W^T, W > 0$, and $V = V^T, V > 0$. Assume that the noise processes $w$ and $v$ are independent; i.e., $E[wv^T] = 0$. Also assume that the initial state $x(0)$ of the plant is a Gaussian random variable of known mean, $E[x(0)] = x_0$, and known covariance, $E[(x(0) - x_0)(x(0) - x_0)^T] = P_{e0}$. Assume also that $x(0)$ is independent of $w$ and $v$. Note that all these are typical assumptions made in practice.

Consider now the estimator (9.57), namely,

$$\dot{\hat{x}} = A\hat{x} + Bu + K(y - C\hat{x}) = (A - KC)\hat{x} + Bu + Ky, \tag{9.71}$$

and let $(A, \Gamma W^{1/2}, C)$ be controllable (-from-the-origin) and observable. It turns out that the error covariance $E[(x - \hat{x})(x - \hat{x})^T]$ is minimized when the filter gain is given by

$$K^* = P_e^* C^T V^{-1}, \tag{9.72}$$

where $P_e^*$ denotes the symmetric, positive definite solution of the quadratic (dual) algebraic Riccati equation

$$P_e A^T + A P_e - P_e C^T V^{-1} C P_e + \Gamma W \Gamma^T = 0. \tag{9.73}$$

Note that $P_e^*$, which is in fact the minimum error covariance, is the positive semidefinite solution of the above Riccati equation if $(A, \Gamma W^{1/2}, C)$ is stabilizable and detectable. The optimal estimator is asymptotically stable.

The above algebraic Riccati equation is the dual to the Riccati equation given in (9.33) for optimal control and can be obtained from (9.33) making use of the substitutions

$$A \to A^T, B \to C^T, M \to \Gamma^T \quad \text{and} \quad R \to V, Q \to W. \tag{9.74}$$

Clearly, methods that are analogous to the ones developed by solving the control Riccati equation (9.33) may be applied to solve the Riccati equation (9.71) in filtering. These methods are not discussed here.

**Example 9.22.** Consider the system $\dot{x} = Ax$, $y = Cx$, where $A = \begin{bmatrix} 0 & 0 \\ 1 & 0 \end{bmatrix}$, $C = [0, 1]$, and let $\varGamma = \begin{bmatrix} 1 \\ 0 \end{bmatrix}$, $V = \rho > 0$, $W = 1$. We wish to derive the optimal filter gain $K^* = P_e^* C^T V^{-1}$ given in (9.72). In this case, the Riccati equation (9.73) is precisely the Riccati equation of the control problem given in Example 9.14. The solution of this equation was determined to be

$$P_e^* = \begin{bmatrix} \sqrt{2\sqrt{\rho}} & \sqrt{\rho} \\ \sqrt{\rho} & \sqrt{2\rho\sqrt{\rho}} \end{bmatrix}.$$

We note that this was expected, since our example was chosen to satisfy (9.74). Therefore,

$$K^* = P_e^* \begin{bmatrix} 0 \\ 1 \end{bmatrix} \frac{1}{\rho} = \begin{bmatrix} \sqrt{\rho} \\ \sqrt{2\rho\sqrt{\rho}} \end{bmatrix} \frac{1}{\rho}.$$

### 9.3.4 Full-Order Observers: Discrete-Time Systems

We consider systems described by equations of the form

$$x(k+1) = Ax(k) + Bu(k), \quad y = Cx(k) + Du(k), \tag{9.75}$$

where $A \in R^{n \times n}$, $B \in R^{n \times m}$, $C \in R^{p \times m}$, and $D \in R^{p \times m}$.

The construction of state estimators for discrete-time systems is mostly analogous to the continuous-time case, and the results that we established above for such systems are valid here as well, subject to obvious adjustments and modifications. There are, however, some notable differences. For example, *in discrete-time systems, it is possible to construct a state estimator that converges to the true value of the state in finite time, instead of infinite time as in the case of asymptotic state estimators.* This is the estimator known as the *deadbeat observer.* Furthermore, in discrete-time systems it is possible to talk about *current state estimators,* in addition to *prediction state estimators.* In what follows, a brief description of the results that are analogous to the continuous-time case is given. Current estimators and deadbeat observers that are unique to the discrete-time case are discussed at greater length.

### Full-State Observers: The Identity Observer

As in the continuous-time case, following (9.57) we consider systems described by equations of the form

$$\hat{x}(k+1) = A\hat{x}(k) + Bu(k) + K[y(k) - \hat{y}(k)], \tag{9.76}$$

where $\hat{y}(k) \triangleq C\hat{x}(k) + Dx(k)$. This can also be written as

$$\hat{x}(k+1) = (A - KC)\hat{x}(k) + [B - KD, K]\begin{bmatrix} u(k) \\ y(k) \end{bmatrix}. \qquad (9.77)$$

It can be shown that the error $e(k) \triangleq x(k) - \hat{x}(k)$ obeys the equation $e(k+1) = (A - KC)e(k)$. Therefore, if the eigenvalues of $A - KC$ are inside the open unit disk of the complex plane, then $e(k) \to 0$ as $k \to \infty$. There exists $K$ so that the eigenvalues of $A - KC$ can be arbitrarily assigned if and only if the pair $(A, C)$ is observable (see Lemma 9.18).

The discussion following Lemma 9.18 for the case when $(A, C)$ is not completely observable, although detectable, is still valid. Also, the remarks on appropriate locations for the eigenvalues of $A - KC$ and noise being the limiting factor in state estimators are also valid in the present case. Note that the latter point should seriously be considered when deciding whether to use the deadbeat observer described next.

To balance the tradeoffs between speed of the estimator response and noise amplification, one may formulate an *optimal estimation problem* as was done in the continuous-time case, the *Linear Quadratic Gaussian (LQG)* design being a common formulation. The Kalman filter (discrete-time case) that is based on the "current estimator" described below is such a quadratic estimator. The LQG optimal estimation problem can be seen to be the dual of the quadratic optimal control problem discussed in the previous section. As in the continuous-time case, optimal estimation in the discrete-time case will be discussed only briefly as follows. First, however, several other related issues are addressed.

*Deadbeat Observer*

If the pair $(A, C)$ is observable, it is possible to select $K$ so that all the eigenvalues of $A - KC$ are at the origin. In this case $e(k) = x(k) - \hat{x}(k) = (A - KC)^k e(0) = 0$, for some $k \le n$; i.e., the error will be identically zero within at most $n$ steps. The minimum value of $k$ for which $(A - KC)^k = 0$ depends on the size of the largest block on the diagonal of the Jordan canonical form of $A - KC$. (Refer to the discussion on the modes of discrete-time systems in Subsection 3.5.5.)

---

**Example 9.23.** Consider the system $x(k+1) = Ax(k)$, $y(k) = Cx(k)$, where

$$A = \begin{bmatrix} 0 & 2 & 1 \\ 1 & -1 & 0 \\ \hline 0 & 0 & 0 \end{bmatrix}, \quad C = \begin{bmatrix} 0 & 1 & 0 \\ 0 & 1 & 1 \end{bmatrix}$$

is in observer form. We wish to design a deadbeat observer. It is rather easy to show (compare with Example 9.11) that

$$K = \begin{bmatrix} A_{dm}^T - \begin{bmatrix} 2 & 1 \\ -1 & 0 \\ 0 & 0 \end{bmatrix} \end{bmatrix} \begin{bmatrix} -1 & 0 \\ 1 & -1 \end{bmatrix},$$

which was determined by taking the dual $A_D = A^T, B_D = C^T$ in controller form, using $F_D = B_m^{-1}[A_{d_m} - A_m]$ and $K = -F_D^T$.

The matrix $A_{d_m}^T$ consists of the second and third columns of a matrix

$$A_d = \begin{bmatrix} 0 & \times & \times \\ 1 & \times & \times \\ 0 & \times & \times \end{bmatrix} \text{ in observer (companion) form with all its eigenvalues at } 0.$$

For $A_{d_1} = \begin{bmatrix} 0 & 0 & 0 \\ 1 & 0 & 0 \\ 0 & 1 & 0 \end{bmatrix}$, we have

$$K_1 = \left[ \begin{bmatrix} 0 & 0 \\ 0 & 0 \\ 1 & 0 \end{bmatrix} - \begin{bmatrix} 2 & 1 \\ -1 & 0 \\ 0 & 0 \end{bmatrix} \right] \begin{bmatrix} -1 & 0 \\ 1 & -1 \end{bmatrix} = \begin{bmatrix} 1 & 1 \\ -1 & 0 \\ -1 & 0 \end{bmatrix},$$

and for $A_{d_2} = \begin{bmatrix} 0 & 0 & 0 \\ 1 & 0 & 0 \\ 0 & 0 & 0 \end{bmatrix}$, we obtain

$$K_2 = \left[ \begin{bmatrix} 0 & 0 \\ 0 & 0 \\ 0 & 0 \end{bmatrix} - \begin{bmatrix} 2 & 1 \\ -1 & 0 \\ 0 & 0 \end{bmatrix} \right] \begin{bmatrix} -1 & 0 \\ 1 & -1 \end{bmatrix} = \begin{bmatrix} 1 & 1 \\ -1 & 0 \\ 0 & 0 \end{bmatrix}.$$

Note that $A - K_1 C = A_{d_1}$, $A_{d_1}^2 = \begin{bmatrix} 0 & 0 & 0 \\ 0 & 0 & 0 \\ 1 & 0 & 0 \end{bmatrix}$, and $A_{d_1}^3 = 0$, and that $A - K_2 C = A_{d_2}$ and $A_{d_2}^2 = 0$. Therefore, for the observer gain $K_1$, the error $e(k)$ in the deadbeat observer will become zero in $n = 3$ steps, since $e(3) = (A - K_1 C)^3 e(0) = 0$. For the observer gain $K_2$, the error $e(k)$ in the deadbeat observer will become zero in $2 < n$ steps, since $e(2) = (A - K_2 C)^2 e(0) = 0$. The reader should determine the Jordan canonical forms of $A_{d_1}$ and $A_{d_2}$ and verify that the dimension of the largest block on the diagonal is 3 and 2, respectively.

---

The comments in the discussion following Lemma 9.18 on taking $K = 0$ are valid in the discrete-time case as well. Also, the approach of determining the state instantaneously in the continuous-time case, using the derivatives of the input and output, corresponds in the discrete-time case to determining the state from current and future input and output values (see Exercise 5.12 in Chapter 5). This approach was in fact used to determine $x(0)$ when studying observability in Section 5.4. The disadvantage of this method is that it requires future measurements to calculate the current state. This issue of using future or past measurements to determine the current state is elaborated upon next.

*Current Estimator*

The estimator (9.76) is called a *prediction estimator*. The state estimate $\hat{x}(k)$ is based on measurements up to and including $y(k - 1)$. It is often of interest in applications to determine the state estimate $\hat{x}(k)$ based on measurements up to and including $y(k)$. This may seem rather odd at first; however, if the computation time required to calculate $\hat{x}(k)$ is short compared with the sample period in a sampled-data system, then it is certainly possible practically to determine the estimate $\hat{x}(k)$ before $x(k+1)$ and $y(k+1)$ are generated by the system. If this state estimate, which is based on current measurements of $y(k)$, is to be used to control the system, then the unavoidable computational delays should be taken into consideration.

Now let $\bar{x}(k)$ denote the current state estimate based on measurements up through $y(k)$. Consider the current estimator

$$\bar{x}(k) = \hat{x}(k) + K_c(y(k) - C\hat{x}(k)),\tag{9.78}$$

where

$$\hat{x}(k) = A\bar{x}(k-1) + Bu(k-1);\tag{9.79}$$

i.e., $\hat{x}(k)$ denotes the estimate based on model prediction from the previous time estimate, $\bar{x}(k-1)$. Note that in (9.78), the error is $y(k) - \hat{y}(k)$, where $\hat{y}(k) = C\hat{x}(k)$ $(D = 0)$, for simplicity.

Combining the above, we obtain

$$\hat{x}(k) = (I - K_cC)A\bar{x}(k-1) + [(I - K_cC)B, -K_c]\begin{bmatrix} u(k-1) \\ y(k) \end{bmatrix}.\tag{9.80}$$

The relation to the prediction estimator (9.76) can be seen by substituting (9.78) into (9.79) to obtain

$$\hat{x}(k+1) = A\hat{x}(k) + Bu(k) + AK_c[y(k) - C\hat{x}(k)].\tag{9.81}$$

Comparison with the prediction estimator (9.76) (with $D = 0$) shows that it is clear that if

$$K = AK_c,\tag{9.82}$$

then (9.81) is indeed the prediction estimator, and the estimate $\hat{x}(k)$ used in the current estimator (9.78) is indeed the prediction state estimate. In view of this, we expect to obtain for the error $\hat{e}(k) = x(k) - \hat{x}(k)$ the difference equation

$$\hat{e}(k+1) = (A - AK_cC)\hat{e}(k).\tag{9.83}$$

To determine the error $\bar{e}(k) = x(k) - \bar{x}(k)$ we note that $\bar{e}(k) = \hat{e}(k) - (\bar{x}(k) - \hat{x}(k))$. Equation (9.78) now implies that $\bar{x}(k) - \hat{x}(k) = K_cCe(k)$. Therefore,

$$\bar{e}(k) = (I - K_cC)\hat{e}(k).\tag{9.84}$$

This establishes the relationship between errors in current and prediction estimators.

Premultiplying (9.81) by $I - K_cC$ (assuming $|I - K_cC| \neq 0$), we obtain

$$\bar{e}(k+1) = (A - K_cCA)\bar{e}(k),\tag{9.85}$$

which is the current estimator error equation. The gain $K_c$ is chosen so that the eigenvalues of $A - K_c CA$ are within the open unit disk of the complex plane. The pair $(A, CA)$ must be observable for arbitrary eigenvalue assignment. Note that the two error equations (9.83) and (9.85) have identical eigenvalues.

---

**Example 9.24.** Consider the system $x(k+1) = Ax(k)$, $y(k) = Cx(k)$, where $A = \begin{bmatrix} 0 & -2 \\ 1 & -2 \end{bmatrix}$, $C = [0, 1]$, which is in observer form (see also Example 9.20). We wish to design a current estimator. In view of the error equation (9.85), we consider

$$\det(sI - (A - K_c CA)) = \det\left(\begin{bmatrix} s & 0 \\ 0 & s \end{bmatrix} - \left(\begin{bmatrix} 0 & -2 \\ 1 & -2 \end{bmatrix} - \begin{bmatrix} k_0 \\ k_1 \end{bmatrix} [1 \ -2]\right)\right)$$

$$= \det \begin{bmatrix} s + k_0 & 2 - 2k_0 \\ k_1 - 1 & s + 2 - 2k_1 \end{bmatrix}$$

$$= s^2 + s(2 - 2k_1 + k_0) + (2 - 2k_1)$$

$$= s^2 + d_1 s + d_0 = \alpha_d(s),$$

a desired polynomial, from which $K_c = [k_0, k_1]^T = [d_1 - d_0, \frac{1}{2}(2 - d_0)]^T$. Note that $AK_c = [d_0 - 2, d_1 - 2]^T = K$, found in Example 9.20, as noted in (9.82).

The current estimator (9.80) is now given by $\bar{x}(k) = (A - K_c CA)\bar{x}(k - 1) - K_c CBu(k - 1) + K_c y(k)$, or

$$\bar{x}(k) = \begin{bmatrix} -k_0 & -2 + 2k_0 \\ 1 - k_1 & -2 + 2k_1 \end{bmatrix} \bar{x}(k - 1) + \begin{bmatrix} k_0 \\ k_1 \end{bmatrix} y(k).$$

---

**Partial or Linear Functional State Observers**

The problem of estimating a linear function of the state, $Tx(k)$, $T \in R^{\tilde{n} \times n}$, where $\tilde{n} \leq n$, using a prediction estimator, is completely analogous to the continuous-time case.

### 9.3.5 Reduced-Order Observers: Discrete-Time Systems

It is possible to estimate the full state $x(k)$ using an estimator of order $n - p$, where $p = \text{rank}\, C$. If a prediction estimator is used for that part of the state that needs to be estimated, then the problem in the discrete-time case is completely analogous to the continuous-time case, discussed before.

### 9.3.6 Optimal State Estimation: Discrete-Time Systems

The formulation of the Kalman filtering problem in discrete-time is analogous to the continuous-time case.

Consider the linear time-invariant system given by

$$x(k+1) = Ax(k) + Bu(k) + \Gamma w(k), \quad y(k) = Cx(k) + v, \qquad (9.86)$$

where the process and measurement noises $w, v$ are white, zero-mean Gaussian stochastic processes; i.e., they are uncorrelated in time with $E[w] = 0$ and $E[v] = 0$. Let the covariances be given by

$$E[ww^T] = W, \quad E[vv^T] = V, \qquad (9.87)$$

where $W = W^T, W > 0$ and $V = V^T, V > 0$. Assume that $w, v$ are independent, that the initial state $x(0)$ is Gaussian of known mean $(E[x(0)] = x_0)$, that $E[(x(0) - x_0)(x(0) - x_0)^T] = P_{e0}$, and that $x(0)$ is independent of $w$ and $v$.

Consider now the current estimator (9.76), namely,

$$\bar{x}(k) = \hat{x}(k) + K_c[y(k) - C\hat{x}(k)],$$

where $\hat{x}(k) = A\bar{x}(k-1) + Bu(k-1)$ and $\hat{x}(k)$ denotes the prior estimate of the state at the time of a measurement.

It turns out that the state error covariance is minimized when the filter gain is

$$K_c^* = P_e^* C^T (C P_e^* C^T + V)^{-1}, \qquad (9.88)$$

where $P_e^*$ is the unique, symmetric, positive definite solution of the Riccati equation

$$P_e = A[P_e - P_e C^T [C P_e C^T + V]^{-1} C P_e] A^T + \Gamma W \Gamma^T. \qquad (9.89)$$

It is assumed here that $(A, \Gamma W^{1/2}, C)$ is reachable and observable. This *algebraic Riccati equation* is the dual to the Riccati equation (9.53) that arose in the discrete-time LQR problem and can be obtained by substituting

$$A \rightarrow A^T, B \rightarrow C^T, M \rightarrow \Gamma^T \quad \text{and} \quad R \rightarrow V, Q \rightarrow W. \qquad (9.90)$$

It is clear that, as in the case of the LQR problem, the solution of the algebraic Riccati equation can be determined using the eigenvectors of the (dual) Hamiltonian.

The filter derived above is called the *discrete-time Kalman filter*. It is based on the current estimator (9.78). Note that $AK_c$ yields the gain $K$ of the prediction estimator [see (9.82)].

## 9.4 Observer-Based Dynamic Controllers

State estimates, derived by the methods described in the previous section, may be used in state feedback control laws to compensate given systems. This section addresses this topic.

In Section 9.2, the linear state feedback control law was introduced. There it was implicitly assumed that the state vector $x(t)$ is available for measurement. The values of the states $x(t)$ for $t \geq t_0$ were fed back and used to generate a control input in accordance with the relation $u(t) = Fx(t) + r(t)$. There are cases, however, when it may be either impossible or impractical to measure the states directly. This has provided the motivation to develop methods for estimating the states. Some of these methods were considered in Section 9.3. A natural question that arises at this time is the following: What would happen to system performance if, in the control law $u = Fx + r$, the state estimate $\hat{x}$ were used in place of $x$ as in Figure 9.4? How much, if any, would the compensated system response deteriorate? What are the difficulties in designing such estimator-(observer-)based linear state feedback controllers? These questions are addressed in this section. Note that observer-based controllers of the type described in the following are widely used.

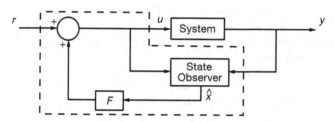

**Figure 9.4.** Observer-based controller

In the remainder of this section we will concentrate primarily on full-state/full-order observers and (static) linear state feedback, as applied to linear time-invariant systems. The analysis of partial-state and/or reduced-order observers with static or dynamic state feedback is analogous; however, it is more complex. In this section, continuous-time systems are addressed. The discrete-time case is completely analogous and will be omitted.

### 9.4.1 State-Space Analysis

We consider systems described by equations of the form

$$\dot{x} = Ax + Bu, \quad y = Cx + Du, \tag{9.91}$$

where $A \in R^{n \times n}$, $B \in R^{n \times m}$, $C \in R^{p \times n}$, and $D \in R^{p \times m}$. For such systems, we determine an estimate $\hat{x}(t) \in R^n$ of the state $x(t)$ via the (full-state/full-order) *state observer* (9.57) given by

$$\dot{\hat{x}} = A\hat{x} + Bu + K(y - \hat{y})$$

$$= (A - KC)\hat{x} + [B - KD, K] \begin{bmatrix} u \\ y \end{bmatrix},$$

$$z = \hat{x}, \tag{9.92}$$

where $\hat{y} = C\hat{x} + Du$. We now compensate the system by *state feedback* using the control law

$$u = F\hat{x} + r, \tag{9.93}$$

where $\hat{x}$ is the output of the state estimator and we wish to analyze the behavior of the compensated system. To this end we first eliminate $y$ in (9.92) to obtain

$$\dot{\hat{x}} = (A - KC)\hat{x} + KCx + Bu. \tag{9.94}$$

The state equations of the compensated system are then given by

$$\dot{x} = Ax + BF\hat{x} + Br,$$
$$\dot{\hat{x}} = KCx + (A - KC + BF)\hat{x} + Br, \tag{9.95}$$

and the output equation assumes the form

$$y = Cx + DF\hat{x} + Dr, \tag{9.96}$$

where $u$ was eliminated from (9.91) and (9.94), using (9.93). Rewriting in matrix form, we have

$$\begin{bmatrix} \dot{x} \\ \dot{\hat{x}} \end{bmatrix} = \begin{bmatrix} A & BF \\ KC & A - KC + BF \end{bmatrix} \begin{bmatrix} x \\ \hat{x} \end{bmatrix} + \begin{bmatrix} B \\ B \end{bmatrix} r,$$
$$y = [C,\ DF] \begin{bmatrix} x \\ \hat{x} \end{bmatrix} + Dr, \tag{9.97}$$

which is a representation of the compensated closed-loop system. Note that (9.97) constitutes a $2n$th-order system. Its properties are more easily studied if an appropriate similarity transformation is used to simplify the representation. Such a transformation is given by

$$P \begin{bmatrix} x \\ \hat{x} \end{bmatrix} = \begin{bmatrix} I & 0 \\ I & -I \end{bmatrix} \begin{bmatrix} x \\ \hat{x} \end{bmatrix} = \begin{bmatrix} x \\ e \end{bmatrix}, \tag{9.98}$$

where the error $e(t) = x(t) - \hat{x}(t)$. Then the equivalent representation is

$$\begin{bmatrix} \dot{x} \\ \dot{e} \end{bmatrix} = \begin{bmatrix} A + BF & -BF \\ 0 & A - KC \end{bmatrix} \begin{bmatrix} x \\ e \end{bmatrix} + \begin{bmatrix} B \\ 0 \end{bmatrix} r,$$
$$y = [C + DF,\ -DF] \begin{bmatrix} x \\ e \end{bmatrix} + Dr. \tag{9.99}$$

It is now quite clear that the closed-loop system is not fully controllable with respect to $r$ (this can be explained in view of Subsection 6.2.1). In fact, $e(t)$ does not depend on $r$ at all. This is of course as it should be, since the error $e(t) = x(t) - \hat{x}(t)$ should converge to zero independently of the externally applied input $r$.

The closed-loop eigenvalues are the roots of the polynomial

$$|sI_n - (A + BF)||sI_n - (A - KC)|. \tag{9.100}$$

Recall that the roots of $|sI_n - (A + BF)|$ are the eigenvalues of $A + BF$ that can arbitrarily be assigned via $F$ provided that the pair $(A, B)$ is controllable. These are in fact the closed-loop eigenvalues of the system when the state $x$ is available and the linear state feedback control law $u = Fx + r$ is used (see Section 9.2). The roots of $|sI_n - (A - KC)|$ are the eigenvalues of $(A - KC)$ that can arbitrarily be assigned via $K$ provided that the pair $(A, C)$ is observable. These are the eigenvalues of the estimator (9.92).

The above discussion points out that the *design of the control law (9.93) can be carried out independently of the design of the estimator (9.92)*. This is referred to as the *Separation Property* and is generally not true for more complex systems. The separation property indicates that the linear state feedback control law may be designed as if the state $x$ were available and the eigenvalues of $A + BF$ are assigned at appropriate locations. The feedback matrix $F$ can also be determined by solving an optimal control problem (LQR). If state measurements are not available for feedback, a state estimator is employed. The eigenvalues of a full-state/full-order estimator are given by the eigenvalues of $A - KC$. These are typically assigned so that the error $e(t) = x(t) - \hat{x}(t)$ becomes adequately small in a short period of time. For this reason, the eigenvalues of $A - KC$ are (empirically) taken to be about 6 to 10 times further away from the imaginary axis (in the complex plane, for continuous-time systems) than the eigenvalues of $A + BF$. The estimator gain $K$ may also be determined by solving an optimal estimation problem (the Kalman filter). In fact, under the assumption of Gaussian noise and initial conditions given earlier (see Section 9.3), $F$ and $K$ can be found by solving, respectively, optimal control and estimation problems with quadratic performance criteria. In particular, the deterministic LQR problem is first solved to determine the optimal control gain $F^*$, and then the stochastic Kalman filtering problem is solved to determine the optimal filter gain $K^*$. The separation property (i.e., Separation Theorem—see any optimal control textbook) guarantees that the overall *(state estimate feedback) Linear Quadratic Gaussian (LQG) control design* is optimal in the sense that the control law $u^*(t) = F^*\hat{x}(t)$ minimizes the quadratic performance index $E[\int_0^\infty (z^T Q z + u^T R u) dt]$. As was discussed in previous sections, the gain matrices $F^*$ and $K^*$ are evaluated in the following manner.

Consider

$$\dot{x} = Ax + Bu + \Gamma w, \quad y = Cx + v, z = Mx \tag{9.101}$$

with $E\{ww^T\} = W > 0$ and $E[vv^T] = V > 0$ and with $Q > 0, R > 0$ denoting the matrix weights in the performance index $E[\int_0^\infty (z^T Q x + u^T R u) dt]$. Assume that both $(A, B, Q^{1/2}M)$ and $(A, \Gamma W^{1/2}, C)$ are controllable and observable. Then the optimal control law is given by

$$u^*(t) = F^*\hat{x}(t) = -R^{-1}B^T P_c^* \hat{x}(t), \tag{9.102}$$

where $P_c^* > 0$ is the solution of the algebraic Riccati equation (9.33) given by

$$A^T P_c + P_c A - P_c B R^{-1} B^T P_c + M^T Q M = 0. \tag{9.103}$$

The estimate $\hat{x}$ is generated by the optimal estimator

$$\dot{\hat{x}} = A\hat{x} + Bu + K^*(y - C\hat{x}), \tag{9.104}$$

where

$$K^* = P_e^* C^T V^{-1}, \tag{9.105}$$

in which $P_e^* > 0$ is the solution to the dual algebraic Riccati equation (9.71) given by

$$P_e A^T + A P_e - P_e C^T V^{-1} C P_e + \Gamma W \Gamma^T = 0. \tag{9.106}$$

Designing observer-based dynamic controllers by the LQG control design method has been quite successful, especially when the plant model is accurately known. In this approach the weight matrices $Q$, $R$ and the covariance matrices $W$, $V$ are used as design parameters. Unfortunately, this method does not necessarily lead to robust designs when uncertainties are present. This limitation has led to an enhancement of this method, called the LQR/LTR (Loop Transfer Recovery) method, where the design parameters $W$ and $V$ are selected (iteratively) so that the robustness properties of the LQR design are recovered.

Finally, as mentioned earlier, the discrete-time case is analogous to the continuous-time case and its discussion will be omitted.

---

**Example 9.25.** Consider the system $\dot{x} = Ax + Bu$, $y = Cx$, where

$$A = \begin{bmatrix} 0 & 1 & 0 \\ 0 & 0 & 1 \\ 0 & 2 & -1 \end{bmatrix}, \quad B = \begin{bmatrix} 0 & 1 \\ 1 & 1 \\ 0 & 0 \end{bmatrix}, \quad C = [1, 0, 0].$$

This is a controllable and observable but unstable system with eigenvalues of $A$ equal to $0, -2, 1$. A linear state feedback control $u = Fx + r$ was derived in Example 9.11 to assign the eigenvalues of $A + BF$ at $-2, -1 \pm j$. An appropriate $F$ to accomplish this was shown to be

$$F = \begin{bmatrix} 2 & -1 & -2 \\ -2 & 0 & 1/2 \end{bmatrix}.$$

If the state $x(t)$ is not available for measurement, then an estimate $\hat{x}(t)$ is used instead; i.e., the control law $u = F\hat{x} + r$ is employed. In Example 9.19, a full-order/full-state observer, given by

$$\dot{\hat{x}} = (A - KC)\hat{x} + [B, K] \begin{bmatrix} u \\ y \end{bmatrix},$$

was derived [see (9.58)] with the eigenvalues of $A - KC$ determined as the roots of the polynomial $\alpha_d(s) = s^3 + d_2 s^2 + d_1 s + d_0$. It was shown that the (unique) $K$ is in this case

$$K = [d_2 - 1, d_1 - d_2 + 3, d_0 - d_1 + 3d_2 - 5]^T,$$

and the observer is given by

$$\dot{\hat{x}} = \begin{bmatrix} 1 - d_2 & 1 & 0 \\ -d_1 + d_2 - 3 & 0 & 1 \\ -d_0 + d_1 - 3d_2 + 5 & 2 & -1 \end{bmatrix} \hat{x} + \begin{bmatrix} 0 & 1 & d_2 - 1 \\ 1 & 1 & d_1 - d_2 + 3 \\ 0 & 0 & d_0 - d_1 + 3d_2 - 5 \end{bmatrix} \begin{bmatrix} u \\ y \end{bmatrix}.$$

Using the estimate $\hat{x}$ in place of the control state $x$ in the feedback control law causes some deterioration in the behavior of the system. This deterioration can be studied experimentally. (See the next subsection for analytical results.) To this end, let the eigenvalues of the observer be at, say, $-10, -10, -10$; let $x(0) = [1, 1, 1]^T$ and $\hat{x}(0) = [0, 0, 0]^T$; plot $x(t)$, $\hat{x}(t)$, and $e(t) = x(t) - \hat{x}(t)$; and compare these with the corresponding plots of Example 9.11, where no observer was used. Repeat the above with observer eigenvalues closer to the eigenvalues of $A + BF$ (say, at $-2, -1 \pm j$) and also further away. In general the faster the observer, the faster $e(t) \to 0$, and the smaller the deterioration of response; however, in this case, care should be taken if noise is present in the system.

---

### 9.4.2 Transfer Function Analysis

For the compensated system (9.99) [or (9.97)], the closed-loop transfer function $T(s)$ between $y$ and $r$ is given by

$$\tilde{y}(s) = T(s)\tilde{r}(s) = [(C + DF)[sI - (A + BF)]^{-1}B + D]\tilde{r}(s), \qquad (9.107)$$

where $\tilde{y}(s)$ and $\tilde{r}(s)$ denote the Laplace transforms of $y(t)$ and $r(t)$, respectively. The function $T(s)$ was found from (9.99), using the fact that the uncontrollable part of the system does not appear in the transfer function (see Section 7.2). Note that $T(s)$ is the transfer function of $\{A+BF, B, C+DF, D\}$; i.e., $T(s)$ is precisely the transfer function of the closed-loop system $H_F(s)$ when no state estimation is present (see Section 9.2). Therefore, the compensated system behaves to the outside world as if there were no estimator present. *Note that this statement is true only after sufficient time has elapsed from the initial time, allowing the transients to become negligible.* (Recall what the transfer function represents in a system.) Specifically, taking Laplace transforms in (9.99) and solving, we obtain

$$\begin{bmatrix} \tilde{x}(s) \\ \tilde{e}(s) \end{bmatrix} = \begin{bmatrix} [sI-(A+BF)]^{-1} & -[sI-(A+BF)]^{-1}BF[sI-(A-KC)]^{-1} \\ 0 & [sI-(A+BF)]^{-1} \end{bmatrix} \begin{bmatrix} x(0) \\ e(0) \end{bmatrix}$$

$$+ \begin{bmatrix} [sI-(A+BF)]^{-1}B \\ 0 \end{bmatrix} \tilde{r}(s),$$

$$\tilde{y}(s) = [C + DF, -DF] \begin{bmatrix} \tilde{x}(s) \\ \tilde{e}(s) \end{bmatrix} + D\tilde{r}(s). \qquad (9.108)$$

Therefore,

$$\tilde{y}(s) = (C + DF)[sI - (A + BF)]^{-1}x(0)$$
$$- [(C + DF)[sI - (A + BF)]^{-1}BF[sI - (A - KC)]^{-1}$$
$$+ DF[sI - (A + BF)]^{-1}]e(0) + T(s)\tilde{r}(s), \qquad (9.109)$$

which indicates the effects of the estimator on the input–output behavior of the closed-loop system. Notice how the initial conditions for the error $e(0) = x(0) - \hat{x}(0)$ influence the response. Specifically, when $e(0) \neq 0$, its effect can be viewed as a disturbance that will become negligible at steady state. The speed by which the effect of $e(0)$ on $y$ will diminish depends on the location of the eigenvalues of $A + BF$ and $A - KC$, as can be easily seen from relation (9.109).

## Two-Input Controller

In the following discussion, we will find it of interest to view the observer-based controller discussed previously as a one-vector output $(u)$ and a two-vector input $(y$ and $r)$ controller. In particular, from $\dot{\hat{x}} = (A - KC)\hat{x} + (B - KD)u + Ky$ given in (9.92) and $u = F\hat{x} + r$ given in (9.93), we obtain the equations

$$\dot{\hat{x}} = (A - KC + BF - KDF)\hat{x} + [K, B - KD]\begin{bmatrix} y \\ r \end{bmatrix},$$

$$u = F\hat{x} + r. \qquad (9.110)$$

This is the description of the ($n$th order) controller shown in Figure 9.4. The state $\hat{x}$ is of course the state of the estimator, and the transfer function between $u$ and $y, r$ is given by

$$\tilde{u}(s) = F[sI - (A - KC + BF - KDF)]^{-1}K\tilde{y}(s)$$
$$+ [F[sI - (A - KC + BF - KDF)]^{-1}(B - KD) + I]\tilde{r}(s). \quad (9.111)$$

If we are interested only in "loop properties," then $r$ can be taken to be zero; in which case, (9.111) (for $r = 0$) yields the output feedback compensator, which accomplishes the same control objectives (that are typically only "loop properties") as the original observer-based controller. This fact is used in the LQG/LTR design approach. When $r \neq 0$, (9.111) is not appropriate for the realization of the controller since the transfer function from $r$, which must be outside the loop, may be unstable. Note that an expression for this controller that leads to a realization of a stable closed-loop system is given by

$$\tilde{u}(s) = [F[sI - (A - KC + BF - KDF)]^{-1}[K, B - KD] + [0, I]]\begin{bmatrix} \tilde{y}(s) \\ \tilde{r}(s) \end{bmatrix} \quad (9.112)$$

(see Figure 9.5). This was also derived from (9.110). The stability of general two-input controllers (with two degrees of freedom) is discussed at length in Chapter 10.

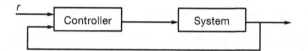

**Figure 9.5.** Two-input controller

At this point, we find it of interest to determine the relationship of the observer-based controller and the conventional block controller configuration of Figure 9.6. Here, the requirement is to maintain the same transfer functions between inputs $y$ and $r$ and output $u$. (For further discussion of stability and attainable response maps in system controlled by output feedback controllers, refer to Chapter 10.) We proceed by considering once more (9.92) and (9.93) and by writing

$$\tilde{u}(s) = F[sI - (A - KC)]^{-1}(B - KD)\tilde{u}(s)$$
$$+ F[sI - (A - KC)]^{-1}K\tilde{y}(s) + \tilde{r}(s) = G_u\tilde{u}(s) + G_y\tilde{y}(s) + \tilde{r}(s).$$

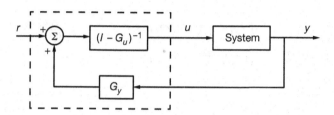

**Figure 9.6.** Conventional block controller configuration

This yields

$$\tilde{u}(s) = (I - G_u)^{-1}[G_y\tilde{y}(s) + \tilde{r}(s)] \tag{9.113}$$

(see Figure 9.6). Notice that

$$G_y = F[sI - (A - KC)]^{-1}K; \tag{9.114}$$

i.e., the controller in the feedback path is stable. The matrix $(I - G_u)^{-1}$ is not necessarily stable; however, it is inside the loop and therefore the internal stability of the compensated system is preserved. Comparing with (9.111), we obtain

$$(I - G_u)^{-1} = F[sI - (A - KC + BF - KDF)]^{-1}(B - KD) + I. \tag{9.115}$$

Also, as expected, we have

$$(I - G_u)^{-1}G_y = F[sI - (A - KC + BF - KDF)]^{-1}K. \tag{9.116}$$

These relations could have been derived directly as well by the use of matrix identities; however, such derivation is quite involved.

**Example 9.26.** For the system $\dot{x} = Ax + Bu$, $y = Cx$ with $A = \begin{bmatrix} 0 & -2 \\ 1 & -2 \end{bmatrix}$, $B = \begin{bmatrix} 0 \\ 1 \end{bmatrix}$, and $C = [0, 1]$, we have $H(s) = C(sI - A)^{-1}B = \frac{s}{s^2+2s+2}$. In Example 9.20, it was shown that the gain matrix $K = [d_0 - 2, d_1 - 2]^T$ assigns the eigenvalues of the asymptotic observer (of $A - KC$) at the roots of $s^2 + d_1 s + d_0$. In fact $sI - (A - KC) = \begin{bmatrix} s & d_0 \\ -1 & s + d_1 \end{bmatrix}$. It is straightforward to show that $F = [\frac{1}{2}a_0 - 1, 2 - a_1]$ will assign the eigenvalues of the closed-loop system (of $A + BF$) at the roots of $s^2 + a_1 s + a_0$. Indeed, $sI - (A + BF) = \begin{bmatrix} s & 2 \\ -\frac{1}{2}a_0 & s + a_1 \end{bmatrix}$. Now in (9.113) we have

$$G_y(s) = F(sI - (A - KC))^{-1}K$$
$$= \frac{s((d_0-2)(\frac{1}{2}a_0-1)+(d_1-2)(2-a_1))+((d_0-d_1)(a_0-2)+(d_0-2)(2-a_1))}{s^2+d_1s+d_0},$$

$$G_u(s) = F(sI - (A - KC))^{-1} B = \frac{s(2-a_1)-d_0(\frac{1}{2}a_0-1)}{s^2+d_1s+d_0},$$

$$(1 - G_u)^{-1} = \frac{s^2+d_1s+d_0}{s^2+s(d_1+a_1-2)+\frac{1}{2}a_0d_0}.$$

## 9.5 Summary and Highlights

*Linear State Feedback*

- Given $\dot{x} = Ax + Bu$, $y = Cx + Du$ and the linear state feedback control law $u = Fx + r$, the closed-loop system is

$$\dot{x} = (A + BF)x + Br, \quad y = (C + DF)x + Dr. \tag{9.3}$$

- If $u$ were implemented via open-loop control, it would be given by

$$\hat{u} = F[sI - (A + BF)]^{-1}x(0) + [I - F(sI - A)^{-1}B]^{-1}\hat{r}. \tag{9.4}$$

- The eigenvalues of $A + BF$ can be assigned to arbitrary real and/or complex conjugate locations by selecting $F$ if and only if the system [or $(A, B)$] is controllable. The uncontrollable eigenvalues of $(A, B)$ cannot be shifted (see Theorem 9.2 and Lemma 9.3).
- Methods to select $F$ to assign the closed-loop eigenvalues in $A + BF$ include
  1. the direct method (see (9.12)), and

2. using controller forms (see (9.20 and (9.24)).
Controller forms are used to derive Ackermann's formula ($m = 1$)

$$F = -e_n^T C^{-1} \alpha_d(A),$$ (9.21)

where $e_n = [0, \ldots, 0, 1]^T$, $\mathcal{C} = [B, \ldots, A^{n-1}B]$ is the controllability matrix and $\alpha_d(s)$ is the desired closed-loop characteristic polynomial (its roots are the desired eigenvalues).

3. Assigning eigenvalues and eigenvectors (see Theorem 9.12).
The flexibility in choosing $F$ that assigns the $n$ closed-loop eignvalues (when $m > 1$) is expressed in terms of desired closed-loop eigenvectors that can be partially assigned,

$$FV = W,$$ (9.29)

where $V \triangleq [M_1 a_1, \ldots, M_n a_n]$ and $W \triangleq [D_1 a_1, \ldots, D_n a_n]$ uniquely specify $F$ as the solution to these $n$ linearly independent equations. When $s_j$ are distinct, the $n$ vectors $M_j a_j$, $j = 1, \ldots, n$, are linearly independent for almost any nonzero $a_j$, and $V$ has full rank.

- Optimal Control Linear Quadratic Regulator. Given $\dot{x} = Ax + Bu$, $z = Mx$, find $u(t)$ that minimizes the quadratic cost

$$J(u) = \int_0^\infty [z^T(t)Qz(t) + u^T(t)Ru(t)]dt.$$ (9.31)

Under controllability and observability conditions, the solution is unique and it is given as a linear state feedback control law

$$u^*(t) = F^* x(t) = -R^{-1}B^T P_c^* x(t),$$ (9.32)

where $P_c^*$ is the symmetric, positive definite solution of the algebraic Riccati equation

$$A^T P_c + P_c A - P_c B R^{-1} B^T P_c + M^T Q M = 0.$$ (9.33)

The corresponding discrete-time case optimal control is described in (9.51), (9.52), and (9.53).

- The closed-loop transfer function $H_f(s)$ is given by

$$\begin{aligned}
H_F(s) &= (C + DF)[sI - (A + BF)]^{-1}B + D \\
&= [C(sI - A)^{-1}B + D][I - F(sI - A)^{-1}B]^{-1} \\
&= H(s)[I - F(sI - A)^{-1}B]^{-1} \\
&= H(s)[F(sI - (A + BF))^{-1}B + I].
\end{aligned}$$

See also (9.42) and (9.45). Also

$$\begin{aligned}
H_F(s) &= N(s)D_F^{-1}(s) = [N(s)D^{-1}(s)][D(s)D_F^{-1}(s)] \\
&= H(s)[D(s)D_F^{-1}(s)].
\end{aligned}$$ (9.44)

*Linear State Observers*

- Given $\dot{x} = Ax + Bu$, $y = Cx + Du$, the Luenberger observer is

$$\dot{\hat{x}} = A\hat{x} + Bu + K(y - \hat{y}),$$   (9.57)

where $\hat{y} = C\hat{y} + D$ or

$$\dot{\hat{x}} = (A - KC)\hat{x} + [B - KD, K]\begin{bmatrix} u \\ y \end{bmatrix},$$   (9.58)

where $K$ is chosen so that all the eigenvalues of $A - KC$ have negative real parts. Then the error $e(t) = x(t) - \hat{x}(t)$ will go to zero asymptotically.
- The eigenvalues of $A - KC$ can be assigned to arbitrary real and/or complex conjugate locations by selecting $K$ if and only if the system [or $(A, C)$] is observable. The unobservable eigenvalues of $(A, C)$ cannot be shifted. This is the dual problem to the control problem of assigning eigenvalues in $A + BF$, and the same methods can be used (see Lemma 9.18).
- Optimal State Estimation. Consider $\dot{x} = Ax + Bu + \Gamma w$, $y = Cx + v$, where $w$, $v$ are process and measurement noise. Let the state estimator be

$$\dot{\hat{x}} = A\hat{x} + Bu + K(y - C\hat{x})$$   (9.117)
$$= (A - KC)\hat{x} + Bu + Ky,$$   (9.71)

and consider minimizing the error covariance $E[(x - \hat{x})(x - \hat{x})^T]$. Under certain controllability and observability conditions, the solution is unique and it is given by

$$K^* = P_e^* C^T V^{-1},$$   (9.72)

where $P_e^*$ is the symmetric, positive definite solution of the quadratic (dual) algebraic Riccati equation

$$P_e A^T + AP_e - P_e C^T V^{-1} C P_e + \Gamma W \Gamma^T = 0.$$   (9.73)

This problem is the dual to the Linear Quadratic Regulator problem.
- The discrete-time case is analogous to the continuous-time case [see (9.76)]. The current estimator is given in (9.78)–(9.80). The optimal current estimator is given by (9.88) and (9.89).

*Observer-Based Dynamic Controllers*

- Given $\dot{x} = Ax + Bu$, $y = Cx + Du$ with the state feedback $u = Fx + r$, if the state is estimated via a Luenberger observer, then the closed-loop system is

$$\begin{bmatrix} \dot{x} \\ \dot{e} \end{bmatrix} = \begin{bmatrix} A + BF & -BF \\ 0 & A - KC \end{bmatrix}\begin{bmatrix} x \\ e \end{bmatrix} + \begin{bmatrix} B \\ 0 \end{bmatrix} r,$$   (9.118)

$$y = [C + DF, \ -DF]\begin{bmatrix} x \\ e \end{bmatrix} + Dr.$$   (9.99)

The error $e = x - \hat{x}$.

- The design of the control law $(F)$ can be carried out independently of the design of the estimator $(K)$ [see (9.101)–(9.106]. (Separation property)
- The compensated system behaves to the outside world as if there were no estimator present—after sufficient time so the transients have become negligible [see (9.109)].
- The observer based dynamic controller is a special case of a two degrees of freedom controller [see (9.113)–(9.115)].

## 9.6 Notes

The fact that if a system is (state) controllable, then all its eigenvalues can arbitrarily be assigned by means of linear state feedback has been known since the 1960s. Original sources include Rissanen [20], Popov [18], and Wonham [24]. (See also remarks in Kailath [11, pp. 187, 195].)

The present approach for eigenvalue assignment via linear state feedback, using the controller form, follows the development in Wolovich [23]. Ackermann's formula first appeared in Ackermann [2].

The development of the eigenvector formulas for the feedback matrix that assign all the closed-loop eigenvalues and (in part) the corresponding eigenvectors follows Moore [17]. The corresponding development that uses $(A, B)$ in controller (companion) form and polynomial matrix descriptions follows Antsaklis [4]. Related results on static output feedback and on polynomial and rational matrix interpolation can be found in Antsaklis and Wolovich [5] and Antsaklis and Gao [6]. Note that the flexibility in assigning the eigenvalues via state feedback in the multi-input case can be used to assign the invariant polynomials of $sI - (A+BF)$; conditions for this are given by Rosenbrock [21].

The Linear Quadratic Regulator (LQR) problem and the Linear Quadratic Gaussian (LQG) problem have been studied extensively, particularly in the 1960s and early 1970s. Sources for these topics include the books by Anderson and Moore [3], Kwakernaak and Sivan [12], Lewis [13], and Dorato et al. [10]. Early optimal control sources include Athans and Falb [7] and Bryson and Ho [9]. A very powerful idea in optimal control is the *Principle of Optimality*, Bellman [8], which can be stated as follows: "An optimal trajectory has the property that at any intermediate point, no matter how it was reached, the remaining part of a trajectory must coincide with an optimal trajectory, computed from the intermediate point as the initial point." For historical remarks on this topic, refer, e.g., to Kailath [11, pp. 240–241].

The most influential work on state observers is the work of Luenberger. Although the asymptotic observer presented here is generally attributed to him, Luenberger's Ph.D. thesis work in 1963 was closer to the reduced-order observer presented above. Original sources on state observers include Luenberger [14], [15], and [16]. For an extensive overview of observers, refer to the book by O'Reilly [19].

When linear quadratic optimal controllers and observers are combined in control design, a procedure called LQG/LTR (Loop Transfer Recovery) is used to enhance the robustness properties of the closed-loop system. For a treatment of this procedure, see Stein and Athans [22] and contemporary textbooks on multivariable control.

# References

1. P.J. Antsaklis and A.N. Michel, *Linear Systems*, Birkhäuser, Boston, MA, 2006.
2. J. Ackermann, "Der Entwurf linearer Regelungssysteme im Zustandsraum," *Regelungstechnik und Prozessdatenverarbeitung*, Vol. 7, pp. 297–300, 1972.
3. B.D.O. Anderson and J.B. Moore, *Optimal Control. Linear Quadratic Methods*, Prentice-Hall, Englewood, Cliffs, NJ, 1990.
4. P.J. Antsaklis, "Some new matrix methods applied to multivariable system analysis and design," Ph.D. Dissertation, Brown University, Providence, RI, May 1976.
5. P.J. Antsaklis and W.A. Wolovich, "Arbitrary pole placement using linear output feedback compensation," *Int. J. Control*, Vol. 25, No. 6, pp. 915–925, 1977.
6. P.J. Antsaklis and Z. Gao, "Polynomial and rational matrix interpolation: theory and control applications," *Int. J. Control*, Vol. 58, No. 2, pp. 349–404, August 1993.
7. M. Athans and P.L. Falb, *Optimal Control*, McGraw-Hill, New York, NY, 1966.
8. R. Bellman, *Dynamic Programming*, Princeton University Press, Princeton, NJ, 1957.
9. A.E. Bryson and Y.C. Ho, *Applied Optimal Control*, Holsted Press, New York, NY, 1968.
10. P. Dorato, C. Abdallah, and V. Cerone, *Linear-Quadratic Control: An Introduction*, Prentice-Hall, Englewood Cliffs, NJ, 1995.
11. T. Kailath, *Linear Systems*, Prentice-Hall, Englewood Cliffs, NJ, 1980.
12. H. Kwakernaak and R. Sivan, *Linear Optimal Control Systems*, Wiley, New York, NY, 1972.
13. F.L. Lewis, *Optimal Control*, Wiley, New York, NY, 1986.
14. D.G. Luenberger, "Observing the state of a linear system," *IEEE Trans. Mil. Electron*, Vol. MIL-8, pp. 74–80, 1964.
15. D.G. Luenberger, "Observers for multivariable systems," *IEEE Trans. Auto. Control*, Vol. AC-11, pp. 190–199, 1966.
16. D.G. Luenberger, "An introduction to observers," *IEEE Trans. Auto. Control*, Vol. AC-16, pp. 596–603, Dec. 1971.
17. B.C. Moore, "On the flexibility offered by state feedback in multivariable systems beyond closed loop eigenvalue assignment," *IEEE Trans. Auto. Control*, Vol. AC-21, pp. 689–692, 1976; see also Vol. AC-22, pp. 140–141, 1977 for the repeated eigenvalue case.
18. V.M. Popov, "Hyperstability and optimality of automatic systems with several control functions," *Rev. Roum. Sci. Tech. Ser. Electrotech Energ.*, Vol. 9, pp. 629–690, 1964. See also V.M. Popov, *Hyperstability of Control Systems*, Springer-Verlag, New York, NY, 1973.
19. J. O'Reilly, *Observers for Linear Systems*, Academic Press, New York, NY, 1983.

20. J. Rissanen, "Control system synthesis by analogue computer based on the generalized linear feedback concept," in *Proc. Symp. Analog Comp. Applied to the Study of Chem. Processes*, pp. 1–13, Intern. Seminar, Brussels, 1960. Presses Académiques Européennes, Bruxelles, 1961.

21. H.H. Rosenbrock, *State-Space and Multivariable Theory*, Wiley, New York, NY, 1970.

22. G. Stein and M. Athans, "The LQG/LTR procedure for multivariable feedback control design," *IEEE Trans. Auto. Control*, Vol. AC-32, pp. 105–114, February, 1987.

23. W.A. Wolovich, *Linear Multivariable Systems*, Springer-Verlag, New York, NY, 1974.

24. W.M. Wonham, "On pole assigment in multi-input controllable linear systems," *IEEE Trans. Auto. Control*, Vol. AC-12, pp. 660–665, 1967.

# Exercises

**9.1.** Consider the system $\dot{x} = Ax + Bu$, where $A = \begin{bmatrix} -0.01 & 0 \\ 0 & -0.02 \end{bmatrix}$ and $B = \begin{bmatrix} 1 & 1 \\ -0.25 & 0.75 \end{bmatrix}$ with $u = Fx$.

(a) Verify that the three different state feedback matrices given by

$$F_1 = \begin{bmatrix} -1.1 & -3.7 \\ 0 & 0 \end{bmatrix}, \quad F_2 = \begin{bmatrix} 0 & 0 \\ -1.1 & 1.2333 \end{bmatrix}, \quad F_3 = \begin{bmatrix} -0.1 & 0 \\ 0 & -0.1 \end{bmatrix}$$

all assign the closed-loop eigenvalues at the same locations, namely, at $-0.1025 \pm j0.04944$. Note that in the first control law $(F_1)$ only the first input is used, whereas in the second law $(F_2)$, only the second input is used. For all three cases, plot $x(t) = [x_1(t), x_2(t)]^T$ when $x(0) = [0,1]^T$ and comment on your results. This example demonstrates how different the responses can be for different designs even though the eigenvalues of the compensated system are at the same locations.

(b) Use the eigenvalue/eigenvector assignment method to characterize all $F$ that assign the closed-loop eigenvalues at $-0.1025 \pm j0.04944$. Show how to select the free parameters to obtain $F_1, F_2$, and $F_3$ above. What are the closed-loop eigenvectors in these cases?

**9.2.** For the system $\dot{x} = Ax + Bu$ with $A \in R^{n \times n}$ and $B \in R^{n \times m}$, where $(A, B)$ is controllable and $m > 1$, choose $u = Fx$ as the feedback control law. It is possible to assign all eigenvalues of $A + BF$ by first reducing this problem to the case of eigenvalue assignment for single-input systems $(m = 1)$. This is accomplished by first reducing the system to a *single-input controllable system*. We proceed as follows.

Let $F = g \cdot f$, where $g \in R^m$ and $f^T \in R^n$ are vectors to be selected. Let $g$ be chosen such that $(A, Bg)$ is controllable. Then $f$ in

$$A + BF = A + (Bg)f$$

can be viewed as the state feedback gain vector for a single-input controllable system $(A, Bg)$, and any of the single-input eigenvalue assignment methods can be used to select $f$ so that the closed-loop eigenvalues are at desired locations.

The only question that remains to be addressed is whether there exists $g$ such that $(A, Bg)$ is controllable. It can be shown that if $(A, B)$ is controllable and $A$ is cyclic, then almost any $g \in R^m$ will make $(A, Bg)$ controllable. (A matrix $A$ is cyclic if and only if its characteristic and minimal polynomials are equal.) In the case when $A$ is not cyclic, it can be shown that if $(A, B, C)$ is controllable and observable, then for almost any real output feedback gain matrix $H$, $A + BHC$ is cyclic. So initially, by an almost arbitrary choice of $H$ or $F = HC$, the matrix $A$ is made cyclic, and then by employing a $g$, $(A, Bg)$ is made controllable. The state feedback vector gain $f$ is then selected so that the eigenvalues are at desired locations.

Note that $F = gf$ is always a rank one matrix, and this restriction on $F$ reduces the applicability of the method when requirements in addition to eigenvalue assignment are to be met.

(a) For $A, B$ as in Exercise 9.4, use the method described above to determine $F$ so that the closed-loop eigenvalues are at $-1 \pm j$ and $-2 \pm j$. Comment on your choice for $g$.

(b) For $A = \begin{bmatrix} 0 & 1 \\ 1 & 1 \end{bmatrix}$ and $B = \begin{bmatrix} 1 & 0 \\ 0 & 1 \end{bmatrix}$, characterize all $g$ such that the closed-loop eigenvalues are at $-1$.

**9.3.** Consider the system $x(k+1) = Ax(k) + Bu(k)$, where

$$A = \begin{bmatrix} 1 & 4 & 0 \\ 2 & -1 & 0 \\ 0 & 0 & 1 \end{bmatrix}, \quad B = \begin{bmatrix} 0 & 0 \\ 1 & 0 \\ -1 & 1 \end{bmatrix}.$$

Determine a linear state feedback control law $u(k) = Fx(k)$ such that all the eigenvalues of $A + BF$ are located at the origin. To accomplish this, use

(a) reduction to a single-input controllable system,
(b) the controller form of $(A, B)$,
(c) $\det(zI - (A + BF))$ and the resulting nonlinear system of equations.

In each case, plot $x(k)$ with $x(0) = [1, 1, 1]^T$ and comment on your results. In how many steps does your compensated system go to the zero state?

**9.4.** For the system $\dot{x} = Ax + Bu$, where

$$A = \begin{bmatrix} 0 & 1 & 0 & 0 \\ 0 & 0 & 1 & 0 \\ 0 & 0 & 0 & 1 \\ 1 & 1 & -3 & 4 \end{bmatrix}, \quad B = \begin{bmatrix} 1 & 0 \\ 0 & 0 \\ 0 & 0 \\ 0 & 1 \end{bmatrix},$$

determine $F$ so that the eigenvalues of $A+BF$ are at $-1 \pm j$ and $-2 \pm j$. Use as many different methods to choose $F$ as you can.

**9.5.** Consider the SISO system $\dot{x}_c = A_c x_c + B_c u, y = C_c x_c + D_c u$, where $(A_c, B_c)$ is in controller form with

$$
A_c = \begin{bmatrix} 0 & 1 & \cdots & 0 \\ \vdots & \vdots & \ddots & \vdots \\ 0 & 0 & \cdots & 1 \\ -\alpha_0 & -\alpha_1 & \cdots & -\alpha_{n-1} \end{bmatrix}, \quad B_c = \begin{bmatrix} 0 \\ \vdots \\ 0 \\ 1 \end{bmatrix}, \quad C_c = [c_0, c_1, \ldots, c_{n-1}],
$$

and let $u = F_c x + r = [f_0, f_1, \ldots, f_{n-1}] x + r$ be the linear state feedback control law. Use the Structure Theorem of Section 6.4 to show that the open-loop transfer function is

$$
H(s) = C_c(sI - A_c)^{-1} B_c + D_c = \frac{c_{n-1} s^{n-1} + \cdots + c_1 s + c_0}{s^n + \alpha_{n-1} s^{n-1} + \cdots + \alpha_1 s + \alpha_0} + D_c
$$

$$
= \frac{n(s)}{d(s)}
$$

and that the closed-loop transfer function is

$$
H_F(s) = (C_c + D_c F_c)[sI - (A_c + B_c F_c)]^{-1} B_c + D_c
$$

$$
= \frac{(c_{n-1} + D_c f_{n-1}) s^{n-1} + \cdots + (c_1 + D_c f_1) s + (c_0 + D_c f_0)}{s^n + (\alpha_{n-1} - f_{n-1}) s^{n-1} + \cdots + (\alpha_1 - f_1) s + (\alpha_0 - f_0)} + D_c
$$

$$
= \frac{n(s)}{d_F(s)}.
$$

Observe that state feedback does not change the numerator $n(s)$ of the transfer function, but it can arbitrarily assign any desired (monic) denominator polynomial $d_F(s) = d(s) - F_c[1, s, \ldots, s^{n-1}]^T$. Thus, state feedback does not (directly) alter the zeros of $H(s)$, but it can arbitrarily assign the poles of $H(s)$. Note that these results generalize to the MIMO case [see (9.43)].

**9.6.** Consider the system $\dot{x} = Ax + Bu, y = Cx$, where

$$
A = \begin{bmatrix} 0 & 1 & 0 \\ 0 & 0 & 1 \\ 1 & 0 & -1 \end{bmatrix}, \quad B = \begin{bmatrix} 0 \\ 0 \\ 1 \end{bmatrix}, \quad C = [1, 2, 0].
$$

(a) Determine an appropriate linear state feedback control law $u = Fx + Gr\ (G \in R)$ so that the closed-loop transfer function is equal to a given desired transfer function

$$
H_m(s) = \frac{1}{s^2 + 3s + 2}.
$$

We note that this is an example of *model matching*, i.e., compensating a given system so that it matches the input–output behavior of a desired model. In the present case, state feedback is used; however, output feedback is more common in model matching.

(b) Is the compensated system in (a) controllable? Is it observable? Explain your answers.

(c) Repeat (a) and (b) by assuming that the state is not available for measurement. Design an appropriate state observer, if possible.

**9.7.** Design an observer for the oscillatory system $\dot{x}(t) = v(t)$, $\dot{v}(t) = -\omega_0^2 x(t)$, using measurements of the velocity $v$. Place both observer poles at $s = -\omega_0$.

**9.8.** Consider the undamped harmonic oscillator $\dot{x}_1(t) = x_2(t)$, $\dot{x}_2(t) = -\omega_0^2 x_1(t) + u(t)$. Using an observation of velocity $y = x_2$, design an observer/state feedback compensator to control the position $x_1$. Place the state feedback controller poles at $s = -\omega_0 \pm j\omega_0$ and both observer poles at $s = -\omega_0$. Plot $x(t)$ for $x(0) = [1, 1]^T$ and $\omega_0 = 2$.

**9.9.** A servomotor that drives a load is described by the equation $\frac{d^2\theta}{dt^2} + \frac{d\theta}{dt} = u$, where $\theta$ is the shaft position (output) and $u$ is the applied voltage. Choose $u$ so that $\theta$ and $\frac{d\theta}{dt}$ will go to zero exponentially (when their initial values are not zero). To accomplish this, proceed as follows.

(a) Derive a state-space representation of the servomotor.

(b) Determine linear state feedback, $u = Fx + r$, so that both closed-loop eigenvalues are at $-1$. Such $F$ is actually optimal since it minimizes $J = \int_0^\infty [\theta^2 + \left(\frac{d\theta}{dt}\right)^2 + u^2] dt$.

(c) Since only $\theta$ and $u$ are available for measurement, design an asymptotic state estimator (with eigenvalues at, say, $-3$) and use the state estimate $\hat{x}$ in the linear state feedback control law. Write the transfer function and the state-space description of the overall system and comment on stability, controllability, and observability.

(d) Plot $\theta$ and $d\theta/dt$ in (b) and (c) for $r = 0$ and initial conditions equal to $[1, 1]^T$.

(e) Repeat (c) and (d), using a reduced-order observer of order 1.

**9.10.** Consider the LQR problem for the system $\dot{x} = Ax + Bu$, where $(A, B)$ is controllable and the performance index is given by

$$\tilde{J}(u) = \int_0^\infty e^{2\alpha t}[x^T(t)Qx(t) + u^T(t)Ru(t)]dt,$$

where $\alpha \in R, \alpha > 0$ and $Q \geq 0, R > 0$.

(a) Show that $u^*$ that minimizes $\tilde{J}(u)$ is a fixed control law with constant gains on the states, even though the weighting matrices $\tilde{Q} = e^{2\alpha t}Q$, $\tilde{R} = e^{2\alpha t}R$ are time varying. Derive the algebraic Riccati matrix equation that characterizes this control law.

(b) The performance index given above has been used to solve the question of relative stability. In light of your solution, how do you explain this?

*Hint:* Reformulate the problem in terms of the transformed variables $\tilde{x} = e^{\alpha t}x, \tilde{u} = e^{\alpha t}u$.

**9.11.** Consider the system $\dot{x} = \begin{bmatrix} 0 & 1 \\ 1 & 1 \end{bmatrix} x + \begin{bmatrix} 1 \\ 0 \end{bmatrix} u$ and the performance indices $J_1, J_2$ given by

$$J_1 = \int_0^\infty (x_1^2 + x_2^2 + u^2)dt \text{ and } J_2 = \int_0^\infty (900(x_1^2 + x_2^2) + u^2)dt.$$

Determine the optimal control laws that minimize $J_1$ and $J_2$. In each case, plot $u(t), x_1(t), x_2(t)$ for $x(0) = [1, 1]^T$ and comment on your results.

**9.12.** Consider the system $\dot{x} = \begin{bmatrix} 0 & 1 \\ 1 & 0 \end{bmatrix} x + \begin{bmatrix} 0 \\ -1 \end{bmatrix} u, \ y = [1, 0]x$.

(a) Use state feedback $u = Fx$ to assign the eigenvalues of $A + BF$ at $-0.5 \pm j0.5$. Plot $x(t) = [x_1(t), x_2(t)]^T$ for the open- and closed-loop system with $x(0) = [-0.6, 0.4]^T$.
(b) Design an identity observer with eigenvalues at $-\alpha \pm j$, where $\alpha > 0$. What is the observer gain $K$ in this case?
(c) Use the state estimate $\hat{x}$ from (b) in the linear feedback control law $u = F\hat{x}$, where $F$ was found in (a). Derive the state-space description of the closed-loop system. If $u = F\hat{x} + r$, what is the transfer function between $y$ and $r$?
(d) For $x(0) = [-0.6, 0.4]^T$ and $\hat{x}(0) = [0, 0]^T$, plot $x(t), \hat{x}(t), y(t)$, and $u(t)$ of the closed-loop system obtained in (c) and comment on your results. Use $\alpha = 1, 2, 5$, and 10, and comment on the effects on the system response.

*Remark:* This exercise illustrates the deterioration of system response when state observers are used to generate the state estimate that is used in the feedback control law.

**9.13.** Consider the system

$$x(k + 1) = Ax(k) + Bu(k) + Eq(k), \quad y(k) = Cx(k),$$

where $q(k) \in R^r$ is some disturbance vector. It is desirable to completely eliminate the effects of $q(k)$ on the output $y(k)$. This can happen only when $E$ satisfies certain conditions. Presently, it is assumed that $q(k)$ is an arbitrary $r \times 1$ vector.

(a) Express the required conditions on $E$ in terms of the observability matrix of the system.
(b) If $A = \begin{bmatrix} 1 & 1 \\ 1 & 1 \end{bmatrix}$, $C = [1, 1]$, characterize all $E$ that satisfy these conditions.

(c) Suppose $E \in R^{n \times 1}$, $C \in R^{p \times 1}$, and $q(k)$ is a step, and let the objective be to asymptotically reduce the effects of $q$ on the output. Note that this specification is not as strict as in (a), and in general it is more easily satisfied. Use $z$-transforms to derive conditions for this to happen.
*Hint*: Express the conditions in terms of poles and zeros of $\{A, E, C\}$.

**9.14.** Consider the system $\dot{x} = Ax + Bu$, $y = Cx + Du$, where

$$A = \begin{bmatrix} 0 & 0 \\ 0 & 0 \end{bmatrix}, \quad B = \begin{bmatrix} 1 & 0 \\ 0 & 1 \end{bmatrix}, \quad C = \begin{bmatrix} 1 & 0 \\ 1 & 2 \end{bmatrix}, \quad D = \begin{bmatrix} 1 & 0 \\ 0 & 1 \end{bmatrix}.$$

Let $u = Fx + r$ be a linear state feedback control law. Determine $F$ so that the eigenvalues of $A + BF$ are $-1, -2$ and are unobservable from $y$. What is the closed-loop transfer function $H_F(s)$ ($\hat{y} = H_F \hat{r}$) in this case?
*Hint*: Select the eigenvalues and eigenvectors of $A + BF$.

**9.15.** Consider the controllable and observable $SISO$ system $\dot{x} = Ax + Bu$, $y = Cx$ with $H(s) = C(sI - A)^{-1}B$.

(a) If $\lambda$ is not an eigenvalue of $A$, show that there exists an initial state $x_0$ such that the response to $u(t) = e^{\lambda t}$, $t \geq 0$, is $y(t) = H(\lambda)e^{\lambda t}$, $t \geq 0$. What happens if $\lambda$ is a zero of $H(s)$?

(b) Assume that $A$ has distinct eigenvalues. Let $\lambda$ be an eigenvalue of $A$ and show that there exists an initial state $x_0$ such that with "no input" ( $u(t) \equiv 0$ ), $y(t) = ke^{\lambda t}$, $t \geq 0$, for some $k \in R$.

# 10

# Feedback Control Systems

## 10.1 Introduction

This chapter focuses on one and two degrees of freedom feedback control systems that have been studied, using Polynomial Matrix (PMD) and Matrix Fractional (MFD) Descriptions. The chapter starts by considering in Section 10.2 interconnected systems and their properties, with emphasis on systems connected via feedback interconnections. Internal stability is central in the development, and all stabilizing feedback controllers are parameterized in Section 10.3. The role of the Diophantine equation is also explained. In Section 10.4 two degrees of freedom controllers are studied at length.

## 10.2 Interconnected Systems

Interconnected systems, connected in parallel, series, and feedback configurations are studied in the present section. It is shown that particular interconnections may introduce uncontrollable, unobservable, or unstable modes into a system; for a more detailed development, see [1, p. 568, Subsection 7.3C]. Feedback configurations, as well as series interconnections, are of particular importance in the control of systems.

### 10.2.1 Systems Connected in Parallel and in Series

*In Parallel*

Consider first systems $S_1$ and $S_2$ connected in parallel as shown in Figure 10.1, and let

$$P_1(q)z_1(t) = Q_1(q)u_1(t), \quad y_1(t) = R_1(q)z_1(t) \tag{10.1}$$

and

$$P_2(q)z_2(t) = Q_2(q)u_2(t), \quad y_2(t) = R_2(q)z_2(t) \tag{10.2}$$

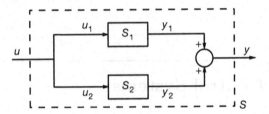

**Figure 10.1.** Systems connected in parallel

be representations (PMDs) for $S_1$ and $S_2$, respectively; see Section 7.5. Since $u(t) = u_1(t) = u_2(t)$ and $y(t) = y_1(t) + y_2(t)$, the overall system description is given by

$$\begin{bmatrix} P_1(q) & 0 \\ 0 & P_2(q) \end{bmatrix} \begin{bmatrix} z_1(q) \\ z_2(q) \end{bmatrix} = \begin{bmatrix} Q_1(q) \\ Q_2(q) \end{bmatrix} u(t), \quad y(t) = [R_1(q), R_2(q)] \begin{bmatrix} z_1(t) \\ z_2(t) \end{bmatrix}.$$
(10.3)

If the systems $S_1$ and $S_2$ are described by the state-space representations $\dot{x}_i = A_i x_i + B_i u_i$, $y_i = C_i x_i + D_i u_i$, $i = 1, 2$, then the overall system state-space description is given by

$$\begin{bmatrix} \dot{x}_1 \\ \dot{x}_2 \end{bmatrix} = \begin{bmatrix} A_1 & 0 \\ 0 & A_2 \end{bmatrix} \begin{bmatrix} x_1 \\ x_2 \end{bmatrix} + \begin{bmatrix} B_1 \\ B_2 \end{bmatrix} u,$$

$$y = [C_1, \ C_2] \begin{bmatrix} x_1 \\ x_2 \end{bmatrix} + [D_1 + D_2] u.$$
(10.4)

If $H_1(s), H_2(s)$ are the transfer function matrices of $S_1$ and $S_2$, respectively, then the overall transfer function can be found from $\hat{y}(s) = \hat{y}_1(s) + \hat{y}_2(s) = H_1(s)\hat{u}_1(s) + H_2(s)\hat{u}_2(s) = [H_1(s) + H_2(s)]\hat{u}(s)$ to be

$$H(s) = H_1(s) + H_2(s).$$
(10.5)

Note that if both $H_1(s)$ and $H_2(s)$ are proper, then $H(s)$ is also proper.

*In Series*

Consider now systems $S_1$ and $S_2$ connected in series, as shown in Figure 10.2, and let (10.1) and (10.2) describe the systems. Here $u_2(t) = y_1(t)$. To derive

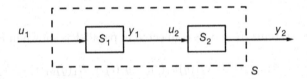

**Figure 10.2.** Systems connected in series

the overall system description, consider $P_2 z_2 = Q_2 u_2 = Q_2 y_1 = Q_2 R_1 z_1$. Then

$$\begin{bmatrix} P_1 & 0 \\ -Q_2 R_1 & P_2 \end{bmatrix} \begin{bmatrix} z_1 \\ z_2 \end{bmatrix} = \begin{bmatrix} Q_1 \\ 0 \end{bmatrix} u_1,$$

$$y_2 = [0, \ R_2] \begin{bmatrix} z_1 \\ z_2 \end{bmatrix}. \tag{10.6}$$

If the systems $S_1, S_2$ are described by the state-space representations $\dot{x}_i = A_i x_i + C_i u_i$, $y_i = C_i x_i + D_i u_i$, $i = 1, 2$, then it can be shown that the overall system state-space description is given by

$$\begin{bmatrix} \dot{x}_1 \\ \dot{x}_2 \end{bmatrix} = \begin{bmatrix} A_1 & 0 \\ B_2 C_1 & A_2 \end{bmatrix} \begin{bmatrix} x_1 \\ x_2 \end{bmatrix} + \begin{bmatrix} B_1 & 0 \\ B_2 D_1 & B_2 \end{bmatrix} \begin{bmatrix} u_1 \\ r_2 \end{bmatrix},$$

$$\begin{bmatrix} y_1 \\ y_2 \end{bmatrix} = \begin{bmatrix} C_1 & 0 \\ D_2 C_1 & C_2 \end{bmatrix} \begin{bmatrix} x_1 \\ x_2 \end{bmatrix} + \begin{bmatrix} D_1 & 0 \\ D_2 D_1 & D_2 \end{bmatrix} \begin{bmatrix} u_1 \\ r_2 \end{bmatrix}. \tag{10.7}$$

If $H_1(s), H_2(s)$ are the transfer function matrices of $S_1$ and $S_2$, then the overall transfer function $\hat{y}_2(s) = H(s)\hat{u}_1(s)$ is

$$H(s) = H_2(s) H_1(s). \tag{10.8}$$

It can be shown that if both $H_1$ and $H_2$ are proper, then $H$ is also proper. Note that poles of $H_1$ and $H_2$ may cancel in the product $H_2 H_1$ and any cancellation implies that there are uncontrollable/unobservable eigenvalues in the overall system internal description.

## 10.2.2 Systems Connected in Feedback Configuration

Consider systems $S_1$ and $S_2$ connected in a feedback configuration as shown in Figure 10.3a, or equivalently as in Figure 10.3b. Let

$$P_1(q) z_1(t) = Q_1(q) u_1(t), \quad y_1(t) = R_1(q) z_1(t) \tag{10.9}$$

and

$$P_2(q) z_2(t) = Q_2(q) u_2(t), \quad y_2(t) = R_2(q) z_2(t) \tag{10.10}$$

be polynomial matrix representations of $S_1$ and $S_2$, respectively. Since

$$u_1(t) = y_2(t) + r_1(t), \tag{10.11}$$

$$u_2(t) = y_1(t) + r_2(t), \tag{10.12}$$

where $r_1$ and $r_2$ are external inputs, the dimensions of the vector inputs and outputs, $u_1$ and $y_2$ and also $u_2$ and $y_1$ must be the same. To derive the overall system description we consider $P_1 z_1 = Q_1 u_1 = Q_1(y_2 + r_1)$ and $P_2 z_2 = Q_2 u_2 = Q_2(y_1 + r_2)$ where $y_1$ and $y_2$ are as above. Then the closed-loop is described by

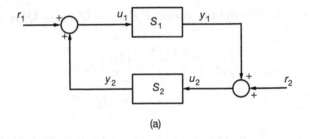

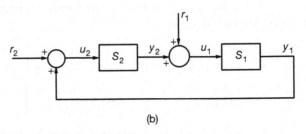

(b)

**Figure 10.3.** Feedback configuration

$$\begin{bmatrix} P_1 & -Q_1R_2 \\ -Q_2R_1 & P_2 \end{bmatrix}\begin{bmatrix} z_1 \\ z_2 \end{bmatrix} = \begin{bmatrix} Q_1 & 0 \\ 0 & Q_2 \end{bmatrix}\begin{bmatrix} r_1 \\ r_2 \end{bmatrix}, \begin{bmatrix} y_1 \\ y_2 \end{bmatrix} = \begin{bmatrix} R_1 & 0 \\ 0 & R_2 \end{bmatrix}\begin{bmatrix} z_1 \\ z_2 \end{bmatrix}.$$
(10.13)

Note that the condition for the closed-loop system to be well defined is that

$$\det\left(\begin{bmatrix} P_1 & -Q_1R_2 \\ -Q_2R_1 & P_2 \end{bmatrix}\right) \neq 0.$$
(10.14)

If this condition is not satisfied, then the closed-loop system cannot be described by the polynomial matrix representations discussed here.

If the systems $S_1$ and $S_2$ are described by the state-space representations $\dot{x}_i = A_i x_i + B_i u_i$, $y_i = C_i x_i + D_i u_i$, $i = 1, 2$, then it can be shown that the closed-loop system state-space description is

$$\begin{bmatrix} \dot{x}_1 \\ \dot{x}_2 \end{bmatrix} = \begin{bmatrix} A_1 + B_1 M_2 D_2 C_1 & B_1 M_2 C_2 \\ B_2 M_1 C_1 & A_2 + B_2 M_1 D_1 C_2 \end{bmatrix}\begin{bmatrix} x_1 \\ x_2 \end{bmatrix} + \begin{bmatrix} B_1 M_2 & B_1 M_2 D_2 \\ B_2 M_1 D_1 & B_2 M_1 \end{bmatrix}\begin{bmatrix} r_1 \\ r_2 \end{bmatrix},$$

$$\begin{bmatrix} y_1 \\ y_2 \end{bmatrix} = \begin{bmatrix} M_1 C_1 & M_1 D_1 C_2 \\ M_2 D_2 C_1 & M_2 C_2 \end{bmatrix}\begin{bmatrix} x_1 \\ x_2 \end{bmatrix} + \begin{bmatrix} M_1 D_1 & M_1 D_1 D_2 \\ M_2 D_2 D_1 & M_2 D_2 \end{bmatrix}\begin{bmatrix} r_1 \\ r_2 \end{bmatrix},$$
(10.15)

where $M_1 = (I - D_1 D_2)^{-1}$ and $M_2 = (I - D_2 D_1)^{-1}$. It is assumed that $\det(I - D_1 D_2) = \det(I - D_2 D_1) \neq 0$.

It is not difficult to see that in the case of state-space representations the conditions for the closed-loop system state-space representation to be well defined is $\det(I - D_1 D_2) \neq 0$. When $D_1 = 0$ and $D_2 = 0$, then (10.15) simplifies to

$$\begin{bmatrix} \dot{x}_1 \\ \dot{x}_2 \end{bmatrix} = \begin{bmatrix} A_1 & B_1 C_2 \\ B_2 C_1 & A_2 \end{bmatrix} \begin{bmatrix} x_1 \\ x_2 \end{bmatrix} + \begin{bmatrix} B_1 & 0 \\ 0 & B_2 \end{bmatrix} \begin{bmatrix} r_1 \\ r_2 \end{bmatrix},$$

$$\begin{bmatrix} y_1 \\ y_2 \end{bmatrix} = \begin{bmatrix} C_1 & 0 \\ 0 & C_2 \end{bmatrix} \begin{bmatrix} x_1 \\ x_2 \end{bmatrix}. \tag{10.16}$$

**Example 10.1.** Consider systems $S_1$ and $S_2$ in a feedback configuration with $H_1(s) = \frac{s}{s+1}$ and $H_2(s) = 1$ and consider the realizations $\{P_1, Q_1, R_1, W_1\} = \{q + 1, q, 1, 0\}$ and $\{P_2, Q_2, R_2, W_2\} = \{1, 1, 1, 0\}$. Then (10.13) becomes

$$\begin{bmatrix} q + 1 & -q \\ -1 & 1 \end{bmatrix} \begin{bmatrix} z_1 \\ z_2 \end{bmatrix} = \begin{bmatrix} q & 0 \\ 0 & 1 \end{bmatrix} \begin{bmatrix} r_1 \\ r_2 \end{bmatrix},$$

$$\begin{bmatrix} y_1 \\ y_2 \end{bmatrix} = \begin{bmatrix} 1 & 0 \\ 0 & 1 \end{bmatrix} \begin{bmatrix} z_1 \\ z_2 \end{bmatrix}.$$

Since $\det\left(\begin{bmatrix} q+1 & -q \\ -1 & 1 \end{bmatrix}\right) = 1 \neq 0$, this is a well-defined polynomial matrix description for the closed-loop system. Note that the transfer function matrix of the closed-loop system is $H(s) = \begin{bmatrix} 1 & 0 \\ 0 & 1 \end{bmatrix} \begin{bmatrix} s+1 & -s \\ -1 & 1 \end{bmatrix}^{-1} \begin{bmatrix} s & 0 \\ 0 & 1 \end{bmatrix} = \begin{bmatrix} s & s \\ s & s+1 \end{bmatrix}$, which is not proper, whereas $H_1$ and $H_2$ were both proper.

Now if state-space realizations of $H_1(s) = \frac{-1}{s+1} + 1$ and $H_2(s) = 1$ are considered, namely $\{A_1, B_1, C_1, D_1\} = \{-1, 1, -1, 1\}$ and $\{A_2, B_2, C_2, D_2\} = \{0, 0, 0, 1\}$, then $1 - D_1 D_2 = 1 - 1 \cdot 1 = 0$; i.e., a state-space description of the closed-loop does not exist. This is to be expected since the closed-loop transfer function is nonproper and as such cannot be represented by a state-space realization $\{A, B, C, D\}$.

---

Next, let $H_1(s)$ and $H_2(s)$ be the transfer function matrices of $S_1$ and $S_2$; i.e., $\hat{y}_1(s) = H_1(s)\hat{u}_1(s)$ and $\hat{y}_2(s) = H_2(s)\hat{u}_2(s)$. In view of $\hat{u}_1 = \hat{y}_2 + \hat{r}_1$ and $\hat{u}_2 = \hat{y}_1 + \hat{r}_2$, we have $\hat{y}_1 = H_1\hat{u}_1 = H_1(\hat{y}_2 + \hat{r}_1) = H_1 H_2 \hat{u}_2 + H_1 \hat{r}_1 = H_1 H_2 \hat{y}_1 + H_1 H_2 \hat{r}_2 + H_1 \hat{r}_1$ or

$$(I - H_1 H_2)\hat{y}_1 = H_1 H_2 \hat{r}_2 + H_1 \hat{r}_1. \tag{10.17}$$

Also, $\hat{y}_2 = H_2\hat{u}_2 = H_2(\hat{y}_1 + \hat{r}_2) = H_2 H_1 \hat{u}_1 + H_2 \hat{r}_2 = H_2 H_1 \hat{y}_2 + H_2 H_1 \hat{r}_1 + H_2 \hat{r}_2$ or

$$(I - H_2 H_1)\hat{y}_2 = H_2 H_1 \hat{r}_1 + H_2 \hat{r}_2. \tag{10.18}$$

Note that $\det(I - H_1 H_2) = \det(I - H_2 H_1)$, and assume that the determinant is nonzero. Then

$$\begin{bmatrix} \hat{y}_1 \\ \hat{y}_2 \end{bmatrix} = \begin{bmatrix} (I - H_1 H_2)^{-1} H_1 & (I - H_1 H_2)^{-1} H_1 H_2 \\ (I - H_2 H_1)^{-1} H_2 H_1 & (I - H_2 H_1)^{-1} H_2 \end{bmatrix} \begin{bmatrix} \hat{r}_1 \\ \hat{r}_2 \end{bmatrix}$$

$$= \begin{bmatrix} H_{11} & H_{12} \\ H_{21} & H_{22} \end{bmatrix} \begin{bmatrix} \hat{r}_1 \\ \hat{r}_2 \end{bmatrix}. \tag{10.19}$$

The significance of the assumption $\det(I - H_1 H_2) \neq 0$ can be seen as follows. Let $\widetilde{D}_1 \tilde{z}_1 = \widetilde{N}_1 u_1$, $y_1 = \tilde{z}_1$ and $D_2 z_2 = u_2$, $y_2 = N_2 z_2$ be representations of the systems $S_1$ and $S_2$. As will be shown below, the closed-loop system description in this case is given by $(\widetilde{D}_1 D_2 - \widetilde{N}_1 N_2) z_2 = \widetilde{N}_1 r_1 + \widetilde{D}_1 r_2$ and $y_1 = D_2 z_2 - r_2$ and $y_2 = N_2 z_2$. Now note that $I - H_1 H_2 = I - \widetilde{D}_1^{-1} \widetilde{N}_1 N_2 D_2^{-1} = \widetilde{D}_1^{-1} (\widetilde{D}_1 D_2 - \widetilde{N}_1 N_2) D_2^{-1}$, which implies that $\det(I - H_1 H_2) \neq 0$ if and only if $\det(\widetilde{D}_1 D_2 - \widetilde{N}_1 N_2) \neq 0$; i.e., if $\det(I - H_1 H_2) = 0$, then the closed-loop system cannot be described by the polynomial matrix representations discussed in this chapter. Thus, the assumption that $\det(I - H_1 H_2) \neq 0$ is essential for the closed-loop system to be well defined.

---

**Example 10.2.** Consider $H_1(s) = \frac{s}{s+1}$ and $H_2(s) = 1$ as in Example 10.1. Here $1 - H_1 H_2 = \frac{1}{s+1} \neq 0$, and therefore, the closed-loop system is well defined. Relation (10.19) assumes in this case the form

$$\begin{bmatrix} \hat{y}_1 \\ \hat{y}_2 \end{bmatrix} = \begin{bmatrix} s & s \\ s & s+1 \end{bmatrix} \begin{bmatrix} \hat{r}_1 \\ \hat{r}_2 \end{bmatrix},$$

a nonproper transfer function that is the transfer function matrix $H(s)$ derived in Example 10.1.

---

For simplicity, assume that both $S_1$ and $S_2$ in Figure 10.3 are controllable and observable and consider the following representations.

For system $S_1$:

$$\text{(1a)} \quad D_1(q) z_1(t) = u_1(t), \quad y_1(t) = N_1(q) z_1(t) \tag{10.20}$$

or

$$\text{(1b)} \quad \widetilde{D}_1(q) \tilde{z}_1(t) = \widetilde{N}_1(q) u_1(t), y_1(t) = \tilde{z}_1(t), \tag{10.21}$$

where $(D_1(q), N_1(q))$ are rc and $(\widetilde{D}_1(q), \widetilde{N}_1(q))$ are lc.

For system $S_2$:

$$\text{(2a)} \quad D_2(q) z_2(t) = u_2(t), \quad y_2(t) = N_2(q) z_2(t) \tag{10.22}$$

or

$$\text{(2b)} \quad \widetilde{D}_2(q) \tilde{z}_2(t) = \widetilde{N}_2(q) u_2(t), \quad y_2(t) = \tilde{z}_2(t), \tag{10.23}$$

where $(D_2(q), N_2(q))$ are rc and $(\widetilde{D}_2(q), \widetilde{N}_2(q))$ are lc.

In view of the connections

$$u_1(t) = y_2(t) + r_1(t), u_2(t) = y_1(t) + r_2(t), \tag{10.24}$$

the closed-loop feedback system of Figure 10.3 can now be characterized as follows [see also (10.13)]:

(i)  Using descriptions (1a) and (2a), and Eqs. 10.20 and 10.21, we have

$$\begin{bmatrix} D_1 & -N_2 \\ -N_1 & D_2 \end{bmatrix} \begin{bmatrix} z_1 \\ z_2 \end{bmatrix} = \begin{bmatrix} I & 0 \\ 0 & I \end{bmatrix} \begin{bmatrix} r_1 \\ r_2 \end{bmatrix}, \begin{bmatrix} y_1 \\ y_2 \end{bmatrix} = \begin{bmatrix} N_1 & 0 \\ 0 & N_2 \end{bmatrix} \begin{bmatrix} z_1 \\ z_2 \end{bmatrix}. \quad (10.25)$$

(ii)  Using descriptions (1b) and (2b), we have

$$\begin{bmatrix} \tilde{D}_1 & -\tilde{N}_1 \\ -\tilde{N}_2 & \tilde{D}_2 \end{bmatrix} \begin{bmatrix} \tilde{z}_1 \\ \tilde{z}_2 \end{bmatrix} = \begin{bmatrix} \tilde{N}_1 & 0 \\ 0 & \tilde{N}_2 \end{bmatrix} \begin{bmatrix} r_1 \\ r_2 \end{bmatrix}, \begin{bmatrix} y_1 \\ y_2 \end{bmatrix} = \begin{bmatrix} I & 0 \\ 0 & I \end{bmatrix} \begin{bmatrix} \tilde{z}_1 \\ \tilde{z}_2 \end{bmatrix}. \quad (10.26)$$

(iii)  Using descriptions (1b) and (2a), we have

$$\begin{bmatrix} \tilde{D}_1 & -\tilde{N}_1 N_2 \\ -I & D_2 \end{bmatrix} \begin{bmatrix} \tilde{z}_1 \\ z_2 \end{bmatrix} = \begin{bmatrix} \tilde{N}_1 & 0 \\ 0 & I \end{bmatrix} \begin{bmatrix} r_1 \\ r_2 \end{bmatrix}, \begin{bmatrix} y_1 \\ y_2 \end{bmatrix} = \begin{bmatrix} I & 0 \\ 0 & N_2 \end{bmatrix} \begin{bmatrix} \tilde{z}_1 \\ z_2 \end{bmatrix}. \quad (10.27)$$

Also, $D_2 z_2 = u_2 = y_1 + r_2 = \tilde{D}_1^{-1} \tilde{N}_1 u_1 + r_2 = \tilde{D}_1^{-1} \tilde{N}_1 (y_2 + r_1) + r_2 = \tilde{D}_1^{-1} \tilde{N}_1 (N_2 z_2 + r_1) + r_2$ and $y_1 = u_2 - r_2 = D_2 z_2 - r_2$, from which we obtain

$$(\tilde{D}_1 D_2 - \tilde{N}_1 N_2) z_2 = [\tilde{N}_1, \tilde{D}_1] \begin{bmatrix} r_1 \\ r_2 \end{bmatrix}, \begin{bmatrix} y_1 \\ y_2 \end{bmatrix} = \begin{bmatrix} D_2 \\ N_2 \end{bmatrix} z_2 + \begin{bmatrix} 0 & -I \\ 0 & 0 \end{bmatrix} \begin{bmatrix} r_1 \\ r_2 \end{bmatrix}. \quad (10.28)$$

(iv)  Using descriptions (1a) and (2b), we have

$$\begin{bmatrix} D_1 & -I \\ -\tilde{N}_2 N_1 & \tilde{D}_2 \end{bmatrix} \begin{bmatrix} z_1 \\ \tilde{z}_2 \end{bmatrix} = \begin{bmatrix} I & 0 \\ 0 & \tilde{N}_2 \end{bmatrix} \begin{bmatrix} r_1 \\ r_2 \end{bmatrix}, \begin{bmatrix} y_1 \\ y_2 \end{bmatrix} = \begin{bmatrix} N_1 & 0 \\ 0 & I \end{bmatrix} \begin{bmatrix} z_1 \\ \tilde{z}_2 \end{bmatrix}. \quad (10.29)$$

Also, $D_1 z_1 = u_1 = y_2 + r_1 = \tilde{D}_2^{-1} \tilde{N}_2 u_2 + r_1 = \tilde{D}_2^{-1} \tilde{N}_2 (y_1 + r_2) + r_1 = \tilde{D}_2^{-1} \tilde{N}_2 (N_1 z_1 + r_2) + r_1$ and $y_2 = u_1 - r_1 = D_1 z_1 - r_1$, from which we obtain

$$(\tilde{D}_2 D_1 - \tilde{N}_2 N_1) z_1 = [\tilde{D}_2, \tilde{N}_2] \begin{bmatrix} r_1 \\ r_2 \end{bmatrix}, \begin{bmatrix} y_1 \\ y_2 \end{bmatrix} = \begin{bmatrix} N_1 \\ D_1 \end{bmatrix} z_1 + \begin{bmatrix} 0 & 0 \\ -I & 0 \end{bmatrix} \begin{bmatrix} r_1 \\ r_2 \end{bmatrix}. \quad (10.30)$$

## Controllability and Observability

The preceding descriptions of the closed-loop system given in (i), (ii), (iii), and (iv) are equivalent and have the same uncontrollable and unobservable modes. The systems $S_1$ and $S_2$ were taken to be controllable and observable, and so the uncontrollability and unobservability discussed below is due to the feedback interconnection only.

*Controllability.* To study controllability, consider the representation (10.25). It can be seen from the matrices $\begin{bmatrix} D_1 & -N_2 \\ -N_1 & D_2 \end{bmatrix}$ and $\begin{bmatrix} I & 0 \\ 0 & I \end{bmatrix}$ that the eigenvalues that are uncontrollable from $r_1$ will be the roots of the determinant of a gcld of $[-N_1, D_2]$ and the eigenvalues that are uncontrollable from $r_2$ will be the roots of a gcld of $[D_1, -N_2]$. The closed-loop system is controllable from $\begin{bmatrix} r_1 \\ r_2 \end{bmatrix}$. Clearly, all possible eigenvalues that are uncontrollable from $r_1$ are eigenvalues of $S_2$. These are the poles of $H_2 = N_2 D_2^{-1}$ that cancel in the product $H_2 N_1$. Similarly, all possible eigenvalues that are uncontrollable from $r_2$ are eigenvalues of $S_1$. These are the poles of $H_1 = N_1 D_1^{-1}$ that cancel in the product $H_1 N_2$.

*Observability.* To study observability, consider the representation (10.26). From the matrices $\begin{bmatrix} \tilde{D}_1 & -\tilde{N}_1 \\ -\tilde{N}_2 & \tilde{D}_2 \end{bmatrix}$ and $\begin{bmatrix} I & 0 \\ 0 & I \end{bmatrix}$, it can be seen that the eigenvalues that are unobservable from $y_1$ will be the roots of the determinant of a gcrd of $\begin{bmatrix} -\tilde{N}_1 \\ \tilde{D}_2 \end{bmatrix}$ and the eigenvalues that are unobservable from $y_2$ will be the roots of the determinant of a gcrd of $\begin{bmatrix} \tilde{D}_1 \\ -\tilde{N}_2 \end{bmatrix}$. The closed-loop system is observable from $\begin{bmatrix} y_1 \\ y_2 \end{bmatrix}$. Clearly, all possible eigenvalues that are unobservable from $y_1$ are eigenvalues of $S_2$. These are the poles of $H_2 = \tilde{D}_2^{-1} \tilde{N}_2$ that cancel in the product $\tilde{N}_1 H_2$. Similarly, all possible eigenvalues that are unobservable from $y_2$ are eigenvalues of $S_1$. These are the poles of $H_1 = \tilde{D}_1^{-1} \tilde{N}_1$ that cancel in the product $\tilde{N}_2 H_1$, $H_2[H_1, I]$.

---

**Example 10.3.** Consider systems $S_1$ and $S_2$ connected in the feedback configuration of Figure 10.3, and let $S_1$ and $S_2$ be described by the transfer functions $H_1(s) = \frac{s+1}{s-1}$, and $H_2(s) = \frac{a_1 s + a_0}{s+b}$. For the closed-loop to be well defined, we must have $1 - H_1 H_2 = 1 - \frac{s+1}{s-1} \frac{a_1 s + a_0}{s+b} = \frac{(1-a_1)s^2 + (b-a_1-a_0-1)s - (b+a_0)}{(s-1)(s+b)} \neq 0$. Note that for $a_1 = 1$, $a_0 = -1$, and $b = 1$, $H_2 = \frac{s-1}{s+1}$ and $1 - H_1 H_2 = 1 - 1 = 0$. Therefore, these values are not allowed for the parameters if the closed-loop system is to be represented by a PMD. If state-space descriptions are to be used, let $D_1 = \lim_{s\to\infty} H_1(s) = 1$ and $D_2 = \lim_{s\to\infty} H_2(s) = a_1$, from which we have $1 - D_1 D_2 = 1 - a_1 \neq 0$ for the closed-loop system to be characterized by a state-space description. Let us assume that $a_1 \neq 1$.

The uncontrollable and unobservable eigenvalues can be determined from a PMD such as (10.28). Alternatively, in view of the discussion just preceding this example, we conclude the following. (i) The eigenvalues that are uncontrollable from $r_1$ are the poles of $H_2$ that cancel in $H_2 N_1 = \frac{a_1 s + a_0}{s+b}(s+1)$; i.e., there is an eigenvalue that is uncontrollable from $r_1$ (at $-1$) only when $b = 1$. If this is the case, $-1$ is also an eigenvalue that is unobservable from $y_1$. (ii) The poles of $H_1$ that cancel in $H_1 N_2 = \frac{s+1}{s-1}(a_1 s + a_0)$ are the eigenvalues that are uncontrollable from $r_2$; i.e., there is an eigenvalue that is uncontrollable from $r_2$ (at $+1$) only when $a_0/a_1 = -1$. If this is the case, $+1$ is also an eigenvalue that is unobservable from $y_2$.

---

## Stability

The closed-loop feedback system is internally stable if and only if all of its eigenvalues have negative real parts. The closed-loop eigenvalues can be determined from the closed-loop descriptions derived above. First recall the identities

$$\det \begin{bmatrix} A & D \\ C & B \end{bmatrix} = \det(A)\det(B - CA^{-1}D) = \det(B)\det(A - DB^{-1}C), \quad (10.31)$$

where in the first expression it was assumed that $\det(A) \neq 0$ and in the second expression it was assumed that $\det(B) \neq 0$. The proof of this result is immediate from the matrix identities $\begin{bmatrix} I & 0 \\ -CA^{-1} & I \end{bmatrix}\begin{bmatrix} A & D \\ C & B \end{bmatrix} = \begin{bmatrix} A & D \\ 0 & B - CA^{-1}D \end{bmatrix}$

and $\begin{bmatrix} I & -DB^{-1} \\ 0 & I \end{bmatrix}\begin{bmatrix} A & D \\ C & B \end{bmatrix} = \begin{bmatrix} A - DB^{-1}C & 0 \\ C & B \end{bmatrix}$.

We now consider the polynomial matrices $\begin{bmatrix} D_1 & -N_2 \\ -N_1 & D_2 \end{bmatrix}$, $\begin{bmatrix} \tilde{D}_1 & -\tilde{N}_1 \\ -\tilde{N}_2 & \tilde{D}_2 \end{bmatrix}$, $(\tilde{D}_1 D_2 - \tilde{N}_1 N_2)$, and $(\tilde{D}_2 D_1 - \tilde{N}_2 N_1)$ from the closed-loop descriptions in (i), (ii), (iii), and (iv). Then

$$\det\left(\begin{bmatrix} D_1 & -N_2 \\ -N_1 & D_2 \end{bmatrix}\right) = \det(D_1)\det(D_2 - N_1 D_1^{-1} N_2)$$

$$= \det(D_1)\det(D_2 - \tilde{D}_1^{-1}\tilde{N}_1 N_2)$$

$$= \det(D_1)\det(\tilde{D}_1^{-1})\det(\tilde{D}_1 D_2 - \tilde{N}_1 N_2)$$

$$= \alpha_1 \det(\tilde{D}_1 D_2 - \tilde{N}_1 N_2), \quad (10.32)$$

where $\alpha_1$ is a nonzero real number. Also

$$\det\left(\begin{bmatrix} D_1 & -N_2 \\ -N_1 & D_2 \end{bmatrix}\right) = \det(D_2)\det(D_1 - N_2 D_2^{-1} N_1)$$

$$= \det(D_2)\det(D_1 - \tilde{D}_2^{-1}\tilde{N}_2 N_1)$$

$$= \det(D_2)\det(\tilde{D}_2^{-1})\det(\tilde{D}_2 D_1 - \tilde{N}_2 N_1)$$

$$= \alpha_2 \det(\tilde{D}_2 D_1 - \tilde{N}_2 N_1), \quad (10.33)$$

where $\alpha_2$ is a nonzero real number.

Similarly,

$$\det\left(\begin{bmatrix} \tilde{D}_1 & -\tilde{N}_1 \\ -\tilde{N}_2 & \tilde{D}_2 \end{bmatrix}\right) = \hat{\alpha}_1 \det(\tilde{D}_2 D_1 - \tilde{N}_2 N_1), \tag{10.34}$$

where $\hat{\alpha}_1 = \det(\tilde{D}_1)\det(D_1^{-1})$ is a nonzero real number, and

$$\det\left(\begin{bmatrix} \tilde{D}_1 & -\tilde{N}_1 \\ -\tilde{N}_2 & \tilde{D}_2 \end{bmatrix}\right) = \hat{\alpha}_2 \det(\tilde{D}_1 D_2 - \tilde{N}_1 N_2), \tag{10.35}$$

where $\hat{\alpha}_2 = \det(\tilde{D}_2)\det(D_2^{-1})$ is a nonzero real number. These computations verify that the equivalent representations given by (i), (ii), (iii), and (iv) have identical eigenvalues.

The following theorem presents conditions for the internal stability of the feedback system of Figure 10.3. These conditions are useful in a variety of circumstances. Assume that the systems $S_1$ and $S_2$ are controllable and observable and that they are described by (10.20)–(10.23) with transfer function matrices given by

$$H_1 = N_1 D_1^{-1} = \tilde{D}_1^{-1}\tilde{N}_1 \tag{10.36}$$

and

$$H_2 = N_2 D_2^{-1} = \tilde{D}_2^{-1}\tilde{N}_2, \tag{10.37}$$

where the $(N_i, D_i)$ are rc and the $(\tilde{N}_i, \tilde{D}_i)$ are lc for $i = 1, 2$. Let $\alpha_1(s)$ and $\alpha_2(s)$ be the pole (characteristic) polynomials of $H_1(s)$ and $H_2(s)$, respectively. Note that $\alpha_i(s) = k_i \det(D_i(s)) = \tilde{k}_i \det(\tilde{D}_i(s))$, $i = 1, 2$, for some nonzero real numbers $k_i, \tilde{k}_i$. Consider the feedback system in Figure 10.3.

**Theorem 10.4.** *The following statements are equivalent:*

*(a) The closed-loop feedback system in Figure 10.3 is internally stable.*
*(b) The polynomial*

*(i)* $\det\left(\begin{bmatrix} D_1 & -N_2 \\ -N_1 & D_2 \end{bmatrix}\right)$, *or*

*(ii)* $\det\left(\begin{bmatrix} \tilde{D}_1 & -\tilde{N}_1 \\ -\tilde{N}_2 & \tilde{D}_2 \end{bmatrix}\right)$, *or*

*(iii)* $\det(\tilde{D}_1 D_2 - \tilde{N}_1 N_2)$, *or*
*(iv)* $\det(\tilde{D}_2 D_1 - \tilde{N}_2 N_1)$
*is Hurwitz; that is, its roots have negative real parts.*
*(c) The polynomial*

$$\alpha_1(s)\alpha_2(s)\det(I - H_1(s)H_2(s)) = \alpha_1(s)\alpha_2(s)\det(I - H_2(s)H_1(s)) \tag{10.38}$$

*is a Hurwitz polynomial.*

*(d) The poles of*

$$\begin{bmatrix} \hat{u}_1 \\ \hat{u}_2 \end{bmatrix} = \begin{bmatrix} I & -H_2 \\ -H_1 & I \end{bmatrix}^{-1} \begin{bmatrix} \hat{r}_1 \\ \hat{r}_2 \end{bmatrix}$$

$$= \begin{bmatrix} (I - H_2 H_1)^{-1} & H_2(I - H_1 H_2)^{-1} \\ H_1(I - H_2 H_1)^{-1} & (I - H_1 H_2)^{-1} \end{bmatrix} \begin{bmatrix} \hat{r}_1 \\ \hat{r}_2 \end{bmatrix} \tag{10.39}$$

*are stable; i.e., they have negative real parts.*
*(e) The poles of*

$$\begin{bmatrix} \hat{y}_1 \\ \hat{y}_2 \end{bmatrix} = \begin{bmatrix} -H_2 & I \\ I & -H_1 \end{bmatrix}^{-1} \begin{bmatrix} 0 & H_2 \\ H_1 & 0 \end{bmatrix} \begin{bmatrix} \hat{r}_1 \\ \hat{r}_2 \end{bmatrix}$$

$$= \begin{bmatrix} (I - H_1 H_2)^{-1} H_1 & (I - H_1 H_2)^{-1} H_1 H_2 \\ (I - H_2 H_1)^{-1} H_2 H_1 & (I - H_2 H_1)^{-1} H_2 \end{bmatrix} \begin{bmatrix} \hat{r}_1 \\ \hat{r}_2 \end{bmatrix} \tag{10.40}$$

*are stable.*

*Proof.* See [1, p. 583, Theorem 3.15]. ∎

*Remarks*

It is important to consider all four entries in the transfer function (10.40) between $\begin{bmatrix} y_1 \\ y_2 \end{bmatrix}$ and $\begin{bmatrix} r_1 \\ r_2 \end{bmatrix}$ [or in (10.39) between $\begin{bmatrix} u_1 \\ u_2 \end{bmatrix}$ and $\begin{bmatrix} r_1 \\ r_2 \end{bmatrix}$] when considering internal stability. Note that the eigenvalues that are uncontrollable from $r_1$ or $r_2$ will not appear in the first or the second column of the transfer matrix, respectively. Similarly, the eigenvalues that are unobservable from $y_1$ or $y_2$ will not appear in the first or the second row of the transfer matrix, respectively. Therefore, consideration of the poles of some of the entries only may lead to erroneous results, since possible uncontrollable or unobservable modes may be omitted from consideration, and these may lead to instabilities.

*Closed-Loop Characteristic Polynomial.* The open-loop characteristic polynomial of the feedback system is $\alpha_1(s)\alpha_2(s)$. The closed-loop characteristic polynomial is a monic polynomial, $\alpha_{cl}(s)$, with roots equal to the closed-loop eigenvalues; i.e., it is equal to any of the polynomials in (b) within a multiplication by a nonzero real number. Then, relation (10.38) implies, in view of (iv), that *the determinant of the return difference matrix $(I - H_1(s)H_2(s))$ is the ratio of the closed-loop characteristic polynomial over the open-loop characteristic polynomial within a multiplication by a nonzero real number.*

---

***Example 10.5.*** Consider the feedback configuration of Figure 10.3 with $H_1 = \frac{s+1}{s-1}$ and $H_2 = \frac{a_1 s + a_0}{s+b}$ the transfer functions of systems $S_1$ and $S_2$, respectively. Let $a_1 \neq 1$ so that the loop is well defined in terms of state-space representations (and all transfer functions are proper). (See Example 10.3.)

All polynomials in (b) of Theorem 10.4 are equal within a multiplication by a nonzero real number, to the closed-loop characteristic polynomial given by $\alpha_{cl}(s) = s^2 + \frac{b-a_1-a_0-1}{1-a_1}s - \frac{b+a_0}{1-a_1}$. This polynomial must be a Hurwitz polynomial for internal stability. If $\alpha_1(s) = s - 1$ and $\alpha_2(s) = s + b$ are the pole polynomials of $H_1$ and $H_2$, then the polynomial in (c) is given by $\alpha_1(s)\alpha_2(s)(1 - H_1(s)H_2(s)) = (1 - a_1)s^2 + (b - a_1 - a_0 - 1)s - (b + a_0) = (1 - a_1)\alpha_{cl}(s)$, which implies that the return difference $1 - H_1(s)H_2(s) = (1 - a_1)\frac{\alpha_{cl}(s)}{\alpha_1(s)\alpha_2(a)}$. Note that $(1 - H_1H_2)^{-1} = (1 - H_2H_1)^{-1} = \frac{(s-1)(s+b)}{\alpha(s)}$ with $\alpha(s) = (1 - \alpha_1)\alpha_{cl}(s)$ and the transfer function matrix in (d) of Theorem 10.4 is given by

$$\begin{bmatrix} \hat{u}_1 \\ \hat{u}_2 \end{bmatrix} = \begin{bmatrix} \frac{(s-1)(s+b)}{\alpha(s)} & \frac{(s-1)(a_1s+a_0)}{\alpha(s)} \\ \frac{(s+1)(s+b)}{\alpha(s)} & \frac{(s-1)(s+b)}{\alpha(s)} \end{bmatrix} \begin{bmatrix} \hat{r}_1 \\ \hat{r}_2 \end{bmatrix}.$$

The polynomial $\alpha(s)$ has a factor $s + 1$ when $b = 1$. Notice that $\alpha(-1) = 2 - 2b = 0$ when $b = 1$. If this is the case ($b = 1$), then

$$\begin{bmatrix} \hat{u}_1 \\ \hat{u}_2 \end{bmatrix} = \begin{bmatrix} \frac{s-1}{\bar{\alpha}(s)} & \frac{(s-1)(a_1s+a_0)}{\alpha(s)} \\ \frac{s+1}{\bar{\alpha}(s)} & \frac{s-1}{\bar{\alpha}(s)} \end{bmatrix} \begin{bmatrix} \hat{r}_1 \\ \hat{r}_2 \end{bmatrix},$$

where $\alpha(s) = (s + 1)\bar{\alpha}(s)$. Notice that three out of four transfer functions do not contain the pole at $-1$ in $\bar{\alpha}(s)$. Recall that when $b = 1$, $-1$ is an eigenvalue that is uncontrollable from the $r_1$ eigenvalue and it cancels in certain transfer functions as expected (see Example 10.3). Similar results can be derived when $a_0/a_1 = -1$. This illustrates the necessity for considering all the transfer functions between $u_1, u_2$ and $r_1, r_2$ when studying the internal stability of the feedback system. Similar results can be derived when considering the transfer functions between $y_1, y_2$ and $r_1, r_2$ in (e).

## 10.3 Parameterization of All Stabilizing Feedback Controllers

In this section, it is shown that all stabilizing feedback controllers can be conveniently parameterized. These parameterizations are very important in control since they are fundamental in methodologies such as the optimal $H^\infty$ approach to control design. Our development builds on the controllability, observability, and particularly the internal stability results introduced in Section 10.2, as well as on Diophantine Equation results [1, Subsection 7.2E]. First, in Subsection 10.3.1, all stabilizing feedback controllers are parameterized, using PMDs. Parameterizations are introduced, using first the polynomial matrix parameters (i) $D_k, N_k$ and $\tilde{D}_k, \tilde{N}_k$ and then the stable rational parameter (ii) $K = N_k D_k^{-1} = \tilde{D}_k^{-1}\tilde{N}_k$. These parameters are very convenient in characterizing stability, but cumbersome when properness of the controller transfer function is to be guaranteed. A parameterization that uses proper and stable MFDs and involves a proper and stable parameter $K'$ is then introduced in Subsection 10.3.2. This is very convenient when properness of $H_2$ is to be guaranteed. The parameter $K'$ is closely related to the parameter $K$ used in the second approach enumerated above. This type of parameterization is useful in certain control design methods such as optimal $H^\infty$ control design. Two degrees of freedom feedback controllers offer additional capabilities in control design and are discussed in Subsection 10.3.2. Control problems are also described in this subsection.

In the following discussion, the term "stable system $S$" is taken to mean that the eigenvalues of the internal description of system $S$ have negative real parts (in the continuous-time case); i.e., the system $S$ is internally stable. Note that when the transfer functions in (10.39) and (10.40) of the feedback system $S$ are proper, internal stability of $S$ implies bounded-input, bounded-output stability of the feedback system, since the poles of the various transfer functions are a subset of the closed-loop eigenvalues.

### 10.3.1 Stabilizing Feedback Controllers Using Polynomial MFDs

Now consider systems $S_1$ and $S_2$ connected in the feedback configuration shown in Figure 10.3. Given $S_1$, it is shown how to parameterize all systems $S_2$ so that the closed-loop feedback system is internally stable. Thus, if $S_1 = S$, called the *plant*, is a given system to be controlled, then $S_2 = S_c$ is viewed as the *feedback controller* that is to be designed. Presently we provide the parameterizations of all stabilizing feedback controllers.

**Theorem 10.6.** *Assume that the system $S_1$ is controllable and observable and is described by the PMD (or PMFD) as (a) $D_1 z_1 = u_1$, $y_1 = N_1 z_1$ given in (10.20), or by (b) $\tilde{D}_1 \tilde{z}_1 = \tilde{N}_1 u_1$, $y_1 = \tilde{z}_1$ given in (10.21). Let the pair $(D_1, N_1)$ and the pair $(\tilde{D}_1, \tilde{N}_1)$ be doubly coprime factorizations of the transfer function matrix $H_1(s) = N_1 D_1^{-1} = \tilde{D}_1^{-1} \tilde{N}_1$. That is,*

$$UU^{-1} = \begin{bmatrix} X_1 & Y_1 \\ -\tilde{N}_1 & \tilde{D}_1 \end{bmatrix} \begin{bmatrix} D_1 & -\tilde{Y}_1 \\ N_1 & \tilde{X}_1 \end{bmatrix} = \begin{bmatrix} I & 0 \\ 0 & I \end{bmatrix}, \qquad (10.41)$$

*where $U$ is a unimodular matrix (i.e., $\det U$ is a nonzero real number) and $X_1, Y_1, \tilde{X}_1, \tilde{Y}_1$ are appropriate matrices. Then all the controllable and observable systems $S_2$ with the property that the closed-loop feedback system eigenvalues are stable (i.e., they have negative real parts) are described by*

$$(a) \quad \tilde{D}_2 \tilde{z}_2 = \tilde{N}_2 u_2, \quad y_2 = \tilde{z}_2, \qquad (10.42)$$

*where $\tilde{D}_2 = \tilde{D}_k X_1 - \tilde{N}_k \tilde{N}_1$ and $\tilde{N}_2 = -(\tilde{D}_k Y_1 + \tilde{N}_k \tilde{D}_1)$ with $X_1, Y_1, \tilde{N}_1, \tilde{D}_1$ given in (10.41) and the parameters $\tilde{D}_k$ and $\tilde{N}_k$ are selected arbitrarily under* *the conditions that $\widetilde{D}_k^{-1}$ exists and is stable, and the pair $(\widetilde{D}_k, \widetilde{N}_k)$ is lc and is such that $\det(\widetilde{D}_k X_1 - \widetilde{N}_k \widetilde{N}_1) \neq 0$.*

*Equivalently, all stabilizing $S_2$ can be described by*

$$(b) \quad D_2 z_2 = u_2, \quad y_2 = N_2 z_2, \tag{10.43}$$

*where $D_2 = \widetilde{X}_1 D_k - N_1 N_k$ and $N_2 = -(\widetilde{Y}_1 D_k + D_1 N_k)$ with $\widetilde{X}_1, \widetilde{Y}_1, \widetilde{N}_1, \widetilde{D}_1$ given in (10.41) and the parameters $D_k$ and $N_k$ are selected arbitrarily under the conditions that $D_k^{-1}$ exists and is stable, and the pair $(D_k, N_k)$ is rc and is such that $\det(\widetilde{X}_1 D_k - N_1 N_k) \neq 0$.*

*Furthermore, the closed-loop eigenvalues are precisely the roots of $\det \widetilde{D}_k$ or of $\det D_k$. In addition, the transfer function matrix of $S_2$ is given by*

$$\begin{aligned} H_2 &= -(\widetilde{D}_k X_1 - \widetilde{N}_k \widetilde{N}_1)^{-1}(\widetilde{D}_k Y_1 + \widetilde{N}_k \widetilde{D}_1) \\ &= -(\widetilde{Y}_1 D_k + D_1 N_k)(\widetilde{X}_1 D_k - N_1 N_k)^{-1}. \end{aligned} \tag{10.44}$$

*Proof.* The closed-loop description in case (a) is given by (10.30) and in case (b) it is given by (10.28). It can be shown [1, Subsection 7.2E] that the expression in (a) and (b) above can also be written as

$$[\widetilde{D}_2, -\widetilde{N}_2] = [\widetilde{D}_k, \widetilde{N}_k] U \tag{10.45}$$

and that

$$\begin{bmatrix} N_2 \\ D_2 \end{bmatrix} = U^{-1} \begin{bmatrix} -N_k \\ D_k \end{bmatrix} \tag{10.46}$$

are parameterizations of all solutions of the *Diophantine equation*

$$\widetilde{D}_2 D_1 - \widetilde{N}_2 N_1 = \widetilde{D}_k \tag{10.47}$$

and

$$\widetilde{D}_1 D_2 - \widetilde{N}_1 N_2 = D_k, \tag{10.48}$$

respectively, where we let $\widetilde{D}_k$ and $D_k$ be desired closed-loop matrices. The fact that $\widetilde{D}_k^{-1}$ (or $D_k^{-1}$) exists and is stable guarantees that all the closed-loop eigenvalues, which are the poles of $\widetilde{D}_k^{-1}$ (or of $D_k^{-1}$), will be stable. The condition $\det(\widetilde{D}_k X_1 - \widetilde{N}_k \widetilde{N}_1) \neq 0$ (or $\det(\widetilde{X}_1 D_k - N_1 N_k) \neq 0$) guarantees that $\det \widetilde{D}_2 \neq 0$ (or $\det D_2 \neq 0$) and therefore the polynomial matrix description for $S_2$ in (10.28) is well defined. Finally, note that the pair $(\widetilde{D}_k, \widetilde{N}_k)$ is lc if and only if the pair $(\widetilde{D}_2, \widetilde{N}_2)$ is lc as can be seen from $[\widetilde{D}_2, -\widetilde{N}_2] = [\widetilde{D}_k, \widetilde{N}_k] U$ given in (10.45) where $U$ unimodular. This then implies that the description $\{\widetilde{D}_2, \widetilde{N}_2, I\}$ for $S_2$ is both controllable and observable. Similarly, the pair $(D_k, N_k)$ is rc, which guarantees that $\{D_2, I, N_2\}$ with $D_2$ and $N_2$ given in (10.46) is also a controllable and observable description for $S_2$. ∎

In place of the polynomial matrix parameters $\widetilde{D}_k, \widetilde{N}_k$ or $D_k, N_k$, it is possible to use a single parameter, a stable rational matrix $K$. This is shown next.

**Theorem 10.7.** *Assume that the system $S_1$ is controllable and observable and is described by its transfer function matrix*

$$H_1 = N_1 D_1^{-1} = \tilde{D}_1^{-1} \tilde{N}_1, \tag{10.49}$$

*where the pairs $(N_1, D_1)$, $(\tilde{D}_1, \tilde{N}_1)$ are doubly coprime factorizations satisfying (10.41). Then all the controllable and observable systems $S_2$ with the property that the closed-loop feedback system eigenvalues are stable (i.e., they have strictly negative real parts) are described by the transfer function matrix*

$$\begin{aligned} H_2 &= -(X_1 - K\tilde{N}_1)^{-1}(Y_1 + K\tilde{D}_1) \\ &= -(\tilde{Y}_1 + D_1 K)(\tilde{X}_1 - N_1 K)^{-1}, \end{aligned} \tag{10.50}$$

*where the parameter $K$ is an arbitrary rational matrix that is stable and is such that $\det(X_1 - K\tilde{N}_1) \neq 0$ or $\det(\tilde{X}_1 - N_1 K) \neq 0$. Furthermore, the poles of $K$ are precisely the closed-loop eigenvalues.*

*Proof.* This is in fact a corollary to Theorem 10.6. It is called a theorem here since it was historically one of the first results established in this area. The parameter $K$ is called the *Youla parameter*.

In Theorem 10.6, descriptions for $H_2$ were given in (10.44) in terms of the parameters $\tilde{D}_k, \tilde{N}_k$ and $D_k, N_k$. Now in view of $-\tilde{D}_k N_k + \tilde{N}_k D_k = 0$, we have

$$\tilde{D}_k^{-1} \tilde{N}_k = N_k D_k^{-1} = K, \tag{10.51}$$

which is a stable rational matrix. Since the pair $(\tilde{D}_k, \tilde{N}_k)$ is lc and the pair $(N_k, D_k)$ is rc, they are coprime factorizations for $K$. Therefore, $H_2$ in (10.50) can be written as the $H_2$ of (10.44) given in the previous theorem, from which the controllable and observable internal descriptions for $S_2$ in (10.42) and (10.43) can immediately be derived. Conversely, (10.50) can immediately be derived from (10.44), using (10.51). Note that the poles of $K$ are the roots of $\det \tilde{D}_k$ or $\det D_k$, which are the closed-loop eigenvalues. ∎

---

***Example 10.8.*** Consider $H_1 = \frac{s+1}{s-1}$. Here $N_1 = \tilde{N}_1 = s + 1$ and $D_1 = \tilde{D}_1 = s - 1$. These are doubly coprime factorizations (a trivial case) since (10.41) is satisfied. We have

$$\begin{aligned} UU^{-1} &= \begin{bmatrix} X_1 & Y_1 \\ -\tilde{N}_1 & \tilde{D}_1 \end{bmatrix} \begin{bmatrix} D_1 & -\tilde{Y}_1 \\ N_1 & \tilde{X}_1 \end{bmatrix} \\ &= \begin{bmatrix} s + \frac{1}{2} & -s + \frac{3}{2} \\ -(s+1) & s - 1 \end{bmatrix} \begin{bmatrix} s - 1, & -(-s + \frac{3}{2}) \\ s + 1, & s + \frac{1}{2} \end{bmatrix} = \begin{bmatrix} 1 & 0 \\ 0 & 1 \end{bmatrix}. \end{aligned}$$

In view of (10.44) and (10.50), all stabilizing controllers $H_2$ are then given by

$$H_2 = -\frac{(-s + \frac{3}{2})d_k + (s-1)n_k}{(s + \frac{1}{2})d_k - (s+1)n_k} = -\frac{(-s + \frac{3}{2}) + (s-1)K}{(s + \frac{1}{2}) - (s+1)K},$$

where $K = n_k/d_k$ is any stable rational function.

---

**Example 10.9.** Consider $H_1(s) = [\frac{1}{s^2}, \frac{s+1}{s^2}] = [1, 0] \begin{bmatrix} s^2 & -(s+1) \\ 0 & 1 \end{bmatrix}^{-1} = N_1 D_1^{-1} = \frac{1}{s^2}[1, s+1] = \tilde{D}_1^{-1} \tilde{N}_1$, which are coprime polynomial MFDs. Relation (10.41) is given by

$$
UU^{-1} = \begin{bmatrix} X_1 & Y_1 \\ -\tilde{N}_1 & \tilde{D}_1 \end{bmatrix} \begin{bmatrix} D_1 & -\tilde{Y}_1 \\ N_1 & \tilde{X}_1 \end{bmatrix}
$$

$$
= \begin{bmatrix} 1 & s+1 & -s^2+1 \\ s & s^2+s+1 & -s^3 \\ -1 & -(s+1) & s^2 \end{bmatrix} \begin{bmatrix} s^2 & -(s+1) & -(s+1) \\ 0 & 1 & s \\ 1 & 0 & 1 \end{bmatrix}
$$

$$
= \begin{bmatrix} 1 & 0 & 0 \\ 0 & 1 & 0 \\ 0 & 0 & 1 \end{bmatrix}.
$$

All stabilizing controllers may then be determined by applying (10.44) or (10.50).

---

*Remark*

In [1, pp. 592–605] a complete treatment of several different parameterizations of all stabilizing controllers is given. The first two parameterizations involving $D_k$ and $K$ were presented here. Another interesting parameterization involves $Q_1$ and $Q_2$ [1, p. 597], which in the case when the plant is stable, it becomes particularly attractive [1, p. 597, Corollary 4.4].

### 10.3.2 Stabilizing Feedback Controllers Using Proper and Stable MFDs

In the above development all systems $S_2$ that internally stabilize the closed-loop feedback system were parametrically characterized. In that development $H_1$, the transfer function of $S_1$ was not necessarily proper and the stabilizing $H_2$ as well as the closed-loop system transfer function were not necessarily proper either. Recall that a system is said to be internally stable when all of its eigenvalues, which are the roots of its characteristic polynomial, have strictly negative real parts. Polynomial matrix descriptions that can easily handle the case of nonproper transfer functions were used to derive the above results and the case of proper $H_1$ and $H_2$ was handled by restricting the parameters used to characterize all stabilizing controllers.

Here we concentrate exclusively on the case of proper transfer functions $H_1$ of $S_1$ and parametrically characterize all proper $H_2$, which internally stabilize the closed-loop system. For this purpose, proper and stable matrix fractional descriptions (MFDs) of $H_1$ and $H_2$ are used. Such MFDs are now described [1, Subsection 7.4C].

Consider $H(s) \in R(s)^{p \times m}$ to be proper, i.e., $\lim_{s \to \infty} H(s) < \infty$, and write the MFD as

$$H(s) = N'(s)D'(s)^{-1}, \tag{10.52}$$

where the $N'(s)$ and $D'(s)$ are proper and stable rational matrices that we denote here as $N'(s) \in RH_\infty^{p \times m}$ and $D'(s) \in RH_\infty^{m \times m}$; that is, they are matrices with elements in $RH_\infty$, the set of all proper and stable rational functions with real coefficients. For instance, if $H(s) = \frac{s-1}{(s-2)(s+1)}$, then $H(s) = \left[ \frac{s-1}{(s+2)(s+3)} \right] \left[ \frac{(s-2)(s+1)}{(s+2)(s+3)} \right]^{-1} = \left[ \frac{s-1}{(s+1)^2} \right] \left[ \frac{s-2}{s+1} \right]^{-1}$ are examples of proper and stable MFDs.

A pair $(N', D') \in RH_\infty$ is called *right coprime (rc) in* $RH_\infty$ if there exists a pair $(X', Y') \in RH_\infty$ such that

$$X'D' + Y'N' = I. \tag{10.53}$$

This is a *Diophantine Equation* over the ring of proper and stable rational functions. It is also called a *Bezout Identity*.

Let $H = N'D'^{-1}$, and write (10.53) as $X' + Y'H = D'^{-1}$. Since the left-hand side is proper, $D'^{-1}$ is also proper; i.e., in the MFD given by $H = N'D'^{-1}$, where the pair $(N', D')$ is rc, $D'$ is biproper ($D'$ and $D'^{-1}$ are both proper).

Note that $X'^{-1}$, where $X'$ satisfies (10.53), does not necessarily exist. If, however, $H$ is strictly proper ($\lim_{s \to \infty} H(s) = 0$), then $\lim_{s \to \infty} X'(s) = \lim_{s \to \infty} D'(s)^{-1}$ is a nonzero real matrix, and in this case $X'^{-1}$ exists and is proper; i.e., in this case $X'$ is biproper.

When the Diophantine Equation (10.53) is used to characterize all stabilizing controllers, it is often desirable to have solutions $(X', Y')$ where $X'$ is biproper. This is always possible. Clearly, when $H$ is strictly proper, this is automatically true, as was shown. When $H$ is not strictly proper, however, care should be exercised in the selection of the solutions of (10.53).

As in the polynomial case, doubly coprime factorizations in $RH_\infty$ of a transfer function matrix $H_1 = N'_1 D'^{-1}_1 = \widetilde{D'}_1^{-1} \widetilde{N'}_1$, where $D'_1, N'_1 \in RH_\infty$ and $\widetilde{D'}_1, \tilde{N'}_1 \in RH_\infty$ are important in obtaining parametric characterizations of all stabilizing controllers. Assume therefore that

$$U'U'^{-1} = \begin{bmatrix} X'_1 & Y'_1 \\ -\widetilde{N'}_1 & \widetilde{D'}_1 \end{bmatrix} \begin{bmatrix} D'_1 & -\tilde{Y'}_1 \\ N'_1 & \tilde{X'}_1 \end{bmatrix} = \begin{bmatrix} I & 0 \\ 0 & I \end{bmatrix}, \tag{10.54}$$

where $U'$ is unimodular in $RH_\infty$, i.e., $U'$ and $U'^{-1} \in RH_\infty$. Also, assume that $X'_1$ and $\widetilde{X'}_1$ have been selected so that $\det X'_1 \neq 0$ and $\det \widetilde{X'}_1 \neq 0$.

## Internal Stability

Consider now the feedback system in Figure 10.3, and let $H_1$ and $H_2$ be the transfer function matrices of $S_1$ and $S_2$, respectively, which are assumed to be controllable and observable. Internal stability of a system can be defined in a variety of equivalent ways in terms of the internal description of the system. For example, in this chapter, polynomial matrix internal descriptions were used and the system was considered as being internally stable when its eigenvalues were stable; i.e., they have negative real parts. In Theorem 10.4, it was shown that the closed-loop feedback system is internally stable if and only if the transfer function between $\begin{bmatrix} u_1 \\ u_2 \end{bmatrix}$ and $\begin{bmatrix} r_1 \\ r_2 \end{bmatrix}$ or $\begin{bmatrix} y_1 \\ y_2 \end{bmatrix}$ and $\begin{bmatrix} r_1 \\ r_2 \end{bmatrix}$ have stable poles, i.e., if and only if the poles of $\begin{bmatrix} I & -H_2 \\ -H_1 & I \end{bmatrix}^{-1}$ or

$\begin{bmatrix} -H_2 & I \\ I & -H_1 \end{bmatrix}^{-1} \begin{bmatrix} 0 & H_1 \\ H_1 & 0 \end{bmatrix}$, respectively, are stable.

In this section we shall regard the feedback system to be internally stable when

$$\begin{bmatrix} I & -H_2 \\ -H_1 & I \end{bmatrix}^{-1} \in RH_\infty, \tag{10.55}$$

i.e., when all the transfer function matrices in (10.55) are proper and stable. In this way, internal stability can be checked without necessarily involving internal descriptions of $S_1$ and $S_2$. This approach to stability has advantages since it can be extended to systems other than linear, time-invariant systems.

**Theorem 10.10.** *Let $H_1 = N_1' D_1'^{-1} = \widetilde{D'}_1^{-1} \widetilde{N'}_1$ be doubly coprime MFDs in $RH_\infty$. Then the closed-loop feedback system is internally stable if and only if $H_2$ has an lc MFD in $RH_\infty$, $H_2 = \widetilde{D'}_2^{-1} \widetilde{N'}_2$, such that*

$$\widetilde{D'}_2 D_1' - \widetilde{N'}_2 N_1' = I, \tag{10.56}$$

*or if and only if $H_2$ has an rc MFD in $RH_\infty$, $H_2 = N_2' D_2'^{-1}$, such that*

$$\widetilde{D'}_1 D_2' - \widetilde{N'}_1 N_2 = I. \tag{10.57}$$

*Proof.* See [1, p. 615, Corollary 4.12]. ∎

In the following discussion, all proper stabilizing controllers are now parameterized.

**Theorem 10.11.** *Let $H_1 = N_1' D_1'^{-1} = \widetilde{D'}_1^{-1} \widetilde{N'}_1$ be doubly coprime MFDs in $RH_\infty$ that satisfy (10.54). Then all $H_2$ that internally stabilize the closed-loop feedback system are given by*

$$H_2 = -(X_1' - K' \widetilde{N'}_1)^{-1}(Y_1' + K' \widetilde{D'}_1) = -(\widetilde{Y}_1' + D_1' K')(\widetilde{X'}_1 - N_1' K')^{-1}, \tag{10.58}$$

*where $K' \in RH_\infty$ is such that $(X_1' - K' \widetilde{N'}_1)^{-1}$ (or $(\widetilde{X}_1 - N_1' K')^{-1}$) exists and is proper.*

*Proof.* It can be shown that all solutions of $\widetilde{D}_2 D_1' - \widetilde{N}_2 N_1' = I$ are given by

$$[\widetilde{D'}_2, -\widetilde{N'}_2] = [I, K'] \begin{bmatrix} X_1' & Y_1' \\ -\widetilde{N'}_1 & \widetilde{D'}_1 \end{bmatrix}, \tag{10.59}$$

where $K' \in RH_\infty$. The proof of this result is similar to the proof of the corresponding result for the polynomial matrix Diophantine Equation. Similarly, all solutions of $\widetilde{D'}_1 D_2' - \widetilde{N'}_1 N_2' = I$ are given by

$$\begin{bmatrix} N_2' \\ D_2' \end{bmatrix} = \begin{bmatrix} D_1' & -\widetilde{Y'}_1 \\ N_1' & \widetilde{X'}_1 \end{bmatrix} \begin{bmatrix} -K' \\ I \end{bmatrix}, \tag{10.60}$$

where $K' \in RH_\infty$. The result follows then directly from Theorem 10.10. ∎

The above theorem is a generalization of the Youla parameterization of Theorem 10.7 over the ring of proper and stable rational functions.

It is interesting to note that in view of (10.54), $H_2$ in (10.58) can be written as follows. Assume that $X_1^{-1}$ and $\widetilde{X}_1^{-1}$ exist. Then

$$\begin{aligned} H_2 &= -(\widetilde{Y'}_1 + X'^{-1}_1(I - Y_1'N_1')K')(\widetilde{X}_1 - N_1'K')^{-1} \\ &= -[\widetilde{Y'}_1\widetilde{X'}_1^{-1}(\widetilde{X'}_1 - N_1'K') + X'^{-1}_1 K'](\widetilde{X}_1 - N_1'K')^{-1} \\ &= -\widetilde{Y'}_1\widetilde{X'}_1^{-1} - X'^{-1}_1 K'(\widetilde{X}_1 - N_1'K')^{-1} = H_{20} + H_{2a}; \end{aligned} \tag{10.61}$$

i.e., any stabilizing controller $H_2$ can be viewed as the sum of an initial stabilizing controller $H_{20} = -\widetilde{Y'}_1\widetilde{X'}_1^{-1}$ and an additional controller $H_{2a}$, which depends on $K'$. When $K' = 0$, then $H_{2a}$, is zero.

---

**Example 10.12.** Let $H_1 = \frac{1}{s-1} = (\frac{1}{s+1})(\frac{s-1}{s+1})^{-1} = N_1'D'^{-1}_1 = (\frac{s-1}{s+a})^{-1}(\frac{1}{s+a}) = \widetilde{D'}^{-1}_1\widetilde{N'}_1$ with $a > 0$, which are doubly coprime factorizations. Note that

$$\begin{bmatrix} X_1' & Y_1' \\ -\widetilde{N'}_1 & \widetilde{D'}_1 \end{bmatrix} \begin{bmatrix} D_1' & -\widetilde{Y'}_1 \\ N_1' & \widetilde{X'}_1 \end{bmatrix} = \begin{bmatrix} \frac{s+3}{s+2} & \frac{s+5}{s+2} \\ -\frac{1}{s+a} & \frac{s-1}{s+a} \end{bmatrix} \begin{bmatrix} \frac{s-1}{s+1} & -\frac{(s+5)(s+a)}{(s+1)(s+2)} \\ \frac{1}{s+1} & \frac{(s+3)(s+a)}{(s+1)(s+2)} \end{bmatrix} = \begin{bmatrix} 1 & 0 \\ 0 & 1 \end{bmatrix}.$$

All stabilizing $H_2$ are parametrically characterized by (10.58).

---

**Example 10.13.** In the above example $H_2 = -(b+1), b > 0$ characterizes all static stabilizing $H_2$. Then for $a = 1$, we have

$$\begin{aligned} K' &= -\left(\frac{s+5}{s+2} - \frac{s+3}{s+2}(b+1)\right)\left(\frac{s-1}{s+1} + \frac{b+1}{s+1}\right)^{-1} \\ &= -\left(\frac{-bs - 3b + 2}{s+2}\right)\left(\frac{s+b}{s+1}\right)^{-1} = \frac{(s+1)(bs + 3b - 2)}{(s+2)(s+b)}, \end{aligned}$$

which will yield the desired $H_2 = -(b+1)$. The closed-loop eigenvalue is in this case at $-b$ as can easily be verified.

---

## Parameterizations Using State-Space Descriptions

Consider $H = N'D'^{-1} = \widetilde{D}'^{-1}\widetilde{N}'$, a doubly coprime factorization in $RH_\infty$; i.e., (10.54) is satisfied. It is possible to express all proper and stable matrices in (10.54) in terms of the matrices of a state-space realization of the transfer function matrix $H(s)$. In particular, we have the following result.

**Lemma 10.14.** *Let* $\{A, B, C, D\}$ *be a stabilizable and detectable realization of* $H(s)$, *i.e.,* $H(s) = C(sI - A)^{-1}B + D$, *which is also denoted by* $H(s) \overset{s}{=} \begin{bmatrix} A & B \\ C & D \end{bmatrix}$, *and with* $(A, B)$ *stabilizable and* $(A, C)$ *detectable. Let* $F$ *be a state feedback gain matrix such that all the eigenvalues of* $A + BF$ *have negative real parts, and let* $K$ *be an observer gain matrix such that all the eigenvalues of* $A - KC$ *have negative real parts. Define*

$$U' = \begin{bmatrix} X' & Y' \\ -\widetilde{N}' & \widetilde{D}' \end{bmatrix} \overset{s}{=} \left[ \begin{array}{c|cc} A - KC & B - KD & K \\ \hline -F & I & 0 \\ -C & -D & I \end{array} \right] \tag{10.62}$$

*and*

$$\widehat{U}' = \begin{bmatrix} \widetilde{D}' & -\widetilde{Y}' \\ N' & X' \end{bmatrix} \overset{s}{=} \left[ \begin{array}{c|cc} A + BF & B & -K \\ \hline F & I & 0 \\ C + DF & D & I \end{array} \right]. \tag{10.63}$$

*Then (10.54) holds and* $H = N'D'^{-1} = \widetilde{D}'^{-1}\widetilde{N}'$ *are coprime factorizations of* $H$.

*Proof.* Relation (10.54) can be shown to be true by direct computation, which it is left to the reader to verify. Clearly, $U', \widehat{U}' \in RH_\infty$. That $N', D'$ and $\widetilde{D}', \widetilde{N}'$ are coprime is a direct consequence of (10.54). That $N'D'^{-1} = \widetilde{D}'^{-1}\widetilde{N}' = H$ can be shown by direct computation and is left to the reader. ∎

In view of Lemma 10.14, $U'$ and $U'^{-1} \in RH_\infty$ in (10.54) can be expressed as

$$U' = \begin{bmatrix} X' & Y' \\ -\widetilde{N}' & \widetilde{D}' \end{bmatrix} = \begin{bmatrix} -F \\ -C \end{bmatrix} [sI - (A - KC)]^{-1}[B - KD, K] + \begin{bmatrix} I & 0 \\ -D & I \end{bmatrix} \tag{10.64}$$

and

$$U'^{-1} = \begin{bmatrix} D' & -\widetilde{Y}' \\ N' & \widetilde{X}' \end{bmatrix} = \begin{bmatrix} F \\ C + DF \end{bmatrix} [sI - (A + BF)]^{-1}[B, -K] + \begin{bmatrix} I & 0 \\ D & I \end{bmatrix}. \tag{10.65}$$

These formulas can be used as follows. A stabilizable and detectable realization $\{A, B, C, D\}$ of $H(s)$ is first determined, and appropriate $F$ and $K$ are found so that $A + BF$ and $A - KC$ have eigenvalues with negative real parts. Then $U'$ and $U'^{-1}$ are calculated from (10.64) and (10.65). Note that appropriate state feedback gain matrices $F$ and observer gain matrices $K$ can be determined, using the methods discussed in Chapter 9. The matrices $F$ and $K$ may be determined, for example, by solving appropriate optimal linear quadratic control and filtering problems. All proper stabilizing controllers $H_2 = N_2'D_2'^{-1} = \widetilde{D}_2'^{-1}\widetilde{N}_2'$ of the plant $H_1$ are then characterized as in Theorem 10.11.

It can now be shown, in view of Lemma 10.14, that all stabilizing controllers are described by

$$\dot{\hat{x}} = (A + BF - K(C + DF))\hat{x} + Ky + (B - KD)r_1,$$
$$u = F\hat{x} + r_1, r_2 = y - (C + DF)\hat{x} - Dr_1, r_1 = K'(q)r_2, \qquad (10.66)$$

which can be rewritten as

$$\dot{\hat{x}} = A\hat{x} + Bu + K(y - (C\hat{x} + Du)),$$
$$u = F\hat{x} + K'(q)(y - (C\hat{x} + Du)). \qquad (10.67)$$

Thus, every stabilizing controller is a combination of an asymptotic (full-state, full-order) estimator or observer and a stabilizing state feedback, plus $K'(q)r_2$ with $r_2 = y - (C\hat{x} + Du)$, the output "error" (see Figure 10.4).

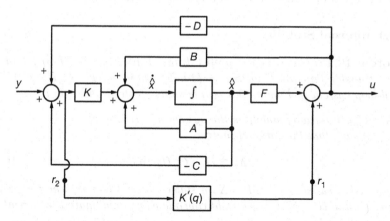

**Figure 10.4.** A state-space representation of all stabilizing controllers

## 10.4 Two Degrees of Freedom Controllers

Consider the two degrees of freedom controller $S_C$ in the feedback configuration of Figure 10.5. Here $S_H$ represents the system to be controlled and is described by its transfer function matrix $H(s)$ so that

$$\hat{y}(s) = H(s)\hat{u}(s). \qquad (10.68)$$

The two degrees of freedom controller $S_C$ is described by its transfer function matrix $C(s)$ in

$$\hat{u}(s) = C(s) \begin{bmatrix} \hat{y}(s) \\ \hat{r}(s) \end{bmatrix} = [C_y(s), C_r(s)] \begin{bmatrix} \hat{y}(s) \\ \hat{r}(s) \end{bmatrix}. \tag{10.69}$$

Since the controller $S_C$ generates the input $u$ to $S_H$ by processing independently $y$, the output of $S_H$, and $r$, it is called a two degrees of freedom controller.

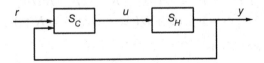

**Figure 10.5.** Two degrees of freedom controller $S_C$

In the following discussion, we shall assume that $H$ is a proper transfer function and we shall determine proper controller transfer functions $C$, which internally stabilize the feedback system in Figure 10.5. The restriction that $H$ and $C$ are proper may easily be removed, if so desired.

### 10.4.1 Internal Stability

**Theorem 10.15.** *Given is the proper transfer function $H$ of $S_H$, and the proper transfer function $C$ of $S_C$ in (10.69) where $\det(I - C_y H) \neq 0$. The closed-loop system in Figure 10.5 is internally stable if and only if*

*(i) $\hat{u} = C_y \hat{y}$ internally stabilizes the system $\hat{y} = H\hat{u}$, and*
*(ii) $C_r$ is such that the rational matrix*

$$M \triangleq (I - C_y H)^{-1} C_r \tag{10.70}$$

*$(u = Mr)$ satisfies $D^{-1}M = X$, a stable rational matrix, where $C_y$ satisfies (i) and $H = ND^{-1}$ is a right coprime polynomial matrix factorization.*

*Proof.* Consider controllable and observable polynomial matrix descriptions (PMDs) for $S_H$, given by

$$Dz = u, \quad y = Nz \tag{10.71}$$

and for $S_C$, given by

$$\tilde{D}_c \tilde{z}_c = [\tilde{N}_y, \tilde{N}_r] \begin{bmatrix} y \\ r \end{bmatrix}, u = \tilde{z}_c, \tag{10.72}$$

where the $N, D$ are rc and the $\tilde{D}_c, [\tilde{N}_y, \tilde{N}_r]$ are lc polynomial matrices. The closed-loop system is then described by

$$(\tilde{D}_c D - \tilde{N}_c N)z = \tilde{N}_r r, y = Nz \qquad (10.73)$$

and is internally stable if the roots of $\det \tilde{D}_o$, where $\tilde{D}_o \triangleq \tilde{D}_c D - \tilde{N}_c N$, have negative real parts.

(*Necessity*) Assume that the closed-loop system is internally stable, i.e., $\tilde{D}_o^{-1}$ is stable. Since $C_y = \tilde{D}_c^{-1}\tilde{N}_y$ is not necessarily a left coprime polynomial factorization, write $[\tilde{D}_c, \tilde{N}_y] = G_L[\tilde{D}_{C_y}, \tilde{N}_{C_y}]$, where $G_L$ is a gcld of the pair $(\tilde{D}_c, \tilde{N}_y)$. Then $\tilde{D}_{C_y} D - \tilde{N}_{C_y} N = G_L^{-1} \tilde{D}_o = \tilde{D}_k$, where $\tilde{D}_k$ is a polynomial matrix, with $\tilde{D}_k^{-1}$ stable; note also that $G_L^{-1}$ is stable. Hence, $u = C_y y = \tilde{D}_{C_y}^{-1} \tilde{N}_{C_y} y$ internally stabilizes $y = Hu = ND^{-1}u$; i.e., part (i) of the theorem is true. To show that (ii) is true, we write $M = (I - C_y H)^{-1} C_r = D \tilde{D}_k^{-1} \tilde{D}_{C_y} (\tilde{D}_c^{-1} \tilde{N}_r) = D \tilde{D}_k^{-1} G_L^{-1} \tilde{N}_r = DX$, where $X \triangleq \tilde{D}_o^{-1} \tilde{N}_r$ is a stable rational matrix. This shows that (ii) is also necessary.

(*Sufficiency*) Let $C$ satisfy (i) and (ii) of the theorem. If $C = \tilde{D}_c^{-1}[\tilde{N}_y, \tilde{N}_r]$ is an lc polynomial MFD and $G_L$ is a gcld of the pair $(\tilde{D}_c, \tilde{N}_y)$, then $[\tilde{D}_c, \tilde{N}_y] = G_L[\tilde{D}_{C_y}, \tilde{N}_{C_y}]$ is true for some lc matrices $\tilde{D}_{C_y}$ and $\tilde{N}_{C_y}$ ($C_y = \tilde{D}_{C_y}^{-1} \tilde{N}_{C_y}$). Because (i) is satisfied, $\tilde{D}_{C_y} D - \tilde{N}_{C_y} N = \tilde{D}_k$, where $\tilde{D}_k^{-1}$ is stable. Premultiplying by $G_L$ we obtain $\tilde{D}_c D - \tilde{N}_y N = G_L \tilde{D}_k$. Now if $G_L^{-1}$ is stable, then $\tilde{D}_o^{-1}$, where $\tilde{D}_o \triangleq \tilde{D}_c D - \tilde{N}_y N = G_L \tilde{D}_k$, will be stable since $\tilde{D}_k^{-1}$ is stable. To show this, write $D^{-1}M = D^{-1}(I - C_y H)^{-1} C_r = \tilde{D}_k^{-1} \tilde{D}_{C_y} (\tilde{D}_c^{-1} \tilde{N}_r) = \tilde{D}_k^{-1} G_L^{-1} \tilde{N}_r$ and note that this is stable, in view of (ii). Observe now that the $G_L, \tilde{N}_r$ are lc; if they were not, then $C = \tilde{D}_c^{-1}[\tilde{N}_y, \tilde{N}_r]$ would not be a coprime factorization. In this case no unstable cancellations take place in $\tilde{D}_k^{-1} G_L^{-1} \tilde{N}_r$ ($\tilde{D}_k^{-1}$ is stable) and therefore, if $D^{-1}M$ is stable, then $(G_L \tilde{D}_k)^{-1} = \tilde{D}_o^{-1}$ is stable or the closed-loop system is internally stable.  ∎

*Remarks*

(i) It is straightforward to show the same results, using proper and stable factorizations of $H$ given by

$$H = N'D'^{-1}, \qquad (10.74)$$

where the pair $(N', D') \in RH_\infty$ and $(N', D')$ is rc, and of

$$C = \tilde{D}'_c{}^{-1}[\tilde{N}'_y, \tilde{N}'_r], \qquad (10.75)$$

where the pair $(\tilde{D}'_c, [\tilde{N}'_y, \tilde{N}'_r]) \in RH_\infty$ and $(\tilde{D}'_c, [\tilde{N}'_y, \tilde{N}'_r])$ is lc. The proof is completely analogous and is left to the reader. The only change in the theorem will be in its part (ii), which will now read as follows: $C_r$ is such that the rational matrix $M \triangleq (I - C_y H)^{-1} C_r$ satisfies $D'^{-1}M = X' \in RH_\infty$, where $C_y$ satisfies (a) and $H = N'D'^{-1}$ is an rc MFD in $RH_\infty$.

(ii) Theorem 10.15 separates the role of $C_y$, the feedback part of the two
degrees of freedom controller $C$, from the role of $C_r$, in achieving internal
stability. Clearly, if only feedback action is considered, then only part (i) of
the theorem is of interest; and if open-loop control is desired, then $C_y = 0$
and (i) implies that for internal stability $H$ must be stable and $C_r = M$
must satisfy part (ii). In (ii) the parameter $M = DX$ appears naturally
and in (i) the way is open to use any desired feedback parameterizations.
In view of Theorem 10.15 it is straightforward to parametrically characterize all internally stabilizing controllers $C$. In the theorem it is clearly
stated [Part (i)] that $C_y$ must be a stabilizing controller. Therefore, any
parametric characterization of the ones developed in the previous subsections, as in [1, Subsection 7.4], can be used for $C_y$. Also, $C_r$ is expressed
in terms of $D^{-1}M = X$ (or $D'^{-1}M = X'$).

**Theorem 10.16.** *Given that* $\hat{y} = H\hat{u}$ *is proper with* $H = ND^{-1} = \tilde{D}^{-1}\tilde{N}$
*doubly coprime polynomial MFDs, all internally stabilizing proper controllers*

$C$ *in* $\hat{u} = C \begin{bmatrix} \hat{y} \\ \hat{r} \end{bmatrix}$ *are given by*

(a) $$C = (I+QH)^{-1}[Q, M] = [(I+LN)D^{-1}]^{-1}[L, X], \qquad (10.76)$$

*where* $Q = DL$ *and* $M = DX$ *are proper with* $L, X$ *and* $D^{-1}(I + QH) = (I + LN)D^{-1}$ *stable, so that* $(I + QH)^{-1}$ *exists and is proper; or*

(b) $$C = (X_1 - K\tilde{N})^{-1}[-(X_2 + K\tilde{D}), X], \qquad (10.77)$$

*where* $K$ *and* $X$ *are stable so that* $(X_1 - K\tilde{N}_1)^{-1}$ *exists and* $C$ *is proper. Also,*
$X_1$ *and* $X_2$ *are determined from* $UU^{-1} = \begin{bmatrix} X_1 & X_2 \\ -\tilde{N} & \tilde{D} \end{bmatrix} \begin{bmatrix} D & -\tilde{X}_2 \\ N & \tilde{X}_1 \end{bmatrix} = \begin{bmatrix} I & 0 \\ 0 & I \end{bmatrix}$ *with*
$U$ *unimodular.*

*If* $H = N'D'^{-1} = \tilde{D}'^{-1}\tilde{N}'$ *are doubly coprime MFDs in* $RH_\infty$, *then all*
*stabilizing proper* $C$ *are given by*

(c) $$C = (X_1' - K'\widetilde{N'})^{-1}[-(X_2' + K'\widetilde{D'}), X'], \qquad (10.78)$$

*where* $K', X' \in RH_\infty$ *so that* $(X_1' - K'\widetilde{N'})^{-1}$ *exists and is proper. Also,*
$U'U'^{-1} = \begin{bmatrix} X_1' & X_2' \\ -\widetilde{N'} & \widetilde{D'} \end{bmatrix} \begin{bmatrix} D' & -\widetilde{X'}_2 \\ N' & \widetilde{X'}_1 \end{bmatrix} = \begin{bmatrix} I & 0 \\ 0 & I \end{bmatrix}$ *with* $U', U'^{-1} \in RH_\infty$.

(d) $$C = (I+QH)^{-1}[Q, M] = [(I+L'N')D'^{-1}]^{-1}[L', X'], \qquad (10.79)$$

*where* $Q = D'L'$, $M = D'X' \in RH_\infty$ *with* $L', X'$ *and* $D'^{-1}(I + QH) = (I + L'N')D'^{-1} \in RH_\infty$ *so that* $(I + QH)^{-1}$ *or* $(I + L'N')^{-1}$ *exists and is*
*proper.*

*Proof.* The proof is based on the parameterizations of Section 10.3. For details, and for additional discussion of the parameters $L$ and $L'$, see [1, p. 624,
Theorem 4.2.2]. ∎

*Remarks*

(a) In [1, pp. 592–605] a complete treatment of different parameterizations of all stabilizing controllers is given. Parameter $K$ in the theorem above was discussed earlier and parameters $Q$, $X$ and $L$ are discussed in [1, pp. 597–605].

(b) Notice that in the above theorem $C_y$ is parameterized by $K$ or $Q$ or $L$, whereas $C_r$ is parameterized by $M$ or $X$.

### 10.4.2 Response Maps

It is straightforward to express the maps between signals of interest of Figure 10.6 in terms of the parameters in Theorem 10.16. For instance, $u = C\begin{bmatrix} y \\ r \end{bmatrix} = [C_y, C_r]\begin{bmatrix} y \\ r \end{bmatrix} = C_y Hu + C_r r$, from which we have $u = (I - C_y H)^{-1}C_r r = Mr$. (In the following discussion, we will use the symbols $u$, $y$, $r$, etc. instead of $\hat{u}$, $\hat{y}$, $\hat{r}$, etc. for convenience.) If expressions in (d) of Theorem 10.16 are used, then

$$u = D'X'r, \quad \text{and} \quad y = Hu = N'D'^{-1}D'X'r = N'X'r \qquad (10.80)$$

in view of $(I - C_y H)^{-1} = D'(I + L'N')D'^{-1}$. Similar results can be derived using the other parameterizations in Theorem 10.16. To determine expressions for other maps of interest in control systems, consider Figure 10.6, where $d_u$ and $d_y$ are assumed to be disturbances at the input and output of the plant $H$, respectively, and $\eta$ denotes measurement noise. Then, $u = [C_y, C_r]\begin{bmatrix} y + d_y + \eta \\ r \end{bmatrix} + d_u$, from which we have $u = (I - C_y H)^{-1}[C_r r + C_y d_y + C_y \eta + d_u]$ and $y = Hu = H(I - C_y H)^{-1}[C_r r + C_y d_y + C_y \eta + d_u]$.

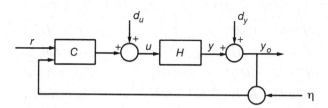

**Figure 10.6.** Two degrees of freedom control configuration

Then, in view of (10.79) in Theorem 10.16, we obtain

$$u = D'X'r + D'L'd_y + D'L'\eta + D'(I + L'N')D'^{-1}d_u$$
$$= Mr + Qd_y + Q\eta + S_i d_u \qquad (10.81)$$

and

$$y = N'X'r + N'L'd_y + N'L'\eta + N'(I + L'N')D'^{-1}d_u$$
$$= Tr + (S_o - I)d_y + HQ\eta + HS_id_u. \qquad (10.82)$$

Notice that $Q = (I - C_yH)^{-1}C_y = D'L'$ is the transfer function between $u$ and $d_y$ or $\eta$. Also,

$$S_i \triangleq (I - C_yH)^{-1} = D'(I + L'N')D'^{-1} = I + QH \qquad (10.83)$$

is the transfer function between $u$ and $d_u$. The matrix $S_i$ is called the *input comparison sensitivity matrix*. Notice also that $y_o = y + d_y = Tr + S_od_y + HQ\eta + HS_id_u$; i.e.,

$$S_o = (I - HC_y)^{-1} = I + HQ \qquad (10.84)$$

is the transfer function between $y_o$ and $d_y$. The matrix $S_o$ is called the *output comparison sensitivity matrix*. The sensitivity matrices $S_i$ and $S_o$ are important quantities in control design. Now

$$S_o - HQ = S_o - N'L' = I \qquad (10.85)$$

since $HQ = H(I - C_yH)^{-1}C_y = HC_y(I - HC_y)^{-1} = -I + (I - HC_y)^{-1} = -I + S_o$, where $S_o$ and $HQ$ are the transfer functions from $y_o$ to $d_y$ and $\eta$, respectively. Equation (10.85) states that disturbance attenuation (or sensitivity reduction) and noise attenuation cannot occur over the same frequency range. This is a fundamental limitation of the feedback loop and occurs also in two degrees of freedom control systems. Similarly we note that

$$S_i - QH = I. \qquad (10.86)$$

We now summarize some of the relations discussed above:

$$T = H(I - C_yH)^{-1}C_r = HM = NX \quad (y = Tr),$$
$$M = (I - C_yH)^{-1}C_r = DX \qquad\quad (u = Mr),$$
$$Q = (I - C_yH)^{-1}C_y = DL \qquad\quad (u = Qd_y),$$
$$S_o = (I - HC_y)^{-1} = I + HQ \qquad\; (y_o = S_od_y),$$
$$S_i = (I - C_yH)^{-1} = I + QH \qquad\; (u = S_id_u),$$

where $y = Tr$ denotes the relation between $y$ and $r$ from (10.82) when all the other signals are zero. Similar expressions hold for the rest of the relations.

### Realizing Desired Responses

The input–output maps attainable from $r$, using an internally stable two degrees of freedom configuration, can be characterized directly. In particular, consider the two maps described by

$$\begin{bmatrix} y \\ u \end{bmatrix} = \begin{bmatrix} T \\ M \end{bmatrix} r, \tag{10.87}$$

i.e., the command/output map $T$ and the command/input map $M$. Let $H = ND^{-1}$ be an rc polynomial MFD.

**Theorem 10.17.** *The stable rational function matrices $T$ and $M$ are realizable with internal stability by means of a two degrees of freedom control configuration [which satisfies (10.87)] if and only if there exists a stable $X$ so that*

$$\begin{bmatrix} T \\ M \end{bmatrix} = \begin{bmatrix} N \\ D \end{bmatrix} X. \tag{10.88}$$

*Proof.* (*Necessity*) Assume that $T$ and $M$ in (4.169) are realizable with internal stability. Then in view of Theorem 10.15, $X \triangleq D^{-1}M$ is stable. Also, $y = Hu = (ND^{-1})(Mr) = NXr$.

(*Sufficiency*) Let (10.88) be satisfied. If $X$ is stable, then $T$ and $M$ are stable. Also, note that $T = HM$. We now show that in this case a controller configuration exists to implement these maps (see Figure 10.7). Note that $u = \widehat{M}r + C_y(\widehat{T}r + y) = [C_y, \widehat{M} + C_y\widehat{T}] \begin{bmatrix} y \\ r \end{bmatrix}$, from which we obtain

$$u = (I + C_yH)^{-1}(\widehat{M} + C_y\widehat{T})r. \tag{10.89}$$

Now if $\widehat{M} = M$ and $\widehat{T} = T$, then in view of $T = HM$, this relation implies that $u = (I + C_yH)^{-1}(I + C_yH)Mr = Mr$ and $y = Hu = HMr = Tr$. Furthermore, $C_y$ is a stabilizing feedback controller, and the system is internally stable since $\widehat{T}$ and $\widehat{M}$ are stable. ∎

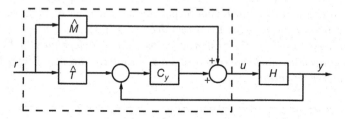

**Figure 10.7.** Feedback realization of $(T, M)$

Note that other (than Figure 10.7), internally stable controller configurations to attain these maps are possible. (The realization of both response maps $T$ and $M$, instead of only $T$ as in the case of the Model Matching Problem, makes the convenient formulation in Theorem 10.17 possible. The realization of both $T$ and $M$ is sometimes referred to as the *Total Synthesis Problem*; see [6], [7] and the references therein.)

The results of Theorem 10.17 can be expressed in terms of $H = N'D'^{-1}$, rc MFDs in $RH_\infty$. In particular, we have the following result.

**Theorem 10.18.** $T, M \in RH_\infty$ *are realizable with internal stability by means of a two degrees of freedom control configuration [which satisfies (10.87)] if and only if there exists $X' \in RH_\infty$ so that*

$$\begin{bmatrix} T \\ M \end{bmatrix} = \begin{bmatrix} N' \\ D' \end{bmatrix} X'. \tag{10.90}$$

*Proof.* The proof is completely analogous to the proof of Theorem 10.17, and it is omitted. ∎

*Remarks*

(i) It is now clear that given any desirable response maps $\begin{bmatrix} y \\ u \end{bmatrix} = \begin{bmatrix} T \\ M \end{bmatrix} r$ such that $\begin{bmatrix} T \\ M \end{bmatrix} = \begin{bmatrix} N' \\ D' \end{bmatrix} X'$, where $X' \in RH_\infty$, the pair $(T, M)$ can be realized with internal stability by using for instance a controller (10.79), $C = [(I + L'N')D'^{-1}]^{-1}[L', X']$, where $[(I + L'N')D'^{-1}, L'] \in RH_\infty$ and $X'$ is given above, as can easily be verified. It is clear that there are many $C$, which realize such $T$ and $M$, and they are all parameterized via the parameter $L' \in RH_\infty$, which for internal stability must satisfy the condition $(I + L'N')D'^{-1} \in RH_\infty$. Other parameterizations such as $K'$ can also be used. In other words, the maps $T, M$ can be realized by a variety of configurations, each with different feedback properties.

(ii) In a two degrees of freedom feedback control configuration, all admissible responses from $r$ under condition of internal stability are characterized in terms of the parameters $X$ (or $M$), whereas all response maps from disturbance and noise inputs that describe feedback properties of the system can be characterized in terms of parameters such as $K$ or $Q$ or $L$. *This is the fundamental property of two degrees of freedom control systems: It is possible to attain the response maps from $r$ independently from feedback properties such as response to disturances and sensitivity to plant parameter variations.*

---

***Example 10.19.*** We consider $H(s) = \frac{(s-1)(s+2)}{(s-2)^2}$ and wish to characterize all proper and stable transfer functions $T(s)$ that can be realized by means of some control configuration with internal stability. Let $H(s) = \frac{s-1}{(s+2)}\left(\frac{(s-2)^2}{(s+2)^2}\right)^{-1} = N'D'^{-1}$ be an rc $MFD$ in $RH_\infty$. Then in view of Theorem 10.18, all such $T$ must satisfy $N'^{-1}T = \frac{s+2}{s-1}T = X' \in RH_\infty$. Therefore, any proper $T$ with a zero at $+1$ can be realized via a two degrees of freedom feedback controller with internal stability. *In general, all unstable zeros of $H$ must appear in $T$ for internal stability to be possible. This shows a fundamental limitation of feedback control.*

Now if a single degree of freedom controller must be used, the class of realizable $T(s)$ under internal stability is restricted. In particular, if the unity feedback configuration $\{I, G_{ff}, I\}$ in Figure 10.10 below is used, then all proper and stable $T$ that are realizable under internal stability are again given by $T = N'X' = \frac{s-1}{s+2}X'$ where $X' = L' \in RH_\infty$ [see 10.100] and in addition $(I + X'N')D'^{-1} = (1 + X'\frac{s-1}{s+2})\frac{(s+2)^2}{(s-2)^2} \in RH_\infty$; i.e., $X' = n_x/d_x$ is proper and stable and should also satisfy $(s+2)d_x + (s-1)n_x = (s-2)^2 p(s)$ for some polynomial $p(s)$. This illustrates the restrictions imposed by the unity feedback controller, as opposed to a two degrees of freedom controller.

---

It is not difficult to prove the following result.

**Theorem 10.20.** $T, M, S \in RH_\infty$ *are realizable with internal stability by a two degrees of freedom control configuration that satisfy (10.87) and (10.85) [$S = S_o$, see Figure 10.6 and (10.81), (10.82)] if and only if there exist $X', L' \in RH_\infty$ so that*

$$\begin{bmatrix} T \\ M \\ S \end{bmatrix} = \begin{bmatrix} N' & 0 \\ D' & 0 \\ 0 & N' \end{bmatrix} \begin{bmatrix} X' \\ L' \end{bmatrix} + \begin{bmatrix} 0 \\ 0 \\ I \end{bmatrix}, \tag{10.91}$$

*where $(I + L'N')D'^{-1} \in RH_\infty$. Similarly, $T, M, Q \in RH_\infty$ are realizable if and only if there exist $X', L' \in RH_\infty$ so that*

$$\begin{bmatrix} T \\ M \\ Q \end{bmatrix} = \begin{bmatrix} N' & 0 \\ D' & 0 \\ 0 & D' \end{bmatrix} \begin{bmatrix} X' \\ L' \end{bmatrix}, \tag{10.92}$$

*where $(I + L'N')D'^{-1} \in RH_\infty$.*

*Proof.* The proof is straightfoward in view of Theorem 10.18. Note that $S$ or $Q$ are selected in such a manner that the feedback loop has desirable feedback characteristics that are expressed in terms of these maps. ∎

### 10.4.3 Controller Implementations

The controller $C = [C_y, C_r]$ may be implemented, for example, as a system $S_c$ as shown in Figure 10.5 and described by (10.72); or as shown in Figure 10.7 with $C = [C_y, M + C_yT]$, where $C_y$ stabilizes $H$ and $T, M$ are desired stable maps that relate $r$ to $y$ and $r$ to $u$; i.e., $y = Tr$ and $u = Mr$. There are also alternative ways of implementing a stabilizing controller $C$. In the following discussion, the common control configuration of Figure 10.8, denoted by $\{R; G_{ff}, G_{fb}\}$, is briefly discussed together with several special cases.

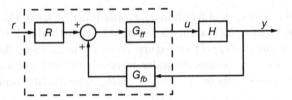

**Figure 10.8.** Two degrees of freedom controller $\{R; G_{ff}, G_{fb}\}$

## $\{R; G_{ff}, G_{fb}\}$ Configuration

Consider the system in Figure 10.8. Note that since

$$u = [C_y, C_r] \begin{bmatrix} y \\ r \end{bmatrix} = [G_{ff}G_{fb}, G_{ff}R] \begin{bmatrix} y \\ r \end{bmatrix}, \tag{10.93}$$

$\{R; G_{ff}, G_{fb}\}$ is a two degrees of freedom control configuration that is as general as the ones discussed before. To see this, let $C = [C_y, C_r] = \widetilde{D'}_c^{-1} [\widetilde{N'}_y, \widetilde{N'}_r]$ be an lc MFD in $RH_\infty$ and let

$$R = \widetilde{N'}_r, G_{ff} = \widetilde{D'}_c^{-1}, G_{fb} = \widetilde{N'}_y. \tag{10.94}$$

Note that $R$ and $G_{fb}$ are always stable; also, $G_{ff}^{-1}$ exists and is stable. Assume now that $C$ was chosen so that

$$\widetilde{D'}_c D' - \widetilde{N'}_y N' = \widetilde{U'}, \tag{10.95}$$

where $\widetilde{U'}, \widetilde{U'}^{-1} \in RH_\infty$. Then the system in Figure 10.8 with $R, G_{ff}$ and $G_{fb}$ given in (10.94) is internally stable. See [1, p. 630] for the proof of this claim.

We shall now discuss briefly some special cases of the $\{R; G_{ff}, G_{fb}\}$ control configuration, which are quite common in practice. Note that the configurations below are simpler; however, they restrict the choices of attainable response maps and so the flexibility offered to the control designer is reduced.

*(i) $\{I; G_{ff}, G_{fb}\}$ Controller*

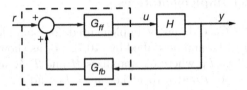

**Figure 10.9.** The $\{I; G_{ff}, G_{fb}\}$ controller

In this case $u = [C_y, C_r] \begin{bmatrix} y \\ r \end{bmatrix} = [G_{ff}G_{fb}, G_{ff}] \begin{bmatrix} y \\ r \end{bmatrix}$; that is,

$$C_y = C_r G_{fb}. \tag{10.96}$$

See Figure 10.9. In view of (10.79) given in Theorem 10.16, this implies that

$$L' = X' G_{fb} \tag{10.97}$$

or that the choice for the parameters $L'$ and $X'$ is not completely independent as in the $\{R; G_{ff}, G_{fb}\}$ case. The $L'$ and $X'$ must of course satisfy $L', X'$ and $(I + L'N')D'^{-1} \in RH_\infty$. In addition, in this case $L'$ and $X'$ must be so that a proper solution $G_{fb}$ of (10.97) exists and no unstable poles cancel in $X'G_{fb}$. Note that these poles will cancel in the product $G_{ff}G_{fb}$ and will lead to an unstable system. Since $L'$ and $X'$ are both stable, we will require that (10.97) has a solution $G_{fb} \in RH_\infty$. This implies that if, for example, $X'^{-1}$ exists, then $X'$ and $L'$ must be such that $X'^{-1}L' \in RH_\infty$; i.e., the $X'$ and $L'$ have the same unstable zeros and $L'$ is "more proper" than $X'$. This provides some guidelines about the conditions $X'$ and $L'$ must satisfy. Also,

$$G_{ff} = [(I + L'N')D'^{-1}]^{-1}X'. \tag{10.98}$$

It should be noted that the state feedback law implemented by a dynamic observer can be represented as a $\{I; G_{ff}, G_{fb}\}$ controller. See [1, Section 7.4B, Figure 7.8].

*(ii)* $\{I; G_{ff}, I\}$ *Controller*

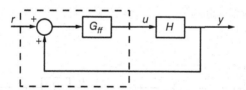

**Figure 10.10.** The $\{I; G_{ff}, I\}$ controller

A special case of (i) is the common unity feedback control configuration; see Figure 10.10. Here $u = [C_y, C_r] \begin{bmatrix} y \\ r \end{bmatrix} = [G_{ff}, G_{ff}] \begin{bmatrix} y \\ r \end{bmatrix}$; that is,

$$C_r = C_y, \tag{10.99}$$

which in view of (10.79) implies that

$$X' = L'. \tag{10.100}$$

In this case the responses between $y$ or $u$ and $r$ (characterized by $X'$) cannot be designed independently of feedback properties such as sensitivity (characterized by $L'$). This is a single degree of freedom controller and is used primarily to attain feedback control specifications. This case is discussed further below.

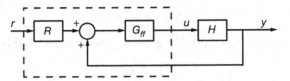

**Figure 10.11.** The $\{R; G_{ff}, I\}$ controller

*(iii) $\{R; G_{ff}, I\}$ Controller*

Here $u = [C_y, C_r] \begin{bmatrix} y \\ r \end{bmatrix} = [G_{ff}, G_{ff}R] \begin{bmatrix} y \\ r \end{bmatrix}$; that is,

$$C_r = C_y R. \tag{10.101}$$

See Figure 10.11. In view of (10.79) given in Theorem 10.16, this implies that

$$X' = L'R. \tag{10.102}$$

The $L'$ and $X'$ must satisfy $L', X', (I + L'N')D'^{-1} \in RH_\infty$. In addition, they must be such that (10.102) has a solution $R \in RH_\infty$. Note that $R$ stable is necessary for internal stability. The reader should refer to the discussion in (i) above for the implications of such assumptions on $X'$ and $L'$. Also,

$$G_{ff} = [(I + L'N')D'^{-1}]^{-1}L'. \tag{10.103}$$

*(iv) $\{R; I, G_{fb}\}$ Controller*

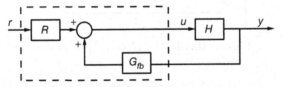

**Figure 10.12.** The $\{R; I, G_{fb}\}$ controller

In this case

$$u = [C_y, C_r] \begin{bmatrix} y \\ r \end{bmatrix} = [G_{fb}, R] \begin{bmatrix} y \\ r \end{bmatrix}. \tag{10.104}$$

See Figure 10.12. For internal stability, $R$ must be stable. In view of (10.79) given in Theorem 10.16, this implies the requirement $[(I + L'N')D'^{-1}]^{-1}X' \in RH_\infty$, in addition to $L', X', (I + L'N')D^{-1} \in RH_\infty$, which imposes significant additional restrictions on $L'$. Here

$$[G_{fb}, R] = [(I + L'N')D'^{-1}]^{-1}[L', X']. \tag{10.105}$$

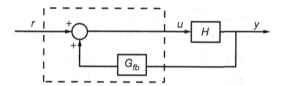

**Figure 10.13.** The $\{I; I, G_{fb}\}$ controller

*(v) $\{I; I, G_{fb}\}$ Controller*

This is a special case of (iv), a single degree of freedom case where $R = I$; see Figure 10.13. Here, $R = I$ implies that

$$X' = (I + L'N')D'^{-1}, \tag{10.106}$$

or that, $X'$ and $L'$ must satisfy additionally the relation

$$D'X' - L'N' = I, \tag{10.107}$$

a (skew) Diophantine Equation. This is in addition to the condition that $L', X', (I + L'N')D^{-1} \in RH_\infty$.

## Unity (Error) Feedback Configuration

Consider the unity feedback (error feedback) control system depicted in Figure 10.14, where $H$ and $C$ are the transfer function matrices of the plant and controller, respectively [see also Figure 10.10 and (10.99), (10.100)]. This configuration is studied further below.

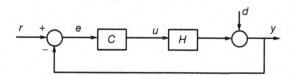

**Figure 10.14.** Unity feedback control system

Assume that $(I + HC)^{-1}$ exists. It is not difficult to verify the relations

$$y = (I + HC)^{-1}HCr + (I + HC)^{-1}d \triangleq Tr + Sd,$$
$$u = (I + CH)^{-1}Cr - (I + CH)^{-1}Cd \triangleq Mr - Md. \tag{10.108}$$

If they are compared with relations (10.81)–(10.86) for the two degrees of freedom controller, then $u = C_y y + C_r r$, $C_y = -C$ and $C_r = C$, since $u = -Cy + Cr$. Hence, for the error feedback system of Figure 10.14, the relations following (10.86) assume the forms

$$M = (I + CH)^{-1}C = DX = -Q = -DL,$$
$$T = H(I + CH)^{-1}C = (I + HC)^{-1}HC = HM = NX,$$
$$S_o = (I + HC)^{-1} = I + HQ = I - HM = I - T,$$
$$S_i = (I + CH)^{-1} = I + QH = I - MH. \tag{10.109}$$

If now Theorem 10.16 is applied to the present error feedback case, then it can be seen that all stabilizing controllers are given by

$$C = [(I - XN)D^{-1}]^{-1}X, \tag{10.110}$$

where $[(I - XN)D^{-1}, X]$ is stable and $(I - XN)^{-1}$ exists. $H = ND^{-1}$ is a right coprime (rc) polynomial matrix factorization.

Similarly, it can be shown by applying Theorem 10.16 that if $H$ is proper and $H = N'D'^{-1}$ is an rc MFD in $RH_\infty$, then all proper stabilizing controllers are given by

$$C = [(I - X'N')D'^{-1}]^{-1}X', \tag{10.111}$$

where $[(I - X'N')D'^{-1}, X'] \in RH_\infty$ and $(I - X'N')^{-1}$ exists and is proper.

*H Square and Nonsingular*

Assume now that $H$ is proper and $H^{-1}$ exists; i.e., $H$ is square and nonsingular. Let $H = ND^{-1}$ be an rc polynomial MFD. If $T$ is the closed-loop transfer function between $y$ and $r$, it can be shown that the system will be internally stable if and only if

$$[N^{-1}(I - T)H, N^{-1}T] \tag{10.112}$$

is stable. Assume that $T \neq I$ in order for the loop to be well defined. Note that if $T$ is proper, then

$$C = H^{-1}T(I - T)^{-1} \tag{10.113}$$

is proper if and only if $H^{-1}T$ is proper and $I - T$ is biproper.

*SISO Case*

If, in addition, it is assumed that $H$ and $T$ are single-input, single-output transfer functions with $H = n/d$, the closed-loop system will be stable if and only if

$$(1 - T)d^{-1} = Sd^{-1} \quad \text{and} \quad Tn^{-1} \tag{10.114}$$

are stable, i.e., if and only if the sensitivity matrix has as zeros all the unstable poles of the plant and the closed-loop transfer function has as zeros all the unstable zeros of the plant.

This is a result that is well known in the classical control literature (refer to the book by J. R. Ragazzini and G. F. Franklin, *Sampled Data Control Systems*, McGraw-Hill, New York, 1958). It is derived here by specializing the more general multi-input, multi-output case results to the single-input, single-output case.

**Example 10.21.** Given $H(s) = \frac{s-1}{(s-2)(s+1)}$, all scalar proper transfer functions $T$ that can be realized via the error feedback configuration, shown in Figure 10.14, under internal stability, are to be characterized. When an error feedback configuration is used, $T$ must satisfy the following conditions: $[(1 - T)d^{-1}, Tn^{-1}] = \left[\frac{d_T - n_T}{(s-2)(s+1)}, \frac{n_T}{d_T(s-1)}\right]$ stable; that is, $T$ must be stable and $d_T - n_T = (s - 2)\hat{d}_T$, $n_T = (s - 1)\hat{n}_T$ ($T$ must have as zero the unstable zero of $H$). The controller $C = \frac{n_T(s-2)(s+1)}{(d_T - n_T)(s-1)} = \frac{\hat{n}_T(s+1)}{\hat{d}_T}$. For $C$ to be proper $H^{-1}T = \frac{(s-2)(s+1)n_T}{(s-1)d_T}$ must be proper and $1 - T = \frac{d_T - n_T}{d_T}$ must be biproper; these are satisfied when $\deg d_T \geq \deg n_T + 1$. The closed-loop eigenvalues are the zeros of $d_c d + n_c n = \hat{d}_T(s - 2)(s + 1) + \hat{n}_T(s + 1)(s - 1) = (s + 1)[(d_T - n_T) + n_T] = (s + 1)d_T$.

If $T$ is to be realized via a two degrees of freedom controller instead, in view of Theorem 10.17, the stability requirement is that $T = NX = (s - 1)X$ with $X$ stable.

It may be of interest to use Theorem 10.18 and proper and stable factorizations. In this case, let $H = \frac{s-1}{(s-2)(s+1)} = \left(\frac{s-1}{(s+1)^2}\right)\left(\frac{s-2}{s+1}\right)^{-1}$. Then $T$ is realizable with internal stability using a two degrees of freedom configuration if and only if $N'^{-1}T = \frac{(s+1)^2 n_T}{(s-1)d_T} = X'$ is proper and stable. That is $n_T = (s - 1)\hat{n}_T$ and $\deg d_T \geq \deg n_T + 1$. In the error feedback case, $[(1 - T)d'^{-1}, Tn'^{-1}] = \left[\frac{(d_T - n_T)(s+1)}{d_T(s-2)}, \frac{n_t(s+1)^2}{d_T(s-1)}\right]$ must be proper and stable, which imply, for stability, that $T$ should be stable, $d_T - n_T = (s - 2)\hat{d}_T, n_T = (s - 1)\hat{n}_T$; and for properness, $\deg d_T \geq \deg n_T + 1$ as before.

## 10.4.4 Some Control Problems

In control problems, design specifications typically include requirements for internal stability or pole placement, low sensitivity to parameter variations, disturbance attenuation, and noise reduction. Also, requirements such as model matching, diagonal decoupling, static decoupling, regulation, and tracking are included in the specifications.

Internal stability has, of course, been a central theme throughout this book, and in this section, all stabilizing controllers were parameterized. Pole placement was also studied in Chapter 9, using state feedback. Sensitivity and disturbance noise reduction are treated by appropriately selecting the feedback controller $C_y$. Methodologies to accomplish these control goals, frequently in an optimal way, are developed in many control books. It should be noted that many important design approaches such as the $H_\infty$ optimal control design method are based on the parameterizations of all feedback stabilizing controllers discussed above. In particular, an appropriate or optimal controller is selected by restricting the parameters used, so that additional control goals are accomplished optimally, while guaranteeing internal stability in the loop.

Our development of the theory of two degrees of freedom controllers can be used directly to study model matching and decoupling, and a brief outline of this approach is given in the following. Note that this does not, by far, constitute a complete treatment of these important control problems, but rather, an illustration of the methodologies introduced in this section.

### Model Matching Problem

In the *model matching* problem, the transfer function of the plant $H(s)$ ($y = Hu$) and a desired transfer function $T(s)$ ($y = Tr$) are given and a transfer function $M(s)$ ($u = Mr$) is sought so that

$$T(s) = H(s)M(s). \tag{10.115}$$

Typically, $H(s)$ is proper, and the proper and stable $T(s)$ is to be obtained from $H(s)$ using a controller under the condition of internal stability. Therefore, $M(s)$ can in general not be implemented as an open-loop controller, but rather, as a two degrees of freedom controller. In view of Theorem 10.18, if $H = N'D'^{-1}$ is an rc MFD in $RH_\infty$, then the pair $(T, M)$ can be realized with internal stability if and only if there exists $X' \in RH_\infty$ so that $\begin{bmatrix} T \\ M \end{bmatrix} = \begin{bmatrix} N' \\ D' \end{bmatrix} X'$. Note that an $M$ that satisfies (10.115) must first be selected (there may be an infinite number of solutions $M$). In the case when $\det H(s) \neq 0$, $T$ can be realized with internal stability by means of a two degrees of freedom control configuration if and only if $N'^{-1}T = X' \in RH_\infty$ (see Example 10.19). In this case $M = D'X'$. Now if the model matching is to be achieved by a more restricted control configuration, then additional conditions are imposed on $T$ for this to happen, which are expressed in terms of $X'$ (see, for instance, Example 10.19 for the case of the unity feedback configuration).

### Decoupling Problem

In the problem of *diagonal decoupling*, $T(s)$ in (10.115) is not completely specified but is required to be diagonal, proper, and stable. In this problem the first input affects only the first output, the second input affects only the second output, and so forth. If $H(s)^{-1}$ exists, then diagonal decoupling under internal stability via a two degrees of freedom control configuration is possible if and only if

$$N'^{-1}T = N'^{-1}\begin{bmatrix} \frac{n_1}{d_1} & & \\ & \ddots & \\ & & \frac{n_m}{d_m} \end{bmatrix} = X' \in RH_\infty, \tag{10.116}$$

where $H = N'D'^{-1}$ is an rc MFD in $RH_\infty$ and $T(s) = \text{diag}[n_i(s)/d_i(s)]$, $i = 1, \ldots, m$. It is clear that if $H(s)$ has only stable zeros, then no additional restrictions are imposed on $T(s)$. Relation (10.116) implies restrictions on the zeros of $n_i(s)$ when $H(s)$ has unstable zeros.

It is straightforward to show that if diagonal decoupling is to be accomplished by means of more restricted control configurations, then additional restrictions will be imposed on $T(s)$ via $X'$. (See Exercise 10.5 below for the case of diagonal decoupling via linear state feedback.) A problem closely related to the diagonal decoupling problem is the problem of the *inverse* of $H(s)$. In this case, $T(s) = I$.

In the problem of *static decoupling*, $T(s) \in RH_\infty m$ is square and also satisfies $T(0) = \Lambda$, a real nonsingular diagonal matrix. An example of such $T(s)$ is $T(s) = \frac{1}{d(s)} \begin{bmatrix} s^2 + 1 & s(s^2 + 2) \\ s(s+2) & s^2 + 3s + 1 \end{bmatrix}$, where $d(s)$ is a Hurwitz polynomial. Note that if $T(0) = \Lambda$, then a step change in the first input $r$ will affect only the first output in $y$ at steady-state and so forth. Here $y = Tr = T\frac{1}{s}$ and $\lim_{s \to 0} sT\frac{1}{s} = T(0) = \Lambda$, which is diagonal and nonsingular. For this to happen, with internal stability when $H(s)$ is nonsingular (see Theorem 10.18), we must have $N'^{-1}T = X' \in RH_\infty$, from which can be seen that static decoupling is possible if and only if $H(s)$ does not have zeros at $s = 0$. If this is the case and if in addition $H(s)$ is stable, static decoupling can be achieved with just a precompensation by a real gain matrix $G$ where $G = H^{-1}(0)\Lambda$. In this case $T(s) = H(s)G = H(s)H^{-1}(0)\Lambda$ from which $T(0) = \Lambda$.

## 10.5 Summary and Highlights

*Interconnected Systems—Feedback*

- Let $y = H_1 u$ and $u = H_2 y + r$, where the plant $H_1 = N_1 D_1^{-1}$, and the controller $H_2 = \widetilde{D}_2^{-1}\widetilde{N}_2$ are both coprime MFD. Then the closed-loop system is stable if and only if $\det(\widetilde{D}_2 D_1 - \widetilde{N}_2 N_1)$ is a Hurwitz polynomial or the poles of $\begin{bmatrix} I & -H_2 \\ -H_1 & I \end{bmatrix}^{-1}$ are stable (see Theorem 10.4).

- The Diophantine Equation

$$\widetilde{D}_2 D_1 - \widetilde{N}_2 N_1 = \widetilde{D}_k$$

is important for feedback systems. The roots of $\det \widetilde{D}_k$ are the closed-loop eigenvalues. See (10.30).

- For interconnected systems in parallel and series, see (10.3)–(10.5) and (10.6)–(10.8).

*Parameterization of All Stabilizing Feedback Controllers*

- Given $H_1 = N_1 D_1^{-1} = \tilde{D}_1^{-1} \tilde{N}_1$, a doubly coprime factorization, all feedback stabilizing controllers $H_2$ are given by

$$\begin{aligned} H_2 &= -(\tilde{D}_k X_1 - \tilde{N}_k \tilde{N}_1)^{-1}(\tilde{D}_k Y_1 + \tilde{N}_k \tilde{D}_1) \\ &= -(\tilde{Y}_1 D_k + D_1 N_k)(\tilde{X}_1 D_k - N_1 N_k)^{-1}, \end{aligned} \qquad (10.44)$$

where

$$UU^{-1} = \begin{bmatrix} X_1 & Y_1 \\ -\tilde{N}_1 & \tilde{D}_1 \end{bmatrix} \begin{bmatrix} D_1 & -\tilde{Y}_1 \\ N_1 & \tilde{X}_1 \end{bmatrix} = \begin{bmatrix} I & 0 \\ 0 & I \end{bmatrix} \qquad (10.41)$$

with $U$ a unimodular matrix. The polynomial matrices $\tilde{N}_k$ and $N_k$ in

$$\tilde{D}_k^{-1} \tilde{N}_k = N_k D_k^{-1} = K \qquad (10.51)$$

are arbitrary and $\tilde{D}_k^{-1}$, $D_k^{-1}$ stable; the closed-loop eigenvalues are the roots of $\det \tilde{D}_k$ or of $\det D_k$ (see Theorem 10.6).
- Equivalently, all stabilizing controllers are given by

$$\begin{aligned} H_2 &= -(X_1 - K\tilde{N}_1)^{-1}(Y_1 + K\tilde{D}_1) \\ &= -(\tilde{Y}_1 + D_1 K)(\tilde{X}_1 - N_1 K)^{-1}, \end{aligned} \qquad (10.50)$$

where the poles of $K$ are the closed-loop eigenvalues (see Theorem 10.7).
- Given $H_1 = N_1' D_1'^{-1} = \widetilde{D'}_1^{-1} \widetilde{N'}_1$, a doubly coprime MFDs in $RH_\infty$, then all proper stabilizing controllers $H_2$ are given by

$$H_2 = -(X_1' - K'\widetilde{N'}_1)^{-1}(Y_1' + K'\widetilde{D'}_1) = -(\widetilde{Y'}_1 + D_1' K')(\widetilde{X'}_1 - N_1' K')^{-1}, \qquad (10.58)$$

where $K' \in RH_\infty$, any rational proper and stable matrix (see Theorem 10.11).
- See (10.67) for all stabilizing controllers in terms of state-space descriptions.

*Two Degrees of Freedom Controllers*

- Given $H = ND^{-1}$ right coprime,

$$\hat{u}(s) = [C_y(s), C_r(s)] \begin{bmatrix} \hat{y}(s) \\ \hat{r}(s) \end{bmatrix} \qquad (10.69)$$

stabilizes $H$ if and only if
(i) $\hat{u} = C_y \hat{y}$ stabilizes $\hat{y} = H\hat{u}$, and
(ii) $C_r$ is such that

$$M \triangleq (I - C_y H)^{-1} C_r \qquad (10.70)$$

$(u = Mr)$ satisfies $D^{-1} M = X$, a stable rational matrix (see Theorem 10.15).

- See Theorem 10.16 for parameterizations of all stabilizing two degrees of freedom controllers.
- See (10.81) and (10.82) for relations between $u$, $y$ and external inputs and disturbances.
- Given $y = ND^{-1}u$, $y = Tr$ and $u = Mr$ are realizable via any control configuration with internal stability if and only if

$$\begin{bmatrix} T \\ M \end{bmatrix} = \begin{bmatrix} N \\ D \end{bmatrix} X,$$

  where $X$ is stable. See Theorems 10.17 and 10.18.
- The cases when a more restricted controller is used are addressed. See (10.96)–(10.107). The error or unity feedback controller is further discussed in (10.108)–(10.114).
- The model matching problem, the diagonal decoupling problem, and the static decoupling problem are discussed. See Subsection 10.4.4.

## 10.6 Notes

Two books that are original sources on the use of polynomial matrix descriptions in Systems and Control are Rosenbrock [18] and Wolovich [22]. In the former, what is now called Rosenbrock's matrix is employed and relations to state-space descriptions are emphasized. In the latter, what are now called Polynomial Matrix Fractional Descriptions are emphasized and the relation to state space is accomplished primarily by using controller forms and the Structure Theorem, which was presented in Chap 6. Good general sources for the polynomial matrix description approach include also the books by Vardulakis [19], Kailath [16], and Chen [9]. A good source for the study of feedback systems using PMDs and MFDs is the book by Callier and Desoer [8].

The development of the properties of interconnected systems, addressed in Section 10.2, which include controllability, observability, and stability of systems in parallel, in series, and in feedback configurations is primarily based on the approach taken in Antsaklis and Sain [7], Antsaklis [3] and [4], and Gonzalez and Antsaklis [14].

Parameterizations of all stabilizing controllers are of course very important in control theory today. Historically, their development appears to have evolved in the following manner (see also the historical remarks on the Diophantine Equation in [1, Subsection 7.2E]): Youla et al. [23] introduced the $K$ parameterization (as in Theorem 10.7 above) in 1976 and used it in the Wiener–Hopf design of optimal controllers. This work is considered to be the seminal contribution in this area. The proofs of the results on the parameterizations in Youla et al. [23] involve transfer functions and their characteristic polynomials. Neither the Diophantine Equation nor PMDs of the system are used (explicitly). It should be recalled that in the middle 1970s most of the control results in the literature concerning MIMO systems involved state-space descriptions and a few employed transfer function matrices. The PMD descriptions of systems were only beginning to make some impact. A version of the linear Diophantine Equation, namely, $AX + YB = C$ polynomial in $z^{-1}$ was used in control design by Kucera in work reported in 1974 and 1975. In that work, parameterizations of all stabilizing controllers were implicit, not explicit, in the sense that the stabilizing controllers were expressed in terms of the general solution of the Diophantine Equation, which in turn can be described parametrically. Explicit parameterizations were reported later in Kucera [17] in 1979. Antsaklis [2] in 1979 introduced the doubly coprime MFDs (used in this book and in the literature) for the first time with the polynomial Diophantine Equation, working over the ring of polynomials, to derive parameterizations of all stabilizing controllers and to prove the results by Youla et al. in an alternative way. In this work, internal system descriptions were connected directly to stabilizing controller parameterizations via the polynomial Diophantine Equation. In Desoer et al. [10] in 1980 parameterizations $K'$ of all stabilizing controllers using coprime MFDs in rings other than polynomial rings (including the ring of proper and stable rational functions) were derived. It should also be noted that proper and stable MFDs had apparently been used earlier by Vidyasagar. In Zames [24] in 1981, a parameterization $Q$ of all stabilizing controllers, but only for stable plants was introduced and used in $H_\infty$ optimal control design. (Similar parameterizations were also used elsewhere, but apparently not to characterize all stabilizing controllers; for example, they were used in the design of the closed-loop transfer function in control systems and in sensitivity studies in the 1950s and 1960s, and also in the "internal model control" studies in chemical process control in the 1980s.) A parameterization $X$ of all stabilizing controllers (where $X$ is closely related to the attainable response in an error feedback control system), valid for unstable plants as well, was introduced in Antsakis and Sain [6]. Parameterizations involving proper and stable MFDs were further developed in the 1980s in connection with optimal control design methodologies, such as $H_\infty$ optimal control, and connections to state-space approaches were derived. Two degrees of freedom controllers were also studied, and the limitations of the different control configurations became better understood. By now, MDFs and PMDs have become important system representations and their study is essential, if optimal control design methodologies are to be well understood. See [1, Subsections 7.2E and Section 7.6] for further discussion of controller parameterizations.

The material on two degrees of freedom controllers in Section 10.4 is based on Antsaklis [4] and Gonzalez and Antsaklis [12], [13], [14], [15]; a good source for this topic is also Vidyasagar [20]. Note that the main stability theorem (Theorem 10.15) first appeared in Antsaklis [4] and Antsaklis and Gonzalez [5]. For additional material on model matching and decoupling, consult Chen [9], Kailath [16], Falb and Wolovich [11], Williams and Antsaklis [21], and the extensive list of references therein.

# References

1. P.J. Antsaklis and A.N. Michel, *Linear Systems*, Birkhäuser, Boston, MA, 2006.
2. P.J. Antsaklis, "Some relations satisfied by prime polynomial matrices and their role in linear multivariable system theory," *IEEE Trans. Auto. Control*, Vol. AC-24, No. 4, pp. 611–616, August 1979.
3. P.J. Antsaklis, *Notes on: Polynomial Matrix Representation of Linear Control Systems*, Pub. No. 80/17, Dept. of Elec. Engr., Imperial College, 1980.
4. P.J. Antsaklis, Lecture Notes of the graduate course, *Feedback Systems*, University of Notre Dame, Spring 1985.
5. P.J. Antsaklis and O.R. Gonzalez, "Stability parameterizations and stable hidden modes in two degrees of freedom control design," *Proc. of the 25th Annual Allerton Conference on Communication, Control and Computing*, pp. 546–555, Monticelo, IL, Sept. 30–Oct. 2, 1987.
6. P.J. Antsaklis and M.K. Sain, "Unity feedback compensation of unstable plants," *Proc. of the 20th IEEE Conf. on Decision and Control*, pp. 305–308, Dec. 1981.
7. P.J. Antsaklis and M.K. Sain, "Feedback controller parameterizations: finite hidden modes and causality," in *Multivariable Control: New Concepts and Tools*, S. G. Tzafestas, Ed., D. Reidel Pub., Dordrecht, Holland, 1984.
8. F.M. Callier and C.A. Desoer, *Multivariable Feedback Systems*, Springer-Verlag, New York, NY, 1982.
9. C.T. Chen, *Linear System Theory and Design*, Holt, Reinehart and Winston, New York, NY, 1984.
10. C.A. Desoer, R.W. Liu, J. Murray, and R. Saeks, "Feedback system design: the fractional approach to analysis and synthesis," *IEEE Trans. Auto. Control*, Vol. AC-25, pp. 399–412, June 1980.
11. P.L. Falb and W.A. Wolovich, "Decoupling in the design of multivariable control systems," *IEEE Trans. Auto. Control*, Vol. AC-12, pp. 651–659, 1967.
12. O.R. Gonzalez and P.J. Antsaklis, "Implementations of two degrees of freedom controllers," *Proc. of the 1989 American Control Conference*, pp. 269–273, Pittsburgh, PA, June 21–23, 1989.
13. O.R. Gonzalez and P.J. Antsaklis, "Sensitivity considerations in the control of generalized plants," *IEEE Trans. Auto. Control*, Vol. 34, No. 8, pp. 885–888, August 1989.
14. O.R. Gonzalez and P.J. Antsaklis, "Hidden modes of two degrees of freedom systems in control design," *IEEE Trans. Auto. Control*, Vol. 35, No. 4, pp. 502–506, April 1990.
15. O.R. Gonzalez and P.J. Antsaklis, "Internal models in regulation, stabilization and tracking," *Int. J. Control*, Vol. 53, No. 2, pp. 411–430, 1991.
16. T. Kailath, *Linear Systems*, Prentice-Hall, Englewood Cliffs, NJ, 1980.
17. V. Kucera, *Discrete Linear Control*, Wiley, New York, NY, 1979.
18. H. H. Rosenbrock, *State-Space and Multivariable Theory*, Wiley, New York, NY, 1970.
19. A.I.G. Vardulakis, *Linear Multivariable Control. Algebraic Analysis and Synthesis Methods*, Wiley, New York, NY, 1991.
20. M. Vidyasagar, *Control System Synthesis. A Factorization Approach*, MIT Press, Cambridge, MA, 1985.
21. T. Williams and P.J. Antsaklis, "Decoupling," *The Control Handbook*, Chapter 50, pp. 745–804, CRC Press and IEEE Press, New York, NY, 1996.

22. W.A. Wolovich, *Linear Multivariable Systems*, Springer-Verlag, New York, NY, 1974.
23. D.C. Youla, H.A. Jabr, and J.J. Bongiorno, Jr., "Modern Wiener–Hopf design of optimal controllers – Part II: The multivariable case," *IEEE Trans. Auto. Control*, Vol. AC-21, pp. 319–338, June 1976.
24. G. Zames, "Feedback and optimal sensitivity: model reference transformations, multiplicative seminorms, and approximate inverses," *IEEE Trans. Auto. Control*, Vol. 26, pp. 301–320, 1981.

## Exercises

**10.1.** Consider the double integrator $H_1 = \frac{1}{s^2}$.

(a) Characterize all stabilizing controllers $H_2$ for $H_1$.
(b) Characterize all proper stabilizing controllers $H_2$ for $H_1$ of order 1.

**10.2.** Consider the double integrator $H_1 = \frac{1}{s^2}$.

(a) Derive a minimal state-space realization for $H_1$, and use Lemma 10.14 to derive doubly coprime factorizations in $RH_\infty$.
(b) Use the polynomial Diophantine Equations to derive factorizations in $RH_\infty$.

**10.3.** Consider $H_1 = [\frac{s^2+1}{s^2}, \frac{s+1}{s^3}]$.

(a) Derive a minimal state-space realization $\{A, B, C, D\}$, and use Lemma 10.14 and Theorem 10.11 to parameterize all stabilizing controllers $H_2$.
(b) Derive a stabilizing controller $H_2$ of order three by appropriately selecting $K'$. What are the closed-loop eigenvalues in this case? Comment on your results.

**10.4.** Consider $H = \begin{bmatrix} \frac{1}{s+1} & \frac{2}{s+3} \\ \frac{1}{s+1} & \frac{1}{s+1} \end{bmatrix}$.

(a) Derive an rc MFD in $RH_\infty, H = N'D'^{-1}$.

(b) Let $T = \begin{bmatrix} \frac{n_1}{d_1} & 0 \\ 0 & \frac{n_2}{d_2} \end{bmatrix}$, and characterize all diagonal $T$ that can be realized under internal stability via a two degrees of freedom control configuration.

**10.5.** In the model matching problem, the transfer function matrices $H \in R^{p \times m}(s)$ of the plant and $T \in R^{p \times m}(s)$ of the model must be found so that $T = HM$. $M$ is to be realized via a feedback control configuration under internal stability. Here we are interested in the *model matching problem via linear state feedback*. For this purpose, let $H = ND^{-1}$ an rc polynomial factorization with $D$ column reduced. Then $Dz = u, y = Nz$ is a minimal realization of $H$. Let the state feedback control law be defined by $u = Fz + Gr$, where $F \in R[s]^{m \times m}, G \in R^{m \times m}$ with $\det G \neq 0$ and $\deg_{c_j} F < \deg_{c_j} D$. To allow additional flexibility, let $r = Kv$ and $K \in R^{m \times q}$. Note that $H_{F,GK} = ND_F^{-1}GK = (ND^{-1})(DD_F^{-1}GK) = (ND^{-1})[D(G^{-1}D_F)^{-1}K] = HM$ where $D_F = D - F$.

In view of the above, solve the model matching problem via linear state feedback, determine $F, G$, and $K$, and comment your results when

(a) $H = \frac{(s+1)(s+2)}{2s^2-3s+2}$,    $T = \frac{s+1}{s+2}$,

(b) $H = \begin{bmatrix} s+1 & 0 \\ \frac{s}{s} & s+2 \\ s & s \end{bmatrix}$,    $T = I_2$,

(c) $H = \begin{bmatrix} \frac{s+2}{s+1} & \frac{s+3}{s+2} \\ \frac{1}{s+1} & 0 \end{bmatrix}$,    $T = \begin{bmatrix} \frac{s+1}{s+4} \\ \frac{-2}{(s+2)(s+4)} \end{bmatrix}$.

Hint: The model matching problem via linear state feedback is quite easy to solve when $p = m$ and rank $H = m$ in view of $(G^{-1}D_F)^{-1}K = D^{-1}M = D^{-1}H^{-1}T = N^{-1}T$.

# A

# Appendix

This appendix consists of nine parts. In the first eight, Sections A.1–A.8, we present results from linear algebra used throughout this book. In the last part, Section A.9, we address some numerical considerations. In all cases, our aim is to present a concise summary of pertinent results and not a full development of the subject on hand. For a more extensive exposition of the materials presented herein, refer to Antsaklis and Michel [1, Section 2.2] and to the other sources cited at the end of this appendix.

## A.1 Vector Spaces

In defining vector space, we require the notion of a field.

### A.1.1 Fields

**Definition A.1.** *Let $F$ be a set containing more than one element, and let there be two operations "$+$" and "$\cdot$" defined on $F$ (i.e., "$+$" and "$\cdot$" are mappings of $F \times F$ into $F$), called addition and multiplication, respectively. Then for each $\alpha, \beta \in F$ there is a unique element $\alpha + \beta \in F$, called the sum of $\alpha$ and $\beta$, and a unique element $\alpha\beta \triangleq \alpha \cdot \beta \in F$, called the product of $\alpha$ and $\beta$. We say that $\{F; +, \cdot\}$ is a field provided that the following axioms are satisfied:*

(i)  *$\alpha + (\beta + \gamma) = (\alpha + \beta) + \gamma$ and $\alpha \cdot (\beta \cdot \gamma) = (\alpha \cdot \beta) \cdot \gamma$ for all $\alpha, \beta, \gamma \in F$ (i.e., "$+$" and "$\cdot$" are associative operations);*

(ii)  *$\alpha + \beta = \beta + \alpha$ and $\alpha \cdot \beta = \beta \cdot \alpha$ for all $\alpha, \beta \in F$ (i.e., "$+$" and "$\cdot$" are commutative operations);*

(iii)  *$\alpha \cdot (\beta + \gamma) = \alpha \cdot \beta + \alpha \cdot \gamma$ for all all $\alpha, \beta, \gamma \in F$ (i.e., "$\cdot$" is distributive over "$+$");*

(iv)  *There exists an element $0_F \in F$ such that $0_F + \alpha = \alpha$ for all $\alpha \in F$ (i.e., $0_F$ is the identity element of $F$ with respect to "$+$");*

*(v)* There exists an element $1_F \in F, 1_F \neq 0_F$, such that $1_F \cdot \alpha = \alpha$ for all $\alpha \in F$ (i.e., $1_F$ is the identity element of $F$ with respect to "."), 

*(vi)* For every $\alpha \in F$, there exists an element $-\alpha \in F$ such that $\alpha + (-\alpha) = 0_F$ (i.e., $-\alpha$ is the additive inverse of $F$);

*(vii)* For any $\alpha \neq 0_F$, there exists an $\alpha^{-1} \in F$ such that $\alpha \cdot (\alpha^{-1}) = 1_F$ (i.e., $\alpha^{-1}$ is the multiplicative inverse of $F$). ∎

In the sequel, we will usually speak of a field $F$ rather than of "a field $\{F; +, \cdot\}$."

Perhaps the most widely known fields are the *field of real numbers* $R$ and the *field of complex numbers* $C$. Another field that we will encounter (see Example A.11) is the *field of rational functions* (i.e., rational fractions over polynomials).

As a third example, we let $F = \{0, 1\}$ and we define on $F$ (binary) addition as $0 + 0 = 0 = 1 + 1$, $1 + 0 = 1 = 0 + 1$ and (binary) multiplication as $1 \cdot 0 = 0 \cdot 1 = 0 \cdot 0 = 0$, $1 \cdot 1 = 1$. It is easily verified that $\{F; +, \cdot\}$ is a field.

As a fourth example, let $P$ denote the set of polynomials with real coefficients and define addition "+" and multiplication "." on $P$ in the usual manner. Then $\{F; +, \cdot\}$ is *not* a field since, e.g., axiom (vii) in Definition A.1 is violated (i.e., the *multiplicative* inverse of a polynomial $p \in P$ is not necessarily a polynomial).

## A.1.2 Vector Spaces

**Definition A.2.** *Let $V$ be a nonempty set, $F$ a field, "+" a mapping of $V \times V$ into $V$, and "." a mapping of $F \times V$ into $V$. Let the members $x \in V$ be called* vectors, *let the elements $\alpha \in F$ be called* scalars, *let the operation "+" defined on $V$ be called* vector addition, *and let the mapping "." be called* scalar multiplication *or* multiplication of vectors by scalars. *Then for each $x, y \in V$, there is a unique element, $x + y \in V$, called the* sum *of $x$ and $y$, and for each $x \in V$ and $\alpha \in F$, there is a unique element, $\alpha x \triangleq \alpha \cdot x \in V$, called the* multiple *of $x$ by $\alpha$. We say that the nonempty set $V$ and the field $F$, along with the two mappings of vector addition and scalar multiplication, constitute a* vector space *or a* linear space *if the following axioms are satisfied:*

*(i)*    $x + y = y + x$ for every $x, y \in V$;

*(ii)*   $x + (y + z) = (x + y) + z$ for every $x, y, z \in V$;

*(iii)*  There is a unique vector in $V$, called the zero vector or the null vector or the origin, that is denoted by $0_V$ and has the property that $0_V + x = x$ for all $x \in V$;

*(iv)*   $\alpha(x + y) = \alpha x + \alpha y$ for all $\alpha \in F$ and for all $x, y \in V$;

*(v)*    $(\alpha + \beta)x = \alpha x + \beta x$ for all $\alpha, \beta \in F$ and for all $x \in V$;

*(vi)*   $(\alpha\beta)x = \alpha(\beta x)$ for all $\alpha, \beta \in F$ and for all $x \in V$;

*(vii)*  $0_F x = 0_V$ for all $x \in V$;

*(viii)* $1_F x = x$ for all $x \in V$. ∎

When the meaning is clear from context, we will write 0 in place of $0_F$, 1 in place of $1_F$, and 0 in place of $0_V$. To indicate the relationship between the set of vectors $V$ and the underlying field $F$, we sometimes refer to a *vector space* $V$ *over the field* $F$, and we signify this by writing $(V, F)$. However, usually, when the field in question is clear from context, we simply speak of a vector space $V$. If $F$ is the field of real numbers $R$, we call the space a *real vector space*. Similarly, if $F$ is the field of complex numbers $C$, we speak of a *complex vector space*.

## Examples of Vector Spaces

---

**Example A.3.** Let $V = F^n$ denote the set of all ordered $n$-tuples of elements from a field $F$. Thus, if $x \in F^n$, then $x = (x_1, \ldots, x_n)^T$, where $x_i \in F$, $i = 1, \ldots, n$. With $x, y \in F^n$ and $\alpha \in F$, let vector addition and scalar multiplication be defined as

$$x + y = (x_1, \ldots, x_n)^T + (y_1, \ldots, y_n)^T$$
$$\triangleq (x_1 + y_1, \ldots, x_n + y_n)^T \tag{A.1}$$

and

$$\alpha x = \alpha(x_1, \ldots, x_n)^T \triangleq (\alpha x_1, \ldots, \alpha x_n)^T. \tag{A.2}$$

In this case the null vector is defined as $0 = (0, \ldots, 0)^T$ and the vector $-x$ is defined as $-x = -(x_1, \ldots, x_n)^T = (-x_1, \ldots, -x_n)^T$. Utilizing the properties of the field $F$, all axioms of Definition A.2 are readily verified, and therefore, $F^n$ is a vector space. We call this space the *space $F^n$ of $n$-tuples of elements of $F$*. If in particular we let $F = R$, we have $R^n$, *the $n$-dimensional real coordinate space*. Similarly, if we let $F = C$, we have $C^n$, *the $n$-dimensional complex coordinate space*.

---

We note that the set of points in $R^2$, $(x_1, x_2)$, that satisfy the linear equation

$$x_1 + x_2 + c = 0, \quad c \neq 0,$$

with addition and multiplication defined as in (A.1) and (A.2), is *not* a vector space.

---

**Example A.4.** Let $V = R^\infty$ denote the set of all infinite sequences of real numbers,

$$x = \{x_1, x_2, \ldots, x_k, \ldots\} \triangleq \{x_i\},$$

let vector addition be defined similarly as in (A.1), and let scalar multiplication be defined similarly as in (A.2). It is again an easy matter to show that this space is a vector space.

On some occasions we will find it convenient to modify $V = R^\infty$ to consist of the set of all real infinite sequences $\{x_i\}$, $i \in Z$.

---

**Example A.5.** Let $1 \le p \le \infty$, and define $V = l_p$ by

$$l_p = \{x \in R^\infty : \sum_{i=1}^{\infty} |x_i|^p < \infty\}, \quad 1 \le p < \infty,$$

$$l_\infty = \{x \in R^\infty : \sup_i\{|x_i|\} < \infty\}. \tag{A.3}$$

Define vector addition and scalar multiplication on $l_p$ as in (A.1) and (A.2), respectively. It can be verified that this space, called the $l_p$-*space*, is a vector space.

---

In proving that $l_p$, $1 \le p \le \infty$, is indeed a vector space, in establishing some properties of norms defined on the $l_p$-spaces, in defining linear transformations on $l_p$-spaces, and in many other applications, we make use of the *Hölder and Minkowski Inequalities for infinite sums*, given below. (These inequalities are of course also valid for *finite* sums.) For proofs of these results, refer, e.g., to Michel and Herget [9, pp. 268–270].

*Hölder's Inequality* states that if $p, q \in R$ are such that $1 < p < \infty$ and $1/p + 1/q = 1$, if $\{x_i\}$ and $\{y_i\}$ are sequences in either $R$ or $C$, and if $\sum_{i=1}^{\infty} |x_i|^p < \infty$ and $\sum_{i=1}^{\infty} |y_i|^q < \infty$, then

$$\sum_{i=1}^{\infty} |x_i y_i| \le (\sum_{i=1}^{\infty} |x_i|^p)^{1/p}(\sum_{i=1}^{\infty} |y_i|^q)^{1/q}. \tag{A.4}$$

*Minkowski's Inequality* states that if $p \in R$, where $1 \le p < \infty$, if $\{x_i\}$ and $\{y_i\}$ are sequences in either $R$ or $C$, and if $\sum_{i=1}^{\infty} |x_i|^p < \infty$ and $\sum_{i=1}^{\infty} |y_i|^p < \infty$, then

$$(\sum_{i=1}^{\infty} |x_i \pm y_i|^p)^{1/p} \le (\sum_{i=1}^{\infty} |x_i|^p)^{1/p} + (\sum_{i=1}^{\infty} |y_i|^p)^{1/p}. \tag{A.5}$$

If in particular $p = q = 2$, then (A.4) reduces to *the Schwarz Inequality for sums*.

---

**Example A.6.** Let $V = C([a, b], R)$. We note that $x = y$ if and only if $x(t) = y(t)$ for all $t \in [a, b]$, and that the null vector is the function that is zero for all $t \in [a, b]$. Let $F$ denote the field of real numbers, let $\alpha \in F$, and let vector addition and scalar multiplication be defined pointwise by

$$(x + y)(t) = x(t) + y(t) \quad \text{for all } t \in [a, b] \tag{A.6}$$

and

$$(\alpha x)(t) = \alpha x(t) \quad \text{for all } t \in [a, b]. \tag{A.7}$$

Then clearly $x + y \in V$ whenever $x, y \in V, \alpha x \in V$, whenever $\alpha \in F$ and $x \in V$, and all the axioms of a vector space are satisfied. We call this space the *space of real-valued continuous functions on* $[a, b]$, and we frequently denote it simply by $C[a, b]$.

---

***Example A.7.*** Let $1 \leq p < \infty$, and let $V$ denote the set of all real-valued functions $x$ on the interval $[a, b]$ such that

$$\int_a^b |x(t)|^p dt < \infty. \tag{A.8}$$

Let $F = R$, and let vector addition and scalar multiplication be defined as in (A.6) and (A.7), respectively. It can be verified that this space is a vector space.

In this book we will usually assume that in (A.8), integration is in the Riemann sense. When integration in (A.8) is in the Lebesgue sense, then the vector space under discussion is called an $L_p$-space (or the space $L_p[a, b]$).

---

In proving that the $L_p$-spaces are indeed vector spaces, in establishing properties of norms defined on $L_p$-spaces, in defining linear transformations on $L_p$-spaces, and in many other applications, we make use of the *Hölder and Minkowski Inequalities for integrals,* given below. (These inequalities are valid when integration is in the Riemann and the Lebesgue senses.) For proofs of these results, refer, e.g., to Michel and Herget [9, pp. 268–270].

Hölder's Inequality states that if $p, q \in R$ are such that $1 < p < \infty$ and $1/p + 1/q = 1$, if $[a, b]$ is an interval on the real line, if $f, g : [a, b] \to R$, and if $\int_a^b |f(t)|^p dt < \infty$ and $\int_a^b |g(t)|^q dt < \infty$, then

$$\int_a^b |f(t)g(t)| dt \leq (\int_a^b |f(t)|^p dt)^{1/p} (\int_a^b |g(t)|^q dt)^{1/q}. \tag{A.9}$$

*Minkowski's Inequality* states that if $p \in R$, where $1 \leq p < \infty$, if $f, g : [a, b] \to R$, and if $\int_a^b |f(t)|^p dt < \infty$ and $\int_a^b |g(t)|^p dt < \infty$, then

$$(\int_a^b |f(t) \pm g(t)|^p dt)^{1/p} \leq (\int_a^b |f(t)|^p dt)^{1/p} + (\int_a^b |g(t)|^p dt)^{1/p}. \tag{A.10}$$

If in particular $p = q = 2$, then (A.9) reduces to *the Schwarz Inequality for integrals.*

---

***Example A.8.*** Let $V$ denote the set of all continuous real-valued functions on the interval $[a, b]$ such that

$$\max_{a \leq t \leq b} |x(t)| < \infty. \tag{A.11}$$

Let $F = R$, and let vector addition and scalar multiplication be defined as in (A.6) and (A.7), respectively. It can readily be verified that this space is a vector space.

In some applications it is necessary to expand the above space to the set of measurable real-valued functions on $[a, b]$ and to replace (A.11) by

$$ess \sup_{a \leq t \leq b} |x(t)| < \infty, \tag{A.12}$$

where $ess$ sup denotes the essential supremum; i.e.,

$$ess \sup_{a \leq t \leq b} |x(t)| = \inf\{M : \mu\{t : |x(t)| > M\} = 0,\},$$

where $\mu$ denotes the Lebesgue measure. In this case, the vector space under discussion is called the $L_\infty$-space.

---

## A.2 Linear Independence and Bases

We now address the important concepts of linear independence of a set of vectors in general and bases in particular. We first require the notion of linear subspace.

### A.2.1 Linear Subspaces

A nonempty subset $W$ of a vector space $V$ is called a *linear subspace* (or a *linear manifold*) in $V$ if (i) $w_1 + w_2$ is in $W$ whenever $w_1$ and $w_2$ are in $W$, and (ii) $\alpha w$ is in $W$ whenever $\alpha \in F$ and $w \in W$. It is an easy matter to verify that a linear subspace $W$ satisfies all the axioms of a vector space and may as such be regarded as a linear space itself.

Two trivial examples of linear subspaces include the null vector (i.e., the set $W = \{0\}$ is a linear subspace of $V$) and the vector space $V$ itself. Another example of a linear subspace is the set of all real-valued polynomials defined on the interval $[a, b]$ that is a linear subspace of the vector space consisting of all real-valued continuous functions defined on the interval $[a, b]$ (refer to Example A.6).

As another example of a linear subspace (of $R^2$), we cite the set of all points on a straight line passing through the origin. On the other hand, a straight line that does not pass through the origin is *not* a linear subspace of $R^2$.

It is an easy matter to show that if $W_1$ and $W_2$ are linear subspaces of a vector space $V$, then $W_1 \cap W_2$, the intersection of $W_1$ and $W_2$, is also a linear subspace of $V$. A similar statement cannot be made, however, for the union of $W_1$ and $W_2$ (prove this). Note that to show that a set $V$ is a vector space, it suffices to show that it is a linear subspace of some vector space.

## A.2.2 Linear Independence

Throughout this section, we let $\{\alpha_1, \ldots, \alpha_n\}, \alpha_i \in F$, denote an indexed set of scalars and we let $\{v^1, \ldots, v^n\}, v^i \in V$, denote an indexed set of vectors.

Now let $W$ be a set in a linear space $V$ ($W$ may be a finite set or an infinite set). We say that a vector $v \in V$ is a *finite linear combination of vectors* in $W$ if there is a finite set of elements $\{w^1, \ldots, w^n\}$ in $W$ and a finite set of scalars $\{\alpha_1, \ldots, \alpha_n\}$ in $F$ such that

$$v = \alpha_1 w^1 + \cdots + \alpha_n w^n.$$

Now let $W$ be a nonempty subset of a linear space $V$ and let $S(W)$ be the set of all finite linear combinations of the vectors from $W$; i.e., $w \in S(W)$ if and only if there is some set of scalars $\{\alpha_1, \ldots, \alpha_m\}$ and some finite subset $\{w^1, \ldots, w^m\}$ of $W$ such that $w = \alpha_1 w^1 + \cdots + \alpha_m w^m$, where $m$ may be any positive integer. Then it is easily shown that $S(W)$ is a linear subspace of $V$, called the *linear subspace generated* by the set $W$.

Now if $U$ is a linear subspace of a vector space $V$ and if there exists a set of vectors $W \subset V$ such that the linear space $S(W)$ generated by $W$ is $U$, then we say that $W$ *spans* $U$. It is easily shown that $S(W)$ is the smallest linear subspace of a vector space $V$ containing the subset $W$ of $V$. Specifically, if $U$ is a linear subspace of $V$ and if $U$ contains $W$, then $U$ also contains $S(W)$.

As an example, in the space $(R^2, R)$, the set $S_1 = \{e^1\} = \{(1,0)^T\}$ spans the set consisting of all vectors of the form $(a, 0)^T, a \in R$, whereas the set $S_2 = \{e^1, e^2\}, e^2 = (0,1)^T$ spans all of $R^2$.

We are now in a position to introduce the notion of linear dependence.

**Definition A.9.** *Let $S = \{v^1, \ldots, v^m\}$ be a finite nonempty set in a linear space $V$. If there exist scalars $\alpha_1, \ldots, \alpha_m$, not all zero, such that*

$$\alpha_1 v^1 + \cdots + \alpha_m v^m = 0, \tag{A.13}$$

*then the set $S$ is said to be linearly dependent (over $F$). If a set is not linearly dependent, then it is said to be linearly independent. In this case relation (A.13) implies that $\alpha_1 = \cdots = \alpha_m = 0$. An infinite set of vectors $W$ in $V$ is said to be linearly independent if every finite subset of $W$ is linearly independent.* ∎

---

***Example A.10.*** Consider the linear space $(R^n, R)$ (see Example A.3), and let $e^1 = (1, 0, \ldots, 0)^T, e^2 = (0, 1, 0, \ldots, 0)^T, \ldots, e^n = (0, \ldots, 0, 1)^T$. Clearly, $\sum_{i=1}^{n} \alpha_i e^i = 0$ implies that $\alpha_i = 0, i = 1, \ldots, n$. Therefore, the set $S = \{e^1, \ldots, e^n\}$ is a linearly independent set of vectors in $R^n$ over the field of real numbers $R$.

---

**Example A.11.** Let $V$ be the set of 2-tuples whose entries are complex-valued rational functions over the field of complex-valued rational functions. Let

$$v^1 = \begin{bmatrix} 1/(s+1) \\ 1/(s+2) \end{bmatrix}, \quad v^2 = \begin{bmatrix} (s+2)/[(s+1)(s+3)] \\ 1/(s+3) \end{bmatrix},$$

and let $\alpha_1 = -1$, $\alpha_2 = (s+3)/(s+2)$. Then $\alpha_1 v^1 + \alpha_2 v^2 = 0$, and therefore, the set $S = \{v^1, v^2\}$ is *linearly dependent over the field of rational functions.* On the other hand, since $\alpha_1 v^1 + \alpha_2 v^2 = 0$ when $\alpha_1, \alpha_2 \in R$ is true if and only if $\alpha_1 = \alpha_2 = 0$, it follows that $S$ is linearly independent over the field of real numbers (which is a subset of the field of rational functions). This shows that *linear dependence* of a set of vectors in $V$ depends on the field $F$.

### A.2.3 Linear Independence of Functions of Time

**Example A.12.** Let $V = C((a,b), R^n)$, let $F = R$, and for $x, y \in V$ and $\alpha \in F$, define addition of elements in $V$ and multiplication of elements in $V$ by elements in $F$ by $(x+y)(t) = x(t)+y(t)$ for all $t \in (a,b)$ and $(\alpha x)(t) = \alpha x(t)$ for all $t \in (a,b)$. Then, as in Example A.6, we can easily show that $(V, F)$ is a vector space. An interesting question that arises is whether for this space, linear dependence (and linear independence) of a set of vectors can be phrased in some testable form. The answer is affirmative. Indeed, it can readily be verified that for the present vector space $(V, F)$, *linear dependence of a set of vectors* $S = \{\phi_1, \ldots, \phi_k\}$ *in* $V = C((a,b), R^n)$ *over* $F = R$ *is equivalent to the requirement that there exist scalars* $\alpha_i \in F$, $i = 1, \ldots, k$, *not all zero, such that*

$$\alpha_1 \phi_1(t) + \cdots + \alpha_k \phi_k(t) = 0 \quad \text{for all } t \in (a,b).$$

Otherwise, $S$ is *linearly independent.*

To see how the above example applies to specific cases, let $V = C((-\infty, \infty), R^2)$, and consider the vectors $\phi_1(t) = [1, t]^T, \phi_2(t) = [1, t^2]^T$. To show that the set $S = \{\phi_1, \phi_2\}$ is linearly independent (over $F = R$), assume for purposes of contradiction that $S$ is linearly dependent. Then there must exist scalars $\alpha_1$ and $\alpha_2$, not both zero, such that $\alpha_1[1, t]^T + \alpha_2[1, t^2]^T = [0, 0]^T$ for all $t \in (-\infty, \infty)$. But in particular, for $t = 2$, the above equation is satisfied if and only if $\alpha_1 = \alpha_2 = 0$, which contradicts the assumption. Therefore, $S = \{\phi_1, \phi_2\}$ is linearly independent.

As another specific case of the above example, let $V = C((-\infty, \infty), R^2)$ and consider the set $S = \{\phi_1, \phi_2, \phi_3, \phi_4\}$, where $\phi_1(t) = [1, t]^T, \phi_2(t) = [1, t^2], \phi_3(t) = [0, 1]^T$, and $\phi_4(t) = [e^{-t}, 0]$. The set $S$ is clearly independent over $R$ since $\alpha_1 \phi_1(t) + \alpha_2 \phi_2(t) + \alpha_3 \phi_3(t) + \alpha_4 \phi_4(t) = 0$ for all $t \in (-\infty, \infty)$ if and only if $\alpha_1 = \alpha_2 = \alpha_3 = \alpha_4 = 0$.

## A.2.4 Bases

We are now in a position to introduce another important concept.

**Definition A.13.** A set $W$ in a linear space $V$ is called a *basis* for $V$ if

*(i) $W$ is linearly independent, and*
*(ii) the span of $W$ is the linear space $V$ itself; i.e., $S(W) = V$.* ∎

An immediate consequence of the above definition is that if $W$ is a linearly independent set in a vector space $V$, then $W$ is a basis for $S(W)$.

To introduce the notion of dimension of a vector space, it is shown that if a linear space $V$ is generated by a finite number of linearly independent elements, then this number of elements must be unique. The following results lead up to this.

Let $\{v^1, \ldots, v^n\}$ be a basis for a linear space $V$. Then it is easily shown that for each vector $v \in V$, there exist *unique* scalars $\alpha_1, \ldots, \alpha_n$ such that

$$v = \alpha_1 v^1 + \cdots + \alpha_n v^n.$$

Furthermore, if $u^1, \ldots, u^m$ is any linearly independent set of vectors in $V$, then $m \leq n$. Moreover, any other basis of $V$ consists of exactly $n$ elements. These facts allow the following definitions.

If a linear space $V$ has a basis consisting of a finite number of vectors, say, $\{v^1, \ldots, v^n\}$, then $V$ is said to be a *finite-dimensional vector space* and the *dimension of* $V$ is $n$, abbreviated $\dim V = n$. In this case we speak of an *n-dimensional vector space*. If $V$ is not a finite-dimensional vector space, it is said to be an *infinite-dimensional vector space*.

By convention, the linear space consisting of the null vector is finite-dimensional with dimension equal to zero.

An alternative to the above definition of dimension of a (finite-dimensional) vector space is given by the following result, which is easily verified: Let $V$ be a vector space that contains $n$ linearly independent vectors. If every set of $n + 1$ vectors in $V$ is linearly dependent, then $V$ is finite-dimensional and $\dim V = n$.

The preceding results enable us now to introduce the concept of coordinates of a vector. We let $\{v^1, \ldots, v^n\}$ be a basis of a vector space $V$, and we let $v \in V$ be represented by

$$v = \xi_1 v^1 + \cdots + \xi_n v^n.$$

The *unique* scalars $\xi_1, \ldots, \xi_n$ are called the *coordinates of $v$ with respect to the basis* $\{v^1, \ldots, v^n\}$.

---

***Example A.14.*** For the linear space $(R^n, R)$, let $S = \{e^1, \ldots, e^n\}$, where the $e^i \in R^n$, $i = 1, \ldots, n$, were defined earlier (in Example A.10). Then

$S$ is clearly a basis for $(R^n, R)$ since it is linearly independent and since given any $v \in R^n$, there exist unique real scalars $\alpha_i, i = 1, \ldots, n$, such that $v = \sum_{i=1}^{n} \alpha_i e^i = (\alpha_1, \ldots, \alpha_n)^T$; i.e., $S$ spans $R^n$. It follows that with every vector $v \in R^n$, we can associate a *unique* $n$-tuple of scalars

$$\begin{bmatrix} \alpha_1 \\ \vdots \\ \alpha_n \end{bmatrix} \quad \text{or} \quad (\alpha_1, \ldots, \alpha_n)$$

relative to the basis $\{e^1, \ldots, e^n\}$, the *coordinate representation* of the vector $v \in R^n$ with respect to the basis $S = \{e^1, \ldots, e^n\}$. Henceforth, we will refer to the basis $S$ of this example as the *natural basis for $R^n$*.

---

***Example A.15.*** We note that the vector space of all (complex-valued) polynomials with real coefficients of degree less than $n$ is an $n$-dimensional vector space over the field of real numbers. A basis for this space is given by $S = \{1, s, \ldots, s^{n-1}\}$ where $s$ is a complex variable. Associated with a given element of this vector space, say $p(s) = \alpha_0 + \alpha_1 s + \cdots + \alpha_{n-1} s^{n-1}$, we have the *unique* $n$-tuple given by $(\alpha_0, \alpha_1, \ldots, \alpha_{n-1})^T$, which constitutes the coordinate representation of $p(s)$ with respect to the basis $S$ given above.

---

***Example A.16.*** We note that the space $(V, R)$, where $V = C([a, b], R)$, given in Example A.6 is an infinite-dimensional vector space.

---

## A.3 Linear Transformations

**Definition A.17.** *A mapping $T$ of a linear space $V$ into a linear space $W$, where $V$ and $W$ are vector spaces over the same field $F$, is called a* linear transformation *or a* linear operator *provided that*

*(L-i)* $\qquad\qquad T(x + y) = T(x) + T(y) \quad$ *for all $x, y \in V$, and*

*(L-ii)* $\qquad\qquad T(\alpha x) = \alpha T(x) \qquad\qquad$ *for all $x \in V$ and $\alpha \in F$.*  ∎

In the following discussion, we consider three specific examples of linear transformations.

---

***Example A.18.*** Let $(V, R) = (R^n, R)$ and $(W, R) = (R^m, R)$ be vector spaces defined as in Example A.3, let $A = [a_{ij}] \in R^{m \times n}$, and let $T : V \to W$ be defined by the equation

$$y = Ax, \quad y \in R^m, \quad x \in R^n,$$

where $Ax$ denotes multiplication of the matrix $A$ and the vector $y$. It is easily verified using the properties of matrices that $T$ is a linear transformation.

***Example A.19.*** Let $(V, R) = (l_p, R)$ be the vector space defined in Example A.5 (modified to consist of sequences $\{x_i\}, i \in Z$, in place of $\{x_i\}$, $i = 1, 2, \dots$). Let $h : Z \times Z \to R$ be a function having the property that for each $x \in V$, the infinite sum

$$\sum_{k=-\infty}^{\infty} h(n, k)x(k)$$

exists and defines a function of $n$ on $Z$. Let $T : V \to V$ be defined by

$$y(n) = \sum_{k=-\infty}^{\infty} h(n, k)x(k).$$

It is easily verified that $T$ is a linear transformation.

The existence of the above sum is ensured under appropriate assumptions. For example, by using the Hölder Inequality, it is readily shown that if, e.g., for fixed $n$, $\{h(n, k)\} \in l_2$ and $\{x(k)\} \in l_2$, then the above sum is well defined. The above sum exists also if, e.g., $\{x(k)\} \in l_\infty$ and $\{h(n, k)\} \in l_1$ for fixed $n$.

***Example A.20.*** Let $(V, R)$ denote the vector space given in Example A.7, and let $k \in C([a, b] \times [a, b], R)$ have the property that for each $x \in V$, the Riemann integral

$$\int_a^b k(s, t)x(t)dt$$

exists and defines a continuous function of $s$ on $[a, b]$. Let $T : V \to V$ be defined by

$$(Tx)(s) = y(s) = \int_a^b k(s, t)x(t)dt.$$

It is readily verified that $T$ is a linear transformation of $V$ into $V$.

Henceforth, if $T$ is a linear transformation from a vector space $V$ (over a field $F$) into a vector space $W$ (over the same field $F$) we will write $T \in L(V, W)$ to express this. In the following discussion, we will identify some of the important properties of linear transformations.

### A.3.1 Linear Equations

With $T \in L(V, W)$ we define the *null space of $T$* as the set

$$\mathcal{N}(T) = \{v \in V : Tv = w = 0\}$$

and the *range space of* $T$ as the set

$$\mathcal{R}(T) = \{w \in W : w = Tv, v \in V\}.$$

Note that since $T0 = 0$, $\mathcal{N}(T)$ and $\mathcal{R}(T)$ are never empty. It is easily verified that $\mathcal{N}(T)$ is a linear subspace of $V$ and that $\mathcal{R}(T)$ is a linear subspace of $W$. If $V$ is finite-dimensional (of dimension $n$), then it is easily shown that $\dim \mathcal{R}(T) \leq n$. Also, if $V$ is finite-dimensional and if $\{w^1, \ldots, w^n\}$ is a basis for $\mathcal{R}(T)$ and $v^i$ is defined by $Tv^i = w^i, i = 1, \ldots, n$, then it is readily proved that the vectors $v^1, \ldots, v^n$ are linearly independent.

One of the important results of linear algebra, called the *fundamental theorem of linear equations*, states that for $T \in L(V, W)$ with $V$ finite-dimensional, we have

$$\dim \mathcal{N}(T) + \dim \mathcal{R}(T) = \dim V.$$

For the proof of this result, refer to any of the references on linear algebra cited at the end of Chapters 4, 6, 8, and 9.

The above result gives rise to the notions of the *rank*, $\rho(T)$, of a linear transformation $T$ of a finite-dimensional vector space $V$ into a vector space $W$, which we define as the dimension of the range space $\mathcal{R}(T)$, and the *nullity*, $\nu(T)$, of $T$, which we define as the dimension of the null space $\mathcal{N}(T)$.

With the above machinery in place, it is now easy to establish the following important results concerning *linear equations*.

Let $T \in L(V, W)$, where $V$ is finite-dimensional, let $s = \dim \mathcal{N}(T)$, and let $\{v^1, \ldots, v^s\}$ be a basis for $\mathcal{N}(T)$. Then it is easily verified that

(i)   a vector $v \in V$ satisfies the equation $Tv = 0$ if and only if $v = \sum_{i=1}^{s} \alpha_i v^i$ for some set of scalars $\{\alpha_1, \ldots, \alpha_s\}$, and furthermore, for each $v \in V$ such that $Tv = 0$ is true, the set of scalars $\{\alpha_1, \ldots, \alpha_s\}$ is unique;

(ii)  if $w^0 \in W$ is a fixed vector, then $Tv = w^0$ holds for at least one vector $v \in V$ (called the *solution* of the equation $Tv = w^0$) if and only if $w^0 \in \mathcal{R}(T)$; and

(iii) if $w^0$ is any fixed vector in $W$ and if $v^0$ is some vector in $V$ such that $Tv^0 = w^0$ (i.e., $v^0$ is a solution of the equation $Tv^0 = w^0$), then a vector $v \in V$ satisfies $Tv = w^0$ if and only if $v = v^0 + \sum_{i=1}^{s} \beta_i v^i$ for some set of scalars $\{\beta_1, \ldots, \beta_s\}$, and furthermore, for each $v \in V$ such that $Tv = w_0$, the set of scalars $\{\beta_1, \ldots, \beta_s\}$ is unique.

## A.3.2 Representation of Linear Transformations by Matrices

In the following discussion, we let $(V, F)$ and $(W, F)$ be vector spaces over the *same* field and we let $\mathcal{A} : V \to W$ denote a linear mapping. We let $\{v^1, \ldots, v^n\}$ be a basis for $V$, and we set $\bar{v}^1 = \mathcal{A}v^1, \ldots, \bar{v}^n = \mathcal{A}v^n$. Then it is an easy matter to show that if $v$ is any vector in $V$ and if $(\alpha_1, \ldots, \alpha_n)$ are the coordinates of $v$ with respect to $\{v^1, \ldots, v^n\}$, then $\mathcal{A}v = \alpha_1 \bar{v}^1 + \cdots + \alpha_n \bar{v}^n$. Indeed, we have $\mathcal{A}v = \mathcal{A}(\alpha_1 v^1 + \cdots + \alpha_n v^n) = \alpha_1 \mathcal{A}v^1 + \cdots + \alpha_n \mathcal{A}v^n = \alpha_1 \bar{v}^1 + \cdots + \alpha_n \bar{v}^n$.

Next, we let $\{\bar{v}^1, \ldots, \bar{v}^n\}$ be any set of vectors in $W$. Then it can be shown that there exists a unique linear transformation $\mathcal{A}$ from $V$ into $W$ such that $\mathcal{A}v^1 = \bar{v}^1, \ldots, \mathcal{A}v^n = \bar{v}^n$. To show this, we first observe that for each $v \in V$ we have unique scalars $\alpha_1, \ldots, \alpha_n$ such that

$$v = \alpha_1 v^1 + \cdots + \alpha_n v^n.$$

Now define a mapping $\mathcal{A} : V \to W$ as

$$\mathcal{A}(v) = \alpha_1 \bar{v}^1 + \cdots + \alpha_n \bar{v}^n.$$

Clearly, $\mathcal{A}(v^i) = \bar{v}^i, i = 1, \ldots, n$. We first must show that $\mathcal{A}$ is linear and, then, that $\mathcal{A}$ is unique. Given $v = \alpha_1 v^1 + \cdots + \alpha_n v^n$ and $w = \beta_1 v^1 + \cdots + \beta_n v^n$, we have $\mathcal{A}(v+w) = \mathcal{A}[(\alpha_1+\beta_1)v^1 + \cdots + (\alpha_n+\beta_n)v^n] = (\alpha_1+\beta_1)\bar{v}^1 + \cdots + (\alpha_n+\beta_n)\bar{v}^n$. On the other hand, $\mathcal{A}(v) = \alpha_1\bar{v}^1 + \cdots + \alpha_n\bar{v}^n$, $\mathcal{A}(w) = \beta_1\bar{v}^1 + \cdots + \beta_n\bar{v}^n$. Thus, $\mathcal{A}(v)+\mathcal{A}(w) = (\alpha_1\bar{v}^1 + \cdots + \alpha_n\bar{v}^n) + (\beta_1\bar{v}^1 + \cdots + \beta_n\bar{v}^n) = (\alpha_1+\beta_1)\bar{v}^1 + \cdots + (\alpha_n+\beta_n)\bar{v}^n = \mathcal{A}(v+w)$. In a similar manner, it is easily established that $\alpha\mathcal{A}(v) = \mathcal{A}(\alpha v)$ for all $\alpha \in F$ and $v \in V$. Therefore, $\mathcal{A}$ is linear. Finally, to show that $\mathcal{A}$ is unique, suppose there exists a linear transformation $\mathcal{B} : V \to W$ such that $\mathcal{B}v^i = \bar{v}^i, i = 1, \ldots, n$. It follows that $(\mathcal{A} - \mathcal{B})v^i = 0, i = 1, \ldots, n$, and, therefore, that $\mathcal{A} = \mathcal{B}$.

These results show that *a linear transformation is completely determined by knowing how it transforms the basis vectors in its domain, and that this linear transformation is uniquely determined in this way.* These results enable us to represent linear transformations defined on finite-dimensional spaces in an unambiguous way by means of matrices. We will use this fact in the following development.

Let $(V, F)$ and $(W, F)$ denote $n$-dimensional and $m$-dimensional vector spaces, respectively, and let $\{v^1, \ldots, v^n\}$ and $\{w^1, \ldots, w^m\}$ be bases for $V$ and $W$, respectively. Let $\mathcal{A} : V \to W$ be a linear transformation, and let $\bar{v}^i = \mathcal{A}v^i, i = 1, \ldots, n$. Since $\{w^1, \ldots, w^m\}$ is a basis for $W$, there are unique scalars $\{a_{ij}\}, i = 1, \ldots, m, j = 1, \ldots, n$, such that

$$
\begin{aligned}
\mathcal{A}v^1 = \bar{v}^1 &= a_{11}w^1 + a_{21}w^2 + \cdots + a_{m1}w^m, \\
\mathcal{A}v^2 = \bar{v}^2 &= a_{12}w^1 + a_{22}w^2 + \cdots + a_{m2}w^m, \\
&\cdots\cdots\cdots\cdots\cdots\cdots\cdots\cdots\cdots\cdots \\
\mathcal{A}v^n = \bar{v}^n &= a_{1n}w^1 + a_{2n}w^2 + \cdots + a_{mn}w^m.
\end{aligned}
\tag{A.14}
$$

Next, let $v \in V$. Then $v$ has the unique representation $v = \alpha_1 v^1 + \alpha_2 v^2 + \cdots + \alpha_n v^n$ with respect to the basis $\{v^1, \ldots, v^n\}$. In view of the result given at the beginning of this subsection, we now have

$$\mathcal{A}v = \alpha_1 \bar{v}^1 + \cdots + \alpha_n \bar{v}^n. \tag{A.15}$$

Since $\mathcal{A}v \in W, \mathcal{A}v$ has a unique representation with respect to the basis $\{w^1, \ldots, w^m\}$, say,

$$\mathcal{A}v = \gamma_1 w^1 + \gamma_2 w^2 + \cdots + \gamma_m w^m. \tag{A.16}$$

Combining (A.14) and (A.16), and rearranging, in view of the uniqueness of the representation in (A.16), we have

$$\gamma_1 = a_{11}\alpha_1 + a_{12}\alpha_2 + \cdots + a_{1n}\alpha_n,$$
$$\gamma_2 = a_{21}\alpha_1 + a_{22}\alpha_2 + \cdots + a_{2n}\alpha_n, \tag{A.17}$$
$$\dots\dots\dots\dots\dots\dots\dots\dots\dots\dots\dots$$
$$\gamma_m = a_{m1}\alpha_1 + a_{m2}\alpha_2 + \cdots + a_{mn}\alpha_n,$$

where $(\alpha_1, \ldots, \alpha_n)^T$ and $(\gamma_1, \ldots, \gamma_m)^T$ are coordinate representations of $v \in V$ and $\mathcal{A}v \in W$ with respect to the bases $\{v^1, \ldots, v^n\}$ of $V$ and $\{w^1, \ldots, w^m\}$ of $W$, respectively. This set of equations enables us to represent the linear transformation $\mathcal{A}$ from the linear space $V$ into the linear space $W$ by the unique scalars $\{a_{ij}\}, i = 1, \ldots, m, j = 1, \ldots, n$. For convenience we let

$$A = [a_{ij}] = \begin{bmatrix} a_{11} & a_{12} & \cdots & a_{1n} \\ a_{21} & a_{22} & \cdots & a_{2n} \\ \cdots & \cdots & \cdots & \cdots \\ a_{m1} & a_{m2} & \cdots & a_{mn} \end{bmatrix}. \tag{A.18}$$

*We see that once the bases $\{v^1, \ldots, v^n\}, \{w^1, \ldots, w^m\}$ are fixed, we can represent the linear transformation $\mathcal{A}$ by the array of scalars in (A.18) that are uniquely determined by (A.14). Note that the $j$th column of $A$ is the coordinate representation of the vector $\mathcal{A}v^j \in W$ with respect to the basis $\{w^1, \ldots, w^m\}$.*

The converse to the preceding statement also holds. Specifically, with the bases for $V$ and $W$ still fixed, the array given in (A.18) is uniquely associated with the linear transformation $\mathcal{A}$ of $V$ into $W$. The above discussion gives rise to the following important definition.

**Definition A.21.** *The array given in (A.18) is called the* matrix $A$ of the *linear transformation $\mathcal{A}$ from a linear space $V$ into a linear space $W$ (over $F$) with respect to the basis $\{v^1, \ldots, v^n\}$ of $V$ and the basis $\{w^1, \ldots, w^m\}$ of $W$.* ∎

If in Definition A.21, $V = W$, and if for both $V$ and $W$ the same basis $\{v^1, \ldots, v^n\}$ is used, then we simply speak of the *matrix $A$ of the linear transformation $\mathcal{A}$ with respect to the basis $\{v^1, \ldots, v^n\}$.*

In (A.18) the scalars $(a_{i1}, a_{i2}, \ldots, a_{in})$ form the $i$th *row* of $A$ and the scalars $(a_{1j}, a_{2j}, \ldots, a_{mj})^T$ form the $j$th *column* of $A$. The scalar $a_{ij}$ refers to that element of matrix $A$ that can be found in the $i$th row and $j$th column of $A$. The array in (A.18) is said to be an $m \times n$ *matrix*. If $m = n$, we speak of a *square matrix*. Consistent with the above discussion, an $n \times 1$ matrix is called a *column vector, column matrix*, or *n-vector*, and a $1 \times n$ matrix is called a *row vector*. Finally, if $A = [a_{ij}]$ and $B = [b_{ij}]$ are two $m \times n$ matrices, then

$A = B$; i.e., $A$ and $B$ are *equal* if and only if $a_{ij} = b_{ij}$ for all $i = 1, \ldots, m$, and for all $j = 1, \ldots, n$. Furthermore, we call $A^T = [a_{ij}]^T = [a_{ji}]$ the *transpose* of $A$.

The preceding discussion shows in particular that if $\mathcal{A}$ is a linear transformation of an $n$-dimensional vector space $V$ into an $m$-dimensional vector space $W$,

$$w = \mathcal{A}v, \tag{A.19}$$

if $\gamma = (\gamma_1, \ldots, \gamma_m)^T$ denotes the coordinate representation of $w$ with respect to the basis $\{w^1, \ldots, w^m\}$, if $\alpha = (\alpha_1, \ldots, \alpha_n)^T$ denotes the coordinate representation of $v$ with respect to the basis $\{v^1, \ldots, v^n\}$, and if $A$ denotes the matrix of $\mathcal{A}$ with respect to the bases $\{v^1, \ldots, v^n\}, \{w^1, \ldots, w^m\}$, then

$$\gamma = A\alpha, \tag{A.20}$$

or equivalently,

$$\gamma_i = \sum_{j=1}^{n} a_{ij}\alpha_j, \quad i = 1, \ldots, m \tag{A.21}$$

which are alternative ways to write (A.17).

### The Rank of a Matrix

Let $A$ denote the matrix representation of a linear transformation $\mathcal{A}$. The rank of $A$, $\rho(A)$, is defined as the rank of $\mathcal{A}$, $\rho(\mathcal{A})$. It can be shown that the *rank $\rho(A)$ of an $m \times n$ matrix $A$* is the largest number of linearly independent columns of $A$. The rank is also equal to the largest numbers of linearly independent rows of $A$. It also equals the dimension of the largest nonzero minor of $A$.

### A.3.3 Solving Linear Algebraic Equations

Now consider the linear system of equations given by

$$A\alpha = \gamma, \tag{A.22}$$

where $A \in R^{m \times n}$ and $\gamma \in R^m$ are given and $\alpha \in R^n$ is to be determined.

1. For a given $\gamma$, a solution $\alpha$ of (A.22) exists (not necessarily unique) if and only if $\gamma \in \mathcal{R}(A)$, or equivalently, if and only if

$$\rho([A, \gamma]) = \rho(A). \tag{A.23}$$

2. Every solution $\alpha$ of (A.22) can be expressed as a sum

$$\alpha = \alpha_p + \alpha_h, \tag{A.24}$$

where $\alpha_p$ is a specific solution of (A.22) and $\alpha_h$ satisfies $A\alpha_h = 0$. This result allows us to span the space of all solutions of (A.22). Note that there are

$$\dim \mathcal{N}(A) = n - \rho(A) \tag{A.25}$$

linearly independent solutions of the system of equations $A\beta = 0$.

3. $A\alpha = \gamma$ has a unique solution if and only if (A.23) is satisfied and

$$\rho(A) = n \le m. \tag{A.26}$$

4. A solution $\alpha$ of (A.22) exists for any $\gamma$ if and only if

$$\rho(A) = m. \tag{A.27}$$

If (A.27) is satisfied, a solution of (A.22) can be found by using the relation

$$\alpha = A^T (AA^T)^{-1}\gamma. \tag{A.28}$$

When in (A.22), $\rho(A) = m = n$, then $A \in R^{n \times n}$ and is nonsingular and the unique solution of (A.28) is given by

$$\alpha = A^{-1}\gamma. \tag{A.29}$$

---

**Example A.22.** Consider

$$A\alpha = \begin{bmatrix} 0 & 0 & 0 \\ 0 & 0 & 1 \\ 0 & 0 & 0 \end{bmatrix} \alpha = \gamma. \tag{A.30}$$

It is easily verified that $\{(0,1,0)^T\}$ is a basis for $\mathcal{R}(A)$. Since a solution of (A.30) exists if and only if $\gamma \in \mathcal{R}(A)$, $\gamma$ must be of the form $\gamma = (0, k, 0)^T$, $k \in R$. Note that

$$\rho(A) = 1 = \rho([A, \gamma]) = \text{rank} \begin{bmatrix} 0 & 0 & 0 & 0 \\ 0 & 0 & 1 & k \\ 0 & 0 & 0 & 0 \end{bmatrix},$$

as expected. To determine all solutions of (A.30), we need to determine an $\alpha_p$ and an $\alpha_h$ [see (A.24)]. In particular, $\alpha_p = (0\,0\,k)^T$ will do. To determine $\alpha_h$, we consider $A\beta = 0$. There are $\dim \mathcal{N}(A) = 2$ linearly independent solutions of $A\beta = 0$. In particular, $\{(1,0,0)^T, (0,1,0)^T\}$ is a basis for $\mathcal{N}(A)$. Therefore, any solution of (A.30) can be expressed as

$$\alpha = \alpha_p + \alpha_h = \begin{bmatrix} 0 \\ 0 \\ k \end{bmatrix} + \begin{bmatrix} 1 & 0 \\ 0 & 1 \\ 0 & 0 \end{bmatrix} \begin{bmatrix} c_1 \\ c_2 \end{bmatrix},$$

where $c_1, c_2$ are appropriately chosen real numbers.

---

# A.4 Equivalence and Similarity

From our previous discussion it is clear that a linear transformation $\mathcal{A}$ of a finite-dimensional vector space $V$ into a finite-dimensional vector space $W$ can be represented by means of different matrices, depending on the particular choice of bases in $V$ and $W$. The choice of bases may in different cases result in matrices that are easy or hard to utilize. Many of the resulting "standard" forms of matrices, called *canonical forms*, arise because of practical considerations. Such canonical forms often exhibit inherent characteristics of the underlying transformation $\mathcal{A}$.

Throughout this section, $V$ and $W$ are finite-dimensional vector spaces over the same field $F$, $\dim V = n$, and $\dim W = m$.

## A.4.1 Change of Bases: Vector Case

Our first aim will be to consider the change of bases in the coordinate representation of vectors. Let $\{v^1, \ldots, v^n\}$ be a basis for $V$, and let $\{\bar{v}^1, \ldots, \bar{v}^n\}$ be a set of vectors in $V$ given by

$$\bar{v}^i = \sum_{j=1}^{n} p_{ji} v^j, \quad i = 1, \ldots, n, \tag{A.31}$$

where $p_{ij} \in F$ for all $i, j = 1, \ldots, n$. It is easily verified that the set $\{\bar{v}^1, \ldots, \bar{v}^n\}$ forms a basis for $V$ if and only if the $n \times n$ matrix $P = [p_{ij}]$ is nonsingular (i.e., $\det P \neq 0$). We call $P$ *the matrix of the basis* $\{\bar{v}^1, \ldots, \bar{v}^n\}$ *with respect to the basis* $\{v^1, \ldots, v^n\}$. Note that the $i$th column of $P$ is the coordinate representation of $\bar{v}^i$ with respect to the basis $\{v^1, \ldots, v^n\}$.

Continuing the above discussion, let $\{v^1, \ldots, v^n\}$ and $\{\bar{v}^1, \ldots, \bar{v}^n\}$ be two bases for $V$ and let $P$ be the matrix of the basis $\{\bar{v}^1, \ldots, \bar{v}^n\}$ with respect to the basis $\{v^1, \ldots, v^n\}$. Then it is easily shown that $P^{-1}$ is the matrix of the basis $\{v^1, \ldots, v^n\}$ with respect to the basis $\{\bar{v}^1, \ldots, \bar{v}^n\}$.

Next, let the sets of vectors $\{v^1, \ldots, v^n\}$, $\{\bar{v}^1, \ldots, \bar{v}^n\}$, and $\{\tilde{v}^1, \ldots, \tilde{v}^n\}$ be bases for $V$. If $P$ is the matrix of the basis $\{\bar{v}^1, \ldots, \bar{v}^n\}$ with respect to the basis $\{v^1, \ldots, v^n\}$ and if $Q$ is the matrix of the basis $\{\tilde{v}^1, \ldots, \tilde{v}^n\}$ with respect to the basis $\{\bar{v}^1, \ldots, \bar{v}^n\}$, then it is easily verified that $PQ$ is the matrix of the basis $\{\tilde{v}^1, \ldots, \tilde{v}^n\}$ with respect to the basis $\{v^1, \ldots, v^n\}$.

Continuing further, let $\{v^1, \ldots, v^n\}$ and $\{\bar{v}^1, \ldots, \bar{v}^n\}$ be two bases for $V$ and let $P$ be the matrix of the basis $\{\bar{v}^1, \ldots, \bar{v}^n\}$ with respect to the basis $\{v^1, \ldots, v^n\}$. Let $a \in V$, and let $\alpha^T = (\alpha_1, \ldots, \alpha_n)$ denote the coordinate representation of $a$ with respect to the basis $\{v^1, \ldots, v^n\}$ (i.e., $a = \sum_{i=1}^{n} \alpha_i v^i$). Let $\bar{\alpha}^T = (\bar{\alpha}_1, \ldots, \bar{\alpha}_n)$ denote the coordinate representation of $a$ with respect to the basis $\{\bar{v}^1, \ldots, \bar{v}^n\}$. Then it is readily verified that

$$P\bar{\alpha} = \alpha.$$

**Example A.23.** Let $V = R^3, F = R$, and let $a = (1,2,3)^T \in R^3$ be given. Let $\{v^1, v^2, v^3\} = \{e^1, e^2, e^3\}$ denote the natural basis for $R^3$; i.e., $e^1 = (1,0,0)^T, e^2 = (0,1,0)^T, e^3 = (0,0,1)^T$. Clearly, the coordinate representation $\alpha$ of $a$ with respect to the natural basis is $(1,2,3)^T$.

Now let $\{\bar{v}^1, \bar{v}^2, \bar{v}^3\}$ be another basis for $R^3$, given by $\bar{v}^1 = (1,0,1)^T, \bar{v}^2 = (0,1,0)^T, \bar{v}^3 = (0,1,1)^T$. From the relation

$$(1,0,1)^T = \bar{v}^1 = p_{11}v^1 + p_{21}v^2 + p_{31}v^3 = p_{11}\begin{bmatrix} 1 \\ 0 \\ 0 \end{bmatrix} + p_{21}\begin{bmatrix} 0 \\ 1 \\ 0 \end{bmatrix} + p_{31}\begin{bmatrix} 0 \\ 0 \\ 1 \end{bmatrix},$$

we conclude that $p_{11} = 1, p_{21} = 0$, and $p_{31} = 1$. Similarly, from

$$(0,1,0)^T = \bar{v}^2 = p_{12}v^1 + p_{22}v^2 + p_{32}v^3 = p_{12}\begin{bmatrix} 1 \\ 0 \\ 0 \end{bmatrix} + p_{22}\begin{bmatrix} 0 \\ 1 \\ 0 \end{bmatrix} + p_{32}\begin{bmatrix} 0 \\ 0 \\ 1 \end{bmatrix},$$

we conclude that $p_{12} = 0, p_{22} = 1$, and $p_{32} = 0$. Finally, from the relation

$$(0,1,1)^T = \bar{v}^3 = p_{13}\begin{bmatrix} 1 \\ 0 \\ 0 \end{bmatrix} + p_{23}\begin{bmatrix} 0 \\ 1 \\ 0 \end{bmatrix} + p_{33}\begin{bmatrix} 0 \\ 0 \\ 1 \end{bmatrix},$$

we obtain that $p_{13} = 0, p_{23} = 1$, and $p_{33} = 1$.

The matrix $P = [p_{ij}]$ of the basis $\{\bar{v}^1, \bar{v}^2, \bar{v}^3\}$ with respect to the basis $\{v^1, v^2, v^3\}$ is therefore determined to be

$$P = \begin{bmatrix} 1 & 0 & 0 \\ 0 & 1 & 1 \\ 1 & 0 & 1 \end{bmatrix},$$

and the coordinate representation of $a$ with respect to the basis $\{\bar{v}^1, \bar{v}^2, \bar{v}^3\}$ is given by $\bar{\alpha} = P^{-1}\alpha$, or

$$\bar{\alpha} = \begin{bmatrix} 1 & 0 & 0 \\ 0 & 1 & 1 \\ 1 & 0 & 1 \end{bmatrix}^{-1} \begin{bmatrix} 1 \\ 2 \\ 3 \end{bmatrix} = \begin{bmatrix} 1 & 0 & 0 \\ 1 & 1 & -1 \\ -1 & 0 & 0 \end{bmatrix} \begin{bmatrix} 1 \\ 2 \\ 3 \end{bmatrix} = \begin{bmatrix} 1 \\ 0 \\ 2 \end{bmatrix}.$$

## A.4.2 Change of Bases: Matrix Case

Having addressed the relationship between the coordinate representations of a given vector with respect to different bases, we next consider the relationship between the matrix representations of a given linear transformation relative to different bases. To this end, let $\mathcal{A} \in L(V, W)$ and let $\{v^1, \ldots, v^n\}$ and

$\{w^1, \ldots, w^m\}$ be bases for $V$ and $W$, respectively. Let $A$ be the matrix of $\mathcal{A}$ with respect to the bases $\{v^1, \ldots, v^n\}$ and $\{w^1, \ldots, w^m\}$. Let $\{\bar{v}^1, \ldots, \bar{v}^n\}$ be another basis for $V$, and let the matrix of $\{\bar{v}^1, \ldots, \bar{v}^n\}$ with respect to $\{v^1, \ldots, v^n\}$ be $P$. Let $\{\bar{w}^1, \ldots, \bar{w}^m\}$ be another basis for $W$, and let $Q$ be the matrix of $\{w^1, \ldots, w^m\}$ with respect to $\{\bar{w}^1, \ldots, \bar{w}^m\}$. Let $\bar{A}$ be the matrix of $\mathcal{A}$ with respect to the bases $\{\bar{v}^1, \ldots, \bar{v}^n\}$ and $\{\bar{w}^1, \ldots, \bar{w}^m\}$. Then it is readily verified that

$$\bar{A} = QAP. \tag{A.32}$$

This result is depicted schematically in Figure A.1.

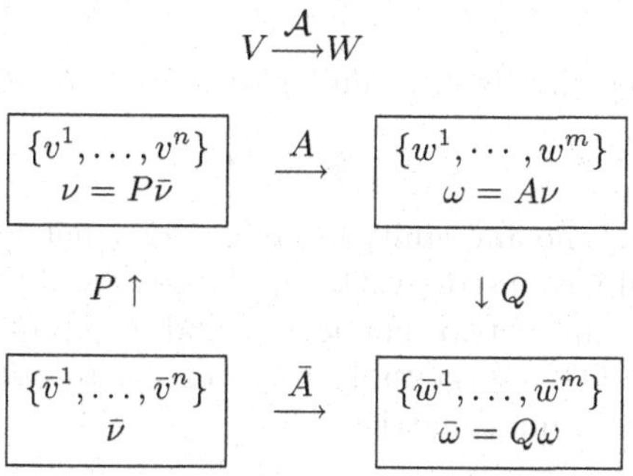

$$V \xrightarrow{\mathcal{A}} W$$

| $\{v^1, \ldots, v^n\}$ $\nu = P\bar{\nu}$ | $\xrightarrow{A}$ | $\{w^1, \cdots, w^m\}$ $\omega = A\nu$ |
|---|---|---|

$$P \uparrow \qquad\qquad \downarrow Q$$

| $\{\bar{v}^1, \ldots, \bar{v}^n\}$ $\bar{\nu}$ | $\xrightarrow{\bar{A}}$ | $\{\bar{w}^1, \ldots, \bar{w}^m\}$ $\bar{\omega} = Q\omega$ |
|---|---|---|

**Figure A.1.** Schematic diagram of the equivalence of two matrices

### A.4.3 Equivalence and Similarity of Matrices

The preceding discussion motivates the following definition.

**Definition A.24.** *An $m \times n$ matrix $\bar{A}$ is said to be* equivalent *to an $m \times n$ matrix $A$ if there exists an $m \times m$ nonsingular matrix $Q$ and an $n \times n$ nonsingular matrix $P$ such that (A.32) is true. If $\bar{A}$ is equivalent to $A$, we write $\bar{A} \sim A$.* ∎

Next, let $V = W$, let $\mathcal{A} \in L(V, V)$, let $\{v^1, \ldots, v^n\}$ be a basis for $V$, and let $A$ be the matrix of $\mathcal{A}$ with respect to $\{v^1, \ldots, v^n\}$. Let $\{\bar{v}^1, \ldots, \bar{v}^n\}$ be another basis for $V$ whose matrix with respect to $\{v^1, \ldots, v^n\}$ is $P$. Let $\bar{A}$ be the matrix of $\mathcal{A}$ with respect to $\{\bar{v}^1, \ldots, \bar{v}^n\}$. Then it follows immediately from (A.32) that

$$\bar{A} = P^{-1}AP. \tag{A.33}$$

The meaning of this result is depicted schematically in Figure A.2. The above discussion motivates the following definition.

**Definition A.25.** *An $n \times n$ matrix $\bar{A}$ is said to be* similar *to an $n \times n$ matrix $A$ if there exists an $(n \times n)$ nonsingular matrix $P$ such that*

$$V \xrightarrow{\mathcal{A}} V$$
$$\{v^1,\ldots,v^n\} \xrightarrow{A} \{v^1,\ldots,v^n\}$$
$$\uparrow P \qquad\qquad \downarrow P^{-1}$$
$$\{\bar{v}^1,\ldots,\bar{v}^n\} \xrightarrow{\bar{A}} \{\bar{v}^1,\ldots,\bar{v}^n\}$$

**Figure A.2.** Schematic diagram of the similarity of two matrices

$$\bar{A} = P^{-1}AP.$$

*If $\bar{A}$ is similar to $A$, we write $\bar{A} \sim A$. We call $P$* a similarity transformation. ∎

It is easily verified that if $\bar{A}$ is similar to $A$ [i.e., (A.33) is true], then $A$ is similar to $\bar{A}$; i.e.,

$$A = P\bar{A}P^{-1}. \tag{A.34}$$

In view of this, there is no ambiguity in saying "two matrices are similar," and we could just as well have used (A.34) [in place of (A.33)] to define similarity of matrices. To sum up, if two matrices $A$ and $\bar{A}$ represent the same linear transformation $\mathcal{A} \in L(V,V)$, possibly with respect to two different bases for $V$, then $A$ and $\bar{A}$ are similar matrices.

## A.5 Eigenvalues and Eigenvectors

**Definition A.26.** *Let $A$ be an $n \times n$ matrix whose elements belong to the field $F$. If there exist $\lambda \in F$ and a nonzero vector $\alpha \in F^n$ such that*

$$A\alpha = \lambda\alpha, \tag{A.35}$$

*then $\lambda$ is called an* eigenvalue *of $A$ and $\alpha$ is called an* eigenvector *of $A$ corresponding to the eigenvalue $\lambda$.* ∎

We note that if $\alpha$ is an eigenvector of $A$, then any nonzero multiple of $\alpha$ is also an eigenvector of $A$.

### A.5.1 Characteristic Polynomial

Let $A \in C^{n \times n}$. Then

$$\det(A - \lambda I) = \alpha_0 + \alpha_1\lambda + \alpha_2\lambda^2 + \cdots + \alpha_n\lambda^n \tag{A.36}$$

[note that $\alpha_0 = \det(A)$ and $\alpha_n = (-1)^n$]. The eigenvalues of $A$ are precisely the roots of the equation

$$\det(A - \lambda I) = \alpha_0 + \alpha_1\lambda + \alpha_2\lambda^2 + \cdots + \alpha_n\lambda^n = 0 \tag{A.37}$$

and $A$ has at most $n$ distinct eigenvalues.

We call (A.36) the *characteristic polynomial of $A$*, and we call (A.37) *the characteristic equation of $A$.*

*Remarks*

The above definition of characteristic polynomial is the one usually used in texts on linear algebra and matrix theory (refer, e.g., to some of the books on this subject cited at the end of this chapter). An alternative to the above definition is given by the expression

$$\alpha(\lambda) \triangleq \det(\lambda I - A) = (-1)^n \det(A - \lambda I).$$

Now consider

$$\det(A - \lambda I) = (\lambda_1 - \lambda)^{m_1} (\lambda_2 - \lambda)^{m_2} \cdots (\lambda_p - \lambda)^{m_p}, \qquad (A.38)$$

where $\lambda_i, i = 1, \ldots, p$, are the distinct roots of (A.37) (i.e., $\lambda_i \neq \lambda_j$, if $i \neq j$). In (A.38), $m_i$ is called the *algebraic multiplicity* of the root $\lambda_i$. The $m_i$ are positive integers, and $\sum_{i=1}^{p} m_i = n$.

The reader should make note of the distinction between the concept of *algebraic multiplicity* $m_i$ of $\lambda_i$, given above, and the *(geometric) multiplicity* $l_i$ of an eigenvalue $\lambda_i$, given by $l_i = n - \rho(\lambda_i I - A)$. In general these need not be the same.

## A.5.2 The Cayley–Hamilton Theorem and Applications

We now state and prove a result that is very important in linear systems theory.

**Theorem A.27.** (Cayley–Hamilton Theorem) *Every square matrix satisfies its characteristic equation. More specifically, if $A$ is an $n \times n$ matrix and $p(\lambda) = \det(A - \lambda I)$ is the characteristic polynomial of $A$, then $p(A) = O$.*

*Proof.* Let the characteristic polynomial for $A$ be $p(\lambda) = \alpha_0 + \alpha_1\lambda + \cdots + \alpha_n\lambda^n$, and let $B(\lambda) = [b_{ij}(\lambda)]$ be the classical adjoint of $(A - \lambda I)$. (For a nonsingular matrix $C$ with inverse $C^{-1} = \frac{1}{\det(C)} \text{adj}(C)$, $\text{adj}(C)$ is called the *classical adjoint of $C$*.) Since the $b_{ij}(\lambda)$ are cofactors of the matrix $A - \lambda I$, they are polynomials in $\lambda$ of degree not more than $n - 1$. Thus, $b_{ij}(\lambda) = \beta_{ij0} + \beta_{ij1}\lambda + \cdots + \beta_{ij(n-1)}\lambda^{n-1}$. Letting $B_k = [\beta_{ijk}]$ for $k = 0, 1, \ldots, n-1$, we have $B(\lambda) = B_0 + \lambda B_1 + \cdots + \lambda^{n-1}B_{n-1}$ and $(A - \lambda I)B(\lambda) = [\det(A - \lambda I)]I$. Thus, $(A - \lambda I)[B_0 + \lambda B_1 + \cdots + \lambda^{n-1}B_{n-1}] = (\alpha_0 + \alpha_1\lambda + \cdots + \alpha_n\lambda^n)I$. Expanding the left-hand side of this equation and equating like powers of $\lambda$, we have $-B_{n-1} = \alpha_n I, AB_{n-1} - B_{n-2} = \alpha_{n-1}I, \ldots, AB_1 - B_0 = \alpha_1 I, AB_0 = \alpha_0 I$. Premultiplying the above matrix equations by $A^n, A^{n-1}, \ldots, A, I$, respectively, we have $-A^n B_{n-1} = \alpha_n A^n, A^n B_{n-1} - A^{n-1}B_{n-2} = \alpha_{n-1}A^{n-1}, \ldots, A^2 B_1 - AB_0 = \alpha_1 A, AB_0 = \alpha_0 I$. Adding these matrix equations, we obtain $O = \alpha_0 I + \alpha_1 A + \cdots + \alpha_n A^n = p(A)$, which was to be shown. ∎

As an immediate consequence of the Cayley–Hamilton Theorem, we have the following results: Let $A$ be an $n \times n$ matrix with characteristic polynomial given by (A.37). Then (i) $A^n = (-1)^{n+1}[\alpha_0 I + \alpha_1 A + \cdots + \alpha_{n-1} A^{n-1}]$; and (ii) if $f(\lambda)$ is any polynomial in $\lambda$, then there exist $\beta_0, \beta_1, \ldots, \beta_{n-1} \in F$ such that

$$f(A) = \beta_0 I + \beta_1 A + \cdots + \beta_{n-1} A^{n-1}. \tag{A.39}$$

Part (i) follows from the Cayley–Hamilton Theorem and from the fact that $\alpha_n = (-1)^n$. To prove part (ii), let $f(\lambda)$ be any polynomial in $\lambda$ and let $p(\lambda)$ denote the characteristic polynomial of $A$. From a result for polynomials (called the *division algorithm*), we know that there exist two unique polynomials $g(\lambda)$ and $r(\lambda)$ such that

$$f(\lambda) = p(\lambda)g(\lambda) + r(\lambda), \tag{A.40}$$

where the degree of $r(\lambda) \leq n - 1$. Now since $p(A) = 0$, we have that $f(A) = r(A)$ and the result follows.

The Cayley–Hamilton Theorem can also be used to express $n \times n$ matrix-valued power series (as well as other kinds of functions) as matrix polynomials of degree $n - 1$. Consider in particular the matrix exponential $e^{At}$ defined by

$$e^{At} = \sum_{k=0}^{\infty} (t^k/k!) A^k, t \in (-a, a). \tag{A.41}$$

In view of the Cayley–Hamilton Theorem, we can write

$$f(A) = e^{At} = \sum_{i=0}^{n-1} \alpha_i(t) A^i. \tag{A.42}$$

In the following discussion, we present a method to determine the coefficients $\alpha_i(t)$ in (A.42) [or $\beta_i$ in (A.39)].

In accordance with (A.38), let $p(\lambda) = \det(A - \lambda I) = \prod_{i=1}^{p} (\lambda_i - \lambda)^{m_i}$ be the characteristic polynomial of $A$. Also, let $f(\lambda)$ and $g(\lambda)$ be two analytic functions. Now if

$$f^{(l)}(\lambda_i) = g^{(l)}(\lambda_i), \quad l = 0, \ldots, m_i - 1, \quad i = 1, \ldots, p, \tag{A.43}$$

where $f^{(l)}(\lambda_i) = \frac{d^l f}{d\lambda^l}(\lambda)|_{\lambda=\lambda_i}$, $\sum_{i=1}^{p} m_i = n$, then $f(A) = g(A)$. To see this, we note that condition (A.43) written as $(f - g)^l(\lambda_i) = 0$ implies that $f(\lambda) - g(\lambda)$ has $p(\lambda)$ as a factor; i.e., $f(\lambda) - g(\lambda) = w(\lambda)p(\lambda)$ for some analytic function $w(\lambda)$. From the Cayley–Hamilton Theorem we have that $p(A) = O$ and therefore $f(A) - g(A) = O$.

---

**Example A.28.** As a specific application of the Cayley–Hamilton Theorem, we evaluate the matrix $A^{37}$, where $A = \begin{bmatrix} 1 & 0 \\ 1 & 2 \end{bmatrix}$. Since $n = 2$, we assume, in view of (A.39), that $A^{37}$ is of the form $A^{37} = \beta_0 I + \beta_1 A$. The characteristic polynomial of $A$ is $p(\lambda) = (1 - \lambda)(2 - \lambda)$, and the eigenvalues of $A$ are $\lambda_1 = 1$ and $\lambda_2 = 2$. In this case, $f(\lambda) = \lambda^{37}$ and $r(\lambda)$ in (A.40) is $r(\lambda) = \beta_0 + \beta_1 \lambda$. To determine $\beta_0$ and $\beta_1$ we use the fact that $p(\lambda_1) = p(\lambda_2) = 0$ to conclude that $f(\lambda_1) = r(\lambda_1)$ and $f(\lambda_2) = r(\lambda_2)$. Therefore, we have that $\beta_0 + \beta_1 = 1^{37} = 1$ and $\beta_0 + 2\beta_1 = 2^{37}$. Hence, $\beta_1 = 2^{37} - 1$ and $\beta_0 = 2 - 2^{37}$. Therefore, $A^{37} = (2 - 2^{37})I + (2^{37} - 1)A$ or $A^{37} = \begin{bmatrix} 1 & 0 \\ 2^{37} - 1 & 2^{37} \end{bmatrix}$.

---

**Example A.29.** Let $A = \begin{bmatrix} -1 & 1 \\ -1 & 1 \end{bmatrix}$, and let $f(A) = e^{At}, f(\lambda) = e^{\lambda t}$, and $g(\lambda) = \alpha_1 \lambda + \alpha_0$. The matrix $A$ has an eigenvalue $\lambda = \lambda_1 = \lambda_2 = 0$ with multiplicity $m_1 = 2$. Conditions (A.43) are given by $f(\lambda_1) = g(\lambda_1) = 1$ and $f^{(1)}(\lambda_1) = g^{(1)}(\lambda_1)$, which imply that $\alpha_0 = 1$ and $\alpha_1 = t$. Therefore,

$$e^{At} = f(A) = g(A) = \alpha_1 A + \alpha_0 I = \begin{bmatrix} -\alpha_1 + \alpha_0 & \alpha_1 \\ -\alpha_1 & \alpha_1 + \alpha_0 \end{bmatrix} = \begin{bmatrix} 1-t & t \\ -t & 1+t \end{bmatrix}.$$

---

### A.5.3 Minimal Polynomials

For purposes of motivation, consider the matrix

$$A = \begin{bmatrix} 1 & 3 & -2 \\ 0 & 4 & -2 \\ 0 & 3 & -1 \end{bmatrix}.$$

The characteristic polynomial of $A$ is $p(\lambda) = (1 - \lambda)^2(2 - \lambda)$, and we know from the Cayley–Hamilton Theorem that

$$p(A) = O. \tag{A.44}$$

Now let us consider the polynomial $m(\lambda) = (1-\lambda)(2-\lambda) = 2 - 3\lambda + \lambda^2$. Then

$$m(A) = 2I - 3A + A^2 = O. \tag{A.45}$$

Thus, matrix $A$ satisfies (A.45), which is of lower degree than (A.44), the characteristic equation of $A$.

More generally, it can be shown that for an $n \times n$ matrix $A$, there exists a unique polynomial $m(\lambda)$ such that (i) $m(A) = O$, (ii) $m(\lambda)$ is *monic* (i.e., if $m$ is an $n$th-order polynomial in $\lambda$, then the coefficient of $\lambda^n$ is unity), and (iii) if $m'(\lambda)$ is any other polynomial such that $m'(A) = O$, then the degree of $m(\lambda)$ is less or equal to the degree of $m'(\lambda)$ [i.e., $m(\lambda)$ is of the lowest degree such that $m(A) = O$]. The polynomial $m(\lambda)$ is called the *minimal polynomial of $A$*.

Let $f(\lambda)$ be any polynomial such that $f(A) = O$ (e.g., the characteristic polynomial). Then it is easily shown that $m(\lambda)$ divides $f(\lambda)$ [i.e., there is a polynomial $q(\lambda)$ such that $f(\lambda) = q(\lambda)m(\lambda)$]. In particular, the minimal polynomial of $A$, $m(\lambda)$, divides the characteristic polynomial of $A$, $p(\lambda)$. Also, it can be shown that $p(\lambda)$ divides $[m(\lambda)]^n$.

Next, let $p(\lambda)$ be given by

$$p(\lambda) = (\lambda_1 - \lambda)^{m_1}(\lambda_2 - \lambda)^{m_2} \cdots (\lambda_p - \lambda)^{m_p}, \tag{A.46}$$

where $m_1, \ldots, m_p$ are the algebraic multiplicities of the distinct eigenvalues $\lambda_1, \ldots, \lambda_p$ of $A$, respectively. It can be shown that

$$m(\lambda) = (\lambda - \lambda_1)^{\mu_1}(\lambda - \lambda_2)^{\mu_2} \cdots (\lambda - \lambda_p)^{\mu_p}, \tag{A.47}$$

where $1 \le \mu_i \le m_i$, $i = 1, \ldots, p$.

It can also be shown that $(\lambda - \lambda_i)^{\mu_i}$ is the minimal polynomial of the $A_i$ diagonal block in the *Jordan canonical form* of $A$, which we discuss in the next section. When $A$ has all $n$ distinct eigenvalues, the Jordan canonical form has $n$ diagonal blocks and, therefore, $\mu_i = 1$ and $p(\lambda) = m(\lambda)$. The Jordan canonical form is described in Section A.6 and in [1, Section 2.2].

## A.6 Diagonal and Jordan Canonical Form of Matrices

Let $A$ be an $n \times n$ matrix $A \in C^{n \times n}$. The following developement follows [10]. To begin with, let us assume that $A$ has *distinct eigenvalues* $\lambda_1, \ldots, \lambda_n$. Let $v_i$ be an eigenvector of $A$ corresponding to $\lambda_i$, $i = 1, \ldots, n$. Then it can be easily shown that the set of vectors $\{v_1, \ldots, v_n\}$ is linearly independent over $C$, and as such, it can be used as a basis for $C^n$. Now let $\tilde{A}$ be the representation of $A$ with respect to the basis $\{v_1, \ldots, v_n\}$. Since the $i$th column of $\tilde{A}$ is the representation of $Av_i = \lambda_i v_i$ with respect to the basis $\{v_1, \ldots, v_n\}$, it follows that

$$\tilde{A} = \begin{bmatrix} \lambda_1 & & & 0 \\ & \lambda_2 & & \\ & & \ddots & \\ 0 & & & \lambda_n \end{bmatrix} \triangleq \mathrm{diag}(\lambda_1, \ldots, \lambda_n). \tag{A.48}$$

Since $A$ and $\tilde{A}$ are matrix representations of the same linear transformation, it follows that $A$ and $\tilde{A}$ are similar matrices. Indeed, this can be checked by computing

$$\tilde{A} = P^{-1}AP, \tag{A.49}$$

where $P = [v_1, \ldots, v_n]$ and where the $v_i$ are eigenvectors corresponding to $\lambda_i$, $i = 1, \ldots, n$. Note that $AP = \tilde{A}P$ is true because the $i$th column of $AP$ is $Av_i$, which equals $\lambda_i v_i$, the $i$th column of $\tilde{A}P$.

When a matrix $\widetilde{A}$ is obtained from a matrix $A$ via a similarity transformation $P$, we say that matrix $A$ *has been diagonalized*. Now if the matrix $A$ has repeated eigenvalues, then it is not always possible to diagonalize it. In generating a "convenient" basis for $C^n$ in this case, we introduce the concept of *generalized eigenvector*. Specifically, a vector $v$ is called a *generalized eigenvector of rank $k$* of $A$, associated with an eigenvalue $\lambda$ if and only if

$$(A - \lambda I_n)^k v = 0 \quad \text{and} \quad (A - \lambda I_n)^{k-1} v \neq 0, \tag{A.50}$$

where $I_n$ denotes the $n \times n$ identity matrix. Note that when $k = 1$, this definition reduces to the preceding defintion of eigenvector.

Now let $v$ be a generalized eigenvector of rank $k$ associated with the eigenvalue $\lambda$. Define

$$
\begin{aligned}
v_k &= v, \\
v_{k-1} &= (A - \lambda I_n)v = (A - \lambda I_n)v_k, \\
v_{k-2} &= (A - \lambda I_n)^2 v = (A - \lambda I_n)v_{k-1}, \\
&\ \ \vdots \\
v_1 &= (A - \lambda I_n)^{k-1} v = (A - \lambda I_n)v_2.
\end{aligned}
\tag{A.51}
$$

Then for each $i$, $1 \le i \le k$, $v_i$ is a generalized eigenvector of rank $i$. We call the set of vectors $\{v_1, \ldots, v_k\}$ a *chain of generalized eigenvectors*.

For generalized eigenvectors, we have the following results:

(i)   The generalized eigenvectors $\{v_1, \ldots, v_k\}$ defined in (A.51) are linearly independent.

(ii)  The generalized eigenvectors of $A$ associated with different eigenvalues are linearly independent.

(iii) If $u$ and $v$ are generalized eigenvectors of rank $k$ and $l$, respectively, associated with the same eigenvalue $\lambda$, and if $u_i$ and $v_j$ are defined by

$$
\begin{aligned}
u_i &= (A - \lambda I_n)^{k-i} u, \quad i = 1, \ldots, k, \\
v_j &= (A - \lambda I_n)^{l-j} v, \quad j = 1, \ldots, l,
\end{aligned}
$$

and if $u_1$ and $v_1$ are linearly independent, then the generalized eigenvectors $u_1, \ldots, u_k, v_1, \ldots, v_l$ are linearly independent.

These results can be used to construct a new basis for $C^n$ such that the matrix representation of $A$ with respect to this new basis is in the *Jordan canonical form* $J$. We characterize $J$ in the following result: For every complex $n \times n$ matrix $A$, there exists a nonsingular matrix $P$ such that the matrix

$$J = P^{-1}AP$$

is in the canonical form

$$J = \begin{bmatrix} J_0 & & & 0 \\ & J_1 & & \\ & & \ddots & \\ 0 & & & J_s \end{bmatrix}, \qquad (A.52)$$

where $J_0$ is a diagonal matrix with diagonal elements $\lambda_1, \ldots, \lambda_k$ (not necessarily distinct), i.e.,

$$J_0 = \text{diag}(\lambda_1, \ldots, \lambda_k),$$

and each $J_p$ is an $n_p \times n_p$ matrix of the form

$$J_p = \begin{bmatrix} \lambda_{k+p} & 1 & 0 & \cdots & 0 \\ 0 & \lambda_{k+p} & 1 & & \vdots \\ \vdots & \vdots & \ddots & \ddots & 1 \\ 0 & 0 & \cdots & & \lambda_{k+p} \end{bmatrix}, \quad p = 1, \ldots, s,$$

where $\lambda_{k+p}$ need not be different from $\lambda_{k+q}$ if $p \neq q$ and $k + n_1 + \cdots + n_s = n$. The numbers $\lambda_i$, $i = 1, \ldots, k + s$, are the eigenvalues of $A$. If $\lambda_i$ is a simple eigenvalue of $A$, it appears in the block $J_0$. The blocks $J_0, J_1, \ldots, J_s$ are called *Jordan blocks*, and $J$ is called the *Jordan canonical form*.

Note that a matrix may be similar to a diagonal matrix without having distinct eigenvalues. The identity matrix $I$ is such an example. Also, it can be shown that any real symmetric matrix $A$ has only real eigenvalues (which may be repeated) and is similar to a diagonal matrix.

We now give a procedure for computing a set of basis vectors that yield the Jordan canonical form $J$ of an $n \times n$ matrix $A$ and the required nonsingular transformation $P$ that relates $A$ to $J$:

1. Compute the eigenvalues of $A$. Let $\lambda_1, \ldots, \lambda_m$ be the distinct eigenvalues of $A$ with multiplicities $n_1, \ldots, n_m$, respectively.
2. Compute $n_1$ linearly independent generalized eigenvectors of $A$ associated with $\lambda_1$ as follows: Compute $(A - \lambda_1 I_n)^i$ for $i = 1, 2, \ldots$ until the rank of $(A - \lambda_1 I_n)^k$ is equal to the rank of $(A - \lambda_1 I_n)^{k+1}$. Find a generalized eigenvector of rank $k$, say $u$. Define $u_i = (A - \lambda_1 I_n)^{k-i}u$, $i = 1, \ldots, k$. If $k = n_1$, proceed to step 3. If $k < n_1$, find another linearly independent generalized eigenvector with the largest possible rank; i.e., try to find another generalized eigenvector with rank $k$. If this is not possible, try $k - 1$, and so forth, until $n_1$ linearly independent generalized eigenvectors are determined. Note that if $\rho(A - \lambda_1 I_n) = r$, then there are totally $(n - r)$ chains of generalized eigenvectors associated with $\lambda_1$.
3. Repeat step 2 for $\lambda_2, \ldots, \lambda_m$.
4. Let $u_1, \ldots, u_k, \ldots$ be the new basis. Observe, from (A.51), that

$$Au_1 = \lambda_1 u_1 = [u_1 u_2 \cdots u_k \cdots][\lambda_1, 0, \ldots, 0]^T,$$
$$Au_2 = u_1 + \lambda_1 u_2 = [u_1 u_2 \cdots u_k \cdots][1, \lambda_1, 0, \ldots, 0]^T,$$
$$\vdots$$
$$Au_k = u_{k-1} + \lambda_1 u_k = [u_1 u_2 \cdots u_k \cdots][0, \ldots, 0, 1, \lambda_1, 0, \ldots, 0]^T,$$

with $\lambda_1$ in the $k$th position, which yields the representation $J$ in (A.52) of $A$ with respect to the new basis, where the $k \times k$ matrix $J_1$ is given by

$$J_1 = \begin{bmatrix} \lambda_1 & 1 & \cdots & 0 \\ 0 & \lambda_1 & & \vdots \\ & & \ddots & 1 \\ 0 & & \cdots & \lambda_1 \end{bmatrix}.$$

Note that each chain of generalized eigenvectors generates a Jordan block whose order equals the length of the chain.

5. The similarity transformation that yields $J = Q^{-1}AQ$ is given by $Q = [u_1, \ldots, u_k, \ldots]$.
6. Rearrange the Jordan blocks in the desired order to yield (A.52) and the corresponding similarity transformation $P$.

---

**Example A.30.** The characteristic equation of the matrix

$$A = \begin{bmatrix} 3 & -1 & 1 & 1 & 0 & 0 \\ 1 & 1 & -1 & -1 & 0 & 0 \\ 0 & 0 & 2 & 0 & 1 & 1 \\ 0 & 0 & 0 & 2 & -1 & -1 \\ 0 & 0 & 0 & 0 & 1 & 1 \\ 0 & 0 & 0 & 0 & 1 & 1 \end{bmatrix}$$

is given by

$$\det(A - \lambda I) = (\lambda - 2)^5 \lambda = 0.$$

Thus, $A$ has eigenvalue $\lambda_2 = 2$ with multiplicity 5 and eigenvalue $\lambda_1 = 0$ with multiplicity 1.

Now compute $(A - \lambda_2 I)^i$, $i = 1, 2, \ldots$, as follows:

$$(A - 2I) = \begin{bmatrix} 1 & -1 & 1 & 1 & 0 & 0 \\ 1 & -1 & -1 & -1 & 0 & 0 \\ 0 & 0 & 0 & 0 & 1 & 1 \\ 0 & 0 & 0 & 0 & -1 & -1 \\ 0 & 0 & 0 & 0 & -1 & 1 \\ 0 & 0 & 0 & 0 & 1 & -1 \end{bmatrix} \quad \text{and} \quad \rho(A - 2I) = 4,$$

$$(A - 2I)^2 = \begin{bmatrix} 0 & 0 & 2 & 2 & 0 & 0 \\ 0 & 0 & 2 & 2 & 0 & 0 \\ 0 & 0 & 0 & 0 & 0 & 0 \\ 0 & 0 & 0 & 0 & 0 & 0 \\ 0 & 0 & 0 & 0 & 2 & -2 \\ 0 & 0 & 0 & 0 & -2 & 2 \end{bmatrix} \quad \text{and} \quad \rho(A - 2I)^2 = 2,$$

$$(A - 2I)^3 = \begin{bmatrix} 0 & 0 & 0 & 0 & 0 & 0 \\ 0 & 0 & 0 & 0 & 0 & 0 \\ 0 & 0 & 0 & 0 & 0 & 0 \\ 0 & 0 & 0 & 0 & 0 & 0 \\ 0 & 0 & 0 & 0 & -4 & 4 \\ 0 & 0 & 0 & 0 & 4 & -4 \end{bmatrix} \quad \text{and} \quad \rho(A - 2I)^3 = 1,$$

$$(A - 2I)^4 = \begin{bmatrix} 0 & 0 & 0 & 0 & 0 & 0 \\ 0 & 0 & 0 & 0 & 0 & 0 \\ 0 & 0 & 0 & 0 & 0 & 0 \\ 0 & 0 & 0 & 0 & 0 & 0 \\ 0 & 0 & 0 & 0 & 8 & -8 \\ 0 & 0 & 0 & 0 & -8 & 8 \end{bmatrix} \quad \text{and} \quad \rho(A - 2I)^4 = 1.$$

Since $\rho(A - 2I)^3 = \rho(A - 2I)^4$, we stop at $(A - 2I)^3$. It can be easily verified that if $u = [0\ 0\ 1\ 0\ 0\ 0]^T$, then $(A-2I)^3 u = 0$ and $(A-2I)^2 u = [2\ 2\ 0\ 0\ 0\ 0]^T \neq 0$. Therefore, $u$ is a generalized eigenvector of rank 3. So we define

$$u_1 \triangleq (A - 2I)^2 u = \begin{bmatrix} 2\ 2\ 0\ 0\ 0\ 0 \end{bmatrix}^T,$$
$$u_2 \triangleq (A - 2I)u = \begin{bmatrix} 1\ -1\ 0\ 0\ 0\ 0 \end{bmatrix}^T,$$
$$u_3 \triangleq u = \begin{bmatrix} 0\ 0\ 1\ 0\ 0\ 0 \end{bmatrix}^T.$$

Since we have only three generalized eigenvectors for $\lambda_2 = 2$ and since the multiplicity of $\lambda_2 = 2$ is five, we have to find two more linearly independent eigenvectors for $\lambda_2 = 2$. So let us try to find a generalized eigenvector of rank 2. Let $v = [0\ 0\ 1\ -1\ 1\ 1]^T$. Then $(A - 2I)v = [0\ 0\ 2\ -2\ 0\ 0]^T \neq 0$ and $(A - 2I)^2 v = 0$. Moreover, $(A - 2I)v$ is linearly independent of $u_1$, and hence, we have another linearly independent generalized eigenvector of rank 2. Define

$$v_2 \triangleq v = \begin{bmatrix} 0\ 0\ 1\ -1\ 1\ 1 \end{bmatrix}^T$$

and

$$v_1 = (A - 2I)v = \begin{bmatrix} 0\ 0\ 2\ -2\ 0\ 0 \end{bmatrix}^T.$$

Next, we compute an eigenvector associated with $\lambda_1 = 0$. Since $w = [0\ 0\ 0\ 0\ 1\ -1]^T$ is a solution of $(A - \lambda_1 I)w = 0$, the vector $w$ will do.

Finally, with respect to the basis $w_1, u_1, u_2, u_3, v_1, v_2$, the Jordan canonical form of $A$ is given by

$$J = \begin{bmatrix} 0 & 0 & 0 & 0 & 0 & 0 \\ 0 & 2 & 1 & 0 & 0 & 0 \\ 0 & 0 & 2 & 1 & 0 & 0 \\ 0 & 0 & 0 & 2 & 0 & 0 \\ 0 & 0 & 0 & 0 & 2 & 1 \\ 0 & 0 & 0 & 0 & 0 & 2 \end{bmatrix} = \begin{bmatrix} \lambda_1 & 0 & 0 & 0 & 0 & 0 \\ 0 & \lambda_2 & 1 & 0 & 0 & 0 \\ 0 & 0 & \lambda_2 & 1 & 0 & 0 \\ 0 & 0 & 0 & \lambda_2 & 0 & 0 \\ 0 & 0 & 0 & 0 & \lambda_2 & 1 \\ 0 & 0 & 0 & 0 & 0 & \lambda_2 \end{bmatrix} \qquad (A.53)$$

and

$$P = \begin{bmatrix} w_1 & u_1 & u_2 & u_3 & v_1 & v_2 \end{bmatrix} = \begin{bmatrix} 0 & 2 & 1 & 0 & 0 & 0 \\ 0 & 2 & -1 & 0 & 0 & 0 \\ 0 & 0 & 0 & 1 & 2 & 1 \\ 0 & 0 & 0 & 0 & -2 & -1 \\ 1 & 0 & 0 & 0 & 0 & 1 \\ -1 & 0 & 0 & 0 & 0 & 1 \end{bmatrix}. \qquad (A.54)$$

The correctness of $P$ is easily checked by computing $PJ = AP$.

## A.7 Normed Linear Spaces

In the following discussion, we require for $(V, F)$ that $F$ be either the field of real numbers $R$ or the field of complex numbers $C$. For such linear spaces we say that a function $\| \cdot \| : V \to R^+$ is a *norm* if

(N-i)   $\| x \| \geq 0$ for every *vector* $x \in V$ and $\| x \| = 0$ if and only if $x$ is the null vector (i.e., $x = 0$);

(N-ii)   for every scalar $\alpha \in F$ and for every vector $x \in V$, $\| \alpha x \| = |\alpha| \| x \|$, where $|\alpha|$ denotes the absolute value of $\alpha$ when $F = R$ and the modulus when $F = C$; and

(N-iii)   for every $x$ and $y$ in $V$, $\| x + y \| \leq \| x \| + \| y \|$. (This inequality is called the *triangle inequality*.)

We call a vector space on which a norm has been defined a *normed vector space* or a *normed linear space*.

---

**Example A.31.** On the linear space $(R^n, R)$, we define for every $x = (x_1, \ldots, x_n)^T$ ,

$$\| x \|_p = (\sum_{i=1}^{n} |x_i|^p)^{1/p}, \quad 1 \leq p < \infty \qquad (A.55)$$

and

$$\| x \|_\infty = \max\{|x_i| : 1 \leq i \leq n\}. \qquad (A.56)$$

Using Minkowski's Inequality for finite sums, see (A.5), it is an easy matter to show that for every $p$, $1 \leq p \leq \infty$, $\| \cdot \|_p$ is a norm on $R^n$. In addition to $\| \cdot \|_\infty$, of particular interest to us will be the cases $p = 1$ and $p = 2$; i.e.,

$$\| x \|_1 = \sum_{i=1}^{n} |x_i| \qquad (A.57)$$

and

$$\| x \|_2 = \left( \sum_{i=1}^{n} |x_i|^2 \right)^{1/2}. \qquad (A.58)$$

The norm $\| \cdot \|_1$ is sometimes referred to as the *taxicab norm* or *Manhattan norm*, whereas, $\| \cdot \|_2$ is called the *Euclidean norm*. The linear space $(R^n, R)$ with norm $\| \cdot \|_2$ is called a *Euclidean vector space*.

The foregoing norms are related by the inequalities

$$\| x \|_\infty \leq \| x \|_1 \leq n \| x \|_\infty, \qquad (A.59)$$

$$\| x \|_\infty \leq \| x \|_2 \leq \sqrt{n} \| x \|_\infty, \qquad (A.60)$$

$$\| x \|_2 \leq \| x \|_1 \leq \sqrt{n} \| x \|_2. \qquad (A.61)$$

Also, for $p = 2$, we obtain from the Hölder Inequality for finite sums, (A.4), the *Schwarz Inequality*

$$|x^T y| = | \sum_{i=1}^{n} x_i y_i | \leq \left( \sum_{i=1}^{n} |x_i|^2 \right)^{1/2} \left( \sum_{i=1}^{n} |y_i|^2 \right)^{1/2} \qquad (A.62)$$

for all $x, y \in R^n$.

---

The assertions made in the above example turn out to be also true for the space $(C^n, C)$. We ask the reader to verify these relations.

---

**Example A.32.** On the space $l_p$ given in Example A.5, let

$$\| x \|_p = \left( \sum_{i=1}^{\infty} |x_i|^p \right)^{1/p}, \quad 1 \leq p < \infty,$$

and

$$\| x \|_\infty = \sup_i |x_i|.$$

Using Minkowski's Inequality for infinite sums, (A.5), it is an easy matter to show that $\| \cdot \|_p$ is a norm for every $p$, $1 \leq p \leq \infty$.

---

**Example A.33.** On the space given in Example A.7, let

$$\| x \|_p = \left( \int_a^b |x(t)|^p dt \right)^{1/p}, \quad 1 \leq p < \infty.$$

Using Minkowski's Inequality for integrals, (A.10), see Example A.7, it can readily be verified that $\| \cdot \|_p$ is a norm for every $p$, $1 \le p < \infty$. Also, on the space of continuous functions given in Example A.8, assume that (A.11) holds. Then

$$\| x \|_\infty = \max_{a \le t \le b} |x(t)|$$

is easily shown to determine a norm. Furthermore, expression (A.12) can also be used to determine a norm.

---

**Example A.34.** We can also define the norm of a matrix. To this end, consider the set of real $m \times n$ matrices, $R^{m \times n} = V$ and $F = R$. It is easily verified that $(V, F) = (R^{m \times n}, R)$ is a vector space, where vector addition is defined as matrix addition and multiplication of vectors by scalars is defined as multiplication of matrices by scalars.

For a given norm $\| \cdot \|_u$ on $R^n$ and a given norm $\| \cdot \|_v$ on $R^m$, we define $\| \cdot \|_{vu} : R^{m \times n} \to R^+$ by

$$\| A \|_{vu} = \sup\{\| Ax \|_v : x \in R^n \text{ with } \| x \|_u = 1\}. \qquad \text{(A.63)}$$

It is easily verified that

(M-i)   $\| Ax \|_v \le \| A \|_{vu} \| x \|_u$ for any $x \in R^n$,

(M-ii)   $\| A + B \|_{vu} \le \| A \|_{vu} + \| B \|_{vu}$,

(M-iii)   $\| \alpha A \|_{vu} = |\alpha| \| A \|_{vu}$ for all $\alpha \in R$,

(M-iv)   $\| A \|_{vu} \ge 0$ and $\| A \|_{vu} = 0$ if and only if $A$ is the zero matrix (i.e., $A = 0$),

(M-v)   $\| A \|_{vu} \le \sum_{i=1}^m \sum_{j=1}^n |a_{ij}|$ for any $p$-vector norms defined on $R^n$ and $R^m$.

Properties (M-ii) to (M-iv) clearly show that $\| \cdot \|_{vu}$ defines a norm on $R^{m \times n}$ and justifies the use of the term *matrix norm*. Since the matrix norm $\| \cdot \|_{vu}$ depends on the choice of the vector norms, $\| \cdot \|_u$, and $\| \cdot \|_v$, defined on $U \triangleq R^n$ and $V \triangleq R^m$, respectively, we say that the matrix norm $\| \cdot \|_{uv}$ is *induced* by the vector norms $\| \cdot \|_u$ and $\| \cdot \|_v$. In particular, if $\| \cdot \|_u = \| \cdot \|_p$ and $\| \cdot \|_v = \| \cdot \|_p$, then the notation $\| A \|_p$ is frequently used to denote the norm of $A$.

As a specific case, let $A = [a_{ij}] \in R^{m \times n}$. Then it is easily verified that

$$\| A \|_1 = \max_j \left( \sum_{i=1}^m |a_{ij}| \right),$$

$$\| A \|_2 = [\max \lambda(A^T A)]^{1/2},$$

where $\max \lambda(A^T A)$ denotes the largest eigenvalue of $A^T A$ and

$$\| A \|_{\infty} = \max_i \left( \sum_{j=1}^{n} |a_{ij}| \right).$$

When it is clear from context which vector spaces and vector norms are being used, the indicated subscripts on the matrix norms are usually not used. For example, if $A \in R^{m \times n}$ and $B \in R^{n \times k}$, it can be shown that

(M-vi) $\| AB \| \leq \| A \| \, \| B \|$.

In (M-vi) we have omitted subscripts on the matrix norms to indicate inducing vector norms.

---

We conclude by noting that it is possible to define norms on $(R^{m \times n}, R)$ that need not be induced by vector norms. Furthermore, the entire discussion given in Example A.34 holds also for norms defined on complex spaces, e.g., $(C^{m \times n}, C)$.

## A.8 Some Facts from Matrix Algebra

### Determinants

We recall that the determinant of a matrix $A = [a_{ij}] \in R^{n \times n}$, det $A$, can be evaluated by the relation

$$\det A = \sum_j a_{ij} d_{ij} \quad \text{for any} \quad i = 1, 2, \ldots, n,$$

where $d_{ij} = (-1)^{i+j} \det A_{ij}$ and $A_{ij}$ is the $(n-1) \times (n-1)$ matrix obtained by deleting the $i$th row and $j$th column of $A$. The term $d_{ij}$ is the *cofactor of $A$ corresponding to $a_{ij}$* and det $A_{ij}$ is the *$ij$th minor of the matrix*. The *principal minors* of $A$ are obtained by letting $i = j$, $i$, $j = 1, \ldots n$.

If any column (or row) of $A$ is multiplied by a scalar $k$, then the determinant of the new matrix is $k \det A$. If every entry is multiplied by $k$, then the determinant of the new matrix is $k^n \det A$. Also,

$$\det A^T = \det A \quad \text{where } A^T \text{ is the transpose of } A.$$

### Determinants of Products

$\det AB = \det A \det B$ when $A$ and $B$ are square matrices, and $\det[I_m - AB] = \det[I_n - BA]$ where $A \in R^{m \times n}$ and $B \in R^{n \times m}$.

### Determinants of Block Matrices

$$\det \begin{bmatrix} \overset{m \times m}{A} & \overset{m \times n}{B} \\ \underset{n \times m}{C} & \underset{n \times n}{D} \end{bmatrix} = \det A \det[D - CA^{-1}B], \quad \det A \neq 0$$

$$= \det D \det[A - BD^{-1}C], \quad \det D \neq 0.$$

## Inverse $A^{-1}$ of $A$

If $A \in R^{n \times n}$ and if $A$ is nonsingular (i.e., $\det A \neq 0$), then $AA^{-1} = A^{-1}A = I$.

$$A^{-1} = \frac{1}{\det A} \operatorname{adj}(A),$$

where $\operatorname{adj}(A) = [d_{ij}]^T$ is the *adjoint* of $A$, where $d_{ij}$ is the cofactor of $A$ corresponding to $a_{ij}$. When

$$A = \begin{bmatrix} a & b \\ c & d \end{bmatrix}$$

is a $2 \times 2$ matrix, then

$$A^{-1} = \frac{1}{ad - cb} \begin{bmatrix} d & -b \\ -c & a \end{bmatrix}.$$

If $A \in R^{m \times m}$ and $C \in R^{n \times n}$, if $A$ and $C$ are nonsingular, and if $B \in R^{m \times n}$ and $D \in R^{n \times m}$, then

$$(A + BCD)^{-1} = A^{-1} - A^{-1}B(DA^{-1}B + C^{-1})^{-1}DA^{-1}.$$

For example

$$\left[I + C(sI - A)^{-1}B\right]^{-1} = I - C(sI - A + BC)^{-1}B.$$

When $A \in R^{m \times n}$ and $B \in R^{n \times m}$, then

$$(I_m + AB)^{-1} = I_m - A(I_n + BA)^{-1}B.$$

*Sylvester Rank Inequality*

If $X \in R^{p \times n}$ and $Y \in R^{n \times m}$, then

$$\operatorname{rank} X + \operatorname{rank} Y - n \leq \operatorname{rank}(xy) \leq \min\{\operatorname{rank} X, \operatorname{rank} Y\}.$$

## A.9 Numerical Considerations

Computing the rank of the controllability matrix $[B, AB, \ldots, A^{n-1}B]$, the eigenvalues of $A$, or the zeros of the system $\{A, B, C, D\}$ typically requires the use of a digital computer. When this is the case, one must deal with the selection of an algorithm and interpret numerical results. In doing so, two issues arise that play important roles in numerical computations using a computer, namely, the *numerical stability or instability* of the computational method used, and how *well or ill conditioned* the problem is numerically.

An example of a problem that can be ill conditioned is the problem of calculating the roots of a polynomial, given its coefficients. This is so because for certain polynomials, small variations in the values of the coefficients, introduced say via round-off errors, can lead to great changes in the roots of the polynomial. That is to say, the roots of a polynomial can be very sensitive to changes in its coefficients. Note that ill conditioning is a property of the problem to be solved and does not depend on the floating-point system used in the computer, nor on the particular algorithm being implemented.

A computational method is numerically stable if it yields a solution that is near the true solution of a problem with slightly changed data. An example of a numerically unstable method to compute the roots of $ax^2 + 2bx + c = 0$ is the formula $(-b \pm \sqrt{(b^2 - ac)})/a$, which for certain parameters $a, b, c$ may give erroneous results in finite arithmetic. This instability is caused by the subtraction of two approximately equal large numbers in the numerator when $b^2 >> ac$. Note that the roots may be calculated in a numerically stable way, using the mathematically equivalent, but numerically very different, expression $c/(-b \mp \sqrt{(b^2 - ac)})$.

We would like of course, to always use numerically stable methods, and we would prefer to have well-conditioned problems. In what follows, we discuss briefly the problem of solving a set of algebraic equations given by $Ax = b$. We will show that a measure of how ill conditioned a given problem is, is the size of the *condition number* (to be defined) of the matrix $A$. There are many algorithms to numerically solve $Ax = b$, and we will briefly discuss numerically stable ones. Singular values, singular value decomposition, and the least-squares problem are also discussed.

### A.9.1 Solving Linear Algebraic Equations

Consider the set of linear algebraic equations given by

$$Ax = b, \tag{A.64}$$

where $A \in R^{m \times n}$, $b \in R^m$ and $x \in R^n$ is to be determined.

### Existence and Uniqueness of Solutions

See also Sec. A.3, (A.22)–(A.29). Given (A.64), for a given $b$, a solution $x$ exists if and only if $b \in \mathcal{R}(A)$, or equivalently, if and only if

$$\rho([A, b]) = \rho(A). \tag{A.65}$$

Every solution of (A.64) can be expressed as a sum

$$x = x_p + x_h, \tag{A.66}$$

where $x_p$ is a specific solution and $x_h$ satisfies $Ax_h = 0$. There are

$$\dim \mathcal{N}(A) = n - \rho(A) \tag{A.67}$$

linearly independent solutions of the systems of equations $Ax = 0$.
$Ax = b$ has a unique solution if and only if (A.65) is satisfied and

$$\rho(A) = n \leq m. \tag{A.68}$$

A solution exists for any $b$ if and only if $\rho(A) = m$. In this case, a solution may be found using

$$x = A^T(AA^T)^{-1}b.$$

When $\rho(A) = m = n$, then $A$ is nonsingular and the unique solution is

$$x = A^{-1}b. \tag{A.69}$$

It is of interest to know the effects of small variations of $A$ and $b$ to the solution $x$ of this system of equations. Note that such variations may be introduced, for example, by rounding errors when calculating a solution or by noisy data.

## Condition Number

Let $A \in R^{n \times n}$ be nonsingular. If $A$ is known exactly and $b$ has some uncertainty $\Delta b$ associated with it, then $A(x + \Delta x) = b + \Delta b$. It can then be shown that the variation in the solution $x$ is bounded by

$$\frac{\| \Delta x \|}{\| x \|} \leq \text{cond}(A)\frac{\| \Delta b \|}{\| b \|}, \tag{A.70}$$

where $\| \cdot \|$ denotes any vector norm (and consistent matrix norm) and $\text{cond}(A)$ denotes the *condition number of A*, where $\text{cond}(A) \triangleq \| A \| \| A^{-1} \|$. Note that

$$\text{cond}(A) = \sigma_{\max}(A)/\sigma_{\min}(A), \tag{A.71}$$

where $\sigma_{\max}(A)$ and $\sigma_{\min}(A)$ are the maximum and minimum singular values of $A$, respectively (see the next section). From the property of matrix norms, $\| AA^{-1} \| \leq \| A \| \| A^{-1} \|$, it follows that $\text{cond}(A) \geq 1$. This also follows from the expression involving singular values. If $\text{cond}(A)$ is small, then $A$ is said to be *well conditioned* with respect to the problem of solving linear equations. If $\text{cond}(A)$ is large, then $A$ is *ill conditioned* with respect to the problem of solving linear equations. In this case the relative uncertainty in the solution $(\| \Delta x \| / \| x \|)$ can be many times the relative uncertainty in $b$ $(\| \Delta b \| / \| b \|)$. This is of course undesirable. Similar results can be derived when variations in both $b$ and $A$ are considered, i.e., when $b$ and $A$ become $b + \Delta b$ and $A + \Delta A$. Note that the conditioning of $A$, and of the given problem, is independent of the algorithm used to determine a solution.

The condition number of $A$ provides a measure of the distance of $A$ to the set of singular (reduced rank) matrices. In particular, if $\| \Delta A \|$ is the norm of the smallest perturbation $\Delta A$ such that $A + \Delta A$ is singular, and is denoted by

$d(A)$, then $dA/\parallel A \parallel= 1/\operatorname{cond}(A)$. Thus, a large condition number indicates a short distance to a singularity and it is not surprising that this implies great sensitivity of the numerical solution $x$ of $Ax = b$ to variations in the problem data.

The condition number of $A$ plays a similar role in the case when $A$ is not square. It can be determined in terms of the singular values of $A$ defined in the next subsection.

## Computational Methods

The system of equations $Ax = b$ is easily solved if $A$ has some special form (e.g., if it is diagonal or triangular). Using the method of *Gaussian elimination*, any nonsingular matrix $A$ can be reduced to an upper triangular matrix $U$. These operations can be represented by premultiplication of $A$ by a sequence of lower triangular matrices. It can then be shown that $A$ can be represented as

$$A = LU, \tag{A.72}$$

where $L$ is a lower triangular matrix with all diagonal elements equal to 1 and $U$ is an upper triangular matrix. The solution of $Ax = b$ is then reduced to the solution of two systems of equations with triangular matrices, $Ly = b$ and $Ux = y$. This method of solving $Ax = b$ is based on the decomposition (A.72) of $A$, which is called the *LU decomposition* of $A$.

If $A$ is a symmetric positive definite matrix, then it may be represented as

$$A = U^T U, \tag{A.73}$$

where $U$ is an upper triangular matrix. This is known as the *Cholesky decomposition* of a positive definite matrix. It can be obtained using a variant of Gaussian elimination. Note that this method requires half of the operations necessary for Gaussian elimination on an arbitrary nonsingular matrix $A$, since $A$ is symmetric.

Now consider the system of equations $Ax = b$, where $A \in R^{m \times n}$, and let rank $A = n(\leq m)$. Then

$$A = Q \begin{bmatrix} R \\ O \end{bmatrix} = [Q_1, Q_2] \begin{bmatrix} R \\ O \end{bmatrix} = Q_1 R, \tag{A.74}$$

where $Q$ is an *orthogonal matrix* $(Q^T = Q^{-1})$ and $R \in R^{n \times n}$ is an upper triangular matrix of full rank $n$. Expression (A.74) is called the *QR decomposition* of $A$. When rank $A = r$, the $QR$ decomposition of $A$ is expressed as

$$AP = Q \begin{bmatrix} R_1 & R_2 \\ 0 & 0 \end{bmatrix}, \tag{A.75}$$

where $Q$ is orthogonal, $R_1 \in R^{r \times r}$ is nonsingular and upper triangular, and $P$ is a permutation matrix that represents the moving of the columns during the reduction (in $Q^T AP$).

The $QR$ decomposition can be used to determine solutions of $Ax = b$. In particular, consider $A \in R^{m \times n}$ with rank $A = n (\leq m)$ and assume that a solution exists. First, determine the $QR$ decomposition of $A$ given in (A.74). Then $Q^T A x = Q^T b$ or $\begin{bmatrix} R \\ 0 \end{bmatrix} x = Q^T b$ (since $Q^T = Q^{-1}$) or $Rx = c$. Solve this system of equations, where $R$ is triangular and $c = [I_n, 0] Q^T b$. In the general case when rank$(A) = r \leq \min(n, m)$, determine the $QR$ decomposition of $A$ (2.7) and assume that a solution exists. The solutions are given by $x = P \begin{bmatrix} R_1^{-1}(c - R_2 y) \\ y \end{bmatrix}$, $c = [I_r, 0] Q^T b$, where $y \in R^{m-r}$ is arbitrary.

A related problem is the *linear least-squares problem* where a solution $x$ of the system of equations $Ax = b$ is to be found that minimizes $\| b - Ax \|_2$. This is a more general problem than simply solving $Ax = b$, since solving it provides the "best" solution in the above sense, even when an exact solution does not exist. The least-squares problem is discussed further in a subsequent subsection.

### A.9.2 Singular Values and Singular Value Decomposition

The singular values of a matrix and the *Singular Value Decomposition Theorem* play a significant role in a number of problems of interest in the area of systems and control, from the computation of solutions of linear systems of equations, to computations of the norm of transfer matrices at specified frequencies, to model reduction, and so forth. In what follows, we provide a brief description of some basic results and we introduce some terminology.

Consider $A \in C^{n \times n}$, and let $A^* = \bar{A}^T$; i.e., the complex conjugate transpose of $A$. $A \in C^{n \times n}$ is said to be *Hermitian* if $A^* = A$. If $A \in R^{n \times n}$, then $A^* = A^T$ and if $A = A^T$, then $A$ is *symmetric*. $A \in C^{n \times n}$ is *unitary* if $A^* = A^{-1}$. In this case $A^* A = AA^* = I_n$. If $A \in R^{n \times n}$, then $A^* = A^T$ and if $A^T = A^{-1}$, i.e., if $A^T A = AA^T = I_n$, then $A$ is *orthogonal*.

### Singular Values

Let $A \in C^{m \times n}$, and consider $AA^* \in C^{m \times m}$. Let $\lambda_i$, $i = 1, \dots, m$ denote the eigenvalues of $AA^*$, and note that these are all real and nonnegative numbers. Assume that $\lambda_1 \geq \lambda_2 \geq \cdots \lambda_r \geq \cdots \geq \lambda_m$. Note that if $r = \text{rank} A = \text{rank}(AA^*)$, then $\lambda_1 \geq \lambda_2 \geq \cdots \geq \lambda_r > 0$ and $\lambda_{r+1} = \cdots = \lambda_m = 0$. The *singular values* $\sigma_i$ *of* $A$ are the positive square roots of $\lambda_i$, $i = 1, \dots, \min(m, n)$. In fact, the nonzero singular values of $A$ are

$$\sigma_i = (\lambda_i)^{1/2}, \quad i = 1, \dots, r, \tag{A.76}$$

where $r = \text{rank} A$, whereas the remaining $(\min(m, n) - r)$ of the singular values are zero. Note that $\sigma_1 \geq \sigma_2 \geq \cdots \geq \sigma_r > 0$, and $\sigma_{r+1} = \sigma_{r+2} = \cdots = \sigma_{\min(m,n)} = 0$. The singular values could also have been found as the square roots of the eigenvalues of $A^*A \in C^{n \times n}$ (instead of $AA^* \in C^{m \times m}$). To see this, consider the following result.

**Lemma A.35.** *Let $m \geq n$. Then*

$$|\lambda I_m - AA^*| = \lambda^{m-n}|\lambda I_n - A^*A|; \tag{A.77}$$

*i.e., all eigenvalues of $A^*A$ are eigenvalues of $AA^*$ that also has $m - n$ additional eigenvalues at zero. Thus $AA^* \in C^{m \times m}$ and $A^*A \in C^{n \times n}$ have precisely the same $r$ nonzero eigenvalues ($r = \text{rank } A$); their remaining eigenvalues, $(m - r)$ for $AA^*$ and $(n - r)$ for $A^*A$, are all at zero. Therefore, either $AA^*$ or $A^*A$ can be used to determine the $r$ nonzero singular values of $A$. All remaining singular values are zero.*

*Proof.* The proof is based on Schur's formula for determinants. In particular, we have

$$\begin{aligned}
D(\lambda) &= \begin{vmatrix} \lambda^{1/2}I_m & A \\ A^* & \lambda^{1/2}I_n \end{vmatrix} = |\lambda^{1/2}I_m| \, |\lambda^{1/2}I_n - A^*\lambda^{-1/2}I_m A| \\
&= |\lambda^{1/2}I_m| \, |\lambda^{-1/2}I_n| \, |\lambda I_n - A^*A| \\
&= \lambda^{\frac{m-n}{2}} \cdot |\lambda I_n - A^*A|,
\end{aligned} \tag{A.78}$$

where Schur's formula was applied to the $(1,1)$ block of the matrix. If it is applied to the $(2,2)$ block, then

$$D(\lambda) = \lambda^{\frac{n-m}{2}} \cdot |\lambda I_m - AA^*|. \tag{A.79}$$

Equating (A.78) and (A.79) we obtain $|\lambda I_m - AA^*| = \lambda^{m-n}|\lambda I_n - A^*A|$, which is (A.78). ∎

---

**Example A.36.** $A = \begin{bmatrix} 2 & 1 & 0 \\ 0 & 0 & 0 \end{bmatrix} \in R^{2 \times 3}$. Here $\text{rank } A = r = 1$, $\lambda_i(AA^*) =$

$$\lambda_i \left( \begin{bmatrix} 2 & 1 & 0 \\ 0 & 0 & 0 \end{bmatrix} \begin{bmatrix} 2 & 0 \\ 1 & 0 \\ 0 & 0 \end{bmatrix} \right) = \lambda_i \left( \begin{bmatrix} 5 & 0 \\ 0 & 0 \end{bmatrix} \right) = \{5, 0\} \text{ and } \lambda_1 = 5, \lambda_2 = 0. \text{ Also,}$$

$$\lambda_i(A^*A) = \lambda_i \left( \begin{bmatrix} 2 & 0 \\ 1 & 0 \\ 0 & 0 \end{bmatrix} \begin{bmatrix} 2 & 1 & 0 \\ 0 & 0 & 0 \end{bmatrix} \right) = \lambda_i \left( \begin{bmatrix} 4 & 2 & 0 \\ 2 & 1 & 0 \\ 0 & 0 & 0 \end{bmatrix} \right), \text{ and } \lambda_1 = 5, \lambda_2 = 0, \text{ and}$$

$\lambda_3 = 0$. The only nonzero singular value is $\sigma_1 = \sqrt{\lambda_1} = +\sqrt{5}$. The remaining singular values are zero.

---

There is an important relation between the singular values of $A$ and its induced Hilbert or 2-norm, also called the spectral norm $\| A \|_2 = \| A \|_s$. In particular,

$$\| A \|_2 (=\| A \|_s) = \sup_{\|x\|_2=1} \| Ax \|_2 = \max_i \{(\lambda_i(A^*A))^{1/2}\} = \bar{\sigma}(A), \quad (A.80)$$

where $\bar{\sigma}(A)$ denotes the largest singular value of $A$. Using the inequalities that are axiomatically true for induced norms, it is possible to establish relations between singular values of various matrices that are useful in MIMO control design. The significance of the singular values of a gain matrix $A(jw)$ is discussed later in this section.

There is an interesting relation between the eigenvalues and the singular values of a (square) matrix. Let $\lambda_i$, $i = 1, \ldots, n$ denote the eigenvalues of $A \in R^{n \times n}$, let $\underline{\lambda}(A) = \min_i |\lambda_i|$, and let $\overline{\lambda}(A) = \max_i |\lambda_i|$. Then

$$\underline{\sigma}(A) \leq \underline{\lambda}(A) \leq \overline{\lambda}(A) \leq \bar{\sigma}(A). \quad (A.81)$$

Note that the ratio $\bar{\sigma}(A)/\underline{\sigma}(A)$, i.e., the ratio of the largest and smallest singular values of $A$, is called the *condition number of $A$*, and is denoted by cond($A$). This is a very useful measure of how well conditioned a system of linear algebraic equations $Ax = b$ is (refer to the discussion of the previous section). The singular values provide a reliable way of determining how far a square matrix is from being singular, or a nonsquare matrix is from losing rank. This is accomplished by examining how close to zero $\underline{\sigma}(A)$ is. In contrast, the eigenvalues of a square matrix are not a good indicator of how far the matrix is from being singular, and a typical example in the literature to illustrate this point is an $n \times n$ lower triangular matrix $A$ with $-1$'s on the diagonal and $+1$'s everywhere else. In this case, $\underline{\sigma}(A)$ behaves as $1/2^n$ and the matrix is nearly singular for large $n$, whereas all of its eigenvalues are at $-1$. In fact, it can be shown that by adding $1/2^{n-1}$ to every element in the first column of $A$ results in an exactly singular matrix (try it for $n = 2$).

## Singular Value Decomposition

Let $A \in C^{m \times n}$ with rank $A = r \leq \min(m, n)$. Let $A^* = \bar{A}^T$, the complex conjugate transpose of $A$.

**Theorem A.37.** *There exist unitary matrices $U \in C^{m \times n}$ and $V \in C^{n \times n}$ such that*

$$A = U \Sigma V^*, \quad (A.82)$$

*where $\Sigma = \begin{bmatrix} \Sigma_r & 0_{r \times (n-r)} \\ 0_{(m-r) \times r} & 0_{(m-r) \times (n-r)} \end{bmatrix}$ with $\Sigma_r = \mathrm{diag}(\sigma_1, \sigma_2, \ldots, \sigma_r) \in R^{r \times r}$ selected so that $\sigma_1 \geq \sigma_2 \geq \cdots \geq \sigma_r > 0$.*

*Proof.* For the proof, see for example, Golub and Van Loan [7], and Patel et al. [11]. ∎

Let $U = [U_1, U_2]$ with $U_1 \in C^{m \times r}, U_2 \in C^{m \times (m-r)}$ and $V = [V_1, V_2]$ with $V_1 \in C^{n \times r}, V_2 \in C^{n \times (n-r)}$. Then

$$A = U\Sigma V^* = U_1\Sigma_r V_1^*. \tag{A.83}$$

Since $U$ and $V$ are unitary, we have

$$U^*U = \begin{bmatrix} U_1^* \\ U_2^* \end{bmatrix} [U_1,\, U_2] = I_m, U_1^*U_1 = I_r \tag{A.84}$$

and

$$V^*V = \begin{bmatrix} V_1^* \\ V_2^* \end{bmatrix} [V_1,\, V_2] = I_n, V_1^*V_1 = I_r. \tag{A.85}$$

Note that the columns of $U_1$ and $V_1$ determine orthonormal bases for $\mathcal{R}(A)$ and $\mathcal{R}(A^*)$, respectively. Now

$$AA^* = (U_1\Sigma_r V_1^*)(V_1\Sigma_r U_1^*) = U_1\Sigma_r^2 U_1^*, \tag{A.86}$$

from which we have

$$AA^*U_1 = U_1\Sigma_r^2 U_1^*U_1 = U_1\Sigma_r^2. \tag{A.87}$$

If $u_i$, $i = 1,\ldots,r$, is the $i$th column of $U_1$, i.e., $U_1 = [u_1, u_2, \ldots, u_r]$, then

$$AA^*u_i = \sigma_i^2 u_i, \quad i = 1,\ldots,r. \tag{A.88}$$

This shows that the $\sigma_i^2$ are the $r$ nonzero eigenvalues of $AA^*$; i.e., $\sigma_i$, $i = 1,\ldots,r$, are the nonzero singular values of $A$. Furthermore, $u_i$, $i = 1,\ldots,r$, are the eigenvectors of $AA^*$ corresponding to $\sigma_i^2$. They are the *left singular vectors of $A$*. Note that the $u_i$ are orthonormal vectors (in view of $U_1^*U_1 = I_r$). Similarly,

$$A^*A = (V_1\Sigma_r U_1^*)(U_1\Sigma_r V_1^*) = V_1\Sigma_r^2 V_1^*, \tag{A.89}$$

from which we obtain

$$A^*AV_1 = V_1\Sigma_r^2 V_1^*V_1 = V_1\Sigma_r^2. \tag{A.90}$$

If $v_i$, $i = 1,\ldots,r$, is the $i$th column of $V_1$, i.e., $V_1 = [v_1, v_2, \ldots, v_r]$, then

$$A^*Av_i = \sigma_i^2 v_i, \quad i = 1, 2, \ldots, r. \tag{A.91}$$

The vectors $v_i$ are the eigenvectors of $A^*A$ corresponding to the eigenvalues $\sigma_i^2$. They are the *right singular vectors* of $A$. Note that the $v_i$ are orthonormal vectors (in view of $V_1^*V_1 = I_r$).

The singular values are unique, whereas the singular vectors are *not*. To see this, consider

$$\widehat{V}_1 = V_1 \operatorname{diag}(e^{j\theta_i}) \quad \text{and} \quad \widehat{U}_1 = U_1 \operatorname{diag}(e^{-j\theta_i}).$$

Their columns are also singular vectors of $A$.

Note also that $A = U_1 \Sigma_r V_1^*$ implies that

$$A = \sum_{i=1}^{r} \sigma_i u_i v_i^*. \tag{A.92}$$

The significance of the singular values of a gain matrix $A(jw)$ is now briefly discussed. This is useful in the control theory of MIMO systems. Consider the relation between signals $y$ and $v$, given by $y = Av$. Then

$$\max_{\|v\|_2 \neq 0} \frac{\| y \|_2}{\| v \|_2} = \max_{\|v\|_2 \neq 0} \frac{\| Av \|_2}{\| v \|_2} = \bar{\sigma}(A)$$

or

$$\max_{\|v\|_2 = 1} \| y \|_2 = \max_{\|v\|_2 = 1} \| Av \|_2 = \bar{\sigma}(A). \tag{A.93}$$

Thus, $\bar{\sigma}(A)$ yields the maximum amplification, in energy terms (2-norm), when the transformation $A$ operates on a signal $v$. Similarly,

$$\min_{\|v\|_2 = 1} \| y \|_2 = \min_{\|v\|_2 = 1} \| Av \|_2 = \underline{\sigma}(A). \tag{A.94}$$

Therefore,

$$\underline{\sigma}(A) \leq \frac{\| Av \|_2}{\| v \|_2} \leq \bar{\sigma}(A), \tag{A.95}$$

where $\| v \|_2 \neq 0$. Thus the gain (energy amplification) is bounded from above and below by $\bar{\sigma}(A)$ and $\underline{\sigma}(A)$, respectively. The exact value depends on the direction of $v$.

To determine the particular directions of vectors $v$ for which these (max and min) gains are achieved, consider (A.92) and write

$$y = Av = \sum_{i=1}^{r} \sigma_i u_i v_i^* v. \tag{A.96}$$

Notice that $|v_i^* v| \leq \|v_i\|\|v\| = \|v\|$, since $\|v_i\| = 1$, with equality holding only when $v = \alpha v_i$, $\alpha \in C$. Therefore, to maximize, consider $v$ along the singular value directions $v_i$ and let $v = \alpha v_i$ with $|\alpha| = 1$ so that $\|v\| = 1$. Then in view of $v_i^* v_j = 0$, $i \neq j$ and $v_i^* v_j = 1$, $i = j$, we have that $y = Av = \alpha Av_i = \alpha \sigma_i u_i$ and $\|y\|_2 = \|Av\|_2 = \sigma_i$, since $\|u_i\|_2 = 1$. Thus, the maximum possible gain is $\sigma_1$; i.e., $\max_{\|v\|_2 = 1} \|y\|_2 = \max_{\|v\|_2 = 1} \|Av\|_2 = \sigma_1 (= \bar{\sigma}(A))$, as was shown above. This maximum gain occurs when $v$ is along the right singular vector $v_1$. Then $Av = Av_1 = \sigma_1 u_1 = y$ in view of (A.92); i.e., the projection is along the left singular vector $u_1$, also of the same singular value $\sigma_1$. Similarly, for the minimum gain, we have $\sigma_r = \underline{\sigma}(A) = \min_{\|v\|_2 = 1} \|y\|_2 = \min_{\|v\|_2 = 1} \|Av\|_2$; in which case, $Av = Av_r = \sigma_r u_r = y$.

Additional interesting properties include

$$\mathcal{R}(A) = \mathcal{R}(U_1) = \text{span}\{u_1, \ldots, u_r\}, \tag{A.97}$$

$$\mathcal{N}(A) = \mathcal{R}(V_2) = \text{span}\{v_{r+1}, \ldots, v_n\}, \tag{A.98}$$

where $U = [u_1, \ldots, u_r, u_{r+1}, \ldots, u_m] = [U_1, U_2]$, $V = [v_1, \ldots, v_r, v_{r+1}, \ldots, v_n]$ $= [V_1, V_2]$.

### A.9.3 Least-Squares Problem

Consider now *the least-squares problem* where a solution $x$ to the system of linear equations $Ax = b$ is to be determined that minimizes $\| b - Ax \|_2$. Write
$$\min_x \| b - Ax \|_2^2 = \min_x (b - Ax)^T (b - Ax) = \min_x (x^T A^T Ax - 2b^T Ax + b^T b).$$
Then $\nabla_x(x^T A^T Ax - 2b^T Ax + b^T b) = 2A^T Ax - 2A^T b = 0$ implies that the $x$, which minimizes $\| b - Ax \|_2$, is a solution of

$$A^T Ax = A^T b. \tag{A.99}$$

Rewrite this as $V_1 \Sigma_r^2 V_1^T x = (U_1 \Sigma_r V_1^T)^T b = V_1 \Sigma_r U_1^T b$ in view of (A.89) and (A.83). Now $x = V_1 \Sigma_r^{-1} U_1^T b$ is a solution. To see this, substitute and note that $V_1^T V_1 = I_r$. In view of the fact that $\mathcal{N}(A^T A) = \mathcal{N}(A) = \mathcal{R}(V_2) = \text{span}\{v_{r+1}, \ldots, v_n\}$, the complete solution is given by

$$x_w = V_1 \Sigma_r^{-1} U_1^T b + V_2 w \tag{A.100}$$

for some $w \in R^{m-r}$. Since $V_1 \Sigma_r^{-1} U_1^T b$ is orthogonal to $V_2 w$ for all $w$,

$$x_0 = V_1 \Sigma_r^{-1} U_1^T b \tag{A.101}$$

is the optimal solution that minimizes $\| b - Ax \|_2$.

The *Moore–Penrose pseudo-inverse* of $A \in R^{m \times n}$ can be shown to be

$$A^+ = V_1 \Sigma_r^{-1} U_1^T. \tag{A.102}$$

We have seen that $x = A^+ b$ is the solution to the least-squares problem. It can be shown that this pseudo-inverse minimizes $\| AA^+ - I_m \|_F$, where $\| A \|_F$ denotes the *Frobenius norm* of $A$, which is equal to the square root of $\text{trace}[AA^T] = \sum_{i=1}^m \lambda_i(AA^T) = \sum_{i=1}^m \sigma_i^2(A)$. It is of interest to note that the Moore–Penrose pseudo-inverse of $A$ is defined as the unique matrix that satisfies the conditions (i) $AA^+ A = A$, (ii) $A^+ AA^+ = A^+$, (iii) $(AA^+)^T = AA^+$, and (iv) $(A^+ A)^T = A^+ A$.

Note that if $\text{rank}\, A = m \leq n$ then it can be shown that $A^+ = A^T(AA^T)^{-1}$; this is, in fact, the *right inverse of* $A$, since $A(A^T(AA^T)^{-1}) = I_m$. Similarly, if $\text{rank}\, A = n \leq m$, then $A^+ = (A^T A)^{-1} A^T$, the *left inverse of* $A$, since $((A^T A)^{-1} A^T)A = I_n$.

Singular values and singular value decomposition are discussed in a number of references. See for example, Golub and Van Loan [7], Patel et al. [11], Petkov et al. [12], and DeCarlo [5].

# A.10 Notes

Standard references on linear algebra and matrix theory include Birkhoff and MacLane [4], Halmos [8], and Gantmacher [6]. Our presentation in this appendix follows Michel and Herget [9]. Conditioning and numerical stability of a problem are key issues in the area of numerical analysis. Our aim in Section A.9 was to make the reader aware that depending on the problem, the numerical considerations in the calculation of a solution may be nontrivial. These issues are discussed at length in many textbooks on numerical analysis. Examples of good books in this area include Golub and Van Loan [7] and Stewart [13] where matrix computations are emphasized. Also, see Petkov et al. [12] and Patel et al. [11] for computational methods with emphasis on system and control problems. For background on the theory of algorithms, optimization algorithms, and their numerical properties, see Bazaraa et al. [2] and Bertsekas and Tsitsiklis [3].

# References

1. P.J. Antsaklis and A.N. Michel, *Linear Systems*, Birkhäuser, Boston, MA, 2006.
2. M.S. Bazaraa, H.D. Sherali, and C.M. Shetty, *Nonlinear Programming Theory and Algorithms*, 2nd edition, Wiley, New York, NY, 1993.
3. D.P. Bertsekas and J.N. Tsitsiklis, *Parallel and Distributed Computation–Numerical Methods*, Prentice Hall, Englewood Cliffs, NJ, 1989.
4. G. Birkhoff and S. MacLane, *A Survey of Modern Algebra*, Macmillan, New York, NY, 1965.
5. R.A. DeCarlo, *Linear Systems*, Prentice Hall, Englewood Cliffs, NJ, 1989.
6. F.R. Gantmacher, *Theory of Matrices*, Vols. I and II, Chelsea Publications, New York, NY, 1959.
7. G.H. Golub and C.F. Van Loan, *Matrix Computations*, The Johns Hopkins University Press, Baltimore, MD, 1983.
8. P.R. Halmos, *Finite Dimensional Vector Spaces*, D. Van Nostrand, Princeton, NJ, 1958.
9. A.N. Michel and C.J. Herget, *Applied Algebra and Functional Analysis*, Dover, New York, NY, 1993.
10. R.K. Miller and A.N. Michel, *Ordinary Differential Equations*, Academic Press, New York, NY, 1982.
11. R.V. Patel, A.J. Laub, and P.M. VanDooren, Eds., *Numerical Linear Algebra Techniques for Systems and Control*, IEEE Press, Piscataway, NJ, 1993.
12. P.Hr. Petkov, N.D. Christov, and M.M. Konstantinov, *Computational Methods for Linear Control Systems*, Prentice Hall-International Series, Englewood Cliffs, NJ, 1991.
13. G.W. Stewart, *Introduction to Matrix Computations*, Academic Press, New York, NY, 1973.

# Solutions to Selected Exercises

## Exercises of Chapter 1

**1.10** $\Delta\dot{x} = (k_1 k_2/L)\,\Delta x + k_2\Delta u$, $\Delta y = (k_1/L)\,\Delta x$ where $L = 2\sqrt{k} - k_1 k_2 t$.

**1.12** (a) $\begin{bmatrix} \Delta\dot{x}_1 \\ \Delta\dot{x}_2 \end{bmatrix} = \begin{bmatrix} 0 & 1 \\ 1 & 1 \end{bmatrix}\begin{bmatrix} \Delta x_1 \\ \Delta x_2 \end{bmatrix}$.

(b) $x_1 = x, x_2 = \dot{x}$

$$\begin{bmatrix} \Delta\dot{x}_1 \\ \Delta\dot{x}_2 \end{bmatrix} = \begin{bmatrix} 0 & 1 \\ 0 & -3 \end{bmatrix}\begin{bmatrix} \Delta x_1 \\ \Delta x_2 \end{bmatrix} + \begin{bmatrix} 0 \\ -1 \end{bmatrix}\Delta u.$$

**1.13** $x_1 = \phi, x_2 = \dot{\phi}, x_3 = s, x_4 = \dot{s}$

$$\begin{bmatrix} \Delta\dot{x}_1 \\ \Delta\dot{x}_2 \\ \Delta\dot{x}_3 \\ \Delta\dot{x}_4 \end{bmatrix} = \begin{bmatrix} 0 & 1 & 0 & 0 \\ \frac{g}{L'} & 0 & 0 & \frac{F}{L'M} \\ 0 & 0 & 0 & 1 \\ 0 & 0 & 0 & -\frac{F}{M} \end{bmatrix}\begin{bmatrix} \Delta x_1 \\ \Delta x_2 \\ \Delta x_3 \\ \Delta x_4 \end{bmatrix} + \begin{bmatrix} 0 \\ -\frac{1}{L'M} \\ 0 \\ \frac{1}{M} \end{bmatrix}\Delta\mu.$$

## Exercises of Chapter 2

**2.4** (b) Not linear; time-invariant; causal.

**2.5** Causal; linear; not time-invariant.

**2.6** Noncausal; linear; time-invariant.

**2.7** Not linear.

**2.8** Noncausal; time-invariant.

**2.9** Noncausal; nonlinear (affine).

**2.10** $y(n) = \sum_{k=-\infty}^{\infty} u(l)[s(n,l) - s(n, l-1)]$, where $s(n,l) = \sum_{l=-\infty}^{\infty} h(n,k)$ $p(k-l)$ is the unit step response of the system.

## Exercises of Chapter 3

**3.1** (a) $(\alpha_1, \alpha_2, \alpha_3) = (1, 0, 5)$. (b) $(\bar{\alpha}_1, \bar{\alpha}_2, \bar{\alpha}_3) = (s, 2, s + 1/s)$.

**3.2** $a = [1, 0, -2]^T$, $\bar{a} = [0, 1/2, 1/2]$.

**3.3** A basis is $\{(1, \alpha)^T\}$, $\alpha \in R$.

**3.4** It is a vector space of dimension $n^2$. The set of nonsingular matrices is not a vector space since closure of matrix addition is violated.

**3.5** Dependent over the field of rational functions; independent over the field of reals.

**3.6** (a) Rank is 1 over complex numbers. (b) 2 over reals. (c) 2 over rational functions. (d) 1 over rational functions.

**3.8** Directly, from the series definition of $e^{At}$ or using system concepts.

**3.9** See also Subsection 6.4.1.

**3.11** $(\lambda_i^k, v^i)$ is an (eigenvalue, eigenvector) pair of $A^k$. Then $f(A)v^i = f(\lambda_i)v^i$.

**3.13** Substitute $x(t) = \Phi(t, t_0)z(t)$ into $\dot{x} = A(t)x + B(t)u$.

**3.14** Take derivatives of both sides of $\Phi(t, \tau)\Phi(\tau, t) = I$ with respect to $\tau$.

**3.19** Verify that $\Phi(t, 0) = e^{At}$ is the solution of $\Phi(t, 0) = A\Phi(t, 0)$, $\Phi(0, 0) = I$.

**3.21** Use Exercise 3.19. $x_1^2 + x_2^2 = 2 = (\sqrt{2})^2$, so trajectory is a circle.

**3.22** $x(0) = [1, 1]^T$ is colinear to the eigenvector of eigenvalue 1, and so $e^t$ is the only mode that appears in the solution.

**3.23** (a) Take $t = 0$ in the expression for $e^{At}$.

**3.25** (a) $x(0) = [-1, 1, 0]^T$.

**3.30** $(I - A)x(0) = Bu(0)$; $u(0) = 2$.

## Exercises of Chapter 4

**4.1** Set of equilibria is $\{(-4v, v, 5v)^T : v \in R\}$.

**4.2** Set of equilibria is $\{(\frac{1}{k\pi}, 0)^T : k \in \mathcal{N}\backslash\{0\}\} \cup \{(0, 0)^T\}$.

**4.3** $x = 0$ is uniformly asymptotically stable; $x = 1$ is unstable.

**4.5** $A > 0$.

**4.7** $x = 0$ is exponentially stable; $x = 1$ is unstable.

**4.9** Uniformly BIBO stable.

**4.10** (a) Set of equilibria $\{(\alpha, -\alpha)^T \in \mathcal{R}^2 : \alpha \geq 0\}$. (b) No equilibrium.

**4.12** $x = 0$ is stable.

**4.13** $x = 0$ is stable.

**4.14** $x = 0$ is not stable.

**4.15** $x = 0$ is not stable.

**4.18** (a) $x = 0$ is stable. (b) $x = 0$ is stable.

**4.21** $x = 0$ is unstable.

**4.22** Not BIBO stable. Theorem cannot be applied.

# Exercises of Chapter 5

**5.2** (a) Controllable from $u$, observable from $y$. (b) when $u_1 = 0$, controllable from $u_2$; when $u_2 = 0$, not controllable from $u_1$. (c) not observable from $y_1$; observable from $y_2$.

**5.3** (a) Use $u(t) = B^T e^{A^T(T-t)} W_r^{-1}(0, T)(x_1 - e^{AT} x_0)$.

**5.4** (a) It can be reached in two steps, with $u(0) = 3$, $u(1) = -1$.
(b) Any $x = (b, a, a)^T$ will be reachable. $a, b \in R$.
(c) $x = (0, 0, a)^T$ unobservable. $a \in R$.

**5.10** (a) $u(0) = -1$, $u(1) = 2$; (b) $y(1) = (1, 2)^T u(0)$.

**5.11** $x_0 = [1\ 0\ \alpha]^T$, $\alpha \in R$.

# Exercises of Chapter 6

**6.2** (b) $\lambda = 3$ uncontrollable (first pair); $\lambda = -1$ uncontrollable (second pair).

**6.3** Controllability indices are 1, 3.

**6.7** Use controller form or Sylvester's Rank Inequality.

**6.13** (a) It is controllable. (b) It is controllable from $f_1$ only. (c) It is observable.

# Exercises of Chapter 7

**7.1** Use the standard form for uncontrollable systems.

**7.5** $\lambda_1 = 1$ is both controllable and observable, $\lambda_2 = -\frac{1}{2}$ is uncontrollable but observable, and $\lambda_3 = -\frac{1}{2}$ is controllable but unobservable.

**7.6** (a) $\lambda = 2$ is uncontrollable and unobservable. (b) $H(s) = \frac{1}{s+1}\begin{bmatrix} 1 & 1 \\ 1 & 0 \end{bmatrix}$.

(c) It is not asymptotically stable, but it is BIBO stable.

**7.7** (a) $p_H(s) = s^2(s+2)$, $m_H(s) = s(s+2)$. (b) $z_H(s) = 1$.

**7.8** (a) $p_H(s) = s^3 = m_H(s)$. (b) $z_H(s) = 1$.

**7.10** (a) They are rc. (b) They are not lc; a glcd is $\begin{bmatrix} s & 0 \\ 0 & 1 \end{bmatrix}$.

**7.13** (a) It is uncontrollable and unobservable.

(b) $H(s) = \frac{2}{s+1}\begin{bmatrix} -1 & 2 \\ 0 & 0 \end{bmatrix}$.

## Exercises of Chapter 8

**8.4** $p_H = s^2 - 1$; McMillan degree is 2.

**8.6** (a) $p_H(s) = s(s+1)(s+3)$; McMillan degree is 3.

(b) $A = \begin{bmatrix} 0 & 0 & 0 \\ 0 & -1 & 0 \\ 0 & 0 & -3 \end{bmatrix}$, $B = \begin{bmatrix} 1 & 0 \\ 0 & 1 \\ 1 & 0 \end{bmatrix}$, $C = \begin{bmatrix} 1 & 2 & 0 \\ 0 & -1 & 1 \end{bmatrix}$, $D = \begin{bmatrix} 0 & 1 \\ 0 & 1 \end{bmatrix}$.

**8.10** (a) $p_H(s) = s^2(s+1)^2$ so the order of any minimal realization is 4.
(b) Take $u_1 = 0$, and find the McMillan degree, which is 2. So in a fourth order realization, system will not be controllable from $u_2$ only. System will be observable from $y_1$.

**8.12** $p_H(s) = s^3$, and so 3 is the order of any minimal realization. A minimal

realization is $A = \begin{bmatrix} 0 & 1 & 0 \\ 0 & 0 & 0 \\ 0 & 0 & 0 \end{bmatrix}$, $B = \begin{bmatrix} 0 & 0 & 0 \\ 0 & 1 & 0 \\ 0 & 0 & 1 \end{bmatrix}$, $C = \begin{bmatrix} 0 & 1 & -1 \\ 1 & 1 & 0 \end{bmatrix}$, $D = \begin{bmatrix} 1 & 0 & 1 \\ 0 & 0 & 0 \end{bmatrix}$.

## Exercises of Chapter 9

**9.4** $F = gf$, $g = (0,1)^T$, $f = (-11 \ -19 \ -12; -10)$ (after reducing the system to single-input controllable).

**9.6** (a) $G = \frac{1}{2}$, $F = [-2, -\frac{7}{2}, -\frac{5}{2}]$. (b) controllable but unobservable.

**9.9** (a) Let $x_1 = \theta$, $x_2 = \dot{\theta}$. Then $A = \begin{bmatrix} 0 & 1 \\ 0 & -1 \end{bmatrix}$, $B = \begin{bmatrix} 0 \\ 1 \end{bmatrix}$, $C = [1 \ 0]$.

(b) $F = [-1, -1]$.

**9.11** $F = [-3.7321, -6, 4641]$ minimizes $J_1$.

**9.12** (a) $F = [\frac{3}{2}, 1]$. (b) $K = [2\alpha, \alpha^2 + 2]^T$.

**9.13** (a) $\mathcal{O}E = 0$. (b) $E = \alpha(1, -1)^T$.

**9.14** $F = - \begin{bmatrix} 1 & 0 \\ 1 & 2 \end{bmatrix}$.

**9.15** (a) $x_0 = -(\lambda I - A)^{-1}B$; If $\lambda$ is a zero of $H(s)$, a pole-zero cancellation will occur. (b) $x_0 = \alpha v$, where $v$ is the eigenvector corresponding to $\lambda$.

## Exercises of Chapter 10

**10.1** (a) For $X_1 = \tilde{X}_1 = 0$, $Y_1 = \tilde{Y}_1 = 1$, $H_2 = (1 + s^2 K)/K$, where $K$ is any stable rational function. Alternatively, for $X_1' = \tilde{X}_1' = (s^2 + 8s + 24)/(s+2)^2$, $Y_1' = \tilde{Y}_1' = (32s + 16)/(s+2)^2$, $N_1' = \tilde{N}_1' = 1/(s+2)^2$, $D_1' = \tilde{D}_1' = s^2/(s+2)^2$, and $H_2 = (32s + 16 - s^2 K')/(s^2 + 8s + 24 - K')$, where $K'$ is any proper and stable rational function.
(b) In general, it is not easy (or may be impossible) to restrict appropriately $K$ or $K'$, so $H_2$ is of specific order. Here let $H_2 = (b_1 s + b_0)/(s + a_0)$ and establish conditions on the parameters so that the closed-loop system is stable.

**10.4** (b) $N'^{-1}T = X'$, a proper and stable function, implies conditions on $n_i, d_i$ of $T$.

**10.5** (a) In view of the hint, $N^{-1}T = 1/(s + 2)^2 = 1/G^{-1}D_F$, from which $G = 2$, $F(s) = FS(s) = [-6, -11][1, s]^T$.

## Exercises of Chapter 10

# Index

A/D, analog-to-digital converter, 117
Abel's formula, 80
Ackermann's formula, 361, 382, 401
Adjoint equation, 132, 133
Adjoint of a matrix, 487
Algebraic multiplicity, 475,
    see also Eigenvalue
Asymptotic behavior, 94, 121,
    see also Mode of system
Asymptotic state estimator, see State
    observer
Automobile suspension system, 140
Autonomous, 10,
    see also Linear ordinary differential
    equation; System
Axioms
    of a field, 455
    of a norm, 483
    of a vector space, 456

Basis, see Vector space
BIBO stable, 170, 174, 187, 189,
    see also Stability
Biproper, 427,
    see also Rational function

Canonical form, 471
    Jordan, 88, 478–481
Canonical Structure Theorem, 245, 269,
    see also Kalman's Decomposition
    Theorem
Cauchy–Peano existence theorem, 18
Cayley–Hamilton Theorem, 91, 127, 475
Characteristic

equation, 474
polynomial, 474,
    see also Matrix; Transfer function
value, see Eigenvalue
vector, see Eigenvector
Cholesky decomposition, 490
Circuit, 14, 279, 286, 291
Closed-loop control, 353, 400
Cofactor, 486,
    see also Matrix
Command input, see Reference input
Companion form matrix, 137
Condition number, 489,
    see also Matrix
Constructibility, 222, 230,
    see also Observability
    continuous-time system, 222–224
    discrete-time system, 202, 228
    Gramian, 223
Continuous function
    at a point, 7
    over an interval, 7
    uniformly, 7
Control problems, 445
Controllability (-to-the-origin), 209,
    210, 230, 303, 307
    continuous-time system, 209, 230
    discrete-time system, 216
    eigenvalue/eigenvector (PBH) test,
    248
    from the origin, see Reachability
    Gramian, 210
    indices, 256

Controllability (*Cont'd*)
  matrix, 197, 206, 230
  output, 234
Controllable (-to-the-origin), 199, 209,
    230
  companion form, *see* Controller form
  eigenvalues, 240
  mode, 240
  single-input, 405
  subspace, 209
Controller
  digital, 116
  feedback, 352, 411
  implementations, 439
  with two degrees of freedom, 431, 448
Controller form, 251, 270
  multi-input, 256, 270
  single-input, 252, 270
Converter(A/D, D/A), 117
Convolution
  integral, 69
  sum, 62
Coprime, 299, 301, 307, 423, 448,
  *see also* Polynomial matrices

D/A, digital-to-analog converter, 117
Deadbeat
  control, 136
  observer, 388
Decoupling
  diagonal, 446, 449
  static, 447, 449
Degree, McMillan, 321, 345
Detectable, 380
Determinant, 486,
  *see also* Matrix
Determinantal divisor, 283
Diagonal decoupling, 446, 449
Difference equations, 51,
  *see also* Solutions of difference
    equations
Differential equations, ordinary,
  *see also* Solutions of differential
    equations
  classification, 10
  first-order, 9
  linear, 20, 21
  linear homogeneous, 11, 12, 28
  linear homogeneous with constant
    coefficients, 11, 12, 78

linear nonhomogeneous, 11, 12, 30, 78
  $n$th-order, 11
  systems of first-order, 10
Digital signal, 117
Diophantine equation, 424, 427, 447
  all solutions, 424, 429
  Bezout identity, 427
Dirac delta distribution, 65, 66, 73, 92
Direct link matrix, 105
Discrete-time impulse, 60
Discrete-time Kalman filter, 391,
  *see also* Kalman filter
Divisor, common, 298
Domain of attraction, 166,
  *see also* Equilibrium
Double integrator, 120, 139
Doubly coprime, 423, 448, 450,
  *see also* Coprime
Dual system, 203, 232

Eigenvalue, 474
  algebraic multiplicity, 475
  controllable, 240
  critical, 152
  geometric multiplicity, 475
  multiple, 475
  observable, 243
Eigenvalue or pole assignment, 357, 401
  direct method, 358, 400
  eigenvector assignment, 364, 401
  using controller form, 359, 401
Eigenvector, 474
  generalized, 479
Equilibrium, 124, 129, 142, 174, 188,
  *see also* Stability
  attractive, 145
  domain of attraction, 145
  qualitative characterization, 144
  trivial solution, 143
Equivalence
  of internal representations, 105, 115,
    303
  of matrices, 473
  zero-input, 107
  zero-state, 107
Estimator, *see* State observer
Euclidean
  norm, 484

Euler
  method, 38
External input, 352

Feedback
  configuration, 413
  control, 411
  gain matrix, 352
  output, 351, 392, 411
  state, 351, 400
Feedback stabilizing controller, 422, 448
  parameterizations, polynomial MFD,
    423, 448
  parameterizations, proper and stable
    MFD, 426, 448
  two degrees of freedom, 434, 449
Field, 455
Frequency response, 138
Function, 7
  continuous, 7
  Hamiltonian, 34
  indefinite, 155
  piecewise continuous, 8, 31
  positive definite, semidefinite, 155
Fundamental matrix, 78–81, 127
Fundamental theorem of linear
    equations, 466

Gaussian elimination, 490
Generalized distance function, 154,
  see also Lyapunov function
Generalized eigenvector, 479
Generalized energy function, 154,
  see also Lyapunov function
Generalized function, 66,
  see also Dirac delta distribution
Geometric multiplicity, 475,
  see also Eigenvalue
Gram
  matrix, 211
Gramian
  constructibility, 223
  controllability, 210
  observability, 220, 231
  reachability, 206, 230

Hamiltonian
  dynamical systems, 34
  function, 34
  matrix, 372

Hamiltonian matrix, 377
Hankel matrix, 322, 345
Harmonic oscillator, 134
Hermite form, 299, 301
Hölder's Inequality, 458
Hurwitz matrix, 165
Hybrid system, 118

Ill conditioned, 489
Impulse response
  continuous-time, 70, 71, 73
  discrete-time, 64, 72
Impulse response matrix, 64, 68, 70, 71
Indices
  controllability, 256
  observability, 265
Infinite series method, 87, 127,
  see also Matrix, exponential
Initial conditions, 5, 6
Initial time, 5, 6
Initial-value problem, 8, 11, 32, 51
  examples, 13
  solutions, 17
Input
  command or reference, 352
  comparison sensitivity matrix, 436
  decoupling zeros, 287, 305
  external, 352
  function observability, 235
  output decoupling zeros, 288, 305
  vector, 5
Input-output description, see System
    representations, descriptions
Instability, 160
Integral equation, 10
Integral representation, 65
Integration, forward rectangular rule,
    38
Internal description, see System
    representations, descriptions
Internal stability, 144–169, 173–186, 188
Invariant factors, polynomials, 283
Invariant property of $(A, B)$, 260
Invariant subspace, 209, 227
Inverted pendulum, 43,
  see also Pendulum

Jacobian matrix, 21, 32
Jordan canonical form, 88, 478–481

Kalman filter
  continuous-time, 385
  discrete-time, 391
Kalman–Bucy filter, 385
Kalman's Decomposition Theorem, 245, 269

Lagrange's equation, 35
Lagrangian, 36
Laplace transform, 92, 127
Least-order realization, 318
Least-squares, 491
Leonhard–Mikhailov stability criterion, 190
Level curve, 157,
  see also Lyapunov function
Lienard equation, 15, 167
Limit cycle, 42
Linear algebraic equation
  fundamental theorem, 466
  solutions, 469, 488
Linear operator, see Linear transformation
Linear ordinary difference equations, 52, 53
Linear ordinary differential equation,
  see also Solutions of differential equations
  autonomous, 10
  homogeneous, 11, 12, 28
  matrix, 80
  nonhomogeneous, 11, 12, 30
  periodic, 10, 12
Linear space, see Vector space
Linear system, 57
Linear transformation, 464
  fundamental theorem, 466
  null space, 465
  nullity, 466
  orthogonal, 491
  principle of superposition, 50
  range space, 466
  representation by a matrix, 466
Linearization, 6, 21, 32, 164, 185, 189
  examples, 24
Linearized equation, 22, 23, 32
Linearly dependent, 461, 462,
  see also Vector
Linearly independent, 210, 461, 462,

  see also Vector
Lipschitz condition, 18, 20
LQG (linear quadratic Gaussian) problem, 385, 402, 403
LQR (linear quadratic regulator) problem
  continuous-time, 369
  discrete-time, 377, 401
LU decomposition, 490
Luenberger observer, 379, 402
Lyapunov function, 154, 183
  construction of, 160
  level curve, 157
Lyapunov matrix equation, 153, 179, 189
Lyapunov stability, 144, 148, 188, 189,
  see also Stability
Lyapunov's Direct or Second Method, 153

Markov parameter, 137, 315, 322, 345
Matrix
  characteristic polynomial, 474
  Cholesky decomposition, 490
  cofactor, 486
  companion form, 137
  condition number, 489
  controllability, 197, 230
  determinant, 486
  diagonal, 478
  equivalent, 473
  exponential, 85, 127
  fundamental, 80
  Gram, 211
  Hamiltonian, 372, 377
  Hankel, 322, 345
  Hermite form, 299, 301
  Hermitian, 491
  Hurwitz, 165
  ill conditioned, 489
  impulse response, 64, 68, 70, 71
  indefinite, 154
  inverse, 487
  Jacobian, 21, 32
  Jordan, 478–481
  LU decomposition, 490
  minimal polynomial, 477
  minor, 155, 284, 307, 486
  Moore–Penrose inverse, 496

Matrix (*Cont'd*)
  negative definite, semidefinite, 154
  nonsingular, 487
  norm, 485
  observability, 220, 231
  orthogonal, 491
  positive definite, semidefinite, 154
  proper rational, 105
  QR decomposition, 490
  rank, 469
  Rosenbrock system, 286, 305
  Schur stable, 178
  similar, 473
  Smith form, 283
  Smith–McMillan form, 284, 307
  state transition, 82, 109, 127
  symmetric, 154, 491
  system, 286, 305
  Toeplitz, 227
  unimodular, 283
  unitary, 491
  well conditioned, 489
Matrix fractional description, 293, 297, 308,
  *see also* System representations, descriptions
McMillan degree, 321, 345
MIMO system multi-input/multi-output, 56
Minimal polynomial, 284, 477
Minkowski's Inequality, 458
Mode of system, 95, 122, 127,
  *see also* System
Model matching problem, 446, 449, 452
Modeling, 2
Moore–Penrose pseudo-inverse, 496

Natural basis, 464,
  *see also* Basis; Vector space
Negative,
  *see also* Function; Matrix
  definite, semidefinite, 154
Nonlinear systems, 4, 147, 164, 185, 189
Norm
  Euclidean, 484
  induced, 485
  Manhattan, 484
  matrix, 485
  taxicab, 484

Observability, 219, 223, 230, 304, 307
  continuous-time system, 219, 221, 230
  discrete-time system, 226
  eigenvalue/eigenvector (PBH) test, 248
  Gramian, 220, 231
  indices, 265
  matrix, 220, 231
  subspace, *see* Unobservable
Observable
  eigenvalue mode, 243
Observer, Luenberger, 379, 402,
  *see also* State observer
Observer form, 263, 271
  multi-output, 265, 271
  single-input, 263, 271
Open-loop control, 353, 400
Operator, linear, *see* Linear transformation
Optimal
  control problem, LQR, 369, 377, 401
  estimation problem, LQG, 385, 391, 402, 404
Optimality principle, 403
Orthogonal, 491
  matrix, 491
Output
  comparison sensitivity matrix, 436
  decoupling zeros, 287, 305
  equation, 5
  function controllability, 235
  reachability, controllability, 234
  vector, 5

Peano–Baker series, 29, 33, 82
Pendulum
  inverted, 43
  simple, 17, 24, 45
Phase
  plane, 134
  portrait, 42
  variable, 134
Picard iterations, 20
  with successive approximations, 28
Pole assignment problem, 357,
  *see also* Eigenvalue or pole assignment
Pole polynomial, 284, 307
Pole, zero relations, 290, 306

Poles at infinity, 283
Poles of a transfer function, 284
Poles of the system, 283, 306,
    see also Eigenvalue
Polynomial
    monic, 477
Polynomial matrices
    coprime, left, right, 299, 301, 307
    division algorithm, 476
    doubly coprime, 423, 448, 450
    greatest common divisors, 298, 300
    Hermite form, 299, 301
    Smith form, 283
    unimodular, 283
Polynomial matrix description, 292,
    see also System representations,
        descriptions
Positive,
    see also Function; Matrix
    definite, indefinite, semidefinite, 155
Prediction estimator, 389
Predictor–corrector method, 40
Proper transfer function, 105

QR decomposition, 490
Quadratic form, 154
Quantization, 117

Rank, 469
    test, 249
Rational function
    biproper, 427
    proper and stable, 308
Rayleigh's dissipation function, 36
Reachability, 205, 214, 216,
    see also Controllability, from the
        origin
    continuous-time system, 205, 230
    discrete-time system, 198, 214, 230
    Gramian, 206, 230
    matrix, see Controllability
    output, 234
    subspace, 206
Reachable, 198, 205, 214, 230,
    see also Controllable; Controllability,
        from the origin
    state, 205
    subspace, 206
Realization algorithms, 324, 345

controller/observer form, 326, 345
matrix $A$ diagonal, 339, 345
singular-value decomposition, 341,
    345
using duality, 324, 345
Realization of systems, 314, 343, 345
    existence and minimality, 316, 345
    least order, irreducible, minimal
        order, 318, 321, 345
Reconstructible, see Constructibility
Reference input, 352
Response, 49, 55, 71
    maps, 435
    total, 100, 111, 128
    zero-input, 100, 111
    zero-state, 100, 111
Return difference matrix, 421, 422
Riccati equation
    continuous-time case, 370, 401
    discrete-time case, 386, 402
Rosenbrock system matrix, 286, 305
Routh–Hurwitz stability criterion, 190
Runge–Kutta method, 39

Sampled data system, 116, 129
Sampling period, rate, 119, 129
Scalar, 456
Schur–Cohn stability criterion, 190
Schwarz Inequality, 458
Semigroup property, 109
Sensitivity matrix, 436
Separation principle, property, 395, 403
Shift operator, 75
Signal, digital, 117
Similarity transformation, 105, 474
Singular
    value, 491
    value decomposition, 493
    vector, left, right, 494
SISO system
    single-input/single-output, 56
Smith form, 283
Smith–McMillan form, 284, 307
Solutions of algebraic equations, 469,
    488
Solutions of difference equations
    particular, 55
    total, 55
Solutions of differential equations, 9,

*see also* Variation of constants
    formula
bounded, 148
continuable, 19
continuation, 18
continuous dependence on initial
    conditions, 20
continuous dependence on parame-
    ters, 19, 20
existence, 18, 27, 33
homogeneous, 28–30, 53
noncontinuable, 18, 19
particular, 31
Peano–Baker series, 29
predictor–corrector method, 40
Runge–Kutta, 39
successive approximations, Picard
    iterations, 20, 28
total, 31
Space
of $n$-tuples, 457
of real-valued continuous functions,
    459
span, 461
Spring, 16
Spring mass system, 139
Stability, 124, 129, 141, 148, 304, 307,
    419, 427
asymptotic, 124, 145, 149, 150, 177,
    178, 185
asymptotic in the large, 146, 149,
    152, 177, 180
attractive equilibrium, 145, 175
bounded-input/bounded-output
    (BIBO), 170, 174, 187, 189
causal, 58, 61, 62, 64, 70, 72, 73
domain of attraction, 145, 175
exponential, 145, 150, 177
exponential in the large, 150, 152
external, 170
global asymptotic, 150
input–output, 170, 186, 189, 281, 306
linear systems, continuous, 148
linear systems, discrete, 173
Lyapunov, 144, 148, 188, 189, 281,
    306
Routh–Hurwitz criterion, 190
Schur–Cohn, 190
Stabilizable, 356

Stable, *see* Stability
Standard form
    Kalman's canonical, 244, 269
    uncontrollable system, 238, 269
    unobservable system, 241
State
    partial, 295
    phase variable, 134
    variables, 5
    vector, 5
State equation, 5
State estimator, *see* State observer
State feedback, 352, 400,
    *see also* Feedback
    eigenvalue assignment, 355, 400
    input–output relations, 372
    optimal, 369, 377, 401
State observer, 378, 402
    current, 389, 402
    deadbeat, 388
    full-order, 378, 387
    identity, 379, 387
    optimal, 385, 391, 402
    partial state, 383, 391
    prediction, 389
    reduced-order, 383, 391
State transition matrix, 82, 109
State unconstructible, 222, 223
State unobservable, 220, 223
Structure theorem
    controllable version, 261, 270
    observable version, 267, 271
Successive approximations, Picard
    iterations, 20, 28, 32,
    *see also* Solutions of differential
        equations
Superposition principle, 50, 57
Sylvester Rank Inequality, 319, 487
System
    at rest, 63, 69
    autonomous, 10
    causal, 58, 72, 73
    conservative, 34
    continuous-time, 5
    discrete-time, 4, 6, 50, 60, 72
    distributed parameter, 4
    dual, 203, 232
    finite-dimensional, 4, 6
    Hamiltonian, 35

System (*Cont'd*)
  hybrid, 118
  infinite-dimensional, 4
  linear, 57
  lumped parameter, 4
  matrix, 286, 305
  memoryless, 57
  mode, 95, 122, 127
  nonanticipative, 58
  nonlinear, 4
  realization, 314, 343, 345
  sampled data, 116, 129
  single input/single output, 56
  time-invariant, 59, 72
  time-varying, 59
  with memory, 57, 72
System interconnections
  feedback, 413, 447
  parallel, 411, 447
  series, or tandem, 412, 447
System representations, descriptions
  balanced, 342, 345
  continuous-time, 48
  controller form, 251, 257, 270
  differential/difference operator, 292,
    307
  discrete-time, 50
  equivalence, of, 105, 115, 128
  external, 6, 56, 72, 293
  input–output, 56, 72, 73
  internal, 6, 293
  matrix fractional, 293, 297, 308
  observer form, 263, 266, 271
  polynomial matrix, 292, 295, 307,
    308, 416
  standard form, uncontrollable,
    unobservable, 238, 239, 241, 242,
    269
  state-space, 5

Time reversibility, 109
Toeplitz matrix, 227
Trajectory, 42, 134
Transfer function
  McMillan degree, 321, 345
  minimal polynomial, 284
  pole polynomial, 284
  strictly proper, 105
Triangle inequality, 483

Truncation operator, 74
Two degrees of freedom controller, 431,
  448

Unconstructible, 227,
  *see also* Constructibility
  subspace, 227
Uncontrollable,
  *see also* Controllability
  eigenvalues, modes, 240, 269
Unimodular matrix, 283
Unit
  pulse, or unit impulse, response, 61
  pulse, or unit sample, 60
  step function, 92
  step sequence, 60
Unit impulse, 66, 73
Unity feedback, 441, 443, 449
Unobservable, 219,
  *see also* Observability
  eigenvalues, modes, 243
  subspace, 219

van der Pol equation, 15, 42
Variation of constants formula, 31, 33,
  132
Vector, 456
  coordinate representation, 464
  linearly dependent, 461, 462
  linearly independent, 461, 462
Vector space, 456
  basis, 463, 472
  dimension, 463
  examples, 457
  finite-dimensional, 463
  normed, 483
  null, 465

Youla parameter, 425, 429, 449

$z$-Transform, 112
Zero, 286, 287, 304, 307
  at infinity, 310
  decoupling input/output, 288, 305
  direction, 292
  invariant, 287, 305, 307
  of transfer functions, 288, 305, 306
  polynomial, 287, 305, 307
  system, 287
  transmission, 288, 305

Zero-input response, 100, 111,
    *see also* Response
Zero-order hold, 117

Zero-state response, 100, 111,
    *see also* Response

# About the Authors

**Panos J. Antsaklis** is the H. Clifford and Evelyn A. Brosey Professor of Electrical Engineering and Concurrent Professor of Computer Science and Engineering at the University of Notre Dame. He served as the Director of the Center for Applied Mathematics of the University of Notre Dame from 1999 to 2005. He is a graduate of the National Technical University of Athens (NTUA), Greece and holds MS and Ph.D. degrees from Brown University.

His research addresses problems of control and automation and examines ways to design engineering systems that will exhibit a high degree of autonomy in performing useful tasks. His recent research focuses on networked embedded systems and addresses problems in the interdisciplinary research area of control, computing and communication networks, and on hybrid and discrete-event dynamical systems.

Dr. Antsaklis has authored a number of publications in journals, conference proceedings, and books, and he has edited six books on intelligent autonomous control, hybrid systems, and networked embedded control systems. In addition, he has co-authored the research monographs *Supervisory Control of Discrete-Event Systems Using Petri Nets* (Kluwer Academic Publishers, 1998, with J. Moody) and *Supervisory Control of Concurrent Systems: A Petri Net Structural Approach* (Birkhäuser, 2006, with M.V. Iordache) as well as the graduate textbook *Linear Systems* (McGraw-Hill, 1997, first printing and Birkhäuser, 2005, second printing, with A.N. Michel).

Dr. Antsaklis has been guest editor of special issues of the *IEEE Transactions of Automatic Control* and the *Proceedings of the IEEE on Hybrid and on Networked Control Systems*. He serves on the editorial boards of several journals, and he currently serves as Associate-Editor-at-Large of the *IEEE Transactions of Automatic Control*.

Dr. Antsaklis has served as program chair and general chair of major systems and control conferences including the Conference on Decision and Control, and he was the 1997 President of the IEEE Control Systems Society (CSS). He has been a plenary and keynote speaker at a number of conferences and research workshops and currently serves as the president of the Mediterranean Control Association.

Dr. Antsaklis serves on the Scientific Advisory Board for the Max-Planck-Institut für Dynamik komplexer technischer Systeme, Magdeburg, Germany. He is currently a member of the subcommittee on Networking and Information Technology of the President's Council of Advisors for Science and Technology (PCAST), which advises the President of the United States on science and technology federal policy issues regarding technology, scientific research priorities, and math and science education.

Dr. Antsaklis is an IEEE Fellow for his contributions to the theory of feedback stabilization and control of linear multivariable systems, a Distinguished Lecturer of the IEEE Control Systems Society, a recipient of the IEEE Distinguished Member Award of the Control Systems Society, and an IEEE Third Millennium Medal recipient. He is the 2006 recipient of the Brown Engineering Alumni Medal from Brown University, Providence, Rhode Island.

**Anthony N. Michel** received the Ph.D. degree in electrical engineering from Marquette University and the D.Sc. in applied mathematics from the Technical University of Graz, Austria. He has extensive industrial and academic experience with interests in control systems, circuit theory, neural networks, and applied mathematics. His most recent work is concerned with stability analysis of finite- and infinite-dimensional discontinuous dynamical systems. He has held faculty positions at Iowa State University and the University of Notre Dame and visiting faculty positions at the Technical University in Vienna, Austria, the Johannes Kepler University in Linz, Austria, and the Ruhr University in Bochum, Germany. He is currently the Frank M. Freimann Professor of Engineering Emeritus and the Matthew H. McCloskey Dean of Engineering Emeritus at the University of Notre Dame.

Dr. Michel has co-authored ten books and a number of publications in journals, conference proceedings, and books. He is a past Editor-in-Chief of the *IEEE Transactions on Circuits and Systems* and has held a variety of positions on the editorial boards of the *IEEE Transactions on Automatic Control*; *IEEE Transactions on Neural Networks*; *Circuits, Systems and Signal Processing*; *International Journal of Hybrid Systems*; *Nonlinear Analysis*; and other journals. He is a past president of the IEEE Circuits and Systems Society and has been a member of the executive committees of several professional organizations.

Dr. Michel is a Life Fellow of the IEEE. He received three prize paper awards from the IEEE Control Systems Society and the IEEE Circuits and Systems Society. He was awarded the IEEE Centennial Medal (1984), the Golden Jubilee Medal of the IEEE Circuits and Systems Society (1999), and the IEEE Third Millennium Medal (2000). He was a Fulbright Scholar at the Technical University of Vienna (1992), and he received the 1995 Technical Achievement Award of the IEEE Circuits and Systems Society, the Alexander von Humboldt Research Award for Senior U.S. Scientists (1997), the Distinguished Member Award of the IEEE Control Systems Society (1998), and the 2005 Distinguished Alumnus Award of the College of Engineering, Marquette University.